The *Arithmetica* of Diophantus

This volume offers an English translation of all ten extant books of Diophantus of Alexandria's *Arithmetica*, along with a comprehensive conceptual, historical, and mathematical commentary.

Before his work became the inspiration for the emerging field of number theory in the seventeenth century, Diophantus (ca. 3rd c. CE) was known primarily as an algebraist. This volume explains how his method of solving arithmetical problems agrees both conceptually and procedurally with the premodern algebra later practiced in Arabic, Latin, and European vernaculars, and how this algebra differs radically from the modern algebra initiated by François Viète and René Descartes. It also discusses other surviving traces of ancient Greek algebra and follows the influence of the *Arithmetica* in medieval Islam, Byzantium, and the European Renaissance down to the 1621 publication of Claude-Gaspard Bachet's edition. After the English translation the book provides a problem-by-problem commentary explaining the solutions in a manner compatible with Diophantus's mode of thought.

The *Arithmetica* of Diophantus provides an invaluable resource for historians of mathematics, science, and technology, as well as those studying ancient Greek, medieval Islamic and Byzantine, and Renaissance history. In addition, the volume is also suitable for mathematicians and mathematics educators.

Jean Christianidis is a professor of history of mathematics at the National and Kapodistrian University of Athens (Greece) and partner member of the Centre Alexandre-Koyré (Paris, France). He has authored or co-authored numerous articles on Diophantus and on Greek and Byzantine mathematics.

Jeffrey Oaks is a mathematics professor at University of Indianapolis (USA). He is co-author of the 2021 book *Al-Hawārī's Essential Commentary: Arabic Arithmetic in the Fourteenth Century*.

Scientific Writings from the Ancient and Medieval World

Series Editor: John Steele

Brown University, USA

Scientific texts provide our main source for understanding the history of science in the ancient and medieval world. The aim of this series is to provide clear and accurate English translations of key scientific texts accompanied by up-to-date commentaries dealing with both textual and scientific aspects of the works and accessible contextual introductions setting the works within the broader history of ancient science. In doing so, the series makes these works accessible to scholars and students in a variety of disciplines, including history of science, the sciences, and history (including Classics, Assyriology, East Asian Studies, Near Eastern Studies, and Indology).

Texts will be included from all branches of early science, including astronomy, mathematics, medicine, biology, and physics, and which are written in a range of languages, including Akkadian, Arabic, Chinese, Greek, Latin, and Sanskrit.

The Babylonian Astronomical Compendium MUL.APIN
Hermann Hunger and John Steele

The Gaṇitatilaka and its Commentary
Two Medieval Sanskrit Mathematical Texts
Alessandra Petrocchi

The Medicina Plinii
Latin Text, Translation, and Commentary
Yvette Hunt

Learning With Spheres
The *golādhyāya* of Nityānanda's *Sarvasiddhāntarāja*
Anuj Misra

The *Arithmetica* of Diophantus
A Complete Translation and Commentary
Jean Christianidis and Jeffrey Oaks

For more information about this series, please visit: www.routledge.com/classicalstudies/series/SWAMW

The *Arithmetica* of Diophantus
A Complete Translation and Commentary

**Jean Christianidis and
Jeffrey Oaks**

Routledge
Taylor & Francis Group

LONDON AND NEW YORK

First published 2023
by Routledge
4 Park Square, Milton Park, Abingdon, Oxon OX14 4RN

and by Routledge
605 Third Avenue, New York, NY 10158

Routledge is an imprint of the Taylor & Francis Group, an informa business

British Library Cataloguing-in-Publication Data
A catalogue record for this book is available from the British Library

ISBN: 978-1-138-04635-1 (hbk)
ISBN: 978-1-032-32445-6 (pbk)
ISBN: 978-1-315-17147-0 (ebk)

DOI: 10.4324/9781315171470

Typeset in Times New Roman
by Apex CoVantage, LLC

Contents

Book V^G

Figures

Tables

Preface

During the ca. 1750 years that have elapsed since its composition, the *Arithmetica* of Diophantus has been read from two major perspectives. From the time of its writing down to the first part of the seventeenth century, it was read and commented on mainly as a book on algebraic problem-solving. To solve arithmetical problems, Diophantus named unknowns and performed operations on those names to set up and solve equations in much the same manner as the algebraists who later wrote in medieval Arabic, Latin, Italian, and sixteenth-century European vernaculars. This was a premodern algebra, in which polynomials were aggregations, and the names of the powers and their abbreviations in notation designated types of numbers and not values.

The seventeenth-century shift in perspective was the result of two roughly simultaneous developments. In the 1630s numerical adaptations of François Viète's new geometrical algebra by René Descartes and others rendered premodern algebra obsolete, though practical textbooks teaching the old algebra continued to be written into the next century. With Viète, for the first time, algebraic expressions were built from operations on letters designating values, thus making it possible to work with undetermined givens and opening the way for applications of algebra beyond problem-solving.[1] At about the same time, beginning with Pierre de Fermat in the early 1640s, the continued relevance of Diophantus was ensured when his book became a major source of inspiration for the emerging field of number theory. Fittingly, it was in Fermat's copy of the *Arithmetica* that he penned his famous last theorem in a marginal note. Since the eighteenth century, through the efforts of Leonhard Euler, Joseph-Louis Lagrange, and countless others, the study of Diophantine equations and number theory in general has blossomed into the vibrant branch of mathematics that we know today. And so Diophantus the algebraist gave way to Diophantus the number theorist.

This book is about Diophantus the algebraist. But just what it means to call him an algebraist has been problematic. This is because the premodern version of his art was entirely forgotten after the eighteenth century, when the last of the practical manuals written in the Renaissance tradition went out of print. It was only about a decade and a half ago that one of us (Oaks) began publishing a series of studies identifying the ways that algebra in medieval Arabic and Italian and also the algebra of sixteenth-century Europe differ both conceptually and procedurally from post-Vietan algebra, and the two of us published an article showing that the same premodern conceptual foundation is also at work in Diophantus. Previously, historians have either implicitly reinterpreted the *Arithmetica* in terms of the concepts and notation of modern algebra, or they have denied that the work contains any algebra at all.

In order to situate the *Arithmetica* as closely as we can to the way its author intended and the way premodern readers understood it, we devote a chapter of our introduction to the vocabulary, concepts, and structure of premodern algebra as practiced by Diophantus and later writers. From there we trace the influence of the *Arithmetica* in ancient Greece, medieval Islam, the Byzantine Empire, and Renaissance Europe down to the publication of Bachet's 1621 edition and Latin translation, stopping just before the rebirth of Diophantus as a number theorist. Of particular interest is the impact the book had when translated into Arabic in the ninth century and into Latin in the sixteenth.

Next, we give a literal English translation of the six books of the *Arithmetica* extant in Greek and the four books preserved only in Arabic translation. The *Arithmetica* is one of the last major Greek mathematical works without an English translation, though Paul ver Eecke's 1926 French translation of the six surviving Greek books is quite good, and the four Arabic books have been duly rendered into both English and French by Jacques Sesiano and Roshdi Rashed, respectively. The most obvious difference between our translation of the Arabic portion and Sesiano's is that he had translated the algebra into modern symbols, while we preserve the rhetorical versions of the expressions.

After the translation we give a mathematical commentary for each problem. Instead of rewriting the solutions in modern algebraic symbols, we have adopted a notation compatible with the notations found in various Greek, Arabic, and pre-1590 European works. In each problem we focus on the methods of assigning algebraic names to the different unknown numbers and to the ways that Diophantus then set up his equations.

The two of us began collaborating on Diophantus about a dozen years ago. In 2012, after we had submitted our joint article on Diophantus to *Historia Mathematica*, we began participating in workshops organized in Paris through the "Séries de problèmes" project supported by the laboratory Histoire et anthropologie des savoirs, des techniques et des croyances (HASTEC). It was around the time of a May 19, 2016, workshop held at Centre Alexandre Koyré that Jean suggested that we write this book, and in the following month we put together our proposal during the international workshop "History of Mathematics through the study of textual sources" held June 10–11, 2016, at National and Kapodistrian University of Athens. The two of us presented portions of our research on Diophantus at subsequent meetings of the Séries de problèmes group in 2017, 2018, and 2019, and both before and after that period we gave lectures individually at meetings in London, Tel Aviv, Marrakesh, Berlin, and Clermont-Ferrand. We thank all of our colleagues who organized or participated in these meetings. Special thanks are due to Alain Bernard and Michalis Sialaros, who invited us repeatedly to present the progress of our research in the meetings they organized.

Special thanks are also due to Nathan Sidoli and Ioanna Skoura, who provided us information about some Byzantine scholars who were involved with Diophantus, and to Konstantinos Dimitrakopoulos, who read the manuscript with patience.

In May of 2018 we were supported in research in pairs programs at Fondation des Treilles in Tourtour, France, and at Centre International de Rencontres Mathématiques (CIRM) in Luminy near Marseilles. We thank the organizations of both institutions. Finally, we thank John Steele, who accepted the publication of our book in the series "Scientific Writings from the Ancient and Medieval World".

Note

1 (Oaks 2021).

Part I

Introduction

1 Diophantus and his work

1.1 What we know about Diophantus

In history of ancient mathematics we often come across mathematicians whose names loom large because of the great works they have authored, and yet for their lives, we have only the most meager information. Such is the case of Diophantus. Although his name was well known throughout late antiquity, virtually no records are extant relating to his life. In the past even the exact spelling of his name, whether it should be Διόφαντος (Diophantos) or Διοφάντης (Diophantēs), was under dispute. Today there is no longer any doubt, the Greek indirect evidence being overwhelmingly in favor of Διόφαντος.[1]

The only allegedly personal information that has been transmitted on the life of Diophantus comes from the collection of the arithmetical epigrams of the Palatine Anthology, attributed to a certain Metrodorus, who lived sometime in late antiquity, in all likelihood not later than the sixth century.[2] One epigram formulates an arithmetical problem which expresses the lifetime in years of Diophantus as a sum of its fractional parts. The epigram reads as follows in Paton's translation (1918, 93–95):

> This tomb holds Diophantus. Ah, how great a marvel! the tomb tells scientifically the measure of his life. God granted him to be a boy for the sixth part of his life, and adding a twelfth part to this, he clothed his cheeks with down; He lit him the light of wedlock after a seventh part, and five years after his marriage He granted him a son. Alas! late-born wretched child; after attaining the measure of half his father's life, chill Fate took him. After consoling his grief by this science of numbers for four years, he ended his life.[3]

---

1 Cf. (Heath 1910, 1). The following is the evidence collected by Tannery in the second volume of Diophantus's opera (Diophantus 1893–95, II): 35.9, 19; 36.24; 38.22–23, 25; 46.15; 60.19; 62.3–4, 24–25; 69.9; 72.11, 14, 15; 73.25–26; 122.2–3; 148.1, 178.25, 198.17, 205.8, 215.1, 218.28, 219.6, 220.6, 224.4, 226.14; 260.24. Another piece of evidence which came to light after Tannery's edition is a mention of Diophantus by the late Neoplatonic scholiast Pseudo-Elias (or Pseudo-David) in his *Lectures on Porphyry's Isagoge* (Westerink 1967, 38–39). On this testimony, see Section 3.2.2.
2 See Section 3.2.3.
3 For the Greek text, see Diophantus (1893–95, II, 60.19–61.2); Paton (1918, 92–94); and Buffière (1970, 91–92).

DOI: 10.4324/9781315171470-2

In modern language the problem is equivalent to the equation $\frac{1}{6}x+\frac{1}{12}x+\frac{1}{7}x+5+\frac{1}{2}x+4=x$, whose solution is 84. The epigram does not refer to any precise date, and the details it relates may be purely anecdotal; yet if regarded as aiming to indicate that Diophantus died at an old age, it should not be dismissed as valueless.[4]

Even the time during which Diophantus lived cannot be pinned down within reasonable limits. The only incontestable evidence we have is confined to the following two facts. In his treatise *On Polygonal Numbers* Diophantus mentions Hypsicles, and he quotes the definition of a polygonal number given by him.[5] Hypsicles was active in Alexandria in the first half of the second century BCE, hence Diophantus must have lived not earlier than, say, 170 BCE.[6] On the other hand, Theon of Alexandria is the earliest scholar we know to have studied the *Arithmetica*, and indeed he mentions Diophantus by name in his commentary on the first book of Ptolemy's *Almagest*.[7] Theon was active in the third quarter of the fourth century CE, which therefore constitutes a lower limit for Diophantus's date. By virtue of these two references, we can say with certainty that Diophantus's date is situated somewhere in a very wide span, sometime between the middle of the first half of the second century BCE and the third quarter of the fourth century CE.

To shrink this wide range, historians beginning with Tannery (1843–1904) have cited a letter written by the Byzantine polymath Michael Psellus (1018–ca. 1081). In this letter Psellus describes "the Egyptian method of numbers, by means of which the problems concerning the analytic are treated",[8] and he gives a list and description of the names of powers up to the ninth, mixing together without

---

4 If it was specifically the life of Diophantus that the epigram wanted to relate, then the choice of 84 for his age cannot be due solely to the fact that that number has many divisors, since many numbers smaller than it also have many divisors. And if the epigram was conceived as giving the fractions of the life of a man who lived a long time, then the choice of Diophantus for the subject implies that he was known for his advanced age at death.

5 See Diophantus (1893–95, I, 470.27–472.4) and Diophantus (2011, 196.23–27). In modern notation Hypsicles's definition says that the nth – agonal number is $\frac{1}{2}n\left[2+(n-1)(a-2)\right]$. On account of this quotation [Bulmer-]Thomas presumes that Hypsicles had written a book on numbers, now lost (1981, 617). [N.B.: References to Tannery's edition of the Greek text of Diophantus will be given henceforth by citing the page and line numbers. Thus, the first citation, shortened as (470.27–472.4), refers to vol. I, p. 470, line 27 through page 472, line 4 of Diophantus (1893–95).]

6 This date is surmised by combining two facts mentioned in the preface to the so-called Book XIV of the *Elements*, which is attributed to Hypsicles. According to the first, Hypsicles's father and Basilides of Tyre studied together a treatise of Apollonius on the dodecahedron and the icosahedron inscribed in the same sphere from an unsatisfactory pre-edition of it. According to the second, Hypsicles himself found later an elaborated version of that treatise. From these facts we may deduce that Hypsicles's father was an elder contemporary of Apollonius. Given that Apollonius died early in the second century BCE it follows that the time of Hypsicles's activity may be placed around the middle of the first half of the second century BCE. See Heath (1885, 4–5), [Bulmer-]Thomas (1981, 616), and Vitrac and Djebbar (2011, 52–53).

7 See Diophantus (1893–95, II, 35.2–36.3) and Theon of Alexandria (1936, 452.21–453.16). We discuss Theon's interest in Diophantus in Section 3.2.1.

8 The text of Psellus is published in Diophantus (1893–95, II, 37–42).

distinction arithmetical terms with algebraic terms.[9] After this, he writes: "Concerning this Egyptian method, Diophantus dealt with it more accurately, while the very learned Anatolius, having selected the most essential parts of the doctrine of that man,[10] most succinctly addressed it to another Diophantus".[11] Tannery identified this Anatolius with the Christian scholar Anatolius of Alexandria (d. ca. 282), who became Bishop of Laodicea ca. 269 (Tannery 1912g, 533, 535–36). What led Tannery to this identification was mainly the mathematical leanings of Anatolius as described in Eusebius's *Ecclesiastical History*:

> He also was by race an Alexandrian, who for his learning, secular education and philosophy had attained the first place among our most illustrious contemporaries; inasmuch as in arithmetic and geometry, in astronomy and other sciences, whether of logic or of physics, and in the arts of rhetoric as well, he had reached the pinnacle. It is recorded that because of these attainments the citizens of the Aristotelian tradition deemed him worthy to establish the school of the Aristotelian tradition at Alexandria.[12]

Eusebius goes on to discuss and quote Anatolius's book on the date of Easter (1932, 235ff),[13] and he also notes "The same person has left behind, also, arithmetical introductions in ten complete treatises and, as well, evidences of his study and deep knowledge of divine things" (1932, 239; translation adapted). This work can be identified with a book on Neopythagorean arithmology titled *On the First Ten Numbers* (Περὶ δεκάδος), partially preserved in Monac. gr. 384 ff. 57v–59r and in quotations in a book misattributed to Iamblichus titled *Theology of Arithmetic* (Θεολογούμενα ἀριθμητικῆς).[14] We will encounter this book again in Section 3.1.6.

But there was another Anatolius who lived at about the same time and who is known to have been proficient in mathematics. In his *Lives of the Philosophers and Sophists* Eunapius names Anatolius as one of the teachers of Iamblichus (Wright 1922, 362), and scholars are split as to whether this is the same person as the bishop Anatolius. The question makes no difference for the dating

---

9 We discuss the differences between arithmetical and algebraic terms in Section 3.1.1, and we discuss the list in Psellus in Section 3.4.1.

10 κατ᾽ ἐκεῖνον. Marwan Rashed (2013, 602) points out that "Le pronom ἐκεῖνος a ici son sens laudatif (« ce grand homme », « cet homme fameux ») et désigne Diophante qui vient d'être nommé".

11 Our translation follows the text carried by the fifteenth-century manuscript Vat. Urb. gr. 78, f. 81r, ll. 23–36. Tannery published the text on the basis of two fourteenth-century manuscripts, Laur. Plut. 58.29 and Scorial. gr. Y.III.12, in (Diophantus 1893–95, II, 37–42), and this passage occurs at pp. 38.22–39.1. Cf. (Diophantus 1893–95, II, xlvii, 38.25; Tannery 1912g, 536). The text does not differ in the three manuscripts, with the exception of the punctuation, which in the Vatican manuscript is executed with greater care.

12 (Eusebius 1932, 229–31), Oulton's translation.

13 A Latin translation titled *De Ratione Paschali* and attributed to Anatolius is extant, but it is not clear how faithful this translation is to the original. For the text, see (McCarthy and Breen 2003); cf. (Mosshammer 2008; McCarthy 1996; Fernández 2011).

14 The Greek text was published in (Heiberg 1901), and Tannery published a French translation in (Tannery 1915, 12–28).

of Diophantus. Even if they are not the same person, they both lived in the latter third century.

Now, if the Anatolius mentioned by Psellus is either the Christian Bishop, as Tannery supposed, or the teacher of Iamblichus, then Diophantus would have been active no later than the beginning of the last quarter of the third century. But Tannery went further than this. He emended the text of Psellus by changing "ἑτέρῳ Διοφάντῳ" ("to another Diophantus") of the manuscripts firstly to "ἑτέρως Διοφάντῳ"[15] ("in a different way from Diophantus") and finally to "⟨τῷ⟩ ἑταίρῳ Διοφάντῳ" ("to the friend Diophantus"), thus making Diophantus to be "en relation d'*hétairie*" and, by implication, contemporary with Anatolius (Diophantus 1893–95, II, xlvii, 38.25; Tannery 1912g, 536). Tannery further posited that Dionysius, the dedicatee of the *Arithmetica*, should be identified with Saint Dionysius of Alexandria, who was a teacher of Anatolius from 247 to 265 and had formerly been the head of the Catechetical school of the city from 231 to 247. By Tannery's account, the *Arithmetica* was written on the request of Dionysius to serve as a textbook for students in that school (1912g, 536–37).

While historians have generally agreed with Tannery's identification of Anatolius, they have been less accepting of his textual emendation and the subsequent identification of Dionysius. In 1903 Friedrich Hultsch (1833–1906) proposed instead Tannery's initial emendation ἑτέρως together with a reading that implies that Diophantus was the dedicatee of the book of Anatolius (1903, col. 1052–53). Thomas Heath (1861–1940) later took up this interpretation and gave his translation of the passage: "but the very learned Anatolius collected the most essential parts of the doctrine as stated by Diophantus in a different way and in the most succinct form, dedicating (προσεφώνησε) his work to Diophantus" (Heath 1910, 2). By this reading Diophantus and Anatolius are still contemporaries, though both Hultsch and Heath dismissed Tannery's conjectures about Dionysius (Heath 1910, 2 n. 3).

In 1934 Jacob Klein (1899–1978) reviewed the evidence and came up with his own proposal for the date of Diophantus (Klein 1968, 244–48 n. 149). He adjusted "ἑτέρῳ Διοφάντῳ" in the passage in Psellus to "ἑτέρως Διοφάντου", so again it means "in a different way from Diophantus". In connection with this, rather than interpret "προσεφώνησε" as meaning "he dedicated", he read the word as meaning "calling a thing something, naming" (1968, 245), and he suggested that the way in which Anatolius differed from Diophantus was in the names they gave to the powers of the unknown. Then, because of similarities between the *Arithmetica* and the *Metrica* of Hero, and that Hero dedicated his *Definitiones* to a Dionysius just as Diophantus did, Klein proposed that Hero and Diophantus were contemporaries and that the two Dionysii were the same person. At the time, *Definitiones* was still regarded as being a genuine work of Hero, and Hero's dates could not reliably be determined better than sometime in the span stretching from ca. 150 BCE to ca. 320 CE. Citing the opinions of A. Stein (1871–1950) and Heiberg

---

15 ἑτέρως Διοφάντου would be a more accurate writing. In fact, as we shall see, this double change (ἑτέρως for ἑτέρῳ and Διοφάντου for Διοφάντῳ) was proposed by Klein (1968, 245).

(1854–1908), Klein settled on the late second century CE for Hero, so he placed Diophantus there as well. He later adjusted this date back a century, based on Otto Neugebauer's (1899–1990) analysis of an eclipse recorded in Hero's *Dioptra* (Neugebauer 1938), and which seemed to securely put Hero in the mid-first century CE. Recently, however, that dating has been shown to be faulty.[16] Regardless the date of Hero, Klein's string of hypotheses is as fragile as Tannery's.

Wilbur Knorr (1945–97) took up the problem of the date of Diophantus in conjunction with his investigation into the authorship of the pseudo-Heronian *Definitiones*. In a 1993 article he criticized the emendations to the text of Psellus by Tannery, Heath, and Klein, and kept the original "ἑτέρῳ Διοφάντῳ". He translated the passage as: "but the very learned Anatolius, having collected the most essential parts of that man's doctrine, to a different Diophantus most succinctly addressed it" (1993, 184). Knorr explained:

> My rendering, if awkward, intends to capture the nuance of the Greek word order: that by juxtaposing the personal references, Psellus is emphasizing the distinction between Diophantus (*ekeinon*) and the dedicatee, his namesake, "a different Diophantus" (*heterôi Diophantôi*). The situation is one that Psellus could hardly have avoided commenting on, had he known of it, since the coincidence of names would be not only inherently confusing, but also tantalizingly apt.
>
> (1993, 184)

As Knorr notes, the "passage is grammatically sound without any revision" (1993, 190 n. 20), and because the text makes sense without any intervention, we concur.[17] But then, supposing that the names of the powers are Diophantus's own invention, he proposed that the δυναμοδύναμις (*dynamodynamis*) in Hero's *Metrica*, and the list of powers in the early third-century CE *Refutation of All Heresies*, came later than the *Arithmetica*, which, presuming Neugebauer's dating of Hero, puts Diophantus in the first century CE at the latest.[18] But, Knorr wrote at the end of his argument (1993, 191), "Admittedly, the originality of Diophantus is a hypothesis that, on present evidence, is no more or less plausible than the alternative, that Diophantus consolidated an arithmetic system initiated by others before him". Indeed, we argue in Chapter 3 that Diophantus did not invent the terms by which the powers are named.

16 It has long been known that because Hero cites Ctesibius, and that he is cited by Pappus, that he must have been active sometime during the period ca. 150 BCE to ca. 320 CE. In 1938, Neugebauer identified a lunar eclipse mentioned in Hero's *Dioptra* as being one that occurred in 62 CE, placing Hero squarely in the first century. Recently, however, it has been shown that Hero did not actually observe this eclipse, so all we can really say is that Hero was active *after* 62 CE. This shrinks the ca. 470-year span down to something under three centuries (Sidoli 2011). (We suspect that there are problems with Masià [2015], who argues that the eclipse recorded by Hero may have been an imaginative example.)

17 Marwan Rashed, also, prefers this reading (M. Rashed 2013, 602), and, more recently, Christianidis and Megremi (2019).

18 We discuss Hero's term in Section 3.1.1 and *Refutation of All Heresies* in Section 3.1.6.

So far, if Psellus was writing about either Anatolius the bishop or Anatolius the teacher of Iamblichus, we can narrow the span of time during which Diophantus can be said to be active to the period ca. 170 BCE to ca. 280 CE. We might be inclined to place Diophantus toward the latter part of this interval for the fact that he is not mentioned by any author across this timespan. Still, we should consider that both identifications might be wrong. A look into Vol. I, Part 2 of *Paulys Realencyclopädie der classischen Altertumswissenschaft* (1894) shows 16 different people named "Anatolius", with one more added in the 1903 Supplement, and the *Prosopography of the Later Roman Empire* (1971, 1980, 1992) shows 33 individuals with this name who lived in the period 260–641 CE. There may have been another Anatolius, perhaps not even among those listed in our reference works, and who may even have lived in Byzantine times, who composed the epitome described by Psellus.

With this uncertainty in mind, it is worth examining the only direct testimony we have for the time of Diophantus, from the Syrian Jacobite scholar Grigōriyōs Abū l-Faraj Bar 'Ebroyo, also known in Arabic as Ibn al-'Ibrī and in Latin as Barhebraeus (1225/26–85).[19] Barhebraeus composed a chronicle in Arabic titled *Mukhtaṣar ta'rīkh al-duwal*, which is known in English as *History of Dynasties*. For his historical studies, Barhebraeus drew on sources in Syriac and Arabic, and most likely he could not read Greek (Takahashi 2010, 25–26). In his chronicle he places Diophantus, together with some other intellectuals, during the reign of the Roman emperor Flavius Claudius Julianus (reigned as sole emperor 361–63 CE). Barhebraeus's chronicle is organized for the Imperial period chronologically by emperor, so in the fourth century it covers Constantius II (d. 361) just before Julian, and after that Jovian, Valentinian, etc. For the emperor Julian Barhebraeus first gives details on the politics of the era, after which he turns to scholarship, reporting the following (Barhebraeus 1890, 139.12–140.6; cf. Barhebraeus 1663, 139–41):

> There was in the time of Julian a writer named Themistius, a famous philosopher of the era who explained most of the books of Aristotle. He wrote a book for Julian on the economics and politics of kingdoms, and another treatise for him embracing the cessation of the persecution of Christians, stating that God Almighty desires to be worshiped by different faces. The philosophers, after all, were themselves divided into three hundred doctrines. His words persuaded him, and he ceased to harm them, so he stopped.
>
> And among the well-known philosophers of this time was Nicolaus, who held precedence in knowledge of wisdom. And among his literary works is a book bearing on the philosophy of Aristotle, and we have a version in Syriac carried out by Ḥunayn ibn Isḥāq; and a book on plants; and a book responding

---

19 See the entry on Barhebraeus in the *Encyclopaedia of Islam* (3rd ed.) by Takahashi (2014). For additional information about Barhebraeus, see Takahashi (2005). For questions on the origin of the name and the familial ancestry this could indicate, see Fathi-Chelhod (2001).

to those who believe that the intellect and the intelligibles are the same thing. Ibn Baṭlān said that he was originally from Laodicea.

And among them was Dorotheus, who was a mathematician. He had decisive influence on the science of astronomy and judicial astrology. His work is famous among the people (who study) the science of nativities and epochs.[20]

And among them was Diophantus, whose book A, B, called algebra (*al-jabr wa-l-muqābala*) is famous, and the observer who engages in it will perceive the vastness of this technique.[21]

Barhebraeus mentions scholars mainly in two clusters, six during the reign of Antoninus Pius, and the four previously listed under Julian. Only three other scholars are cited for the Roman imperial period, all in different reigns.[22] It thus seems that we should regard the reign as the general time during which a particular scholar flourished and not as the precise period of his activity. So if his testimony is accurate, Diophantus would have been active around the middle of the fourth century.

Barhebraeus likely began this passage with Themistius because the philosopher had written treatises for Julian, and from there he decided to list other thinkers of that general era. Themistius (ca. 317–ca. 385) indeed worked during the time of Julian, and our best estimate for the date of Nicolaus of Laodicea comes from this reference (Fazzo 2008). But this Dorotheus is certainly the astrologer Dorotheus of Sidon, who is known to have been active in the first century CE. Thus Barhebraeus, or his source, got his date wrong by three centuries (unless we wish to posit the existence of yet another astrologer named Dorotheus!). With the exception of Theon of Alexandria, other intellectuals from Roman times are placed correctly. So we cannot quite trust Barhebraeus for the date of Diophantus.

In his monumental *Histoire des Mathematiques* Jean-Étienne Montucla (1725–99) cited the testimony of Barhebraeus to date Diophantus to ca. 365 CE, though he added "Ce témoignage, à la vérité, n'est pas entiérement décisif" (Montucla 1758, 315). Later, in 1797, D. Pietro Cossali (1748–1815) gave a justification for doubting Barhebraeus: that Barhebraeus must have confused Diophantus the

---

20 The word translated as "epochs" is *adwār*, which Pococke translates as "Circulis". It means something like "rotations".

21 The word translated as "technique" is *nauʿ*. This is also the most common word for the "species" (powers) in algebra. It is singular here. Wehr (1979, 1185) gives the meanings of this "kind, sort, type, species; variety; way, manner, mode, fashion; form; nature, character, quality, grade. . . ."

22 Under Antoninus Puis (reigned 138–61) Barhebraeus lists the theologians Valentinus and Marcion and the astronomer Ptolemy, who are known to have been active during this time. He also lists Galen, who was only beginning his career at that time. Alexander of Aphrodisias is tacked on the end of this reign probably because, as Barhebraeus mentions, "he and Galen had many exchanges" (Barhebraeus 1890, 124.5). Outside the reigns of Antoninus Pius and Julian, Barhebraeus correctly places the theologian Cerinthus under Claudius, Apollonius of Tyana under Domitian, and Porphyry under Diocletian. But, he errs when he makes Theon, the commentator of the *Almagest*, a contemporary of Ptolemy (Barhebraeus 1890, 123.17).

algebraist with another Diophantus, "the teacher of rhetoric to Libanius, Julian the Apostate's favorite orator".[23] Eunapius (4th–5th c.) calls this teacher "Diophantus of Arabia"[24] and, as Wright reports (1922, 330–31), professed rhetoric at Athens at least from 336, when Libanius came there as a student, to 367, when the sophist Proheraesius died, for it was he who delivered the funeral oration.

To discredit the date given by Barhebraeus, Cossali had simply found another Diophantus who lived in the time of Julian and claimed that one was confused for the other. But Barhebraeus was well versed in the sciences, and he clearly knew that he was writing about the mathematician. One could entertain the possibility that the two Diophantii – the mathematician and the rhetorician – were the same person, since we do find within the *Arithmetica* terms and expressions with a rhetorical hue.[25] But according to Tannery (1912a, 66), what makes identification impossible is that the Diophantus mentioned by Eunapius was an Arabian, not an Alexandrian, as the mathematician Diophantus allegedly was, and he taught at Athens. We should also consider that there is nothing to prevent there from having been two Diophantii working about the same time, one on algebra and the other teaching rhetoric.

To conclude, we know for certain that Diophantus worked sometime in the more than five-century span from ca. 170 BCE to ca. 370 CE. If the Anatolius mentioned by Psellus is either the bishop or the teacher of Iamblichus, then Diophantus could not have been active later than ca. 280 CE. But this identification is by no means secure, given that the name Anatolius was not uncommon. The passage in Barhebraeus, which states that Diophantus worked in the middle of the fourth century, is problematic but cannot be entirely disregarded. Apart from Theon and Dorotheus, he correctly dated the other scholars he mentioned.

1.2 The works of Diophantus

Two works of Diophantus have come down to us, the *Arithmetica* and the short treatise *On Polygonal Numbers*. In their present state both are incomplete. The topic of polygonal numbers can be traced back to the Pythagoreans, and several later Greek mathematicians and philosophers worked on them, including

---

23 "il maestro di eloquenza a Libanio favorito oratore di Giuliano apostata" (Cossali 1797, 65).

24 In Wright (1922, 514–17). Cf. (Tannery 1912a, 66).

25 Such terms are, for example, παρισότης, meaning "approximate equality", and its cognate πάρισος, meaning "approximatively equal", employed in Problems 11 and 14 of the fifth Greek book. In ancient rhetoric the term παρίσωσις refers to a specific "figure of speech", and it has the meaning of "the succession of two or more coordinate clauses, which tend to have the same construction and length (measured by number of words or syllables)" (Rowe 1997, 137). The famous phrase "Veni vidi vici" (I came, I saw, I conquered), attributed to Julius Caesar, is an example of παρίσωσις. Another example of a word with rhetorical overtone is the verb πλέκειν, "to weave", employed by Diophantus in a phrase of the introduction, the literal translation of which would be "most arithmetical problems are woven". The related substantive πλοκή, "plot", is very common in ancient rhetorical writings. The word ἀντίδοσις, employed in three problems (I.22–23 and II.18), is another example. On this word, see the note on Problem I.22. Also, the word εὕρεσις, "invention", the very first non-grammatical word of the introduction. For this word, too, see the corresponding note.

Philippus of Opus (fl. ca. 350 BCE), Speusippus (ca. 407–339 BCE), Hypsicles (fl. ca. 170 BCE), Theon of Smyrna (fl. ca. 115–40 CE), and Nicomachus of Gerasa (fl. ca. 100 CE).[26] Diophantus's treatise is a work typical of Greek demonstrative mathematics. Rather than solve arithmetical problems by converting them to equations, as is the case with *Arithmetica*, this work consists of propositions and their proofs. In its present form it comprises a preface and five propositions. The first four propositions contain preliminary material on arithmetical progressions, and the main result, in modern notation, is the relation

$$8 \cdot P_n^a \cdot (a-2) + (a-4)^2 = \left[(a-2) \cdot (2n-1) + 2 \right]^2,$$

where a indicates the number of vertices and n the rank (called by Diophantus the "side") of a polygonal number P (cf. Heath 1921, II, 515; Vogel 1971, 116; Acerbi 2011, 550). From this relation Diophantus derives two procedures, one the inverse of the other, namely, given the value of a how to calculate the rank n of a polygonal number P, and vice versa (Acerbi 2011, 550). The treatise breaks off in treating a proposition which investigates how many ways a number P can be a polygonal number. Restorations of the missing part of that proposition have been proposed by Wertheim (1897), Heath (1910), and Acerbi (2011; in Diophantus 2011, 147–51).

Several suggestions have been made regarding *On Polygonal Numbers*. Tannery proposed that the last, incomplete proposition is not genuine but was added by some later editor (Diophantus 1893–95, II, 477 n. 1), while others have even doubted that Diophantus is the author of the work (Bossut 1802, 26; Allard 1984, 317 n. 2; Rashed and Houzel 2013, 4). The transmission of the treatise together with the *Arithmetica*,[27] and the didactic concern that both works display,[28] are hardly compatible with the latter opinion. Still others have proposed that the treatise on polygonal numbers was one of the lost books of the *Arithmetica*. But this, too, cannot be supported. As Tannery wrote, its characteristics "comme fonds, comme form et comme composition" speak against such a proposition (1912d, 86).

The work of Diophantus for which he is best known, and which has had an enormous impact on the subsequent development of mathematics, is the *Arithmetica*. Originally this work consisted of 13 books, as we are told by Diophantus himself at the very end of the introduction, where he writes of the problems he

---

26 Cf. Heath (1921, II, 514–16), Ver Eecke in (Diophantus 1959, xlv–l), Acerbi in (Diophantus 2011, 39–40).

27 In the four Byzantine manuscripts that contain the complete text of the Greek *Arithmetica*, and from which all other manuscripts derive (see Section 1.3), the treatise *On polygonal numbers* immediately follows the text of the *Arithmetica*; see Acerbi in (Diophantus 2011, 118).

28 The didactic aspect of the *Arithmetica* is discussed in this introduction (see Section 5). The didactic concern of the treatise *On polygonal numbers* is manifested in the phrase "We will describe ⟨this⟩ in a more didactic way, even for those who want what they are looking for through procedures" (Diophantus 1893–95, I, 474.10–11; Diophantus 2011, 197.16–17); our English translation of this passage is aligned with Acerbi's Italian translation in (Diophantus 2011, 211).

is about to present: "As for their complete treatment, it has been done in thirteen books". Of these, six books are preserved in Greek and four more books, discovered only in 1968,[29] survive in a medieval Arabic translation. During the time that only the six Greek books were known, much speculation circulated about the contents of the missing parts, as well as about the place of the lost books (at the time, seven) within the whole work. After the discovery of the Arabic translation, many of the opinions that had been expressed on these matters in the previous two centuries were rendered partly or wholly obsolete.[30] Today, the state of the art of the issues bearing on the status, the structure, and the scope of the work as a whole can be described as follows.

From Arabic sources (Ibn al-Nadīm, Ibn al-Qiftī, Ibn Abī Uṣaybī'a, Barhebraeus) it is attested that the *Arithmetica* were translated into Arabic; that the translator was Qusṭā ibn Lūqā (d. ca. 910), a Christian polymath of Greek origin;[31] that at least three Arabic authors wrote commentaries on it;[32] that the treatise of Diophantus was designated in Arabic by several appellations: "Treatise on algebra" (where "algebra" is either *al-jabr wa-l-muqābala* or simply *al-jabr*), "Treatise on arithmetical problems", "Treatise on problems of algebra (*al-jabr*)", and "The art of algebra (*al-jabr*)";[33] that the translation comprised at least the first seven books of the *Arithmetica*; as for the date of the translation, it is placed, according to Sesiano (1982, 3), "around or after the middle of the ninth century", while for Rashed (Diophantus 1984, 3, xxii) it is located in the period between 860 and 890 CE. Further, it is generally admitted that the preserved Arabic text of books IV–VII stems from Qusṭā's translation, this being explicitly stated in the incipit of the unique manuscript which carries the text and which is kept in the library Astān Quds, in Mashhad, Iran.[34] However, Sesiano leaves room for doubt as to whether this manuscript, which was copied in 595H/1198 CE,[35] is a direct copy of Qusṭā's translation or a copy of an apograph, which, if nothing else, was interspersed with extraneous material deriving from annotations by previous readers or copyists (1982, 29–35).[36] More than that, there is reason to believe that the Arabic text we possess is a recension of Qusṭā's translation made by an unknown editor. The additions it has, aiming to complete or clarify the text, are indicative of this (Sesiano 1982, 29–33). The most striking indication that it is a recension is the appearance of a new preface to Book IV, and of another short preface to Book VII. That these prefaces were not included in the text

---

29 See Sesiano (2004, 262 n. 4). For the public controversy which erupted in the 1980s about the discovery of the Arabic manuscript, we refer the reader to Section 1.4. See Allard and Rashed (1984); Toomer (1985); Hogendijk (1985); Chemla, Morelon, and Allard (1986); Allard (1987); Jaouiche (1987, 1990); Morelon (1990); and Rashed in (Diophantus 1984).

30 (Sesiano 1982, 76). The several views are summarized in (Heath 1910).

31 Sesiano (1982, 8–9) and Rashed in (Diophantus 1984, 3, xff; xvii).

32 Sesiano (1982, 8ff) and Rashed in (Diophantus 1984, 3, x–xiii).

33 Sesiano (1982, 13) and Rashed in (Diophantus 1984, 3, xiv–xvi).

34 Sesiano (1982, 21–22) and Rashed in (Diophantus 1984, 3, lxii).

35 Sesiano (1982, 22) and Rashed in (Diophantus 1984, 3, lxviii).

36 On the contrary, for Rashed (Diophantus 1984, 3, lxx) "la traduction dont nous disposons ne comporte aucun élément qui ne se remonte à Ibn Lūqā et à son époque. Rien n'autorise à déceler des gloses de lecteurs tardifs incorporées dans le manuscrit de Meshed".

of Diophantus is clear from the fact that the preface to Book IV repeats the rules for operating on the species that are given in the introduction to the Greek text.

More importantly, Sesiano (1982, 68–75) conjectures that the Arabic translation stems not directly from Diophantus's treatise but from a "Major Commentary" on it, most likely the commentary that the Byzantine encyclopedia of *Souda* ascribes to Hypatia, the renowned teacher of philosophy and daughter of Theon of Alexandria.[37] For, as Sesiano writes (1982, 71),

> the Arabic text, unlike the extant Greek text, possesses all the characteristics of a commentary made at about the time of the decline of Greek mathematics. Thus, the idea that our text might be (part of) Hypatia's commentary arises quite naturally.

The same conjecture was expressed, before Sesiano, by I. G. Bashmakova, E. I. Slavutin, and B. A. Rozenfeld.[38] Sesiano writes that the Major Commentary "consists, in fact, in a *rewriting* of the entire text" (1982, 68; his emphasis). But a "rewriting of the entire text" is not a characteristic of a commentary. As Acerbi points out, the Arabic version is in fact "a new recension of the treatise, not a commentary, and normally a commentary does not interfere with the text in such a way" (2008, 436; in Diophantus 2011, 116–17).

In the Arabic text the technical algebraic terms of Diophantus have been translated into the corresponding Arabic terms, where "number" (ἀριθμός, *arithmos*) is now "thing" (*shay'*) and "power" (δύναμις, *dynamis*) is "*māl*".[39] Moreover, the Arabic terms "restore" (from *al-jabr*) and "confront" (from *al-Muqābala*) are employed in place of the Greek clauses "let the lacking be added in common" and "let likes be subtracted from likes", respectively. Also, numbers are expressed in full words, as was usual in Arabic algebra (Sesiano 1982, 37), while none of the stylistic forms employed in the Greek manuscripts for writing fractions[40] is adopted in the Arabic translation. Clearly all these changes are due to Qusṭā ibn Lūqā, who quite naturally decided to translate the text of Diophantus, a text in itself of algebraic character, into the language of the Arabic algebra of his time. Furthermore, the possibility that the introductory and concluding phrases in the beginning and the end of each book are also due to Qusṭā ibn Lūqā, should not be ruled out.

The problems in the Arabic text are generally more verbose than those in the Greek text. Contrary to the problems preserved in Greek, the Arabic problems include the following: (i) most often after the setting-out (instantiation) the enunciation of the instantiated problem is stated; (ii) in the final stage of the solution,

---

37 See *Suidae lexicon, s.v.* Ὑπατία (Adler 1935, 644–46). The entry ascribes to Hypatia "a commentary (ὑπόμνημα) on Diophantus". Cf. the critical discussion of *Souda*'s entry by Tannery (1912c).

38 See Bashmakova, Slavutin, and Rozenfeld (1978). A concise English version of this article was published in 1981.

39 On the technical vocabulary employed in both Greek and Arabic books of the *Arithmetica*, see Section 4.2.1.

40 For the ways fractions are presented in *Arithmetica*, see Section 4.2.7. Cf. (Heath 1910, 44–47).

just before the calculations of the numerical values, the assignments through which names were given to the sought-after numbers are restated; (iii) the calculations of the numerical values of the sought-after numbers are pursued in every detail; (iv) while in the Greek text most often the proof is merely announced with clauses like "and the proof is obvious", or "and they fulfill the problem", in the Arabic text all calculations are performed with great care; and (v) quite often the final statement (conclusion) is appended. These additions, together with some other minor interventions (references to previous problems, statement of banal identities or theorems used in the course of the solutions, etc.)[41] make the problems of the Arabic text more prolix than the Greek problems.[42]

The manuscript of the Arabic translation is not lacking any leaves. It contains both a title page at the beginning and an explicit at the end announcing the termination of the copyist's work. The title to the work is given as "The fourth book of the work of Diophantus", which of course should not be the title of even this abridgement. Most likely a copyist with an incomplete manuscript mistook the heading for Book IV for the title of the whole treatise. We know that the four books are complete because at the end of each one is a statement "End of the fourth book . . .", "End of the fifth book . . .", etc., and which give the number of problems in that book. The explicit begins "The work is finished", and the copyist then thanks God and gives the date of completion.

These four books are not the only books of Diophantus's treatise that were translated. Our main evidence comes from the [*Book of*] *al-Fakhrī on the Art of Algebra* (*al-Fakhrī fī ṣinā'a al-jabr wa-l-muqābala*, henceforth *al-Fakhrī*), written in the early eleventh century CE by the Persian mathematician Fakhr al-Dīn Abū Bakr Muḥammad b. al-Ḥasan (or al-Ḥusayn) al-Karajī (or al-Karkhī), henceforth al-Karajī. Among the 255 worked-out problems in this book are about 100 problems taken, usually in order, from the first three Greek books and Book IV of the Arabic. Also, there are some other Arabic authors who reproduce some of Diophantus's problems from Book I, including 'Alī al-Sulamī (tenth century) and al-Samaw'al (ca. 1180).

Thus, at least seven books were translated. It is not true, as Rashed and Sesiano have both claimed, that Qusṭā translated no more than these first seven books.[43] The fact that our single manuscript stops at the end of Book VII is not an indication of the end of Qusṭā's work since the manuscript also omits the first three books, which are known to have been translated. More important, we have the evidence from al-Karajī regarding the solutions to three-term equations, which are not explained in any of the extant books. In his introduction, Diophantus promised to explain the solution to equations in which "two species are left equal to one", and in the Greek

---

41 See Sesiano (1982, 69).

42 For the parts of a Diophantine problem, see Section 4.1.

43 Sesiano in (1982, 9) and Rashed in (Diophantus 1984, 3, xv n. 17). However, the possibility that the Arabic translation covered more books is not excluded by Sesiano, as we gather from the following statement: "My assertion in the *Méthodes . . . chez Abū Kāmil*, p. 90, that Qusṭā's translation 'n'a très certainement jamais contenu plus que les livres I à VII' may thus be too absolute" (Sesiano 1982, 9 n. 21).

Books IV and VI he solves such equations. In his *al-Fakhrī* al-Karajī derives the rules for such equations by what he calls the "method of Diophantus" that must have come from a lost portion of the *Arithmetica*.[44] Similarly, al-Samaw'al, who wrote a lost commentary on the *Arithmetica*, attributes to Diophantus a proposition that is not found in the extant books (see Section 3.3.9). There is no foundation, then, for arguing that Qusṭā translated only some of the 13 books.

That the four Arabic books are indeed Books IV–VII of Diophantus's original composition is borne out through internal references, and from al-Karajī's copying in order from the first four books. The problem of the placement of the last three Greek books remains unresolved. Diophantus's lost explanation of solutions to three-term equations must have been given after the Arabic Book VII, and either before the fourth Greek book or in a lost prolegomena to that book. We can offer no further suggestions as to the nature of the contents of the lost books of the *Arithmetica*.[45]

Two other works which are attributed to Diophantus do not survive. The first is a collection of propositions under the title *Porisms*. Diophantus makes explicit mention of it in three problems of his Greek Book V (3, 5, and 16), and he quotes the corresponding theorems (in fact, two theorems and a problem) referring for their proofs "in the *Porisms*".[46] Tannery was of the opinion that such theorems formed a part of the text of the *Arithmetica* itself, as lemmas to the corresponding problems, and that they have been lost in the course of the transmission of the text. For the three porisms at issue he thinks that they were attached to Problems III.10, III.15, and IV.1–2 (Diophantus 1893–95, II, xix). Most historians, however, following Hultsch (1903, col. 1071) and Heath (1910, 8–10, 99–101), assume that the *Porisms* were a separate work. Indeed, by assuming – as seems more plausible – that the theorems were accompanied by the respective proofs, then such material could be part not of the original text of the *Arithmetica* but of an independent work, *mutatis mutandis*, like the homonymous work of Euclid,[47] with an orientation more theoretical than the *Arithmetica*.

The other work attributed to Diophantus, *Moriastica* (*On Parts*), is mentioned only once, in a Neoplatonic scholium to Iamblichus' treatise *To Nicomachus's arithmetic* (i.e., introduction to Nicomachus's arithmetic). The scholium, which has been transmitted, together with three other anonymous Neoplatonic scholia through the fourteenth-century codex Plut. 86.3 of the Biblioteca Medicea Laurenziana (Florence), is printed in Diophantus (1893–95, II, 72). Commenting on the phrase of Iamblichus "Some of the Pythagoreans said that 'the unit is the borderline between number and parts'" (Iamblichus 1975, 11.9–11), the commentator writes: "So Diophantus (writes) in *On Parts*; for parts (involve) progress in diminution carried to infinity".[48] Tannery conjectures that *Moriastica* was not

---

44 See Oaks (2018b).
45 For another suggestion, see Sesiano (2004).
46 See Problems V^G.3, 5, and 16.
47 Cf. (Proclus 1873, 212.13, 302.11–13; Pappus 1986, 85.16–17, 94–105, 260–295).
48 Translation slightly adapted from Klein (1968, 137).

a separate work by Diophantus but a collection of scholia to the corresponding part in the introduction where Diophantus defines the reciprocal powers of the unknown (Diophantus 1893–95, II, 72 n. 2). And Heath (1910, 3–4) thinks it is possible that the scholium refers specifically to those definitions. But, as Hultsch points out (1903, col. 1071), the very title, *Moriastica*, indicates that the scholiast refers to an independent work, the subject of which was calculation with fractions. A token of such calculations is preserved in the body of the *Arithmetica*, in Problem 36 of the Greek Book IV, where the addition of the fractions that in modern notation can be written as $\frac{3x}{1x-3}$ and $\frac{4x}{1x-4}$ is explained in detail (288.1–8), while addition of common fractions is implied also in Problem III.1. After all, the composition of a work predating the *Arithmetica* on fractions (teaching operations with fractions) fits well with the solutions Diophantus gives in the *Arithmetica*.

Diophantus is sometimes mentioned as the author of texts which are in fact compilations written in later periods. Thus, Tannery published in the second volume of his edition three texts under the heading "Diophantus pseudepigraphus" (Diophantus 1893–95, II, 1–31). The first, titled "From the Arithmetic of Diophantus", is an excerpt from the introductory paragraph of a text exposing two methods of extracting the square root of a square number. The fact that the first method requires writing the number according to the decimal place-value system, which did not reach the Greek speaking world before the Byzantine period, shows that the text has nothing to do with Diophantus.[49] The second text is a fragment titled "Useful methods for the multiplications of the degrees in the astronomical table, which preserve, more than the other methods, all the accuracy".[50] We are now in position to know that this text belongs to a late antique compilation which is known as "Prolegomena to the *Almagest*", constituting part 5 of its 17 parts (cf. Acerbi, Vinel, and Vitrac 2010, 57). The third text, beginning with a fragment bearing the heading "Diophantus's plane measurement", is extracted from Byzantine compilations transmitted under the name of Hero.[51]

Recently a question has been raised about whether a lost treatise on *Arithmetic Elements* (Ἀριθμητικὴ στοιχείωσις) should be ascribed to Diophantus, on the basis of a reference to such a title made in a Neoplatonic scholium to Iamblichus's *To Nicomachus's arithmetic*, which also refers to Problem I.39 of the *Arithmetica* for details on the harmonic mean.[52] The question was posed by Christianidis (1991), and negative answers to it were given first by Waterhouse (1993), then by Rashed (1994a). In between, Knorr (1993) suggested that such work did exist. It constituted one of a pair of works, *Preliminaries to the Arithmetic Elements* and *Preliminaries to the Geometric Elements*. Knorr proposed that the former is lost and that the latter is extant but is now known by the alternative title *Definitions*,

---

49 The fragment is also printed by Heiberg in Hero (1912, xv.7–17).
50 The text had been edited before Tannery by Charles Henry (1879) as *Opusculum de multiplication et divisione sexagesimalibus Diophanto vel Pappo attribuendum*.
51 Heiberg in Hero (1912, xvii–xxiii) offers a comparison of the text published by Tannery from codex Par. Gr. 2448 with the text carried by the codex Constantinopolitanus Palatii veteris 1.
52 See Section 3.2.2.

which was misattributed to Hero of Alexandria. Knorr's suggestion, which as we have seen is paired with a conclusion about the time period in which Diophantus lived, is highly hypothetical.

1.3 The text of the *Arithmetica* and its history

In this section, we review the extant manuscripts containing Diophantus's *Arithmetica*, first those in Greek, followed by the Arabic manuscript. The manuscript tradition of the Greek books is rich, including 33 complete or fragmentary witnesses, dated from the second half of the eleventh century to the early seventeenth century. The Arabic books have been transmitted by a unique manuscript of the end of the twelfth century. The complete list is shown in Table 1.1.[53]

There is a consensus among scholars who have studied the manuscript tradition of the works of Diophantus, from Tannery (1893–95) to Allard (1980) to Acerbi (2011), that the Greek manuscripts fall into two main classes, conventionally called "Planudean" and "non-Planudean". The Planudean class, which includes far more witnesses, stems from Marcianus gr. 308, the most direct descendant of Mediolanensis Ambrosianus Et 157 sup., the latter being an incomplete autograph of Maximus Planudes, written in 1292 or 1293 and containing only 23 sheets. The most ancient manuscript of the non-Planudean class, on which all other non-Planudean manuscripts depend, is Matritensis Bibl. Nat. 4678, previously believed to be of the thirteenth century but now dated in the second half of the eleventh century (Pérez Martín 2006).

According to Tannery (Diophantus 1893–95, II, xviii), the whole tradition of the Greek text descends from a progenitor, now lost, written in the eighth or ninth century. This was the era of Leo the Mathematician, during which, to quote Kurt Vogel (1888–1985), "most of the manuscripts forming the vital link in the line of descent from antiquity were written" (Vogel 1967). This hypothesis, although quite plausible, lacks any positive evidence to support it (Pérez Martín 2006, 451 n. 59). The studies of Allard (1982–83), and later of Acerbi (Diophantus 2011), have improved the picture, in some cases adding details and in others modifying Tannery's conclusions. We now have a more accurate picture of the tradition, which, as Acerbi points out, is still "only partially under control" (in Diophantus 2011, 113). Both scholars arrived at the same basic conclusion. According to Acerbi,

> the rich tradition of the *Arithmetica* . . . can readily be reduced to no more than four independent manuscripts: Matrit. 4678, Vat. gr. 191 and 304, and Marc. gr. 308, the latter in fact containing a recension made by the renowned scholar Maximus Planudes (d. 1305).
>
> (2013a, 712)

---

53 The list of the Greek manuscripts has been compiled from Tannery (in Diophantus 1893–95, II, xxii–xxxiv), Allard (1982–83, 1983a), Acerbi (in Diophantus 2011, 113–21), Vitrac (2018), and from the electronic database Pinakes available at http://pinakes.irht.cnrs.fr/ (accessed September 5, 2022).

Table 1.1 Inventory of the manuscripts of the *Arithmetica*.

GREEK MANUSCRIPTS	
1	Matrit. Bibl. Nat. 4678 (ff. 58r–130v)	XI (2nd half)
	http://bdh.bne.es/bnesearch/detalle/bdh0000099980	siglum **A**
2	Ambros. Et 157 sup. (ff. 13, 14, 8, 18, 20, 15, 9, 16, 17, 19)*	1292/1293
3	Vat. gr. 191 (ff. 360r–390r)	1296–1298
	https://digi.vatlib.it/view/MSS_Vat.gr.191	siglum **V**
4	Vat. gr. 304 (ff. 77r–118r)	XIV (1st half)
	https://digi.vatlib.it/view/MSS_Vat.gr.304	siglum **T**
5	Marc. gr. 308 (ff. 50v–263r)	XIII (end) siglum **B**
6	Fir. Laur. Acquisti e Doni 163 + 164 (163: ff. 2r–80v; 164: ff. 1–66)	XVI
7	Guelferbytanus Herzog Aug. Bibl. Gudianus gr. 1 (ff. 1–80v)	1560–1565
8	Krakov. Bibl. Jagiel. 544 (ff. 1–203r)	XVI
9	Leiden. Bibl. Rijksuniv. BPG 74 G (ff. 48–49r)*	XVI (1st half)
	https://digitalcollections.universiteitleiden.nl/view/item/2030897#page/1/mode/1up	
10	Ambros. A 91 sup. (ff. 1–122r)	XV
11	Ambros. E 5 inf. (ff. 1–98r)	XVI (1st half)
12	Ambros. Q 121 sup. (ff. 44–69)*	XVI (1st half)
13	Neap. Borb. gr. 275 (ff. 1–75r)	XVI (end)
14	Oxon. Bodl. Lib. Barocci 166 (ff. 244r–248r)*	ca. 1570 (or shortly
	https://digital.bodleian.ox.ac.uk/objects/60e08c28-9bde-4916-bd3a-ba4ef57d5cd0/	after)
15	Oxon. Bodl. Lib. Savile 6 (ff. 1–83v)	1582
16	Paris. gr. 2378 (ff. 1–63r)	XVI
	https://gallica.bnf.fr/ark:/12148/btv1b10723750z	
17	Paris. gr. 2379 (pp. 1–132)	XVI
	https://gallica.bnf.fr/ark:/12148/btv1b10722381s	
18	Paris. gr. 2380 (ff. 1–192v)	XVI (1st half)/
	https://gallica.bnf.fr/ark:/12148/btv1b10723637w	XVII
19	Paris. gr. 2485 (ff. 1–194r)	XVI
	https://gallica.bnf.fr/ark:/12148/btv1b10722263g	
20	Paris. Bibl. Arsenal 8406 (pp. 1–280)	XVI
21	Scorial. gr. P II 3 (ff. 1–235r)	XVI (end)
22	Scorial. gr. P III 18 (ff. 1–148v)	XVI (end)
23	Scorial. gr. T I 11 (ff. 1–143v)	1545
24	Scorial. gr. Ω I 15 (ff. 1–153v)	ca 1555
25	Taurin. Bibl. Naz. Univ. C I 4 (ff. 1–71v)	XVI
26	Upsaliens. UB gr. 53 (ff. 124r–218r)	XVI (end)/XVII
	www.manuscripta.se/ms/100055	(early)
27	Urbin. Bibl.Univ. Fondo dell'Università 102 (ff. 8r–65v)	XVI (mid)
28	Vat. Barb. gr. 267 (ff. 268r–272r)*	XVI
	https://digi.vatlib.it/view/MSS_Barb.gr.267	
29	Vat. gr. 200 (ff. 1–194r)	XV
30	Vat. Palat. gr. 391 (ff. 1–122v)	1571–1579
	https://digi.ub.uni-heidelberg.de/diglit/bav_pal_gr_391/0001/thumbs	
31	Vat. Regin. gr. 128 (ff. 1–77)	1571
32	Vat. Urbin. gr. 74 (ff. 9r–75r)	ca. 1555
	https://digi.vatlib.it/view/MSS_Urb.gr.74	
ARABIC MANUSCRIPT	
33	Codex 295 (Astān Quds library, Mashhad, Iran)	1198

An asterisk (*) denotes an incomplete manuscript.

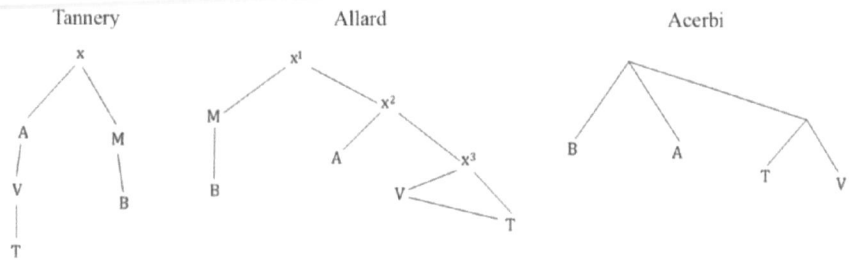

Figure 1.1 Diagrams showing the relationships between the main Greek manuscripts of the *Arithmetica*.

Similarly, in Allard's terms (1983b, 664–65),

> tous les manuscrits connus, complets ou fragmentaires, qui contiennent la partie conservée en grec des *Arithmétiques*, sont issus des manuscrits suivants:
>
> A: Matritensis Bibl. Nat. 4678 (olim N 48) (XIIIe s.)[54]
> V: Vaticanus gr. 191 (XIIIe s.)
> T: Vaticanus gr. 304 (XIVe s.)
> M: Mediolanensis Ambrosianus Et 157 sup. (= gr. 780) (fragmentaire, autographe de Maxime Planude) (XIIIe s.)
> B: Venetus Marcianus gr. 308 (XIIIe s.).

Apart from this basic conclusion shared by all, the stemmata proposed by the three scholars showing the relationships between the basic manuscripts differ in some details, as is seen in Figure 1.1.

A few words about these important manuscripts are in order at this point. The following are gathered from the fundamental studies of Tannery (Diophantus 1893–95, II, xxii–xxxiv),[55] Allard (1982–83), and Acerbi (in Diophantus 2011, 119–21), and occasionally from other sources mentioned in the references. For detailed descriptions of all manuscripts which carry the Greek text of the *Arithmetica* the reader is referred to these studies.

Mediolanensis Ambrosianus Et 157 sup. is the archetype of the Planudean branch of the tradition. It is an autograph of Maximus Planudes copied in 1292 or 1293 (Turyn 1972, I, 78–81, and table 57; Allard 1979). Only 23 folios are extant, bound in the wrong order, 10 of which (ff. 13, 14, 8, 18, 20, 15, 9, 16, 17, 19) contain fragments from the first two books of the *Arithmetica* (Turyn 1972, I, 78–79; Pérez Martín 2006, 435 n. 8), corresponding approximately to half of the two books (Tannery 1912f, 274). The manuscript also contains fragments of Planudes's scholia to these two books, of his *Great calculation according to*

---

54 In fact, the part of the manuscript which carries the text of Diophantus is of the eleventh century.
55 Cf. also the studies of Tannery (1912e, 1912f, 1912i), first published prior to his edition of 1893–95.

the Indians, of the pseudo-Iamblichean *Theologumena arithmetica*, and of a text attributed to Michael Psellus.

Marcianus gr. 308 consists of two parts. The first part, occupying ff. 1–49, was copied in the second half of the fourteenth century, and the older part, occupying ff. 50–284, dates from the late thirteenth century like Ambrosianus Et. 157 sup. It is this older part that gives the *Arithmetica*, written in two columns on sheets 50v–263r, together with the scholia of Planudes to the first two books, written in one column. The older part continues with Diophantus's tract *On Polygonal Numbers* (ff. 263r–272v) and Planudes's fragment on the *Great calculation according to the Indians* (ff. 273r–284r). The more recent, first part contains the two books of the *Circular Theory of Heavens* of Cleomedes, the pseudo-Aristotelian text *On the Universe*, and some astronomical fragments, including an anonymous text on the astrolabe. The manuscript once belonged to Cardinal Bessarion (d. 1472), who donated it to the Library of San Marco in Venice in 1468. It is described in the inventory of that year as "Item Cleomedes et Diophantes arithmeticus, in papyro" (Labowsky 1979, 166; quoted by Acerbi in Diophantus 2011, 114).

Vaticanus gr. 191, composed of vi + 397 (+ 11) sheets, is a true compendium of the exact sciences, comprising works on arithmetic, geometry, astronomy, astrology, geography, and harmonics, penned by 17 different copyists, four of whom were involved with copying the *Arithmetica* (De Biase 2011–12, 64–66). The manuscript consists of three parts, which were assembled into one codex between 1296 and 1298, apparently by a single individual who seems to have been in charge of revising the miscellany. As for the identification of the revisor, several hypotheses have been advanced. According to Alexander Turyn (1964, 89–97), the manuscript was produced by a circle of scribes and scholars close to Maximus Planudes. David Pingree (1983, 532–33) believed that the revisor was Gregory Chioniades (d. ca. 1320), and Inmaculada Pérez Martín (2010) identified him with John Pothos Pediasimus (b. ca. 1250). More recently Emanuele De Biase (2011–12, 2016) put forward the hypothesis that the manuscript was produced by a circle of scholars distinct from that of Planudes, under the supervision of Manuel Bryennius (fl. ca. 1300). The manuscript bears witness, as Alexander Turyn points out (1964, 89), of the blossoming of scientific studies in Byzantium under the Emperor Andronicus II Palaeologus (r. 1282–1328).[56] During the papacy of Paul III (1464–71) the manuscript was acquired by the papal library as a legacy of Cardinal Isidore of Russia (d. 1463). It is described in several inventories, the earliest of which was compiled in 1475 (Devreesse 1965, 42, 60; Allard 1982–83, 69; Acerbi in Diophantus 2011, 115). The *Arithmetica*, in six books, occupies sheets 360r–390r, followed by the fragment *On Polygonal Numbers* (ff. 390r–392v), the whole corpus of Diophantus being written by five hands. The first book of the *Arithmetica* is accompanied by scholia written in the margins, which have been published by Allard (1983b). The manuscript also contains, among other texts, the *Data* of Euclid, the *Commentary* of Eutocius to the *Conica* of Apollonius,

56 Cf. Gaul (2016).

the *Geography* and the *Harmonics* of Ptolemy, two musical treatises attributed to Euclid, the *Sectio Canonis* and the *Introductio Harmonica*, and various minor texts of astronomical, astrological, and musical content.[57]

Vaticanus gr. 304 is made up of v + 256 sheets. According to Acerbi (in Diophantus 2011, 114), the codex in its present form probably derives from the joining of two distinct manuscripts, the second part starting on f. 77. This part, which opens with the *Arithmetica*, entered the papal library during the papacy of the bibliophile Nicholas V (1447–55), and is registered in several inventories, the earliest of which was compiled by Cosmas of Montserrat in 1455 (Devreesse 1965, 36; Allard 1982–83, 70; Acerbi in Diophantus 2011, 114). The *Arithmetica*, in six books, occupy sheets 77r–118r, and is followed by the tract *On Polygonal Numbers* (ff. 118v–121r), the two being copied by one hand. Several scholia are written in the margins of the first book and have been published by Allard (1983b). The manuscript also contains a paraphrase of the Aristotelian *Posterior Analytics* attributed to Theodoros Prodromos (twelfth century) (ff. 1r–24v), two works of Theon of Alexandria – the first two books of his *Commentary to the Almagest* (ff. 25r–76r) and the *Short Commentary* to Ptolemy's *Handy Tables* (ff. 122r–134r) – the *Commentary* of Stephanos of Alexandria to the same work of Ptolemy (ff. 135r–171v), and some astronomical texts, including the treatise *On the Astrolabe* by John Philoponus (ca. 490–ca. 570) (ff. 176r–180v).

Matritensis 4678, housed in the National Library of Madrid, is a very important manuscript for the textual tradition of the *Arithmetica*. It is the oldest witness we possess of the Greek text of Diophantus, and for this reason it served as the main source for Tannery's edition (Sesiano 1982, 18; De Biase 2016, 360).[58] It consists of 145 sheets of two parts, the older of which is dated to the second half of the eleventh century, and a restored part which was added in the Palaeologan era. The older part was written by two hands, the first one dated, according to the recent paleographic study by Inmaculada Pérez Martín, to the years 1060–80. It was possibly copied under the behest and with the support of Michael Psellus, the leading intellectual figure of the time, whose interest on Diophantus is well known (Pérez Martín 2006, 439, 451; see below Section 3.4.1). The *Arithmetica* occupies sheets 58r–130v, all belonging to the older portion of the manuscript, among which only the sheets from 58r to 62r, line 26, are the work of the first copyist. The manuscript also contains the tract *On Polygonal Numbers* (ff. 130v–135v); the *Arithmetical Introduction* of Nicomachus (ff. 4r–57v, whose ff. 4–8 were added between 1260 and 1280); the *Introductio Harmonica* of Cleonides (attributed in the manuscript to Zosimus); the two musical treatises attributed to Euclid, *Introductio Harmonica* and *Sectio Canonis*; a fragment of the *Almagest*, and some incomplete astronomical tables. Traces showing that this manuscript was studied are detected in Byzantine sources. In a letter of Maximus Planudes dated to

---

57 For more details on Vat. Gr. 191, see the doctoral dissertation of Emanuele De Biase (2011–12).

58 Cf. Allard's statement that "l'édition de P. Tannery est basée quasi exclusivement sur le Matritensis Bib. Nat. 1678 et l'édition de Bachet" (1984, 318 n. 4).

ca. 1292/93 and addressed to Theodore Mouzalōn, a high-ranking official who was granted the title of "*prōtovestiarios*", we learn that the latter had asked Planudes for a recently repaired and bound codex of Diophantus, which was kept in the imperial library of his monastery (Pérez Martín 2006, 437).[59] The repaired manuscript has been determined to be the actual Matritensis 4678 (Wendel 1940, 414–17; cf. Pérez Martín 2006). Besides Maximus Planudes, John Chortasmenos (d. before 1439) also studied and annotated this codex. Among his many marginal or interlinear annotations[60] is the famous imprecation written in the bottom of f. 74r (Wilson 1996, 279).[61] After the fall of Constantinople (1453), the codex was taken to Messina by Constantine Lascaris (1434–1501), and having subsequently passed through the hands of several noblemen it was confiscated in the beginning of eighteenth century by Philip V, king of Spain, and finally entered the Royal Library, now Biblioteca Nacional de España.

The four books of the *Arithmetica* which are preserved in Arabic translation are carried by the unique MS Mashhad, Astān Quds 295 of 80 leaves numbered in recent times as pages. The Central Library of Astān Quds Raḍawī, attached to the shrine of Imam Rezā at Mashhad, Iran, acquired the manuscript in 1932. The manuscript was copied in the year 595 of the hegira, corresponding to the year 1198 CE, by two copyists. The first, who began the work, was a certain Muḥammad ibn Abī Bakr ibn Ḥākīr, who completed only the first seven folios. As already mentioned, the manuscript contains some material not belonging to the translation of Qusṭā ibn Lūqā but which was incorporated into the text from marginal or supralineal annotations by earlier copyists (Sesiano 1982, 29–37).

1.4 Editions and translations of Diophantus

Even if several Greek manuscripts of the works of Diophantus could be found in Italian and other European collections since the mid-1400s, it was only in 1575 that a Latin translation was published. This was the work of the German humanist Wilhelm Holzmann (1532–76), who wrote under his Hellenicized name

---

59 The letter is published under the number 67 in Planudes (1890, 81–85).
60 Published partly by Allard (1983b); cf. Pérez Martín's remarks in (2006, 450 n. 57).
61 "Diophantus, may your soul be with Satan because of the difficulty of the other theorems of yours, and in particular of the present theorem". Acerbi (2013b) claimed that the problem whose difficulty provoked the imprecation was not Problem II.8, as Tannery and other scholars after him asserted, but Problem II.7, of which the "exasperated" John Chortasmenos was supposedly unable to understand the phrase "by a given number greater than in ratio", which "is never met elsewhere in the *Arithmetica*" except Problem II.7 (2013b, 387). But the relation described by this phrase also appears in Problem I.3. In contrast, Problem II.8 is the first indeterminate problem that Diophantus solves and the first to make use of an *al-istiqrā'* assignment, which indeed makes it a difficult problem. (For the technique *al-istiqrā'*, see Section 2.7.2 and Appendix 2.) The fact that more than a century before Chortasmenos, another renowned Byzantine scholar, Maximus Planudes, devoted seven full pages in Tannery's edition to comment on Problem II.8 (see Section 3.4.3), in sharp contrast to the only one page he devotes to the Problems II.6–7 together, shows what the truly difficult, and mathematically interesting, problem is.

Xylander,[62] based in a manuscript lent to him by the Hungarian humanist and diplomat Andreas Dudith (1533–89). Xylander not only translated the six Greek books of the *Arithmetica*, the commentary of Maximus Planudes on the first two books, and the tract *On Polygonal Numbers*, but he accompanied them with his own commentary. Xylander's intention of also publishing the Greek text of the *Arithmetica* was never realized, for he died on February 10, 1576, shortly after the Latin translation appeared.

Although the algebra in the *Arithmetica* added little to what Europeans had already been practicing, its publication gave further support for the trend toward regarding algebra as a proper, scientific area of study. The *Arithmetica* also gave algebraists their first real exposure to indeterminate problems about squares and cubes, together with *al-istiqrā'* and the double equality, the two main methods for solving them (see Appendix 2). Guillaume Gosselin singled out these techniques in his *De Arte Magna* (1577), and large numbers of problems were adapted into the books on algebra of several authors of the time, including Rafael Bombelli (1572), Simon Stevin (1585), François Viète (1593d), and Christoph Clavius (1608), the first and last of these consulting manuscripts directly.

The Greek text was first published in 1621, together with a facing Latin translation and a comprehensive commentary, in the meritorious edition of Claude Gaspart Bachet, sieur de Méziriac (1581–1638) (Diophantus 1621). Bachet based his edition on a manuscript which he designated "Codex Regius" and which has been identified by Tannery as the sixteenth-century Par. gr. 2379 (1912f, 291 n. 1). For his work Bachet also made use of other supporting texts he had at his disposal. He consulted a partial copy of the codex Vat. gr. 304, which Jacques Sirmond (1559–1651) transcribed for him, and also some notes taken for him by Claude Saumaise (1588–1653) from Xylander's copy of the manuscript of Dudith (Ver Eecke in Diophantus 1959, lxxx–lxxxi). For his Latin translation Bachet took Xylander as his starting point, and to make sense of the text he also made use of Bombelli. Besides the text of Diophantus, Bachet's edition adds after the last problem of Greek Book V 45 arithmetical epigrams of the Palatine Anthology, also published for the first time, on the basis of a copy made for him by Saumaise (Ver Eecke in Diophantus 1959, lxxxii). All of this material is augmented with Bachet's personal notes bearing on both mathematical and textual matters, among which is a conjectural reconstruction of the lost *Porisms* ("In Diophantum Porismatum") in three books, which contain 24, 21, and 19 propositions, respectively.

It is Bachet's edition that was studied by Pierre de Fermat (1607–65). Fermat inscribed his famous notes in his personal copy of the book, and after Fermat's death his son Clément-Samuel (1634–97) collected these notes and other results of Fermat's investigations gathered from his correspondence with the

---

62 In Greek Ξύλανδρος (*Xylandros*), from ξύλον = Holz, and ἀνήρ = mann. Turning their names into Greek was not uncommon for the Hellenists of the period. Thus, Johannes Reuchlin (1455–1552) turned his name into Καπνίων (*Kapniōn*), from καπνός = Rauch. Also, the great theologian Phillipp Schwarzerd (1497–1560) is better known with the Greek equivalent of his name Μελάγχθων (*Melanchthōn*), meaning "black earth". Cf. (Pfeiffer 1976, 87–88).

mathematician Jacques de Billy (1602–79), and to this material he added a reprint of Bachet's edition, all published in a single volume in 1670 (Diophantus 1670). The value of this edition lies only with the additions due to Fermat. In particular, the reprinted Greek text is much inferior to that of 1621 (Heath 1910, 29).

Between 1670 and the 1890s, when Paul Tannery published his critical edition with a Latin translation of the Greek text of the *Arithmetica*, translations or adaptations of the work appeared in several languages. In 1585 the Dutch-Flemish mathematician and engineer Simon Stevin (1548–1620) included in his *L'Arithmétique* a French version of the first four books of the *Arithmetica*, based on Xylander's translation. This work was reprinted in 1625 by the Dutch mathematician Albert Girard (1595–1632), who added Books V and VI to the four published by Stevin. In 1822 the German philologist Johann Otto Leopold Schulz (1782–1849) published a commented German translation, based on the reprint of Bachet's edition of 1670. In it he included the German translation of the tract *On Polygonal Numbers* which had previously been published in 1810 by Friedrich Theodor Poselger (1771–1838). A new German translation of Diophantus's works was published by Gustav Wertheim (1843–1902) in 1890. This work also includes translations of Fermat's notes, 46 arithmetical epigrams of the Palatine Anthology, and the cattle problem attributed to Archimedes. Finally, in 1885 Thomas Little Heath (1861–1940) published, just before Tannery's critical edition was issued, a modern transcription of the solutions of the problems of *Arithmetica*. A new revised and augmented edition of this work, which appeared in 1910, has proven highly influential for most of the past century.

In 1893–95 Tannery published his critical edition in two volumes, which remains the standard edition of the works of Diophantus. The first volume contains the Greek text with a new Latin translation, and the second volume gives a collection of late antique and Byzantine testimonies, scholia, and other secondary texts related to Diophantus, as well as 44 arithmetical epigrams from the Palatine Anthology with their scholia. Several translations appeared during the twentieth century based on Tannery's edition. In 1926 a literal French translation was published by the Belgian mathematician, engineer, and historian of mathematics Paul Ver Eecke (1867–1959). A new German translation by Arthur Czwalina (1884–1964) appeared in 1952. A Modern Greek translation, together with the ancient text of Tannery's edition, was published in 1963 by Evangelos Stamatis (1898–1990). A Russian translation of the works of Diophantus by Ivan Nicolaevitch Vesselovski (1892–1977), accompanied with a rich mathematical and historical commentary by Isabella Grigoryevna Bashmakova (1921–2005), was published in 1974. The most recent translation of Diophantus is in Spanish, published in 2007. It contains the Greek and Arabic books of the *Arithmetica* (Diophantus 2007).

In 1980 a new edition with a new French translation of the Greek books of the *Arithmetica* was given by André Allard (1937–2014) in his doctoral dissertation at the Université Catholique de Louvain (Allard 1980). Allard's investigation was principally motivated by the discovery of the four Arabic books of the *Arithmetica*, which, as he believed, had made necessary a reexamination of the Greek books with a view to preparing a new critical edition of Diophantus, which could only be possible if both the Greek and Arabic manuscripts were taken into account. The greatest merit of Allard's edition, which has never been published, lies in the critical apparatus which is richer than that of Tannery.

Although the Astān Quds Raḍawī Library acquired their manuscript of the Arabic translation in 1932, it remained unknown to historians of mathematics until 1968, when it was discovered by Fuat Sezgin (1924–2018) during a trip to Mashhad.[63] The library catalog was published soon after.[64] Jacques Sesiano produced an edition of the text with English translation and commentary in his Brown University PhD dissertation in 1975. This was later published, with minor modifications, in Sesiano (1982). Roshdi Rashed published the Arabic text in Cairo in 1975 (Diophantus 1975), and later, with corrections together with a French translation and commentary (written in conjunction with the mathematician Gilles Lachaud (1946–2018)), in Diophantus (1984).

Finally, the following piece of information is provided in an obituary for Abigail Lousada (1763–1833), published in *The Gentlemen's Magazine* (April 1833, 377–78):

> She was endowed with superior talents; and her favourite studies were History, ancient and modern, Natural History, and Mathematics. When young she compiled a chronological history from the Creation to the American War, with maps; and among her papers is a translation by her of Diophantus, which has been consigned at her own request to the care of an eminent Mathematician, and, after a proper examination, it is expected will be made public. Nothing of hers is as yet extant, except a few papers in a Mathematical miscellany.

This translation was never published and probably has not survived.

---

63 Sezgin wrote to Khalil Jaouiche (1925–2002) on July 17, 1986:

> 1. J'ai découvert le manuscrit en question parmi d'autres livres importants lors de ma première visite à la bibliothéque de Meshed en 1968. Comme ce voyage a été financé par le *Deutsch Forschungsgemeinschaft*, j'ai mentionné ce fait dans mon rapport addressé à cette organisation.
>
> 2. J'ai proposé à M. Jensen, un danois, qui a fait, si je me souviens bien, des recherches en 1969 dans notre institut l'édition de ce manuscrit comme thèse. Lorsqu'il n'avait pas pu obtenir ce microfilm, il en a pu recevoir un grâce à mes efforts en 1970.
>
> Vous pouvez conclure de ces faits, que ce n'est pas moi qu'a appris par M. Rashed, en 1973, de l'existence de ce manuscrit, mais que la verité est bien le contraire.

A facsimile of the letter is included in the lecture given by Ahmed Djebbar at "The First International Prof. Dr. Fuat Sezgin History of Science in Islam Symposium" held in Istanbul in 2019. That the manuscript was discovered by Sezgin in 1968, and not by Rashed in 1971 (Diophantus 1984, 3, lx), is confirmed by Jan Hogendijk (1985, 82), Ivor Bulmer-Thomas (1985, 256), Gerald Toomer (1985, 240), and Khalil Jaouiche (1987, 308).

64 The catalog was officially published in 1973 (Gulchīn-i Maʿānī 1973), though Toomer had seen a draft in 1972 (Toomer 1985, 237).

2 Numbers, problem solving, and algebra

2.1 Introduction

If we trace the history of algebra backwards in time through its texts, we pass from modern algebra to the works of Descartes and Viète, then through sixteenth-century European books to Luca Pacioli's 1494 *Summa*, back through two centuries of the writings of medieval Italian abacus masters, then to books in medieval Latin (of which Fibonacci's *Liber Abaci* of 1228 is the best known), and from there across the Pyrenees and the Mediterranean to the Arabic-speaking world. The trail of extant texts eventually comes to an end with the *Book of Algebra* (*Kitāb al-jabr wa-l-muqābala*) of Muḥammad ibn Mūsā al-Khwārazmī and a fragment of a book of the same title by Ibn Turk, both composed in Baghdad in the early ninth century CE. The continuity of this tradition is evident even in the name given to the art, which was transliterated from the Arabic *al-jabr* to the Latin and Italian *algebra* (with many variations), to the modern algebra/algèbre/álgebra etc. that we employ today.

But the algebra practiced in the ninth century is not the same as algebra today. Through a series of studies one of us has shown that medieval Arabic, Latin, and Italian algebra, as well as the algebra of sixteenth-century Europe, operated on a fundamentally different conceptual basis from the algebra initiated by François Viète in 1591 and which saw its first full expression in Descartes's *La Geometrie* of 1637.[1] Evidence for this conceptual difference is found in the ways that algebraists before ca. 1600 worked out problems. They structured their solutions differently, they worded certain operations in ways that make no sense or seem convoluted in modern notation, they shunned irrational coefficients, and the various notations from Ibn al-Yāsamīn to Stevin and Clavius show deliberate differences that do not agree with our Cartesian notation. These anomalies, taken together, are testimony to a different way of conceiving of monomials, polynomials, and equations. We call this pre-Vietan algebra "premodern algebra" to distinguish it from modern elementary algebra.

---

1 Premodern algebra in medieval Arabic, Latin, and Italian is explained in Oaks (2009, 2010a, 2010b, 2012a, 2017, 2021).

DOI: 10.4324/9781315171470-3

The question then arises regarding the mathematical practices of still earlier times: Do any of them conform to either the premodern or modern conceptual foundation for algebra? A look into the method exhibited in the *Brāhma-sputa-siddhānta* of Brahmagupta and other works from India (seventh century CE and later) suggests that it is of a different nature than Arabic and later algebra. It certainly warrants being called algebra, in that unknowns are given names, operations are performed on these names, and equations are set up and solved.[2] In some respects, it is more sophisticated than what we see later in Arabic authors, and it seems to have had no influence on Arabic algebra. These texts have not yet been examined on a conceptual level, so we cannot say how closely Indian algebra matches our premodern algebra, or perhaps even modern elementary algebra.

Babylonian "algebra" – we preserve Høyrup's (2002, 1) quotation marks – is fundamentally different from our premodern algebra. Mesopotamian cuneiform tablets represent known and unknown numerical values with lines and rectangles that are then manipulated, but there is no notion of an equation that can be simplified to a standard form. If anything, Babylonian "algebra" can be viewed as a precursor to the premodern algebra we describe later.

The *Arithmetica* of Diophantus, on the other hand, shows the same kind of nomenclature for the powers of the unknown, the same underlying concepts of monomial, polynomial, and equation, and the same basic procedure for setting up, simplifying, and solving equations, all of which differ from those of modern elementary algebra.[3] Even with hardly any trace of a textual link between Diophantus's book and Arabic algebra, the agreement between their concepts and methods leaves no doubt that there is some historical connection between them. Thus, when we speak of "premodern algebra" we include also Diophantus.

It is the task of the present chapter to describe this premodern algebra and to explain how it differs from modern elementary algebra. We draw most of our examples from Diophantus and Arabic algebra, with occasional references to the algebra of medieval and Renaissance Europe.

2.2 The arithmetical foundation of algebra

The knowns and unknowns in premodern algebra are numbers, and the concepts of polynomial and equation we briefly alluded to earlier derive from the ways that numbers were conceived and represented in arithmetic. To put premodern algebra in its proper context, then, we must first describe the premodern numbers to which they refer. There are three aspects to numbers in Greek, Arabic, and ca. pre-1600 Europe that we cover in this section: first, that numbers admit multiplicity; second, that numbers come in different kinds, or species; and third, that numbers are often expressed as aggregations or differences of two or more numbers. All three

---

2 See Plofker (2009, 191–96).

3 We explain how Diophantus's method conforms to Arabic and medieval European algebra in Christianidis and Oaks (2013). See also Christianidis (2007, 2018a, 2018b).

of these aspects were carried over to algebraic expressions and form the basis for the premodern understanding of those expressions.

2.2.1 Multiplicity, species, and aggregations

At least as far back as the time of Plato a distinction was made in Greece between ἀριθμητική (*arithmētikē*), "arithmetic", and λογιστική (*logistikē*), "logistic".[4] Arithmetic dealt with what we call elementary number theory, such as the even and the odd and their subspecies, polygonal numbers, perfect and amicable numbers, etc. Our main sources for this arithmetic are Books VII–IX of Euclid's *Elements*, Theon of Smyrna's *On Mathematics Useful for the Understanding of Plato*, Nicomachus of Gerasa's *Arithmetical Introduction*, and a number of late antique works composed in the tradition of Nicomachus.

Because the unit in Greek arithmetic was regarded as being indivisible,[5] numbers were restricted to positive integers. Euclid defined the unit as "that by virtue of which each of the things that exist is called one", and he defined number as "a multitude composed of units" (Heath 1926, II, 277). One consequence of these definitions is that numbers are not unique. As Ian Mueller observed, "In Greek arithmetic there are indefinitely many units and indefinitely many ways of combining them into multitudes. Clearly, then, there is no unique 2 or 3; any pair of units is a 2, for example" (Mueller 1981, 59). And because numbers are multitudes, they must be positive. Zero and negative numbers would have made no sense in Greek arithmetic.

Greek logistic is concerned with calculations on numbers: their addition, subtraction, multiplication, division, and root extraction. We learn from Plato's dialogues that by his time a distinction was made between theoretical and practical logistic.[6] Practical logistic dealt with numbered material things, such as cows and apples, as well as measures in particular units of weight, length, time, etc. Although one would not likely wish to speak of two-thirds of a cow, fractions were a necessary part of calculations in units that naturally admit continuous measure. For example, a papyrus from ca. 300 BCE records the number of daytime and nighttime hours, one example being "The night is 13 12′ 45′ hours, the day 10 3″ 5′ 30′ 90‴" (Fowler 1999, 230).[7] These are expressed with unit fractions, the latter number being the sum of $10, \frac{2}{3}, \frac{1}{5}, \frac{1}{30},$ and $\frac{1}{90}$, which we

---

4 See, for example, Heath (1921, I, 13–16), [Bulmer-]Thomas (1939, 3ff), and Klein (1968). For the classifications of mathematical sciences in antiquity, see Vitrac (2005a).

5 See, for example, Plato's *Republic* 525e, *Parmenides* 143a, *Sophist* 245a. Cf. the statement of Eutocius, in his commentary to Archimedes's *On the Sphere and the Cylinder*: "in the case of superparticulars (and) superpartients, it is no longer possible for the quantity to be taken with the unit remaining undivided; so that in these cases the unit must be divided – which, even if this does not belong to what is proper in arithmetic, yet it does belong to what is proper in logistic" (Netz's translation [2004, 313], slightly adapted).

6 For a discussion, see Klein (1968).

7 See Fowler and Turner (1983) for a facsimile of the papyrus.

would write as $10\frac{26}{49}$. Another example, a potsherd from Egypt from the time of Augustus (30 BCE–14 CE), records the measurements of fields. The lengths of the four sides of one particular field are given as 2′ 8′, 2′ 4′ 8′, 2′ 4′, and 2′ *schoinia* (lit. "measuring chords"), and the area of the field is found to be 4′ 8′ 16′ *arouras*, where an *aroura* is one square *schoinion*.[8] In our notation the four sides are $\frac{5}{8}$, $\frac{7}{8}$, $\frac{3}{4}$, and $\frac{1}{2}$ *schoinia* and the area is $\frac{7}{16}$ *arouras*. Theoretical logistic differs from the practical in that it deals with calculations on abstract, intelligible (yet still divisible) units instead of sensible objects. Again, because there are indefinitely many units, numbers admit multiplicity in both kinds of logistic. Any pair of apples or units is a "two", for example. And because the numbers in logistic count or measure something, they, too, cannot be zero or negative.

It is from a single passage in the pseudo-Heronian book *Definitiones* that we know that at least some Greek authors worked with irrational numbers. The author of that work cites another, similar work that deals with arithmetic:

> What ⟨numbers⟩ are irrational (ἄλογοι) and incommensurable (ἀσύμμετροι), and what are rational (ῥητοί) and commensurable (σύμμετροι) – this has been said in the Preliminaries of the *Arithmêtikê stoicheiôsis*. But now, following Euclid, the "Elementator" (*Stoicheiôtês*, sc. author of the *Elements*), in the case of magnitudes, we say that commensurable magnitudes are defined as those measured by the same measures while incommensurable ones are those for which there can be no common measure.[9]

The word Knorr translates as "rational" (ῥητοί) appears often in the *Arithmetica*, where we translate it more literally as "expressible". Diophantus writes "inexpressible" (οὐ ῥητός) instead of ἄλογος to mean "irrational". We do not know if irrational numbers, being inexpressible, were operated on as they were later in Arabic books. But they were acknowledged, and Hero for one routinely approximated them with rational numbers. If we look to Arabic mathematics for context, irrational numbers would have fallen within the purview of architects and surveyors, who naturally encountered irrational lengths in mensuration. Given how little practical Greek mathematics survives, it is not surprising to find only this one reference to irrational numbers in the literature.

Numbers in Greek arithmetic and logistic possess a kind or species. In practical logistic a five, for example, is five *of something*. It can be five cows, five bowls, five *minae*, or five hours, while in arithmetic and in theoretical logistic it is five (abstract) units. The idea of different species of number is present in sexagesimal calculation, too. Theon of Alexandria calls degrees (μοῖραι), minutes

---

8　See Fowler (1999, 231ff). His translations.
9　(Hero 1912, 84.17–22); translated in Knorr (1993, 181; cf. also 189 n. 6). We added the Greek terms in square brackets.

(πρῶτα ἑξηκοστά), seconds (δεύτερα ἑξηκοστά), etc. different "species" (εἴδη) in his commentary on Book I of Ptolemy's *Almagest*.[10] So a five can also be five degrees, five minutes, five seconds, or five of any other sexagesimal place. In another setting the astronomer Geminus gives a nice example when he writes "ἡμερῶν τξε ἐννεακαιδεκάτων ε" lit. "days 365 ninteenths 5" for "$365\frac{5}{19}$ days". The species of 5 is nineteenths, and the species of $365\frac{5}{19}$ is days.[11] Jacob Klein (1968, 7) expressed this idea of numbers possessing species by writing that a number (ἀριθμός) in Greek arithmetic and logistic "*never* means anything other than 'a definite amount of definite objects'".[12]

Theoretical logistic was largely abandoned by the first century BCE, and afterward practical logistic came to be distinguished from arithmetic not so much because it was concerned with calculations, but because it dealt with sensible objects as opposed to intelligible units. It is this shift that accounts for the fact that Diophantus's book, which is unquestionably a book on logistic, bears the title *Arithmetica*. Thomas Heath explained (1921, I, 16):

> Problems of the Diophantine type, like those of the arithmetical epigrams, had previously been enunciated of concrete numbers (numbers of apples, bowls, &c.), and one of Diophantus' problems (V. 30) is actually in epigram form, and is about measures of wine with prices in drachmas. Diophantus then probably saw that there was no reason why such problems should refer to numbers of any one particular thing rather than another, but that they might more conveniently take the form of finding numbers *in the abstract* with certain properties, alone or in combination, and therefore that they might claim to be part of arithmetic, the abstract science or theory of numbers.

In Arabic mathematics, too, a distinction was made between arithmetic (number theory) and logistic. The former was developed in connection with Euclid's *Elements* Books VII–IX and Nicomachus's *Arithmetical Introduction*, both of which had been translated into Arabic by the early ninth century CE. The Arabic version of practical logistic was called *ḥisāb*, which can be translated as "calculation" or "reckoning". It did not derive from Greek books but was cultivated by practitioners who transmitted their knowledge orally. Just as in Greek, the numbers in Arabic number theory are restricted to whole numbers, while those of *ḥisāb* include (positive) fractions and irrational roots. As in Greek, the numbers in Arabic number theory and in *ḥisāb* are not unique. To pick just one example, the tenth-century

---

10 See Section 3.2.1.
11 (Geminus 1898, 122.15). The "5 nineteenths" is thus not an unresolved division, nor does it express the taking of a nineteenth of 5. Even if Geminus may not have calculated with common fractions, he, and presumably others of his time, could state them as values. Later, Ptolemy routinely expresses fifths similarly, such as μοίρας ς καὶ γ πέμπτα, lit. "degrees 6 and 3 fifths", for $6\frac{3}{5}°$. See Ptolemy (1898–1903, II, 299.21–22) and Ptolemy (1998, 471.8).
12 Translation adjusted in Hopkins (2011, 96).

algebraist ʿAlī al-Sulamī solves a problem in which he arrives at "eleven roots of ten". This is a collection of eleven $\sqrt{10}$s, and he writes "So make that the root of one number", and he then converts it to $\sqrt{1210}$.[13]

As in Greek ἀριθμητική and λογιστική, numbers in Arabic come in kinds or species. A five can be five men, five dirhams (a denomination of silver coin), five mithqals (a measure of weight), five sevenths ($\frac{5}{7}$), five thousands (5,000), five roots of three (a quintet of $\sqrt{3}$s) or five units. In his *Lifting the Veil*, Ibn al-Bannāʾ (1301) wrote about the two aspects of a number, its "meaning" (*maʿnan*) and its "term" (*lafẓ*) (1994, 255.3).[14] For "three men", the meaning is "three" and the term is "men". The term was of course disregarded when working through operations.

Often an amount is expressed as a collection of different numbers. This is more common in Arabic than in Greek in part because of the scarcity of our sources for practical Greek calculation and also because writers in Arabic expressed their fractions differently and they worked extensively with irrational roots, both of which provide many examples. Thus, we begin with examples from Arabic.

Arabic authors worked with common fractions such as "three fifths" ($\frac{3}{5}$) and "four parts of thirteen parts of a dirham" ($\frac{4}{13}$). These fractions were subject to various kinds of combinations, like al-Ḥaṣṣār's "eight ninths and two fourths of a ninth" for our $\frac{17}{18}$, and which he wrote in notation as $\frac{2\ 8}{4\ 9}$,[15] and Ibn al-Bannāʾ's "five sixths and four fifths", our $\frac{49}{30}$, which is showed in notation by al-Hawārī as "$\frac{4\ 5}{5\ 6}$" (Abdeljaouad and Oaks 2021, 78). One could choose based on convenience which form to write a fraction, and calculators could easily shift from one form to another. In one of his inheritance problems al-Khwārazmī wrote, "two sevenths of a share and two thirds of a seventh of a share, and that is eight parts of twenty-one parts of a share", thus converting $\frac{2\ 2}{3\ 7}$ to $\frac{8}{21}$ (al-Khwārazmī 2009, 249.17), and in another problem he starts with "the eighth and the tenth (of the estate)", and later writes it as "thirty-one parts of forty parts of the estate" (al-Khwārazmī 2009, 253.6).

Ibn al-Bannāʾ's "five sixths and four fifths" is a fraction expressed as an aggregation, with "two names": sixths and fifths. There is no word or sign in notation between the $\frac{5}{6}$ and $\frac{4}{5}$ to indicate addition, and the "and" (*wa*) in Ibn al-Bannāʾ's verbal formulation is the common conjunction that is also used in stating whole numbers, like "four and twenty" for 24. Ibn al-Bannāʾ expresses his fraction the same way we would say "five bottles and four cans."

Sometimes a fraction was expressed as the difference of two fractions. Al-Qalaṣādī (d. 1486) works with several examples, such as "seven eighths and half an eighth less a fourth" (our $\frac{11}{16}$), which is shown in notation as "$\frac{1\ 7}{2\ 8}$ less $\frac{1}{4}$" (al-Qalaṣādī 1999 Arabic p. 162.8). The "less" here is the Arabic word *illā*,

---

13 ʿAlī al-Sulamī, MS Vatican, Sbath 5, f. 30r, l. 5.

14 Ibn al-Bannāʾ defines two types of multiplication based on whether the "term" changes. See also his two types of division, explained at Ibn al-Bannāʾ (1994, 263.4).

15 Al-Ḥaṣṣār, MS LJS 293 (copied 1194), University of Pennsylvania, f. 21r, l. 2–4.

which is the negative counterpart to "and". It indicates that what follows it is missing or lacking from what precedes it. The meaning is similar to when we say, "five minutes shy of an hour" or "two cents short of a dollar". The word *illā* takes the same meaning in Arabic grammar, too, like in the phrase "all the students except (*illā*) Stacy". Here "all the students" is called *mustathnā minhu*, lit. "excluded from it" or "diminished", and "Stacy" is called *mustathnā*, "excluded" or "absent". The word *illā* does not intend the operation of subtraction, which in any case was represented in notation in Arabic arithmetic with its preposition *min* ("from").

Quadratic binomials were necessarily written as aggregations. For example, al-Karajī expressed the number we would write as $3 + \sqrt{8}$ by "three and a root of eight" (al-Karajī 1964, 42.24), with the same *wa* ("and") as in the examples with fractions and whole numbers. When Arabic authors wrote this number in notation it appeared again without any word or sign between them (Abdeljaouad and Oaks 2021, 183). The corresponding apotome $3 - \sqrt{8}$ was expressed by al-Karajī as "three less a root of eight", with the same "less" we saw earlier in al-Qalaṣādī's fraction. As with fractions, this number was regarded as being a diminished 3, which is lacking $\sqrt{8}$.

What few examples we have from Greek logistic are consistent with these Arabic examples. We find aggregations in both sexagesimal notation and in the Egyptian, or unit, fractions commonly found in many authors. We saw earlier that Theon of Alexandria characterized the places of a sexagesimal number as "species", so a number like Ptolemy's 93° 15′ 27″ for the length of a chord of an arc of 102° is an aggregation of three numbers whose species are degrees, minutes, and seconds. (For sexagesimal numbers, ′ indicates minutes [sixtieths] and ″ indicates seconds [sixtieths of sixtieths].) Likewise, Egyptian or unit fractions are expressed as collections of the reciprocals of positive integers. For example, Hero writes at one point "$\bar{\iota}\,\gamma'\,\iota\beta'''$", which we can translate as "$10\frac{1}{3}\frac{1}{12}$" (Hero 2014, 356.10) and which is the same as our $10\frac{5}{12}$. The entire "$\bar{\iota}\,\gamma'\,\iota\beta'''$" is expressed as the accumulation of three parts, which together are meant to represent a single number. It does not intend two additions like $10 + \frac{1}{3} + \frac{1}{12}$. We do this ourselves with mixed fractions, so that $10\frac{5}{12}$ is read as one number and not as the addition problem "$10 + \frac{5}{12}$". Sometimes the parts are joined by the word καί ("and"), as in this example from Ptolemy's *Almagest*: "$\bar{\delta}$ καὶ ∠′ καὶ γ′ καὶ ιε′″" for "4 and 2′ and 3′ and 15′″" (Ptolemy 1898–1903, II, 214.20; Ptolemy 1998, 424.12), which is our $4\frac{9}{10}$. An Egyptian contract from 72 CE written in Greek for dividing a property speaks of "a fourth part and a third of a twelfth part, which is a thirty-sixth part" of a property (Winter 1936, 205.7; 207). And as a last example, the enunciations of two epigrams of the Palatine Anthology show fractions as aggregations. Epigram 121 speaks of "one eighth and the twelfth part of one tenth" $\left(\frac{2}{15}\right)$, and epigram 140 has "twice two sixths and twice one seventh" $\left(\frac{20}{21}\right)$ (Paton 1918, 89; 103) (for the latter, see Section 3.2.3).

Hero provides an example of a diminished number in his *Metrica*. He writes "οδ ⳱ ιδ′″" to mean "74 lacking $\frac{1}{14}$" (Hero 2014, 324.20; 21). The sign ⳱ abbreviates the word λεῖψις (*leipsis*), which means "lacking", or "less", like the

Arabic *illā*.[16] We will see later that this sign is routinely used by Diophantus in an algebraic context with the same meaning. Heath (1910, 41–44) explains that the sign most likely originated as an "I" superimposed on a "Λ", which are the first letters in λιπ, the stem for the word λεῖψις.[17] Ptolemy uses this word to express numbers lacking from other numbers in a few places. For example, he expresses a number of revolutions as "8 lacking 2 4′ degrees" where "lacking" is the word λειπούσαις. As in Arabic, this word is the counterpart to the word meaning "and". Just after this, Ptolemy gives a number of revolutions as "46 and 1 degree."[18]

2.2.2 Operations

This practice of expressing numbers as aggregations or differences of two or more numbers contrasts with the way arithmetical operations were stated.[19] Operations were expressed with two verbs, one for the operation and one to announce the outcome. For example, one of many arithmetical calculations from Hero's *Metrica* is: "Add together the 16 and the 6, they become 22".[20] The verb for the operation is σύνθες ("add"), and the verb for the outcome is γίνονται ("become"). The operation is an act that is performed in time which produces a particular result. Today we sometimes still state operations this way, but more often we write them in a way that is alien to premodern mathematics. For this example, we would write "16 + 6 = 22", or verbally, "sixteen plus six equals twenty-two". Here the "plus" functions in its quasi-prepositional role to indicate the operation of addition, while the one verb in the sentence is "equals". Unlike the premodern operation, the verb here is symmetric in the sense that we could also say "twenty-two equals sixteen plus six". With our "plus" we conflate the operation and its outcome, while the two were consistently differentiated in premodern mathematics. An example of a subtraction from the same book is: "Subtract the 6 from the 11, there remains 5" (Hero 2014, 172.18).[21] Today we usually say instead "eleven minus six equals five" where "minus" is the negative counterpart to "plus", and the one verb in the sentence is again the bidirectional "equals". Other Greek operations are worded similarly. For example, Hero squares 13 by writing: "and the 13 (multiplied) by

16 See Bruins (1963, I, 201 [f. 103r, l. 5, 6]).

17 For further discussion on this sign, see Fowler (1999, 232 n. 28).

18 See Ptolemy (1898–1903, II, 215.8–9,12) and Ptolemy (1998, 424.19,22). For other examples of "lacking", see Ptolemy (1898–1903, I, 204.17, 205.11, 375.2, 464.8; II, 214–15) and Ptolemy (1998, 138.11,21, 230.24, 276.–6, 424), for three more examples, and with four examples using "and", Ptolemy (1898–1903, II, 388.7, 391.5) and Ptolemy (1998, 523.6, 524.–6).

19 For the ways operations are stated by Diophantus, see Section 4.2.4.

20 (Hero 2014, 174.21). Acerbi and Vitrac translate this as "Compose (σύνθες) les 16 et les 6 : en résulte 22".

21 Acerbi and Vitrac translate this as "Soustrais les 6 des 11 : il en résulte 5 restants". Similarly, in the thirteenth problem of the papyrus of Akhmîm (6th or 7th c.) we read "subtract 17 from the 221, it is left 204" (Baillet 1892, 70); the verb announcing the result is here λείπεται (*leipetai*).

themselves become 169" (Hero 2014, 174.19).[22] Contrast this with the modern version, "13 squared equals 169".

Ibn Ghāzī (1483) gives a simple Arabic example of addition in his breakdown of the steps to add 4043 to 2685: "Add 5 to 3 to get 8" (Ibn Ghāzī 1983, 26.8). Here, too, the operation is an act that is performed in time to produce a particular outcome. The verb for the operation is *ijma'*, "add", and the verb to announce the result is *yakūn*, "to get". Al-Ḥaṣṣār (late 12th c.) gives this simple example of subtraction: "Subtract eight from nine, leaving one".[23] Again, the operation occurs in time, with an outcome. The verb for the operation is *iṭraḥ*, "subtract", and the result is stated with the verb *baqiya*, "to leave".

Aggregations are not additions, but they are often the result of an addition. Al-Qalaṣādī writes: "Add a root of six to a root of five. You say the result is a root of six and a root of five" (al-Qalaṣādī 1999, 214.10). In our notation this seems to accomplish nothing: $\sqrt{6} + \sqrt{5} = \sqrt{6} + \sqrt{5}$. The operation should be read for what it says: the result of adding together the two roots is the single number $\sqrt{6}\,\sqrt{5}$, which can only be expressed as an aggregation. Today aggregations and the operation of addition are subsumed under our sign "+", but the two were always distinguished in premodern arithmetic.

The same distinction exists between the operation of subtraction and apotomes/ diminished numbers. Al-Qalaṣādī gives this example: "Subtract a root of five from a root of eight. You say the remainder is a root of eight less a root of five" (al-Qalaṣādī 1999, 215.9). Again, in modern notation the operation seems to do nothing: $\sqrt{8} - \sqrt{5} = \sqrt{8} - \sqrt{5}$. So again, where premodern arithmeticians differentiated between the operation of subtraction and diminished amounts, we apply our sign "−" to both. There were no words in Greek, Arabic, Latin, Italian, or even in sixteenth-century European languages with the meanings of our modern "plus" and "minus".[24] Addition and subtraction were operations that produced results, and sometimes those results were expressed with more than one "name", like "five sixths and four fifths" or "three less a root of eight". Our modern notation allows us to conveniently blur these distinctions.

2.2.3 Methods of calculation

Calculations could be performed in a variety of ways in premodern arithmetic.[25] One common method practiced in Græco-Roman antiquity, the Islamic world, and probably in many other parts and periods of the Old World, was

22 The 13 is plural because they are 13 units.

23 (Al-Ḥaṣṣār, MS LJS 293, f. 9v, l. 15).

24 For the meanings of "plus" and "minus" in sixteenth-century European mathematics, see Oaks (2018c, 253–54). Likewise, there was no word in premodern mathematics with the meaning of "times", as in "eight times two equals sixteen". What we translate as, for example, "four times two units" (Problem VII.1) is literally "four copies of (*amthāl*) two units". This is a repeated duplication, not the operation of multiplication.

25 General surveys can be found, for example, in Menninger (1969), Ifrah (1994), and Schärlig (2001).

finger reckoning (Smith 1958, II, 196ff; Pellat 1977; Schärlig 2006). In this system operations were performed mentally, with intermediate results being stored by positioning the fingers in particular ways. Another system was the sexagesimal arithmetic of the astronomers, which originated around 4,000 years ago in Mesopotamia. This is a base sixty place-value system that spread to Greece, India, the Islamic world, and later to medieval Europe. In Greek and Arabic, the numbers were written using their alphabetic numerals for 1 through 59 together with a special sign for an empty place. Greeks and Romans also calculated with the abacus, which consisted of a board or table marked with lines designating units, fives, tens, fifties, hundreds, etc. (or sometimes other divisions) in one direction, and with lines for fractions in the other direction. Pebbles or some other kind of counter were placed on these lines to represent numbers, and operations were performed by shifting them. The abacus is not known to have been used by medieval Muslims, but they did appropriate from India the base ten system with the nine figures 1, 2, 3, 4, 5, 6, 7, 8, 9, and the 0 for an empty place. We call these "Arabic" numerals because Europeans learned them from Arabic sources, but in Arabic they were called "Indian numerals", or "dust figures", after the dust-board on which calculations were often performed. Reckoning with Indian numerals was unknown in Greek antiquity. The figures had reached Syria by 662 CE, but it may be that rules for calculating with them were learned by Muslims only in the mid-eighth century (Kunitzsch 2003).

The Arabic method of calculating that is most relevant to algebra is finger reckoning. In Islamic countries this was the method preferred by government secretaries, merchants, and others who performed calculations in their jobs, and it is in this environment that the numerical problem-solving we are about to describe was practiced.

2.3 Numerical problem-solving

A distinction was made within the Arabic finger-reckoning tradition between problems asking for the result of a calculation on known numbers, like "multiply three fourths of eight by three fifths of ten", from Ibn al-Yāsamīn's (d. 1204) *Grafting of Opinions of the Work on Dust Figures*, and problems of finding an unknown number, like "a quantity: you add its third and a dirham and its sixth and two dirhams to get fifteen. How much is the quantity?", from the same book (Zemouli 1993, 154.19, 203.13). Today a question asking for an unknown number is regarded as belonging to algebra, and we would work out this one by setting up and solving the equation $\frac{1}{3}x+1+\frac{1}{6}x+2=15$. But in Ibn al-Yāsamīn's time, and across the span of premodern practice, several methods were available for solving such problems. Ibn al-Yāsamīn, in fact, solves this problem three different ways: first by single false position, then by algebra, and finally by double false position. It was only roughly around the turn of the nineteenth century that algebra emerged as nearly the *only* method taught in Western countries for finding unknown numbers.

Single false position applies the rule of three, or as it was called in Arabic, "the four proportional numbers". One makes a convenient, but probably incorrect, guess for the answer, and the value of the unknown is found by a calculation on the guess and its error. Single false position is employed to solve a problem in the Rhind Papyrus, written in Egypt ca. 1650 BCE; a version of the method was practiced in the Old Babylonian period, ca. 1700–1600 BCE; the Palatine Anthology, written in Greek in late antiquity, contains some problems solved this way; and the method was applied in India before the rise of Islam.[26]

Double false position, called in Arabic the method of "the two errors" (*al-khaṭa'ān*) or the method of "scales" (*kiffāt*), works similarly, but with two guesses and their respective errors. The only known pre-Islamic examples of this method are found in the *Nine Chapters*, a Chinese text from around the first century CE. Single and double false position were likely practiced in other parts of the ancient world as well. We just have no evidence one way or the other.

Algebra (*al-jabr wa-l-muqābala*, or often simply *al-jabr*) functions differently. Where in other problem-solving methods one operates exclusively on known numbers, in algebra one or more unknowns are named, and operations are performed on these names to set up an equation, which is then simplified and solved. This method is more sophisticated than the others, and it can be applied to a larger class of problems.

Here is our translation of Ibn al-Yāsamīn's problem, showing his solutions by the three methods just described.

(*Enunciation*)

Problem. A quantity:[27] you add its third and a dirham and its sixth and two dirhams to get fifteen. How much is the quantity?

(*Solution by single false position*)

To work out this problem you add up the appended dirhams and you remove them from fifteen, leaving twelve. It is as if (someone) had said to you, a quantity: you add its third and its sixth to get twelve.[28] So you see what denominates the third and the sixth, and that is eighteen. You add its third and its sixth to get nine, so it is the head which you will divide by. Then multiply the twelve by the denominator to get two hundred sixteen. Divide it by the head, resulting for you in twenty-four. It is the sought-after (quantity).

---

26 For Egypt: (Imhausen 2003, 37, 51; Gillings 1972, 154ff). For Babylonia: (Høyrup 2002, 59–60, 102, 311–13). For Sanskrit: (Hayashi 1995, 396–99). One of eight examples from the Palatine Anthology is translated and discussed in Christianidis and Oaks (2013, 129–30).

27 The word "quantity" is translated from the Arabic *māl*. This is also the word designating the second power of the unknown in Arabic algebra, and it was also often used to mean "dividend" in a division problem. In practical arithmetic, as here, the word took the meaning of a common noun meaning "quantity" or "amount of money" (Oaks and Alkhateeb 2005).

28 This reformulation of the problem is necessary for the solution by single false position to work.

And another way is that you divide the denominator by the head, resulting for you in two. Then multiply it by the twelve, giving twenty-four. It is the unknown quantity. And another way, by analogy (*al-qiyās*), is that you work out how much you must restore nine so that it becomes eighteen, and that is by itself. So restore the twelve by itself to get twenty-four.

(*Solution by algebra*)

And another way, by algebra (*al-jabr*), is that you make your quantity a thing (*like our x*). You take its third and a dirham and its sixth and two dirhams to get that half a thing and three dirhams equal fifteen dirhams (*this gives the equation we write as* $\frac{1}{2}x + 3 = 15$). You subtract the three dirhams from the fifteen, leaving half a thing equals twelve dirhams ($\frac{1}{2}x = 12$). Then you see by how much you must restore[29] half a thing so that it gives one (thing), and that is by multiplying it by two. You multiply twelve by two, so the thing is twenty-four, which is what you wanted.

(*Solution by double false position*)

And another way, by the scales (*kiffāt*), is that you take (the first) scale to be thirty-six, and you add its third and a dirham and its sixth and two dirhams, to get twenty-one. This exceeds the fifteen by six, which is the error of the first scale. Then you take the second scale to be forty-eight. You add its third and a dirham and its sixth and two dirhams, amounting to twenty-seven. This exceeds the fifteen by twelve dirhams, which is the error of the second scale.

Then you multiply the error of the first scale by the second scale, and the error of the second scale by the first scale, and you subtract the smaller from the greater, leaving one hundred forty-four. Divide it by the difference between the two errors, which is six, resulting in twenty-four, which is the quantity.[30]

Apart from Diophantus's *Arithmetica*, the only notable collection of worked-out arithmetical problems from ancient Greece is found in the Palatine Anthology, where both algebra and single false position are applied. We have translated one

---

29 The word "restore" here, as in the previous solution, is conjugated from *al-jabr*. Meanings of this word will be explained later.

30 In other problems, Ibn al-Yāsamīn shows the diagram that is drawn for the solution by double false position. He does not show it for this problem, but it would have looked like this:

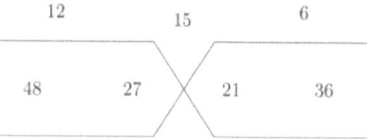

solution by single false position in Christianidis and Oaks (2013, 129–30), and we discuss the algebraic solutions from this anthology in Section 3.2.3.[31]

It is important to note that with solutions by three different methods, Ibn al-Yāsamīn's enunciation cannot be considered to be an algebra question. Like other questions in premodern texts, it is an arithmetic question that can, if one chooses, be solved *by* algebra.[32] Premodern algebra was just one of several techniques of numerical problem-solving. In contrast with the looser and more abstract notions that the words "algebra" and "algebraic" convey today, premodern algebra was not used to state arithmetical identities like $(a + b)^2 = a^2 + b^2 + 2ab$, describe curves like $y = 3x^2 - 1$, state formulas or physical laws like $A = \pi r^2$ and $d = \frac{1}{2}gt^2$, and it made only very rare appearances in proofs. We may be inclined to identify certain arguments in premodern mathematics as being "algebraic", but to those who wrote these texts, the term *al-jabr wa-l-muqābala* designated a specific problem-solving method for finding unknown numbers with its own technical vocabulary.

The earliest extant Arabic descriptions of the purpose of algebra attest that it was a method of solving problems. Al-Khwārazmī wrote in the introduction to his early ninth-century *Book of Algebra* that the caliph al-Ma'mūn

> has encouraged me to write a concise book on algebraic calculation which encompasses the fine and important parts of its calculations that people constantly need in cases of their inheritances, their wills, their partnerships, their legal decisions, and their trade, and in all their dealings with one another, such as the surveying of land, the digging of canals, architecture, and other procedures and kinds of work.
>
> (al-Khwārazmī 2009, 95.11)

In the latter part of the tenth century, the lexicographer Muḥammad ibn Aḥmad al-Khwārazmī wrote: "Algebra (*al-jabr wa-l-muqābala*) is the art of arts of calculation (*ḥisāb*) and the beautiful method of solving difficult problems in wills, inheritance, commercial transactions, and exchanges" (al-Khwārazmī 1895, 199.14). Soon after, ca. 1011/12 CE, al-Karajī gave this brief characterization before going on to explain the process in more detail: "Know that the objective of algebra (*al-jabr wa-l-muqābala*) is to find unknowns from known quantities". And 'Alī al-Sulamī (MS Sbath 5, f. 2v.14), writing around the same time, wrote: "The goal of the science of algebra (*al-jabr wa-l-muqābala*) is to find unknown numbers".

31 See also Christianidis and Megremi (2019).
32 There are, of course, questions posed already in algebraic terms, but these are found in chapters explaining algebra. For example, in the section on multiplying algebraic expressions al-Karajī asks and then solves the problem "If (someone) said to you, multiply ten dirhams and a Thing by ten dirhams and a Thing" (Saidan 1986, 108.14).

From another angle, algebra was classified by the early tenth-century philosopher al-Fārābī not as a part of arithmetic or geometry, but as a "technique" (*ḥīla*), along with other techniques such as the construction of astronomical and musical instruments and the making of military weapons.[33] Apart from its rare use in proofs[34] and the occasional mathematician exploring the workings of polynomials for their own sake (al-Karajī, al-Samaw'al), premodern algebra never lost this focus on problem-solving. Rather than being a theory of equations centered on the solutions and geometric proofs for quadratic equations that could, if one wished, be applied to work out problems, algebra consisted of a collection of rules, occasionally with geometrical or arithmetical proofs, devised specifically for finding unknown numbers.

We should stress the numerical nature of premodern algebra. As al-Khwārazmī noted in his introduction, "When I considered what people need in calculation (*ḥisāb*) I found that it is always a number" (al-Khwārazmī 2009, 97.1). Algebra could be used to solve problems in arithmetic, geometry, metrology, and even horology, as long as the parameters are given numerical measure, from which the solution is then found as a number. Algebra was not used, for instance, to find a geometric magnitude in itself, as in geometrical analysis,[35] but only its numerical size.

2.4 Premodern algebra: vocabulary and structure

This Graeco-Arabic-Latin-Italian-Renaissance algebra which we call premodern algebra differs from post-Vietan algebra in key respects. The underlying concept of what constitutes a monomial rests on an entirely different basis from ours, and by consequence also what constitutes a polynomial and an equation. These concepts affected the ways they framed their operations, the ways they approached the setting-up of the equation, and the ways they represented expressions in notation. In this section we give the basic vocabulary and structure of premodern algebraic solutions, after which we investigate the premodern concepts.

2.4.1 The names of the powers

The powers of the unknown were given individual names in premodern algebra. The names commonly found in Greek (Diophantus), Arabic, and medieval Italian algebra up to the fourth power are given in the following table:[36]

---

33 From *Enumeration of the Sciences* (*Iḥṣā' al-'ulūm*), first half of the tenth century (al-Fārābī 2015, 156ff).

34 Most notably in Abū Kāmil and al-Fārisī. See Oaks (2011, 2019).

35 Cf. Section 4.1.

36 For the complete overview of the Greek terms, see Section 4.2.1

Table 2.1 Names for the powers in Greek, Arabic, and Italian up to the fourth.

Greek	Arabic	Italian	Modern
μονάς (unit)	*ahad* (unit)	*numero* (number)	1
	dirham	*dragma*	
ἀριθμός (number)	*shay'* (thing)	*cosa* (thing)	x
	jidhr (root)		
δύναμις (power)	*māl* (amount of money)	*censo* (amount of money)	x^2
κύβος (cube)	*ka'b* (cube)	*cubo* (cube)	x^3
δυναμοδύναμις	*māl māl*	*censo di censo*	x^4

Powers above the fourth in all three languages were usually written as combinations of the names for the second and third powers, such as δυναμόκυβος for the fifth power in Diophantus, or *māl māl māl māl* for the eighth power in Abū Kāmil. Diophantus also gives names for the reciprocal powers corresponding to our x^{-1}, x^{-2}, etc. These are called ἀριθμοστόν, δυναμοστόν, etc., which we translate as "inverse number", "inverse power", etc. Inspired by Diophantus, al-Karajī was the first Arabic algebraist we know to work with reciprocal powers, and we also find them in medieval Italian and later texts. Some Arabic and later authors occasionally solved indeterminate problems by introducing other names for independent unknowns. Two common Arabic names for this purpose were *dīnār* and *fals*, denominations of gold and copper coin, respectively. These are used in the example we translate in Section 2.7.1.

Some of the names in both Greek and Arabic could take other meanings in arithmetic. In Greek, the word ἀριθμός (*arithmos*) is also the common word for "number", and κύβος (*kybos*) for a cubic number,[37] like 27 or $\frac{8}{125}$. In Arabic *jidhr* and *māl* can take the meanings of "square root" and "quantity/amount of money", respectively, and *ka'b* meant "cube root" or sometimes "cube". For all of these terms, which meaning is intended is clear from the context. To help the reader we capitalize the names of the algebraic powers in our translation. For example, for the Greek text we translate "Number" for ἀριθμός when it is the first power of the unknown, and "number" when it takes its ordinary meaning.

Diophantus calls the names of the powers the "species" (εἴδη) of his "arithmetical theory" (ἀριθμητικὴ θεωρία). Words with meanings varying like the English "type", "species", "kind", and "sort" were used to describe the names of the powers in Arabic: *ḍurūb, jins, nauʿ*, and *aṣl*, and the names were called *spezie/spetie* in medieval Italian. These are all words commonly used for any kind of classification and are not restricted to algebra or arithmetic, nor are they tied to specifically philosophical usage. The powers are also sometimes called by words with meanings close to "name" or "denomination" in Greek (ἐπωνυμίαι), Arabic (*ism, lafẓ*, and sometimes *laqab*), and Italian (*nome, denominatione*). Also, the

---

37 See LSJ (= Liddell, Scott, and Jones 1996), *s.v.* κύβος, II.

powers are often called "ranks" (*martaba*) in Arabic, reflecting the increasing powers. Finally, in the late antique Greek-speaking world and in Byzantium the algebraic powers were known as "the Diophantine numbers", and the problem-solving through them as "the method of the Diophantine numbers" (see Sections 3.2.1 and 3.4.4).

It is with these names, or ranks, that equations are formed. Al-Khayyām writes: "solutions in algebra (*al-jabr*) are accomplished by means of the equation (*mu'ādala*), I mean the equating of these ranks with one another" (Rashed and Vahabzadeh 1999, 123.8). A sample equation from Abū Kāmil's *Book on Algebra* is: "sixty-two Things and half a Thing less four *Māl*s and a fourth of a *Māl* equal a hundred dirhams and two *Māl*s" (Abū Kāmil 2012, 339.9), which corresponds to our $62\frac{1}{2}x - 4\frac{1}{4}x^2 = 100 + 2x^2$. Equations in Arabic were written with the same "and" (*wa*) and "less" (*illā*) that we saw for numbers. A sample equation from Problem IV^G.24 in Diophantus is: "8 Cubes, 4 Numbers lacking 12 Powers. These are equal to 6 Numbers lacking 1 Power", which corresponds to our $8x^3 + 4x - 12x^2 = 6x - x^2$. The "lacking" here is the Greek word λείψει (dative of λεῖψις), which in this context takes the same meaning as the Arabic *illā*.

2.4.2 The structure of algebraic solutions

The structure of a solution to a problem in premodern algebra consists of the following basic stages:

1   One or more unknowns are named in terms of the preassigned names of the powers, and the operations or conditions called for in the enunciation are applied to set up (ideally) a polynomial equation.
2   Next, the equation is simplified. In premodern algebra, a polynomial equation is simplified when (i) it contains no diminished terms (i.e., no terms of the form "*A* lacking/less *B*") and (ii) there is at most one term for each power in the equation. There are thus two basic steps to simplify a polynomial equation: (i) Terms of the form "*A* lacking *B*" should be restored to a full *A* (this is the Arabic *al-jabr*) and the *B* should be added to the other side; and (ii) like terms on opposite sides should be confronted, leaving their difference on the side of the greater (this is the Arabic *al-muqābala*). These steps will be explained later.
3   The simplified equation is solved using the prescribed rule. If the equation has one term on each side, then the solution is trivial. If the equation contains three terms of consecutive powers, two on one side and one on the other, then a rule equivalent to our quadratic formula is applied.
4   Once the value of the ἀριθμός/*shay'*/*cosa* is found, the unknowns in the problem are calculated.

Some solutions deviate a little from this scheme. Occasionally in Arabic and later algebra the equation set up in stage 1 will include a division of a polynomial, or it will have square roots of algebraic expressions. In these cases, the expected multiplications are performed to simplify it. In some indeterminate problems, again in Arabic and later, equations are set up with two or more independent unknowns, so the equation simplifies to one in which one unknown is expressed in terms of the others. An example from Abū Kāmil is translated in Section 2.7.1.

We illustrate these stages in the context of two simple determinate examples, Problem I.11 from Diophantus and Problem 1 from Abū Kāmil's ninth-century *Book on Algebra* (Abū Kāmil 2012, 335.12). First, the enunciations:

Diophantus	*Abū Kāmil*
Given two numbers, to add the first to, and subtract the second from the same number, and make the produced (numbers) have to one another a given ratio. Let it be proposed to add 20 to, and subtract 100 from the same number, and make the greater the triple of the smaller.	Ten: you divide it into two parts. You multiply each part by itself, and you cast away the smaller from the greater, leaving eighty.

Diophantus and Abū Kāmil begin stage (1) of their solutions by naming their unknowns:

Stage 1	*Stage 1*
Let the sought-after (number) be 1 Number.	Its rule is that you make the smaller part a Thing, and the greater ten less a Thing.

Next, the conditions of the enunciation are applied to these names to set up an equation:

And if we add to it 20 units, it becomes 1 Number, 20 units; while, if 100 units are subtracted from it, there remains 1 Number lacking 100 units. And the greater should be the triple of the smaller. Thrice the smaller, therefore, are equal to the greater. But thrice the smaller makes 3 Numbers lacking 300 units. Thus, 3 Numbers lacking 300 units are equal to 1 Number, 20 units.	Multiply ten less a Thing by itself to get a hundred dirhams and a *Māl* less twenty Things. Then multiply a Thing by itself to get a *Māl*. You subtract it from a hundred dirhams and a *Māl* less twenty Things, leaving a hundred dirhams less twenty Things equal eighty dirhams.

Translated into modern notation, Diophantus's equation becomes $3x - 300 = x + 20$ and Abū Kāmil's becomes $100 - 20x = 80$.

In stage (2) the equation is simplified. The two steps to be performed are to restore diminished terms and to confront like terms on opposite sides:

Stage 2	Stage 2
Let the lacking be added in common and let likes be subtracted from likes. Thus, 320 units are equal to 2 Numbers,	Restore the hundred dirhams by the twenty Things and add it to the eighty, to get: twenty Things and eighty dirhams equal a hundred dirhams. Cast away eighty from a hundred, leaving twenty dirhams equal twenty Things.

Diophantus words this succinctly and obtains what in our notation is $320 = 2x$. Abū Kāmil spells out the operations and obtains the simplified equation that in modern notation is written as $20 = 20x$.

The solutions to both equations in stage (3) are trivial:

Stage 3	Stage 3
and the Number becomes 160 units.	So the Thing is one,

In stage (4) Diophantus then calculates the unknowns from the value of 1 Number, and Abū Kāmil identifies the parts of ten. Diophantus marks stage 4 with the phrase "To the hypostases."[38] This stage is generally not singled out in Arabic algebra since the value of the thing is usually the sought-after unknown.

Stage 4	Stage 4
To the hypostases: thus, the greater will be 180 units, and the smaller 60 units. And it is fulfilled that the greater is the triple of the smaller.	which is the smaller part, and the greater is nine, which is what is left over from the ten.

We will investigate these stages more thoroughly in Section 2.6. Until then there are some other features of premodern algebra that need to be addressed.

2.4.3 The shift to algebraic language

Today, when confronted with a problem such as "Find the dimensions of a rectangle with perimeter 22 and area 21", we might call the sides x and y and use the first condition to find that $y = 11 - x$, from which we set up the equation $x(11 - x) = 21$.

---

38 Hypostasis (ὑπόστασις) has here the meaning of "numerical value." For a discussion, see Christianidis (2015).

What we have done here is to translate the enunciation, which is a problem of metrical geometry, into an equation, which is a problem in algebra. Similarly, in the two problems of Diophantus and Abū Kāmil translated above, an arithmetic problem is translated into an algebraic equation. In modern algebra, the shift is more noticeable due to our notation, but it is no less present in premodern algebra. As with the problem of Ibn al-Yāsamīn translated in Section 2.3, the enunciations to the Greek and Arabic problems are not questions in algebra. No unknown has yet been named in terms of the names of the powers, and no equation has been stated. These steps are taken by our authors in stage 1 of their solutions.

The questions Diophantus poses in his enunciations ask for unknown numbers (ἀριθμοί, sing. ἀριθμός), and often a number is required to be a perfect square (τετράγωνος, *tetragōnos*) or a perfect cube (κύβος, *kybos*). In the solutions these are translated into terms of "the arithmetical theory" (ἡ ἀριθμητικὴ θεωρία) (4.14), i.e., into expressions formed from the names of the powers ἀριθμός, δύναμις, κύβος, etc. The terms in the two series carry very different meanings, even if two of them are the same word. In the problem just translated, Diophantus writes about three numbers (ἀριθμοί), two of them to be given as 20 and 100 in the instantiation, and the other to be found in the solution. Here the word ἀριθμός is a common noun and not the name of a specific number. After all, there are three of them, and they are all different. In the beginning of the solution "the sought-after (ἀριθμός)" is named "1 ἀριθμός" (in our translation "1 Number"). This "1 ἀριθμός" is the name given to a specific amount whose value will be found at the end of the solution. By naming "the sought-after (ἀριθμός)" as "1 ἀριθμός" Diophantus gives it a designation that can be operated on. From "1 ἀριθμός" he constructs "1 ἀριθμός, 20 units" and "1 ἀριθμός lacking 100 units" by an addition and a subtraction, and he triples the latter to get the expression "3 ἀριθμοί lacking 300 units", which leads to the equation and its solution. These operations are fairly simple, so it would be possible to work through them with unnamed amounts, but in more complicated problems such an approach would be intractable. For example, in *Arithmetica* Problem IV^G.24, where Diophantus subtracts "2 Numbers lacking 1 unit" (like our $2x - 1$) from its cube to get "8 Cubes, 4 Numbers lacking 12 Powers" (like our $8x^3 + 4x - 12x^2$). This is a calculation performed on numbers expressed in terms of the names of the powers. In other words, the calculations are performed on *algebraic expressions*.

The difference between the second-degree terms is more evident, since the words are different. The name of the second power, δύναμις, never appears in an enunciation, and its counterpart τετράγωνος (perfect square) is never the name of the second-degree unknown in a solution. This is illustrated in Problem II.12, whose instantiation reads: "Now, let it be proposed to subtract from 9 and 21 the same number and make each of the remainders a τετράγωνος". The first assignment in the solution is: "So, let the τετράγωνος which is subtracted from the 9 units be 1 δύναμις". The word τετράγωνος is clearly a common noun by virtue of the fact that there are two of them, and they are not equal. The "1 δύναμις", on the other hand, is the name, or proper noun, given to one of these τετράγωνοι, and being a name, it can be operated on to yield meaningful results. First, it is subtracted from 9 units to obtain "9 units lacking 1 δύναμις"; this is then subtracted from 21 to get "1 δύναμις lacking 12 units", which is then equated with another named amount expressed in terms of the powers of the same unknown, "1 δύναμις, 16 units lacking 8 Numbers". The value of 1 Number is then found to be "4 8ths", which makes the value of 1 δύναμις $\frac{1}{4}$.

The same differences can be observed for the two meanings of the word κύβος (pl. κύβοι). As with the two meanings of ἀριθμός and τετράγωνος/δύναμις, the κύβος (cubic number) in the enunciation differs from the κύβος (Cube) through which it is named much like the word "city" differs from "City", the district of London. The first is a category of number, and the second is the name given to a specific cube. Also, sometimes a quantity is named with a different corresponding power, as in Problem V.17, where three numbers (ἀριθμοί) are named "2 Cubes", "9 Cubes", and "28 Cubes".

In Arabic algebra, too, enunciations are stated in terms of common nouns which are given algebraic names in the solution. In Abū Kāmil's problem translated previously, the common nouns are the two "parts" of ten, which are named "a Thing" and "ten less a Thing" in the solution. These names are then operated on to obtain the equation. It is context that determines the meanings of the potentially ambiguous words *māl* and *jidhr*. Many Arabic enunciations are framed in terms of an unknown *māl* (quantity, or sum of money), like the problem of Ibn al-Yāsamīn translated in Section 2.3. Sometimes a problem also calls on the *jidhr* (square root) of this *māl*. These words then will appear in an algebraic solution as the names of the powers. The same kinds of indications are present to distinguish the two meanings of *māl* and *jidhr*, as we see in Diophantus for ἀριθμός and κύβος.[39]

2.5 Monomials, polynomials, and equations in premodern algebra

Before examining the four stages of an algebraic solution in more detail, it is necessary to describe the nature of premodern monomials, polynomials, and equations and to explain how they differ from their modern counterparts.

If we avoid the temptation to gloss over the curious and consistent differences between the ways premodern algebraists solved problems and the way we solve them today, we discover that our idea of what constitutes a monomial, a polynomial, and even an equation is inconsistent with premodern wording, procedure, and notations. For example, to simplify an equation that we would express as $4x = 10 - x$, al-Khwārazmī writes, "Restore the ten by the Thing and add it to the four Things" (al-Khwārazmī 2009, 149.12, our translation), while we would just add the x to the other side. What, we might ask, does the ten have to do with this step? And algebraists from Diophantus to Clavius (1538–1612) routinely worked

39 See Oaks and Alkhateeb (2005).

out all operations called for in the enunciation before setting up their equations, even in cases in which it is clear (to us) that it is far easier to set up the equation first. Also, medieval algebraists forbid coefficients to be irrational, even if irrational numbers permeate their works. And last, in premodern notation, whether in Greek, Arabic, Latin, Italian, or in any European algebra text written before Viète's 1591 *Isagoge*, a "coefficient" of 1 was always written. We, instead, write x or x^2 and not $1x$ or $1x^2$.

2.5.1 Monomials in premodern algebra

The features of premodern numbers described earlier – that they admit multiplicity, that they come in species, and that they are sometimes expressed as aggregations or differences – also apply to premodern algebraic expressions. Some authors state explicitly that the names of the powers are kinds or species of number. Al-Khwārazmī situates the powers in the beginning of his *Book of Algebra*: "I found that the numbers which are necessary for calculation in algebra are of three kinds (*ḍurūb*), which are roots, *māl*s, and simple numbers unrelated to a root or a *māl*" (al-Khwārazmī 2009, 97.9). Now, in addition to being counted in men, dirhams, mithqals, or units, a number can be (algebraic) Roots, *Māl*s, and extending to higher powers, Cubes, *Māl Māl*s, etc. These species are still ways of measuring intelligible units, so the units themselves are now called "simple numbers". Diophantus, for his part, calls the names of the powers "denominated numbers" (ἐπονομασθέντες ἀριθμοί), and Theon of Alexandria, in his commentary on Book XIII of Ptolemy's *Almagest*, speaks of terms expressed by the names of the powers as being "Diophantine numbers" in his algebraic solution to a mensuration problem.[40] It is worth noting, too, that in his commentary on Book I Theon explains the multiplication of numbers expressed in sexagesimal notation by explicitly comparing the species (εἴδη) of units (degrees) and the reciprocal powers of sixty (minutes, seconds, etc.) with the species of units and the reciprocal powers of the unknown in Diophantus.[41]

Another indication that the powers are kinds of number is the way terms are worded in the texts. For example, when there is one of a power it is stated in the singular, and when there is more than one it is plural. Diophantus, for example, writes "1 δύναμις" ("1 Power") and "25 δυνάμεις" ("25 Powers"). These "25 Powers" are expressed the same way we would say "25 cats" or "25 degrees". The "25 Powers" is a 25 of a particular kind. Contrast this with the way we say "twenty-five ex-squared" for $25x^2$.

---

40 See Christianidis and Skoura (2013). The same expression ("Diophantine numbers") is also employed in the fourteenth century by the Byzantine astronomer and mathematician Isaac Argyros; cf. Sections 3.2.1 and 3.4.4.

41 See Skoura and Christianidis (2014), Skoura (2016), and Christianidis and Megremi (2019); cf. Section 3.2.1.

What we call the "coefficient" was often called the "multitude" (*plēthos*, πλῆθος) of a term in Diophantus, and the "number" (*'adad*) of a term in Arabic. In premodern algebra, the multitude/number designates how many of that object are present, keeping in mind that numbers are numbers of something, and not scalar multiples like our modern coefficients. Consequently, the "number" of a term was not allowed to be irrational. One can have five of something or one and a fourth of something, but it makes no sense to have a square root of five ($\sqrt{5}$) of something. Throughout medieval Arabic, Latin, and Italian algebra, and stretching into the sixteenth century, irrational numbers of a term were avoided by taking a root of the whole term. 'Alī al-Sulamī explains:

> If we want to multiply some of the unknowns by a root of a known number, like multiplying a Thing by a root of five, you multiply the Thing by itself to get a *Māl*, then by five to get five *Māl*s. Then a root of five *Māl*s is the product of a Thing by a root of five.

> ('Alī al-Sulamī, MS Sbath 5, f. 22v, l. 10)

In modern notation, he multiplies x by $\sqrt{5}$ to get $\sqrt{5x^2}$, where we instead write $\sqrt{5}x$. In 'Alī al-Sulamī's answer the "number" (coefficient), 5, of *Māl*s remains rational, while the root of the term makes sense as a (potentially) irrational amount. It was only in notation, and then only in some algebraists, that irrational coefficients were first admitted in sixteenth-century Europe (Oaks 2017). Our word "coefficient" made its debut in François Viète's 1591 *In Artem Analyticem Isagoge*.

For these reasons, the relationship between the coefficient and the power in a term like $7x^3$ is radically different from the relationship between the multitude/number and the species in a premodern term like "seven Cubes". In our $7x^3$, the 7 and the x^3 are both numbers, one known and the other unknown, that are understood to be multiplied together. In the term "seven Cubes", the 7 is the multitude/number, while the Cube is the kind or species, or in Ibn al-Bannā''s words, "7" is the meaning and "Cubes" is the term. They are two different aspects of a single number. Just as there is no scalar multiplication involved in the phrase "seven tomatoes", there is also none in the phrase "seven Cubes". While it is true that we sometimes think of a term like $7x^3$ as being "seven cubes", the premodern understanding becomes intractable when we move to terms like πr^2 or $3xy$.

This accounts for the fact that the word "multitude" or "number" is often omitted in Diophantus and in Arabic algebra. In Problem IV^G.19, for example, Diophantus has formed the polynomial "9 Powers, 24 Numbers, 13 units", an amount consisting of numbers of three different kinds. He then writes what is literally translated as "I have the Powers (to be) square", by which he means that its multitude, 9, is a square. The "9 Powers" is composed of Powers, and to identify this term in the expression it suffices to refer to it as "the Powers". Or, when al-Khwārazmī begins the solution to his famous equation "a *māl* and ten roots equal thirty-nine dirhams", he writes "you halve the roots" instead of "you halve the number of roots".[42] Think of al-Khwārazmī's equation as consisting of three groups of coins of different denominations, like "a peso and ten euros equal thirty-nine pounds". The "ten euros" is à pile of ten one-euro coins, so to single out this ten in the equation it suffices to refer to it as "the euros", and its half is five. Several premodern algebraists, in fact, compared different algebraic powers with different denominations of coins to clarify aspects of the art: al-Bīrūnī (Arabic, eleventh c.; 1934, 37–38), Luca Pacioli (Italian; 1494, f. 112r, l. 42), Michael Stifel (Latin; 1544, f. 231r), Marco Aurel (Spanish; 1552, f. 71r), and Jacques Peletier (French; 1554, 22). The analogy is helped in Arabic by the fact that "dirham" is already a denomination of silver coin. The multitude/number and the species were two aspects of the same "pile", not two separate entities brought together to form the term. In our translation, we sometimes add "(multitude of)" before the name of a power to make the reading easier, but this clarification was not necessary for the ancient or medieval reader.

We have already mentioned that Diophantus calls the names of the powers (ἀριθμός, δύναμις, κύβος, δυναμοδύναμις, etc.) "species" (εἴδη), while, on the other hand, he calls the three terms through which enunciations are framed (ἀριθμός, τετράγωνος, and κύβος) "sought-after numbers" (ζητούμενοι ἀριθμοί), a phrase often shortened to just οἱ ζητούμενοι, the substantive ἀριθμοί being implied, as we saw in Problem I.11. Diophantus never calls the terms in this latter series εἴδη. His use of the word εἶδος is thus different from that of other authors, such as Nicomachus of Gerasa. Nicomachus follows Aristotle by classifying the numbers of ἀριθμητική into the εἴδη "even" and "odd", and further dividing these into sub-εἴδη, and he uses the term for other kinds of classification, too. In the language of Ibn al-Bannā', Nicomachus applies the word εἶδος to classify the "meanings" of numbers, while Diophantus applies the term to classify the "terms" of numbers. For Diophantus, "two δυνάμεις" and "four δυνάμεις" are the same species because they are both δυνάμεις, while for Nicomachus they are the same species because "two" and "four" are both even.

2.5.2 *Polynomials in premodern algebra*

Adding three apples to two apples gives five apples, but adding three apples to two pears can only give the aggregation "three apples and two pears". In reckoning, one can add together commensurable numbers to get a result expressed with "one name", such as adding "two roots of two" (a pair of $\sqrt{2}$s) to "three roots of eight" (a trio of $\sqrt{8}$s) to get "a root of one hundred twenty-eight" ($\sqrt{128}$).[43] But as we saw earlier, when the two numbers are incommensurable, or in the

---

42 (al-Khwārazmī 2009, 101.5). Sometimes a text will read "you halve the number of roots", but it is more common to read "you halve the roots".

43 From a 1387 commentary by Ibn al-Hā'im (Abdeljaouad 2003, 179.1). Premodern arithmeticians preferred not to work with multiple roots like these, and in fact Ibn al-Hā'im converts the addends to "a root of eight" and "a root of seventy-two" before adding them.

case of fractions when it happens to be convenient, the result is expressed as an aggregation of the addends. Adding three to a root of eight gives "three and a root of eight", for example.

The same holds for the addition of algebraic terms. Diophantus writes in Problem IV^G.8: "Therefore, if 1 Number is added to 2 Numbers, they become 3 Numbers, while if (it is added) to 8 Cubes, they become 8 Cubes, 1 Number". The "1 Number" and "2 Numbers", being of the same species, can be added to make the unitary "3 Numbers". But "1 Number" added to "8 Cubes" must be expressed as the aggregation "8 Cubes, 1 Number". This last amount is a collection of nine objects of two different kinds, 8 of one and 1 of the other. It is written as an inventory, like "8 chairs, 1 table", or like the Egyptian fraction 2′ 18′, which is an aggregation equal to our $\frac{5}{9}$. Or again, it is written like Theon of Alexandria's sexagesimal number 37° 4′ 55″, which is expressed in terms of the different species (εἴδη) of degrees, minutes, and seconds. In Diophantus's operations, the "3 Numbers" and "8 Cubes, 1 Number" are the *results* of the additions. Neither is expressed using the operation of addition. Again, the "and" which sometimes joins the species is the common conjunction.

In Arabic, the "and" (*wa*) joins algebraic terms gathered together through addition just as it does for fractions and incommensurable roots. In the first worked-out problem in his *al-Fakhrī*, al-Karajī has: "two Things. And you add to it the five dirhams, so it becomes two Things and five dirhams" (Saidan 1986, 170.5).[44] This result is a collection of seven objects of two different kinds, and is expressed the same way we would say "two apples and five pears". In modern notation, the operation seems to accomplish nothing: $2x + 5 = 2x + 5$. As in arithmetic, our notation distorts the premodern operation in two ways. First, we use the "+" for both the operation "add" and the conjunction "and", and second, our "=" removes any sense that the right side is the outcome of the operation on the left.

Subtractions in Arabic algebra also follow the pattern we see in *ḥisāb*. In Problem II.19 al-Karajī writes, "If you cast away the Thing from fourteen, it leaves fourteen less a Thing" (Saidan 1986, 193.15). Again, in modern notation, this seems to do nothing: $14 - x = 14 - x$. Like the apotome "three less a root of eight", the result of the subtraction is a single amount that can only be expressed using two names. Diophantus words his subtractions the same way. In Problem I.16 he writes: "if I subtract 20 units from 1 Number, I shall have . . . 1 Number lacking 20 units". The "lacking", as already mentioned, is the Greek word λεῖψις (more precisely, the dative λείψει), which, as we saw with numerical calculation in Hero and Ptolemy, takes the same meaning as the Arabic *illā*. It indicates that what follows is lacking, or has been removed, from what precedes it.

If we have placed ample attention on how premodern polynomials were conceived, it is because modern polynomials are fundamental in algebra today. The premodern texts in fact rarely speak of polynomials as objects in themselves,

---

44 While "two Things" is plural, sometimes an Arabic algebraist would consider it as one term and regard it as being singular, as in this example.

and this lack of attention is consistent with the premodern conception. Diophantus has no word that can be translated as "polynomial", and in Arabic only al-Karajī and some later algebraists influenced by him differentiating between "simple numbers" ('*adad mufrad*)[45] and "composite numbers" ('*adad murakkab*), which roughly correspond to our "monomial" and "polynomial".[46] These terms are borrowed from the arithmetic of finger reckoning. Al-Karajī introduced them in his arithmetic book *al-Kāfī fī l-ḥisāb* to explain how to multiply whole numbers. He first covers "simple numbers", which are numbers with a single non-zero place, like 3, 20, or 7,000. In his first example, he multiplies 200,000,000 by 30,000 by multiplying 2 by 3 to get 6, and then adding the zeros to get 6,000,000,000,000. Composite numbers have more than one nonzero place, like 12, 405, and 3,559. Al-Karajī gives the example of multiplying 555 by 444, which entails multiplying each rank of the multiplicand by each rank of the multiplier and adding the results. In other words, one adds the results of multiplying the simple numbers of which the composite numbers are composed (al-Karajī 1986, 38ff).

It is natural, then, that when he wrote the section on the multiplication of algebraic expressions in his *al-Fakhrī*, al-Karajī would borrow the terms "simple numbers" and "composite numbers" from arithmetic. After all, a composite number like "a hundred and five and twenty" (Saidan 1986, 41.8) (125) was spoken much like "a *Māl* and five dirhams and two Things" (corresponding to our $x^2 + 5 + 2x$). Simple numbers in the setting of algebra are what we call monomials, like "thirteen Things" ($13x$) or "half of a *Māl Māl*" $\left(\frac{1}{2}x^4\right)$. The rule for multiplying them is like the rule for integers: multiply the "numbers" of the terms, and then add the powers. In one example, al-Karajī multiplies "five Cubes" ($5x^3$) by "five *Māl Māls*" ($5x^4$) to get "twenty-five *Māl Māl* Cubes" ($25x^7$). Then he explains that to multiply "composite numbers", which are made up of more than one power, one multiplies all the terms in the multiplicand by all the terms in the multiplier, just like in arithmetic. One of his examples is to "multiply five Cubes and three *Māl*s and four Things by four numbers and five *Māl*s and three Things" (corresponding to our $5x^3 + 3x^2 + 4x$ by $4 + 5x^2 + 3x$), to eventually get, after adding the nine products, "twenty-five *Māl* Cubes and thirty *Māl Māl*s and forty-nine Cubes and twenty-four *Māl*s and sixteen Things" (our $25x^5 + 30x^4 + 49x^3 + 24x^2 + 16x$).

Recall that it was in the context of explaining the multiplication of numbers in sexagesimal notation that Theon of Alexandria made the comparison between the species in sexagesimal arithmetic with the species in algebra. Just as al-Karajī did over six centuries earlier, Theon first gave the rule for multiplying single species by multiplying the multitudes and then adding the powers. He then gave

---

45 This is a different use of the phrase '*adad mufrad* than we find al-Khwārazmī's list of powers of the unknown.

46 These later algebraists include al-Samaw'al (d. ca. 1175), Ibn Fallūs (1194–1239), Ibn al-Khawwām (1277), and Ibn al-Hā'im (1387).

the rule for multiplying numbers composed of more than one species, which he illustrated with the example of multiplying 37° 4′ 55″ by itself.

Returning to al-Karajī, it is in the subsection on multiplying simple (algebraic) numbers in *al-Fakhrī* that he explains why expressions of the form "*A* less *B*" should be considered to be simple numbers, i.e., monomials, and not composite numbers. Just after multiplying ten less a Thing by ten to get "a hundred less ten Things" $(100 - 10x)$, he writes (Saidan 1986, 105.24):

> And some people believe that this number is composite, since it is of two types (i.e., units and Things). But this is not so, because in saying "ten less a Thing" you denote one number of the rank of units. But if there were in its place "ten and a Thing", that would be composite.

Al-Karajī's "a hundred less ten Things" is a deficient hundred, which is lacking ten Things. It is not composed of units and Things, but only of units.[47] Based on what we read in al-Karajī, a "composite (algebraic) number" is a collection of terms of at least two different powers, including the reciprocal powers, which may or may not be diminished (with "less") by other powers. A "simple (algebraic) number" is a term of a single power, which also may or may not be diminished (with "less") by other powers. It is not worth attempting to pin down these definitions any further by considering square roots or unresolved divisions, which we will encounter later, since the idea of a polynomial was only of marginal importance in algebra before the sixteenth century. It is only in the context of outlining the rules for operating on algebraic terms that al-Karajī and some of his successors use the terms "simple number" and "composite number". Like quadratic binomials and fractions such as "five sixths and four fifths", what we see as a polynomial was regarded as being a single number that happens to be expressed with more than one name.[48]

2.5.3 Equations in premodern algebra

Taking into account the relationship between the multitude/number and the species of a term, and the ways terms are combined to make polynomials with words meaning "and" and "lacking", it becomes clear that premodern polynomials contain no operations. Consider two examples, "6 4′ inverse Powers, 25 Powers lacking 9 units" from Diophantus's Problem V^{G}.27, and "a *Māl Māl* and twelve *Māl*s and a fourth of a *Māl* less seven Cubes" from Abū Kāmil's Problem 51 (Abū Kāmil 2012, 445.8).[49] In modern notation, these are $6\frac{1}{4}x^{-1} + 25x^2 - 9$ and $x^4 + 12\frac{1}{4}x^2 - 7x^3$, respectively. Where the modern versions are built from the

---

47 Later, al-Samaw'al recognized the convenience of regarding quantites of the form "*A* less *B*" as being composite for the purpose of multiplication and root extraction.

48 Consider also that the word "polynomial", meaning "many names", was coined in the context of premodern arithmetic and algebra.

49 This is in Problem 57 by Rashed's numbering.

operations of exponentiation, scalar multiplication, addition, and subtraction, the premodern polynomials are simply collections of the different kinds of algebraic number. Diophantus's polynomial, for instance, is the gathering of $6\frac{1}{4}$ objects of one kind with 25 objects of another, and which are lacking 9 objects of a third kind. Note that although the polynomials themselves are not expressed with operations, one can perform operations on them. By analogy, the number 4 is not expressed with any operation, but one can add 3 to it, divide it by 11, etc.

The Arabic word for "equal" in the statement of an algebraic equation is the otherwise unusual verb ʿadala ("to be equal") (Oaks 2010a). Outside algebra, equating is almost always expressed using some other word. To equate numbers, geometric magnitudes, angles, and ratios, one finds instead the words *sawiya*, *mithl*, the prefix *ka-*, and the implied verb "to be". (Neither ʿadala nor any of these other words is used to announce the result of an operation in arithmetic or algebra.) This makes it easy to distinguish the statement of an algebraic equation from other kinds of equating that might occur in a text. From now on, and in our translation, we write "Equal" with a capital "E" for conjugations of ʿadala, and "equal" in lower case for other words, such as *sawiya* and *mithl*.

The word "equal" for stating equations in Greek is the adjective ἴσος (*isos*). Unlike the Arabic case, this is the common word used to equate any kind of object in Greek mathematics. Diophantus uses it frequently in enunciations, for example, to express the equality of two or more amounts. The associated noun ἴσωσις (*isōsis*) sometimes takes the meaning of "(algebraic) equation", as in this example from Problem IV^G.24: "And if the (multitudes of) Numbers in either part of the *equation* were equal, it would be reduced to equate Cubes equal to Powers". Other times ἴσωσις means the action of equalizing, that is, the "setting-up of an equation", or the "making equal", as in this example from Problem III.10: "Therefore, 52 Powers together with 12 units must make a square. And if the multitude of the 13 units of the first number were a square, the *setting-up of the equation* would be easy". Another related noun, ἰσότης (*isotēs*), "equality", is used with the meaning of "(algebraic) equation" in Problem IV^G.8 ("Therefore, 19 Powers are equal to 1 unit. And the one unit is a square; if 19, the multitude of the Powers, were also a square, the *equality* would be solvable"), while three times (Problems IV^G.32, VI^G.12, VI^G.14) it is used to denote forthcoming equations (see next paragraph). Considering the use of both nouns in the equivalent expressions "double equation" and "double equality" the two words, ἴσωσις and ἰσότης, appear to be interchangeable. We write "equality" for ἰσότης only to distinguish between the two Greek words.

Many times, Diophantus states what we call a forthcoming equation. This is an equation that has not yet been fully established. In nearly all instances one side of the equation has been cast in terms of the names of the powers, but the other side has not yet been named. Here is an example from Problem II.33: "16 Powers, 7 Numbers. These are equal to a square". The "16 Powers, 7 Numbers" is stated in algebraic terms, but the assignment for the "square" still has to be determined. In his introduction, Heath (1910, 67ff) rewrites these incomplete, forthcoming equations in modern notation and calls them "indeterminate equations". For example, he writes the equation just mentioned as "$16x^2 + 7x = y^2$" (Heath 1910, 68). In Heath's version both sides are expressed in algebraic notation, and the equation is stated in terms of two independent unknowns x and y. It is Heath who has completed the forthcoming equation by naming the unnamed square as y^2. In Diophantus the equation is only partially complete in one unknown, and in the next step he writes: "I form the square from 5 Numbers. Thus, 25 Powers are equal to 16 Powers, 7 Numbers", from which the solution is found. Apart from two accountable exceptions to be discussed in Section 2.7.1, Diophantus does not set up any indeterminate equations.

In some forthcoming equations in Problems V.1, V.2, and V.3, the one side is still only partially completed, as in "one *Māl Māl*, which, together with twelve cubes, Equal a square" from V.2. Here the "one *Māl Māl*" is an algebraic name, while the "twelve cubes" and the "square" still wait to be expressed in algebraic terms. Problem V.1 is unusual in stating an equation in which no part of either side of the forthcoming equation has yet been put into algebraic terms. See our commentary on that problem for an explanation.

The word for "equation" in Arabic algebra is *mu'ādala*, which derives from *'adala*. This word does not appear in al-Khwārazmī or Abū Kāmil. We first encounter it in Qusṭā's translation of the *Arithmetica*, so he may have coined the term himself as a translation of Diophantus's ἴσωσις or ἰσότης. As with the related verb *'adala*, from now on we capitalize "Equation" in our translation from the Arabic.[50] We have seen *mu'ādala* once in Section 2.4.1, where al-Khayyām explained equations: "solutions in algebra (*al-jabr*) are accomplished by means of the Equation (*mu'ādala*), I mean the Equating of these ranks with one another".[51]

2.6 The stages of an algebraic solution

2.6.1 Stage 1: setting up the equation

In stage 1 of an algebraic solution the conditions of a problem are converted into an equation. This process is begun by naming one or more unknowns in terms of the names of the powers. Al-Karajī begins his description of the steps of solving problems by algebra in his book *al-Kāfī* with what is needed for stage 1 (al-Karajī 1986, 169.6):

> We know that problems are solved by means of three things. The first is a search for the way to deal with the problem dictated by its conditions, and second is the givens which you are given in the problem, and third is multiplication, division, duplication, halving, gathering, partitioning, adding, and

50 Earlier authors, like al-Khwārazmī or Abū Kāmil, call equations by the more generic term "problems" (*masā'il*), and this word continued to be used by most later algebraists. On the uses of the words *masā'il* and *mu'ādala*, see Oaks (2010a).
51 For further discussion of Diophantus's language for stating equations, see Section 4.2.5.

subtracting, to bring the problem to two Equated (*mut'ādalīn*, from *'adala*) expressions.[52]

Al-Karajī then continues by explaining the simplification of the equation in stage 2 by *al-jabr* and *al-muqābala*, after which he covers stage 3 by classifying and solving the six simplified equations of degree one and two, with arithmetical proofs. What should be noted here is that the working-out of operations leads to the setting up of the equation, which in modern algebra is often done in reverse order.

Al-Fārisī, about two centuries later, described this process in more practical terms. His explanation of stage 1 even reads like it may have been written with Diophantine-type indeterminate problems in mind. Like al-Karajī, al-Fārisī solves some problems that ultimately come from Diophantus.

> The general explanation of the solving of problems by means of the methods of this science is that you assign the unknown to be a type of the types originating from the Thing[53] appropriate for it, according to how the questioner described it. If there is a square, (make it) a *Māl*, and if there is a cube, (make it) a Cube, and if it is not related to a type of the unknown, then assign it to be a Thing or something composed of two types or more, as an aggregation or an exclusion.[54] Then work out the operations according to how the questioner characterized the unknown, and you carry out the computation to where it leads by means of the light of your intuition and the clarity of your talent, until you reach an Equation (*mu'ādala*) of one type or more to one type or more.
>
> (al-Fārisī 1994, 463.1)

Like al-Karajī, al-Fārisī continues by describing stages 2 and 3. In both authors it is the working out of the operations that leads to a polynomial equation "of one type or more to one type or more".

Diophantus organizes his introduction following the steps one takes in solving problems (Christianidis and Oaks 2013, 131–34; cf. Section 4.1), and he, too, covers operations before he brings up equations. After giving the rules for operating on monomials and polynomials, and just before his first mention of species equal to other species, he summarizes stage 1:

> Then, it is good that you who are beginning this study should have acquired practice in addition, subtraction, and the multiplications concerning the

---

52 For a complete translation of the passage, see Oaks (forthcoming).
53 Here "the types originating from the Thing" are the names of the powers: Thing, *Māl*, Cube, *Māl Māl*, etc. They all originate from the Thing through mutiplication.
54 For example, Abū Kāmil solves some problems by naming the parts of ten "five and a Thing" and "five less a Thing", these being an aggregation and an exclusion, respectively. In one indeterminate problem, al-Fārisī names an unknown as the exclusion "a *Māl* less ten" so that when increased by 10 it is a square.

species; and how to add extant and lacking species, not of the same multitude, to other species, themselves either extant or likewise extant and lacking, and how from extant and other lacking species to subtract other (species), either extant or likewise extant and lacking.

These three descriptions of stage 1 are in agreement with the way premodern algebraists solved their problems: after naming unknowns, the operations are worked out in terms of these names based on the conditions of the enunciation, and this leads to the setting-up of an equation. With the operations taken care of first, the two sides of the equation should both be polynomials, free of any unresolved operations. By contrast, because our modern equations are built from operations, we can set up an equation first and then continue to work out the operations contained in them.

In most cases setting up a polynomial equation comes easily, since the addition, subtraction, and multiplication of polynomials always gives another polynomial. This is not true for division and roots, however. So when these operations are called for in the enunciation, some way around them must be found. Abū Kāmil's Problem 25 is a good example showing the trouble algebraists could take to ensure that the equation has polynomials on both sides.[55] The enunciation reads (Abū Kāmil 2012, 399.6):

> And if (someone) said to you, ten: you divide it into two parts. You divide the greater by the smaller, then you add what results from the division to the ten. Then you multiply what is gathered by the greater part, to get sixty-nine dirhams.

Abū Kāmil begins the first of his two solutions by naming the smaller part "a Thing" and the greater part "ten less a Thing", but he does not set up a rhetorical version of the equation $\left(\frac{10-x}{x}+10\right)(10-x)=69$. Instead, he proceeds to reason his way through the operations, first multiplying the 10 by the $10-x$, and subtracting it from the 69 to get "ten Things less thirty-one dirhams" ($10x-31$). He then has to address the $\left(\frac{10-x}{x}\right)(10-x)$, which he does by multiplying the quotient by the deleted "Thing" from the "ten less a Thing" to get "ten less a Thing", which is added to the $10x-31$, resulting in "nine Things less twenty-one" ($9x-21$). Now he is left with the $\left(\frac{10-x}{x}\right)10$, which he expresses as "the product of what results from dividing ten less a Thing by a Thing, by ten". He divides the $9x-21$ by the 10, and then he multiplies the result by "a Thing" to finally set up the equation "nine tenths of a *Māl* less two Things and a tenth of a Thing Equal ten less a Thing", which corresponds to our equation $\frac{9}{10}x^2-2\frac{1}{10}x=10-x$.

Diophantus also works his operations first. His enunciations in the preserved part of the *Arithmetica* do not call for divisions, so the difference with modern practice is not as evident. But it can be detected in some problems, such as IV[G].1.

Its enunciation and instantiation read: "To divide a given number into two cubes, whose sides (together) are given. Now, let it be (proposed) to divide the number 370 into two cubes whose sides (together) are 10 units". Diophantus begins his solution by naming one of the sides "1 Number, 5 units" (our $x + 5$) and the other "5 units less 1 Number" (our $5 - x$). But he does not set up the premodern equivalent of the equation $(x + 5)^3 + (5 - x)^3 = 370$. Instead, he expands and adds the cubes first, and only then does he establish the polynomial equation "30 Powers, 250 units. These are equal to 370 units" (our $30x^2 + 250 = 370$).

2.6.2 *Divisions and roots in equations*

There are many examples in the history of mathematics in which people have disregarded the accepted norms of mathematical practice to allow for what should be impossible objects. In the sixteenth century Christoff Rudolff (1525), followed by Cardano (1539) and others, began to admit irrational "coefficients" in their notation, even if it conflicted with the meaning of the "number" of a term as a multitude. For another example from the same century, Michael Stifel explained why irrational numbers should not exist but then went ahead to show how to calculate with them anyway because of "the results that follow from their use" (Stifel 1544, f. 103r; translated in Nunn 1914, 412). Other mathematicians of that century, notably Cardano (1545) and Bombelli (1572), began to work with negative and complex numbers, and François Viète (1593a) manipulated four dimensional geometric magnitudes like "Quadrato-quadratum ex BC", where BC is a line shown in a diagram, in two propositions in geometry (Oaks 2017, 2018c). In a similar vein, Diophantus and some Arabic algebraists relaxed the ban on operations in algebraic expressions for divisions and roots.

Although the enunciations in Diophantus do not call for division, there are occasions in stage 1 in which he finds himself in need of dividing an expression by a polynomial, that is, in forming a rational expression.[56] In Problem IV$^\mathrm{G}$.36, for instance, he constructs "3 Numbers in a part of 1 Number lacking 3 units" and "4 Numbers in a part of 1 Number lacking 4 units", and he adds them to get "7 Powers lacking 24 Numbers of a part of 1 Power, 12 units lacking 7 Numbers". In modern notation, he forms $\frac{3x}{x-3}$ and $\frac{4x}{x-4}$, and he adds them to get $\frac{7x^2-24}{x^2+12-7x}$. These are not written as divisions, but as algebraic versions of common fractions, like his "65 of a part of 12,678 units" for $\frac{65}{12,678}$, from Problem III.19. In one problem, VI$^\mathrm{G}$.21, these expressions also involve the unresolved squares and cubes of binomials. There we find expressions like "2 Power-Powers in a part of the cube of 1 Power less 2 units", which in our notation is $\frac{2x^4}{\left(x^2-2\right)^3}$. Diophantus performs multiplications on such terms to eliminate the denominators before he states his equation so that all equations in Diophantus are polynomial equations. So although Diophantus might occasionally bend the rules by manipulating terms that are not strictly

---

[56] In the Greek books this occurs in the lemma to IV$^\mathrm{G}$.36, and in Problems IV$^\mathrm{G}$.36, V$^\mathrm{G}$.10, VI$^\mathrm{G}$.12, VI$^\mathrm{G}$.13, VI$^\mathrm{G}$.14, VI$^\mathrm{G}$.19, and VI$^\mathrm{G}$.21. Rational expressions are not found in the Arabic books IV–VII.

aggregations and differences of the powers, he nevertheless resolves these terms into polynomials before writing his equation.

The rules were bent in Arabic algebra, too. Enunciations in Arabic texts frequently call for divisions, and algebraists routinely worked to find ways of setting up equations without them, as we saw in the example from Abū Kāmil. Sometimes, though, an Arabic algebraist would admit an unresolved division to an equation. One problem that is given across several authors, including al-Khwārazmī, Abū Kāmil, al-Karajī, and Ibn al-Bannā', is particularly instructive on this point. The enunciation in Abū Kāmil is (Abū Kāmil 2012, 337.17):

> If (someone) said to you, ten: you divide it into two parts. You divide each of them by the other, so they result in four and a fourth.

If we name the two parts x and y, then the enunciation can be translated into the modern algebraic system of equations $x + y = 10$ and $\frac{x}{y} + \frac{y}{x} = 4\frac{1}{4}$. (Some authors make the sum $2\frac{1}{6}$ instead of $4\frac{1}{4}$.) Al-Khwārazmī solves this problem by stating an arithmetical (not algebraic) rule equivalent to our $a^2 + b^2 = \left(\frac{a}{b} + \frac{b}{a}\right)ab$, and then using it to set up a polynomial equation. Abū Kāmil solves the problem five different ways, each time using some rule or artifice to ensure that the equation he sets up has polynomials on both sides. Ibn al-Bannā' solves the problem five different ways, too, setting up a polynomial equation each time. Al-Karajī works out the problem four different ways. In the first three he sets up polynomial equations, but in his last solution he admits divisions to his equation. Here is how this solution begins (Saidan 1986, 212.25):

> And if you want to work out this problem another way, make one of the parts a Thing and the other ten less a Thing. Divide each of them by the other, to get: ten less a Thing divided by a Thing and a Thing divided by ten less a Thing, and that Equals two dirhams and a sixth.

Here we have the Arabic version of $\frac{x}{10-x} + \frac{10-x}{x} = 2\frac{1}{6}$, with the divisions intact. Unlike Diophantus, al-Karajī includes these algebraic fractions in his equation, and he uses the phrase "divided by" (*maqsūm 'alā*) to express them, thus explicitly citing the operation. He simplifies the equation by multiplying everything by the denominators to get a polynomial equation, which he then simplifies and solves by the standard rules.

It is not that al-Khwārazmī, Abū Kāmil, and Ibn al-Bannā' could not have worked with rational expressions. They all, in one other problem each, include an unresolved division in an equation. Al-Karajī is just more liberal with divisions. He solves 14 problems that call for the operation of division in the enunciation, and in 11 of them he works with rational expressions. But although some later authors continued to work with expressions of the form "A divided by B", such expressions never became common in Arabic algebra. We encounter rational expressions much more frequently in medieval Italian algebra, where they were made more palatable by writing them in the running text in notation with the division bar, and this practice continued into the Renaissance.[57]

Square roots of even-power monomials are common in Arabic and later algebra. For example, in Problem 40 Abū Kāmil sets up the equation "two Things and a root of two *Māl Māl*s Equal thirty dirhams" (Abū Kāmil 2012, 425.19), which in our notation is $2x + \sqrt{2x^4} = 30$. We occasionally find a square root of an odd power term or of a polynomial, too. In Problem 53 Abū Kāmil sets up and solves the equation "a Thing and a root of forty-nine dirhams Equal a root of thirty-three Things and a third of a Thing" (Abū Kāmil 2012, 449.10) $\left(x + \sqrt{49} = \sqrt{33\frac{1}{3}x}\right)$, while in Problem 61 he works with equations containing terms like "a root of twenty and four Things" $\left(\sqrt{20 + 4x}\right)$ and "a root of one thousand six hundred less sixty-four *Māl*s" $\left(\sqrt{1600 - 64x^2}\right)$ (Abū Kāmil 2012, 491.19; 493.8).

Unresolved divisions and roots in Arabic algebra were worked on and manipulated as if they were ordinary monomials until it came time to perform the appropriate multiplication to eliminate the denominator or the square root. This occasional admission of divisions and roots did not affect the ban on other operations in equations, however. Multiplications, additions, and subtractions continued to be worked out before the setting-up of the equation in premodern algebra down to the end of the sixteenth century. Equations remained ideally polynomial equations, with the premodern understanding of the term. Admitting divisions and roots was a somewhat illicit patch to overcome computational difficulties. Algebraists permitted them, with reservations, because they worked.

2.6.3 *Stage 2:* al-jabr *and* al-muqābala

In stage 2 the equation is simplified to a standard form. In modern elementary algebra, quadratic polynomial equations simplify to the single form $ax^2 + bx + c = 0$, and the rule for solving this equation is $x = \frac{-b \pm \sqrt{b^2 - 4ac}}{2a}$. In Arabic and subsequent premodern algebra, the rules for solving simplified equations in stage 3 similarly use the numbers of the terms. Because these numbers cannot be zero or negative, there are six types of simplified equation of degree 1 and 2. Table 2.2 displays how Al-Khwārazmī (2009, 97.18, 101.3) lists them.

Each type is solved by its own rule. The first three equations require only a division to obtain the Root or the *Māl*, while the last three are solved by a rule equivalent to our quadratic formula.

Recall that in Arabic and subsequent premodern algebra, a polynomial equation is simplified, that is, it requires no further manipulation before calculating the solution, when it satisfies the following two conditions: (i) there are no diminished terms (of the form "*A* less *B*"), and (ii) there do not appear two

---

Table 2.2 The six equations of al-Khwārazmī.

Simple equations[1]	
1 "some *Māl*s Equal some Roots"	$(ax^2 = bx)$
2 "some *Māl*s Equal a number"	$(ax^2 = c)$
3 "some Roots Equal a number"	$(bx = c)$
Composite equations	
4 "some *Māl*s and some Roots Equal a number"	$(ax^2 + bx = c)$
5 "some *Māl*s and a number Equal some Roots"	$(ax^2 + c = bx)$
6 "some Roots and a number Equal some *Māl*s"	$(bx + c = ax^2)$

[1] We label these with al-Karajī's terms "Simple equations" and "Composite equations".

terms of the same power in the equation. Thus, only two steps are required to simplify a polynomial equation, and they are called in Arabic *al-jabr* and *al-muqābala*.

The word *al-jabr* takes the meaning of "restoration" of an incomplete amount. To rectify the presence of a term of the form "*A* less *B*" two steps must be performed. First, the "*A* less *B*", being a deficient or incomplete "*A*", must be restored to a full "*A*". Then, to account for this increase, a "*B*" must be added to the other side of the equation. The step is explained well in this example from Abū Kāmil's Problem T1 (2012, 321.11):

fifteen Things less a *Māl* and a half Equal a *Māl* $\left(15x - 1\frac{1}{2}x^2 = x^2\right)$. Restore (*ajbir*) the fifteen by the *Māl* and a half so that it is equivalent to fifteen Things. Then add the *Māl* and a half to the *Māl* to get: two *Māl*s and half a *Māl* Equal fifteen Things.[58]

Most authors word this more succinctly, as al-Khwārazmī does in this example: "four Things Equal . . . ten less a Thing. Restore the ten by the Thing and add it to the four Things to get: five Things Equal ten" (al-Khwārazmī 2009, 149.12).[59] Some later Arabic authors from the Maghreb call both of these steps a restoration. One example from Ibn al-Yāsamīn is: "a hundred less twenty Things Equal forty. Restore the hundred less twenty Things and also restore the forty, so you get: a hundred Equals forty and twenty Things" (Zemouli 1993, 225.5).[60] The wording in Fibonacci is similar, but not quite the same. He writes in one problem "40 radices, minus 4 censibus, que equantur censui (like our $40x - 4x^2 = x^2$). Restaura ergo 4 census ab utraque parte, erunt 5 census, qui equantur 40 radicibus" (Fibonacci 1857, 410.9, 2002, 558.27–28). Translating, he writes "Then restore 4 *census* on

58 That the "fifteen" is not stated as "fifteen Things" is not important. The term is an example of the number 15, so to identify it there is no need to designate its species.
59 See also stage 2 of the problem of Abū Kāmil translated in Section 2.4.2.
60 The text mistakenly has "a hundred and a *Māl* less twenty Things Equal".

either side", where now it is the lacking part that is restored. There was no particular way to word this step in medieval Italian algebra.

Al-Fārisī's description of the process of solving problems by algebra continues with his description of stage 2 (al-Fārisī 1994, 463.6):

> Then restore (*ajbir*) what is excluded from one of the two sides by adding the amount of the excluded to it and add its same to the other side, which is *al-jabr*. Then concentrate on confronting several of the types Equated on the two sides, by which you cast away the repeated types (that are) equal on the two sides, and this is *al-muqābala*.

The verb *qabila*, from which *al-muqābala* derives, is rarely used when this latter step is performed. The one verb in the step is usually *laqiya*, "to cast away", for the subtraction. An example from al-Khwārazmī is: "seven Things Equal two and half a Thing. Cast away half a Thing from seven Things, leaving six Things and a half Equal two dirhams" (al-Khwārazmī 2009, 169.11).[61]

Several Arabic authors use the phrase *al-jabr wa-l-muqābala* (restoration and confrontation) to mean *al-jabr* and/or *al-muqābala*, to cover all of stage 2, even if only one of the steps is performed. Ibn al-Yāsamīn gives this example which covers both steps (Zemouli 1993, 227.8): "A hundred and a *Māl* less twenty Things Equal two *Māl*s and a fourth of a *Māl*. Then restore and confront, so you get: a *Māl* and a fourth of a *Māl* and twenty Things Equal a hundred dirhams". This phrase, *al-jabr wa-l-muqābala*, is also the name given to the art of algebra in Arabic, and with this meaning it was often shortened to just *al-jabr*. This naming suggests that what Arabic practitioners recognized as the core element of algebra was the transformation of an equation into one of the six types, or said differently, the way that the equations that are set up in any of the myriad of problems that can be posed will simplify to one of a small number of types whose solutions are provided in advance.

The words *al-jabr* and *al-muqābala* also took on a number of other roles in the solutions of problems by algebra,[62] and like the terms *Māl* and *Jidhr*, they are found outside algebra as well. In arithmetic, *al-jabr* is the restoration, or increase, of a number to a desired number in proportional calculations, and it sometimes means "to round up" to the next whole number. The verb *qabila*, from which *al-muqābala* derives, is used in solutions by double false position to "confront" a calculated value with the desired value, resulting in their difference.

The rules Diophantus gives for simplifying equations correspond to these two Arabic steps. He explains them in his introduction with the aim of obtaining a two-term equation:

> if there result from a problem certain species equal to the same species, but not of the same multitude, it will be necessary to subtract likes from likes

---

61 For another example, see the problem of Abū Kāmil translated in Section 2.4.2.
62 See Oaks (forthcoming).

from either side, until it results in one species equal to one species. But if by chance there be on either (side) or on both lacking species, it will be necessary to add the lacking species on both sides, until the species on each side become extant, and again to subtract likes from likes until one species is left on each side.

An example of the first of these steps, from Problem I.3, is worded like this: "Thus, 4 Numbers and 4 units are equal to 80 units. And I subtract likes from likes. Then, there remain 76 units equal to 4 Numbers". Lacking species are taken care of in this example from the second solution to I.18: "2 Numbers less 5 units, therefore, are equal to 65 units. Let the lacking be added in common. Thus, 2 Numbers are equal to 70 units". When both steps need to be performed, Diophantus will often write something like "Let the lacking be added in common and likes from likes".

Diophantus does not give any classification of equations in the extant parts of the *Arithmetica*. Because he sought rational solutions, and because he routinely worked with higher powers, he would not have found it useful to distinguish the three simple equations that we find in Arabic algebra. The majority of his equations simplify to the form $ax^{n+1} = bx^n$, up to $n = 8$ in several problems. Thus, for him, the equation types 1 and 3 are subcases of his preferred general case. The procedure for solving these equations may have seemed too easy to warrant an explanation, so none is given. In the rare cases where Diophantus sets up an equation that simplifies to the form $ax^{n+2} = bx^n$, he ensures ahead of time that the quotient $\frac{b}{a}$ is a perfect square.[63] Here, too, the solution is trivial and does not require any explanation.

Diophantus mentions composite equations in his introduction, where he promises to explain "how, when two species are left equal to one, such a case is solved". That part of the *Arithmetica* is not extant, but he does solve composite equations in six problems in Books IV[G] and VI[G]. We will examine them next.

2.6.4 Stage 3: solving the simplified equation

Finding the solutions to two-term simplified equations is easy. The large majority of equations in Diophantus reduce to the form $ax^{n+1} = bx^n$, or more rarely to the form $ax^{n+2} = bx^n$, and many problems in Arabic algebra simplify to one of the three simple types as well. But sometimes in Diophantus, and quite often in Arabic and later algebra, the simplified equation has three terms of consecutive powers, and its solution requires a rule equivalent to our quadratic formula. Diophantus solves such equations six times in six problems, IV[G].22, IV[G].31, and VI[G].6 through VI[G].9, but he does not explain how he arrived at his solutions.

---

63 He does this in Problem IV[G].8, for example. He first set up an equation that led to $19x^3 = x$, but since 19 is not a square, he retraced his steps and adjusted the numbers so that the equation instead simplifies to $169x^3 = x$.

For example, in Problem IV^G.31 he has: "3 Numbers, 18 units less 1 Power are equal to a square. Now I assign (the square) to be 324 Powers, and the Number becomes 78 325ths, that is 6 25ths". The equation he sets up, corresponding to our $3x + 18 - x^2 = 324x^2$, simplifies to $325x^2 = 3x + 18$, and the solution $\frac{78}{325}$ is given immediately. Although Diophantus does not show his steps in these six problems, in two others he explains the steps in the context of an inequality treated as if it were an equation. In Problem IV^G.39 he has arrived at $2x^2 > 6x + 18$:

> But when we solve such an equation, we multiply half of the Numbers by itself; it becomes 9; and the 2 Powers, by the 18 units; they become 36; add (this) to 9; they become 45, of which (we take) the side; it is not smaller than 7 units;[64] add half of the Numbers; ⟨it becomes not smaller than 10 units; and divide by the (multitude of) Powers;⟩ it becomes not smaller than 5 units.

Later, in Problem V^G.10, he does the same for the inequality $72x > 17x^2 + 17$:

> (Multiply) half the Numbers by itself; it becomes 1296. Subtract the Powers (multiplied) by the units, that is 289; so, the remainder is 1007. (Take) the side of these; it is not greater than 31.[65] Add the half of the Numbers; it becomes not greater than 67. Divide by the multitude of the Powers; the Number becomes ⟨not greater than⟩ 67 17ths.

In two other problems, VI^G.6 and VI^G.22, the discriminant is calculated. In VI^G.6 Diophantus arrives at the equation we write as $6x^2 + 3x = 7$. He then notes, "And it is necessary to add to half the Numbers multiplied by itself (the product of) the Powers ⟨by the units⟩, and make a square". In Problem VI^G.22, from $172x = 336x^2 + 24$, he writes "But this is not always possible, unless half of the Numbers (multiplied) by itself, when it lacks (the product of) the Powers by the units, makes a square".

From these examples, the rule for solving equations of the type $ax^2 = bx + c$, contorted into modern notation, is:

$$x = \frac{\sqrt{\left(\tfrac{1}{2}b\right)^2 + ac} + \tfrac{1}{2}b}{a},$$

and the rule for solving equations of the form $bx = ax^2 + c$ is:

$$x = \frac{\sqrt{\left(\tfrac{1}{2}b\right)^2 - ac} + \tfrac{1}{2}b}{a}.$$

64 Diophantus is calculating a lower bound for his approximation. The square root of the discriminant of the solution for the true value of 1 Number will be greater than the discriminant found here, which is $\sqrt{45}$. So he sets a bound on the discriminant at 7.

65 In fact, the square root of 1,007 lies between 31 and 32; the solution proposed by Diophantus adopts 31 as an upper limit.

Evidence from al-Karajī, given next, suggests that Diophantus knew that this equation type can also yield another positive solution,

$$x = \frac{\frac{1}{2}b - \sqrt{\left(\frac{1}{2}b\right)^2 - ac}}{a}.$$

From the calculation of the discriminant of Problem VI^G.6, the rule for solving the third composite type $c = ax^2 + bx$ should then have been

$$x = \frac{\sqrt{\left(\frac{1}{2}b\right)^2 + ac} - \frac{1}{2}b}{a}.$$

The rules followed by Arabic algebraists to solve composite equations are nearly the same. The only difference is that Arabic authors first normalize the equation, that is, they set the number (coefficient) of the *Māl*s to one before applying the rule. Here is an example from ʿAlī al-Sulamī (MS Sbath 5, f. 7v, l. 8):

> six *Māl*s and twenty dirhams Equal twenty-two Roots. Return that to one *Māl*, by which you divide everything you have by six, to get: a *Māl* and three dirhams and a third[66] Equals three Roots and two thirds. Halve the number of Roots to get one and a half and a third. Multiply that by itself to get three and a third and a fourth of a ninth. Cast away the dirhams from it, leaving a fourth of a ninth. Take its root, which is a sixth, then add it to half the Roots, to get two. This is a root of the *Māl*, and the *Māl* is four. And if you subtract it from half the Roots, it leaves a dirham and two thirds, which is the Root, and the *Māl* is two and seven ninths.

The Arabic rule for solving the normalized equation of the form $x^2 + c = bx$, in modern notation, is:

$$x = \frac{1}{2}b \pm \sqrt{\left(\frac{1}{2}b\right)^2 - c}.$$

The rules for the other equations in Arabic follow this pattern. First, the number of *Māl*s is set to one, then the rule is applied to obtain the solution.

We are fortunate that al-Karajī gives us what is probably Diophantus's own derivation for his rules. In the algebra textbook *al-Fakhrī*, al-Karajī gives three rules each for solving each of the three composite equations. First, he gives the standard Arabic rule which begins by normalizing the equation. Next, he solves the equation like Diophantus, without first setting the number of *Māl*s to one. In his third solution, he gives Abū Kāmil's rule for finding the *Māl* directly. These solutions are followed by geometric proofs for each of them, and after the proofs he gives a derivation that he attributes to Diophantus. For the sample type 4

---

66 The text mistakenly has "two thirds".

equation "a *Māl* and ten Roots Equal thirty-nine units" he writes (Saidan 1986, 154.13):

> And if you want to find the root of the *Māl* in the manner followed by Dio-phantus, you search for a number which, if added to a *Māl* and ten Things, has a root. It is nothing but twenty-five, which added to a *Māl* and ten Things has a root that is a Thing and five dirhams. And you knew that a *Māl* and ten Things are thirty-nine units, so if you remove the *Māl* and ten Things, and you put in its place thirty-nine units, they become sixty-four units. Its root is eight, and that Equals a Thing and five dirhams. So the Thing Equals three dirhams, which is the root of the *Māl*.

Transforming this to modern algebra, adding 25 to $x^2 + 10x$ gives the polynomial $x^2 + 10x + 25$, which is the square of $x + 5$. Substituting the $x^2 + 10x$ with the 39 from the equation, we have that $39 + 25$, or 64, is the square of $x + 5$. Their square roots must then be equal, so $x + 5$ is 8. This gives the answer $x = 3$. The work takes place within the context of the squaring of $x + 5$, and the equation serves to substitute the $x^2 + 10x$ with the 39.

For the sample equation $x^2 + 21 = 10x$, the derivation is (Saidan 1986, 159.17):

> And if you want to solve this problem according to the method of Diophan-tus, you look for a square which, if you subtract from it (and the *Māl*) ten Roots which are equal to the *Māl* and twenty-one units, then the remainder is a square. So make that square from a side of a Thing less five, or five less a Thing, and each of them leads to a quantity of units, and that quantity is a *Māl* and twenty-five units less ten Roots. Replace ten Roots with a *Māl* and twenty-one units, since it equals it, leaving four in number. And a root of that is two, so (if) five less a Thing is two, then the Thing is three. And (if) a Thing less five is two, then the Thing is seven.

Completing the square for $x^2 - 10x$ gives $x^2 + 25 - 10x$, which is the square of $x - 5$ or $5 - x$ (depending on whether x is greater or smaller than 5). Replacing the $10x$ with $x^2 + 21$ from the equation, we find that the remainder is 4, which is equal to the square of $x - 5$ or $5 - x$, and the two solutions follow. Again, the work takes place in the context of the squaring of a binomial, and the equation serves to make a substitution.

Because of a lacuna in our manuscripts of al-Karajī's book, the method of Dio-phantus is lacking for the third composite type. But we are again fortunate because two other works repeat these methods of Diophantus from *al-Fakhrī* for all three composite equation types. In a little-studied work titled *Causes of Calculation* al-Karajī reworks the three methods of Diophantus into proofs for the standard Arabic rules. For equations of the form $bx + c = ax^2$ al-Karajī works through the arguments in the context of the equation $5 + 4x = x^2$. He completes the square for $x^2 - 4x$ to get $x^2 + 4 - 4x$, which is the square of $x - 2$. Substituting the x^2 with the

$5 + 4x$ from the equation, we have that 9 is the square of $x - 2$, and the solution then falls out. The other work that copies these methods is the book *Guiding Light on Algebra* (*Nūr al-dalāla fī l-jabr wa-l-muqābala*) by the thirteenth-century Persian polymath Fakhr al-Dīn al-Ḥilātī, an associate of Naṣīr al-Dīn al-Ṭūsī at the Maragha observatory. Among the several rules and derivations that he gives for the three composite types are the Diophantine derivations, though neither Diophantus nor al-Karajī is mentioned. The wording of these derivations for the first two types matches very closely those in *al-Fakhrī*, though al-Ḥilātī works with different sample equations. His third derivation runs like this:

> A *Māl* Equals four Things and five numbers. Subtract from each of the two sides four Things, leaving: a *Māl* less four Things Equal five in number. Then you look for a number such that if you add it to a *Māl* less four Things, the result is rooted (i.e., it has a rational square root). You find (it to be) four dirhams. So you add it to it, so it becomes: a *Māl* and four numbers less four Things, and its root is a Thing less two dirhams. But the *Māl* less the four Things are five numbers, so you add it to four dirhams to substitute for the *Māl* less four Things. The outcome is nine, and its root is three, and that is a Thing less two dirhams. So your Root is five.
>
> (al-Ḥilātī, Tehran MS 4409, p. 28.10)

Here the argument is slightly different from what we see in al-Karajī's *Causes of Calculation*. Instead of substituting x^2 with $4x + 5$, al-Ḥilātī changes the equation to $x^2 - 4x = 5$ and completes the square for $x^2 - 4x$. Then, knowing that $x^2 + 4 - 4x$ is the square of $x - 2$, he replaces the $x^2 - 4x$ with 5, so that 9 is then the square of $x - 2$, and the solution follows.

Given the solutions we find in Diophantus and the rules and derivations/proofs in al-Karajī and al-Ḥilātī, it seems that in a lost portion of the *Arithmetica*, between Book VII and the fourth Greek book, Diophantus derived his rules for solving composite equations by the methods reported in our Arabic sources. Al-Karajī presumed that the simplified equation is normalized, but the derivation does not require it. Given an equation of the form $ax^2 + bx = c$, for example, one can multiply by a to get $(ax)^2 + abx = ac$, and then complete the square for $(ax)^2 + abx$ by adding the square of half of b. This adjustment will give the rules followed by Diophantus. Instead, dividing each term by a will give the Arabic rules.

While Diophantus most likely justified his rules through these derivations, several Arabic books instead justify them with proofs. Some give geometrical proofs, beginning with the earliest books of al-Khwārazmī and Ibn Turk. Later, starting with al-Karajī's *Causes of Calculation* and *al-Kāfī*, other authors give arithmetical proofs based in one of several types of argument.[67] In the late ninth century Thābit ibn Qurra gave what amounts to geometrical derivations based in Euclid's

---

67 These foundations can be completing the square, arithmetical versions of Euclid's *Elements* Propositions II.5 and II.6, a particular finger-reckoning rule for multiplying numbers, or treating the

propositions II.5 and II.6 in the style of the propositions in Euclid's *Data*, which he calls both "proofs" (*barāhīn*) in the title and "solutions" (sing. *istikhrāj*) in the text. Many books give no proofs or derivations at all, including those of ʿAlī al-Sulamī and Ibn Badr.

Stage 4, calculating the unknowns from the value of the Number/Thing requires no separate subsection. The algebraist returns to the initial assignments to determine their values.

2.7 Enunciations vs. equations and the assignments of names

2.7.1 Indeterminate problems or indeterminate equations?

Today we make equations out of just about any kind of mathematical statement. Take, for example, this rhetorical rule:

> If you want to add the consecutive numbers from one to whatever number you wish . . ., the way to do it is to multiply the upper number by itself and add to its outcome the same as that upper (number). Then half the outcome is the desired amount.

We can naturally transform this into the algebraic formula

$$\sum_{i=1}^{n} i = \frac{n^2 + n}{2}.$$

But while the verbal (procedural) and notational (formulaic) versions may be regarded as being identical in the eyes of those well-versed in algebra, this process of making a formula out of a verbally stated procedure remains a daunting task for many students. The statement and the equation, in fact, belong to two different realms. The former resides in arithmetic, and the latter is its translation into modern algebraic terms. The difference between the arithmetical and algebraic versions is not due to the fact that the former is expressed in words while the latter is in notation. Premodern algebra is usually expressed in books rhetorically, too. The verbal statement of the identity belongs to arithmetic because no number has been given a name. The word "number" remains a common noun, and the rule is presented as a sequence of operations to perform on it. The formula, instead, is written with algebraic expressions on named numbers (*i* and *n*). While many today may not even notice this distinction and see the two as being identical,[68] premodern authors consistently and deliberately differentiated between arithmetical and

---

equation as a problem and setting up a new equation that simplifies to a simple type (Oaks 2018a; al-Ḥilātī, MS 4409).

68 As does B. L. van der Waerden, in his response to Unguru, when he writes "$(a + b)^2 = a^2 + b^2 + 2ab$ can be stated in words thus: 'The square of a sum is the sum of the squares of the

algebraic language. This particular statement is translated from the chapter on finger reckoning in the *Completion of Arithmetic* of al-Baghdādī (d. 1038 CE) (1985, 179), a book that contains no algebra at all. In medieval Arabic, such rules were always expressed in terms of operations on unnamed amounts, with no connection to the named unknowns of algebra and the equations formed from them. If this distinction has gone unnoticed by many historians it is because the range of what is considered to be "algebra" has expanded greatly since the seventeenth century so that any statement involving unspecified numbers can now be called "algebraic". But in medieval Arabic, Latin, Italian, and in sixteenth-century Europe, algebra was strictly a method of problem solving using named unknowns, and it did not trespass into the realm of arithmetical identities.

It is not just arithmetical identities that are often restated in modern algebraic form by modern historians. The enunciations of most problems, too, can easily be converted into modern equations, and often historians provide such equations to explain to the modern reader what is being asked. We have done this ourselves in the examples in this chapter. Diophantus's Problem II.20, to pick another example, asks, "To find two numbers such that the square of each, if it receives in addition the other, makes a square". This has been rendered by different historians as:

$x_1^2 + x_2 = \square$, $x_2^2 + x_1 = \square$	Tannery (Diophantus 1893–95, II, 290)
$x^2 + y = u^2$, $y^2 + x = v^2$	Heath (1910, 262)
$X^2 + Y = \alpha^2$, $Y^2 + X = \beta^2$	Ver Eecke (Diophantus 1959, 68 n. 1)
$\begin{cases} a^2 + b = \square, \\ b^2 + a = \square' \end{cases}$	Sesiano (1982, 465)
$x^2 + y = z_1^2$, $y^2 + x = z_2^2$	Rashed and Houzel (2013, 196)

and in our commentary[69] we show it as

$$\begin{cases} x^2 + y = \square, \\ y^2 + x = \square' \end{cases}$$

While these notational versions may help us understand what is being asked in the problem, the differences between the two should be kept in mind. We historians have taken the steps of naming the unknowns as x_1 and x_2 / X and Y / a and b / x and y, and of constructing equations from the operations stated in the enunciation. Premodern authors, however, did not regard enunciations as being equations, nor as belonging to algebra. The naming of unknowns takes place in the beginning of an algebraic solution, and an equation is only set up after operations have been performed. In this problem, Diophantus names the two

---

terms and twice their product.' The statement in words says exactly the same thing as the formula" (Van der Waerden 1976, 200).

69 For our notation, see our commentary (Part 3) and the conspectus in Appendix 4.

numbers as "1 Number" and "1 unit, 2 Numbers", which, after also naming the second square, eventually leads to the equation "4 Powers, 5 Numbers, 1 unit. These are equal to . . . 4 Powers, 4 units lacking 8 Numbers", which, in modern notation, would be $4x^2 + 5x + 1 = 4x^2 + 4 - 8x$ (keeping in mind that the premodern polynomials contain no operations). The enunciations in the *Arithmetica* do not express equations. It is in the process of solving the problems that equations are established.

Not only are enunciations in the *Arithmetica* not equations, but with two related exceptions to be discussed presently, the equations in all 10 extant books are determinate in one unknown.[70] Diophantus solves his indeterminate problems by making assignments that satisfy some of the conditions of the enunciation and then setting up a single determinate equation to satisfy the one remaining condition. He sets up indeterminate equations only in Problems IV^G.16 and IV^G.17. These equations are designed to facilitate a renaming of the desired unknowns, not to solve the problem at hand. In both problems, after making initial assignments to satisfy two of the four conditions, he wants "13 Numbers" to be a square. So, he equates this amount to "169 Powers", where this "Power" is unrelated to his "Number". In modern notation, we would write the equation as $13x = 169y^2$. Solving for "1 Number" allows Diophantus to re-express his assignments by substituting each "Number" with "13 Powers", and with this renaming a third condition is satisfied. Then, to satisfy the fourth condition, a determinate equation is set up using these new names and is solved. In modern terms, the indeterminate equation allows for a "change of variables", though such a phrase is anachronistic in that premodern algebraists worked with unknowns, not variables. Setting aside these two equations, it is Diophantus's problems, not his equations, that are indeterminate.

Indeterminate equations can be found in other premodern authors, however. The solutions to most indeterminate problems in Arabic books take the approach of Diophantus by making assignments that satisfy conditions, and then setting up an equation in one unknown. But some problems are solved by setting up an indeterminate equation in two or more independent unknowns and then choosing values for the "free variables". Two names common for these other unknowns are *dīnār* and *fals*, denominations of gold and copper coin, respectively, and which we translate below with capital letters as "Dinar" and "Fals". In Problem 3 of Abū Kāmil's *Book of Birds*, for example, 100 birds of 4 different species and prices are presented, and one is asked to find how many of each species can be purchased for a total of 100 dirhams. What follows is a translation of the enunciation and the first solution. He sets up an equation between the three independent unknowns – "Thing", "Dinar", and "Fals" – which is

---

70 Zeuthen (1902, 210), Heath (1910, 67), and others following them call an "indeterminate equation" in Diophantus what we call a "forthcoming equation". This is an incomplete equation, which we discuss in Section 2.5.3 and in Appendix 2. This issue is fully covered in (Christianidis 2018a, 2018b).

simplified so that the "Thing" is expressed in terms of the other two (Abū Kāmil 2012, 741–43):

> If someone gave you a hundred dirhams and said to you, buy with them a hundred birds of four kinds: ducks, chickens, pigeons, and starlings. A duck is four dirhams, starlings are a tenth of a dirham, pigeons are two for a dirham, and chickens are a dirham.

(*Stage 1*)

Its rule is that you buy a Thing of the ducks for four Things in dirhams, and a Dinar of starlings for tenth of a Dinar in dirhams, and a Fals of the pigeons for half a Fals in dirhams. What is left of the hundred dirhams is a hundred dirhams less four Things and less a tenth of a Dinar and less half a Fals, and of the chickens a hundred less a Thing and less a Dinar and less a Fals. You buy them from figuring a chicken for a dirham, so it is the same as its number, which is a hundred dirhams less a Thing and less a Dinar and less a Fals. So it Equals what remains of the dirhams, which is a hundred dirhams less four Things and less a tenth of a Fals and less half a Dinar $(100 - x - y - z = 100 - 4x - \frac{1}{10}y - \frac{1}{2}z)$.

(*Stage 2*)

So you confront it,[71] to get that the Thing, after the confrontation, Equals three tenths of a Dinar and a sixth of a Fals $(x = \frac{3}{10}y + \frac{1}{6}z)$.

(*Stage 3*)

Make the Dinar something that has a tenth, such as ten, which are the starlings, and the Fals something which has a sixth, such as six, which are the pigeons, and the Things are four, which are the ducks, since they are three tenths of the starlings and a sixth of the pigeons. The chickens are what remains to complete the hundred, which are eighty chickens. This is one correct (solution).

In some problems in his *Book on Algebra*, Abū Kāmil sets up an equation in four independent unknowns, solves for one of them, then uses it to make a substitution in the setup of the next equation, after which the process is repeated.[72] This is close to working with a system of equations in more than one unknown, but the equations are not stated together. But we do find a system of equations set up

---

71 By "confront it", he means to simplify the equation. Abū Kāmil is unusual in using the word "confront" (*qabila*) where others would write "restore and confront".

72 See Problems 39–43 in the third part of the treatise, especially the second solution to Problem 40 (Abū Kāmil 2012, 655–79).

and solved in al-Samaw'al's second solution to *Arithmetica* (determinate) Problem I.16, translated in Section 3.3.9. He sets up three equations in three unknowns and adds the three equations to get a new equation, which he uses together with each of the original three to find his solution. Another medieval example of a system of equations is set up and solved in an anonymous fourteenth-century Italian *Trattato d'algibra*. This problem concerns three people who want to buy a goose instead of the usual horse. In the solution the money of the first is called a "thing" (*chosa*), and the value of the "goose" (*ocha*) is treated as an independent unknown. Two equations are set up, which in modern notation would be $7g = 13t + 4$ and $4g = 2t + 176$, where g is the value of the goose and t is the value of the thing. This system is then solved by double false position, positing first that the value of the goose is 40, and then 80 (Anonimo 1988, 149).

Systems of linear equations in more than one unknown are not rare in premodern algebra, and they become even more common in sixteenth-century Europe. For example, in one problem in his 1559 *Logistica*, Jean Borrel sets up a system of four linear equations with four unknowns that he calls A, B, C, and D, and which he solves by repeatedly subtracting an equation from a multiple of another equation to eliminate unknowns until he is left with a single equation in one unknown.[73] And looking back again to ancient Greece, an early second-century CE papyrus found in Egypt shows the elimination of an unknown by subtracting a multiple of one "equation" from another in an arithmetical context for a problem similar in structure to Abū Kāmil's birds problem.[74] Perhaps Diophantus just never found it expedient to set up indeterminate equations the way Abū Kāmil, al-Samaw'al, and other later algebraists did.

It is in part because of the expanded reach of the term "algebra" today that many historians refer to the enunciations in the *Arithmetica* as equations, even if they recognize the role of the names of the powers ἀριθμός, δύναμις, and κύβος in the solutions. Heath (1910, 66) writes of the "[s]imultaneous equations involving quadratics" that characterize some enunciations, while in the majority of his analysis he correctly focuses on the equations set up in the solutions.[75] Jacques Sesiano, too, can write on a single page of the equations comprising the enunciation and the equations established in the solution (2004, 260). This ambivalence regarding the nature of equations may account for the lack of attention given in the secondary literature to the techniques Diophantus employs in assigning algebraic names to unknowns in stage 1 of his solutions. We address these techniques now.[76]

---

73 (Borrel 1559, 193–94). For another example, see p. 357. Heeffer (2010) discusses independent unknowns in Renaissance Europe.

74 Problem III.iv, edited and translated in (Winter 1936, 39, 44–45), with commentary on pp. 50–51.

75 See also p. 114 for his claim that there are indeterminate equations in the Palatine Anthology.

76 In our commentary we have taken care to distinguish problems from equations and to draw attention to Diophantus's dexterity in assigning names for the unnamed sought-after numbers in each problem.

2.7.2 *Techniques of naming unknowns*

The transition from the unnamed quantities and the operations to which they are subjected in the enunciation to the named quantities and their aggregations and differences in the solution is initiated by the naming of unknowns in terms of the algebraic powers. How one chooses the names affects the kind of equation that is set up. Take, for example, the first two solutions Abū Kāmil gives for his Problem 2, first mentioned in Section 2.6.2. The enunciation asks: "Ten: you divide it into two parts. You divide each of them by the other, resulting in four and a fourth". (Converted into a modern system of equations, this becomes: $x + y = 10$; $\frac{x}{y} + \frac{y}{x} = 4\frac{1}{4}$.) In the first solution, Abū Kāmil names one of the parts "a Thing" (corresponding to our x), making the other "ten less a Thing" (our $10 - x$), and the resulting equation translates into modern notation as $42\frac{1}{2}x - 4\frac{1}{4}x^2 = 100 + 2x^2 - 20x$. In the second solution, he names one part "five and a Thing" ($5 + x$) making the other "five less a Thing" ($5 - x$), and the equation is set up as the medieval version of $106\frac{1}{4} - 4\frac{1}{4}x^2 = 50 + 2x^2$.[77] The second equation is easier to solve since it simplifies to two terms.

Diophantus requires his solutions to be rational as well as positive, so he chooses his names carefully in order to set up, when possible, a two-term equation with consecutive powers, and when this is not possible, of ensuring in some other way that the answer is "expressible" (ῥητός). To complicate matters, many problems in the *Arithmetica* ask for several unknowns, and thus, several namings are required. Here new namings must be intermingled with prior calculations so that we witness a virtuoso, sequential handling of the different unknowns with their sometimes tentative named assignments and half-built expressions that culminates in the establishment of the desired two-term equation. With this complexity comes different ways of making assignments.

An assignment can be direct or it can be derivative. A simple example of this difference is found in *Arithmetica* Problem I.15. The enunciation and instantiation read:

> To find two numbers such that each one, if it receives from the other a proposed number, has to the remainder a proposed ratio. Now, let it be proposed that the first, if it receives 30 units from the second, becomes its double (i.e., the double of 30 less than the second), while the second, if it receives 50 units from the first, becomes its triple (i.e., the triple of 50 less than the first).

The solution begins with a direct assignment of the second number and a derivative assignment of the first number: "Let the second be assigned to be 1 Number and the 30 units that it gives. The first, therefore, will be 2 Numbers lacking 30 units, so that, after receiving 30 units from the second, it will become its double".

---

77 (Abū Kāmil 2012, 337.19, 339.18). Our translation of the first equation is given at the end of Section 2.4.1.

A direct assignment is made explicitly and (to some extent) arbitrarily. Diophantus could have named the second number "1 Number" or "1 Number, 2 units" or anything else he wanted, but such assignments would have led to a more difficult solution. A derivative, or calculated, assignment is one that arises as a consequence of former assignments. In this example, once the second number is named, by the first condition the first number must be "2 Numbers lacking 30 units" (our $2x - 30$), so that if the first receives 30 units from the second, it becomes double. Abū Kāmil's Problem 2, cited earlier, also shows the two kinds of assignment. If one of the parts is named "a Thing", then the other is necessarily "ten less a Thing", or if it is "five and a Thing", then the other must be "five less a Thing".

Direct assignments of more than one unknown are made, for example, in Problem II.19, the enunciation and instantiation of which read:

> To find three square (numbers) such that the difference of the greatest and the middle has to the difference of the middle and the least a given ratio. Now, let it be proposed that the difference is the triple of the difference.

Diophantus assigns the smaller square to be "1 Power" (our x^2) and the middle "1 Power, 2 Numbers, 1 unit, that is, from a side of 1 Number, 1 unit" (our $x^2 + 2x + 1$), and he derives the name of the greater square, which is "1 Power, 8 Numbers, 4 units" (our $x^2 + 8x + 4$). Here we have two direct assignments and one derivative.

Often it is not one of the desired unknowns that is directly named, but some combination of them. Diophantus's Problem I.19, for instance, asks for four numbers, and it is their sum that is named "2 Numbers" at the start of the solution. From this the four sought-after numbers are given derivative assignments of "1 Number lacking 15 units", "1 Number lacking 20 units", "1 Number lacking 25 units", and "1 Number lacking 10 units", and the equation is formed by equating the sum of these expressions to the original 2 Numbers.

For indeterminate problems involving squares and/or cubes, direct assignments should respect the type, so, for example, one should not name an unknown square as "1 Number" or "2 Powers". The former would make the problem difficult, and the second would make it impossible. Thus, Diophantus will name a square as "1 Power", "4 Powers", "1 Power, 1 unit lacking 2 Numbers", or some other square expression, and a cube as "1 Cube" or "8 Cubes", etc. Diophantus finds ways of assigning names to make a square or a cube in more complicated circumstances, too. In Problem II.26 he needs to give names to two numbers, let us call them A and B, such that $AB + A$ is a square. So, he names A "1 Number" and B "4 Numbers lacking 1 unit", so that that $AB + A$ is 4 Powers, which is a square. Diophantus explains how he found the two names as follows:

> Since, if there are two numbers the greater of which is the quadruple of the smaller except one unit, then their product makes a square if it receives in addition the smaller, I assign the smaller to be 1 Number, and the greater, 4 Numbers lacking 1 unit. And it comes out, similarly, that their product makes a square if it receives in addition the smaller.

Here Diophantus cites the rule that we would write today as $m \cdot (k^2 m - 1) + m = (km)^2$, which is structurally similar with the condition of the problem, and he assigns names modeled on this rule.[78] (Recall that arithmetical identities were not expressed using algebraic language.) A variant of this technique is the conversion of a relation between concrete numbers into a structurally similar relation but this time between algebraic terms, Numbers or Powers. For example, in Problem II.25, which asks to find two numbers such that the square of the sum less either of them makes a square, Diophantus gives the following heuristic explanation for the assignments he is going to make:

> I first take a certain square from which, if I remove two numbers whatever, I leave a square. Let it be 16. Indeed, if this lacks 12 units, it becomes a square, and again if (it lacks) 7 units, it becomes a square. So, I assign them . . . to be in (terms of a) Power – so that the one is 12 Powers, the other 7 Powers, and the square of the sum 16 Powers – and it is established that the square of the sum lacking either makes a square.

The two relations, $4^2 - 12 = 2^2$ and $4^2 - 7 = 3^2$, serve as models for the conditions of the problem, so the one sought-after number is posited to be "12 Powers" ($12x^2$), the other "7 Powers" ($7x^2$), and the square of their sum "16 Powers" ($16x^2$).

In many problems, Diophantus establishes one side of an equation in terms of the names of the powers, and the other side must be a square or a cube. To name this square or cube for one such equation, he chooses its side (i.e., its square or cube root) in such a way that one of the terms on both sides will drop out to give a two-term equation. For example, in Problem II.32 the expression (in modern notation) $16x^2 + 25x + 9$ must be made equal to a square. He assigns the square to be "from a side of 4 Numbers lacking 4 units" so that the square itself is (in our notation) $16x^2 - 32x + 16$. In the resulting equation the $16x^2$s cancel, which gives $57x = 7$. In other cases it is the number of units that cancels. In these situations, not only should the square be made a square expression, but it should be done in such a way as to achieve a two-term equation. This technique was called *al-istiqrā'* by Arabic algebraists, and al-Karajī gives a thorough investigation of it in his *al-Badī' fī l-ḥisāb* (*Marvelous [Book] on Calculation*). Many times Diophantus sets up two such incomplete (forthcoming) equations. The techniques for naming the unnamed squares or cubes in forthcoming equations are described in Appendix 2, where we also give lists of each instance for single and double equations.

In most problems of the *Arithmetica* a combination of several techniques of naming is applied. Problem II.32 illustrates this very well. In it three numbers are requested such that the square of any one of them, if it receives in addition the next one, makes a square. Using modern symbols, the problem asks for x, y, and z such that $x^2 + y = \Box$, $y^2 + z = \Box'$, and $z^2 + x = \Box''$. Diophantus begins the solution

---

78 One might say that the rule serves as a "simulator" for the condition, hence the description of this technique for assigning names as "simulation" in Bernard and Christianidis (2012).

by a direct assignment: he assigns the first number (x) to be "1 Number". Then he finds the names of the other two sought-after numbers on the basis of the following heuristic explanation:

> And since, if a number is one unit greater than the double of a number, the square of the smaller, if it receives the greater, makes a square, let the second be assigned to be one unit greater than the double of the first; so, it is clear that it will be 2 Numbers, 1 unit; and moreover, the third, one unit greater than the double of that; it will be 4 Numbers, 3 units. And it comes out that the square of the first, if it receives in addition the second, becomes a square, (namely) 1 Power, 2 Numbers, 1 unit; and similarly, the square of the second, if it receives the third, makes a square, (namely) 4 Powers, 8 Numbers, 4 units.

Diophantus cites a rule that we would write today as $m^2 + (2m + 1) = (m + 1)^2$, from which two "simulators" are generated, $a^2 + (2a + 1) = (a + 1)^2$ and $(2a + 1)^2 + (4a + 3) = (2a + 2)^2$, which are structurally similar to the last two conditions of the problem. So, the assignments for the remaining two sought-after numbers result immediately: the second number (y) is posited to be "2 Numbers, 1 unit", and the third (z) to be "4 Numbers, 3 units". The solution continues by stating that the square of the third number, if added to the first, must be a square. The sum is "16 Powers, 25 Numbers, 9 units", and this must be made equal to a square ($\square''$). This is a standard *al-istiqrā'* case, and Diophantus assigns the side of this square to be "4 Numbers less 4 units", and the solution follows easily.[79]

2.8 Notation[80]

In the part of the introduction in which he presents the names of the powers, Diophantus also instructs his readers on how to abbreviate these names. For example, he writes "So, the square is called a Power, and its sign is Δ marked with the sign Υ, i.e. Δ^{Υ}, δύναμις". The sign for the units (μονάς, *monas*) is "$\overset{\bullet\bullet}{M}$", a ligature of the first two letters of the word; for Numbers it is "$\mathcal{S}$"; for Cubes it is "K^{Υ}", for κύβος, followed by "$\Delta^{\Upsilon}\Delta$", "ΔK^{Υ}", and "$K^{\Upsilon}K$" for the fourth, fifth, and sixth powers. It has been supposed that the sign "$\mathcal{S}$" for the first power is the final sigma, "ς". Heath (1910, 33ff) criticizes this explanation and proposes that the sign began as a ligature of the first two letters in the word ἀριθμός and that it eventually took on the form we see in the manuscripts.

The sign for ἀριθμός is frequently duplicated in the manuscripts to express the plural, and it is also often given case endings. There is no consistency in these practices, and often the case ending is wrong. According to Tannery, these

---

79 For a thorough discussion of the techniques of assigning names employed in the first three books of the *Arithmetica* we refer the reader to Bernard and Christianidis (2012).

80 Some variations on the notation are shown in Section 4.2.1.

are Byzantine practices, and there is no evidence that Diophantus himself wrote them. He points to the fact that the other signs, for the unit, Powers, etc., are never doubled, and it is the later manuscripts that regularly show the case endings (Diophantus 1893–95, II, xxxiv–xxxix; cf. Heath 1910, 33).

Among the worked-out problems, the names of the powers are sometimes written out in full in the manuscripts, and other times the abbreviations are used. For ἀριθμός and κύβος, the abbreviations even occasionally appear when the word takes its meaning as a common noun. Despite what might be shown in manuscripts, Tannery wrote the common nouns ἀριθμός or κύβος in full, and where the same words are the algebraic names he wrote them in notation. Diophantus also used the sign for λεῖψις (*leipsis*, "lacking" or "less") that we saw in the example from Hero: "the sign of the lacking is a truncated Ψ turned upside down, i.e. "⋔"". That these signs originate with Diophantus himself and are not scribal additions should be clear by the fact that they are presented in the introduction. A scribe would have simply written them without any instructions to the reader on how to form them.

Next is an example of an equation written in notation as it appears in the manuscript Matrit. 4678 (f. 123v), copied in the second half of the eleventh century. It is from Problem VI^G.8, and we expanded it in our translation to "1 Power, 12 4′ units less 7 Numbers, (which is) equal to 1 Power, 1 unit". (The equation is broken across two lines in the manuscript. We show the two pieces in one line.) Note that a line over letters indicates that the letters are to be read as numbers:

In this instance, the "ἀριθμοί" and "ἴσοι" are written in full, though they are often abbreviated in the manuscripts. Also, the word, or the sign, for "lacking" (λεῖψις) is missing. These manuscripts were written in minuscule (lower-case) script, which was introduced during the ninth and tenth centuries. Diophantus would have written it all with capital letters, and it is likely that he would have used abbreviations for the two words,[81] so we can guess that it would have appeared something like this in his autograph:

$$\Delta^Y\overline{A} \; \overset{\bullet\bullet}{\iota\iota} \; \overline{IB} \; \Delta^o \; ⋔ \; \wedge\!\wedge \; \overline{Z} I \Sigma \Delta^Y \overline{A} \; \overset{\bullet\bullet}{\iota\iota}\overline{A}.$$

Spaces between the terms will help the reader make sense of the signs. Also, the numbers are grasped more easily for the modern reader if we replace the Greek alphabetic numerals with Arabic numerals, and the "ΙΣ"/"ἴσοι" with the English word "equal":

$$\Delta^Y 1 \; \overset{\bullet\bullet}{\iota\iota} 12 \; 4′ \; ⋔ \; \wedge\!\wedge 7 \; \text{equal} \; \Delta^Y 1 \; \overset{\bullet\bullet}{\iota\iota} 1.$$

---

81 Cf. what is said in Section 4.2.1 about the different stages in the preparation of a work in antiquity.

Translating it fully into modern notation gives this equation:

$$x^2 + 12\tfrac{1}{4} - 7x = x^2 + 1.$$

But now we have gone too far, having obliterated many of the premodern elements with our modern "+", "−", and exponential notation. So let us return to the penultimate version. Like the rhetorical forms of algebraic expressions, there is no sign in notation for addition because terms were joined together as aggregations. The expression on the right side of the equation, for example, is simply a collection of two numbers of two different kinds. The sign " ⟰ " naturally takes the same meaning as the word it stands for, λεῖψις, to mean "lacking" or "less". Another important difference is that where there is one of a term, the multitude is always written. Where we write x^2 without a coefficient, Diophantus will write "$\Delta^Y 1$" rather than just "Δ^Y". This is done with all the powers consistently throughout the manuscripts. This practice is not due simply to convention. A multitude/number of 1 was always included in notation throughout the premodern period in every language. Let us briefly look into examples of Arabic, Italian, and sixteenth-century European notations to put it in context.

Today we work out problems with the same notation that we print in books. But in medieval Arabic, solutions were not communicated the same way they were worked out. To solve a problem one would perform the calculations mentally via finger reckoning, or one would work it out in notation on a dust-board or other temporary surface. Then, to communicate the solution to others in a book, a rhetorical version was composed. This is because medieval Arabic books were regarded as being transcriptions of spoken language. Books were recited, and in the case of textbooks they were often memorized and recited back. Notation serves no purpose to a listener, so it was not copied into manuscripts as part of the text. This is why we even find numbers written out in words in most books.[82] Al-Khwārazmī and other authors who wrote books explaining calculation with Indian (i.e., Arabic) numerals would still write the numbers in words and show the notation only as an illustration of what one should put down on the dust-board. It is because al-Khwārazmī was not explaining methods of writing and calculating with Indian numerals in his *Book of Algebra* that that text is entirely rhetorical.

An Arabic notation particular to algebra is shown in some textbooks from the Maghreb and al-Andalus, the earliest being Ibn al-Yāsamīn's (d. 1204) *Grafting of Opinions of the Work on Dust Figures*. In this notation abbreviations for the names of the powers are placed above their numbers. *Shay'* ("Thing") is shown as the letter *shīn* (ش), though sometimes it appears as only the three dots of the letter, and other times as the letter without the dots. A *mīm* (م, but with a horizontal tail), the first letter in *Māl*, is placed above the number of the second power, and

---

82 A few authors wrote some numbers in their books in Indian notation, but in a way that does not interfere with recitation.

a *kaf* (ك), the first letter in *ka'b*, stands for "Cube". Here is an example shown in Ibn al-Qunfūdh's 1370 book *Lowering the Veil from the Faces of Arithmetical Operations*:[83]

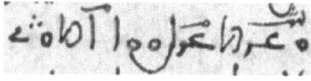

Starting on the right is the *shīn* (ش) above a 40, followed by a stylized *illā* (الا) in black ink and a *mīm* (م) over a 4. The sign for "equal" was a large (sometimes elongated) *lām* (ل), the last letter in *ta'dīl* ("equal"). Next is a 100, which looks like our version of the same number, then a *mīm* (م) over a 1, the stylized "less" in black again, and finally the three dots from a *shīn* (ش) over a 20. Next we write the equation left to right, substituting "t" for "Things", "m" for "*Māls*", "ℓ" for "less", and an "=" for the *lām*, and we write the modern version on the right:

$$\begin{matrix} t & & m & & m & & t \\ & \ell & & = 100 & & \ell & \\ 40 & & 4 & & 1 & & 20 \end{matrix} \qquad 40x - 4x^2 = 100 + x^2 - 20x.$$

As in Diophantus, no sign is shown to "add" the hundred and the *Māl* since they are gathered as an aggregation.

The Arabic notation was designed for performing algebraic calculations on a dust-board or other surface. It is an extension of the notation of Indian numerals, which served the same purpose for arithmetical calculations. In books that show algebraic notation, the calculations are still expressed rhetorically, and the notation appears intermittently only to show the student how to write it on the board. The equation of Ibn al-Qunfūdh shown earlier is stated rhetorically just before as "forty Things less four *Māls* equal a hundred in number and a *Māl* less twenty Things". The algebraic notation was not regarded as being significant for the development of algebra itself but was only a tool for working out problems. It is not mentioned or shown in more advanced books on algebra from the time, including Ibn al-Bannā''s *Book on the Fundamentals and Preliminaries in Algebra* (ca. 1300) and Ibn al-Hā'im's *Commentary on the Poem of al-Yāsamīn* (1387).

We will describe two other notations among the many that were proposed in medieval and Renaissance Europe, one Italian in origin and the other German. In his 1494 *Summa de Arithmetica Geometria Proportioni & Proportionalita* Luca Pacioli adopts the abbreviations "co", "ce", and "cu" for *cosa*, *censo*, and *cubo*, respectively, which are ultimately translations of the Arabic terms for the first three powers *shay'*, *māl*, and *ka'b*. One equation taken from this book is "400. m̃ .40.co. p̃ .1.ce. e sira equale a.1½.ce" (Pacioli 1494, f. 96v, l.-9). In modern notation, it would be $400 - 40x + x^2 = 1\frac{1}{2}x^2$. The m̃ abbreviates *meno*, and p̃ *più*,

83 Ibn al-Qunfūdh, *Ḥaṭṭ al-niqāb 'an wujūh a'māl al-ḥisāb*. Lawrence J. Schoenberg collection, MS ljs 464, copied 11 Sha'bān 849 (November 1445), p. 294, l. 20.

which take the meanings of "less" and "more", respectively. These words had not yet assumed meanings like the modern English "minus" and "plus".

Another fairly common notation first appeared in Germany around the time of Pacioli, employing these special signs for the first three powers: $2\!\varrho$, $\mathcal{Z}$, and $c\!\varrho$. Michael Stifel (1544, f. 242r), for example, writes the equation "$1\mathcal{Z} -22\varrho + 24$ æquata 6 $2\varrho + 72$ " for what is in modern notation $x^2 - 2x + 24 = 6x + 72$. Here the signs "+" and "−" still retain their premodern meanings of "more" and "less". While this equation may look tantalizingly modern, what Stifel writes about his knowns and unknowns, and the presence of the coefficient of 1 on the square term, show that he, too, was still thinking like a premodern algebraist.

In the notations just reviewed, as with the countless other notations that were devised down to the latter part of the sixteenth century, a "1" is always included when there is one of a term. This is a consequence of the relationship between the multitude/number and the power, described in Section 2.5.1. Consider one snippet from the equation of Diophantus: "$\Delta^Y 1$ ⚖︎$12\ 4$"' (our $x^2 + 12\frac{1}{4}x$). If the units were pounds (lb) and ounces (oz) instead of δυνάμεις (Powers) and ἀριθμοί (Numbers), this would read like "lb 1, oz $12\frac{1}{4}$" (1 pound, $12\frac{1}{4}$ ounces). Without the 1, the number of pounds would be unspecified: "lb, oz $12\frac{1}{4}$" ("pounds, $12\frac{1}{4}$ ounces"). "Pound" ("lb") is a type or species of weight, while "1 pound" is a specific weight. Similarly, δύναμις ("Δ^Y") is a type or species of number, while "1 δύναμις" denotes a specific value. Our x^2, instead, already represents a value, so it does not need a coefficient. To reiterate, in a term like our $3x^2$ the "3" and the "x^2" are both numbers that are multiplied together, while in a term like "$\Delta^Y 1$" the "Δ^Y" is the type or species, and the "1" is how many are present. The premodern term cannot be interpreted as a multiplication because "Δ^Y" is not a number, but a type of number. It is for this reason that a "1" is never omitted in premodern notation.

What we know about the Greek notation is consistent with the purpose of the notation in Arabic. Diophantus instructs Dionysius on how to abbreviate the powers, just as Ibn al-Qunfūdh and other Arabic authors explain the Arabic abbreviations to their readers, while they all present their solutions more or less rhetorically. As we shall see in the next chapter, there are other texts showing solutions by algebra from Greek antiquity, and these also show some version of the notation. These include a papyrus of the early second century CE; Theon of Alexandria's commentary on Ptolemy's *Almagest* (Christianidis and Skoura 2013; Skoura and Christianidis 2014); and Par. suppl. gr. 384, the manuscript that conveys the late antique scholia to the arithmetical epigrams of the Palatine Anthology. In this last book the notation is part of the rhetorical text, like that we find in several medieval Italian abacus texts. In Arabic, by contrast, the notation is shown only in figures or illustrations to explain to the student how to write it.

Ever since G. H. F. Nesselmann (1811–71) identified three stages of historical development in algebra in his 1842 book *Versuch Kritischen Geschichte der Algebra* (*Critical Essay on the History of Algebra*) (1842, 301ff), the notation in Diophantus's *Arithmetica* has been classified as a kind of middle ground in the development toward symbolic algebra. Nesselmann calls the most primitive stage "rhetorical algebra", where all calculations are written verbally. Here he placed the *Book of Algebra* of al-Khwārazmī, which shows no symbols at all. Next is "syncopated algebra", which was still essentially rhetorical but with some recurring abbreviations. It is in this category that Nesselmann placed Diophantus. The third stage is "symbolic algebra", where calculations are represented in a language independent of oral presentation.

There are two major flaws with this trichotomy. First, the language written in books is not always the language in which problems were worked out. In Arabic, problems were often solved in notation on a dust-board or some other temporary surface, and then for inclusion in a book a rhetorical version was composed. Also, because of the two-dimensional character of the Arabic notation, it would have been written and read visually, independent of real or imagined speech. It thus fits nicely into Nesselmann's "symbolic" category. The rhetorical version of the same work, on the other hand, was categorized as being "rhetorical". These two ways of writing algebra do not reflect two stages of the development of algebra but are different ways of expressing the same ideas. Second, Nesselmann was unaware of the conceptual differences between premodern and modern algebra, and thus, he could not have appreciated the leap made in the time of Viète and Descartes that included a radical shift in how notation was interpreted. Nesselmann's three stages classify superficial differences in the few books for which he had access. Instead, close study of the ways that algebra is presented in books shows that the development of algebra is divided into two main periods characterized not by notation but by the concepts behind the rhetorical and notational forms.

3   History

3.1 Evidence for the practice of algebra before Diophantus

Diophantus's *Arithmetica* is the earliest preserved work that systematically presents the way to solve arithmetical problems by algebra. But despite claims expressed in the past by mathematicians and historians that Diophantus was, to quote J. L. Lagrange (1736–1813), "l'inventeur de l'Algèbre" (1877, 219),[1] we know that the art of problem solving by algebra was practiced before his time. In this section, we review the evidence for this, which is both varied and fragmentary. We first examine the testimony of the Arabic bio-bibliographer Ibn al-Nadīm (d. 995 or 998), who wrote that the astronomer Hipparchus of Nicaea (d. after 127 BCE) wrote a book on algebra, that this book was translated into Arabic, and that Abū l-Wafā' (940–98 CE) wrote a commentary on it. There is also a recently published papyrus from Hipparchus's century that shows abbreviations for the names of powers up to the eighth, together with the corresponding powers of two. Our most important evidence is a fragment of papyrus from the early second century CE, kept in the collection of the University of Michigan, which preserves three arithmetical problems worked out by algebra. And finally, a Christian theological text from the 220s CE lists the names of the powers up to the sixth in the context of Neopythagorean arithmology.

But before presenting this evidence we first give two preliminary sections on the names of the powers of the unknown, the first being an investigation into non-algebraic meanings of the words in Greek and Arabic, and the second into the ways that the algebraic powers were named in Arabic and Renaissance Europe. These sections provide a necessary background for interpreting the names in Greek texts, both ancient and Byzantine.

---

1 This phrase comes from one of Lagrange's courses in the Ecole Normale, in the year 1795. The passage is the following: "Diophante peut être regardé comme l'inventeur de l'Algèbre; en effet, par un mot de sa préface, ou plutôt de son épître d'envoi (car les anciens Géomètres envoyaient leurs Ouvrages à quelques-uns de leurs amis, comme on le voit aussi par les préfaces des Ouvrages d'Apollonius et d'Archimède); par un mot, dis-je, de sa préface, on voit qu'il a été le premier à s'occuper de cette partie de l'Arithmétique qui a été nommée depuis Algèbre. Son Ouvrage contient les premiers éléments de cette science".

DOI: 10.4324/9781315171470-4

3.1.1 Δύναμις, κύβος, and related terms in Greek and Arabic geometry and arithmetic

We shall see here, based on several independent testimonies, that algebra had been practiced in Greece before Diophantus. The names given to the powers, then, were not his own invention. In fact, as Tannery noted, Diophantus indicates as much in his introduction when he writes that "it is an established fact (ἐδοκιμάσθη) that each of these numbers, once it has been given an abbreviated denomination, is an element of the arithmetical theory".[2] These same terms, sometimes up to the sixth power, are often found in mathematics outside algebra. We presently describe their uses and compare them with the arithmetical terms πλευρά (*pleura*, "side") and τετράγωνος (*tetragōnos*, "square"). This will not only give us some clues regarding the origin of the algebraic names, but it will also serve as a background for our assessment of the lists of the names of the powers found in later Greek texts.

We begin with the second degree. The word τετράγωνος was the common word for a geometrical or arithmetical "square" in Greek mathematics. Euclid, for instance, explains how "to describe a square on a given straight line" in *Elements* Proposition I.46, and he defines a "square number" as "equal multiplied by equal, or a number which is contained by two equal numbers" in Definition 18 of Book VII. Hero, to pick just one other example, also gives examples of the word τετράγωνος in both senses. He writes of "the square on (ἀπό) ΒΑ" in one place, and "the closest square to 720 is 729" in another (Hero 2014, 312.7, 164.10). It is in the numerical sense of the word that Diophantus speaks of τετράγωνος ("square") in the enunciations of his problems.

According to Liddell and Scott, the common meaning of δύναμις (*dynamis*) is "power", "might", "strength", "ability to do anything", with a secondary meaning of "worth" or "value" of money.[3] The word is found in a geometrical setting in many Greek books, where it means "square", but in a different way than τετράγωνος. Most notably, δύναμις plays a critical role in Plato's *Theaetetus* 147c7–148d7, where the young Theaetetus expounds on the notion of the incommensurability of the sides of certain squares. Because of the importance of this part of the dialogue for history of mathematics, and because of the seemingly contradictory ways that δύναμις is used there, the mathematical meaning of the word has been the topic of studies by several scholars.[4] Fortunately, there is more agreement than disagreement among them, so we are safe in the following characterization.

---

2 Tannery noted "notre auteur reproduit une tradition consacrée" (1912i, 68).

3 See LSJ (= Liddell, Scott, and Jones 1996), *s.v.* δύναμις.

4 Some of those taking part in this extended debate include Á. Szabó (1969), W. R. Knorr (1975), M. F. Burnyeat (1978), C. M. Taisbak (1980), J. Høyrup (1990), and B. Vitrac (2008); cf. Knorr (1979) and Burnyeat (1979). See Burnyeat (1978) and Høyrup (1990) for references to others. Szabó's thesis is repeated in English in Szabó (1978).

In geometry, a δύναμις is a characteristic of a line. One did not make reference to a δύναμις directly by naming its opposite vertexes, but rather to a side in respect of δύναμις. As Jens Høyrup proposed, a δύναμις is "a square identified with its side" or "a line seen under the aspect of a square" (1990, 210). The word often occurs in the dative form, δυνάμει ("in power", or "in square"), as in this passage of Archimedes:

> And since the rectilinear ⟨figures⟩ inscribed in the circles A, B are similar, they have the same ratio to each other, which their radii ⟨have⟩ in square. But the triangles ΚΤΔ, ΖΡΛ also have to each other ⟨the⟩ ratio, which the radii of the circles ⟨have⟩ in square; therefore, the rectilinear ⟨figure⟩ inscribed in the circle A to the rectilinear ⟨figure⟩ inscribed in the ⟨circle⟩ B, and the triangle ΚΤΔ to the triangle ΛΖΡ, have the same ratio.
>
> (Archimedes 2004, 86)

Wilbur Knorr (1978), and following him Høyrup (1990, 204–05), explain that the use of δύναμις in geometry was common until about the time of Euclid. Afterward, people preferred τετράγωνος, preserving δύναμις only in traditional, formulaic expressions.

The way the word δύναμις was applied suggests that it was somehow associated with measurement or calculation. Because one never spoke of a δύναμις given in position, the term necessarily intended only its relative size with respect to other planar magnitudes. This is consistent with the frequent juxtaposition of the terms δύναμις and μῆκος (*mēkos*, "length"). We find the pair in Plato's *Theaetetus*, in the works of Archimedes (Knorr 1978, 264), and Eudemus writes regarding the work of Hippocrates, "the diameter being twice the side in length is its quadruple in square".[5] Euclid's only use of δύναμις occurs in Book X, where lines "μήκει σύμμετροι" ("commensurable in length") are juxtaposed with lines "δυνάμει σύμμετροι" ("commensurable in square"). Hero takes the terms into the realm of numerical measure by comparing ratios δυνάμει with ratios μήκει in Problem I.15: "there remains 25 in square (δυνάμει), which are 5 in length (μήκει)" (Hero 2014, 184.22). Based on examples like these, Knorr suggests that δύναμις took the meaning of a "quantity of two-dimensional extension" (1975, 67). In geometry, it designated specifically the size of a two-dimensional surface and not a particular square.

Both Árpád Szabó and Jens Høyrup have suggested that the mathematical δύναμις was understood to take a meaning related to "value" or "worth". Szabó (1978, 39) writes that "the term δύναμις meant 'value' in general in the language of finance, so in geometry it came to mean 'the value of the square of a rectangle', then 'the value of the square' and finally just 'square'".[6] Høyrup (1990, 208) com-

---

5 ([Bulmer-]Thomas 1939, 251.4). Translation adapted from those of Thomas (1939) and Høyrup (1990, 203).
6 But he misunderstood the word to refer to a rectangle considered as a square.

pares δύναμις and its related verb δύνασθαι with the Babylonian verb *maḫārum*, which means "to stand up against, to encounter, to receive [an antagonist, an equivalent, a peer]". The related noun *mitḫartum* meant "geometric square", or more literally, the "confrontation of equivalents", while another related term, *maḫirum*, meant "countervalue", or "commercial rate". He notes that "[a] number of texts show that the *mitḫartum*, when a number is ascribed to it, *is* the length of the side and *possesses* an area" (1990, 209; his emphases). "So, the linking of 'square,' 'side of square,' 'commercial rate,' 'equivalence,' and 'confrontation of force,' so puzzling in Greek mathematics, is shared with the mathematics of the old eastern neighbor" (1990, 208). Høyrup is careful to only suggest that the Babylonian term *might* be the source for the Greek term since "[b]oth concepts could have developed independently on the basis of analogous or shared measuring *practices*" (1990, 211; his emphasis).

The parallel between the Babylonian and Greek terms invites a comparison with *māl*, the Arabic term for the second power in algebra. The ordinary meaning of *māl* is "property", "wealth", or, as Lane notes, originally "what one possesses of gold and silver" (1863–93, 3026). The word was apparently adopted for use in algebra via practical arithmetic problems asking for an unknown *māl* ("amount of money"), like the problem of Ibn al-Yāsamīn translated in Section 2.3. But this connection with Arabic arithmetic does not explain why *māl* is the name of the *square* of the first-degree unknown. Because the Babylonian *mitḫartum*, the Greek δύναμις, and the Arabic *māl* share the common meaning of "possession/wealth", they may all have come to denote the square that a side or a number "possesses". Without any further textual evidence, we cannot suggest more than that the Arabic *māl* might be a claque of either *mitḫartum* or of δύναμις, or that its meanings followed the same path as the Babylonian and Greek terms.

So much for the meaning of δύναμις in geometry. Høyrup (1990, 206–07) also reviews the evidence for an arithmetical meaning of the term, beginning with Plato's *Republic* 587d, and he finds that in logistic one spoke of a number δυνάμει ("in power/value/square") in much the same way that one spoke of a line δυνάμει. There is, then, a difference in use between τετράγωνος and δύναμις in logistic that is conveyed in the passage in Plato's *Republic*. The former is a category of number, which includes 4, 9, $\frac{25}{16}$, etc., while the latter is the square *of* a number. For example, the number 9 is a τετράγωνος, but 9 δυνάμει is 81.[7] Said another way, while any number either is or is not a τετράγωνος, a δύναμις is the value *in square* of any particular number.

Greek algebraists gave δύναμις still a different meaning as the name of the square of the unknown ἀριθμός. Although the algebraic δύναμις is generated from the ἀριθμός through multiplication, one did not need to cite the ἀριθμός when working with a δύναμις. The δύναμις thus gained an independence from its "side", much like the independence granted to the Arabic *jidhr* ("root") in algebra:

7 See, for example, Plato's *Republic* 587d, discussed in Høyrup (1990, 206).

in arithmetic, a *jidhr* is always the root of some number, while in algebra it is simply "the root".

The word κύβος (*kybos*, "cube") can take a geometrical or arithmetical meaning in Greek mathematics just as τετράγωνος does, and again both meanings are found in Euclid's *Elements*. Book XI, Definition 25 reads, "A cube is a solid figure contained by six equal squares", and Book VII, Definition 19 states, "And a cube is equal multiplied by equal and again by equal, or a number which is contained by three equal numbers". But this time there is no counterpart like δύναμις to signify the measure or value of the cube on a side, so the term κύβος serves also as the name of the algebraic third power. This suggests that the algebraic terms for the first three powers were borrowed from arithmetic and not the other way around.[8]

In both the geometrical and arithmetical senses of κύβος and τετράγωνος, the one-dimensional root is called its πλευρά, "side". Contrast this with δύναμις: where one speaks of the πλευρά of a τετράγωνος, one would speak of the δύναμις of an ἀριθμός.[9] The references are reversed, and this may account for how the terms ἀριθμός and δύναμις were adopted for their technical meanings in Greek algebra: the powers are understood to be generated through multiplication from the ἀριθμός, as Diophantus explains in his introduction.

The word δυναμοδύναμις for the fourth power in an arithmetical sense is found in Problem I.17 of Hero's *Metrica*. In four instances, Hero writes the phrase "δυναμοδύναμις on (ἀπό) ΒΓ" to mean the fourth power on the line ΒΓ, whose size is given from the outset.[10] To arrive at this formulation, he establishes a proportion based in a lettered geometric diagram: the square (τετράγωνον) on ΒΓ is to the square on ΑΔ as 4 is to 3. Then, he multiplies both squares by the square on ΒΓ to get four-dimensional amounts, the stated δυναμοδύναμις and the "(square) on ΒΓ by (ἐπί) the (square) on ΑΔ" (Hero 2014, 190.8–9). These may make no sense as geometric magnitudes, but the goal is to compute with them numerically.[11] The language of Hero in this passage is a mix of arithmetic and geometry. The preposition ἀπό ("on") and the letters from the diagram are geometrical, while the preposition ἐπί and the term δυναμοδύναμις are arithmetical. Ἐπί ("by") implies the multiplication of numbers, and δυναμοδύναμις would be meaningless in ancient geometry.

In one instance, Diophantus uses the term δυναμοδύναμις in an arithmetical sense to mean the fourth power of a number. In the first lemma to Problem VI[G].12 he writes: "But the area of the triangle is the sextuple of the δυναμοδύναμις of the

---

8 Both Klein (1968, 141) and Sesiano (1982, 43 n. 33) make the same suggestion but without specifying any evidence.

9 It is probably only because of a lack of texts on practical arithmetic that we can point to no instance of a phrase such as "3 δυνάμει" for 9. Practitioners tended not to write books.

10 (Hero 1903, 48.11,19,21), where line 5 has simply "δύναμις" in error; (Hero 2014, 190.7,10,12,13). The word is sometimes stated with the definite article.

11 These magnitudes designate sizes, just like the four-dimensional magnitudes that François Viète would build in Propositions XIIII and XV in his 1593 work *Effectionum Geometricarum* (Oaks 2018c).

⟨smaller⟩ number, while the number for the smaller of the perpendiculars is 3 of the squares (τετραγώνων) of the smaller". Note that for the second power he uses the arithmetical term τετράγωνος.

The names of powers up to the sixth, transliterated from Greek, appear in an arithmetical context in a fragment preserved in the *Corpus agrimensorum*, a compilation of ancient Latin works on surveying.[12] The earliest manuscripts of the *Corpus*, Codex Arcerianus A and B, Wolfenbüttel Aug. f. 36, 23, date from the sixth century CE (Thulin 1911, 10). The text of interest occupies 12 pages of Nicolaus Bubnov's edition (Gerbert of Aurillac 1899, 494–503). Bubnov reasonably attributes it to M. Terentius Varro (116–27 BCE). It consists of solved mensuration problems, some of which call for higher powers, like Problem 21:

> There is a field which has a width of 112 feet and a half, but also, the number of *jugera* it has, I multiply *in dynamokybum*, and it makes the length. To seek the *jugera* and the length.
>
> I seek (the solution) this way: As was shown above, I note that a *jugerum* of a field has 28,800 (square) feet. I divide this by the width, 112 feet and a half, which makes 256. Then I take the fourth root (*latus dynamodynami*), which makes 4. This is how many *jugera* there are. To find the length, I multiply the 4 *jugera* just found *in dynamokybum*, which makes 1024. This is how many feet long the field is.
>
> (Gerbert of Aurillac 1899, 502.16)

To multiply (*multipico*) *in dynamokybum*, then, means to take the fifth power of a number. Other problems of the same kind, nos. 14, 19, 20, 22, and 23, call for multiplication *in dynamum* (for the square), *in kybum* (cube), *in dynamodynamum* (fourth power), and *in kybokybum* (sixth power). Roots are expressed similarly. The text has *latus quadratum* for the square root, *latus kybi* for the cube root, *latus dynamodynami* for the fourth root, and *latus dynamokybi* for the fifth root. Writing *in dynamum* for the squaring of a number is not found in any other Latin text. The usual way was to multiply a number *in se* ("by itself"), as in the passages we will translate below.[13] This way was common in Greek, too, where, for example, Hero writes, "the 13 by themselves results in 169" (τὰ ιγ ἐφ' ἑαυτά· γίγνεται ρξθ) (Hero 2014, 158.13).

Another text in the *Corpus*, a compilation of the writings of two authors, Epaphroditus and Vitruvius Rufus, poses and solves mensuration and arithmetic problems (Gerbert of Aurillac 1899, 516–51). Bubnov dates Epaphroditus to no later than the second century CE, and this Vitruvius may be the same person as the author of *De Architectura* (1st c. BCE). Near the end of the work the fourth power and the cube of 10 are found:

---

12 The Corpus is edited in Gerbert of Aurillac (Pope Sylvester II) (1899). See Folkerts (1992) for an overview of the work.

13 Many examples are found in the texts in Gerbert of Aurillac (1899).

Ten *in* ⟨*dynamo*⟩*dynamum*. I make 10 by itself, which makes 100, and a hundred by itself, which makes 10,000. It will be ten *in* ⟨*dynamo*⟩*dynamum*.[14]

Ten *in quibo*. You will make 10 by itself, which makes 100. This you carry out tenfold, which makes 1000. It will be 10 *in quibo*.

(Gerbert of Aurillac 1899; Balbus *et al.* 1996)

The word *quibo* appears to be another transliteration of κύβος since we would expect that a Latin verb "to cube" would be related to the noun *cubus* ("cube"). In the next problem, the sum of the cubes from 1^3 to 10^3 is found using the same phrase *in quibo*: "Also, from 10 to 1 *in quibo*".[15]

The names of the powers took on arithmetical meanings in Arabic, too. Recall the names in Arabic algebra: *shay'* ("thing") or *jidhr* ("root") for the first power, then *māl*, *ka'b* ("cube"), and *māl māl* for the second, third, and fourth powers. Judging from our extant texts, it was al-Karajī who standardized the terms for the higher powers in Arabic (see Section 3.1.2). Starting from the fifth power and extending to the ninth he calls them *māl ka'b*, *ka'b ka'b*, *māl māl ka'b*, *māl ka'b ka'b*, and *ka'b ka'b ka'b*. With some differences for the first three powers, these terms were also used to mean "perfect square", "perfect cube", etc. in arithmetic. For example, Naṣīr al-Dīn al-Ṭūsī, in a 1265 CE arithmetic book that contains no algebra, applies the terms up to the ninth power in his description of expanding the powers of a binomial (Saidan 1967, 138ff).

The differences between the algebraic and arithmetical series are much like the differences between their Greek counterparts. The usual Arabic term for a geometrical square or a square number is *murabba'*, and a geometric or arithmetical cube is called a *muka''ab* ("cube", with the same root as the algebraic term *ka'b*). These correspond nicely with the meanings of τετράγωνος and κύβος. The word for the "side" of a *murabba'* or a *muka''ab* is *ḍil'*, the Arabic version of πλευρά. It was natural, then, that these three terms, *murabba'*, *muka''ab*, and *ḍil'*, were chosen to translate their Greek equivalents in the geometrical and the arithmetical books of Euclid's *Elements*. Neither *murabba'* nor *ḍil'* is used as an algebraic unknown in Arabic, just like their Greek counterparts τετράγωνος and πλευρά.[16] And with two known exceptions, Arabic algebraists respected the distinction between the arithmetical *muka''ab* and the algebraic *ka'b*. The exceptions are Sinān ibn al-Fatḥ, who writes *muka''ab* for the algebraic term, and al-Karajī, who sometimes has *muka''ab* instead of *ka'b* in algebra. (In Arabic script *ka'b* and *muka''ab* differ by a single letter.) Another word indicating that a number is a

---

14 We follow Guillaumin in restoring *dynamum* to ⟨*dynamo*⟩*dynamum* (Balbus et al. 1996, 194).

15 "Item *X* ab asse in quibo" (Gerbert of Aurillac 1899, 550.14). It is later written "ab asse usque *X* in quibo" (line 18).

16 There is one exception to this rule, in Ibn Fallūs. For first- and second-degree equations he calls the first-degree term a "root" (*jidhr*), but when the cube is present he calls it a "side" (*ḍil'*). This is a borrowing from arithmetic, where *jidhr* is the term for "square root" and *ḍil'* is a term for "cube root" or any higher root (Ibn Fallūs, MS Landberg 199, ff. 13rff).

perfect square, and which has no counterpart in our Greek texts, is *majdhūr*. This word derives from *jidhr* and literally means "rooted", or "has a rational square root". We saw this word in our translation of the passage from al-Ḥilātī in Section 2.6.4.

Al-Karajī gives this example of arithmetical uses of *ḍilʿ*, *murabbaʿ*, and *māl māl* in his *al-Badīʿ*: "And every odd number, if you add one to it and you find that its half is a square (*murabbaʿ*), then if you take that odd number with the sum of the odd numbers before it, then it is a *māl māl*. Its *ḍilʿ* is a root of half the sum of the larger number with one".[17] Here is a different arithmetical use of *mukaʿʿab* and *māl māl* from al-Samawʾal, in his explanation of the multiplication of $\sqrt[3]{10}$ by $\sqrt[4]{2}$: "we multiply a *māl māl* of the ten by a *mukaʿʿab* of 2. The result of the multiplication is eighty thousand" (Rashed 2021, 155.8). Here the words take on a role similar to δύναμις in Greek arithmetic/logistic, to indicate the power of a number. So a number can be a *māl māl* just as a number can be a square or a cube, and one can also speak of the *māl māl* of a number. The first of these is a category of number that includes 16, 81, $\frac{625}{10,000}$, etc., while the second is a way to indicate the fourth power of a particular number. This second way is close to the algebraic meaning of the term, but in algebra one never wrote anything like "three *māl māl*s of a thing". The term in algebra is always simply a "*māl māl*", and as such, it functions as a name. Arithmetical uses of the terms as categories of numbers in Arabic date back at least to the tenth century, for Ibn al-Nadīm reports in his *Fihrist* that Abū l-Wafāʾ wrote a *Book on extracting a side of the cube and the* māl māl *and what is composed of these terms*.[18]

Table 3.1 Arithmetical and algebraic powers in Greek and Arabic.

Degree	Greek arithmetic	Arabic arithmetic	Greek algebra	Arabic algebra
1	πλευρά (side)	*ḍilʿ* (side) *jidhr* (square root)	ἀριθμός (number)	*jidhr* (root) *shayʾ* (thing)
2	τετράγωνος (square) δύναμις	*murabbaʿ* (square) *māl* *majdūr* (rooted)	δύναμις	*māl*
3	κύβος (cube)	*mukaʿʿab* (cube)	κύβος (cube)	*kaʿb* (cube)
4 etc.	δυναμοδύναμις	*māl māl*	δυναμοδύναμις	*māl māl*

Note in particular that where we find terms meaning "side" (πλευρά/*ḍilʿ*) and "square" (τετράγωνος/ *murabbaʿ*), the setting is arithmetic, and when the first-degree term is called an ἀριθμός or *shayʾ*, the setting is algebra. Other terms, such as δύναμις and *māl*, belong to both.

17 I.e., $1 + 3 + 5 + \ldots + 2n^2 - 1$ is a fourth power (n^4). Its "side" (fourth root) is $\sqrt{\frac{1}{2}\left(\left(2n^2 - 1\right) + 1\right)}$ (al-Karajī 1964, 18.17).

18 "What is composed of these terms" are the higher powers expressed in terms of *māl* and *kaʿb*.

Table 3.1 shows the terms used in each series for Greek and Arabic. Corresponding terms are shown on the same row. The "Arabic arithmetic" column shows the terms listed by Naṣīr al-Dīn al-Ṭūsī (Saidan 1967, 138ff).[19] These terms are found in several other books as well. The "Arabic algebra" column shows the terms used by every extant author. Beginning with al-Karajī, the arithmetical and algebraic series are identical from the fourth power on.

3.1.2 *A survey of ways of naming powers in Arabic, Italian, etc.*

In order to provide some context for assessing the different ways of naming the powers that are preserved in different Greek texts, we should present some of the variations on naming that are found in Arabic, Italian, and sixteenth-century European books.

In Arabic texts that give such powers, the second algebraic power is always called a *māl*, the third power a *ka'b* (or rarely *muka''ab*), and the fourth power a *māl māl* (we refrain from capitalizing the algebraic names in this section). Powers above the fourth are usually expressed with the words *māl* and *ka'b*, and differences first appear in the fifth power. Schemes can generally be divided into two categories, "additive" and "multiplicative". In an additive scheme the powers 2 and 3 of *māl* and *ka'b* are added so that a *māl ka'b* is the fifth power. Diophantus's names are additive in this way. In a multiplicative scheme the powers are multiplied so that a *māl ka'b* would be the sixth power. In multiplicative schemes special names must be given to the fifth, seventh, and other powers that are not products of the powers of two and three.

The earliest extant book explaining algebra in Arabic that extends beyond the second power is Qusṭā's translation of Diophantus, from the mid to late ninth century. Like the Greek *Arithmetica*, the translation employs an additive scheme. The fifth power exhibits variations even within the same problem. In Problems IV.6 and IV.7 it is called both a "cube *māl*" and a "*māl* cube", and in Problems VI.4 through VI.7 the fifth power is repeatedly called a "cube multiplied by a *māl*" or a "*māl* multiplied by a cube". There is also variation in the eighth power, which is called a "cube cube *māl*" in Problems IV.29–33, IV.42–44, and V.4–6, but is a "*māl māl māl māl*" in Problems VI.17 and VII.1.

There are three extant Arabic texts from the late ninth to tenth centuries that are apparently uninfluenced by the way Diophantus names the higher powers. The earliest is Abū Kāmil's late ninth century *Book on Algebra*. In his introduction Abū Kāmil describes only "simple numbers", "thing", and "*māl*", but among his worked-out problems he calculates with the third, fourth, fifth, sixth, and eighth powers more or less following an additive scheme. He calls the fifth power a "*māl māl* multiplied by a thing", the sixth power a "*ka'b ka'b*", and the eighth power a "*māl māl māl māl*". In one place a fifth power term is written as "eighty-eight

---

19 Al-Ṭūsī also includes *ka'b* in his list of arithmetical terms, though in arithmetic this word more commonly meant "cube root".

cubes multiplied by a *māl*" (Abū Kāmil 1986, 101.18). This way of writing one name "multiplied by" another name in Qusṭā's translation and in Abū Kāmil may have been a way to make clear what power is intended, since "*māl* cube" might be understood as the sixth power.

Another text, this one dating from the first half of the tenth century, is the *Book on the Cube, the* Māl, *and Proportional Numbers* by Sinān Ibn al-Fatḥ (MS Cairo, Riyāḍāt 260/4, ff. 95r–104v). This author presents a multiplicative scheme. The names of the powers starting from the third are *muka*ʿʿ*ab* (3rd), *māl māl* (4th), *midād* (5th), *māl muka*ʿʿ*ab* (6th), eighth proportion (*nisba al-thāmin*) (7th), and ninth proportion (*nisba al-tāsiʿ*) (8th). The name for the fifth power is unusual and we do not know its origin.[20] The seventh power is called an "eighth proportion" because it is the eighth term in the series of continued proportions of the powers starting with the unit, and similarly for the name of the eighth power. No other algebraist is known to have used these names for powers above the fourth.

ʿAlī al-Sulamī gives an interesting hybrid of additive and multiplicative schemes, in which the *māl*s multiply while *ka*ʿ*b*s add. He lists the first 20 powers in the introduction to his *Sufficient Introduction to Calculation by Algebra and What One Can Learn from Its Examples* (henceforth *Sufficient Introduction*). Table 3.2 shows his powers starting from the fifth (ʿAlī al-Sulamī, MS Vatican, Sbath 5, ff. 13r–14r). For example, the 14th power is a *ka*ʿ*b ka*ʿ*b māl māl māl* because $3 + 3 + 2 \cdot 2 = 14$. No other algebraist is known to have used this scheme.

Table 3.2 The first 20 powers in ʿAlī al-Sulamī.

*ka*ʿ*b māl* (5th)	*ka*ʿ*b ka*ʿ*b ka*ʿ*b māl māl* (13th)
*ka*ʿ*b ka*ʿ*b* (6th)	*ka*ʿ*b ka*ʿ*b māl māl māl* (14th)
*ka*ʿ*b māl māl* (7th)	*ka*ʿ*b ka*ʿ*b ka*ʿ*b ka*ʿ*b ka*ʿ*b* (15th)
māl māl māl (8th)	*māl māl māl māl* (16th)
*ka*ʿ*b ka*ʿ*b ka*ʿ*b* (9th)	*ka*ʿ*b ka*ʿ*b ka*ʿ*b māl māl māl* (17th)
*ka*ʿ*b ka*ʿ*b māl māl* (10th)	*ka*ʿ*b ka*ʿ*b ka*ʿ*b ka*ʿ*b ka*ʿ*b ka*ʿ*b* (18th)
*ka*ʿ*b māl māl māl* (11th)	*ka*ʿ*b māl māl māl māl* (19th)
*ka*ʿ*b ka*ʿ*b ka*ʿ*b ka*ʿ*b* (12th)	*ka*ʿ*b ka*ʿ*b ka*ʿ*b ka*ʿ*b ka*ʿ*b ka*ʿ*b māl* (20th)

It is with al-Karajī that things settle down. Al-Karajī's two main inspirations for his algebra book *al-Fakhrī* were Diophantus and Abū Kāmil. He took the scheme of naming the powers from Diophantus, listing in his introduction the powers up to the ninth. As we mentioned earlier, starting from the fifth they are: *māl ka*ʿ*b* (5th), *ka*ʿ*b ka*ʿ*b* (6th), *māl māl ka*ʿ*b* (7th), *māl ka*ʿ*b ka*ʿ*b* (8th), and *ka*ʿ*b ka*ʿ*b ka*ʿ*b* (9th). After al-Karajī, every Arabic text we have examined

---

20 Rashed (1994b, 20 n. 8) writes: "If the term *midād* is Arabic in origin, it must derive from the root *mdd*, which signifies the elongation of something, or the extension of one thing by another. It can also indicate the plural of *mudd*, a kind of measurement which originally signified the holding out of hand for food. The reason for such a choice to signify x^5, or the sixth position, is not clear. It is not impossible that it was borrowed from Persian to indicate the sixth position".

that shows powers above the fourth follows this additive scheme, including al-Khayyām, al-Samaw'al, Ibn Fallūs, Ibn Badr, Fakhr al-Dīn al-Ḥilātī, Ibn al-Yāsamīn and his commentators, Ibn al-Khawwām, al-Fārisī, Ibn al-Bannāʾ and his commentators, al-Kāshī, Sibṭ al-Māridīnī, and Muḥammad ibn Sālim Ḥifnī.

Nine of these authors[21] write that the powers can be expressed with different names, such as writing *māl māl māl* in place of *kaʿb kaʿb*. Al-Hawārī (1305), for one, explains:

> We can also work this out by separating the powers by twos or threes or by grouping them: first the *māl*, and then the cube, and then the *māl* and the cube. Suppose someone said, for example, "What is a term for eight?" We would say a *māl māl māl māl*. Or if we wish, we can say a cube *māl* cube, or a cube cube *māl*, or a *māl* cube cube, or anything else that is allowable.
>
> (Abdeljaouad and Oaks 2021, 117 [226.3])

Even Sinān Ibn al-Fatḥ (MS Cairo, Riyadāt 260/4, f. 96r, l. 2), author of the multiplicative scheme described earlier, writes that other ways of naming the powers are acceptable. We see this in practice, too. In al-Karajī, a *māl kaʿb kaʿb* is sometimes called a *kaʿb kaʿb māl* (Saidan 1986, 255.18, 298.23,24; BnF 2459 f. 83v, l. 16, f. 104v, l. 1,2), and ʿAlī al-Sulamī calls the 10th power a *māl māl kaʿb kaʿb* when he first introduces it, but it appears later in his table as a *kaʿb kaʿb māl māl* (MS Cairo, Riyadāt 260/4, f. 9v, l. 4). We have no examples of this flexibility in the problems of the Greek books of the *Arithmetica*. The only power above the fourth that appears there is the cube-cube (6th), in Problems IV^G.18, V^G.18, 21, 22, and VI^G.21.

The names of the first four powers in medieval Latin algebra are translated from Arabic, usually as *res, census, cubus*, and *census census*, and authors rarely ventured to include higher powers. In a couple of problems in his 1228 *Liber Abaci* Fibonacci works with the sixth power "censum census census", which he writes is the same as "cuborum cubi", and his eighth power is "censum census census census" (Fibonacci 1857, 446–47). These are additive names inspired by Abū Kāmil, since they occur in the solutions to problems copied from him.[22] But Jean de Murs, in his 1343 *Liber Quadripartitum*, gives multiplicative names for powers in a multiplication table. His fifth, sixth, and eighth powers are called *altera parte longius, census cubi*, and *census census census* (L'Huillier 1990, 464).

---

21 Ibn Badr, Fakhr al-Dīn al-Ḥilātī, al-Fārisī, Ibn al-Bannāʾ, al-Hawārī, Ibn al-Qunfūdh, al-Qalaṣādī, Ibn Ghāzī, and Muḥammad ibn Sālim Ḥifnī.

22 Problems 57 and 59 (Rashed's numbers 63 and 65), which begin in the Latin translation at lines 2867 and 2920 of Sesiano's edition. Higher powers do not occur in the solution to Problem 57 in either Arabic or Latin, and in Problem 59 Abū Kāmil does not identify his "cube cube" with a *māl māl māl* as Fibonacci does (Abū Kāmil 2012, 473.14, 479.1; Sesiano 1993, 403–04).

The names of the first three powers in Italian abacus books of the fourteenth and fifteenth centuries likewise derive from Arabic: *cosa* ("thing", 1st), *censo* ("amount of money", 2nd, sometimes written as *zenso*), and *chubo* ("cube", 3rd), and the fourth power, *censo di censo*, is also common. In Italian, too, higher powers are not encountered often. Giovanni di Davizzo, in a 1339 work preserved in a 1424 manuscript, follows an additive scheme when he writes "*chubo* (multiplied) by *chubo* makes *chubo di chubo*" and "*zenso* by *chubo* makes *zenso* di *chubo*".[23] Antonio de' Mazzinghi (d. 1383), quoted in codex Palatino 543, lists the powers as "*chosa, censo, chubo, censo di censo, chubo relato, chubo di chubo*, etc." (Arrighi 2004, 191). The sixth power is additive, and the fifth is curious. Høyrup observes that *chubo relato* may have been inspired by the term *radice relata* for the fifth root. Antonio explains in another work:

> as we say, the *radice relata* of a number is that which is multiplied by itself, and that which is made again by itself, and the multiplication again multiplied by the stated root, such as the *radice relata* of 32 is 2, and of 243 is 3.[24]

Some other authors who show additive names for the fifth and/or the sixth power are the anonymous author of Ottobon. lat. 3307, Maestro Dardi, Michael of Rhodes, and Benedetto da Firenze (1463). Benedetto, in particular, adopts Antonio's *chubo relato* for the fifth power and *chubo di chubo* for the sixth (Høyrup 2019; Michael of Rhodes 2009).

Two anonymous manuscripts mix additive and multiplicative terms. One is a *Trattato d'Algibra* from ca. 1390, which calls the fifth power *chubo di censi* and the sixth power by both *censo di chubo* and *chubo di chubo* (Franci and Pancanti 1988, 4–5). Another treatise, titled *On the Roots of Numbers and a Method of Finding Them* (*Della Radice de' Numeri e Metodo di Trovarla*), probably copied in the last quarter of the fifteenth century, is at least consistent. Here the powers are listed as: *numero* (constant), *chossa* (1st), *zenso* (2nd), *Qubo* (3rd), *zenso di zenso* (4th), *cossa di zenso di zenso* (5th), *zenso di Qubo* (6th), *cossa di zenso di Qubo* (7th), *zenso de zenso di zenso* (8th), and *Qubo de Qubo* (9th). The fifth and seventh powers are additive, while the sixth, eighth, and ninth are multiplicative (Anonimo 1986, 20).

And last, during the latter fifteenth century one Giovanni del Sodo adopted a multiplicative scheme that was later reported in Francesco Ghaligai's 1521 *Pratica d'Arithmetica*: Numero, Cosa (1st), Censo (2nd), Cubo (3rd), Relato (5th), Pronico (7th), Tromico (11th), and Dromico (13th). Judging by the notation that is introduced here, the names of the composite powers are: *censo di censo* (4th), *cubo di censo* (6th), *censo di censo di censo* (8th), *cubo di cubo* (9th), *relato di*

---

23 "chubo via chubo fa chubo di chubo", and "zenso via chubo fa zenso di chubo" (Høyrup 2007, 479).

24 "chome diremo, radice ralata d'uno numero è quella che in sè multiplichata et quel che fa anchora in sè et la multiplichatione anchora multplichata per la detta radice, chome radice relata di 32 è 2 et di 243 è 3" (Arrighi 1967, 38).

censo (10th), *cubo di censo di censo* (12th), *pronico di censo* (14th), and *cubo relato* (15th) (Ghaligai 1521, ff. 71r–71v). Giovanni's student Raffaello Canacci composed a *Ragionamenti d'Algebra* ca. 1495, in which he adopts Giovanni's names. For the sixth power he lists *relato*, but elsewhere he calls it a *chubo relato* (Procissi 1954, 443, 482). If these different Italian schemes do not appear to be well thought out, we should understand that powers above the fourth were almost never encountered in practice.

It is starting with Luca Pacioli (1494) that higher powers become common in Italian algebra. In the margin of fol. 67v of his 1494 *Summa de Arithmetica, Geometria, Proportioni et Proportionalita* he takes the names all the way up to the 30th power in a multiplicative scheme. His powers up to the fourth are the same as we find in earlier Italian works. Starting with the fifth power they are: "*primo relato*" ("first related", 5th), "*censo de cubo* and also *cubo de censo*" (6th), "*secondo relato*" ("second related", 7th), "*censo de censo de censo*" (8th), "*cubo de cubo*" (9th), "*censo de primo relato*" (10th), "*terzo relato*" ("third related", 11th), "*cubo de censo de censo* or you can say the reverse" (12th), etc. Note that Pacioli allows the same flexibility in the names that we see in Arabic, Latin, and earlier Italian books.

Multiplicative schemes remained the norm in sixteenth-century Europe. Christoff Rudolff, to pick an early example, gave this list of names, mostly originating in Latin/Italian, in his 1525 *Coss*: "*dragma oder numerus*" ("drachma or number" for the constant), "radix" (1st degree), "zensus" (from the Latin *census*, 2nd), "cubus" (3rd), "zens de zens" (4th), "sursolidum" (5th), "zensi cubus" (6th), "bissursolidum" (7th), "zens zenz de zens" (8th), "cubus de cubo" (9th) (Rudolff 1525, f. Diiiv; Kaunzner and Röttel 2006, 174). Here, too, the fifth and seventh powers get special names.[25]

At the end of Section 3.4.1, we argue based in several Greek texts from antiquity and Byzantine times that Diophantus probably also allowed for variations in the names of his powers. The terms *arithmos*, *dynamis*, and *kybos* served as technical names for the first three powers, but above that one could form the names any way "that is allowable", as al-Hawārī wrote.

3.1.3 The testimonies about Hipparchus

Ibn al-Nadīm completed his famous bio-bibliographic work *Kitāb al-Fihrist* (*The Catalog*) in Baghdad in 987 CE.[26] In two different sections he mentions a book on algebra written by the astronomer Hipparchus of Nicaea (active 147–127 BCE).

---

25 Other authors who used a multiplicative scheme include Francesco Galigai (1548), Michael Stifel (1544), Johann Scheubel (1550), Jacques Peletier (1554), M. Valentin Mennher (1556), Petrus Ramus (1560), Niccolò Tartaglia (1560), Pedro Nuñez (1567), Rafael Bombelli (1572), Juan Pérez de Moya (1573), and Guillaume Gosselin (1577). The only additive schemes we know from the century are Xylander's translation of Diophantus (1575) and François Viète's adoption of the same for his *logistice numerosa* (1591).

26 His full name was Abū l-Faraj Muḥammad ibn Isḥāq ibn Muḥammad ibn Isḥāq ibn al-Nadīm (Stewart 2014, 167). He is called simply "al-Nadīm" in error in some sources. The text was first edited in Flügel (Ibn al-Nadīm 1871–72) and most recently in Sayyid (2009).

Because this testimony has been questioned by some historians of mathematics, we give a detailed account of it here and explain how recent scholarship shows that their objections are no longer valid.

In his chapter on the mathematical sciences Ibn al-Nadīm lists the works of 30 Greek, 4 Indian, and over 80 "recent" scholars writing in Arabic. The entries for Hipparchus and Diophantus come one after the other:

Hipparchus *al-Zafanī*   Among his books are

Book_____

Art of Algebra (*al-jabr*) also known as "definitions". This book was translated, and Abu l-Wafā' Muḥammad ibn Muḥammad the arithmetician adjusted (*ṣalaḥa*) the book and also explained it and (gave) its causes by means of geometrical proofs

Book_____

    Division of Numbers

         Diophantus the Greek,   Among his books is
           of Alexandria

Book_____

    Art of algebra (*al-jabr*).[27]

The entry for Abū l-Wafā', a contemporary of Ibn al-Nadīm, is given a few pages later. The following is taken from the list of books he wrote:

Book_____ _____	Book_____
Commentary on the book of al-Khwārazmī on algebra (*al-jabr wa-l-muqābala*)	Commentary on the book of Diophantus on algebra (*al-jabr*)
Book_____	Book_____
Commentary on the book of Hipparchus on algebra (*al-jabr*)	Introduction to the *Arithmetica* (in one) chapter
Book_____	Book_____
On what one should keep in mind when approaching the book (of) *Arithmetica*	The proofs to the assertions (*qaḍā*) that Diophantus employs in his book, and to what he (Abu' l-Wafā') employs in the commentary.[28]

---

27 (Sayyid 2009, II/1, 219; Ibn al-Nadīm 1871–72, II, 269). We arrange the text as it appears in MS Istanbul, Şehit Ali Paşa 1934, f. 116v, a manuscript one removed from Ibn al-Nadīm's autograph (Stewart 2014, 173). Stewart (2014, 174) notes that prior scholars argue that this "copy was intended to imitate the original very closely in script and format, including the size of the book, the pagination, and the layout of pages, something extraordinary in the history of Islamic books".

28 (Sayyid 2009, II/1, 259.11; Ibn al-Nadīm 1871–72, II, 283.14).

Among all these works, only Diophantus's *Arithmetica* ("on the art of algebra") is extant, if only partially.

In 1851 Franz Woepcke (1826–64) translated the entry for Hipparchus in the introduction to his *L'Algèbre d'Omar Alkhayyâmî* and made this remark after also calling attention to the entry on Abū l-Wafā':

> The testimony of these passages, which attribute to Hipparchus works apart from those which have illustrated him as an astronomer, is corroborated by the following words of Plutarch: "Chrysippus is refuted by all the arithmeticians, among them Hipparchus himself".[29]

Later, in 1855, he brought up these passages again:

> I insist on this point, because these two passages of *Kitāb al-Fihrist*, and especially the last one, which, if necessary, would suffice alone, contain the direct and formal confirmation of an important fact for the history of algebra, and moreover, almost self-evident, to know that Diophantus is not the inventor of the science which he developed so prodigiously, and that algebra was cultivated, and even more invented, long before a time when the brilliance of Greek science had begun to fade.[30]

Woepcke's enthusiasm was not shared by Gustav Flügel, editor of the *Fihrist*. In his notes to the entry on Hipparchus, Flügel found reason to doubt the authenticity of the attribution of these books to the Greek astronomer. Not only is there no mention in any other source that Hipparchus authored works on these topics, but the *nisba* attributed to him, *al-Zafanī*, indicates a region near Emesa in Syria, and not Nicaea in Bithynia, his true place of origin. To make sense of this *nisba*, Flügel compared the entry on Hipparchus in the *Fihrist* with entries in another bio-bibliographic work, an abridgement of *Information on Scholars according to Learned (Men)* of Ibn al-Qifṭī (d. 646/1248) made in 647/1249 by al-Zawzanī, and retitled *Ta'rīkh al-ḥukamā'* (*History of Learned [Men]*).[31] Ibn al-Qifṭī copied much of his material word for word from the *Fihrist*, but he also included informa-

---

29 "Le témoignage de ces passages, qui attribuent à Hipparque des travaux en dehors de ceux qui l'ont illustré comme astronome, est corroboré par les mots suivant de Plutarque: 'Χρύσιππον δὲ πάντες ἐλέγχουσιν οἱ ἀριθμητικοί, ὧν καὶ Ἵππαρχός ἐστιν'" (Woepcke 1851, xi). The quotation of Plutarch is from *On Stoic Self-Contradictions*, 1047D. Its translation here is taken from (Plutarch 1976, 527).

30 "J'insiste sur ce point, parce que ces deux passages du *Qitâb Alfihrist*, et surtout le dernier, qui, au besoin, suffirait seul, contiennent la confirmation directe et formelle d'un fait important pour l'histoire de l'algèbre, et d'ailleurs presque évident par lui-même, savoir que Diophante n'est pas l'inventeur de la science qu'il développa d'une manière si prodigieuse, et que l'algèbre fut cultivée, et à plus forte raison inventée, longtemps avant une époque où l'éclat de la science grecque commençait déjà à pâlir" (Woepcke 1855, 35).

31 The Arabic title of Ibn al-Qifṭī's book is *Kitāb ikhbār al-ʿulamā' bi-akhbār al-ḥukamā'*. The abridgement *Ta'rīkh al-ḥukamā'* is edited in Lippert (1903). The full names of the authors are

tion and titles from other sources. And instead of arranging the material by topic and then chronologically as Ibn al-Nadīm had done, Ibn al-Qifṭī listed his names alphabetically. Three consecutive entries in the abridgement cover "Hipparchus", "Hipparchus the poet",[32] and "Aristippus".[33] The entry for Hipparchus has nothing in common with Ibn al-Nadīm's. It describes Hipparchus's accomplishments in astronomy and lists only one book, an apocryphal work on astrology titled *Secrets of the Stars* (*Asrār al-nujūm*). Curiously, the two works attributed to Hipparchus in the *Fihrist* are now found tacked on the end of the entry for the philosopher Aristippus, a person Flügel knew could not have authored them. Further, although it is stated that Aristippus hailed from Cyrene, there is mention of the place *al-Zafanī* as well in that entry.[34]

So it appeared to Flügel, and also later to Heinrich Suter (1848–1922) (1892, 54–55) and Moritz Steinschneider (1816–1907) (1897, 349–50), that the *nisba* *al-Zafanī* and the two works on algebra and numbers are misattributed in both books, to Hipparchus in the *Fihrist*, and to Aristippus in *Ta'rīkh al-ḥukamā'*. They must, then, belong to some other author. Suter proposed that the works were originally listed for Diophantus, though he later changed his mind by suggesting instead the ninth-century CE mathematician Muḥammad ibn Yaḥyā ibn Aktham, author of a book on numerical problems mentioned later in the *Fihrist*.[35] Regarding the commentary on Hipparchus listed for Abū l-Wafā', he writes: "Of course, this could have been included in the entry for Abū l-Wafā' by later copyists who recalled the passage in the already-corrupt entry for Hipparchus".[36] Steinschneider, for his part, believed that *al-Zafanī* was somehow a corruption for "Bithynia", and the two books were added in error. He suggested in particular that the "art of algebra" belongs instead to Diophantus.

Two pieces of evidence gave rise to these different conjectures: first, that no other text ascribes any book on arithmetic or algebra to Hipparchus, and second, the incorrect *nisba* *al-Zafanī*. Recently Devin Stewart has cleared up the confusion surrounding *al-Zafanī*. He observes that "a number of copyist's errors in the *Fihrist* can only be explained by positing that a *wāw* has been connected to a following letter, contrary to the usual practice in ordinary handwriting, though it occurs often when scribes write quickly" (2016, 149). Stewart identifies this error in seven instances, including the apparent *al-Zafanī*, after the names of both Hipparchus and Simplicius, and which he corrects to *al-rūmī*. This reading makes sense. In texts of the period *rūmī* was an alternative word to *yūnānī* to mean

---

Jamāl al-Dīn Abū al-Ḥasan ʿAlī ibn Yūsuf ibn al-Qifṭī and Muḥammad ibn ʿAlī ibn Muḥammad al-Khatibī al-Zawzanī.

32 Hipparchus the poet, d. 514 BCE, was a tyrant of Athens. For his introduction of Homeric performance to Athens, see Nagy (1996, 70).

33 Aristippus was a philosopher and a student of Socrates who lived ca. 435–ca. 356 BCE.

34 (Lippert 1903, 70). Lippert's edition shows the nisba as *al-Rafanī*.

35 (Suter 1900, 213; Ibn al-Nadīm 1871–72, II, 282.1; Sayyid 2009, 255.5).

36 "Dies könnte freilich auch in den Artikel über Abu'l-Wafa durch spätere Abschreiber hineingefügt worden sein, die sich der Stelle in dem schon verdorbenen Artikel über Hipparchos erinnerten".

"Greek", or sometimes more specifically "Greek from Anatolia". Ibn al-Nadīm also gives the *nisba* of *al-rūmī* to Vettius Valens (2nd c. CE), but for this author the word was read correctly by later copyists.[37] Now, with no internal inconsistency in the *Fihrist* on Hipparchus, there is no reason to suppose that the text is corrupt for these entries. According to Ibn al-Nadīm himself, Hipparchus the Greek wrote a book on algebra and a book on numbers, the former was translated into Arabic, and Abū l-Wafā' wrote a commentary on it.

Still, Ibn al-Nadīm might have been wrong, so the veracity of these claims might be doubted without support from some other source. We are fortunate, then, that we have such a source, by none other than the author of the commentary. Abū l-Wafā' wrote his *Book of What is Necessary for Scribes, Dealers, and Others from the Science of Arithmetic* in the period 961–76 CE, and the Arabic text was published by A. S. Saidan in 1971.[38] Speaking about the characterization of division as the inverse of multiplication, he writes: "And we have mentioned it ourselves in our commentary on the book of Hipparchus the Bithynian on the elements of numbers (*uṣūl al-aʿdād*)".[39] This leaves us with a small quandary. On the one hand, the topic of characterizing division and the "title" *uṣūl al-aʿdād* point to the book "Division of Numbers" mentioned by Ibn al-Nadīm. But on the other hand, it is the book on algebra that is said to have been commented upon by Abū l-Wafā'. But the difference in the descriptions "art of algebra" and "elements of numbers" cannot be regarded as significant. Ibn al-Nadīm, for example, called Diophantus's *Arithmetica* a "book on algebra" in one place, and a "book on arithmetical problems" in another place.[40] Regardless which book Abū l-Wafā' intended, his passage gives more support for Hipparchus's authorship of books on arithmetic and algebra, and there should no longer be any ground for regarding the works listed for Hipparchus in the *Fihrist* as errors. It is Ibn al-Qifṭī, his editor al-Zawzanī, or some copyist who later mistakenly transferred the books of Hipparchus over to the entry on Aristippus.

Even knowing that there was an Arabic translation of a book on algebra attributed to Hipparchus to which Abū l-Wafā' wrote a commentary, there is still

---

37 (Stewart 2014, 198, 2016, 149). Flügel read the mistaken *al-Zafanī* after Simplicius's name correctly as *al-rūmī*, with no mention of it in his notes.

38 Arabic title: *Kitāb fīmā yaḥtāju ilayhi al-kuttāb wa l-ʿummāl wa ghayruhum min ʿilm al-ḥisāb.*

39 (Saidan 1971, 126.7). Saidan makes the connection between this passage and the entries in the *Fihrist* in an endnote (p. 417). He brought all this up again, this time in English, in comments to his translation of al-Uqlīdisī's *Chapters on Indian Arithmetic* (Saidan 1978, 405). We wonder why, then, in that same book, he wrote "Abū al-Wafā' . . . states that he also translated a text on number by Hipparchus. We have no evidence that Hipparchus the Bithynian wrote any such book" (Saidan 1978, 15).

40 For "book on algebra", see the entries for Diophantus and Abū l-Wafā' translated earlier. The *Arithmetica* is called a "book on arithmetical problems" in the entry for Qusṭā ibn Lūqā (Ibn al-Nadīm 1871–72, II, 295.23; Sayyid 2009, II/1, 294.5). Also, the "Arithmetica" (*Arithmāṭīqī*) mentioned in other books listed for Abū l-Wafā' may refer to the *Arithmetical Introduction* of Nicomachus of Gerasa.

another objection that should be addressed. Fuat Sezgin gave this assessment in 1974, after the publication of Abū l-Wafā''s arithmetic book:

> On the basis of the fact that not all writings allegedly originating from Hero are actually genuine or that they partly represent later editions, and further, that no Greek source for an algebra of Hipparchus is attested, I am inclined to see in the Arabic titles the translations of pseudo-Hipparchean writings. These were probably written in late antiquity by the same scholars who are also the authors of these pseudo-Heronean writings.[41]

Arguing from a lack of Greek sources is very tenuous, given what we know about the scale of our loss of ancient writings. In the case of Hipparchus, we can point to a Greek source that attests to a remarkable achievement of his in combinatorics, one that Sezgin could not have appreciated in 1974. That passage is the one from Plutarch already mentioned by Woepcke in 1851. Plutarch's passage asserts that Hipparchus corrected the calculation of "the number of conjunctions produced from ten assertibles" that the Stoic philosopher Chrysippus (3rd c. BCE) had given as being greater than one million, by proving that "affirmation gives 103,049 conjoined assertibles and negation 310,952" (Acerbi 2003, 466). Until very recently historians could not make sense of these numbers. Heath, for instance, wrote in 1921, "it seems impossible to make anything of these figures" (1921, II, 256). It was only in the 1990s that their significance came to be recognized. In 1994 Richard Hough, a mathematics graduate student, discovered that the first number coincides with the tenth Schröder number, and in 1998 Habsieger, Kazarian, and Lando found that 310,952 is half the sum of the 10th and 11th Schröder numbers, provided that the last digit is corrected to be 4. Today we know different ways of generating these numbers. Friedrich Wilhelm Karl Ernst Schröder (1841–1902) himself (1870) calculated them as the number of ways to surround n identical letters with parentheses, such as the example $x(xx)((xxx)x)xxx$ for $n = 10$. Schröder numbers also arise in the context of plane trees and in polygonal dissections, and their investigations in the mathematical literature date back to the mid-eighteenth century. Calculating these "bracketing numbers" is not a trivial matter and speaks to a sophistication in combinatorics well beyond what anyone thought Greeks had accomplished. In particular, it reveals Hipparchus's competence in and attention to arithmetical calculations and justifies the designation "arithmetician" (ἀριθμητικός) assigned to him by Plutarch.[42]

---

41 "Von den Tatsachen ausgehend, daß nicht alle angeblich von Heron stammenden Schriften tatsächlich echt sind, bzw. daß sie z. T. spätere Bearbeitungen darstellen und daß ferner von keiner griechischen Quelle eine Algebra von Hipparch bezeugt ist, bin ich geneigt, in den arabischen Titeln die Übersetzungen von Pseudo-Hipparch-Schriften zu sehen. Diese sind wohl in der Spätantike von denselben Gelehrten verfaßt worden, die auch die Urheber der Pseudo-Heron-Schriften sind" (Sezgin 1974, 147).

42 (Stanley 1997; Habsieger, Kazarian, and Lando 1998; Acerbi 2003). We still do not know through what process or even through what particular problem Hipparchus investigated these numbers.

Greek combinatorics is an instructive example illustrating the perils of arguing from lack of evidence. The accepted view before the 1990s is well stated by N. L. Biggs in his 1979 article "The roots of combinatorics", where, after reviewing the passage from Plutarch, he writes: "The lack of any other significant reference to such calculations points to the conclusion that the Greeks took no interest in these matters" (Biggs 1979, 114). Now that the numbers cited by Plutarch have been identified, Fabio Acerbi can more correctly observe: "we are forced to conclude that the vagaries of textual tradition have (almost) annihilated the field" of Greek combinatorics (2003, 466). The selection of texts to copy combined with the intermittent destruction of books over the course of succeeding centuries has robbed us of an entire chapter of Greek mathematics. The same can be said for Greek algebra before Diophantus, where in place of Plutarch's single glimpse we are left with only the meager sources discussed in this chapter.

So far, we have addressed arguments by historians who deny that Hipparchus wrote the two books listed in the *Fihrist*. By contrast, three prominent historians in the first part of the twentieth century accepted Hipparchus's authorship: Moritz Cantor (1829–1920) (1907, 362–63), Johannes Tropfke (1866–1939) (1934, 40–41), and Thomas Heath (1931, 398, 530). Heath had deferred to Suter's opinion in his influential *History of Greek Mathematics* (1921, II, 256), but later, in his lesser-known *Manual of Greek Mathematics*, he, and later Tropfke, also saw a possible link between the Babylonian algebra recently identified by Neugebauer and Sachs and Greek algebra. Hipparchus was seen as being the ideal intermediary due to his known appropriation of elements of Babylonian astronomy. Such reasoning is still compelling, but lack of any positive evidence of Babylonian influence on Greek algebra will keep this hypothesis in the realm of speculation.

Despite the opinions of Cantor, Heath, and Tropfke, scholars since the middle of the twentieth century have generally regarded the testimony on Hipparchus in the *Fihrist* as being spurious.[43] The one dissenting view we have found is Fabio Acerbi's. Writing in the wake of the discovery of Hipparchus's calculation of Schröder numbers he states: "it is in my opinion unreasonable to deny the Hipparchian authorship of the treatises" (Acerbi 2003, 496 n. 95).

So, we know that Hipparchus wrote a book that was recognized by Arabic scholars as one on algebra. But what specifically was this book about? Interpretation of Ibn al-Nadīm's description "art of algebra (*al-jabr*)" is straightforward. In Arabic mathematics, the term *al-jabr wa-l-muqābala* and its common shortened form *al-jabr* meant the problem-solving technique of algebra, distinct from other techniques, such as double false position and the rule of three, and distinct from any method of calculating on known numbers, such as finger reckoning, base ten

---

43 See, for example, the entries on Hipparchus in the *Dictionary of Scientific Biography* (Toomer 1970), the *New Dictionary of Scientific Biography* (A. Jones 2008), the *Biographical Encyclopedia of Astronomers* (Kwan 2014), and the *Encyclopedia of Ancient Natural Scientists* (Lehoux 2008).

(Indian arithmetic), or sexagesimal arithmetic. Every medieval Arabic book or chapter on *al-jabr* dealt in some way with some or all of the following aspects of this art: the names given to the powers of the unknown, the assignment of names in the beginning of a solution, operations on monomials and polynomials, the setting up and simplification of equations, and the solutions to those equations (sometimes with proofs). Diophantus's *Arithmetica* is such a book, while, for example, Nicomachus's *Arithmetical Introduction* is not. This is why the former, but not the latter, was called a book on *al-jabr* by Arabic authors. Because Hipparchus's book was called a book on the "art of algebra", we must accept it as such, for the term *al-jabr* could not have meant anything else. This is made even more evident by the way the commentary on Hipparchus's book is grouped together with the commentaries on the algebra books of al-Khwārazmī and Diophantus in Ibn al-Nadīm's list for Abū l-Wafā'.

Ibn al-Nadīm gives us a clue about the contents of Hipparchus's book on algebra by writing that it is also known as "definitions" (*ḥudūd*). This designation recalls the "definitions" tradition in medieval Arabic scholarship. Our earliest extant source for this tradition is the *Treatise on the definitions of things and their descriptions* (*Risāala fī ḥudūd al-ashyā' wa rūsumihā*) by the philosopher al-Kindī, written in the mid-ninth century CE. This book consists entirely of the definitions of ca. 100 technical terms in philosophy and mathematics. Ibn Sīnā (early 11th c.) also wrote a *Kitāb al-ḥudūd* (*Book of definitions*) in which he, too, defines around 100 philosophical terms. Other books on definitions, of similar structure, are known in the areas of *kalām* (theology) and *fiqh* (law). These books are typically quite short. The two works on philosophy do not run to more than 20 pages each. Because Hipparchus's book was known also as "definitions", we can surmise that it, too, consisted at least mainly in the definitions of terms. This tells us that it probably did not consist largely in the solutions to problems, as we find in Diophantus.

Ibn al-Nadīm writes that Abū l-Wafā' gave "its causes by means of geometrical proofs". One might be tempted to read this as meaning the proofs to the rules for solving three-term equations, like those we find in many Arabic books. But it is also possible that they were geometrical proofs for rules for multiplying and dividing polynomials. Both Abū Kāmil, in his *Book on Algebra*, and Maximus Planudes, in his commentary on the *Arithmetica*, give such proofs. Both authors justify the rules of calculation in terms of specific examples, such as Abū Kāmil's multiplication of 10u ℓ1T by 10u 1T, which gives 100u ℓ 1M (see our commentary for an explanation of this notation), or, in modern notation, $(10 - x) \cdot (10 + x) = 100 - x^2$, and Maximus Planudes's multiplication of 1N 3u by 4u ℓ1N, which gives 1N12u ℓ1P (in modern notation, $(x + 3) \cdot (4 - x) = x + 12 - x^2$) (see Section 3.4.3 for the proof of another example in Planudes).[44]

To summarize, Hipparchus was the author of a lost book on "the art of algebra". Based on the few remarks in Ibn al-Nadīm, this would have been a brief book

---

44 (Abū Kāmil 2012; Diophantus 1893–95, II, 144.3–145.12).

giving definitions of the technical vocabulary of Greek algebra. We cannot say anything about the contents or form of the other book attributed to Hipparchus, on "division of numbers" (*qisma al-a'dād*), except what we might glean from its title.

3.1.4 The Cairo papyrus

Cairo Museum S.R. 3069 is a recently published Greek papyrus from the second century BCE (Aish 2016). It shows two texts on each side. The recto begins with a mainly illegible financial account, and below it in a second hand is a geometric exercise with diagrams of three triangles. The verso shows abbreviations for names of the powers together with the powers of 2 up to 256, and below that, again in another hand, is an account listing various sums of money. The papyrus measures 7.5 cm wide by 14.5 cm tall.

Aish's transcription of the portion showing the list of powers reads:

$$] [$$

ὁ (πρῶτος) γνώ(μων) ρ(ιθμῶν) ἐστίν	
ὁ δὲ ρ(ιθμὸς) μο(νάδων?)	β
ϱικυ( ) (γίνεται) δύ(ναμις)	δ
ἐπὶ ἀρ(ιθμὸν) (γίνεται) κύ(βος)	η
ἐπὶ ἀρ(ιθμὸν) (γίνεται) δυ(ναμο)δύ(ναμις)?	ιϛ
ἐπὶ ἀρ(ιθμὸν) (γίνεται) δύ(ναμις)ᵉ	λβ
ἐπὶ ἀρ(ιθμὸν) (γίνεται) δύ(ναμις)ˢ	ξδ
ἐπὶ ἀρ(ιθμὸν) (γίνεται) δύ(ναμις)ˢ	ρκη
ἐπὶ ἀρ(ιθμὸν) (γίνεται) δύ(ναμις)ⁿ	σνϛ

The word δύναμις (power) is abbreviated as δύ. For the fifth through the eighth power this abbreviation has a superscript in Greek alphabetic numerals: "δύ°" for "5th power", "δύˢ" for "6th power", etc. Other words are also abbreviated: γνώμων (gnomon) as γνώ; ἀριθμός (number) as ἀρ; μονάς (unit) as μ̣; κύβος (cube) as κύ; and γίνεται (becomes) is shown as a vertical line. The "by" (ἐπί) is the preposition for multiplication in arithmetic. Here is our translation of this transcription:

the (first) gnomon of numbers is	⟨1?⟩
and the number of units	2
⟨by a number⟩ becomes a power	4
by a number becomes a cube	8
by a number becomes a power-power	16
by a number becomes a 5th power	32
by a number becomes a 6th power	64
by a number becomes a 7th power	128
by a number becomes an 8th power	256

The repeated phrase "by a number" suggests an algebraic reading of ἀριθμός. If this were merely a list of powers, it should have been phrased as "by two" or something similar. Also, δύναμις appears as a noun, so rather than writing that 2 *in power* is 4, we have the power *is* 4.

The presence of specific values 2, 4, 8, etc., for the powers does not imply an arithmetical reading for these abbreviated names. In fact, in Arabic and later algebra many texts illustrate how the names of the powers function by associating them with the powers of specific numbers. In the beginning of his algebra book *al-Fakhrī* al-Karajī introduces the names of the powers of the unknown up to the ninth via multiplication, he explains that they are in continued proportion, and he observes that the first three powers can be associated with the three geometric dimensions. He then gives this example by supposing the value of the root is two, and then giving the corresponding values of the other powers up to the *māl* cube cube:

> If someone said, the root is two. How much is the *māl*? You should say four, since two by two is four. And the cube is eight, and the *māl māl* is sixteen, and the *māl* cube is thirty-two, and the cube cube is sixty-four, since it comes from multiplying the cube by the cube. And the *māl māl* cube is one hundred twenty-eight, since it comes from multiplying the root by the cube cube, or from multiplying eight by sixteen. And the *māl* cube cube is two hundred fifty-six, since it comes from multiplying a *māl* cube by a cube.
>
> (Saidan 1986, 99.17)

Other algebraists show a table with the names of the powers together with the powers of specific numbers. ʿAlī al-Sulamī gives a long exposition of the names in his *Sufficient Introduction,* including the example of forming the name for the 45th degree term. He then draws up a table listing the powers up to the 20th degree that extends across three pages of the Vatican manuscript. Here is a translation of the first of these pages (MS Vatican, Sbath 5, f. 13r) (read three vertical cells at a time):

first degree	second degree
root	*māl*
two	four
third degree	fourth degree
cube	*māl māl*
eight	sixteen
fifth degree	sixth degree
cube *māl*	cube cube
thirty-two	sixty-four
seventh degree	eighth degree
cube *māl māl*	*māl māl māl*
one hundred twenty-eight	two hundred fifty-six

Some others who give lists or tables associating the names of the algebraic powers with the powers of specific numbers include Sinān Ibn al-Fath (Arabic, 10th c., with powers of 10; see MS Cairo, Riyadāt 260/4, f. 96r), Michael Psellus (Greek, 11th c., text translated in Section 3.4.1, powers of 2), Maximus Planudes (powers of 3; see Diophantus 1893–95, II, 126.2–9), Fakhr al-Dīn al-Ḥilātī

(Arabic, 13th c., with powers of 2 and 3; see Tehran MS 4409, p. 4), Dionigi Gori (Italian, 1544, with powers of 2 and 3; see Gori 1984, 4), Christoff Rudolff (German, 1525, with powers of 2, 3, 4, and 2/3; see Kaunzner and Röttel 2006, 174), and Girolamo Cardano (Latin, 1539, with powers of 2; see Cardano 1539, f. 6r).

That the names of the powers in the papyrus diverge beginning with the fifth power from those of Diophantus is not important, given the variability we witness of the names in Arabic algebra.

3.1.5 The Michigan papyrus

Papyrus Michigan 620 is of unknown origin and authorship. It is dated to the early second century CE. It was purchased by F. W. Kelsey in Egypt in 1921, and later that year it was donated to the University of Michigan. It was first published by F. E. Robbins (1929), and its definitive edition was published seven years later by J. G. Winter (1936). In the meantime, Kurt Vogel (1930) published important comments and corrections. The obverse of the papyrus contains portions of three worked-out arithmetical problems, specifically the last part of the first, the end of the second, and the beginning of the third problem, written in two columns.[45] After the text of each of the first two problems, a tabular setting of the solution is added, much like the tabular settings of the solutions later drawn by Maximus Planudes in his commentary to the first two books of the *Arithmetica*, or the similar layout of Theon's solution of a mensuration problem from Book XIII of Ptolemy's *Almagest*.[46] (We transliterate examples from these texts in Sections 3.4.3 and 3.2.1, respectively.) According to its first editor, the papyrus "is most probably a schoolbook of some sort, and perhaps from it or others like it Diophantus may have derived ideas which served as a basis for his mathematical methods" (Robbins 1929, 329).

The importance of the papyrus lies with the fact that the problems it contains are worked out by algebra, thus making it our earliest evidence of the practice of algebra that has come down to us. But there is more to it than that. There are some special signs in the papyrus that also appear later in the Byzantine manuscripts of Diophantus. One is the sign ς, which appears in our papyrus in two forms: with a diagonal stroke crossing its upper tip downward from left to right (ϛ̓), and without a stroke (cf. Winter 1936, 32). With the stroke it stands for the word δραχμή (drachma), but what interests us is the other form, which is used in the same way as the similar sign that appears in some of the earliest manuscripts of Diophantus (Marc. gr. 308, Vat. gr. 304, Vat. gr. 191) for denoting the name of the first-degree algebraic unknown.[47] Another sign common in manuscripts of Diophantus, in

---

45 The papyrus is available online at https://quod.lib.umich.edu/a/apis/x-2698/620R.TIF (accessed June 24, 20202).

46 For details about the manuscript carrying Theon's commentary and the possible involvement of Planudes in its production, see Christianidis and Skoura (2013, 42).

47 In the most ancient Diophantine manuscript, the Matrit. 4678, the sign used for the unknown takes the form ; however, the other form, which looks like an ς, also appears in it, in the sentence of

this papyrus and in many other papyri, including the Cairo papyrus discussed in the previous section, is the abbreviation for μονάς, "unit", which consists of the Greek letter M with a small O above it (ᴹ̇). In addition, we find the word ὑπόστασις (hypostasis) in the papyrus, with the meaning of the numerical value of the unknown *arithmos* (cf. Christianidis 2015), which is common in Diophantus but is unusual in Greek mathematics generally.

The first problem

The preserved part of the text of the first problem is the following:

> Again (multiply) the 150 by the 30 ⟨N⟩umbers[48] of the fourth: 4500. And the 600 drachmas in its hypostasis: 5100. This much is the fourth. Then add the four, 1050 and 1200 and 2550 and 5100: 9900. Proof: Since it says "let the second exceed the first by a seventh part", take the seventh of the 1050 (drachmas) of the first: 150. Add these to the 1050: 1200, as much as the second. Again, since it says "let the third exceed the two by 300 drachmas", add the first and the second: 2250. And add the 300 drachmas of the excess: 2550, as much as the third. And since it says "let the fourth exceed the three by 300 drachmas", add the three: 4800. And the 300 drachmas of the excess: 5100, as much as the fourth. The total is 9900.[49]

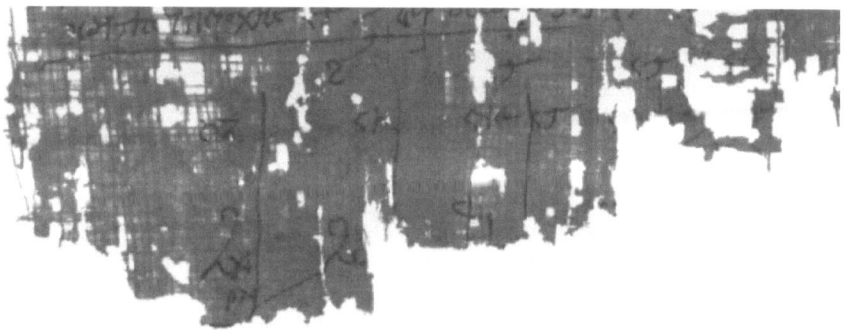

Figure 3.1 Facsimile of the table for the first problem of papyrus Michigan 620.

---

the preface with which Diophantus introduces the technical term ἀριθμός (Number). Heath points out that in the manuscripts of Diophantus the sign "occurs for ἀριθμός in the ordinary as well as for ἀριθμός in the technical sense" (1921, II, 457). Furthermore, he remarks that the sign is by no means confined to Diophantus. Indeed, in several more or less similar forms of the sign are found in the famous ninth-century manuscript of Euclid Bodleianus, D'Orville, X, 1 inf. 2,30, in the manuscripts of the *Sand-reckoner* of Archimedes, in a manuscript of Theon of Smyrna, and in manuscripts that carry treatises of the Heronian corpus, among which is the Constantinopolitanus Palatii veteris 1 (G.I.1) (second half of the 10th c.), which carries the *Metrica*. In all cases the sign, which is often completed with case terminations written above and to the right of the sign, like modern exponents, is used to denote the word ἀριθμός in the non-algebraic meaning of "number".

48 That is "by 30, the (multitude) of the ⟨N⟩umbers."

49 For the Greek text, see Winter (1936, 29.1–10).

	ϛ′		ϛτ		ϛτ		ϛΘτ	
ϛϛ		ϛη	˙ϛιε	ϛτ	[ϛ]λ	ϛχ		
						ϛ̣ξ	[ϛτ]	
						ρ[ν]		
Αν		Ασ	Βφν		Ε̣[ρ]	[Θτ]		
ρν								

Figure 3.2 Edition of the table for the first problem of papyrus Michigan 620.[50]

	7′		D300		D300		D9900
N7		N8	N15D300	30D600			
						N60[D900]	
					1[50]		
1050		1200	2550	5[100]		[9900]	
150							

Figure 3.3 Our transliteration of the table.[51]

The text is followed by an abbreviated solution in tabular form. We show this table in Figure 3.1, followed by the edited version from (Winter 1936, 29.11–16) in Figure 3.2 and then our transliteration in Figure 3.3.

Given the preserved part of the problem, which contains the final computations of the sought-after numbers, the proof, and the abbreviated solution in the end, we can easily reconstruct the missing enunciation and the solution of the problem. The enunciation should have been stated something like this:

> Given 9900 drachmas, let it be divided into four parts; let the second be greater than the first by a seventh part of the first, let the third exceed the first two by 300 drachmas, and let the fourth exceed the (first) three by 300 drachmas. To find the numbers.[52]

50 Winter wrote in his transcription the alphabetic numerals for 1–999 in lower case. To denote thousands the scribe drew a large hook above the letter, and these Winter indicated using capital letters. Thus "α" would be 1, while "A" is 1000.

51 We write "D" in place of the abbreviation for "drachma" and "N" for the abbreviation of the sign for the algebraic "Number".

52 Reconstruction proposed by Winter (1936, 30), slightly adapted.

So, the problem, stated in modern notation, is to find w, x, y, and z such that

$$w + x + y + z = 9900,$$
$$x = w + \tfrac{1}{7} w,$$
$$y = w + x + 300,$$
$$z = w + x + y + 300.$$

The solution can be followed in the table, where the first four vertical lines correspond to the four unknowns. The three excesses are written in the top row ($\frac{1}{7}$, 300, and 300), followed by the 9900 that the four numbers add up to. In the second row, the first number is assigned to be "7 Numbers", making the second number "8 Numbers". Adding these together with the 300 drachmas give the third number as 15 Numbers, 300 drachmas. Moving to the right, the scribe did not bother writing the sign for "Numbers" in designating the fourth number as 30 Numbers, 600 drachmas. To the right of that, and a little lower, is their sum: 60 Numbers, 900 drachmas. These must equal the 9900, and the value of 1 Number obtained from solving the equation is written in the fourth column: 150. Moving to the left, the first number is calculated to be 1050. A seventh of that, 150, is written below it and is added to the 1050 to get the second number, 1200. Then the other numbers are calculated, and it is checked that they add to 9900.

Notwithstanding that the main part of the solution is missing, and despite the condensed nature of the text, the algebraic character of the solution is easily recognized: names are assigned to the unknowns, operations are performed on the named terms, and an equation is established in the setting of the named terms. Table 3.3 shows the stages of the solution up to the statement of the equation. Here we write 7N for N7, etc., to make the reading easier.

Table 3.3 The set up of the equation in the first problem of papyrus Michigan 620.

Assignment of names	Operations with names	Equation
1st := 7N		
	$\frac{1}{7}$ of 7N →1N	
	Add 7N and 1N →8N	
2nd := 8N		
	Add 7N and 8N → 15N	
	Add 15N and 300D → 15N 300D	
3rd := 15N 300D		
	Add 7N, 8N, and 15N 300D → 30N 300D	
	Add 30N 300D and 300D → 30N 600D	
4th := 30N 600D		
	Add 7N, 8N, 15N 300D, and 30N 600D → 60N 900D	
		60N 900D = 9900D

Second problem

As with the previous problem, the preserved part of the text corresponds to the last part of the solution:

> (Since) the second is quadruple the first, make the 42 (units) four times larger: 168. And the 12 (units) of the excess: 180. This much is the second. Proof: Take the sixth of the second: 30. And 12 besides: 42, as much as the first. And make the 42 (units) four times: 168. And 12 besides: 180, as much as the second.[53]

After the text comes the tabular presentation of the solution, for which we again show the scan (Figure 3.4), the edited version (Figure 3.5), and then our transliteration (Figure 3.6).[54]

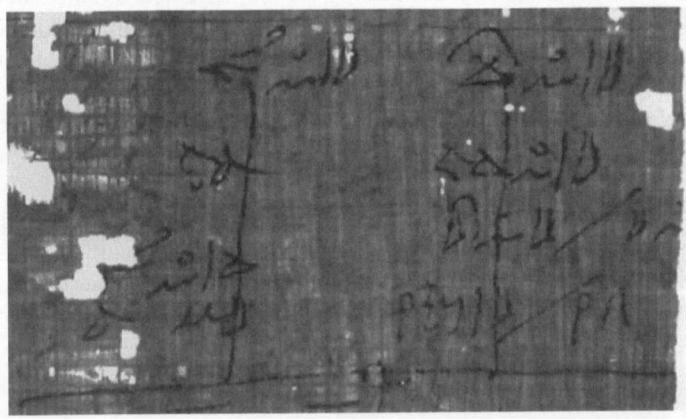

Figure 3.4 Facsimile of the table for the second problem of papyrus Michigan 620.

	ϛʹ	μ̈	ιβ			δ̂	μ̈	ιβ		
	ϛα					ϛδ	μ̈	ιβ		
						β̂μ	β	| β μ̈	[ιδ]	
γʹ	μ̈	ιδ								
	α	μβ			ρξη	ιβ	| ρπ			

Figure 3.5 Edition of the table for the second problem of papyrus Michigan 620.

53 For the Greek text, see Winter (1936, 30.1–5).
54 See Winter (1936, 30.6–10).

$$6'u12 \qquad\qquad \hat{4}u12$$
$$N1 \qquad\qquad N4u12$$
$$3'\ u14 \qquad\qquad \hat{2}u2\ /\ 2u[14]$$
$$1\ 42 \qquad\qquad 168\ 12\ /\ 180$$

Figure 3.6 Our transliteration of the table.

In our transliteration we write "U" as the abbreviation for "units" (μονάδες), and again "N" for "Numbers" (ἀριθμοί). Here the $\hat{4}$ means "quadruple", while $\hat{2}$ means "2/3".

The enunciation of the problem is missing, yet it should have been framed somewhat like this: "Two numbers. The first exceeds the sixth of the second by 12 units, and the second exceeds the quadruple of the first by 12 units".[55] So the problem proposes, in modern notation, to find two numbers x and y, such that $x = \frac{1}{6}y + 12$ and $y = 4x + 12$.

The preserved part of the text begins with the computation of the second number, which, being the quadruple of the first increased by 12 units, is 180. This is followed by the proof, that is, the verification that the numbers satisfy the conditions of the problem.

The solution in the table unfolds in this order:

6'U12	1/6 of the second number, and 12 units is the first number.
$\hat{4}$U12	Quadruple the first number, and 12 units is the second number.
N1	Assign the first number to be "1 Number".
N4U12	The second number is then 4 Numbers, 12 units.
$\hat{2}$U2 / 2U[14]	Taking a sixth of this gives 2/3 Numbers, 2 units, then adding 12 units gives 2/3 Numbers, 14 units. (The second 2 [B] should have a hat [$\hat{2}$].)
3' U14	Equating this to 1 Number, and subtracting 2/3 Number from both sides, gives the equation 1/3 Number = 14 units.
1 42	Thus 1 Number = 42 units, which is the first number.
168 12 / 180	Multiplying by 4 gives 168, then adding 12 gives 180, which is the second number.

There are two unknowns, thus two columns. The vertical lines do not separate columns, but simply line up the calculations, which sometimes run over the lines. In the equations in the bottom two rows on the left there is no sign for "=", and the sign for "Numbers" is not written. Because this work is intended only for immediate use, there is no need for such symbols. This economy of notation is also sometimes present in Arabic algebra. For example, one equation in a manuscript of Ibn

55 Reconstruction proposed by Winter (1936, 33), slightly adapted.

Ghāzī shows "1 2 = 99", for what we would write as "$x^2 + 2x = 99$".[56] No signs for the powers are needed because the person writing it knows which number corresponds to which power. The notation, after all, was not intended to be read and understood by others. In the papyrus, the diagonal slash "/" indicates the outcome of operations. It is probable that had the original papyrus been larger in width, the equations in the first column would have been written in a third column.

As with the previous problem, this one exhibits a solution by algebra of a numerical problem. The solution is shown in Table 3.4.

Third problem

In contrast with the two previous problems, in this one the enunciation and the beginning of the solution are preserved in a very mutilated condition. Most of the solution, the proof, and the tabular setting of the solution are lost. Our translation of the text, as Winter restored it (1936, 30.11–16), is the following:

> Three numbers. All three (sum up to) 5300. Le⟨t the first and the second⟩ be 24 times the third, ⟨and let the second and the thir⟩d be quintuple the first. ⟨The three numbers⟩ are found in this manner. Since the first and the se⟨cond are 24 ti⟩mes ⟨the third⟩; therefore, all three are ⟨25 times the third. Divide 5300 by t⟩he 25: 212; this ⟨much is the third⟩.

Translated into modern algebraic notation, the problem asks to find three numbers x, y, and z, such that $x + y + z = 5{,}300$, $x + y = 24z$, and $y + z = 5x$".[57] It is thus similar to Diophantus's Problem I.20, which asks to find three numbers such that their sum is a given number, the sum of the first two is a given multiple of the third, and

Table 3.4 The set up of the equation in the second problem of papyrus Michigan 620.

Assignment of names	Operations with names	Equation
$x := 1\mathrm{N}$		
	Quadruple 1N → 4N	
	Add 4N, 12U → 4N 12U	
$y := 4\mathrm{N}\ 12\mathrm{U}$		
	$\frac{1}{6}$ of 4N 12U → $\frac{2}{3}$N 2U	
	Add 12, and $\frac{2}{3}$N 2U → $\frac{2}{3}$N 14U	
		$\frac{2}{3}$N 14U = 1N

56 The numbers are written with Arabic numerals, and the "=" is shown as an elongated letter *lām*, the last letter in *taʿdil*, "equal" (Oaks 2012a).

57 An alternative reconstruction that has been considered, according to which the problem would be $x + y + z = 5{,}300$, $x + y = 24z$, and $y = 5x$, is not tenable for grammatical reasons (Winter 1936, 34; Robbins 1929, 328).

the sum of second and third is a given multiple of the first. Instead of 5300, 24, and 5, the given numbers in Diophantus are 100, 3, and 4 (46.9–14).

The solution begins with arithmetical, not algebraic, calculations to find that the third number is 212: since the first and the second are 24 times the third, the sum of the first, second, and third will be 25 times the third. But it is given that the sum of the three numbers is 5300. Therefore, the third number is 212. So far no unknown has been given a name, and no equation has been set up. Diophantus also begins his solution by finding the third number, but he does so by naming the third number as "1 Number" and working it out by algebra.

Now, the problem is reduced to finding two numbers such that the first and second sum to 5088 and the second together with 212 is five times the first. If the lost part of the solution followed the same reasoning as in Diophantus, the first would be assigned to be 1N. Then the second would be 5N ℓ 212U . This leads to the equation 6N = 5300U. Solving this, the first number is found to be $883\frac{1}{3}$, and the second to be $4204\frac{2}{3}$. It is unlikely that this would have been worked out arithmetically, as with the finding of the third number, since the calculations are not so trivial.

In the beginning of this section we pointed out the textual parallels between this papyrus and the text of Diophantus by indicating notation and vocabulary they have in common. This is consistent with the fact that both exhibit solutions by premodern algebra. There is, however, a stylistic difference between the problems in the papyrus and the problems worked out in the Greek text of the *Arithmetica*. In the papyrus all computations involved in the determination of the numerical values of the sought-after numbers and the verification at the end of the solution are carried out, in contrast with the Greek Diophantus, in which quite often this task is skipped altogether (one might suggest that the calculations are left to the reader). However, this is not the case with the Arabic books of Diophantus, in which such calculations are meticulously executed. From this perspective, our papyrus is closer to the Arabic than to the Greek Diophantus. It has been suggested that the prolix character of the Arabic version of the *Arithmetica* is due either to some later editor who, writing in Arabic, added this material to the translation, or it was already present in the Greek manuscript from which Qusṭā ibn Lūqā made the translation – the latter supposedly being a recension of the original text of Diophantus made either by Hypatia or by some late antique Greek editor.[58] In either case, the preserved Greek text, being much terser than the Arabic, is considered as representing the original text of Diophantus. If the problems in the papyrus Michigan 620 are representative of the way algebraic problem solving was taught in the time just before Diophantus, then such a claim should be moderated, and the possibility that the Arabic text more faithfully represents the Greek original should not be ruled out. On the other hand, a reasonable explanation for the fact that the final stages of the resolution are not fully pursued in the Greek text would be that the

58 See Sections 1.2 and 3.2.1.

aim of Diophantus, the very reason why he wrote his work, is to teach what in premodern algebra was the most demanding task of a solution, that is, the setting up of the equation. Considering the evidence we have, we cannot lean in one direction or the other on this issue.

3.1.6 *The lists of powers in* Refutation of All Heresies

In the first half of the third century CE an anonymous Christian bishop composed a treatise in Greek titled *Refutation of All Heresies*, henceforth *Refutation*.[59] This is not the kind of book in which one would expect to find, twice in different chapters no less, lists of the names of the algebraic powers up to the Cube-Cube. Algebra and theology are linked in this case through Neopythagorean arithmology. To understand how it all fits together we should first provide some background, beginning with a description of the book.

Until recently the author of *Refutation* was thought to be the Roman bishop Hippolytus. We know now, thanks to the work of several scholars since the 1940s, that this attribution is wrong. Indeed, we have no clue regarding the name of our author, though we do know that he was a rival bishop in Rome and that he probably wrote his book shortly after 222 CE.[60] The purpose of the book was to show that the Christians whom the author regarded as heretics plagiarized from pagan Greek sources, mainly Greek philosophy (Litwa 2016, xliiff). This philosophy is distinctly Neopythagorean, which situates Pythagoras as the first philosopher, from whose teachings the works of later philosophers, Plato in particular, derived.

Refutation was composed in 10 Books. Book 1 consists of a survey of Greek philosophy in standard Neopythagorean fashion, centered on the figures of Pythagoras and Plato. It is in the chapter on Pythagoras that the first list of powers is given. Books 2 and 3 are lost. The remaining books address the views of the heretical plagiarists, and the other list of powers appears in Book 4, where it is repeated nearly word for word for the purpose of exposing the theologians Simon and Valentinus. The lists in both chapters are given to round out the author's brief explanation of what is otherwise a fairly standard account of the arithmetical foundation of Neopythagorean philosophy. This explanation, but without the algebraic names, is deliberately presented for a third time in Book 6 for the purpose of an immediate comparison with another aspect of the theology of Valentinus (Litwa 2016, 395 [6.23.1–2]). This explanation is particularly concise and clear:

> Pythagoras, then, announced that the first principle of the universe is the unborn Monad. The Dyad and all the other numbers are generated. Moreover, he says that the Monad is the Father of the Dyad, and that the Dyad is the mother of all generated beings, as one generated produces those generated. . . . The numbers three to ten arise, in turn, from the Dyad, as Pythagoras says. It is the number ten that Pythagoras deems the only perfect number. For

---

59 The text has recently been edited, with English translation and commentary, in Litwa (2016).
60 Litwa (2016, xxxiiff) reviews the history of attempts to identify the author.

eleven and twelve are an addition and a reiteration of the decad, and generation arises from no other number.

The decad produces every solid body from incorporeal elements. The indivisible point is the building block and source of both corporeal and incorporeal entities. A point gives rise to a line, and a line to a plane. When a plane becomes three-dimensional, a solid body is formed. Therefore there is even an oath among Pythagoreans consisting of the harmony of the four elements. They swear as follows:

> Yea, by him who delivered to our mind the Tetraktys,
> Source possessing the roots of ever-flowing nature.[61]

The idea of identifying the "unborn Monad" as the "first principle of the universe" in fact originates with Plato, not with Pythagoras (Zhmud 2016, 329). The word τετρακτύς (*tetraktys*) to which the Pythagorean oath is directed literally means a "collection of four things", or a "quaternity" (Delatte 1915, 254). The Pythagorean τετρακτύς is described in several places in *Refutation*, such as in Book 4, just after describing the monad, dyad, triad, and tetrad:

> So all the numbers arose, stemming from four classes – number itself being an undefined class –, from which the perfect number was composed: the decad. For the one, two, three, four becomes ten, if the proper (according to its essence) denomination be preserved for each of the numbers. Pythagoras called this ⟨decad⟩ "the holy tetraktys".[62]

Neopythagoreans were fond of listing attributes of the numbers from 1 to 10. Some of these attributes are mathematical, others not. Certain numbers held special importance, notably 4, associated with the τετρακτύς and the perfect number 10, and 7, the hebdomad.

Anatolius, bishop of Laodicea in the second half of the third century, gives a characteristic treatment of the numbers in his book *On the First Ten Numbers* (cf. Section 1.1). Here is the beginning of his chapter "On the four" (Heiberg 1901, 31–32):

> The four is called justice, because of the square which comes from an area equal to its perimeter, while for preceding numbers the perimeter of the square is greater than the area, and for those that follow, the perimeter is less than the area. It is the first square, and ⟨the first⟩ among the evens. This is the first *tetraktys*, since the ⟨numbers⟩ from 1 to 4 make 10, which is said to be the perfect number. This is the first ⟨number⟩ that exhibits the nature of a solid; for ⟨one has first⟩ the point, then the line, then the surface, then the solid, which is a body. ⟨We see⟩ this in the game of building pyramids with nuts.

---

61 (Litwa 2016, 395 [6.23.1]), his translation.
62 (Litwa 2016, 185 [4.51.6]), our translation.

He then goes on to mention that there are four elements, four seasons, that four is the first evenly even number, four is the first number that admits a sesquitertian ratio with another number, there are four astrological points, and that the universe is ordered according to four principles: substance, shape, form, principle, etc.

It was probably in the 30s CE that the Jewish theologian Philo of Alexandria wrote his *On the Creation of the Cosmos According to Moses*, known today also by the Latin title *De Opificio Mundi* (Philo 2001, 1, 4). In part of this book Philo runs through the seven days of creation, listing the properties of the number of each day in classic Neopythagorean fashion. For the fourth day, for example, he observes that the sum of 1, 2, 3, and 4 is 10, the "all-perfect number", and he associates the sequence with the musical ratios 3:4, 2:2, and 1:2; then with point, line, surface, and solid; and also with the four elements and the four seasons (Philo 2001, 57–58 [§45]). The chapter on the seventh day is the most important in the book and takes up nearly a quarter of the whole treatise. "I do not know whether anyone can give sufficient praise to the nature of the seven" (2001, 70 [§90]). He begins by noting two ways of speaking of the seven, either "confined within the ten" or "outside the ten". The former is associated with addition from 1 and terminating with 7, and the latter with doubling or tripling from 1 and terminating with either 64 or 729:

> The seven is spoken of in two ways. In the former case it is confined within the ten. It is measured by taking the unit on its own seven times and consists of seven units. In the latter case it falls outside the ten, a number whose starting-point in each case is the unit in accordance with the double or the triple or with numbers generally corresponding to these, such as 64 and 729, the former attained by doubling (seven times) from the unit, the latter by tripling. . . . The second kind possesses a distinction which is very apparent. Every time that a number, beginning from the unit, is doubled or tripled or generally multiplied analogously, the resultant seventh number is both a cube and a square. This number contains the kinds of both incorporeal and corporeal being, the former through the surface produced by squares, the latter through the solidity produced by cubes.
>
> (Philo 2001, 71 [§91–92])

Among the other 33 attributes of the hebdomad that Philo presents are the seven stages of life that he attributes to Hippocrates: "child, boy, youth, young man, man, older man, old man" (2001, 75 [§105]); the seven nonrational parts of the soul: "the five senses, the organ of speech, and finally the reproductive part" (2001, 78 [§117]); and the seven bodily secretions:

> Through the eyes tears flow forth, through the nostrils the filtrations of the head, through the mouth the saliva that is spat out. There are also two discharges for drawing off what remains after the digestion of food and drink, one at the front and one at the back. The sixth occurs all over the body, a

profusion in the form of sweat, while the seventh is the emission of seed through the reproductive parts, which is very much a part of nature.

(2001, 78 [§123])

Many extant texts from the first few centuries CE present similar accounts of the first 10 numbers.[63] Such texts constitute an entire genre of Greek literature that since Delatte (1915) has been called "arithmological". They are part of a tradition that traces back to a single, lost source from the early first century BCE (Robbins 1921; Zhmud 2016). The attributes that these texts list are more or less repeated from one author to the next, with occasional original additions. Though this tradition claims to stem from Pythagoras himself, in fact it is more recent. Kalvesmaki (2013, 8) points out that "Pythagoras, who had flourished seven centuries earlier, left behind no writings, and his followers had disappeared by the time of Aristotle". The lists of the attributes of numbers in Neopythagorean texts were a new development in post-Aristotelian Pythagorean philosophy (Thesleff 1961, 1965; Zhmud 2016).

Returning to arithmetic, Anatolius, like Philo, notes that the hebdomads 1, 2, 4, 8, 16, 32, 64 and 1, 3, 9, 27, 81, 243, 729 terminate in a number that is both a square and a cube.[64] This hebdomad was sometimes presented in reduced form as a τετρακτύς. Theon of Smyrna, in his expansive list of τετρακτύες, notes, as Philo does, that numbers can be generated through addition, as in 1, 2, 3, 4, or through multiplication, as in 1, 2, 4, 8 and 1, 3, 9, 27 (1892, 154–56; 1979, 62–63). These same two quaternaries consisting of the powers of two and three are also found in three other works, by Philo, Anatolius, and Macrobius (ca. 400 CE), and they also make an appearance in *Refutation*.[65] Apart from *Refutation*, where there is evidently a lacuna, all of these works cite Plato's markedly Pythagorean dialogue *Timaeus* as the source (35b–36a). F. M. Cornford notes regarding Plato's numbers, "the unit had been held by the Pythagoreans to contain within itself both the 'elements' of number, the even (or 'unlimited') and the odd ('limited' or 'limit')" (Cornford 1937, 66). For Plato, as for the Neopythagoreans who reproduced the series of powers, whether as a τετρακτύς or a hebdomad, the numbers 2 and 3 serve as stand-ins for the even and the odd. Thus, they are to be interpreted as embodying the general notion of numbers generated through multiplication.

---

63 Runia utilized for his study of Philo the following texts, arranged chronologically: Philo of Alexandria, Nicomachus of Gerasa, Theon of Smyrna, Alexander of Aphrodisias, Censorinus, Anatolius, Calcidus, Ps.-Iamblichus, Macrobius, Martianus Capella, and John Lydus (Philo 2001, 28–29). He did not intend for this list to be complete.

64 (Heiberg 1901, 35). French translation by Tannery, ibid., 50; English translation in (Philo 2001, 304–05).

65 They are found in: an Armenian translation of a lost work of Philo (Terian 1984); Anatolius, in his chapter "On the decad" (Heiberg 1901, 39; French translation by Tannery, ibid., 54); Macrobius's *Commentary on the Dream of Scipio* VI.46 (Macrobius 1990, 109); and *Refutation*, in a section on astronomical magnitudes in Book 4 (Litwa 2016, 115 [4.8.6]).

We are now finally in a position to examine the lists of the algebraic powers in *Refutation*. In Books 1 and 4, after recounting the origin from the unborn monad, the generation of the dyad, triad, and tetrad, that they add to 10 to form the holy τετρακτύς, and how the 10 is the perfect number, our author continues by citing another τετρακτύς, one that is formed by the first four algebraic powers, and then extending this to a hebdomad:

> There are what are called "four components" of the decad, the perfect number: ἀριθμός, μονάς, δύναμις, and κύβος, the combinations and mingling of which are conducted for the generation of growth, bringing to natural completion the productive number. For a δύναμις multiplied by a δύναμις becomes a δυναμοδύναμις; a δύναμις by a κύβος becomes a δυναμόκυβος; and a κύβος by a κύβος becomes a κυβόκυβος. Accordingly, all the numbers from which come the origin of all generated beings are seven: ἀριθμός, μονάς, δύναμις, κύβος, δυναμοδύναμις, δυναμόκυβος, κυβόκυβος.[66]

Immediately after this, in Book 4, Simon and Valentinus are accused of changing the names of this hebdomad:

> Simon and Valentinus recounted wondrous myths about this hebdomad (changing out the names), to improvise a basic plan for themselves. Simon calls it "Mind, Thought, Name, Voice, Reasoning, Conception, and the One Who Stood, Stands, and Will Stand". Valentinus calls the hebdomad "Mind, Truth, Word, Life, Human, Church" – and added to them the Father. He did so according to the same principles as those who practice the philosophy of arithmetic.[67]

These two hebdomads were indeed composed along the same lines as those in other authors, and they are not known from other sources. They were likely original with Simon and Valentinus.

But what about the τετρακτύς and hebdomad of our author? After mentioning the sequence 1, 2, 3, 4, he cites only one τετρακτύς – ἀριθμός, μονάς, δύναμις, κύβος – and then a single hebdomad that includes the powers through the κυβόκυβος. These are not found in any other extant source either, so they, too, may not have been part of the Neopythagorean textual tradition. But neither should we definitively attribute them to the author of *Refutation*, since he might have copied them from some other, lost book. Whoever came up with them seems to have wanted to consider numbers generated through multiplication and perhaps

---

66 Translation adapted from Litwa (2016, 17 [1.2.9–10], 187 [4.51.7–8]). The texts of the two passages are almost identical. In transliterated form these names are *arithmos, monas, dynamis, kybos, dynamodynamis, dynamokybos,* and *kybokybos.*

67 (Litwa 2016, 187 [4.51.9]), his translation, but we replaced his "theorem" with "plan". The *Lexicon* of Dēmētrakos (1958) translates ὑπόθεσιν as "goal, purpose, plan, pretext, occasion, allegation".

found the names of the powers to be more suitable than listing the powers of 2 and 3 that we find in other authors. Whatever his motive, he did not understand the names. Ἀριθμός (as the name of the first power) should come after μονάς (the unit) in the series, as he would have known had he practiced algebra. One clue for the reversal of these terms might be found in what our author wrote shortly before this in Book 4, after laying out the monad through tetrad: "So all the numbers arose, stemming from four classes (i.e., 1, 2, 3, and 4) – number itself being an undefined class".[68] If ἀριθμός is taken in its ordinary meaning as this undefined class and not the name given to the first power, then it is of a distinct type from the others, and the sequence 1, 2, 3 could then naturally be identified with μονάς, δύναμις, and κύβος.

In this context it was an appropriate choice to list algebraic powers, with μονάς and ἀριθμός instead of the arithmetical terms πλευρά and τετράγωνος. The algebraic terms are generated through multiplication just like the powers of 2 and 3, while the arithmetical terms are simply classes of numbers. And more important for the present study is that the two passages in *Refutation* show that by the 220s CE people were already working with the same names of the powers up to the sixth that we find in Diophantus.

3.2 Readers and writers on Diophantus in late antiquity

In the decades and centuries after his time, Diophantus's mathematical practice remained alive in the Greek-speaking world before the same practice blossomed in the new social and cultural environment of the Islamic world. In the present section we provide a list of all known testimonies on Diophantus that are preserved in Greek sources dated in or referring to the period extending from the fourth to the seventh century CE. The testimonies gathered and discussed show that there existed erudite circles which not only recognized Diophantus as an authority in arithmetics and occasionally consulted his text, but more importantly, kept alive and continued practicing the art of solving problems via the method taught in the *Arithmetica*. The latter, in particular, is witnessed by Theon's commentary on Ptolemy's *Almagest*, as well as by a series of late antique scholia to a number of arithmetical epigrams contained in Book XIV of the so-called Palatine Anthology.

3.2.1 Theon and Hypatia

The earliest and most important piece of evidence showing familiarity with at least the first book of Diophantus's treatise by an ancient scholar comes from Theon, the famous Alexandrian commentator of Ptolemy and editor of Euclid and Ptolemy. In Tannery's time Theon's expressed interest in Diophantus was only partially known. A short paragraph from Theon's commentary on the first book of Ptolemy's *Almagest* mentions the name of Diophantus twice and quotes verbatim

---

68 (Litwa 2016, 185 [4.51.6]), our translation.

two excerpts from the introductory chapter of the *Arithmetica* that refer to the ways the unit and the parts of the unit function in multiplications. This paragraph is the first selection Tannery included in his "Testimonia Vetera de Diophanto" in the second volume of his edition (Diophantus 1893–95, II, 35.1–36.3). It has been valuable to historians of mathematics mainly because it provides a limiting date for Diophantus.[69] Here we will read it for what it can tell us about the reception of Diophantus's mathematical practices after his time.

The passage published by Tannery begins:

> According to what Diophantus says, "Because the unit, being immutable and always constant, any species multiplied by it will remain the same species", in the same manner, the degree, whatever the species by which it is multiplied might be, preserves the same species. Therefore, a degree multiplied by degrees will make degrees; by minutes, minutes; by seconds, seconds; by thirds, thirds; and so on. On the other hand, the situation is different for the parts of the degree, as we shall show in the following. For, again, as in Diophantus, when multiplying the parts of the unit the species become different – indeed a part of a Number, (such as) a third, multiplied by itself makes a part of a Power, (such as) a ninth, and it changes the species – in the same manner, here, the parts of the degree cause a change in species.
>
> (Diophantus 1893–95, II, 35.8–36.1)

This is developed into a thorough exposition of the multiplication between sexagesimal numbers that covers seven full pages in Adolph Rome's edition (Theon of Alexandria 1936, 452.1–459.17). Theon's objective is to show that the multiplication of sexagesimal numbers can be made in the same way Diophantus multiplies polynomials, that is, according to the premodern meaning of the term, as aggregates of the unit and the reciprocal powers of the unknown. After explaining this connection, Theon gives a geometric proof for the rule and then restates Ptolemy's arithmetical explanation in terms of proportions. Then he explains the corresponding rule for division, again with a proof by geometry. Next are examples of each operation. The first is the multiplication of 37° 4′ 55″ by itself performed

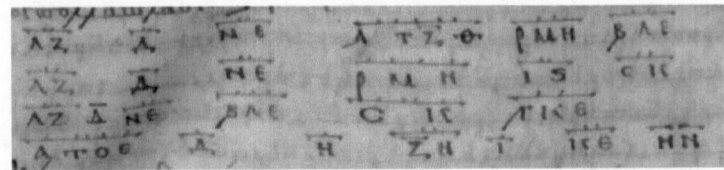

Figure 3.7 Theon's multiplication of 37° 4′ 55″ by itself, from codex Laur. Plut. 28.18, f. 35r.

term by term in the same manner as a multiplication between polynomials consisting of units (degrees) and the reciprocal powers (minutes and seconds).[70] Theon explains that to multiply two terms one finds the product of the numbers, and for the species, using our notation, one adds the primes. For example, the result of multiplying 4′ by 55″ is 220‴. After explaining verbally each of the nine multiplications, the calculations are shown in tabular form in the manuscripts, such as in this image from codex Laur. Plut. 28.18 (first half 9th c.), f. 35r:

Here is the translation:[71]

37°	4′	55″	1369°	148′		2035″		
37°	4′	55″	148′	16″		220‴		
37° 4′ 55″		2035″	220‴	3025⁗				
1375°	4′		6°	68′	10‴	25⁗	8″	50‴

Theon's exposition of the multiplication and division of sexagesimal numbers reveals a familiarity with operations on aggregations of species, which is an important prerequisite for the calculations involved in problem-solving via the method that Diophantus teaches in the *Arithmetica*.

There is another passage in the commentary in which Theon cites Diophantus, but which was unknown to Tannery. It is found in the commentary to Book XIII, which remains unedited.[72] In it, Theon solves a mensuration problem "by the process of the Diophantine numbers", which shows that algebra was recognized by Theon as a particular, identifiable method, distinguished from other problem-solving methods practiced at the time. The problem asks to find three angles, say the angles φ, ω, σ, such that $φ : ω :: 5 : 9$, $φ = 4\frac{1}{3} - σ$, and $ω = 7 - σ$. It corresponds to Diophantus's Problem I.9, which stipulates "to subtract the same number from two given numbers and make the remainders have to one another a given ratio". The difference between the two problems lies not just in the numerical values of the given data but also in the context within which they are formulated. The Diophantine problem is arithmetical, while the Theonine is a problem of mensuration, which originated in astronomy. What makes the algebra of Diophantus applicable for mensuration is that the angles are presumed to have numerical measures, and calculations are performed in the realm of arithmetic.

The passage is the following:[73]

70　See Theon of Alexandria (1936, 458.15–459.6).

71　The scribe did not understand what he wrote, for he misplaced the partial results, he wrote the multiplicand three times, and he has an error in the last line when he put an $\overline{\text{H}}$ (8) instead of $\overline{\text{ϛ}}$ (6). These facts prove that the table existed in the late antique prototype and was not prepared by the Byzantine scribe of the first half of the ninth century. Cf. Rome's remarks in the criticus apparatus and in Theon of Alexandria (1936, 580–81). For the history of the codex Laur. Plut. 28.18, see Rome (1938).

72　For a thorough discussion of Theon's interest in Diophantus, see Skoura and Christianidis (2014) and Skoura (2016, 3–44).

73　Translation slightly adapted from Christianidis and Skoura (2013, 47–48).

If there are two given numbers, and some equal (numbers) are subtracted from them, and the ratio of the remainders is given, those, the ratio of which was given, will also be given, and the others, that is, the equal ones, ⟨will be given too⟩. Let us work this out methodically by the process of the Diophantine numbers in the case of the proposed numbers, the $4\frac{1}{3}$ and the 7, and the ratio 5 to 9.

Let the (amount) subtracted from each of the two, the $4\frac{1}{3}$ and the 7, be 1 Number. Now, when subtracted from $4\frac{1}{3}$ units, the remainder will be $4\frac{1}{3}$ units lacking 1 Number. And when from 7 units, the remainder will be 7 units lacking 1 Number. Therefore, $4\frac{1}{3}$ units lacking 1 Number must have to 7 units lacking 1 Number the ratio that 5 has to 9. But the 5 falls short of the 9 by 4 fifths of themselves. Therefore, $4\frac{1}{3}$ units lacking 1 Number, likewise, falls short of 7 units less 1 Number by four fifths of themselves. If, therefore, we add to $4\frac{1}{3}$ units lacking 1 Number 4 fifths of themselves they will be equal to 7 units lacking 1 Number. But $4\frac{1}{3}$ units lacking 1 Number, when they receive in addition the 4 fifths of themselves, become 117 fifteenths of a unit lacking 9 fifths of a Number, that is to say 27 fifteenths, as we will show below.

Therefore, 117 15ths of a unit lacking 27 15ths of a Number are equal to 7 units lacking 1 Number; that is to say 105 fifteenths of a unit lacking 15 fifteenths of a Number.

Let the lacking 42 15ths of a Number be added in common; therefore, 117 15ths of a unit and 15 15ths of a Number are equal to 105 15ths of a unit and 27 15ths of a Number. And ⟨let⟩ likes ⟨be subtracted⟩ from likes; it remains, therefore, that 12 15ths of a unit are equal to 12 15ths of a Number. And all fifteen times. Therefore, the Number will be 1 unit.

To the hypostases: I assigned one of those having the given ratio to be $4\frac{1}{3}$ units lacking 1 Number; it will be $3\frac{1}{3}$ units. And the other, 7 units lacking 1 Number; it will be 6 units. And the rest, that is to say each of the equal ones, which is what is subtracted from each of the two, will be the 1 unit of the Number.

Now, the fact that $4\frac{1}{3}$ units lacking 1 Number, if they receive in addition the four fifths of themselves become 117 15ths of a unit lacking 27 fifteenths of a Number, is made manifest in this manner: for, since the 5th of the $4\frac{1}{3}$ becomes 52 sixtieths, then the 4 5ths will be 208 sixtieths, that is 52 15ths; the $4\frac{1}{3}$ units, on the other hand, are 65 15ths. Therefore, the $4\frac{1}{3}$ ⟨units⟩, if they receive in addition the 4 5ths of themselves, sum up to 117 15ths. There is also the lacking ⟨part⟩ of the Number together with the 4 5ths of itself, which makes 9 fifths of a Number less, that is to say, 27 15ths ⟨of a Number⟩. Therefore, $4\frac{1}{3}$ units lacking 1 Number, if they receive in addition 4 5ths of themselves, become 117 15ths of a unit falling short by 27 15ths of a Number.

In Theon's solution we recognize the main features of an algebraic solution. He gives the name "1 Number" ($\sigma := 1N$) to the subtracted angle, he subtracts "1 Number" from the $4\frac{1}{3}$ and the 7 to get "$4\frac{1}{3}$u ℓ 1N" and "7u ℓ 1N", which are, in the setting of the technical algebraic language, the names of the numerical values of the angles φ and ω, respectively, (the letter ℓ in the previous expressions stands for the word "lacking"). Because they have the ratio that 5 has to 9, adding $\frac{4}{5}$ of $4\frac{1}{3}$u ℓ 1N to itself will become equal to 7u ℓ 1N. Once the operations are worked out, this results in the equation $\frac{117}{15}$u ℓ $\frac{27}{15}$N = 7u ℓ 1N. This is simplified by applying "restoration and confrontation" to get $\frac{12}{15}$u = $\frac{12}{15}$N and then by multiplying everything by 15. Theon arrives at the simplified equation 12N = 12u (which is not stated explicitly), whose solution is 1. Finally, by substituting this 1 in the expressions of the names, he finds the remaining two angles to be $\varphi = 3\frac{1}{3}$ and ω = 6. Theon's solution shows that his study of the treatise of Diophantus goes beyond the introductory chapter of the *Arithmetica* and that he was proficient in the basic method of algebraic problem solving.

Theon was certainly not the only person of his time familiar with the *Arithmetica*. Indeed, from the heading of Book 3 of Theon's commentary on Ptolemy's *Almagest*, which reads, in the edition of A. Rome (1943, 807), "Commentary of Theon of Alexandria on the third of Ptolemy's *Mathematical Syntaxis*, the edition having been proof-read by the philosopher Hypatia, my daughter",[74] we infer that Theon and Hypatia (d. 415 CE) pursued scientific activities in collaboration. Our testimony for Hypatia's interest in Diophantus is *Souda*, the popular Byzantine lexicon of the late tenth century, whose entry for Hypatia reports that "She wrote a commentary on Diophantus, ⟨on⟩[75] the astronomical canon, a commentary on Apollonius' *Conics*" (Adler 1935, 644.3–5). It is not clear what the second of the aforementioned works might be. Acerbi has pointed out that the wording at this particular point "does not allow the reader to decide whether Hypatia wrote an astronomical table or a commentary on such a work (in this case almost surely Ptolemy's *Handy Tables*)" (Acerbi 2008, 436). The other two works are described as ὑπομνήματα, commentaries, and none of them has survived.

Concerning Hypatia's commentary on Diophantus – certainly the *Arithmetica* is intended – several hypotheses have been raised. One of them, rooted in Tannery's investigations, claims that the whole Greek textual tradition of the *Arithmetica* stems from her reworking of the Diophantine text.[76] Along with this Tannery suggested that the fact that only six books survive in Greek can be explained by assuming that Hypatia's commentary did not extend beyond the sixth book (Tannery 1912d). This view is no longer tenable given that the Arabic books are placed immediately after the third Greek book and before the Greek books transmitted as

---

74 Translation proposed by Acerbi (2008, 436). Alternative translations are attested in the literature, especially because of the uncertainties concerning the exact meaning of the participle παραναγνωσθείσης, translated here as "proof-read", for which other translations have been proposed, like "prepared" (Knorr 1989, 756), "checked" (Cameron 1990, 107), and "revised" (Cameron and Long 1993). Because of that it is unclear what the precise role of Hypatia in the collaborative undertaking with Theon was.

75 Emendation proposed by Tannery (1912c, 77).

76 See Tannery's diagram showing the relation of the manuscripts of the *Arithmetica* in (Diophantus 1893–95, II, xxiii).

being Books IV–VI, without duplicating any of them. Tannery's proposal is also problematic since it entails that Hypatia prepared in fact a revised edition of the whole text, and not a commentary on it.

From another standpoint, Hypatia's commentary has been called on to account for stylistic features, which distinguish the verbosity of the Arabic version from the austerely terse wording employed in the Greek books (Knorr 1985, 151). In this regard some scholars have gone so far as to claim that the Arabic text derives from the text commented by Hypatia. Thus, Sesiano (1982, 68–75) conjectures that the Arabic translation stems not directly from Diophantus's treatise but from a "major commentary" on it – consisting "in a rewriting of the entire text" (1982, 68) – most likely the commentary of Hypatia. For, as he writes,

> the Arabic text, unlike the extant Greek text, possesses all the characteristics of a commentary made at about the time of the decline of Greek mathematics. Thus the idea that our text might be (part of) Hypatia's commentary arises quite naturally.
>
> (1982, 71)

A similar conjecture had been expressed earlier by I. G. Bashmakova, E. I. Slavutin, and B. A. Rozenfeld.[77] Unfortunately, "rewriting the entire text" is not a characteristic of a commentary. As Acerbi points out, the Arabic version is in fact "a new recension of the treatise, not a commentary, and normally a commentary does not interfere with the text in such a way" (Acerbi 2008, 436; in Diophantus 2011, 116–17). Besides, judging from the Diophantine solution presented in Theon's commentary, which is expressed in strict conformity with the austere style of the Greek text,[78] we find no reason for ascribing to Hypatia the pedantic additions that the Arabic text displays. Thus, we propose that this commentary of Hypatia's on Diophantus is lost and is not connected with the Arabic translation.

3.2.2 Some Neoplatonic scholia[79]

Another set of testimonies comes from the intellectual activity of the Neoplatonic circles of late antique Alexandria. To this category belong, to start with, four anonymous scholia to Iamblichus's *To Nicomachus's Arithmetic*, and which provide terminological explanations and other remarks on Diophantus. They have

77 "It is only natural to suppose that she (i.e., Hypatia) was the author of the books whose Arabic translation we are now discussing. To be more exact, the Arabic manuscript is the translation of part of Diophantus' work commented on by Hypatia, the original problems of Diophantus being interspersed with her own comments" (Bashmakova, Slavutin, and Rozenfeld 1981, 159).

78 Even the characteristic of the Greek text clause "To the hypostases", which is absent from the Arabic version, is present in Theon's solution.

79 The content of this and the next section is almost exclusively based on the study (Christianidis and Megremi 2019).

been transmitted through the fourteenth-century codex Plut. 86.3 of the Biblioteca Medicea Laurenziana in Florence and are printed in Tannery's edition of Diophantus (1893–95, II, 72) and more recently by U. Klein in Iamblichus (1975). These are:

1   Commenting on the phrase of Iamblichus, "Some of the Pythagoreans said that 'unit is the boundary between number and parts'" (1975, 11.9–11), the commentator writes: "So Diophantus (writes) in *Moriastica* (*On Parts*); for parts (involve) progress in diminution carried on indefinitely".[80]

2–3 Commenting on the words "square" and "cube" in Iamblichus's phrase "Again, as from two squares multiplied by one another becomes a square, so from two cubes (becomes) a cube" (1975, 98.2–4), the commentator writes, respectively: "Diophantus calls it a Power-Power", "Diophantus calls it a Cube-Cube". In the first comment one can recognize a misunderstanding of the manner in which Diophantus uses the terms "square" (τετράγωνος) and "Power" (δύναμις), which probably stems from the fact that the commentator's copy of the *Arithmetica* was contaminated, with the terms "Power-Power"(δυναμοδύναμις), "Power-Cube"(δυναμόκυβος), and "Cube-Cube" (κυβόκυβος) having already been introduced in the place of the introduction where Diophantus describes how the arithmetical problems are formulated.[81]

4   Commenting on Iamblichus's discussion on the harmonic mean (1975, 110.7 ff.), the commentator makes the following remark: "We shall learn the properties of the harmonic mean more completely in the last theorem of the first book of Diophantus's *Arithmetical Elements* (Ἀριθμητικὴ στοιχείωσις), and the diligent ought to read these things there".[82]

These scholia make clear that at the time of their composition the works of Diophantus, including the lost *Moriastica*, were not only circulating but were also studied in the erudite circles of Alexandria.[83]

Another Neoplatonic testimony comes from one of the late antique commentators of Porphyry's *Isagoge*, who is referred to in the literature as Pseudo-Elias (or Pseudo-David), and who lived and worked in the school of Alexandria in the second half of the sixth century under Olympiodorus (Wildberg 2016). In one place the commentator presents the four quadrivial sciences (arithmetic, geometry, music, astronomy), which concern the immaterial and universal objects, and the four disciplines associated with them (logistic, geodesy,

---

80 Translation slightly adapted from Klein (1968, 137).

81 The misunderstanding in question has permeated the historical studies on Diophantus. For a detailed discussion of the matter, see Ruska (1917, 68–69), Klein (1968, 251 n. 177), Christianidis (2007, 295–96), and Christianidis and Oaks (2013, 135–36).

82 Cf. Section 1.2.

83 M. Rashed (2013, 596) has characterized these scholia "célèbres" and "fruit de l'érudition alexandrine tardive".

music in matter, and spherics), which concern material and physical objects. The author then cites the names of the highest authorities in each field. The text runs as follows:

> Let us mention successively the inventors of these sciences, who have distinguished themselves in them. So, Nicomachus occupies the first place in arithmetic, while Diophantus in logistic; again, the Pythagoreans occupy the first place in music, ⟨while the Aristoxeneans in the music in matter; and Euclid in geometry,⟩ while Hero in geodesy; and Paul in astronomy, while Theodosius in spherics.[84]

This testimony was unknown to Tannery, so it does not appear among the "Testimonia Vetera" in his edition of Diophantus. It confirms that the topic of Diophantus's *Arithmetica* is logistic, and not arithmetic in the traditional sense as we find in Euclid and Nicomachus.

The juxtaposition of Nicomachus and Diophantus as experts in the respective fields of arithmetic and logistic would persist well into Byzantine times and is present even in casual remarks made in hagiographic works (see Section 3.2.5). We find it again in another testimony, undoubtedly also of Neoplatonic origin. The unknown author of the prolegomena to the *Arithmetical Introduction* of Nicomachus, published by Tannery from codex Par. gr. 2372 (1893–95, II, 73.25–28), runs as follows: "For Diophantus, in the thirteen books of his Arithmetic, treats of the measured number (μετρούμενον ἀριθμόν), while the objective of Nicomachus is to treat the measuring number (μετροῦντα ἀριθμόν)". This fragment has been cited as evidence that arithmetic is the "abstract" science of number, in contradistinction with logistic, which is the "concrete" study of number (Klein 1968, 128).

Returning to the text of Pseudo-Elias (or Pseudo-David), if we take its account literally we infer that arithmetical problem solving was considered to be a part of logistic at the time. Now, the association of arithmetical problem solving with logistic is well documented in the history of Greek mathematics, and there is hardly a need to invoke this text to establish it. Among the most important pieces of evidence attesting to this are: (i) the fragment entitled "On logistic", usually ascribed to Geminus, which is transmitted through the pseudo-Heronian treatise *Definitiones* (Hero 1912, 98.12–23) and as an anonymous scholium to passage 165e6 of Plato's *Charmides* (Cufalo 2007, 173); and (ii) a passage from the fourth lecture of Olympiodorus's commentary to Plato's *Gorgias* (Olympiodorus 1970, 4.8.43–49). Both sources confirm that logistic was concerned with arithmetical problems. The "On logistic" fragment mentions the so-called cattle problem, which is transmitted under the name of Archimedes, and the problems about "sheep-numbers" and "bowl-numbers", which refer to the passage 819b–c

---

84 For the Greek text, see Westerink (1967, 38–39).

of the seventh book of Plato's *Laws*. Even more interesting is the commentary of Olympiodorus, from which we quote:[85]

> But (logistic) also instructs some subtleties, e.g. about the statues, on (the first of) which was written, "I have the next and a third of the third", and on the second, "I have the next and a third of the first", and on the third, "I have eight minae and a third of the middle". And about the streams of the lion pouring into the tank, and about other methods.

The fragment indicates two arithmetical problems of the kind that logistic deals with. Both problems are included in the series of problems which are known as the arithmetical epigrams of the Palatine Anthology. The first problem coincides with the epigram 51, with the minor difference that the latter substitutes 10 for 8. The epigram is:[86]

> I have what the second has and the third of what the third has.
> I have what the third has and the third of what the first has.
> And I have ten minae and the third of what the second has.

The epigram, as transmitted by the unique manuscript which carries the XIVth Book of the Anthology, Paris. suppl. gr. 384, makes no reference to statues, as Olympiodorus's scholion does. However, the same epigram is also preserved in codex Par. gr. 1630 (14th c.), one of the manuscripts comprising one of the so-called syllogæ minores, where it is accompanied by a lemma (f. 195r) explaining that the epigram refers "to three statues" (εἰς ἀνδριάντας τρεῖς). Now, abstractly stated, that is, without any reference to statues and weights, the same problem is found in Diophantus. It is Problem I.21, which begins

> To find three numbers such that the greatest exceeds the middle by a given part of the least, the middle exceeds the least by a given part of the greatest, and the least exceeds a given part of the middle by a given number;

the numerical values that Diophantus assigns in the instantiated version of the problem are the same as those of epigram 51.

The other problem to which Olympiodorus refers is epigram 7:[87]

> I am a brazen lion; my spouts are my two eyes,
> my mouth, and the flat of my right foot.

---

85 The English translation is adapted from Olympiodorus (1998, 89), with a major correction on the point where the translators ascribe the two problems the passage speaks about to Nicomachus.

86 For the Greek text, see Diophantus (1893–95, II, 53.1–4), Paton (1918, 50), and Buffière (1970, 64). The English translation is Paton's. The manuscript has in the margin the answer, 45, 37 ½, and 22 ½.

87 For the Greek text, see Diophantus (1893–95, II, 46.20–47.4), Paton (1918, 30), and Buffière (1970, 54). The English translation is Paton's.

My right eye fills a jar in two days,
my left eye in three, and my foot in four.
My mouth is capable of filling it in six hours;
tell me how long all four together will take to fill it.

There is no problem in the *Arithmetica* corresponding to this epigram.

3.2.3 The scholia to the arithmetical epigrams of the Palatine Anthology

Our last piece of evidence for the persistence of Diophantus's heritage in the late antique world are some scholia for arithmetical epigrams of the Palatine Anthology, in which problems are solved by algebra and the name of Diophantus is explicitly mentioned. The Palatine Anthology is a collection of epigrams, that is, short texts expressed in verse. They are preserved in a single manuscript, the famous codex Palatinus gr. 23, which is composed of 15 books. Until the early nineteenth century it was hosted in the Palatine Library at Heidelberg, at which time it was divided into two parts, the Palatinus Heidelbergensis gr. 23 and the Paris. suppl. gr. 384, the former comprising Books I–XIII of the anthology, the latter Books XIV–XV.[88] The Palatinus is largely, but not exclusively, based on an earlier anthology (now lost) compiled by the Byzantine Constantine Cephalas, who, according to current scholarly consensus, was active in Constantinople around the end of the ninth and the beginning of the tenth century (Cameron 1993, 254ff). However, there is no consensus among specialists concerning the dating of the Palatinus, the extent to which it is based on the lost anthology of Cephalas, the precise sources of both, and the paleographic details concerning the manuscript.[89]

The book of the Palatinus that interests us is the XIVth, which is composed of a mix of 150 oracles and riddles. Forty-four of these riddles pose problems in arithmetic, and 32 of them are accompanied by scholia that give solutions, either written in the margins of the manuscript or integrated into the text.[90] Of the 32 scholia, 8 give solutions by single false position, the first of these citing Euclid's *Elements* Proposition VII.39 to find the initial guess. Nine other epigrams give solutions by algebra, sometimes mentioning Diophantus by name. The algebraic notation appears in the solutions, although not consistently, as the sign for the technical word "Number" is sometimes used for the non-technical "number". The

---

88 For details on the history of the Palatinus 23, see the introduction of Pierre Waltz in (Waltz 1960, xxxviii–xlviii) and (Waltz 1928). See also the references mentioned in the next note.

89 There is a rich literature on these topics, for a survey of which the interested reader is referred to the works (Waltz 1928, 1960; Aubreton 1968, 1969; Cameron 1970, 1993; Irigoin 1997; Taub 2017) and to the bibliography provided there.

90 According to Buffière (1970, 30–32), the 14th book was not included in the anthology of Cephalas. As already said, it is carried by only one manuscript, Par. suppl. gr. 384; there is, however, a number of epigrams of it which are also transmitted through other less important collections, which are known as "syllogæ minores".

remaining problems are solved arithmetically by reasoning through the conditions.[91] We will review the algebraic solutions next.

Scholars in the past century have generally endorsed Tannery's assessment of the history of the arithmetical epigrams (Tannery 1912g, 1912h; Diophantus 1893–95, II, x–xii), specifically that: (i) They can be classified into two series, of which the first could be designated by the name of Socrates, according to the lemma Σωκράτους ("of Socrates"), which follows directly after the title of the book, and the second by the name of Metrodorus, on the basis of the inscription Μητροδώρου ἐπιγράμματα ἀριθμητικά ("Arithmetical epigrams of Metrodorus"), which is written in the margin next to the first epigram of that series. All but four of the arithmetical epigrams belong to the Metrodorus series, and among these, scholia are appended to all but the last three. (ii) The two designations do not tell us, in fact, anything about who the authors of the epigrams might have been in reality. (iii) The "series of Metrodorus" is accompanied by systematic scholia revealing a scholar well acquainted with (at least parts of) some classical works of ancient Greek mathematics, such as the *Elements* and the *Data* of Euclid (Vitrac 2018, 225) and the *Arithmetica* of Diophantus. Tannery proposed that this Metrodorus compiled the epigrams from several sources available at the time[92] and composed the scholia himself. (iv) Concerning Socrates and Metrodorus and their respective dates, one cannot avoid resorting to speculations due to the lack of reliable information. For Socrates we know almost nothing except that Diogenes Laertius (3rd c. CE) mentions an author of epigrams with this name (Diogenes Laertius 2013, 179). As for Metrodorus, besides the epigrams of Book XIV there are two more epigrams transmitted under this name, namely, epigrams IX.360 and IX.712, the latter being attributed according to the Palatine manuscript to Μητρόδωρος ὁ γραμματικός ("Metrodorus grammaticus"), who was active in the time of Constantine the Great (Duffière 1970, 36). In the second volume of his edition of Diophantus Tannery reports several other people with this name, mentioned by authors like Servius, Pliny the Elder, Ptolemy, and Fabricius, who might be linked to the epigrams (Diophantus 1893–95, II, xii). However, in a later publication he advanced another suggestion, according to which Metrodorus was a grammarian of the sixth century CE, brother of the mathematician Anthemius of

---

91 The eight problems solved by single false position are numbered 1, 2, 6–9, 11, and 16 in Tannery, following the order in Metrodorus, and as 2, 116, 119–122, 124, and 138 in Anthology. Some problems solved arithmetically are given more than one solution. Besides algebra other arithmetical methods of solution are used in the scholia, a fact showing that algebra was considered one of several problem-solving techniques. Epigram 7 (19th of the collection of Metrodorus) is the most characteristic in this respect, being accompanied by seven solutions, none of them by algebra. That problem is of the type of spouts that fill a tank. See Diophantus (1893–95, II, 46.19–51.6). The arithmetical epigrams and their scholia are published in the second volume of Tannery's edition, among the "Testimonia Vetera de Diophanto" (Diophantus 1893–95, II, 43–72).

92 In Tannery's words, "son rôle a été surtout, sinon exclusivement, celui d'un compilateur" (1912h, 445).

Tralles (1912g, 531).[93] Notwithstanding the fragile elements upon which rests any attempt to identify the Metrodorus to whom the arithmetical epigrams of Book XIV of Palatinus are linked, it seems that since Tannery there is a consensus among the scholars that he was a grammarian who lived sometime in the period of late antiquity not later than the sixth century CE.[94]

Next we discuss the nine epigrams together with the accompanying scholia that refer to Diophantus. The numbering of the epigrams is given in accordance with the original collection of Metrodorus, and the numbers of the actual Palatinus manuscript (Par. suppl. gr. 384) are given in parenthesis. For the correspondence of the two series of numbers, see (Buffière 1970, 35).

Epigram 15 (128)

"What violence my brother has done me, dividing our father's fortune of five talents unjustly! Poor tearful I have this fifth part of the seven-elevenths of my brother's share. Zeus, thou sleepest sound" (Paton 1918, 95). This is accompanied with the following detailed scholium:[95]

> It is supposed that someone is complaining because he has found himself being wronged by his brother on the partition of the paternal inheritance. The brothers were two in total, one of whom was the more powerful. And the total paternal inheritance was 5 talents. The problem is solved according to the second (problem) of the first Book of Diophantus: to divide a proposed number, as in the present case the 5, into two numbers such that the one be the fifth part of the 7 11ths of the other. So let the smaller be 7 Numbers. Then the 7 11ths of the other will be 35 Numbers; thus, the 1 11th will be 5 Numbers; the entire greater, therefore, will be 55 Numbers. But the smaller was 7 Numbers. Therefore, both together will be 62 Numbers; but, also, 5 talents. And the talent becomes 12 3′ 15′ Numbers. If we multiply the Numbers by five in order to get a whole number, they will be 62 Numbers equal to 1 talent; and the Number becomes 1 62nd. The smaller will be 35 62nds, while the greater, 275 62nds. And the 7 11ths of this greater are 175 62nds, and the problem is solved.

The problem asks to find two shares whose sum is 5 and such that the smaller is $\frac{1}{5}\left(\frac{7}{11}\right)$ of the greater. The scholiast recognized the problem type as being the

---

93　Tannery's conclusion relies upon information provided by the Byzantine historiographer Agathias (6th c.), according to which "Anthemius was absolutely outstanding in his field and was a first-rate mathematician enjoying a similar pre-eminence to that enjoyed, mutatis mutandis, by his brother Metrodorus in the sphere of grammar. Their mother was, I think, particularly fortunate to have given birth to such talented children" (Agathias 1975, V.6.4.4–5).

94　See, for example, Heath (1910, 113), Vogel (1967, 269), Hunger (1978, II, 231), Herrin (2000, 30), Albiani (2006), and Taub (2017, 33). According to Alan Cameron, Metrodorus was an author "of uncertain date but hardly later than the fourth century" (1970, 341).

95　For the Greek text, see Diophantus (1893–95, II, 61.26–62.15).

same as that of *Arithmetica* I.2, so he mentions that he will follow that method. He introduces the technical term Number by positing the smaller share to be 7N. The assignment for the greater share is found in a sequence of steps: $\frac{7}{11}$ of it must be 35N, so $\frac{1}{11}$ of it is 5N, and consequently the greater share must be 55N. The two shares together, therefore, are the sum of 7N and 55N, which is 62N ("both together will be 62 Numbers"). But they are given to be 5 talents. Thus, the equation, in notation, is set up as 62N = 5t, where we write "t" for talents. Therefore, 1 talent will be $\frac{62}{5}$ N, or, as the scholiast writes, "12 3' 15' Numbers". At this point, the scholiast replaces 1N with 5M, that is, he introduces a new unknown, which is one-fifth of the "Number". This new unknown is also called a Number, and it is written in the manuscript by the same ϛ-like sign. Then, 1 talent will be 62M; that is the equation 62M = 1t is set up (the corresponding phrase in the text is "62 Numbers equal to 1 talent"), from which the new Number is found to be $\frac{1}{62}$. Therefore, the smaller share is $\frac{35}{62}$ talents and the greater share is $\frac{275}{62}$ talents. The scholiast concludes by verifying that the values found satisfy the requirements of the problem: $\frac{7}{11}$ of $\frac{275}{62}$ is $\frac{175}{62}$, and the problem is satisfied since $\frac{1}{5}$ of it makes $\frac{35}{62}$.

We recognize in this solution the critical features of an algebraic solution: assigning names to the unknowns, operating with the names, setting up an equation framed in terms of the names, and the final solution of the equation.

Epigram 16 (129)

> A traveler, ploughing with his ship the broad gulf of the Adriatic, said to the captain, "How much sea have we still to traverse?" And he answered him, "Voyager, between Cretan Ram's Head and Sicilian Peloris are six thousand stades, and twice two-fifths of the distance we have traversed remains till the Sicilian strait".
>
> (Paton 1918, 95)

In the *Palatinus* manuscript the epigram is followed by this comment:[96]

> This too is similar to the 15th and it is solved through the second problem of the first (Book) of the *Elements* of Diophantus: to divide the 6000 ⟨stades⟩ into two numbers such that one part is sesquiquartan of the other. So the Number becomes 666 3" units, and the greater number becomes 3333 3' units, while the smaller becomes 2666 3".

The problem asks to divide the number 6,000 into two numbers such that one of them is $\frac{4}{5}$ ("twice two-fifths") of the other. Note that the second condition is stated in the scholium as its reciprocal "one and one fourth of", the latter being what is

---

96 For the Greek text, see Diophantus (1893–95, II, 62.23–63.2)." Here 3" means ⅔.

known as the "sesquiquartan of". Thus, the scholiast has reversed the order of the numbers. The solution is again conducted by means of the introduction of the technical term "Number", the first-degree algebraic unknown. The solution from that point is not given, but it can be reconstructed easily: let the fourth part of the smaller sought-after number be assigned to be 1 Number, so the smaller is 4N. Then, the other sought-after number will be 5 Numbers, or 5N. Both together, therefore, are 9N. But they are given to be 6,000. Therefore, 9N = 6,000. This is the equation that is set up, from which we deduce the value of the Number to be $666\frac{2}{3}$. The numerical values of the sought-after numbers are then determined, the greater number being $3,333\frac{1}{3}$, and the smaller $2,666\frac{2}{3}$.

Epigram 17 (130)

"Of the four spouts one filled the whole tank in a day, the second in two days, the third in three days, and the fourth in four days. What time will all four take to fill it?" (Paton 1918, 97). This epigram refers to the well-known problem of a number of spouts that fill a tank in different times, requiring the time in which the tank will be filled if all spouts are run together. The answer is given as a "sub-double-twelfth of the day",[97] but the scholiast finds it more convenient to interpret this cumbersome expression in terms of hours, so he writes: "Thus it is led to dividing the 12 hours of the day into two parts such that the one is one and a twelfth of the other, and the problem is solved".[98] This problem is of the sort that can be solved according to Problem I.2 of the *Arithmetica*. The reference is implicit; however, the scholiast mentions neither the problem nor the name of Diophantus.

Epigram 27 (139)

Diodorus, great glory of dial-makers,[99] tell me the hour since when the golden wheels of the sun leapt up from the east to the pole. Quadruple three-fifths of the distance he has traversed remain until he sinks to the western sea.[100]

The scholium to the epigram is the following:[101]

This is solved methodically according to the second (problem) of the first Book of the *Elements* of Diophantus. For it is required to divide the number

---

97 Following the manuscript, Tannery writes ὑποδιπλασιοδωδεκάτῳ, although the correct word should be ὑποδιπλασιεπιδωδεκάτῳ. It seems that the mistake was noticed by some reader of the Palatinus codex, for there is a line drawn just below the letters ασι (ὑποδιπλασιοδωδεκάτῳ). See Par. suppl. gr. 384, f. 12r, l. 16 from bottom.

98 For the Greek text, see Diophantus (1893–95, II, 63.16–18).

99 Γνωμονικοί (*gnōmonikoi*), i.e., experts in the mathematical science of telling time by the diurnal motion of the sun. For the art of gnomonics in antiquity, see Evans (2005).

100 Translation adjusted slightly from Paton (1918, 101).

101 For the Greek text, see Diophantus (1893–95, II, 69.8–13).

12 into two numbers bearing the ratio that 5 has to 12. And the Number becomes 12 17ths. Thus the parts of the day that have gone by are 60 17ths, while the ones remaining, 144 17ths, and the problem is solved.

Here it is required to divide the number 12 into two numbers such that the one is "quadruple three-fifths" of the other, or, in the commentator's reformulation, "in the ratio that 5 has to 12", the problem is to find two numbers such that the second number is $4 \cdot \left(\frac{3}{5} \right) = \frac{12}{5}$ of the first, or equivalently, they are in the ratio of 5:12. The working out of the solution can be reconstructed as follows: let a fifth of the first number be 1N; therefore, the first number is 5N, so the second number is 12N. The two together, therefore, are 17N. But they were stated to be 12. Thus, the equation is 17N = 12u, and the Number becomes $\frac{12}{17}$. Therefore, the first number will be $\frac{60}{17}$, and the second will be $\frac{144}{17}$.[102]

Epigram 28 (6)

Epigram 28 of the Metrodorus collection, which coincides with the sixth epigram of the Socrates collection, is: "'Best of clocks, how much of the day is past?' There remain twice two-thirds of what is gone" (Paton 1918, 31). The scholium is the following:[103]

> This, as the 27th, is also worked out methodically by the 2nd (problem) of the 1st Book of Diophantus. One must divide the (number) 12 in sesquiter-tan ratio, and the Number becomes 12 7ths. So the (part of) the day that has passed will be 36 7ths, while the remaining (part), 48 7ths.

Here the "twice two-thirds" in the epigram is restated in the scholium as "ses-quitertan ratio", or one and a third. The solution can be reconstructed as follows: let one third of the smaller sought-after number be assigned to be 1 Number. Then, the smaller is 3N, and thus, the greater is 4N. Both together, therefore, are 7N. But they have been given to be 12. Therefore, 7N = 12u, and the Number becomes $\frac{12}{7}$. So, the smaller sought-after number is $\frac{36}{7}$, and the greater is $\frac{48}{7}$.

---

102 Problems requiring one to find how much of a day or a night has passed were common in pre-modern traditions. Here is an example of such a problem from al-Karajī's book *al-Fakhrī* (Saidan 1986, 185): "So if someone said, what remains of the night is a fourth of what has passed and half of what remains. How much remains and how much has passed? You knew that it is as if he had said, the remainder is half of what remains and a fourth of what has passed. So half of the remainder is a fourth of what has passed. So make what has passed a Thing, so the remainder is half a Thing, ⟨and add them⟩. This equals twelve hours $\left[1\frac{1}{2}x=12 \right]$. So the Thing equals eight hours, which is what has passed, and the remainder is four hours". As in the Greek epigrams, the elliptical arithmetical relationship expressed in the enunciation is restated in more understandable terms in the beginning of the algebraic solution.

103 For the Greek text, see Diophantus (1893–95, II, 46.14–18).

Epigram 29 (140)

> Blessed Zeus, are these deeds pleasing in thy sight that the Thessalian women do in play? The eye of the moon is blighted by mortals; I saw it myself. The night still wanted till morning twice two-sixths and twice one-seventh of what was past.
>
> (Paton 1918, 103)

The scholium accompanying the epigram is:[104]

> This too is similar to the 28th and (the) 27th. For it is required to divide the (number) 12 in the ratio of one and one twentieth, that is the (ratio) that 21 has to 20, and the Number becomes 12 41sts. So the (part) of the night that has past will be 252 41sts, while the rest to come, 240 41sts.

Once again, the scholiast transforms the ratio stated in the enunciation of the epigram into a ratio, and here that ratio is stated two ways. The $2 \cdot \frac{2}{6} + 2 \cdot \frac{1}{7}$ of the epigram is reformulated as the ratio $1\frac{1}{20}$, or 21:20. The resolutory procedure can be reconstructed as follows: let one-twentieth of the smaller sought-after number be assigned to be 1 Number. Therefore, the smaller is 20N, and thus, the greater is 21N. Both together, therefore, are 41N. But they were stated to be 12. Thus, 41N = 12u, and the Number becomes $\frac{12}{41}$. So, the smaller sought-after number will be $\frac{240}{41}$, and the greater will be $\frac{252}{41}$.

Epigram 30 (141)

> Tell me the transits of the fixed stars and planets when my wife gave birth to a child yesterday. It was day, and till the sun set in the western sea it wanted sextuple twice a seventh of the time since dawn.[105]

The commentary to this epigram is:[106]

> This too is similar to the 29th. For it is required to divide the number 12 into two numbers in the ratio of one and five-sevenths, and the Number becomes 12 19ths. So the (part of) the day that has past will be 84 19ths and the rest to come 144 19ths.

Here the "sextuple twice a seventh" is translated into $1\frac{5}{7}$ by the scholiast. The procedure for solving this, which is not presented, is that the smaller sought-after number be assigned to be 7 Numbers. Then, the greater will be 12N, and both

---

104 For the Greek text, see Diophantus (1893–95, II, 69.19–23).
105 Translation adjusted slightly from Paton (1918, 103).
106 For the Greek text, see Diophantus (1893–95, II, 70.6–9).

together, therefore, are 19N. But they were stated to be 12. Thus, $19N = 12u$, and the Number becomes $\frac{12}{19}$. So, the smaller will be $\frac{84}{19}$, and the greater will be $\frac{144}{19}$.

Epigram 32 (142)

"Arise, work-women, it is past dawn; a fifth part of three-eighths of what remains has gone by" (Paton 1918, 103, slightly modified). This is accompanied with the following scholium:[107]

> This too is similar to the one before it. For it is required to divide the number 12 into two numbers such that the one is a fifth part of three eighths of the other, that is they have to one another the ratio of thirteen and a third. So the Number becomes 12 43rds. So the (part) of the day that has past becomes 36 43rds, while the rest, 480 43rds.

After again adjusting the stated ratio, the scholiast would have solved this problem as follows: let one-third of the smaller number be assigned to be 1 Number. Therefore, the smaller is 3N, and thus, the greater is 40N. Both together, therefore, are 43N. But they have been given to be 12. Therefore, $43N = 12u$, and the Number becomes $\frac{12}{43}$. So, the smaller will be $\frac{36}{43}$, and the greater will be $\frac{480}{43}$.

Epigram 33 (143)

> The father perished in the shoals of the Syrtis, and this, the eldest of the brothers, came back from that voyage with five talents. To me he gave twice two-thirds of his share, on our mother he bestowed two-eighths of my share, nor did he sin against divine justice.
>
> (Paton 1918, 103)

The accompanying scholium is:[108]

> This too is similar to the 32nd. For it is required to divide the number 5 into two numbers in the ratio of one and one third, and again to divide the greater part in quadruple ratio. So the first division will yield the Number 5 7ths, that is, the one (number) becomes 15 7ths, and the other, 20 7ths. And if we divide the 20 7ths in the quadruple ratio, the Number becomes 5 7ths. And the greater part of it will be ⟨15 7ths⟩, the lesser ⟨5 7ths⟩, and the other, the lesser from the first division, 15 7ths.

The problem asks, first, to divide the five talents into two shares such that the one is twice two-thirds of the other, or, as the scholiast says, they are "in the ratio of one and one third". Moreover, it is required to divide anew the greater share into

---

107 For the Greek text, see Diophantus (1893–95, II, 70.13–18).
108 For the Greek text, see Diophantus (1893–95, II, 71.1–9).

two numbers such that the one is two-eighths of that share, that is, as the scholiast says, in "the quadruple ratio". The solution can be reconstructed as follows: Let a third of the smaller share be 1N. Then the smaller share is 3N, and the greater is 4N. The equation becomes 7N = 5u, from which the Number is found to be $\frac{5}{7}$. So, the smaller share is $\frac{15}{7}$ talents and the greater is $\frac{20}{7}$ talents. It remains now to divide the $\frac{20}{7}$ into quadruple ratio, that is, to separate out one-fourth of it. The two parts will be in the ratio 3:1, and again solving the problem by algebra by naming the parts 1N and 3N, the scholiast finds that "the Number becomes 5 7ths". Therefore, the three shares are: $\frac{15}{7}$ talents for each brother and $\frac{5}{7}$ talent for the mother.

The scholia to the arithmetical epigrams of the Palatine Anthology examined earlier constitute one of the earliest examples of the practice of algebra in the period immediately following Diophantus's time. Regardless of whether the scholia should be ascribed to Metrodorus or to some other late antique author, they portray a scholar acquainted with algebraic problem solving, who had access to a manuscript of the *Arithmetica*. Besides, he was a scholar familiar with the broader arithmetical culture of his time, as we infer from the remarkable dexterity with which he switches from numerical expressions with ratios to their equivalent fractional expressions, of one or another form, and vice versa.[109]

3.2.4 A reference by John Philoponus

One last reference to Diophantus in late antiquity, dating from 517 CE, comes from the commentary on Aristotle's *Physics* by the Christian philosopher John Philoponus (ca. 490–ca. 570), a student of Ammonius who worked in Alexandria (ca. 435/445–517/526). In this passage we again find the *Arithmetica* treating number in the abstract:

> The highest part of mathematics is easy to distinguish and separated from the study of nature – examples are the *Sphaerica* of Theodosius, the 13 books of Euclid, the *Arithmetica*, for in these there is absolutely no mention of matter – but its lowly (part) is in a way close to the study of nature.[110]

---

109 For details, see Christianidis and Megremi (2019). The distinction between ratios and fractions is sufficiently emphasized in the historical writings on ancient Greek mathematics. This distinction was fundamental for both mathematicians and mathematically minded philosophers, who worked in the "scientific" fields of ancient mathematical science and who aimed to endow numbers with a rigorous, ontological foundation. Plato, Aristotle, Euclid, and Nicomachus all worked in an arithmetic that posited an indivisible unit, and thus restricted the definition of number to positive integers. This is not the case with writers and practitioners who worked in the "subscientific" areas of ancient science (the term is borrowed from Høyrup 1989), within which fractions had always been in common use, along with ratios. Engineering, surveying, architecture, trade, astronomical calculation, and, above all, accounting (λογιστική) are such areas. Because the sort of problems illustrated by the arithmetical epigrams of the Palatine Anthology belong to the latter, it should not come as a surprise that the scholiast of the epigrams does not trouble himself with theoretical concerns about ratios and fractions.

110 (Philoponus 1887, 220.14–17). Translation adapted from Philoponus (2014, 33.15–19). We thank Nathan Sidoli for calling our attention to this passage.

3.2.5 Early Byzantine hagiography

The earliest known mention of Diophantus in Byzantium is by Ignatios the Deacon, in a passage of his *Life* of Tarasios, Patriarch of Constantinople in 784–806. The *Life* was composed in the 840s, but clearly after 843, the year of the restoration of the icons of veneration. Referring to the food that Tarasios distributed every day to the poor, Ignatios writes:

> Yet, which Diophantus, the calculator (λογιστής), could measure out, or which Nicomachus count up by multiplication, the distribution that he would make every day into the palms to the poor by his own generous hand, more abundant than sand and comforting to the belly of the needy?
>
> (Ignatios 1998, 180, translation slightly adapted)

So even in the second half of the ninth century Diophantus and Nicomachus continued to be the recognized sages in logistic and arithmetic, respectively.

Another hagiographic work citing ancient mathematicians, this time Pythagoras, Diophantus, and Euclid, is the *Life of John of Damascus* (Halkin 1957, No 884), attributed to a certain "Patriarch John". John of Damascus was born ca. 655 in Damascus into the wealthy Arabic-Christian family of Manṣūr,[111] some members of whom served as high functionaries in the central administration of the caliphate during the first decades of the Umayyad period.[112] John apparently also served in the fiscal administration of the caliphate for the first period of his life. Then, at some point in the early eighth century, he left his position to become a monk in Palestine, where he composed a number of important theological works. He died some time before 745 (Kontouma 2015, 28). From the high level of his writings it is clear that John had received a solid education along classical Greek lines in addition to religious instruction.

The *Life* was written over two centuries later, in the late tenth or early eleventh century, and was based on two earlier works, a *Life of the Methodists* and an Arabic *Life* of John (Kontouma 2015, 23). As was common in such hagiographical biographies, the details of this *Life* are notoriously untrustworthy.[113] The *Life* recounts that John and his adopted brother Cosmas were taught the classical disciplines of the trivium and the quadrivium and also theology from a Greek-Italian teacher who was also named Cosmas. It further states that the youths

---

111 In all probability John was the name he received after he entered monastic life.

112 For John's family, see Nasrallah (1950, 57–69), Le Coz (1992, 43–49), and Griffith (2016, 29–51). In recent decades scholars have pointed out the Arabic lineage of his family. Thus, in Griffith's statement, "John of Damascus belonged to the world of Islam" (2001, 18); similarly, according to Louth, "The background of Umayyad caliphate is more relevant to John's life than one might expect" (2002, 5). Cf., however, the more balanced view of Kontouma: "Bien qu'il (i.e. JD) évoluât dans une société dominée par l'Islam, il fut vraisemblablement plus marqué par les milieux chrétiens chalcédoniens et hellénophones fortement représentés à Damas, au sein desquels il reçut sans doute son éducation" (Kontouma 2000, 1002); updated and augmented English version in Kontouma (2015).

113 On the unreliability of the *Life*, see Kazhdan and Gero (1989), Auzépy (1994, 193), Louth (2002, 2, 4, 16), and Kontouma (2015, 23).

were practicing arithmetic proportions as skillfully as Pythagoras or Diophantus. Also, in geometry, they were trained to such an extent in making proofs, that they were held to be something like Euclid and others like him. With respect to harmonics they became so excellent that they appeared to the wise men like those who composed divine melodies. With respect to astronomy, as regards intervals, and configurations, and proportions of the distances, John dealt with them, albeit briefly, providing a tiny knowledge to the laymen, as it is illustrated from his writings. The same also holds true for Cosmas.

(Diophantus 1893–95, II, 36.8–19)

Although the story is apocryphal, here, too, the reader was expected to be familiar with the names of Pythagoras, Diophantus, and Euclid.

3.3 Diophantus in medieval Arabic

To properly assess the influence and place of Diophantus in Arabic mathematics we should first say a few words about Arabic algebra, into which the *Arithmetica* was classified by Arabic authors. It is well known that algebra had been practiced in Arabic before Qusṭā translated the *Arithmetica* in the mid- to late ninth century, so we begin this section by reviewing the contents of the two earliest complete Arabic algebra books, by al-Khwārazmī and Abū Kāmil.

3.3.1 Arabic algebra before Qusṭā's translation: al-Khwārazmī and Abū Kāmil

The oldest extant books on algebra in Arabic, by al-Khwārazmī and Ibn Turk, were written in the period 813–33 CE and show no sign of Diophantine influence. Only a short part of Ibn Turk's book has come down to us, but we have the *Book of Algebra* (*Kitāb al-jabr wa-l-Muqābala*) of Muḥammad ibn Mūsā al-Khwārazmī complete. This book is divided into three main parts, the first being what we call the "algebra proper", which covers the vocabulary and rules of algebraic problem solving followed by a collection of determinate worked-out problems. The second part covers the rule of three and mensuration. The rule of three is, of course, a method distinct from algebra, and among the rules and problems of finding areas and volumes in the section on mensuration, algebra is applied in only two cases. This inclusion of non-algebraic, utilitarian material is a reflection of the fact that algebra in medieval Arabic was a part of *ḥisāb*, or practical calculation. As Saidan has noted, books on *ḥisāb* cover not just rules of calculation, but usually also the rule of three, false position, algebra, and mensuration (Saidan 1978, 20–21). The third and final part of the *Book of Algebra*, taking up about half of the work, consists of a collection of inheritance problems, most of which are solved by algebra.[114]

Al-Khwārazmī's algebra proper begins with an explanation of the names of the powers *Māl*, Root, and simple number, which leads to his classification and rules

---

114   For an overview of the structure of early Arabic algebra books, see Oaks (2015).

for solving the six canonical equations of first and second degree. Then come the proofs for these rules, followed by some rules for operating on polynomials. The worked-out problems given after this are borrowed and adapted from oral tradition (Oaks 2012b). The first six are framed carefully, and the equations set up in their solutions simplify in order to the six types of equation. Among the 32 assorted problems that follow, 22 are solved by algebra, while the remaining 10 are solved by simply reasoning through the solution to a problem, a technique we have called "arithmetical reasoning".[115] In those simple problems, it is not worth the trouble to name an unknown as "a thing" and then set up and solve an equation. Compared with the *Arithmetica*, the introductory material in al-Khwārazmī is organized differently, the worked-out problems are unlike those in Diophantus, and the terminology does not derive from the Greek text. If there was any influence of Greek algebra on ninth-century Arabic algebra, it did not come directly from Diophantus.

As we can discern from examining the wording and structure of al-Khwārazmī's problems, algebra had been transmitted orally among professionals before he set down the technique in his book (Oaks 2012b). This is in fact what is explained by Muḥammad al-Khuzāʿī, a Yemeni scholar from the thirteenth century who wrote a commentary on al-Khwārazmī's *Algebra*. This commentary relates that before the time of al-Khwārazmī, "the people transmitted this knowledge (of algebra) orally without it being recorded in any book" (King 1988, 26, his translation). While the details of al-Khuzāʿī's story, such as names and dates, may be fabricated, the idea that oral practical knowledge was first written down in books in early ʿAbbāsid times is consistent with the similar pattern we see for finger reckoning, double false position, mensuration, and folk astronomy. They, too, were practiced before they were first explained in books in the ninth century. This oral tradition of algebra did not come to an end with al-Khwārazmī, of course. It is evident from the practical problems we find in some later authors that they also drew from this rich source of practitioner's knowledge.

While al-Khwārazmī solved only determinate problems, indeterminate problems also circulated among Arabic practitioners before Qusṭā made his translation, including problems about squares and cubes that were solved in the same manner as we find in Diophantus. The earliest witness to the Arabic tradition of indeterminate problems is the third part of the *Book on Algebra* (*Kitāb fī l-jabr wa-l-Muqābala*) by Shujāʿ ibn Aslam, better known today as Abū Kāmil. This book was written sometime in the latter ninth century CE, perhaps a decade or two after Qusṭā's translation. Jacques Sesiano is correct when he writes,

> There is nothing to suggest that the Egyptian Abū Kāmil had any direct (or even indirect) knowledge of Diophantus' *Arithmetica*, although the problems in his *Algebra* dealing with indeterminate analysis are perfectly Diophantine in form and the basic methods are attested to in the *Arithmetica*.[116]

---

115 For the differences between algebra and "arithmetical reasoning", see Oaks and Alkhateeb (2005).
116 (Sesiano 1982, 9). See Sesiano (1977b) for details of the argument.

Abū Kāmil modeled his book on the *Algebra* of al-Khwārazmī, and he supplied many proofs for propositions in arithmetic and algebra in the style of those in Euclid's *Elements*. The book takes up 111 folios in the Istanbul manuscript[117] and is divided into three main parts. The first part is his "algebra proper", covering the vocabulary and rules of algebra together with 74 worked-out determinate problems, which are nearly all solved by algebra. Like in al-Khwārazmī, numbers can be any positive amount, including fractions and irrational roots. The second part of the treatise, titled "On the pentagon and decagon", solves 20 geometry problems by algebra. In the third part Abū Kāmil solves still more problems in arithmetic, beginning with 43 indeterminate problems solved by algebra.[118] Forty-one determinate problems follow, sometimes solved by algebra, sometimes not, and the book concludes with a description of the chessboard problem.

There are two basic types of indeterminate problem in Arabic mathematics, and both are found in the third part of Abū Kāmil's treatise. One is of the type common in Diophantus, in which some of the quantities called for in the enunciation are specified to be squares or cubes. This necessitates a restriction of the solution to rational numbers, since otherwise any positive number(s) could serve as a solution. An example is Abū Kāmil's Problem 17, whose enunciation reads: "If (someone) said to you, a *māl*: if you add it to twenty dirhams, then (the result) has a root, and if you add it to thirty dirhams, then (the result) has a root. This (problem) is also indeterminate" (Abū Kāmil 1986, 172.8, 2012, 615.9). The *māl* here takes the meaning of "quantity" or "amount". It is not intended to be a square number, nor is it the algebraic name of the second power of the unknown. By "has a root" Abū Kāmil means that the result has a rational square root. The wording of these Arabic problems differs from those in Diophantus in that they usually ask for a "*māl*" instead of a "number" ('*adad*), and when this amount is intended to be a square, they add that it "has a root" instead of writing that it "is a square".

In the other type of indeterminate problem, the enunciation simply does not give enough constraints to make the problem determinate, such as Abū Kāmil's Problem 39:

> If (someone) said to you: four individuals get together to buy a horse.[119] The first says to the (other) three: if you give me half of what you have, then I will have the price of the horse. The second says to the (other) three: if you give me a third of what you have, then I will have the price of the horse. The third says to the (other) three: if you give me a fourth of what you have, then I will have the price of the horse. And the fourth says to the (other) three: if you give me a fifth of what you have, then I will have the price of the horse. How much does each one have?
>
> (Abū Kāmil 2012, 655)

---

117 Beyazit, Kara Muṣṭafa Paşa 370 (new number: 19046).

118 In some problems no unknown is named, but these problems reduce to problems previously solved by algebra.

119 The word *dābba* can mean "animal, beast; riding animal (horse, mule, donkey)" (Wehr 1979, 312). We translate it as "horse" to conform with the many problems of this type in other languages.

Had one more constraint been given, such as the price of the horse, then the problem would have been determinate. There is no implicit restriction to rational solutions for problems like these, though rational solutions are encouraged by the typically quotidian terms in which the problems are framed. The first 38 problems in the third part of Abū Kāmil's treatise are of the Diophantine type calling for certain numbers to be squares, while Problems 39 through 43 are of the latter, practical type. From here on, all indeterminate problems we speak about will be of the former, Diophantine type.

Abū Kāmil solves nearly all of his indeterminate Problems 1–38 by algebra. Recall that the two main techniques Diophantus applies to complete forthcoming equations are *al-istiqrāʾ* and the double equality (see Appendix 2). Like other algebraists working in the Arabic tradition, Abū Kāmil routinely completes his forthcoming equations by *al-istiqrāʾ*, but not by the double equality. With one exception, none of his problems is close enough to a problem in Diophantus to raise the possibility of dependence, and the style and wording of the enunciations and solutions show consistent differences with the *Arithmetica*. Here we translate a simple example, Problem 4 (Abū Kāmil 2012, 585.9): "If (someone) said to you, a *māl*: it has a root. If you subtract from it six of its roots, then that has a root. This problem is also indeterminate". In the enunciation, the word *māl* again takes its arithmetical meaning as "quantity" or "amount". Unlike the *māl* in Problem 17, this particular *māl* must be a square, so Abū Kāmil specifies that it "has a root". The phrase at the end, *ghayr maḥdūda*, literally means "not determinate".

The solution begins with the naming of the unknown *māl*: "Its rule is that you make your *māl* a *Māl*". The first instance of *māl* in this line is the "quantity" from the enunciation, and the second instance is the algebraic second power. This is like saying, in modern algebra, "you make your quantity x^2". It is important to note that this is a naming, and not a change of variables. If the word *māl* took only its algebraic meaning, there would be no use in renaming the *māl* as a *māl*, as Abū Kāmil does in 21 of his 38 problems.[120]

"You cast away from it six of its roots, leaving a *Māl* less six Roots". The first instance of "roots" means "square roots", while in the second instance "Roots" is the name of the first degree unknown. Next, Abū Kāmil switches to using "Thing" instead of "Root", and he explains how to assign the side of the square that will be set equal to the algebraic expression "a *Māl* less six Roots" (1M ℓ 6T). Abū Kāmil has no word to designate this technique of naming the side of this square, but it is called *al-istiqrāʾ* by later algebraists:

> Its (square) root is smaller than a Thing, so we make it a Thing less four dirhams, or less five dirhams, or less three and a half, or anything you wish in number, as long as it is greater than half of the six Things excluded from the *Māl*. So we make it a Thing less four dirhams, and we multiply it by

120 Problems 1, 4–6, 8–11, and 26–38.

itself to get: a *Māl* and sixteen dirhams less eight Things equal a *Māl* less six Things.

Thus, the equation is set up as 1M 16d ℓ 8T = 1M ℓ 6T.

"Confront them, so the Thing is eight and the *Māl* is sixty-four. Casting away from it six of its roots, which are forty-eight, leaves sixteen, and its root is four". Abū Kāmil often used the word "confront" (*qabila*, related to *muqābala*) to mean "simplify and solve the equation".[121] The two instances of "root" here mean "square root".

Table 3.5 provides a list in modern notation of the enunciations to Abū Kāmil's indeterminate Problems 1–38. Keep in mind that these are all questions in arithmetic with no named amounts, so any modern notational rendition of them will necessarily be anachronistic. However, if one keeps in mind the ways that enunciations are framed, such a list is a handy way to summarize the questions. Key: M

Table 3.5 Indeterminate problems in Abū Kāmil.

(1) $M = \square$; $M + 5 \rightarrow \square'$	(23) $M = \square$; $M + 2\sqrt{M} \rightarrow \square'$; $M - 3\sqrt{M} \rightarrow \square''$
(2) $M = \square$; $M - 10 \rightarrow \square'$	(24) $M = \square$; $(10\sqrt{M} \,\ell\, 8) - M \rightarrow \square'$
(3) $M = \square$; $M + 3\sqrt{M} \rightarrow \square'$	(25) $M = \square$; $(260 \,\ell\, 6\sqrt{M}) - M \rightarrow \square'$
(4) $M = \square$; $M - 6\sqrt{M} \rightarrow \square'$	(26) $M = \square$; $\underset{\wedge}{M + 2\sqrt{M}} \rightarrow \square'$; $A + 3\sqrt{A} \rightarrow \square''$
(5) $M = \square$; $M + (10\sqrt{M} \,\&\, 20) \rightarrow \square'$	
(6) $M = \square$; $M - (8\sqrt{M} \,\&\, 30) \rightarrow \square'$	(27) $M = \square$; $\underset{\wedge}{M + 3\sqrt{M}} \rightarrow \square'$; $A + 6\sqrt{A} \rightarrow \square''$
(7) $M = \square$; $M + \sqrt{M} \rightarrow \square'$; $M + 2\sqrt{M} \rightarrow \square''$	
(8) $M = \square$; $M + \sqrt{M} \rightarrow \square'$; $M + 3\sqrt{M} \rightarrow \square''$	(28) $M = \square$; $\underset{\wedge}{M + 2\sqrt{M}} \rightarrow \square'$; $A + \sqrt{A} \rightarrow \square''$
(9) $M = \square$; $M - 2\sqrt{M} \rightarrow \square'$; $M - 3\sqrt{M} \rightarrow \square''$	
(10) $M = \square$; $\sqrt{M} - M \rightarrow \square'$	(29) $M = \square$; $\underset{\wedge}{M + 4\sqrt{M}} \rightarrow \square'$; $A + 2\sqrt{A} \rightarrow \square''$
(11) $M = \square$; $M + \sqrt{M} \rightarrow \square'$; $\sqrt{M} - M \rightarrow \square''$	
(12) $1^2 + 2^2 = 5$; Divide 5 into two other squares	(30) $[M = \square]$; $\underset{\wedge}{M - 4\sqrt{M}} \rightarrow \square'$; $A - 2\sqrt{A} \rightarrow \square''$
(13) $20 + P_1 = \square$; $50 - P_2 = \square$	(31) $M = \square$; $M + \sqrt{M} \rightarrow \square'$; $M + 1 \rightarrow \square''$
(14) $10 + P_1 = \sqrt{2}$; $50 + P_2 = \square$	(32) $M = \square$; $\underset{\wedge}{M - 5} \rightarrow \square'$; $A + \sqrt{A} \rightarrow \square''$
(15) $3 - M \rightarrow \square$; $2 + M \rightarrow \square'$	
(16) $10 - M \rightarrow \square$; $20 - M \rightarrow \square'$	(33) $M = \square$; $M + 4\sqrt{M} \rightarrow \square'$; $M - (2\sqrt{M} \,\&\, 1) \rightarrow \square''$
(17) $20 + M \rightarrow \square$; $30 + M \rightarrow \square'$	(34) $M = \square$; $\underset{\wedge}{M - 2\sqrt{M}} \rightarrow \square'$; $A + \sqrt{A} \rightarrow \square''$
(18) $10 + M \rightarrow \square$; $10 - M \rightarrow \square'$	
(19) $(8\sqrt{M} \,\&\, 109) - M \rightarrow \square$	(35) $M = \square$; $M + (3\sqrt{M} \,\&\, 1) \rightarrow \square'$; $M - (3\sqrt{M} \,\ell\, 2) \rightarrow \square''$
(20) $M + 8\sqrt{M} \rightarrow \square$; $2\sqrt{M} - M \rightarrow \square'$	(36) $M = \square$; $M + (1 \,\ell\, \sqrt{M}) \rightarrow \square'$; $M - (1 \,\ell\, \sqrt{M}) \rightarrow \square''$
(21) $(2\sqrt{M} \,\&\, 49) - M \rightarrow \square$	(37) $M = \square$; $M + (2 \,\ell\, \sqrt{M}) \rightarrow \square'$; $M - (3 \,\ell\, \sqrt{M}) \rightarrow \square''$
(22) $M = \square$; $M + \sqrt{M} \rightarrow \square'$; $M - \sqrt{M} \rightarrow \square''$	(38) $M = \square$; $M + (\sqrt{M} \,\&\, 1) \rightarrow \square'$; $M + (2\sqrt{M} \,\&\, 2) \rightarrow \square''$

---

121 Other algebraists used the term differently. See Oaks and Alkhateeb (2007, 56) for different uses of *muqābala* in Arabic algebra.

stands for *māl*, meaning "quantity" or "amount", and P_1 and P_2 are two parts of ten, or $P_1 + P_2 = 10$. The signs "+" and "−" stand for the operations of addition and subtraction, and not for "and" and "less". Aggregations are juxtaposed so that, for example, "10$\sqrt{M}$ 20" in Problem 5 stands for "ten of its roots and twenty dirhams".

Problem 12 is the only problem that is also found in the *Arithmetica*, as Problem II.9. Abū Kāmil prefaces this problem with a general claim that we will encounter later in al-Khāzin: "Know that for any number divided into two parts such that each part has a root, you can divide it into two other parts so that each part has a root, and again into two other parts, indefinitely" (Abū Kāmil 2012, 605.3).

3.3.2 *Qusṭā ibn Lūqā, translator of the* Arithmetica

It was against this backdrop of a vibrant algebraic tradition in Arabic that Qusṭā ibn Lūqā made his translation of the *Arithmetica* of Diophantus. Qusṭā was a Christian of Greek origin from Baalbek who translated several books from Greek into Arabic and also wrote a number of original treatises in Arabic.

Qusṭā recognized the algebraic nature of the solutions in Diophantus, so he translated the Greek algebraic terms into the corresponding terms in Arabic. There was no need for any reinterpretation because the basic method of problem solving by Diophantus – of naming unknowns, of operating on these named unknowns, then setting up and solving equations – was already being practiced by Arabic algebraists. But there was not an exact match of terms, so we find Qusṭā introducing the new word *mu'ādala* ("[algebraic] equation") as a translation of the Greek ἴσωσις or ἰσότης. Previously equations had been called simply "problems" (sing. *mas'ala*) in Arabic (Oaks 2010a). He also introduced the names of the reciprocal powers, rendering the terms ἀριθμοστόν, δυναμοστόν, etc., as "a part of a Thing", "a part of a *Māl*", etc., after the way that common fractions were expressed in Arabic arithmetic. And the additive scheme employed by Diophantus for naming powers above the fourth became the Arabic standard after it was adopted by al-Karajī in the early eleventh century.

Two other, connected differences have to do with numbers and problems. For Diophantus, all numbers must be rational, while in Arabic algebra irrational solutions are both accepted and common; and the focus of the problems in the *Arithmetica* lies with indeterminate problems, while in Arabic books determinate problems are far more numerous. These differences account for a difference in emphasis between Diophantus and Arabic algebra. In Diophantus careful attention must be paid to the ways of naming unknowns to ensure that a two-term equation of consecutive powers is set up, in order to guarantee a rational solution. Arabic algebraists generally had no reason to avoid three-term equations since the answers to their questions are allowed to be irrational. Only when solving indeterminate problems does an Arabic author require a rational solution, and there the techniques of assignment found in Diophantus are applied so that the simplified equation has only two terms. Thus, al-Khwārazmī and later Arabic algebraists gave the rules for solving three-term equations in the beginnings of their books and applied them routinely in their worked-out problems, while Diophantus waited until sometime after Book VII to show the way to solve them, and even then he used the rules only sparingly. This is not a difference in method, but a difference in emphasis.

Titles of books in medieval Arabic were sometimes not so much official names, but served the purpose of describing its contents, or of differentiating one book from another. It is not surprising, then, to find that the "title" of Diophantus's book is given different designations even within a single work. In different places in his *Kitāb al-Fihrist* Ibn al-Nadīm writes of the "book of Diophantus on arithmetical problems", the "book of Diophantus on algebra (*al-jabr*)", and "book on the art of algebra (*al-jabr*)". Both Sesiano (1982, 13) and Rashed (Diophantus 1984, 3, xiv) note that among Arabic authors, the *Arithmetica* has been described as a book on algebra (either *al-jabr* or *al-jabr wa-l-muqābala*), on problems of algebra (*masā'il al-jabr*), on the art of algebra (*ṣinā'at al-jabr*), on arithmetical problems (*masā'il 'adadīyya*), and on calculation (*ḥisāb*). In the beginning of Book V of the manuscript of Qusṭā's translation the *Arithmetica* is called "the work of Diophantus the Alexandrian on arithmetical problems", and at the end of Book VII it is "the work of Diophantus the Alexandrian on algebra (*al-jabr wa-l-muqābala*)". There is no contradiction between the designations "arithmetical problems" and "algebra". The enunciations to the problems are indeed problems in arithmetic, while their solutions belong to algebra. One could even describe the book more fully as "the work of Diophantus the Alexandrian on arithmetical problems solved by algebra".

Ibn al-Nadīm completed his *Kitāb al-Fihrist* (*The Catalog*) in Baghdad in 987 CE. Under Qusṭā, the last work mentioned is "his commentary on three and a half books of the work of Diophantus on arithmetical problems" (Ibn al-Nadīm 1871–72, II, 295.22; Sayyid 2009, II, 294.5). This was later repeated by Ibn Abī Uṣaybī'a (Savage-Smith, Swain, and van Gelder 2020, [10.44.5]). We have no other testimony regarding this lost commentary.

3.3.3 Al-Khāzin

Probably the earliest Arabic mathematician we know who cites to Diophantus by name, and the only one we know before Fermat who read him primarily as a number theorist, is Abū Ja'far al-Khāzin (d. ca. 971). Al-Khāzin was a Persian astronomer and mathematician who, according to al-Khayyām, was the first to solve an irreducible cubic equation by means of conic sections.[122] Two short works of his on number theory, both dealing with Pythagorean triples, are preserved in a single codex, Paris BnF 2457. It is in the second of these works,

---

122 On al-Khāzin, see Dold-Samplonius (1970) and Anbouba (1978b).

covering folios 204r–215r, that Diophantus is mentioned.[123] This treatise comprises three main parts: the first part covers the formation of Pythagorean triples, the second part is devoted to solving a particular indeterminate problem from the Arabic tradition, and the third part gives a sequence of results that clarify the procedure employed in *Arithmetica* Problem III.19, including a "lemma" stated but not proven by Diophantus in the beginning of the problem.[124] None of the problems or lemmas are worked out by algebra in either of al-Khāzin's treatises.

The theorems in the first part of the treatise are stated and proven in the setting of Greek number theory, with proofs citing Euclid's *Elements* and sometimes showing the numbers as labeled line segments like we find in *Elements* Books VII–IX. Numbers here are presumed to be positive integers. The propositions in the second part of the treatise lead to the solution to the following indeterminate problem from the Arabic tradition: "One seeks in the art of algebra (*al-jabr*) for a *māl* that has a root, and if twenty is added to it, the outcome has a root, and if twenty is subtracted from it, then the remainder has a root" (Anbouba 1979, 162.15). This problem is not posed by Diophantus in his extant books, though he does find three squares with equal differences as part of the solutions to Problems III.7 and VII.8–10. Neither Abū Kāmil nor al-Sulamī solve this problem either, but it is found in later medieval works.[125] It is in this part of his treatise that al-Khāzin mentions in passing *al-istiqrā'*. Referring to his method of obtaining a solution to this problem, he writes "this method is not limited, and it is similar to *al-istiqrā'* since one obtains square numbers without limit" (Anbouba 1979, 162.20). It is not likely that *al-istiqrā'* here refers to the technique of assigning the side of an unknown square to complete a forthcoming equation, which was also called *al-istiqrā'* by al-Karajī and later algebraists (see Appendix 2). Al-Karajī later solved this problem, evidently by trial and error, in his book *al-Badī'*, to be discussed below.

Al-Khāzin begins the third part of his treatise with:

> Related to what we have mentioned, we present a collection of properties of numbers for which each of them is divided into two square numbers, then multiplying by a number divided into two square numbers for which one of them is a square, or each of them a square, or neither one is a square. This will clarify the lemma that Diophantus introduced in the nineteenth problem

123 The text is edited in Anbouba (1979). This article and Rashed (1979), give mathematical commentaries.

124 Using Anbouba's numbering of passages, the first part consists of passages 1–21, the second part 22–34, and the third part 35–42.

125 Al-Khāzin covers it again in his other number theory treatise in the same manuscript (ff. 86v–92v). The enunciation there is stated in nearly identical terms at f. 91v, 1.10, except that in place of *māl* he has *'adad* ("number"). Al-Karajī solves this problem for the constant 5 in his *al-Badī'* (al-Karajī 1964, 77.8), Fakhr al-Dīn al-Ḥilātī solves it for the constant 6 in his algebra book (MS Tehran 4409, p. 53, 1.12), Ibn al-Khawwām lists it as an impossible problem when 10 replaces 20 (al-Fārisī 1994, 604.15), and Fibonacci treats the problem at length, beginning with Proposition 14 in his *Book of Squares* (Fibonacci 1987, 53ff).

of the third Book of his treatise on algebra (*al-jabr*), and which is useful for other problems.

(Anbouba 1979, 161.12)

That "lemma" is a rule stated but not proven by Diophantus in the first part of the solution to Problem III.19:

> Further, the 65 is divided naturally into (two) squares in two ways: into 16 and 49, but also, 64 and a unit. This is due to the fact that the number 65 is contained by the numbers 13 and 5, each of which is divided into two squares.

Al-Khāzin then proceeds to state and explain seven rules exploring the number of different ways a number can be divided into two squares. The first rule is that if a number can be expressed as the sum of two squares, then its square can also be expressed as the sum of two squares. In modern notation, if $x = a^2 + b^2$, then $x^2 = (2ab)^2 + (b^2 - a^2)^2$. It is the fourth of these rules that addresses the lemma of Diophantus, using the same number 65:

> And if we multiply a number (that is) divided into two square numbers one time by a number (that is) divided into two square numbers one time, (then) the number composed of them[126] is divided into two square numbers two times. For example, five is composed of one and four, and thirteen is composed of four and nine, and the product of one of them by the other is sixty-five. This is divided into two square numbers two times, since it is clear that five by thirteen is five by four and five by nine, and that five by four is four by four and one by four, and five by nine is four by nine and one by nine, since the five is divided into four and one, and thirteen is divided into four and nine. Thus, the sides of these squares are two and four and three and six. And since the ratio of the two to four is as the ratio of three to six, the product of two by six is equal to the product of four by three, and the product of three by four two times is equal to the product of two by six two times.
>
> And the product of two by six two times with the square of the difference between them is equal to the sum of the squares of two and six. But the product of two by six two times is equal to the product of three by four two times, and the product of three by four two times with the sum of the squares of three and four is equal to the square of the sum of three and four.
>
> So, the square of the difference between two and six with the square of the sum of three and four is equal to the sum of the squares of two and three and four and six, which is sixty-five. Therefore, sixty-five is divided into two

---

126 Here "composed" (*murakkab*) means their product. In the next instances, *murakkaba* means their sum.

squares, a side of one of them being the sum of three and four, and a side of the other being the difference between two and six, the first time, and the other time (it is) divided into two squares, a side of one of them being the difference between three and four, and a side of the other the sum of two and six. So we (have) divided sixty-five the first time into forty-nine and sixteen, and the other time into one and sixty-four. And one (can) divide similarly the product of any of two numbers, each of them being divided into two square numbers, one of them by the other.

(Anbouba 1979, 160.6)

In modern notation, if $x = a^2 + b^2$ and $y = c^2 + d^2$, then xy equals $(ac + bd)^2 + (ad - bc)^2$, and also $(ac - bd)^2 + (ad + bc)^2$.

Al-Khāzin's last rule explains why the square of 65 can be divided into two squares in four different ways, a result necessary for the solution of Diophantus's problem. This is shown by combining the first and fourth rules described earlier. Specifically, since $65 = 8^2 + 1^2$, by the first rule

$$65^2 = \left(2 \cdot 8 \cdot 1\right)^2 + \left(8^2 - 1^2\right)^2 = 16^2 + 63^2,$$

and since $65 = 7^2 + 4^2$, again by the first rule

$$65^2 = \left(2 \cdot 7 \cdot 4\right)^2 + \left(7^2 - 4^2\right)^2 = 56^2 + 33^2.$$

But also, since $65^2 = (8^2 + 1^2) \cdot (7^2 + 4^2)$, by the fourth rule,

$$65^2 = \left(8 \cdot 7 + 1 \cdot 4\right)^2 + \left(8 \cdot 4 - 1 \cdot 7\right)^2 - 60^2 + 25^2,$$

and also

$$65^2 = \left(8 \cdot 7 - 1 \cdot 4\right)^2 + \left(8 \cdot 4 + 1 \cdot 7\right)^2 = 52^2 + 39^2.$$

At the end of this third part of the treatise al-Khāzin returns to Diophantus's Problem III.19, where he suggests that the lemma can be used to solve another indeterminate problem that Diophantus does not pose in his extant books:

The square of sixty-five, along with what it is composed of, is what Diophantus introduced in the problem that we mentioned, which is to find four numbers such that if one adds each of them to a square of their sum, then the outcome has a root, and if one subtracts each one of them from it, then the remainder has a root. This lemma makes possible a way to find four different numbers such that their sum is a square, and the sum of any two of them is a square.

Fibonacci also states and proves this in Proposition 6 in his *Book of Squares*, but without mentioning any sources. His proof is in the style of those in Euclid's number theory books (Fibonacci 1987, 23ff).

3.3.4 *Al-Nīsābūrī and ʿAlī al-Sulamī*

Two authors also from the tenth century show a modest influence of Diophantus. According to Boris Rosenfeld, Yūsuf ibn Aḥmad al-Nīsābūrī, author of a treatise on calculation with Indian (i.e., Arabic) numerals, was active in the tenth century.[127] In the beginning of Chapter 1 in this work al-Nīsābūrī writes:

> So we say the philosophers of numbers from Greece and India, [—][128] the Indian and Pythagoras the arithmetician and Nicomachus of Gerasa and Diophantus the algebraist[129] and others among them, describe the science of numbers and are the best-known in philosophy and geometry, and establish proofs mentioned in their books (on) properties of their ranks.
>
> (MS Leiden Or. 780, p. 6, l. 13)

Here we seem to have the standard names of the nearly mythical experts in arithmetic and logistic, much like we saw in contemporary Byzantine hagiographic works in Section 3.2.5. Pythagoras is truly mythical, and al-Nīsābūrī shows no sign in his book of any familiarity with Diophantus's work. But he does note on the next page that "the philosopher Nicomachus" was "author of the *Arithmetica*", and just before that, he writes "the origin of every number is the one, which is included in all of them and is not a number with them" (MS, p. 7.1).

Probably the earliest extant Arabic text to show direct influence from Diophantus, but without any mention of his name, is ʿAlī al-Sulamī's *Sufficient Introduction*. We have no information on this author other than what can be inferred from the one surviving manuscript, which was copied in 1211 CE.[130] Matvievskaya and Rosenfeld (1983, II, 183) date him also to the tenth century, and we see no evidence to the contrary.[131] The main influences on al-Sulamī were the algebra books of al-Khwārazmī and Abū Kāmil.

Among the 78 worked-out problems in al-Sulamī's book, 9 of them are indeterminate problems that belong to the Arabic tradition. The wording and the kinds of questions asked are similar to the problems in Abū Kāmil, including the absence of the word *al-istiqrāʾ*, and none of the problems are posed in either the extant *Arithmetica* or in Abū Kāmil. Al-Sulamī's problems are thus

---

127 The treatise is titled *Attainment of Students on Truths in the Science of Calculation (Bulūgh al-ṭilāb bi l-ḥaqāʾ iq fī ʿilm al-ḥisāb)*. It is extant in a single manuscript, Leiden Or. 780, copied in 843 H/1439–40 CE. Matvievskaya and Rosenfeld (1983, II, 160) date al-Nīsābūrī to the tenth century. In his book al-Nīsābūrī mentions Ibn al-Fatḥ, who, judging by the latter's position in Ibn al-Nadīm's *Fihrist*, probably worked in the first half of the tenth century.

128 There is a blank space in the manuscript where the Indian's name should have been written.

129 We read *al-j-y-l-y* as being miscopied from *al-j-b-r-y* which, adding the vowels, would mean "algebraist."

130 Arabic title: *al-Muqaddima al-kāfiyya fī ḥisāb al-jabr wa-l-muqābala wa mā yuʿ rafu bihi qiyāsuhū min al-amthila* (ʿAlī al-Sulamī, MS Sbath 5).

131 Rashed (1983, 98) claims that al-Sulamī was a successor of al-Karajī, but we see no evidence that al-Sulamī knew any work of al-Karajī. This is especially evident in al-Sulamī's curious system of naming the powers.

an independent testimony to the Arabic oral tradition in this area from which Abū Kāmil had drawn earlier (Oaks 2015). Here is one of his enunciations: "If (someone) said, a *māl*: it has a root. If you add to it its root, then it has a root, and if you add to it two of its roots, it has a root" ('Alī al-Sulamī, MS Sbath 5, f. 52v, l. 11).

However, al-Sulamī does work out one problem that ultimately comes from Diophantus, determinate Problem I.27 from the *Arithmetica*. As is common in medieval mathematics, the source for the problem is not mentioned. The enunciation in al-Sulamī reads: "If (someone) said, two different *māls*. If you add them it gives twenty, and if you multiply one of them by the other it gives ninety-six". He solves this the same way as Diophantus, by naming the two parts of 20 as "ten and a Thing" and "ten less a Thing". It is probably because Diophantus gave the necessary condition for this problem to have a solution that al-Sulamī, after solving it, observes that if the 96 were instead 100, then the problem would be impossible. After this, he notes that it is possible for 80, and he solves it again, this time getting an irrational solution. Here are his solutions ('Alī al-Sulamī, MS Sbath 5, f. 64v, l. 8):

> Make one of the parts ten and a Thing and the other ten less a Thing, so their sum is twenty. Then multiply ten and a Thing by ten less[132] a Thing to get: a hundred dirhams less a *Māl* equals ninety-six. Restore[133] the hundred by the *Māl* and add it to ninety-six, and cast away ninety-six by its same, leaving: a *Māl* equals four. Its root is two, which is the Thing. So you add it to ten to get twelve, and the other (part) is eight.
>
> If (someone) said, if it were a hundred, then it would be impossible. If (someone) said, if it were eighty, then it would be possible. You multiply ten and a Thing by ten less[134] a Thing to get: a hundred less a *Māl* equals eighty. So restore and subtract the two sides, leaving: a *Māl* equals twenty. The Thing is a root of twenty, and one of the parts is ten and a root of twenty, and the other is ten less a root of twenty. If you multiply one of them by the other it gives a hundred less twenty, and that is eighty.

We have also found this problem solved in the book *Magnificent Benefits of the Rules of Calculation* (*al-Fawā'id al-bahā'iyya fī l-qawā'id al-ḥisābiyya*) by the Persian physician and mathematician Ibn al-Khawwām. This book, completed in 1277, is a guide to practical calculation, with chapters on algebra at the end. Ibn al-Khawwām poses the problem with the same given numbers 20 and 96, but he solves it by naming the parts "a thing" and "twenty less a thing". Sometime in the period 1284–1301 his student Kamāl al-Dīn al-Fārisī wrote a commentary on this book, and he included this problem along with the others that Ibn al-Khawwām solved (al-Fārisī 1994, 551.8).

---

132 The text has *wa* ("and") in error.
133 The text has *ā-ḍ-r*, which is visually close to *ajbir* ("restore").
134 The text again has "and" in error.

3.3.5 Abū l-Wafā'

Abū l-Wafā' al-Būzjānī was a tenth-century Persian mathematician and astronomer who worked in Baghdad. Unfortunately, his books relating to algebra are not extant. Ibn al-Nadīm lists the following lost works by him:

1   Commentary on the book of al-Khwārazmī on algebra (*al-jabr wa-l-muqābala*)[135]
2   Commentary on the book of Diophantus on algebra (*al-jabr*)
3   Commentary on the book of Hipparchus on algebra (*al-jabr*)
4   Introduction to the *Arithmetica* (in one) chapter
5   On what one should keep in mind when approaching the book (of) *Arithmetica*
6   The proofs to the assertions (*qaḍā*) that Diophantus employs in his book and to what he (Abu' l-Wafā') employs in the commentary

The "*Arithmetica*" in either or both of 4 and 5 might refer to the *Arithmetical Introduction* of Nicomachus and not the *Arithmetica* of Diophantus. Later Arabic sources add nothing new to this list (Sesiano 1982, 10).

3.3.6 Al-Karajī

Diophantus was a major influence on the Persian mathematician and engineer al-Karajī, who wrote his mathematics books in Baghdad in the early eleventh century CE. Three of al-Karajī's works are of interest to us: his *[Book of] al-Fakhrī on the Art of Algebra* (*al-Fakhrī fī ṣinā'at al-jabr wa-l-muqābala*), henceforth *al-Fakhrī*, composed 1011/12 CE; his *Causes of Calculation in Algebra and Its Proof* ('*Ilal ḥisāb al-jabr wa-l-muqābala wa l-burhān 'alayhi*), henceforth *Causes of Calculation*; and *Marvelous [Book] of Arithmetic* (*al-Badī' fī l-ḥisāb*), henceforth *al-Badī'*. The last two were both written after *al-Fakhrī*.[136]

Al-Fakhrī is al-Karajī's introduction to algebra. Like Abū Kāmil and al-Sulamī, he modeled it on earlier works, which in his case are the *Book on Algebra* of Abū Kāmil and the *Arithmetica* of Diophantus. Diophantus and Abū Kāmil had both presented the rules of algebra followed by a collection of worked-out problems, and al-Karajī did the same. But where Abū Kāmil had defined only the first two powers of the unknown in his introduction and then turned to the classification, solutions, and proofs to the six equations, al-Karajī took the lead of Diophantus by defining also higher powers, up to the ninth in his case, together with their reciprocals. And like Diophantus, he then explained their multiplications and divisions, covering the six equations only later.[137] Even al-Karajī's names for the

---

135 Abū l-Wafā' himself mentions this commentary in his arithmetic book (Saidan 1971, 132.6).
136 Both *al-Fakhrī* and *Causes of Calculation* are published in Saidan (1986), and *al-Badī'* is published in al-Karajī (1964).
137 For his part, 'Alī al-Sulamī begins by introducing the powers up to the *māl māl*, then he covers the six equations and their solutions, and after that he describes powers up to the 20th, but with a system that differs from that in Diophantus and al-Karajī. See Section 3.1.2.

powers above the fourth follow the scheme in Diophantus, and not the schemes or examples in the earlier texts of Abū Kāmil, Sinān Ibn al-Fatḥ, or ʿAlī al-Sulamī.

The treatment of the six equations in *al-Fakhrī* shows influence from both Abū Kāmil and Diophantus. For each of the three types of composite equation Abū Kāmil had given the standard Arabic rule for finding the "Thing", which begins with the normalization of the equation, that is, setting the number of *Māl*s to 1. He then added another rule for finding the *Māl* directly. Al-Karajī gives three rules for each composite equation: first, the standard Arabic rule for finding the "Thing"; then, the rule of Diophantus for finding the "Thing" which does not require normalization; and finally, the rule of Abū Kāmil for finding the *Māl* directly. Neither Abū Kāmil nor Diophantus is explicitly named in this section of the book. Al-Karajī then gives proofs to these rules based on Euclid's *Elements* propositions II.5 and II.6 (Oaks 2018b).

After the proofs for the first two of the three composite equation types, al-Karajī gives a derivation for the standard Arabic rule that he ascribes to Diophantus. It is because of a lacuna in our manuscripts of *al-Fakhrī* that we lack the corresponding derivation for the third type. We have already translated and analyzed these derivations in Section 2.6.4, but we should repeat them here to illustrate the link with his later *Causes of Calculation*. For the sample equation "a *Māl* and ten Roots Equal thirty-nine units" al-Karajī writes (Saidan 1986, 154.13):

> And if you want to find the root of the *Māl* in the manner followed by Diophantus, you search for a number which, if added to a *Māl* and ten Things, has a root. It is nothing but twenty-five, which added to a *Māl* and ten Things, has a root that is a Thing and five dirhams. And you knew that a *Māl* and ten Things are thirty-nine units, so if you remove the *Māl* and ten Things, and you put in its place thirty-nine units, they become sixty-four units. Its root is eight, and that Equals a Thing and five dirhams. So the Thing Equals three dirhams, which is the root of the *Māl*.

Then, for the sample equation $x^2 + 21 = 10x$, the derivation is (Saidan 1986, 159.17):

> And if you want to solve this problem according to the method of Diophantus, you look for a square which, if you subtract from it (and the *Māl*) ten Roots which are equal to the *Māl* and twenty-one units, then the remainder is a square. So make that square from a side of a Thing less five, or five less a Thing, and each of them leads to a quantity of units, and that quantity is a *Māl* and twenty-five units less ten Roots. Replace ten Roots with a *Māl* and twenty-one units, since it equals it, leaving four in number. And a root of that is two, so (if) five less a Thing is two, then the Thing is three. And (if) a Thing less five is two, then the Thing is seven.

Al-Karajī diverged from Abū Kāmil by generally avoiding proofs. Where Abū Kāmil gave over 50 proofs to various rules in arithmetic and algebra, most of them based in one way or another on Euclid's *Elements*, al-Karajī restricted his proofs nearly exclusively to the rules for solving the three composite equations. He wrote just before giving his geometric proofs:

> I made a determination in this book to strip it of proofs, lengthy explanations, and numerous examples. But I cannot avoid giving a brief summary of the proofs (*burhān*) for the connected problems[138] and the cause ('*illa*) of halving the roots and what is associated with it.[139]
>
> (Saidan 1986, 151.8)

One reason, perhaps the main reason, for the omission of proofs is made clear in *Causes of Calculation*. Many readers of *al-Fakhrī* were trained in *ḥisāb*, specifically practical calculation by finger reckoning, and had no knowledge of Euclid. He thus decided to write this short tract in order to give arithmetical proofs to the rules for solving equations (Saidan 1986, 354.6):

> I aspired to establish the proof(s) for what was prescribed in halving the roots, etc., establishing the proof(s) by means of lines and figures. The knowledge coming from this is clear from evidence that cannot be refuted and does not rely on any other (knowledge). Then I saw that people seeking knowledge of calculation found it difficult to understand the correctness (of the proofs) by means of those lines and figures, given their understanding by means of the tongue and hand.[140] I discovered that many people found it very difficult when reading them (i.e., the proofs) in books.
>
> And I decided because of this to make the proofs in this book easier, because the foundation (1) that eliminates the uncertainty of (both) everything that was drawn and the arguments that make up each cause, and (2) that is independent of those drawings, is instead proofs by arithmetic, using algebra and number, (which) puts an end to the incomprehensibility of the realm of lines and figures. Instead of having to piece together the validity of the proof(s) from what you drew, everything for learning the science of arithmetic can be seen in this book.

For these new, arithmetical proofs al-Karajī adapted the arguments from the "method of Diophantus". Here is the proof for the type 4 equation, now performed in the context of the equation "a *Māl* and ten Roots equal twenty-four dirhams":[141]

---

138 I.e., the composite equations.
139 The "what is associated with it" is the rest of the rule, after taking half of the roots.
140 Geometry was viewed as a visual science, while arithmetic was traditionally regarded as a spoken science. This is partly due to the prevalence of finger reckoning, in which intermediate results of calculations are stored by positioning the fingers in particular ways. This method of calculation involved the tongue (for speaking) and the hand (for storing numbers).
141 (Saidan 1986, 359.23). Translations of all three proofs are given in Oaks (2018b, 289–91).

If you want to show the reason for saying, in the first problem, which is the *Māl* and Roots Equal a number, that the way (to solve) it is that you halve the Roots, and you multiply (it) by itself, and add to the total the number Equal to the *Māl* and Roots, then take the (square) root of the sum, then subtract from it half the Roots, so the remainder is the (square) root of the *Māl*, we say:

Suppose a *Māl* and ten Roots Equal twenty-four dirhams (1M 10R = 24). We want the agreement between the Roots and the number. We add to the Root the same as the number of half the Roots, which is five, since it is appended in this first problem,[142] as we explained before,[143] to get a Root and five in number (1T 5). So we multiply that by itself to get a *Māl* and ten Roots and twenty-five dirhams (1M 10R 25). And we already had the *Māl* and ten Roots in the problem Equal to twenty-four dirhams. So if we put the twenty-four in this multiplication in place of a *Māl* and ten Roots, then the multiplication of a Root and five by itself is forty-nine in number. So the root of forty-nine, which is seven, is a Root and five. You subtract five from it, leaving the Root to be two, which is the square root of the *Māl*, and the *Māl* is four, and ten Roots are twenty. If you add them, it gives twenty-four. So this rule is true analogously in the realm of number.[144]

Al-Karajī also later gave a different set of arithmetical proofs for these rules in his arithmetic book *The Sufficient [Book] on Arithmetic* (*al-Kāfī fī l-ḥisāb*). Several later algebraists, including Ibn al-Yāsamīn, Ibn al-Bannā', Ibn al-Hā'im, and al-Fārisī, took al-Karajī's lead by devising other proofs based in arithmetic rather than geometry (Oaks 2018a).

The influence of Diophantus on *al-Fakhrī* is just as evident among the 255 worked-out problems that al-Karajī gives after explaining the vocabulary and rules. These problems are divided into five "categories" (*ṭabaqāt*), which, since Woepcke's transcription of 1853, have been numbered I–V. Problems are borrowed from three main sources: the Arabic oral tradition from which earlier Arabic authors had also drawn; Abū Kāmil's *Algebra*, both determinate and indeterminate problems; and Diophantus's *Arithmetica*, again, both determinate and indeterminate problems. It is tenuous to identify problems as coming directly from the oral tradition, since al-Karajī may have copied or adapted some of them from a now lost book by another author. Either way, they ultimately come from Arabic practical arithmetic. These include all 51 problems in Category I and the first 21 problems in Category II. The remaining 183 problems in Categories II–V are nearly all taken from Abū Kāmil or Diophantus, usually with 10 or more problems in a row from one author, then a switch to the other

---

142 The roots are appended to the *māl* rather than on the opposite side of the equation with the number.

143 He had explained the completion of the square extensively just before these proofs.

144 Following this he works it out again for the example $x^2 + 10x = 30$, where the root of the sum is not a perfect square, to get the solution $x = \sqrt{55} - 5$.

author. From Abū Kāmil we have identified 32 determinate and 23 indeterminate problems that were borrowed or adapted by al-Karajī. From Diophantus al-Karajī borrowed 16 determinate problems and 85 indeterminate problems, all of them from Books I–IV. Table 3.6 provides a list of the problems taken from the *Arithmetica*.[145]

Al-Karajī usually preserved the distinction between the wording of indeterminate problems taken from Abū Kāmil and those taken from Diophantus, keeping the term *māl* and the phrase "has a root" from the former, and using "number" and "is a square" for the latter. One change he introduced to problems from both sources is that he occasionally uses the term *al-istiqrā'* in his

Table 3.6 Problems of al-Karajī taken from Diophantus.

Karajī	Dioph						
		III.40	II.11	IV.47	III.7	V.13	IV.15c
II.45	I.4	III.41	II.12	IV.48	III.8	V.14	IV.16
II.46	I.8	III.42	II.13	IV.49	III.9	V.15	IV.17
II.47	I.9	III.43	II.14	IV.50	III.10	V.16	IV.18
II.48	I.10	III.44	II.15	IV.51	III.11	V.17	IV.19
II.50	II.22	III.45	II.16	IV.52	III.12	V.19	IV.20
III.1	II.20	IV.1	II.22	IV.53	III.13	V.20	IV.22
III.2	II.21	IV.2	II.23	IV.54	III.14	V.21	IV.23
III.3	II.8*	IV.3	II.24	IV.55	III.15	V.22	IV.24
III.4	II.20*	IV.4	II.25	IV.56	III.16	V.28	IV.27
III.7	I.12	IV.5	II.26	IV.57	III.17	V.29	IV.28
III.20	I.13	IV.6	II.27	IV.58	III.18	V.30	IV.29
III.24	I.16	IV.7	II.28	IV.59	III.21	V.31	IV.30
III.25	I.17	IV.8	II.29	IV.60	III.20	V.32	IV.31
III.26	I.24	IV.9	II.30	IV.61	III.19	V.33	IV.32
III.27	I.25	IV.10	II.31	V.1	IV.1	V.34	IV.33
III.28	I.39	IV.11	II.32	V.2	IV.2	V.35	IV.34
III.29	I.18	IV.12	II.33	V.3	IV.3	V.36	IV.35
III.30	I.19	IV.13	II.34	V.4	IV.4	V.37	IV.36
III.31	I.20	IV.14	II.35	V.5	IV.5	V.38	IV.37
III.32	I.21	IV.40	II.18	V.6	IV.6	V.39	IV.38
III.33	I.22*	IV.41	II.19	V.7	IV.7	V.40	IV.39
III.34	I.22	IV.42	III.1	V.8	IV.8	V.41	IV.40
III.35	I.23	IV.43	III.2	V.9	IV.9	V.42	IV.41
III.36	II.8	IV.44	III.3	V.10	IV.10	V.43	IV.20
III.37	II.9	IV.45	III.5	V.11	IV.11		
III.38	II.10	IV.46	III.6	V.12	IV.14		

145 Numerical parameters are often different from those in Diophantus. Problems with an asterisk are minor variations on the Diophantine problem. Of these, al-Karajī's III.3 and III.4 ask for two *māl*s and not two numbers, so these problems may not have been borrowed from the *Arithmetica*. Al-Karajī's Problem V.13 is the corollary to *Arithmetica* Problem IV.15. Some other problems appear to be problems invented by al-Karajī in imitation of those in Diopantus: II.43, III.39, V.18, and V.23–V.27. Also, al-Karajī's Problem V.43 is the same as his V.19, with the solutions worded differently.

solutions. Here is Problem II.28, adapted from Abū Kāmil's indeterminate
Problem 11:

> If (someone) said: a *māl*, it has a root. If you add to it its root, then it has a
> root, and if you subtract the *māl* from its root, the remainder has a root.
>
> The method for this problem is that you look for a number such that if you
> add it to a rooted number, then the outcome is rooted, and if you subtract the
> square number form it, the remainder is rooted. The way to do this is that you
> make the sought-after square a *Māl*, and you know that if you add to it two of
> its roots and a dirham, then the outcome is rooted. So subtract the square from
> two Roots and a dirham, leaving two Things and a dirham less a *Māl*. Make
> a root of that, by *al-istiqrā'*, a dirham less a Thing, and multiply it by itself
> to get a *Māl* and a dirham less two Things, and that Equals two Things and
> a dirham less a *Māl*. So if you restore and confront, you find that the Thing
> Equals two dirhams, and the *Māl* is four dirhams, which is the square number.
>
> Since we made the other number two things and a dirham, it is five dir-
> hams. When you add it to the four, it becomes nine, which is rooted. And if
> you subtract the four from it, it leaves one, which is a square.
>
> Once you have found these two numbers, divide four by five to get four
> fifths, which is a root of the *Māl*, and the *Māl* is sixteen (parts) of twenty-five
> (parts). It is the *māl* to which you add to its root to get (a sum that is) rooted,
> and if you subtract it from its root, the remainder is rooted.
>
> (Saidan 1986, 195)

We also translate two problems that come from Diophantus, one for which the
Greek version is extant, and one for which the Arabic translation is extant. First
is al-Karajī's Problem IV.47, corresponding to *Arithmetica* Problem III.7 (Saidan
1986, 271):

> If (someone) said, three numbers: the excess of the first over the second is
> equal to the excess of the second over the third, and the sum of any two is a
> square.
>
> Its rule is that you look for three squares of equal difference such that the
> sum of any two is greater than the third. So you make the first a *Māl* and the
> second a *Māl* and two Things and one, and the third a *Māl* and four Things
> and two. It is necessary that this third also be a square. So make its root
> something that leads to the *Māl* that results being greater than two of its roots
> and one, so that the sum of the *Māl* with the *Māl* and two things and one is
> greater than the *Māl* and four Things and two. If you make it a Thing less
> eight dirhams, then its square turns out to be: a *Māl* and sixty-four dirhams
> less sixteen things Equal a *Māl* and four things and two dirhams.
>
> So if you restore and confront, it results in the Thing is thirty-one parts of
> ten parts of a dirham, and the *Māl* is nine hundred sixty-one parts of a hundred
> parts, which is the first number, and the second is one thousand six hundred
> eighty-one parts of a hundred parts, since we made it a *Māl* and two Things

and one. And because we made the third a *Māl* and four Things and two dirhams, it is two thousand four hundred one (parts) of a hundred parts of a unit.

So you make the sought-after numbers, all of them, one Thing, and you make the first and the second nine hundred sixty-one dirhams, leaving the third as a Thing less nine hundred sixty-one. And you make the second and the third one thousand six hundred eighty-one, leaving the first as a Thing less one thousand six hundred eighty-one. And you make the third and the first two thousand four hundred one. The sum of these three is three things less five thousand forty-three units, and that Equals a Thing. So the one Thing is two thousand five hundred twenty-one dirhams and a half.

Then cast away from it one thousand six hundred eighty-one, leaving the first as eight hundred forty dirhams and a half. Then cast away from it two thousand four hundred one, leaving the second as one hundred twenty and a half. So make this the first, and make the second eight hundred forty and a half, and the third one thousand five hundred sixty and a half.

The second problem is *al-Fakhrī* V.37, which is al-Karajī's version of *Arithmetica* IV.36 (Saidan 1986, 303.4):

If (someone) said, a cubic number: if you add to it four times the *māl*[146] which comes from its side, it gives a square, and if you subtract from it five times the *māl* which comes from its side, then the remainder is a square.

Its rule is that you make the cube a Cube, and you add to it four *Māl*s, and you subtract from it five *Māl*s. One of them becomes a Cube less five *Māl*s, and the other a Cube and four *Māl*s. And each one of these two aggregations Equal a square. So the difference between them is nine *Māl*s. And this cannot be worked out by the double equality.

If you wish, look for two squares such that the difference between them is nine. So you find that they are twenty-five and sixteen. If you wish, confront twenty-five *Māl*s with a Cube and four *Māl*s, resulting in the Thing being twenty-one. And if you wish, confront sixteen *Māl*s with a Cube less five *Māl*s, resulting in the thing also being twenty-one. After that, the *Māl* is then four hundred forty-one, and the Cube is nine thousand two hundred sixty-one.

Al-Karajī wrote his *Marvelous [Book] of Calculation* (*al-Badīʿ fī l-ḥisāb*) after *al-Fakhrī*. In this work he sets about to explain the foundations and techniques of problem solving by algebra, including the techniques employed by Diophantus. A basic understanding of algebra, such as is presented in *al-Fakhrī*, is presumed. Hebesein writes of this work, "Si certains passages sont écrits hâtivement, d'autres au contraire sont approfondis et détaillés" (Hebesein 2009, 3). Indeed, we will see evidence of this haste at the end of al-Karajī's investigations into *al-istiqrāʾ*.

---

146 The meaning of the word *māl* here and in the next instance is the "square" of the number, like the way al-Samawʾal applied the term *māl māl* described in Section 3.1.2. Each instance of *māl* after that in this problem is the name of the algebraic power.

Al-Badīʿ is divided into five parts, as explained in the introduction.[147] The first part, *On the Elements*, covers number theory and arithmetic and is based mainly in Euclid's *Elements*. Al-Karajī begins by giving most of the definitions from Euclid's Books VII and V, and he then states well over half of the propositions from the number theory Books VII–IX, as well as many propositions reinterpreted arithmetically from Books II and X. It is in this part that al-Karajī states and proves the Diophantine rule for solving problems by the double equality:

> For any two square numbers, if you divide the difference between them by the difference between the two roots, then you add the result of the division and the divisor, then half of that is a root of the greater of the two squares; and if you cast away the smaller of the two amounts from the greater, then half of the remainder is a root of the smaller of the two squares.

In modern notation, if $x = \frac{b^2-a^2}{b-a} + (b-a)$, then $b = \frac{1}{2}x$; and if $y = \frac{b^2-a^2}{b-a} - (b-a)$ then $a = \frac{1}{2}y$. He continues:

> This is because the difference between the two squares is equal to the product of the difference between the two roots by the smaller of the roots two times and (the product) by itself one time ($b^2 - a^2 = 2(b - a)a + (b - a)^2$). So if you divide that difference by the difference between the roots, the result of the division is the root of the smaller two times and the difference between them one time. And if you cast away from that the difference between the two roots, then half of the remainder is the smaller root, and if you add to it the difference between the two roots, then half of the outcome is the root of the smaller and the difference between the two roots, and that is equal to the greater root, and that is what we wanted to show.
>
>   This is called the double equality, which is a fundamental element for solving many of the problems, and we will point out its great usefulness when we encounter it.
>
> (al-Karajī 1964, 20.12)

The second part of *al-Badīʿ*, *On unknowns*, is devoted to ways of taking the square root of a polynomial that is a perfect square. Here al-Karajī expands a chapter on the topic that he had already written about in *al-Fakhrī*.

The third part, *On al-istiqrāʾ*, consists of a long exposé on the algebraic technique of *al-istiqrāʾ* as applied by Diophantus and Arabic algebraists when faced with the task of equating a non-square polynomial to a square, or sometimes a cube, in order to obtain a rational solution. Here, too, al-Karajī expands a chapter that he had already written about in *al-Fakhrī*. That earlier explanation takes up three pages of the published edition, and he mentions at the end, "I have already

---

147 On the five-part structure of the book, see Sesiano (1977a, 298–300).

written a book that gives a thorough account of *al-istiqrā*".[148] So the long treatment he gives the topic in *al-Badīʿ* is at least his third time presenting it in a book.

Where Book II was concerned with finding, for example, that the square root of the polynomial we write as $x^8 + 2x^6 + 11x^4 + 10x^2 + 25$ is "a *Māl Māl* and a *Māl* and five units" (al-Karajī 1964, 49.– 5), Book III is concerned with

> looking for roots of quantities which have a root in power,[149] like a *Māl* and five Things, and four *Māls* and ten Things and ten units, and what is similar[150] to them. Although these and similar quantities are irrational with respect to what is indicated in their formulation,[151] it is possible (for example) for a *Māl* with five of its roots to be a square. And for quantities which do not have a root according to what is indicated in their formulation, (it might be) possible or impossible.
>
> (al-Karajī 1964, 62.3)

Al-Karajī then spends the next 10 pages of Anbouba's edition running through different polynomial types and explaining how to make the *al-istiqrā'* assignment in each of them. Here is an example:

> For a quantity consisting of three types Equal to a square in which one of the types is a square, you confront it with a square which has in it the same as the quantity that has a root, in order for you to succeed with the confrontation.[152] For example, a *Māl* and five Things and ten dirhams Equal a square. You confront that with a *Māl* and twenty-five units less ten Things.
>
> (al-Karajī 1964, 65.11)

The goal is for the equation to simplify to two equated terms of consecutive powers, thus guaranteeing a positive rational solution. In this example, the equation is 1M 5T 10u = 1M 25u ℓ 10T, so the 1Ms cancel, and the value of the Thing can be found to be $\frac{5}{3}$ u. In all, al-Karajī covers over two dozen different types of polynomials across five chapters (the descriptions are ours):

Chapter 1: Monomials equal to a square or a cube
Chapter 2: Binomials equal to a square, with at least one appended term being a square
Chapter 3: Trinomials of consecutive powers equal to a square, with either the highest or lowest power term being appended and a square

---

148 (Saidan 1986, 165.19–168.14). These pages are translated into French in Woepcke (1853, 72–74).
149 We infer that "quantities which have a root in power" are polynomials that have an appended (i.e., not deleted, or in modern terms, not subtracted) term that is a perfect square.
150 For example, "four *māls* and twenty things" is similar to (*shabh*) "a *māl* and five things" since their ratio is a square. This does not extend the possibilities. The word comes ultimately from Definition 21 in Book VII of Euclid's *Elements*.
151 In modern terms, they are not the squares of other polynomials with rational coefficients.
152 I.e., to succeed in obtaining a rational solution after solving the equation.

Chapter 4: Binomials two (or four) ranks apart equal to a square, with no
appended term a square

Chapter 5: Quadratic three-term polynomials in which no term is a square

In the first three chapters, whenever the polynomial is not a single term or two terms
of consecutive powers, then the highest- or lowest-power term is an appended perfect square, so one can always find a rational solution. The last two chapters cover
types that may or may not have a solution, and these types are rarely encountered
when solving problems. Diophantus completes and solves forthcoming equations
following the rules described in *al-Badī* 160 times in the extant books of the *Arithmetica*, and only one of those times, in Problem IV^G.31, does he encounter an
equation type from the last two chapters. Al-Karajī does not include the rules from
Chapters 4 or 5 in the survey of *al-istiqrā'* in *al-Fakhrī*, and he only applies techniques from the first three chapters in his problems in that book. Also, we have
seen no other instance of the application of a rule from Chapters 4 and 5 in the
works of other Arabic algebraists. We discuss this further in Appendix 2.

The rules described in Chapters 4 and 5 fall into three basic types. In the first
type, a similarity condition on the coefficients allows him to reverse the roles
of the *Māl* and the square to obtain a solution. In the second type one arrives at
a three-term composite equation, so one needs to make sure that the discriminant is a perfect square in order to obtain a rational solution. In the third type
al-Karajī attempts to use the method he had given for the first type, but he com-
mits an error.

In his first example in Chapter 4, al-Karajī equates "any *Māl*s with a number
excluded from it that is similar to its number" to a square. These are, in modern
notation, polynomials of the form $ax^2 - b$ for which the ratio of a to b is a square.
His rule is to solve for the square, then reverse the rules of the *Māl* and the square
to obtain a new forthcoming equation. Here he works it out in the context of the
forthcoming equation "three *Māl*s less twelve units Equal a square" (al-Karajī
1964, 66.26):

> If you take a third of everything and you restore, you get: the *Māl* is Equal to
> a third of a square and four units. If you confront it with a *Māl* and four units
> less four Things, then the one Thing is six. And since we confronted a third
> of a square and four units with the square, its root is a Thing less two, so the
> root of that is four, and the *Māl* is sixteen. And if you take it three times and
> you subtract twelve units from it, the remainder is thirty-six.

Starting with 3M ℓ 12 = $\Box$, he solves for 1M to get 1M = $\frac{1}{3}\Box$ 4. Reversing the
roles of the square and the *Māl* gives $\frac{1}{3}$M 4 = $\Box$. Equating the right side with the
square 1M 4 ℓ 4T gives 1T = 6, which makes the $\Box$ 16, which is the original *Māl*.

For the second polynomial type, one needs to choose the square so that the discriminant of the resulting composite equation is a perfect square. To do this, the
rule for solving three-term equations must be applied. Diophantus finds himself in
this situation in *Arithmetica* Problem 31 of the Greek Book IV, where 3N 18u ℓ 1P

must be made a square: "I am reduced to looking for a square which, if it receives in addition 1 unit, and has been taken 18 times, and it receives in addition 2 4' units, makes a square". In modern notation, Diophantus wants to equate $3x + 18 - x^2$ with $(ax)^2$ for some a. The discriminant in this equation will be $2\frac{1}{4}+\left(a^2+1\right)18$, which must be a square. He then sets up a new equation, where the unknown a^2 is named "1 Power":

> Let the square be 1 Power. This, together with 1 unit, and having been taken 18 times, and having received in addition 2 4' units, ⟨makes⟩ 18 Powers, 20 4' units, (which must be) equal to a square. Everything 4 times. 72 Powers, 81 units become equal to a square.

Now he has a new forthcoming equation that can be solved easily, since 81 is a square. This way he finds a^2 to be 324, and the discriminant is the square of $76\frac{1}{2}$.

Al-Karajī explains this process of examining the discriminant for various two- and three-term polynomials in *al-Badī'*. He writes for his first example (al-Karajī 1964, 69.6):

> If (someone) said, three Cubes and five Things equal a square. You confront it with *Māl*s (so that) if you halve its number and you multiply it by itself, and you cast away from the outcome what you get from multiplying the number of cubes by the number of things, the remainder is a square.

In modern notation, equating $3x^3 + 5x$ to $(ax)^2$ and then dividing by x (al-Karajī does not mention the reduction in power) gives a composite quadratic equation. He gives the rule for the discriminant not from the standard Arabic rule for solving that equation, but from Diophantus's rule that does not call for normalization. He also uses Diophantus's rule and not the Arabic rule in the other two instances in which he states the procedure for calculating the discriminant: for the forthcoming equations $3x^3 - 3\frac{1}{3}x = \left(ax\right)^2$ and $3x^2 + 3x + 10 = (ax)^2$ (al-Karajī 1964, 69.13, 70.3). This suggests that al-Karajī did not devise this method himself but that he took it from the *Arithmetica*. After all, apart from his three examples introducing the composite equations in *al-Fakhrī*, in every other instance in which he solves a three-term equation across all of his published texts, he follows the standard Arabic rule of first setting the number of *Māl*s to one. We already knew that the "method of Diophantus" that he gives in *al-Fakhrī* must have been the derivation that Diophantus himself gave for these equations in a lost part of the *Arithmetica*, so this method of assigning the side of a square in the part on *al-istiqrā'* very likely also comes from the *Arithmetica*, perhaps the same lost part where Diophantus derived his rules, or maybe from the solutions to particular problems. This, then, suggests that perhaps even the first method al-Karajī gives in Chapter 4, where he reverses the roles of the *māl* and the square, may also have originated in the *Arithmetica*. But without any further evidence, these will remain conjectures. It is still entirely possible that these rules were devised by al-Karajī from his own handling of indeterminate problems.

In the third rule al-Karajī is faced with a trinomial that must be made equal to a square. He attempts to apply the first rule by reversing the roles of the *Māl* and the square, but he neglects to also convert the Things in the equation to roots of the square, so he does not get a rational solution. He was aware of his mistake, for he wrote: "Truly, it is necessary in this problem that all the *Māl*s in the problem be equal, and likewise the Things, for if they were not equal, the solution would be wrong, as it is wrong in this problem" (al-Karajī 1964, 71.6).

In the last two parts of *al-Badīʿ* al-Karajī explains different approaches to solving problems, both determinate and indeterminate. Several example problems are given, many of them taken from *al-Fakhrī*, and three of these originate in the *Arithmetica*: Problems III.32/I.21, III.40/II.11, and III.1/II.20 (al-Karajī 1964, 73.10, 75.15, 78.3). Similarly, the one problem of Diophantus that is solved in al-Karajī's arithmetic book *al-Kāfī fī l-ḥisāb*, also written after *al-Fakhrī*, is *Arithmetica* Problem I.16, which had also appeared in *al-Fakhrī* as Problem III.24 (al-Karajī 1986, 182).

In the centuries that followed, the influence of Diophantus in the Islamicate world persisted in part indirectly through al-Karajī's impact on other authors. Many people condensed, commented on, reworked, and improved on different aspects of *al-Fakhrī* and *al-Badīʿ*, and some of their works were in turn read and commented on by still more authors. The rule of Diophantus for solving composite equations was later given, along with the standard Arabic rule and other rules, by al-Samawʾal, Ibn al-Yāsamīn,[153] Ibn Fallūs, Ibn al-Bannāʾ, Fakhr al-Dīn al-Ḥilātī, Ibn al-Hāʾim, and Ibn al-Majdī, and Ibn Ghāzī.[154] Fakhr al-Dīn al-Ḥilātī and Ibn Fallūs also reproduced the "method of Diophantus" for deriving these rules, but without naming their source.[155] Problems of Diophantus that al-Karajī had included in his *al-Fakhrī* were re-copied by other authors, too. The most notable aurthor is Ibn al-Hāʾim, who, in his commentary on the poem of Ibn al-Yāsamīn, reproduced the following problems from *al-Fakhrī*, all originating in the *Arithmetica*: II.50; III.24,25,41; IV.12,46; and V.3,37 (Ibn al-Hāʾim 2003, French 80–84). However, the vast majority of problems, both determinate and indeterminate, that we find in later Arabic books come from the Arabic tradition, and not from Diophantus.

153 Ibn al-Yāsamīn explained the Diophantine rules in his *Grafting of Opinions* (Zemouli 1993, 242.1, etc.), but more important are the two lines he gave them in his popular *Poem on Algebra*, which spawned the explanations in the many commentaries that that work inspired: "Or multiply the *māl*s by the numbers / and proceed in the same way; then divide the corresponding root from before by / the number of *māl*s, and take from all this the value" (Ibn al-Hāʾim 2003, 52.38–39). These commentaries include the works of Ibn al-Hāʾim and Ibn al-Majdī cited in the next footnote.

154 (Rashed 2021, 68.16); (Ibn Fallūs, MS Landberg 199, f. 12v, l. 1); (al-Ḥilātī MS, p. 23, l. 2); Ibn al-Bannāʾ: his *Algebra* (Saidan 1986, 549.11ff); (Ibn al-Hāʾim 2003, 107); (Ibn al-Majdī MS, f. 159r, l. -2); (Ibn Ghāzī 1983, 264.7). See the previous footnote for Ibn al-Yāsamīn.

155 Fakhr al-Dīn al-Ḥilātī was discussed in Section 2.6.4, and we discuss Ibn Fallūs in Section 3.3.10.

And last, although the technique of *al-istiqrā'* had been routinely applied to complete forthcoming equations before the translation of the *Arithmetica*, the many later authors who wrote sections inspired by *al-Badī'* explaining the technique can be said to have spread the influence of Diophantus. These include al-Samaw'al, Ibn Fallūs, Fakhr al-Dīn al-Ḥilātī, Ibn al-Bannā', Ibn al-Khawwām, al-Fārisī, Ibn al-Hā'im, and Ibn al-Majdī.[156] Al-Samaw'al's coverage is the most comprehensive, totaling 18 pages in the published edition (Rashed 2021, 78–95).

3.3.7 Al-Zanjānī

'Izz al-Dīn al-Zanjānī was a Persian scholar active in the middle of the thirteenth century who worked in a number of areas, including jurisprudence, language, poetry, and mathematics. One of his books, titled *Balance of the Equation in the Science of Algebra*,[157] is heavily indebted to both Abū Kāmil and al-Karajī. After eight chapters explaining the rules of algebra, he gives two long chapters of worked-out problems. In Chapter 9 al-Zanjānī solves 189 determinate problems, and in Chapter 10 he solves 126 indeterminate problems. Many of the problems in Chapter 10 come from Diophantus via al-Karajī. We have not seen al-Zanjānī's text, but Sammarchi (2019) translates Problem 4 from Chapter 10, which asks for a number such that if it is subtracted from 4 and from 5, then the remainders are squares. This problem corresponds to *Arithmetica* II.12, where the given numbers are 9 and 21, and to al-Karajī's *al-Fakhrī* III.41, where the given numbers are 5 and 3. It is also stated but not solved in al-Karajī's *al-Badī'* (al-Karajī 1964, 75.28), where the numbers are again 4 and 5. Diophantus solves the problem by *al-istiqrā'*, while al-Karajī works it in *al-Fakhrī* by the double equality. Al-Zanjānī solves it by *al-istiqrā*, which might suggest that he took his solution directly from the translation of the *Arithmetica*. But because his given numbers are 4 and 5, and because his wording of the enunciation is very close to the enunciation as stated in *al-Badī'*, it seems more likely that he wrote his solution based in the instructions for *al-istiqrā'* given in that book.

3.3.8 Ibn al-Haytham

It was also around the middle of the thirteenth century that the Syrian physician Ibn Abī Uṣaibī'a wrote his *Lives of the Physicians* ('*Uyūn al-anbā*'). At the end of his

---

156 (Rashed 2021, 78ff); Ibn Fallūs (590/1194–637/1239): *Kitāb niṣāb al-ḥabr fī ḥisāb al-jabr* (Berlin, MS Landberg 199, f. 10v, l. 8); Fakhr al-Dīn al-Ḥilātī (1197–1282): (Tehran, MS 4409, p. 32, l. 4); Ibn al-Bannā': his *Algebra* (Saidan 1986, 556.13); Ibn al-Khawwām (1245–1325): *Al-Risāla al-shamsiyya fī l-qawā'id al-ḥisābiyya* (Paris, MS BnF 2470, f. 66r, l. 7); al-Fārisī (1266/7–1319): (al-Fārisī 1994, 558.15), Ibn al-Hā'im (ca. 1355–1412): (Ibn al-Hā'im 2003, French 69, Arabic 209.5); and Ibn al-Majdī's *Ḥāwī al-lubāb wa-sharḥ talkhīṣ a'māl al-ḥisāb* (1431) (British Library, Add MS 7469, f. 163v, l. 7).

157 *Qisṭās al-mu'ādala fī 'ilm al-jabr wa'l-muqābala*. For al-Zanjānī's biography, see Yadegari (1980), and for the structure of al-Zanjānī's book, see Sammarchi (2019).

biography of Ibn al-Haytham he writes: "There is also a catalogue (*fihrist*), which I found, of the books of Ibn al-Haytham (that he wrote) up to the end of the year 429 (i.e., October 1038). This includes:" followed by a list of 92 works, the penultimate being: "Notes recorded by the physician Isḥāq ibn Yūnūs in Egypt from Ibn al-Haytham concerning Diophantus' book *On Problems of Algebra* (*al-jabr*)".[158] Sesiano adds that Isḥāq ibn Yūnūs was a student of Ibn al-Haytham (Sesiano 1982, 11 n. 26).

3.3.9 Al-Samaw'al

Al-Samaw'al ibn Yaḥyā al-Maghribī, a Jewish native of Baghdad who later converted to Islam, is best known for his *Dazzling Book on the Science of Calculation* (*Kitāb al-Bāhir fī 'ilm al-ḥisāb*),[159] composed ca. 1150 CE when he was just 19 years old. *Al-Bāhir* is largely his answer to al-Karajī's *al-Badī'*. As Anbouba noted (1978a, 93), "Al-Samaw'al's intention in writing the *al-Bāhir* was to compensate for the deficiencies that he found in al-Karajī's work and to provide for algebra the same sort of systematization that the *Elements* gave to geometry". Al-Samaw'al comments not only on al-Karajī's *al-Badī'* but on the works of several other authors, including Abū Kāmil and Diophantus. It is from this book that we know that al-Samaw'al also wrote a commentary, now lost, on the *Arithmetica*.[160] He writes in the very last remark he makes in the book:

> And as for those who want to practice the process and the investigation regarding the nature of the strategies for (solving) the problems in light of the diversity of their formulations, we have commented on its practice on the book of Diophantus the Alexandrian. It is comprehensive (*muḥīṭ*) in (this) part of the workings of this art.
>
> (Rashed 2021, 240.7)

Al-Samaw'al mentions Diophantus in three other places in the book. In the first instance, he states the arithmetical proposition underlying the rule of the double equality, but with a variation on the version given by al-Karajī:

> Statement of Diophantus and proof of al-Samaw'al. If one divides the difference between each of two squares by whatever number, and one adds the outcome of the division and the divisor, then half of that is a root of the greater of the two squares, and if one takes half of the difference between the divisor and the outcome of the division, then it gives a root of the smaller square.

---

158 (Savage-Smith, Swain, and van Gelder 2020, [14.22.5.2]), their translation.

159 The title is taken from the title page of MS Aya Sofia 2718.

160 This does not mean that al-Samaw'al wrote the commentary before he wrote *al-Bāhir*. The latter was probably revised by the author at different points in his career, and the passage translated is most likely a later addition (Anbouba 1978a).

> Al-Samaw'al said: We have found that Diophantus used this idea in his treatise and later algebraists used it in this art, and none of them gave a proof of it in what has come down to us of their writings.
>
> (Rashed 2021, 103.1)

He was apparently not satisfied with the explanation al-Karajī had given, translated above on page 153. Al-Samaw'al then gives an arithmetical proof in the style of those in Euclid's Books VII–IX and which relies on *Elements* II.6. He makes clear in the proof that this "whatever number" is either the sum of the roots of the two squares, or the difference between the roots.

Later in the book we find: "Diophantus said: for any number or quantity that is composed of two square numbers, its double is composed of two square numbers and its half is composed of two square numbers" (Rashed 2021, 137.16). Here a number "composed of two square numbers" is the sum of those squares. This proposition was known to al-Khāzin, at the end of the first section of his three-part treatise we discussed earlier (Anbouba 1979, 147–48; Sesiano 1982, 12). It is not found, however, in the extant *Arithmetica*. Al-Samaw'al would have been familiar with the contents of Diophantus's work since he wrote a commentary on it, so most likely this proposition was given in a now lost part of the work.

Later in the book al-Samaw'al reviews different kinds of arithmetical problems and their solutions, together with proofs. Two of these come from Diophantus, the first being Problem I.26, but whose solution is given and proven without the use of algebra, i.e., without naming an unknown and setting up and solving an equation as Diophantus did:

> An example that has one answer:[161] We want to find a number such that if we multiply it by two given numbers, gives from its multiplication by one of them a square number, and from its multiplication by the other a side of that square.
>
> Let the two numbers be 5 and 200. We want to find a number such that if we multiply it by 200 it results in a square, and if we multiply it by five it results in a side of that square.[162] Let us divide the two hundred by a square of the five. The result of the division is eight, which is the desired number.
>
> A proof of that is that the eight multiplied by 25 results in two hundred, since dividing the two hundred by 25 results in eight. The product of 8 by 200 is equal to the product of 25 by 8, then by 8. But the product of 25 by 8, then by 8 is equal to the product of 8 by 8, then by 25. So the product of the eight by the two hundred is equal to the product of a square of the eight by a square of the five. But the product of a square of 8 by a square of the five is equal to

---

161　That is, as opposed to an indeterminate problem.

162　This is the instantiated reformulation of the enunciation (διορισμός, "definition of goal"). Because it does not exist in the preserved Greek text, Sesiano concludes that the Arabic problem derives from a reworking of the original text of Diophantus, which, according to him, happened in late antiquity and resulted from a "major commentary". See Section 3.2.1 for a discussion of this "major commentary".

a square of the product of the eight by the five. So the product of the eight by the five is equal to a root of the product of the eight by the two hundred, and that is what we wanted to show.

(Rashed 2021, 215.8)

Later, Fakhr al-Dīn al-Ḥilātī would state and solve this problem the same way, but in its uninstatiated form. He may have taken it from al-Samaw'al (al-Ḥilātī, Tehran, MS 4409, f. 37v, l. 2).

This next problem is common in Arabic algebra and is not in the extant *Arithmetica*. It is unlikely to have been in Diophantus's treatise since it is a simple determinate problem that would have been included in Book I. Instantiated versions of this problem are common in Arabic algebra, so al-Samaw'al's version can be read as a practical Arabic problem transformed into the style of those in the *Arithmetica*.

> And an example of what has a requirement on its conditions: we want to find two numbers such that the sum of their squares is equal to a given number, and the product of one of them by the other is equal to another given number. In this problem there is a necessary condition, which is that that the number equal to the sum of their squares (must) exceed double the surface which is contained by them, and this is clear from proposition 7 in the second Book of the work of Euclid.
>
> (Rashed 2021, 215.18)

The next problem is *Arithmetica* I.16, which al-Samaw'al solves three different ways, all by algebra. The second solution is interesting in that al-Samaw'al works with three named independent unknowns: a Thing (*shay'*), a Number (*'adad*), and an Unknown (*majhūl*), and he sets up a system of three equations. The third solution is the one found in the *Arithmetica* and is explicitly attributed to Diophantus.

> And a second example: we want to find three numbers such that the sum of any two of them is equal to a given number. It is necessary that half the sum of the three given numbers be greater than each one of them.
>
> Let us assign the first and the second, if added, to be 25; and the second and the third to be 35; and the third with the first to be 30. It is clear that half the sum of these three numbers, which is 45, is greater than each of them.
>
> We make the first a Thing, thus, the second is 25 less a Thing. And since the second and the third are 35, if we subtract the second, which is 25 less a Thing, from 35, it leaves ten and a Thing, and that is the third number. Then we add the first and the third to get ten units and two Things, and that Equals 30 units. If we confront,[163] it gives the one Thing is ten, so it is ten. The sec-

---

163 By "confront" he means simplify and solve the equation.

ond is 25 less a Thing, so it is fifteen, and the third is ten units and a Thing, so it is 20. So we have found the sought-after numbers.

If we wish, we can make the first a Thing, and the second a Number, and the third an Unknown, and we add the first and the second to get a Thing and a Number Equal 25. And we add the second and the third to get a Number and an Unknown Equal 35. And we add the third and the first to get an Unknown and a Thing Equal 30. These three quantities (1T 1N, 1N 1U, 1U 1T) are equal to three other quantities (25, 35, 30) and if equals are added to equals, they all become equal. So the sum of the three unknown quantities is equal to the sum of the three quantities Equated to them. Thus, two Things and two Numbers and two Unknowns Equal 90 units (2T 2N 2U = 90). And if we return each one to its half, then a Thing and a Number and an Unknown Equal 45 units (1T 1N 1U = 45). So we have worked out that the sum of the three numbers is 45. We cast away from it the sum of the second and the third, which is 35, leaving ten, which is the first quantity. We cast it away from the sum of the first and the second, which is twenty-five, leaving 15, which is the second number. We cast it away from the sum of the second and the third, which is 35, leaving the third (as) twenty.

And if we wish, we can work it out as Diophantus did: we make the sum of the three numbers a Thing. The first is a Thing less thirty-five, since the second and the third are 35. And the second is a Thing less 30, since the first and the third are 30. And the third is a Thing less twenty-five, since the first and the second are 25. We add the three numbers to get: three Things less 90 units Equal one Thing. So if we restore and confront, it gives two Things Equal 90. So the one Thing is 45, which is the sum of the three numbers, since we assigned it to be a Thing. (Now) complete the work as I described.

(Rashed 2021, 216.3)

Al-Samaw'al then poses a more complex version of the same problem: find 10 numbers given the sums of every combination of 6 of them. He lists these 210 given numbers in a long table, and he solves the problem, finding the numbers to be 1, 4, 9, 16, 25, 10, 15, 20, 25, and 5.

3.3.10 *Ibn Fallūs*

Ibn Fallūs[164] (590/1194–637/1239), a mathematician who worked in Cairo and Damascus, mentions Diophantus twice in his short book *Preparation for Writing on*

---

164 His full name is given in Rosenfeld and İhsanoğlu (2003, 206) as Shams al-Dīn Abū'l-Ṭāhir Ismāʿīl ibn Ibrāhīm ibn Ghāzī al-Māridīnī. See Brentjes (1990, 240) for more on his name. Rashed in Diophantus (1984, 3, xiii) calls him Shams al-Dīn al-Māridīnī.

Calculation in Algebra (*Kitāb niṣāb al-ḥabr fī ḥisāb al-jabr*).[165] The first instance occurs where he gives the solutions to the three composite equations. For each equation, Ibn Fallūs first states the standard Arabic rule, and then a solution that he attributes to Diophantus that is a modification of the "method of Diophantus" that we saw in al-Karajī's *al-Fakhrī*. Here is his treatment of the first composite equation:

> *Māl*s and Things Equal a number. The method is that you add the number to a square of half the number of Roots, and you subtract from a root of the outcome half the number of Roots, so what remains is the Thing. And if you wish, work it out by the method mentioned by Qāḍī al-Māristān as being taken from the books of Diophantus, which is that you add to the *Māl* and the Roots something so that the sum has a root, and likewise to the number. Then you confront, so the Root emerges.

For the method attributed to Diophantus, starting with $x^2 + ax = b$, add $\left(\frac{1}{2}a\right)^2$ to the $x^2 + ax$ so that it is the square $\left(x + \frac{1}{2}a\right)^2$. Add this same $\left(\frac{1}{2}a\right)^2$ to b. One then "confronts", i.e., simplifies, the equation $x^2 + ax + \left(\frac{1}{2}a\right)^2 = \left(\frac{1}{2}a\right)^2 + b$, by first equating the square roots of both sides, and then subtracting the $\frac{1}{2}a$. This differs from the rule stated by al-Karajī and Fakhr al-Dīn al-Ḥilātī, who, knowing that $x^2 + ax + \left(\frac{1}{2}a\right)^2$ is the square of $x + \frac{1}{2}a$, substitute the $x^2 + ax$ with b, so that $b + \left(\frac{1}{2}a\right)^2$ is the square of $x + \frac{1}{2}a$, from which the solution follows. So where al-Karajī and al-Ḥilātī work within the context of the square of $x + \frac{1}{2}a$ and use the equation to make a substitution, Ibn Fallūs adds the $\left(\frac{1}{2}a\right)^2$ to the two sides of the equation and solves it in that context.

Here is how the second equation type is explained by Ibn Fallūs:

> And the second: Roots Equal *Māl*s and numbers. And its method is that you subtract the number from a square of half the number of Roots, and you take a root of the remainder. That you add to half the number of Roots, or subtract from it, so the Root emerges. And if you wish, work it out by the other above-mentioned (method), which is that you subtract from the two Equated parts a number (consisting of) the Roots less a number so that the remainder is a square, then you confront, so the Root emerges.

For the second method, starting with $ax = x^2 + b$, subtract $ax - \left(\left(\frac{1}{2}a\right)^2 - b\right)$ from the $x^2 + b$ to get $x^2 - ax + \left(\frac{1}{2}a\right)^2$, which is the square $\left(x - \frac{1}{2}a\right)^2$. Subtract likewise $ax - \left(\left(\frac{1}{2}a\right)^2 - b\right)$ from the other side of the equation, ax, to get $\left(\frac{1}{2}a\right)^2 - b$. From here the equation is solved by taking square roots and restoring the x. Again,

---

165 See Brentjes (1990) for bio-bibliographical information on Ibn Fallūs.

where al-Karajī and al-Ḥilātī worked in the context of the square of $x - \frac{1}{2}a$, Ibn Fallūs works in the context of the equation.

> And third: *Māl*s Equal Roots and numbers. And its method is that you add a square of half the number of Roots to the number and you take a root of the outcome. You add it to half the number of Roots, so what comes out is the Root. And if you wish, work it out by the above-mentioned method, which is that you subtract from the *Māl* the Roots and an excluded number so that it literally becomes a square, and you confront its same on the other side, so the Root emerges.

Again, in the second method, starting with $x^2 = ax + b$, subtract $ax - \left(\frac{1}{2}a\right)^2$ from the x^2 so that it becomes the square $\left(x - \frac{1}{2}a\right)^2$. Likewise, subtract $ax - \left(\frac{1}{2}a\right)^2$ from the $ax + b$ to get $b + \left(\frac{1}{2}a\right)^2$. The solution then follows by equating their roots and then restoring the x.

These same derivations/solutions involving completing the square are explained more fully in Ibn al-Bannā''s late thirteenth century book on algebra,[166] with the only minor variation that for the last type, Ibn al-Bannā' subtracts the ax from x^2 first and then adds the $\left(\frac{1}{2}a\right)^2$, where Ibn Fallūs does this in one step (Oaks 2018a). The derivation for the second composite type is not obvious, so we are led to suspect that Ibn al-Bannā may have learned it directly or indirectly from Ibn Fallūs.

Qāḍī al-Māristān (1050–1141) is one of the names by which Abū Bakr Muḥammad ibn ʿAbd al-Bāqī al-Baghdādī (or al-Mawṣilī) al-Faraḍī was known. He was a jurist mathematician and astronomer who became known in medieval Europe for the Latin translation of his commentary on Book X of Euclid's *Elements*. He is also author of a treatise on algebra preserved in a manuscript in Damascus, which might contain some reference to Diophantus (Rosenfeld and İhsanoğlu 2003, 170–71). Al-Māristān may have learned of this method from al-Karajī's *al-Fakhrī*, and that somewhere in the transmission to Ibn Fallūs the technique was modified so that the calculations take place in the context of the equation.

Later, regarding cubic equations, Ibn Fallūs wrote:

> Some of them can be solved by means of the six well-known (equations), and those which cannot be solved by means of them must be carried out by the method of ʿUmar al-Khayyām (which is) taken from the books of Diophantus. If it still cannot be solved, then it must be solved by the table devised by the Imām Sharaf al-Dīn al-Muẓaffar ibn Muḥammad al-Ṭūsī and and its associated solution.
>
> (Ibn Fallūs, MS Landberg 199, f. 13v, l. 9)

---

166 *Kitāb al-uṣūl wa-l-muqaddimāt fī l-jabr wa-l-muqābala* (*Book on the Fundamentals and Preliminaries in Algebra*).

The "method of 'Umar al-Khayyām" here apparently means the reduction of the degree of a three-term equation that has no constant term, like reducing an equation of the type "a Cube and *Māl*s Equal Roots" to the quadratic type "a *Māl* and Roots equal a number", which is one of the six equations. This kind of reduction is routinely performed by Diophantus, and it is demonstrated geometrically by al-Khayyām. Al-Khayyām does not mention Diophantus in his extant works, though he seems to point to Diophantus in his remark that no solution to irreducible cubic equations was known from the works of "the ancients".[167] Sharaf al-Dīn al-Ṭūsī (d. 1213) is the author of the treatise *Problems of Algebra (Masā' il al-jabr wa-l-muqābala)*, where, like al-Khayyām, he classifies and solves the 25 equations of degree three or less. In addition to geometric solutions, he solves irreducible cubic equations numerically by means of tables.[168]

3.3.11 Ibn al-Qifṭī and al-Nūayīrī

It is worth translating what the bio-bibliographer Ibn al-Qifṭī (d. 646/1248) wrote about Diophantus. In al-Zawzanī's 647/1249 reworking of Ibn al-Qifṭī's book, titled *History of Learned (Men)*, we find:

> Diophantus, a Greek of Alexandria. Excellent, consummate, well-known in his time. His composition is *Art of Algebra (al-jabr)*, a famous, celebrated book, rendered into Arabic, and the people of this art work according to it (*'alāhu*). And when one studies it deeply, one discerns him to be a man of vast knowledge in this branch (of learning).
>
> (Lippert 1903, 184.11)

And last, Diophantus is also mentioned by the historian of literature al-Nūayīrī (1321), who calls the *Arithmetica* a *Book on Calculation (Kitāb al-ḥisāb)* (Diophantus 1984, 3, xiii–xiv).

3.4 Diophantus in Byzantium

In welcome contrast to the meager references to Diophantus in Byzantine sources of the ninth and tenth centuries, Greek interest in our algebraist becomes abruptly abundant in the eleventh century, beginning with Michael Psellus.

3.4.1 Michael Psellus

In an eleventh-century letter to an anonymous recipient, the Byzantine polymath Michael Psellus cites the names of powers of the "Egyptian method", explaining

---

167 (R. Rashed and Vahabzadeh 1999, 147.6, 117.8). See Brentjes (1990, 246) for a French translation of the passage in Ibn Fallūs.

168 For his numerical solutions, Sharaf presumes that his equations have positive integer answers, though his technique would work for aproximations, too. His book is edited and translated in Sharaf al-Dīn al-Ṭūsī (1986).

their ranks as he goes (Diophantus 1893–95, II, 37).[169] He begins with the Euclidean definition of "unit" (μονάς), followed by both the Diophantine (algebraic) and the Euclidean definitions of "number" (ἀριθμός), and which he also identifies with the arithmetical term "side" (πλευρά). This mixing of sources continues with the higher powers (we refrain from capitalizing the algebraic names from here to the end of this section):

A most elegant service in the organization in relation to the numbers offers also the Egyptian method of numbers,[170] by means of which the problems concerning the analytic are treated. First you must understand the names of the numbers (given) by them, and what power each one possesses. According to them, as well as by us, a unit (μονάς) is (that) according to which each of the beings is called one. A number (ἀριθμός) specifically is called by them that which on the one hand does not possess any characteristic, and on the other has in itself an indefinite multitude of units; this number is also called by them a side (πλευρά). A power (δύναμις) is when a number is multiplied by itself; this is also called a square number (τετράγωνος ἀριθμός). Thus, if we assume the number (to be) 2 units, the power will be 4 units. A cube (κύβος) is when a number is multiplied by the power; for example, if we assume the number (to be) 2 units, its power (is) the 4; if (this) is multiplied by the side, the 2, the number 8 will be produced, which is clearly a cube. A power-power (δυναμοδύναμις) is when the power is multiplied by itself; for example, the 4 by itself, and it becomes the (number) 16. A power-cube (δυναμόκυβος) is when the power is multiplied by a cube, as for instance the 4 by the 8, and it becomes 32. That which is called a first unspoken (ἄλογος πρῶτος), for it is neither a square (τετράγωνυς) nor a cube (κύβος), is also a fifth number (ἀριθμὸς πέμπτος). For, first (is the) number simply, second (the) power, third (the) cube, fourth (the) power-power, and fifth this one, the power-cube. A cube-cube (κυβόκυβος) is when a cube, multiplied by itself, will produce (some) number. And a second unspoken (ἄλογος δεύτερος) number is when a power is multiplied by a first unspoken; for the power being 4 units, as was said, and the first unspoken 32 units, their product will be 128 units, which is called second unspoken. This is also called a seventh number (ἀριθμὸς ἕβδομος). A quadruple power (τετραπλῆ δύναμις) is when a power is multiplied by a cube-cube. And an evolved cube (κύβος ἐξελικτός) is when a power is multiplied by a second unspoken. And the homonymous parts of these numbers will be called after these. Of the number,

---

169 The letter has also a second part dealing with measurement of solid shapes. The content of this part is very much like the collection "De mensuris", which has been transmitted under the name of Hero of Alexandria (Hero 1914, 164ff, 1864, 188ff).

170 ἡ κατ' Αἰγυπτίους τῶν ἀριθμῶν μέθοδος. Cf. the corresponding expression of Theon, who refers to the problem-solving method taught in the *Arithmetica* by calling it "the method of the Diophantine numbers" (see Section 3.2.1).

Table 3.7 Names of powers in Michael Psellus.

	Algebraic	Alternate algebraic	Arithmetical
0	unit		
1	number		side
2	power		square number
3	cube		
4	power-power		
5	power-cube	first unspoken/fifth number	
6	cube-cube		
7		second unspoken/seventh number	
8	quadruple power		
9	evolved cube		

inverse number[171] (ἀριθμοστόν); of the power, inverse power (δυναμοστόν); of the cube, inverse cube (κυβοστόν); of the power-power, inverse power-power (δυναμοδυναμοστόν); of the power-cube, inverse power-cube (δυναμοκυβοστόν); of the cube-cube, inverse cube-cube (κυβοκυβοστόν).

(Diophantus 1893–95, II, 37–38)

To sort out all of these names, we organize them in Table 3.7.[172]

The terms "first unspoken" (ἄλογος πρῶτος) and "second unspoken" (ἄλογος δεύτερος) correspond well with Luca Pacioli's *primo relato* and *secundo relato*, and Rudolff's *sursolidum* and *bissursolidum* that we described in Section 3.1.2, so they surely originate in a multiplicative scheme for the naming of powers. If this multiplicative system was like those in Pacioli and later European algebra, then the first four powers would coincide with the standard additive system, and the sixth power would be a "power-cube" (δυναμόκυβος). Perhaps Psellus did not list δυναμόκυβος for the sixth power because he did not want to readers to confuse it with the additive fifth power of the same name. Similarly, the multiplicative eighth power would be a "triple power" (τριπλῆ δύναμις) or something equivalent, but this would be another name for the sixth power in the additive system.

If the fifth power can be called a "fifth number" (ἀριθμὸς πέμπτος) and the seventh a "seventh number" (ἀριθμὸς ἕβδομος), then it stands to reason that such names could be formed starting with the first power, which would be a "first number" (ἀριθμὸς πρῶτος), followed by "second number" (ἀριθμὸς δεύτερος), etc. But such names may never have been coined. Recall the seventh and eighth powers in Ibn al-Fath, which are called "eighth proportional" and "ninth proportional". These names seem to have been devised for powers whose names would

---

171 Just as we construct the word "seventh" from "seven", In Diophantus and others constructed ἀριθμοστόν ("numberth") from ἀριθμός ("number"), and similarly for the higher powers. We translate these as "inverse number", "inverse power", etc.

172 Compare this with the reconstruction given by Tannery (1920b, 279–80), who attributes all names not found in Diophantus, including the arithmetical terms, to Anatolius. Tannery also extended the "fifth number" and "seventh number" all the way back to "first number".

otherwise have been more complicated in his multiplicative scheme. Likewise, the terms "fifth number" and "seventh number" seem to belong to a multiplicative system like "first unspoken" and "second unspoken", or perhaps the two pairs were valid alternatives in the same scheme.

The extant Greek books of Diophantus's *Arithmetica* list the powers up to the sixth, and the books of the Arabic translation use the eighth and ninth powers. What is lacking, and very well may have been lacking in all 13 books of Diophantus's autograph, is a name for the seventh power. Psellus may have been in possession of a text or texts that show all of Diophantus's names, and which gave the eighth and ninth powers as "quadruple power" and "evolved cube". The former is additive and would not fit with the multiplicative names of the fifth and seventh powers.

Because Psellus simply listed names of powers from different sources, he seems to have been playing the role of lexicographer. Rather than present the powers as an introduction to their use, he gives all the names that he has seen. This would account for (i) the inclusion of the Euclidean definitions and the arithmetical "side" and "square number"; (ii) the lack of any attempt to differentiate between the additive and multiplicative series of names; and (iii) the omission of potentially confusing multiplicative powers ("power-cube", "triple power"). Last, it may be that the term ἐξελικτός ("evolved") should be understood as meaning "folded", so that κύβος ἐξελικτός is the folding of one cube on another, and again on another (like the way one folds a towel or a piece of paper).

As already said,[173] Tannery published the letter of Michael Psellus on the basis of two manuscripts, Laur. Plut. 58.29 and Scorial. gr. Y.III.12, both dated to the fourteenth century (Diophantus 1893–95, II, 37–42). In addition to these two there is a third manuscript carrying the same text, Vat. Urbin. gr. 78, dated to the fifteenth century (Moore 2005, 311).[174] Tannery was unaware of this codex which, according to the incipit, contains the treatise *On Music* of Manuel Bryennius, the commentary of Porphyry on the *Harmonics* of Ptolemy, and a number of the minor philosophical works (μικρὰ τινὰ συγγράμματα) of Michael Psellus. The text from the Vatican codex is essentially the same as the text transmitted by the manuscripts of Florence and Escorial, with the exception that the punctuation in the Vatican manuscript is executed with great care. We mention this because the punctuation helps clarify the meaning of the passage that immediately follows the list of powers, and which we translated previously in Section 1.1:

> Concerning this Egyptian method, Diophantus dealt with it more accurately, while the very learned Anatolius, having selected the most essential parts of the doctrine of that man, most succinctly addressed it to another Diophantus.[175] And if someone knows the methods therefrom, it is quite

---

173 See footnote 11.
174 Cf. M. Rashed (2013, 602 n. 14).
175 See Section 1.1 for Tannery's successive attempts to amend the text.

certain that he would solve some of the arithmetical problems, which are proposed in the metrical epigrams. Indeed, some of them are resolved by this theorem (θεώρημα) of the Egyptian analysis, while others by other. For it is required to divide the proposed number either in sesquitertian ratio,[176] or in sesquiquartan,[177] or in another like this. And from such a division, what is proposed will be made easily seen. And that ⟨is enough⟩ for you so far.

(f. 81r)

Although the Greek text presents several difficulties,[178] some conclusions can be drawn with relative safety. This Anatolius drew from Diophantus's work, from which he selected "the most essential parts" and presented it in a most succinct form. Also, it seems that Anatolius addressed his work to some other Diophantus – not to be identified with the author of the *Arithmetica*. On the identification of Psellus's Anatolius, see our discussion in Section 1.1.

This "Egyptian analysis" is the method taught by Diophantus, that is, premodern algebra. When an epigram is stripped of its quotidian context and read as a problem in numbers, it often corresponds to a problem in the *Arithmetica*. The sentence "some of them (i.e., metrical epigrams) are solved by this theorem of the Egyptian analysis, while others by other" explains that to solve an epigram one follows the procedure in the abstract version presented by Diophantus. For example, the epigrams discussed in Section 3.2.3 are solved by the procedure employed in Diophantus's Problem I.2, while the epigram we saw in Section 3.2.2 is solved by the procedure employed in Problem I.21. Moreover, the fact that such a correlation is mentioned in a text that concerns specifically Anatolius gives room for advancing the hypothesis that, besides the introductory chapter with the denominations of the powers of the unknown, the work of Anatolius contained also arithmetical problems in the form of metrical epigrams that he solved by algebra.[179]

The letter of Psellus should be considered in connection with a fragment carried by three manuscripts (Laur. Plut. 58.29, ff. 202r–v; Vat. Urbin. gr. 78, f. 85v; and Laur. Plut. 28.11, f. 67v), which preserves terms for the powers up to the 11th. Although all three manuscripts contain several works of Psellus, nowhere does any of them ascribe the fragment to him, and Paul Moore classifies it as

---

176 Two numbers a and b are in sesquitertian ratio if $a = 1\frac{1}{3}b$.

177 Two numbers a and b are in sesquiquartan ratio if $a = 1\frac{1}{3}b$.

178 The passage of Psellus is discussed, especially with regard to the question of who this Anatolius might be, among others, in Diophantus (1893–95, II, xlvii), Tannery (1912h, 535–37), Hultsch (1903, col. 1052–53), Heath (1910, 2), Klein (1968, 245–46 n. 149), Knorr (1993, 184), and M. Rashed (2013, 600–03). Cf. also O'Meara (1989, 23–25), Goulet (1994), and Bernard (2008).

179 On the relationship between problems stated as epigrams and the abstractly stated problems in Diophantus, see Christianidis and Megremi (2019), Christianidis and Sialaros (2022), and Christianidis (2008).

spurious in his monumental *Iter Psellianum* (2005, 554). The fragment is the following:

> Every side multiplied by itself is named a power; and by the power, is called a cube; and the power by itself, is called a power-power; and by the cube, a power-cube; and the cube by itself, is called a cube-cube; and by the power-power, is called a cube-power-power; and by the power-cube, a cube-power-cube; and the power-power multiplied by itself, is called a ⟨power-power-⟩power-power, while by the power-cube, a ⟨power-⟩power-power-cube; and the cube-cube by itself makes a cube-cube-cube-cube, while by the cube-power-power, a cube-cube-cube-power⟨-power⟩. And by proceeding one after the other in order you will find accurately the designation of each.[180]

The term "side" is arithmetical, while the term "power" is typically algebraic. We do not know if the author of this passage intended the terms to be algebraic, but in any case, the connection with algebra would have been impossible to overlook by readers familiar with the method.

It is important to note that the variety of designations for the powers above the fourth in the Cairo papyrus, in Diophantus and *Refutation*, in Psellus, and in the fragment just translated, shows that, just as in Arabic, Italian, and Renaissance European algebra, there were no set names for those powers. For example, the fifth power in Psellus is given three designations: power-cube, first unspoken, and fifth number. More telling is the eighth power, which in Psellus is a quadruple power, but in the fragment just translated is a cube-power-cube, or a ⟨power-power⟩-power-power. We see this also in Qusṭā's translation of Diophantus, where the eighth power is called a *kaʿb kaʿb māl* in Book 4 and a *māl māl māl māl* in Books 6 and 7. So the fact that Diophantus calls the fifth power of the unknown a power-cube in his introduction does not mean that he would not call it by some other name, such as a cube-power, somewhere else. After all, al-Karajī, who was deeply inspired by Diophantus, did just that. The technical terms in Greek algebra are "unit", "number", "power", and "cube". Beyond that one could construct the names of the powers in whatever way was convenient.

3.4.2 From Psellus to the Palaeologan Renaissance

Michael Psellus was not the only person of his time in Constantinople studying Diophantus. A close reading of the marginal annotations in the oldest

---

180 Πᾶσα πλευρὰ ἐφ' ἑαυτὴν πολλαπλασιασθεῖσα, καλεῖται δύναμις. ἐπὶ δὲ τῇ δυνάμει, λέγεται κύβος. ἡ δὲ δύναμις ἐφ' ἑαυτὴν, λέγεται δυναμοδύναμις. ἐπὶ δὲ τῷ κύβῳ, δυναμόκυβος. ὁ δὲ κύβος ἐφ' ἑαυτὸν, λέγεται κυβόκυβος. ἐπὶ δὲ τῇ δυναμοδυνάμει, λέγεται κυβοδυναμοδύναμις. ἐπὶ δὲ τῷ δυναμοκύβῳ, κυβοδυναμόκυβος. ἡ δὲ δυναμοδύναμις ἐφ' ἑαυτὴν μὲν πολλαπλασιαζομένη, λέγεται δυναμοδύναμις (sic!), ἐπὶ δὲ τῷ δυναμοκύβῳ, δυναμοδυναμόκυβος (sic!). ὁ δὲ κυβόκυβος, ἐφ' ἑαυτὸν μὲν ποιεῖ κυβοκυβοκυβόκυβον, ἐπὶ δὲ τῇ κυβοδυναμοδυνάμει, κυβοκυβοκυβοδύναμιν (sic!). καὶ καθεξῆς ποιῶν ἀκολούθως, εὑρήσεις ἀκριβῶς τὴν ἑκάστου ὀνομασίαν.

manuscript, Matritensis 4678, reveals that others in the imperial court in the latter part of the eleventh century had also indulged in study of the *Arithmetica* (Pérez Martín 2006, 451).[181] Moving forward a century, Ilias Nesseris (2014) has shown that two learned men of the late eleventh and early twelfth century may have had some knowledge of the work of Diophantus. Michael Pantechnēs, physician of the emperor Alexios I Komnēnos (r. 1081–1118), was once a student of Michael Italikos (d. before 1157), a teacher of philosophy, rhetoric, and medicine. Italikos wrote of his former student: "he had not neglected (to be acquainted with) the Egyptian art, which deals with additions, subtractions, and divisions of numbers".[182] If by "Egyptian art" is meant what Psellus calles the "Egyptian method" in the passage discussed earlier, then we have here an indirect testimony on the kind of mathematics practiced by Diophantus. And last, we have a report by Theophylact, Archbishop of Ohris (d. after 1107), that Theodōros Chryselios "knew about cubes, and explained to the ignorants about cube-cubes".[183]

We are better acquainted with the mathematics taught at the end of the twelfth century by Constantine Kaloēthēs in the Patriarchal School of Constantinople. We will see that the intellectual lineage of Kaloēthēs extends through four generations to George Pachymeres and that Diophantus regularly figures into this lineage over the course of the thirteenth century. To start, we know from a letter written by the intellectual Manuel Sarantēnos (or Karantēnos) to Kaloēthēs that the latter had taught him advanced mathematics, mentioning Diophantus explicitly.[184] We also know that Kaloēthēs taught mathematics, including chapters stemming directly or indirectly from the *Arithmetica*, by the fact that some 20 years later another student of his, Prodromos of Skamandros, taught mathematics to Nicephoros Blemmydes (Nikēphoros Blemmydēs, 1197–1269) through the works of Nicomachus, Diophantus, Euclid, and others. It is safe, therefore, to assume that Kaloēthēs possessed, or had access to, a manuscript of Diophantus.

The polymath Blemmydes was the leading scholar of the Empire of Nicaea[185] in the first half of the thirteenth century. After receiving his general education in

---

181 We have no positive evidence concerning the archetype of Matritensis. Tannery in Diophantus (1893–95, II, xviii), followed, among others, by Heath (1910), Vogel (1967), Herrin (2000, 29), and Folkerts (2006), suggested that it was a codex copied in the eighth or ninth century, probably in connection with the activity of Leo the Mathematician. Cf. Pérez Martín (2006, 451 n. 59).

182 Greek text quoted by Nesseris (2014, 291 n. 49) from Italikos (1972).

183 Greek text quoted by Nesseris (2014, 291–92 n. 50) from Theophylacte d'Achrida (1986).

184 "I can add numbers, and raise them high in a cube and a pyramid, and excel in the art of Diophantus (διοφαντήσω γενναίως), and find the power and the power-power, and the immutable unit, and multiply a deficit by a deficit and produce a strangest existence, of which you, highest of the wise (men), taught me". Greek text quoted by Nesseris (2014, 292 n. 51) from Criscuolo (1977).

185 The Empire of Nicaea was the Greek state founded in the western part of Asia Minor after the conquest of Constantinople by the Fourth Crusade, and which flourished during the Latin occupation of the capital (1204–61).

Nicaea, he wanted to continue his studies with more advanced subjects. Because he could not find a teacher, he crossed the border into the Latin area, where he was taught mathematics at the monastery of Skamandros by Prodromos. Blemmydes says in his *Autobiography* that this teacher instructed him in the quadrivial sciences of arithmetic, basic plane and solid geometry, advanced geometry, and astronomy, as well as optics, syllogistics, and physics (Blemmydes 1896, 5.1–7, 55.10–11). Blemmydes reports that he learned arithmetic specifically from the works of Nicomachus and Diophantus, the latter not in its totality, but what his teacher had mastered of it (1896, 5.1–4). Constantinides (1982, 8, 137) remarks that the curriculum described by Blemmydes cannot be explained without presupposing that a rich library was available at Prodromos's school in the 1220s. The possibility that a manuscript of the *Arithmetica* existed in that library (the manuscript of Kaloēthēs?) seems very likely.

3.4.3 Pachymeres, Planudes, and the Palaeologan Renaissance

Byzantine interest in the work of Diophantus found its culmination in the late thirteenth century, a period of literary flourishing referred to by historians as the "Palaeologan Renaissance". The peak of this renaissance occurred during the reign of Andronicus II (r. 1282–1328), the second emperor of the Palaeologan dynasty.

It was during this period, at the end of the thirteenth century, that real use of the work of Diophantus was made by two of the most distinguished scholars of the time, George Pachymeres and Maximus Planudes. Pachymeres (Geōrgios Pachymerēs, 1242–ca. 1310) was a polymath, prolific writer, and church official who had been educated by George Akropolites (1217–1282),[186] a student of Blemmydes. Around the year 1296 Pachymeres composed a voluminous work, the full title of which is *Treatise on the Four Mathematical Sciences: Arithmetic, Music, Geometry, and Astronomy*; however, it is best known by the concise title *Quadrivium*.[187] *Quadrivium* is a true encyclopedia of the mathematical sciences, for which Pachymeres drew from a variety of ancient sources. Rather than slavishly repeat the contents of what he learned, Pachymeres developed the material in creative ways, as Herbert Hunger has pointed out (1978, II, 246). According to Constantinides (1982, 157) the main sources from which Pachymeres drew are, for the book on arithmetic, Nicomachus, Diophantus, and Euclid; for music, Ptolemy and Porphyry; for geometry, Euclid (not only the *Elements* but the *Optics* as well); and for astronomy, Aratos, Archimedes, Aristotle, Cleomedes, Euclid, Ptolemy, and Theon. His relative independence from these authoritative sources is illustrated in the way Pachymeres uses

---

186 Cf. Golitsis (2007, 53).

187 The "Quadrivium" is preserved in at least 14 manuscripts – the most important of which, Angelicus 38 (C. 3. 7), has been identified as an autograph of Pachymeres himself (Harlfinger 1971, 357 n. 3) – and its complete edition was prepared by P. Tannery but published posthumously by E. Stéphanou (Pachymeres 1940). For the date of composition, see Golitsis (2007, 63–64).

Euclid's *Elements* in the part of the work dealing with arithmetic. While one would expect him to draw only from the arithmetical books of the *Elements*, Pachymeres includes also propositions from Book II, which he interprets arithmetically with integer numbers.[188]

The book on arithmetic of the *Quadrivium* consists of 74 chapters. The first five chapters together form a philosophical preface. Chapters 6–24 constitute the main "Nicomachean" part of the book. Next comes the "Diophantine" part, in Chapters 25–44. After two more chapters dealing with square and oblong numbers, the book concludes with chapters 47–74, which form its "Euclidean" part.

The "Diophantine" part of the book thus covers 20 chapters.[189] The first of these chapters presents the technical terms of the theory and operations on these terms that Diophantus lays out in his preface. In Chapters 26–44 Pachymeres poses and solves a total of 78 arithmetical problems combining two sources: problems selected from the first 11 problems of the first Book of the *Arithmetica* and the ratio theory of Nicomachus's *Arithmetical Introduction*. The problems are divided into nine types, corresponding to *Arithmetica* Problems I.1, I.2, I.3, x, I.4, I.8, I.9, I.10, and I.11, where "x" stands in place of a type not found in Diophantus. The second through the fifth types call for a ratio between two values, so for them Pachymeres poses different enunciations in which the instantiated ratios are taken in order from Nicomachus. For example, for the second type the ratios chosen are multiple, superparticular, superpartient, multiple superparticular, and multiple superpartient. Pachymeres does not solve these problems by algebra as Diophantus does. Instead, he reasons his way to the solution as Nicomachus might have, by considering the ratios of the unknown numbers. Only in the last two problems does he mention Diophantus, where he sketches the algebraic solution as a demonstration of his solution via ratios.

Let us see how this is carried out in a concrete example.[190] The second problem of Diophantus asks for a given number to be divided into two numbers having a given ratio. This generic enunciation is followed in Diophantus by a single instantiation, in which the given number is 60 and the given ratio is triple. Pachymeres discusses problems modeled on this problem in six chapters of his book (27–32). The division in chapters is made on the basis of the kind of the ratio, and in each chapter he solves more than one problem. He solves three problems with "multiple" ratio in Chapter 27, six problems with "superparticular" ratio in Chapter 28, three problems with "superpartient" ratio in Chapter 29, eight problems with

---

188 Pachymeres was not the first to give Book II a numerical reading in the context of Euclidean/Nichomachean number theory. The fifth volume of Euclid's *Elements* in the edition of Heiberg-Stamatis provides ample evidence of late antique scholia, which do the same. In the tenth century the Ikhwān al-Ṣafāʾ (Brethren of Purity) translated *Elements* II.1–10 into arithmetical terms in the same way (Goldstein 1964, 154–57), and in the early eleventh century al-Karajī did the same in his *al-Badīʿ*, for II.5–10, giving his own arithmetical proofs (al-Karajī 1964, 18ff).

189 This part was first published by Tannery in the second volume of Diophantus's works (1893–95, II, 78–122).

190 For a more detailed study, see Megremi and Christianidis (2015).

"multiple superparticular" ratio in Chapters 30 and 31, and five problems with "multiple superpartient" ratio in Chapter 32.[191] In the following chart we reproduce the introductory sentence and the first of the three problems of Chapter 27 of Pachymeres, next to the corresponding Problem I.2 of Diophantus:

Pachymeres's Chapter 27, first problem *(Pachymeres 1940, 50.12–21)*	*Diophantus's Problem I.2* *(16.24–18.6)*
Well, the arithmetical problems in multiplications are as follows: if we are asked to divide the given number in ratio either double, or triple, or whatever, such that the part has to the part this ratio. So, if we are asked to divide in double ratio, we must take the sub-triple of the whole, and make it the lesser term, of which we take the double, that is, the remainder, (which is) the greater term. And the problem is made. For example, if we are asked to divide the number 24 into (two numbers having) double ratio, we are looking for its sub-triple, and it is 8. Its double, that is, the remainder, 16, becomes the greater term, and the problem is satisfied. Indeed, 16 is the double of 8, and 16 and 8 (make) 24.	It is required to divide a proposed number into two numbers having a given ratio. Now, let it be proposed to divide 60 into two numbers having a triple ratio. Let the smaller be assigned to be 1 Number. Therefore, the greater will be 3 Numbers, so the greater is triple the smaller. Then the two must be equal to 60 units. But the two, when added together, are 4 Numbers. Thus, 4 Numbers are equal to 60 units; therefore, the Number is 15 units. Thus, the smaller will be 15 units, and the greater 45 units.

In Diophantus's solution names are assigned to the sought-after numbers, operations are performed on the names, and an equation framed in the language of the names is produced. None of these features is present in Pachymeres's solution: no names are applied, and no equation is created. Diophantus solves the problem by algebra, while Pachymeres's solution is arithmetical. This fact in itself confirms Hunger's claim, mentioned earlier, that Pachymeres does not imitate blindly his sources and, at the same time, discredits the widespread view, since Tannery, that Pachymeres's text is a mere "paraphrasis" of Diophantus.

The most important Byzantine commentator of Diophantus was Maximus Planudes (1255–1305 or 1260–1310). His work, writes Sesiano, "represents the farthest-reaching commentary on the *Arithmetica* made in Byzantine times, and,

---

191 The meaning of the terms denoting ratios is easily grasped by using modern symbolism. Thus, in the following definitions, the letters A, B, k, m, and n denote positive integers. A ratio $A:B$ is called a "multiple" if $A = nB$; "submultiple" if $B = nA$; "superparticular" if $A = B + \frac{1}{n} \cdot B$; "subsuperparticular" if $B = A + \frac{1}{n} \cdot A$; "superpartient" if $A = B + \frac{m}{m+n} \cdot B$; "subsuperpartient" if $B = A + \frac{m}{m+n} \cdot A$; "multiple superparticular" if $A = kB + \frac{1}{n} \cdot B$; "submultiple superparticular" if $B = kA + \frac{1}{n} \cdot A$; "multiple superpartient" if $A = kB + \frac{m}{m+n} \cdot B$; and, finally, "submultiple superpartient" if $B = kA + \frac{m}{m+n} \cdot A$.

though limited in length and content, it is particularly noteworthy coming from a man renowned as one of the foremost Byzantine humanists" (1982, 17). This assessment, when it is supplemented with a reference to the crucial role he played in the preservation, restoration, and transmission of the Diophantine text, portrays accurately the importance of Planudes's occupation with Diophantus.

Alexander Turyn (1972, 80) has shown that it was in the years 1292–93 that Planudes undertook the project of establishing a reliable text of the *Arithmetica*. In fact, his autograph, the codex Mediolanensis Ambrosianus Et 157 sup., is the archetype of an entire branch of the tradition (see Section 1.3). The project is outlined in two letters of Planudes, from which we can infer the following information.[192] In the 1290s there existed in Constantinople at least three manuscripts of the *Arithmetica*.[193] One manuscript belonged either to the high official Theodore Mouzalōn (d. 1294), or to the library of Planudes's monastery, the text is not so clear at this point. According to Turyn, Planudes had borrowed the manuscript from some library under Mouzalōn's jurisdiction and now he was returning it to Mouzalōn (1972, 80), while, according to Pérez Martín, the manuscript "n'appartenait pas à Mouzalon mais à la bibliothèque du monastère impérial dont était chargé Planude" (2006, 437), and Mouzalōn "(l')avait sollicité, nous ignorons pourquoi" (2006, 436). Before sending the manuscript to Mouzalōn, Planudes repaired it. Here, too, there is an ambiguity, whether Planudes's work consisted in repairing the codex or in restoring the text. The latter is the viewpoint advanced by Pérez Martín, according to which Planudes transcribed the Diophantine text in order to produce a new copy, which might be either his autograph, the manuscript of the Biblioteca Ambrosiana, or a working copy prior to the Milan copy. Be that as it may, it is important to note that the manuscript at issue has been identified with the Matritensis 4678 (Wendel 1940, 414–17). Another manuscript belonged to Manuel Bryennius, who was an active astronomer and teacher of mathematics at that time in Constantinople (Constantinides 1982, 93). Planudes asked Bryennios to lend him the manuscript, for he wanted to collate it with his own.

Planudes commented on the first two books of the *Arithmetica*. His commentary is full of arithmetical examples that illustrate the rules and properties stated or presupposed in the text of Diophantus, and of explanations of difficult terms and passages. One such explanation, we will see, stands out, by explaining the relationship between the arbitrary multitude of Numbers in the *al-istiqrāʾ* assignment with the given numbers in the problem. Other noteworthy features of Planudes's commentary are the abbreviated presentations of the solutions, in tabular format, in the beginning of each problem, and the geometrical justifications he

---

192 The letters at issue are published by M. Treu in Planudes (1890) under the numbers 33 (pp. 53–55) and 67 (pp. 81–85). Cf. the more recent edition by P. L. Leone (Planudes 1991). The references to Diophantus are given in Treu's edition in pp. 53.7–10 and 82.31–36. The corresponding pages in Leone's edition are 66.14–17 and 99.24–29. Cf. also the recently published French translation of Planudes's letters by J. Schneider in Planudes (2020); the two letters appear in pp. 138–40 and 202–07.

193 The details reported here are gathered from several sources, among which are Turyn (1972), Allard (1979), Constantinides (1982), and Pérez Martín (2006).

appends to the rules stated in the preface as well as to a couple of problems of the first book.

For an example of the latter we show how he justifies "by lines" the rule for multiplying together diminished quantities in the context of the squaring of "5 lacking 1 Number"[194] (Diophantus 1893–95, II, 140.9–142.12): Let two straight lines, AB and BΓ, be set out at right angles, each of 5 units, to which a lacking amount of 1 Number is applied, and we suppose that the Number is 2 units, and let the lacking amount on AB be AE, 1 Number, i.e., 2 units; EB, therefore, will be 3 units; and let the lacking amount on BΓ be ZΓ, also 1 Number, i.e., 2 units; BZ, therefore, will be 3 units. Now, he writes,

> since there are two straight lines, AB and BΓ, each of 5 units lacking 1 Number, and they must be multiplied with each other, so that to show, on the one hand, that a lacking (amount) multiplied by a lacking (amount) makes an extant (amount), and, on the other hand, that (a lacking amount) multiplied by an extant (amount) makes a lacking (amount), it is necessary, according to the modus operandi of the multiplication, first for the extant (amount) of units to be multiplied by itself; then, for the same extant (amount) of units by the lacking Number; and again for the lacking Number by the extant (amount) of units; and finally for the lacking Number by itself, that is, by the lacking (Number), and the required will have been shown.
>
> (Diophantus 1893–95, II, 140.20–141.4)

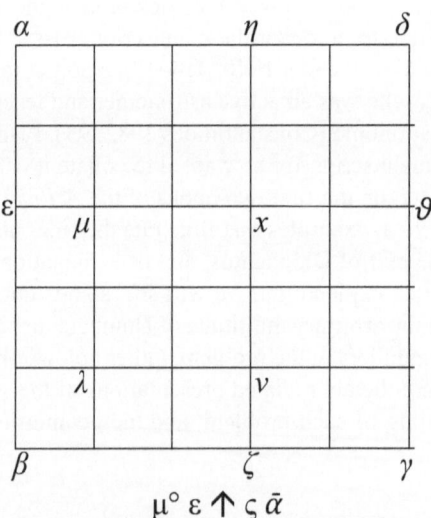

Figure 3.8 Diagram accompanying Planudes's proof of the rule that "a lacking (amount) multiplied by a lacking (amount) makes an extant (amount)".

194 Including why "a lacking (amount) multiplied by a lacking (amount) makes an extant (amount)".

Next, Planudes performs the multiplications. First, he multiplies AB by BΓ, resulting in the square ABΓΔ, which is 25 units. Then he draws all the 25 units by drawing parallel lines as in the diagram. Then he multiplies AB, the extant 5 units, by ZΓ, the lacking ⟨1⟩ Number in the straight line BΓ; and since units multiplied by Numbers produce Numbers, and an extant amount multiplied by a lacking amount makes a lacking amount, the rectangle ZΔ, which is the lacking 5 Numbers, i.e., 10 units, will be removed from the square ABΓΔ. The remainder is the rectangle AZ, which is 15 units. Again, he multiplies AE, the lacking 1 Number, by BΓ, the extant 5 units, resulting again in a lacking amount of 5 Numbers, and this lacking amount is the rectangle AΘ. But since the square HΘ was already removed, we remove the rectangle AK, of 3 Numbers, and the square KΛ, whose side is 2 Numbers, besides, so that the lacking amount, which is the area AHKNΛME, becomes again 5 Numbers, i.e., 10 units. The remainder is the gnomon BEMΛNZ, which is 5 units. But since, after subtracting the lacking amounts AE and ZΓ, each of the remainders EB and BZ are 3 units, and the square contained by them is 9 units; but the remaining gnomon is only 5 units; therefore, in order to produce the square on the 3 units, 4 units must be added. But the lacking AE of 1 Number multiplied by the lacking ZΓ of 1 Number produces an extant amount of 1 Power, which is 4 units. So, the Power is the square KΛ of 4 units, which, having been subtracted before, is now added to the gnomon BEMΛNZ, that is, to the 5 units, to produce a square of 9 units. The result is the same as what would be obtained if EB alone were multiplied by BZ, that is, the 3 units by the 3 units, without taking the lacking amounts into account. That is in the square EBZK, which is 1 Power, 25 units lacking 10 Numbers, that is, 29 units lacking 20 units,[195] i.e., 9 units.

To show how Planudes organizes the solutions to problems in tabular form, we give the example of Problem II.26. The problem asks: "To find two numbers such that their product, if it receives in addition either, makes a square, while the sides of the squares, if added together, make a proposed number". In the instantiation the proposed number is 6. Diophantus solves the problem by positing the smaller number to be 1 Number (1N), and the greater to be 4 Numbers lacking 1 unit (4N ℓ 1u). These assignments satisfy the stipulation that their product together with the smaller makes a square. The remaining stipulation is that the product of the two numbers (4P ℓ 1N), together with the greater (4N ℓ 1u), must be the square. Their sum, 4P 3N ℓ 1, is set equal to the square of 6u ℓ 2N to arrive at the equation 4P 3N ℓ 1u = 4P 36u ℓ 24N. From this the Number is found to be $\frac{37}{27}$.

Planudes outlines the solution as shown in Figure 3.9, followed by the same table in Tannery's edition in Figure 2.10 (Diophantus 1893–95, II, 243.2–9) and our translation in Table 3.8.

The shortened words in the left column designate tasks performed in the solution. More precisely: ἔκθ. (*ekth.*), "setting-out", refers to the assignment of names

---

195 Here the subtraction is denoted in Tannery's text with the sign for "lacking": $\mu°\overline{\kappa\theta}$ ⋀ $\mu°\overline{\kappa}$. Marc. gr. 308, the most direct descended of Planudes's autograph, Ambros. Et 157 sup., has the full word λείψει (f. 62v, l. 1).

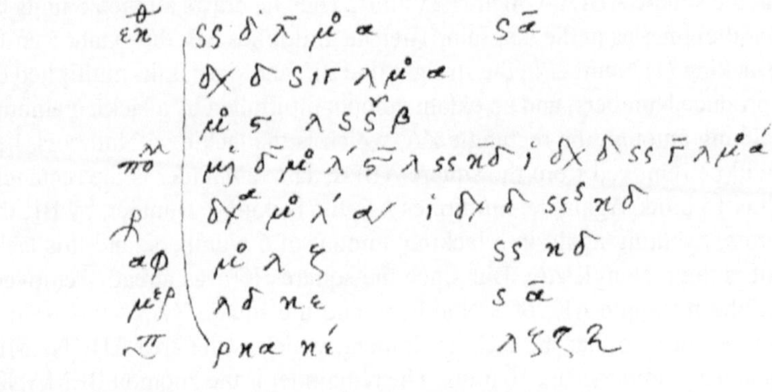

Figure 3.9 Facsimile of Planudes's tabular setting of the solution to II.26, from Par. gr. 2485 (f. 100r).[196]

ἐκθ.	ςς δ̄ ⋔ μ° ᾱ		ς ᾱ
	Δ^Υ δ̄ ςς γ̄ ⋔ μ° ᾱ		
	μ° ς̄ ⋔ ςς β̄		
πολλ.	Δ^Υ δ̄ μ° λς̄ ⋔ ςς κδ̄	ἴσ.	Δ^Υ δ̄ ςς γ̄ ⋔ μ° ᾱ
πρ.	Δ^Υ δ̄ μ° λζ̄	ἴσ.	Δ^Υ δ̄ ςς κζ̄
ἀφ.	μ° λζ̄		ςς κζ̄
μερ.	λζ̄ κζ^α		ς ᾱ
ὑπ.	ρκα κζ^α		λζ̄ κζ^α.

Figure 3.10 The same table in Tannery's edition.

Table 3.8 Translation of Planudes's tabular setting of the solution to II.26.

setting-out	4N ℓ 1u		1N
	4P 3N ℓ 1u		
mult.	6u ℓ 2N	*eq.*	4P 3N ℓ 1
add.	4P 36u	*eq.*	4P 27N
subt.	37u		27N
div.	$\frac{37}{27}$		1N
hypost.	$\frac{121}{27}$		$\frac{37}{27}$u

196 There are several scribal errors in the image: (i) In the second line, after the sign for Number, there is the number ΙΓ (13) instead of Γ (3); (ii) in the fifth and sixth lines the numbers at right are written as κδ (24), while the correct is κζ (27); (iii) in the penultimate line the number κε (25) must be κζ (27); (iv) same for the last line: the κε (25) must be κζ (27). Also, the abbreviation for "equal" has only the initial letter ι (without the letter σ, the second letter of the Greek ἴσος).

to the unknowns; πολλ. (*poll.*), "multiplication", refers to the squaring of 6u ℓ 2N; πρ. (*pr.*), "addition", refers to the task announced by the sentence "let the lacking be added in common"; ἀφ. (*aph.*), "subtraction", refers to the task announced by "let likes be subtracted from likes"; μερ. (*mer.*), "division", refers to the division by 27 of the two sides of the equation 37u = 27N so as to get the normalized equation 1N = $\frac{37}{27}$ u ; ὑπ. (*hyp.*), "hypostases", refers to the task announced by the clause "to the hypostases", that is, the calculation of the sought-after numbers; and, ἰσ. (*eq.*), is short for "equal".

The practice of producing notational versions of solutions that show only the calculations necessary for the solution has its origin in antiquity. These solutions omit all didactic, heuristic, and even stylistic ornaments and additions that an edited work would normally include, especially for a work intended for teaching purposes. We saw such solutions in the Michigan papyrus in Section 3.1.5. The abbreviations described in the introduction to the *Arithmetica* are discussed in Section 4.2, where we advance the idea that they are aimed primarily at facilitating the working mathematician, the *problem solver*, in his problem-solving endeavor, on the one hand, and the mathematical author who stores the notational versions of worked-out problems in view of a future publication, on the other, and they were not intended to be systematically used in the final edition of the treatise. So to solve a problem one would work out the calculations in notation on a wax tablet or other surface, and to communicate this solution in a book a rhetorical version would be created. In Planudes we see the recreation of the notational version from the rhetorical version, presumably to show the reader how it is that one goes about solving problems.

In his commentary Planudes proposes terminological and mathematical explanations to a number of abstruse points of the Diophantine text, some of which have been proven to be notoriously puzzling for modern historians of mathematics. For example, he asserts that the term πλασματικόν (*plasmatikon*), employed for the first time in Problem I.27, qualifies the problem which needs a determination, and not the determination itself. In conformity with this explanation we translate this term as "arising with the formation (of the problem)".[197]

Mathematically the most interesting of the comments of Planudes on the first two books of the *Arithmetica* is that on Problem II.8. This is the first indeterminate problem in the treatise, and it is the first that is solved by setting up a forthcoming equation and assigning an algebraic name for the unnamed square via *al-istiqrā'*. The commentary on Problem II.8 covers seven full pages in Tannery's edition, a good part of which is devoted to the *al-istiqrā'* assignment. The problem asks to divide a given square, which is 16 in the instantiation, into two squares. Diophantus assigns one square to be 1P, so the other square will be 16u ℓ 1P. This expression must be a square number, so Diophantus assigns the side of that square to be 2N ℓ 4u, noting that the multitude of the Numbers (here 2) can be

---

197 See the note (Christianidis 1996). For a comprehensive overview and critical discussion of the interpretations proposed since Xylander and Bachet for the term πλασματικόν, see Acerbi (2009).

any number whatever, while the deficit should be the same as the side of the given 16 units. The equation then becomes 4P 16u ℓ 16N = 16u ℓ 1P. The units cancel, and the equation simplifies to 5P = 16N (90.14–15).

Planudes begins his remarks on the *al-istiqrā'* assignment by observing that had Diophantus made the side of the square 4N ℓ 4u instead of 2N ℓ 4u, it would have led him to a different solution. He next explains that the multitude of Numbers must be greater than 1, and then he lays out how the choice of this multitude affects the solutions that are obtained. He derives the nature of this relationship in purely arithmetical terms, without any algebraic language, first in general and then for numerical examples. He then relates it to the setting of the algebraic assignment (we have added the bracketed letters A to E in the first paragraph for reference):

> In general, (A) for all squares which are divided into two squares, (B) the side of the divided (square), together with the side of either of the (squares) from the division, has with respect to the side of the other a certain ratio. (C) So, whatever the subtracted among the (squares) from the division be, (D) the side of the other will be of so many units, as it was, less the side of the divided, (E) the side of the subtracted (square) taken so many times[198] as was the side of the other, together with the side of the divided, of the side of the subtracted.
>
> For example, since 25 is composed of 16 and 9, and is divided into these, and the side of the 25, i.e., 5, together with the side of the 9, i.e., 3, is double the side of the 16, i.e., 4, if I subtract the 16, the side of the 9 will be, because of the double ratio, twice the side of the 16 less the side of the 25, that is, 8 units except 5, which is 3. Again, since the side of the 25, i.e., 5, together with the side of the 16, i.e., 4, is triple the side of the 9, i.e., 3, if I subtract the 9, the side of the 16 will be, because of the triple ratio, thrice the side of the 9 less the side of the 25, that is, 9 units except 5 units, which is 4 units.
>
> (Diophantus 1893–95, II, 217.18–218.9)

We give a commentary to this arithmetical derivation using modern algebraic notation. Let us write the division of the given square into two squares as $a^2 = x^2 + y^2$, with the idea that in the algebraic solution to the problem, x will be the side of the square assigned to be "1 Power", and the side y will be assigned to be "2 Numbers lacking 4 units", or in modern, general terms, $y := mx - a$ for some m. Planudes starts from (A) $a^2 = x^2 + y^2$. He states that (B) $(a + y) : x$ is a certain ratio, whose numerical manifestation we will call m. Then (C) supposing x is the subtracted side, i.e., it is the side of the square named "1P" that will be subtracted from 16 (= a^2), (D) "the side of the other (y) will be of so many units (mx), as it was, less the side of the divided (a)", or $y = mx - a$, and he explains what this "so

---

198 Marc. gr. 308 (f. 130r, l. 13) has here the word ὁσόλογος, which can be translated "as the ratio" or "as great as the ratio". This word does not appear in the other basic manuscripts, and Tannery put it in the *criticus apparatus*.

many units" is: "the side of the subtracted square (x) taken so many times (m) as was also the side of the other (y), together with the side of the divided (a), of (i.e., "with respect to") the side of the subtracted (x)", or $m = (y + a) : x$.

In the second paragraph, Planudes illustrates this derivation by means of a numerical example.[199] In it, from the equality $25 = 16 + 9$, he notes that $5 + 3 = 2 \cdot 4$, from which, $3 = 2 \cdot 4 - 5$, or, as in the text, "the side of the 9 will be, because of the double ratio, twice the side of the 16 lacking the side of the 25, that is, 8 units except 5, which is 3". Then, he states it again, reversing the roles of 3 and 4: $5 + 4 = 3 \cdot 3$, from which, $4 = 3 \cdot 3 - 5$, or, as in the text, "the side of the 16 will be, because of the triple ratio, thrice the side of the 9 lacking the side of the 25, that is, 9 units except 5, which is 4 units".

After two more examples using the Pythagorean triple 25, 144, 169, he ties it in with the assignment: "Thus, since the sides can be with each other in any ratio whatever, but always lacking the side of the divided (square), that is why he (Diophantus) says 'from however many Numbers lacking the side of the divided'". This kind of analysis is not given for any other problem. It is possible that Planudes did not investigate other *al-istiqrā'* solutions because in the next three that use it, Problems II.11, 12, and 13, the value of 1N found in the solution would make the assigned side of the square negative.[200]

3.4.4 Later Byzantine testimonies

The work of Diophantus remained accessible to scholarly circles in Constantinople during the fourteenth and fifteenth centuries, constituting for some Byzantine humanists part of their education. Theodore Metochites (1270–1332), for example, was a prolific author, statesman, and collector of books of the early fourteenth century. In Book II, Chapter 3 of his *Introduction to Astronomy* (Στοιχείωσις ἐπὶ τῇ ἀστρονομικῇ ἐπιστήμῃ), where he discusses the multiplications between sexagesimal numbers, he mentions Diophantus and quotes the passage from the preface to the *Arithmetica*, saying that "the unit, because it is immutable and always constant, any species multiplied by it will remain the same species"; while, "in the multiplications of the parts of unit, says again Diophantus, the species change".[201]

In the mid-fourteenth century Nicolaos Artabasdos Rhabdas, in two letters of arithmetical content – one addressed to George Chatzykes, the other, from 1341, to Theodore Tzabouches – reproduces almost *verbatim* the beginning of the preface to the *Arithmetica*.[202]

---

199 This example is followed by a second example, with the squares 25, 144, and 169 (Diophantus 1893–95, II, 218.10–21).

200 Such cases are not uncommon in the *Arithmetica*. See Appendix 2 for a discussion.

201 See Vat. gr. 1087 (13th/14th c.), f. 195v, Vat. gr. 1365 (14th c.), f. 248r. We thank Dr. Ioanna Skoura for this information about Metochites's mention of Diophantus.

202 The letters have been published, with French translation, by Tannery (1920a). Recently Ioanna Skoura discovered a third, hitherto unknown letter of Rhabdas, in which he also reproduces

Barely a generation later Isaac Argyros (1300/10–1371/75), the leading figure of Ptolemaic astronomy and mathematics in Byzantium in the fourteenth century, composed a revised commented edition of Ptolemy's *Geography*, based on a previous work of his teacher Nicephorus Gregoras (1293/94–1358/61). In one scholium to *Geography* I.24 Argyros solves an arithmetical problem motivated by Ptolemy's text, by using the method of the "Diophantine" numbers, a designation we encountered in Theon of Alexandria (see Section 3.2.1). The exact phrase is: "He works out this methodically by means of the Diophantine ⟨numbers⟩". Next to this phrase Vat. Urb. gr. 80 has the phrase "Demonstration according to Diophantus of Alexandria, the arithmetician", written, as the editor of the scholium points out, by John Chortasmenos (Tsiotras 2006, 427.22–23). The problem, which is indeed solved by algebra, is the following: "He seeks which number, if added to 63 – the difference of their latitudes – makes the sum to be with respect to the assumed ⟨number⟩ in such a ratio"[203] (Tsiotras 2006, 427.20–22). The text of the scholium is published by Tsiotras (2006), and the solution to the problem runs as follows:

> He adds to 63, 1 Number. Therefore, 1 Number and 63 units have with respect to the added number a double-sesquiquintan ratio. But 2 and 5′ Numbers have also with respect to 1 Number a double-sesquiquintan ratio. Therefore, 2 and 5′ Numbers are equal to 1 Number and 63 units. And the equal being subtracted from equal, 63 units, therefore, are equal to 1 and 5′ Number. And the quintuples; therefore, 315 units are equal to 6 Numbers, and the Number becomes approximatively 52.[204]

> (Tsiotras 2006, 427.24–428.30)

In the second half of the fourteenth century Demetrius Kydones (ca. 1324–ca. 1398) was able to recognize in a book, for which he does not give further specifications, "some of the theorems (θεωρήματα)" of Diophantus, which were "passed unnoticed" by the author or, most probably, the compiler of the book. Kydones supplied the "theorems" with "proofs", as he had already done for some theorems of the arithmetical books of Euclid's *Elements*, "for Diophantus had gotten away with only proposing *questiones* (ζητήματα, i.e., enunciations of problems)", and sent all this to a non-specified correspondent.[205] Acerbi rightly remarks that "this is not a description of the *Arithmetica* of Diophantus, unless we suppose that Kydones had come across an epitome for which there is no longer any trace" (2017, 176). He

---

almost *verbatim* the introductory paragraph of Diophantus's preface. This letter is about *computus ecclesiasticus*, it is addressed to a certain Demetrius Myrsiniotes, and is dated 1342. Skoura announced her discovery in a paper presented at the 1st Conference of the International Academy of the History of Science (Athens, September 14, 2019). The paper is published in *Neusis* (Skoura 2019–20).

203 Namely, a "double-sesquiquintan" ratio, that is, the sum is $2\frac{1}{5}$ times as great as the number.
204 We are grateful to Ioanna Skoura, to whom we owe the information about Argyros's scholium.
205 See Cydonès Démétrius (1956–60, II, 287.3–10).

further suggests that the book to which Kydones refers could be a collection of problems formulated as we read in *Rechenbücher*,[206] or in the form of the epigrams of the Palatine Anthology, which Kydones had found attributed to Diophantus (2017, 176–77). Going one step further we could associate that book with the kind of literature that the letter of Michael Psellus writes about (see Section 3.4.1), and which he associates with the names of Diophantus and especially Anatolius.

The last Byzantine intellectual we know to have undertaken study of the *Arithmetica* was John Chortasmenos (b. ca. 1370, d. before 1439), a collector of books and a copyist, described by Pérez Martín as "un 'détective' reconnu des bibliothèques constantinopolitaines" (2006, 452), who added many annotations in Matrit. 4678, some of which go beyond the first two books of the treatise (2006, 450).

3.5 Diophantus in the Renaissance

3.5.1 Bessarion and Regiomontanus

It was at the end of the fourteenth century that the attention of Italian humanists began to turn to the collection, study, and translation of ancient Greek texts. One event decisive in initiating this shift was the three-year tenure in 1397–1400 of the Byzantine diplomat Manuel Chrysoloras (1355–1415) in Florence, who resurrected the study of the Greek language in Italy through lectures to eager and talented students. Chrysoloras also taught Italian students in Constantinople, both before and after this period. With this awakening came the push to acquire and study Greek manuscripts. Hundreds of Greek codices were purchased and brought to Italy from Byzantine territory by Jacopo d'Angelo da Scarperia (1396), Guarino Veronese (1408–10), Giovanni Aurispa (1423), Francesco Filefo (1427), Ciriaco d'Ancona (various trips), and others. Aurispa alone returned with 238 codices.[207] Additionally, many educated Byzantines fleeing the crumbling Eastern Empire emigrated to Italy, bringing with them manuscripts, and still more manuscripts were acquired from Greek monasteries in Italy and other parts of Europe.

The primary interest of the humanists may have lain with the poetry, history, and philosophy of classical antiquity, but this enthusiasm brought with it a renewed respect for the quadrivium. Pier Paolo Vergerio (1370–1444/45), for example, restored the honored place of the mathematical sciences in the highly influential educational curriculum he presented in his *De Ingenuis Moribus et Liberalibus Studiis* (1400–02) (Rose 1975, 12–13, 27). Indeed, among the authors listed in the inventories of manuscripts acquired in this period we find

---

206 Such Byzantine collections have been published by Hunger and Vogel (1963) and Vogel (1968).
207 For the codices of Aurispa, see Spoerri (1980, 17–19). N. G. Wilson (1992, 25–26) points out that Figure 238 is usually quoted to refer to the non-patristic texts of Aurispa's collection.

Euclid, Apollonius, Archimedes, Hero, Ptolemy, Pappus, Diophantus, Theon, and Proclus.

We know of five manuscripts of Diophantus that were brought to Italy, all of which are now in European libraries:[208]

> Vat. gr. 304 (ff. 77r–118r), copied in the first half of fourteenth century, entered the Papal library after 1447, and is listed in the 1455 inventory.
>
>   (Devreesse 1965, 36; Rose 1975, 37)

> Matrit. Bibl. Nat. 4678 (ff. 58r–130v), copied in the second half of the eleventh century, was taken to Messina by Constantine Lascaris after the fall of Constantinople in 1453.

> Ambros. Et 157 sup., fragments of the autograph of Maximus Planudes, copied 1292/1293.

> Vat. gr. 191 (ff. 360r–390r), copied 1296–98, was acquired by the Vatican sometime in the period 1464–71 from the estate of the humanist Cardinal Isidore of Russia. It appears in the 1475 inventory.
>
>   (Mercati 1926, 68, 86;
>   Devreesse 1965, 42, 60; Rose 1975, 38;
>   Acerbi in Diophantus 2011, 115)

> Marc. gr. 308 (ff. 50v–263r), copied at the end of the thirteenth century, belonged to Cardinal Bessarion. The Biblioteca Marciana was founded when he donated his extensive library to Venice in 1468, and MS gr. 308 appears in the initial inventory of that year.
>
>   (Omont 1894, 158)

Cardinal Bessarion began his career as a Byzantine monk and diplomat. He was sent to Italy in 1437 by the emperor John VIII Palaeologus to garner support for the defense of Constantinople against the Turks. He converted to the Latin faith, and from 1440 until his death he remained an important figure attached to the Roman curia. In conjunction with his continued efforts to secure defense for Constantinople, and then after 1453 to liberate the city, Bessarion was deeply concerned with preserving Greek learning.

Because the demand for manuscripts by both private and public libraries exceeded the number arriving from abroad, many copies were made on Italian soil. Not only did individuals routinely copy treatises for their own use (Rose 1975, 45–47), but larger collections were built through the hiring of professional copyists. To establish the Libreria Medicea Privata in the 1440s, Cosimo de' Medici employed 45 scribes who copied 200 manuscripts in 22 months, and pope Nicholas V (1447–55) was responsible for initiating the humanist collection of the Vatican, employed several copyists and also translators (Rose 1975,

---

208 Cf. Section 1.3.

34, 36). Only two manuscripts of the works of Diophantus are known to have been copied in Europe by the end of the century. Ambros. A 91 sup. was copied from Bessarion's Marc. gr. 308, and Vat. gr. 200 was in turn copied from Ambros. A 91 sup. (Allard 1984, 323) This Vatican manuscript is listed in the 1481 inventory (Devreesse 1965, 93; Allard 1984, 323), but not in the 1475 inventory.[209]

With still few Italians able to read Greek, Latin translations were needed to make the ancient works accessible to a wide readership. Yet despite the influx of mathematical manuscripts and the deep interest of both mathematicians and humanists, by the dawn of the sixteenth century the only newly discovered works with any mathematical content that were translated into Latin are two geographies, one of Ptolemy and the other of Strabo. The translation of Ptolemy's *Geography* was begun by Chrysoloras and completed by 1410 by his student Jacopo d'Angelo (Swerdlow 1993, 160), and Strabo's was begun by Guarino Guarini, another student of Chrysoloras, and completed by Gregorio Tifernate by 1456 (Morse 1981, 31–32). These were strictly humanist endeavors: in neither case did the translators understand the mathematics. Two other Greek mathematical works were rendered into Latin by 1500 for which medieval translations were already in circulation. Jacobus Cremonensis translated most of the extant works of Archimedes around 1450, but although this translation was important, William of Moerbeke's 1269 translation, which had been made from the same manuscript Jacobus used, ultimately proved to be more influential for Renaissance mathematicians. And George of Trebizond (1396–1472) managed in the span of only eight months in 1451 to translate all of Ptolemy's *Almagest* and to write an extensive yet flawed commentary. The inadequacy of his translation ensured that Gerard of Cremona's twelfth-century translation from Arabic would remain in use, aided immensely by the *Epitome* of Ptolemy by Peurbach and Regiomontanus, completed ca. 1463. Throughout the fifteenth century the older medieval translations, often made from Arabic, served the needs of those who wanted to read Euclid, Ptolemy, and Archimedes.

We are aware of only one person who took note of Diophantus's *Arithmetica* during the 1400s. Johannes Müller of Königsberg (1436–76), known as Regiomontanus,[210] famously announced his discovery of a manuscript of the *Arithmetica* in a lecture delivered in Padua in 1464. Regiomontanus was at the time the most competent astronomer in Europe. He had been educated at the University of Vienna by the humanist astronomer Georg Peurbach (1423–61), and from 1457 he served on the faculty as Peurbach's colleague and friend. Peurbach had traveled in Italy in the early 1450s, where, among other activities, he had lectured in Ferrara and Padua (Rose 1975, 92).

---

209 Oxon. Bodl. Lib. Barocci 166 is mistakenly dated to the fifteenth century in Coxe (1969, 281). Watermarks show that it was written in or shortly after 1566 Constantinides and Browning (1993, 380).

210 "Joannes de Regio monte" (Königsberg means "King's Mountain").

Although algebra was not yet commonly practiced in Germany, Regiomontanus had already become familiar with the art in his student days. He owned a copy of the *Quadripartitum Numerorum* of Jean de Murs, a book that incorporates nearly all of the algebra proper of al-Khwārazmī, as well as many worked-out problems from Abū Kāmil, mainly via Fibonacci. Regiomontanus also possessed a copy of Gerard of Cremona's twelfth-century Latin translation of al-Khwārazmī's *Algebra* that he probably copied himself, and in 1456 he copied in Latin a brief introduction to algebra that originated in the medieval Italian tradition. This work comprises an introduction to the vocabulary and rules together with 64 problems.[211] In it Regiomontanus gives the names the powers up to the sixth in the context of their multiplications in an additive scheme, keeping the Italian terms: *numero, cosa, censo, cubo, censo de censum, duplex cubo,* and *cubo de cubum.*[212]

It was a diplomatic visit to Vienna by Bessarion in May of 1460 that changed the course of Regiomontanus's life. Bessarion was keenly interested in astronomy, so he sought out Peurbach during his stay and asked him to produce a Latin summary of the *Almagest* to make that difficult work more accessible. Peurbach began at once, but he had not quite finished covering the first six books when he died on April 8, 1461. Regiomontanus then accompanied Bessarion back to Italy, where he would remain for seven years. He completed Peurbach's *Epitome in Almagestum Ptolemaei* by the end of April 1463 (Zinner 1990, 52), and it soon became the main work through which Greek astronomy was disseminated in the Renaissance (Swerdlow 1993, 150). It was in the company of his patron Bessarion that Regiomontanus learned to read Greek, and he had Bessarion's vast library of Greek astronomical and mathematical manuscripts at his disposal, giving him a firsthand overview of books that had been inaccessible to medieval European mathematicians.

In 1463–64 Regiomontanus corresponded with the astronomer Giovanni Bianchini of Ferrara (ca. 1400–ca. 1470). Their letters dealt mainly with astronomy, but the two also posed numerical problems that the other sometimes solved by algebra, or, as Regiomontanus says in his second letter, by "the art of the thing and *census* (called algebra in Arabic)".[213] In his third letter, written in late

---

211 Currently these three works are bound together with other treatises in a single codex that belonged to Regiomonanus: MS Columbia University, Plimpton 188 (Folkerts 2002, 414). The "title" of the algebra of Italian origin is given at the top of f. 85r: "Regule de cosa et censo sex sunt capitula". A scan of this page is online: http://ds.lib.berkeley.edu/PlimptonMS188_20 (accessed June 24, 2022).

212 These names are given on f. 85r (see the previous footnote). We know of one other work that names the fifth power as *duplex cubo*. Fridericus Amman (d. 1464/65), a monk working near Regensburg, copied several short mathematical treatises that are now bound in MS München, CLM 14908. One of these (ff. 136r–146v), copied in 1461 and written in German, carries the largely Italian title "Regule delacose secundum 6 capitula". There, the first six powers are the same as those in Regiomontanus but with the German *Ding* replacing *cosa* (Curtze 1895, 52). Amman did not copy Regiomontanus (Høyrup 2019, 339), but the two manuscripts are clearly related.

213 "Artem rei et census (vocant arabice algebram)" (Curtze 1902, 216.6, 305.2; von Murr 1786, 93).

February 1464 (Zinner 1990, 64–65), Regiomontanus solves a problem posed by Bianchini involving compound interest, and in his response he mentions that algebraic solutions to problems of this type generally involve an equation of the form "rem, censum, cubum, censum de censu, cubum de censu, et cubum cubi equari numero", or, as we would write in modern notation, $ax + bx^2 + cx^3 + dx^4 + ex^5 + fx^6 = g$. These names for the first six powers come from neither al-Khwārazmī nor from Jean de Murs. They are Latinized forms of Italian names and differ only slightly from the names Regiomontanus had copied in his 1456 notebook. It is at this point that Regiomontanus mentions the manuscript of Diophantus:

> I'll tell you that recently in Venice I discovered Diophantus, a Greek arithmetician not yet translated into Latin. In the introduction, where he defines the terms of this art, he ascends to the cube of the cube. Thus, the first term he calls "number", which we call "thing"; the second he calls "power", where we say "*census*"; then "cube"; then "power of the power", where we say "*census* of the *census*"; again "cube of the *census*"; and finally, "cube of the cube". Yet I do not know if he followed out all the combinations, for not more than six books are found, though in the preface he promises thirteen. If this book, which is really most wonderful and most difficult, could be found complete, I should like to translate it into Latin, for the knowledge of Greek which I have acquired while staying with my most revered master (Bessarion) would suffice for this. Take care, please, to see if you can find the complete book somewhere. There are in your city some who are experts in Greek letters, those among them who are wont because of their ability to be involved with books of this kind. In the meantime, if you urge me to do so, I will translate the six books into Latin, so that Latinity will not be without this new and precious work.[214]

Regiomontanus had noticed the correspondence between the Diophantine names "number" (*numerus*, from ἀριθμός), "power" (*potentiam*, from δύναμις), "cube" (*cubum*, from κύβος), and "power of the power" (*potentiam potentie*, from δυναμοδύναμις), and the Latin names thing (*res*), *censum*, *cubum*, and *censum de censu*. For the fifth and sixth powers, he merely gives the Latin names, *cubum de censu* and *cubum cubi*, that he had just written for the general sixth degree equation.

Because Regiomontanus observes that this book "is really most wonderful and most difficult" (*revera pulcerrimus est et difficillimus*), we can surmise that he had read past the introduction to engage with the solutions to problems. So it is both with the correspondence between the names *and* the process of solving problems with them that Regiomontanus identified the *Arithmetica* as a work on algebra.

---

214 (Murr 1786, 135–36; Curtze 1902, 256–57). English translation taken mostly from Morse (1981, 60), modified slightly, and Heath (1910, 20).

Regiomontanus and Bessarion had arrived in Venice in late July 1463 (Rose 1975, 95), so Regiomontanus discovered this Diophantus manuscript sometime between then and February 1464, when he wrote to Bianchini about it. Undoubtedly, this manuscript is Marcianus gr. 308, which was later given to the city of Venice by Bessarion.[215]

Soon after writing this letter Regiomontanus traveled to Padua to deliver a series of lectures on the Arabic astronomer Alfraganus (al-Farghānī, d. after 861).[216] Only the inaugural lecture, delivered some time before March 1, 1464, survives.[217] It is divided into two main parts. First, Regiomontanus recounts briefly the important figures in the development of the various mathematical sciences: geometry, arithmetic, astronomy, music, and optics. In the second part he praises mathematics at the expense of scholastic philosophy, and he concludes with a panegyric on astrology. Considered in its cultural context, Noel Swerdlow describes it as

> an oration in praise of the mathematical sciences, a species of *epideictic*, that is *demonstrative*, oratory which is devoted to the praise or blame (*laus vel vituperatio*) of someone or something and showing why the subject of the oration is worthy of praise or blame.
>
> (Swerdlow 1993, 141)

The first part of the oration, then, is not intended to be taken as a short history of mathematics but as a genealogy designed to impart a respect for mathematics. The mathematicians Regiomontanus mentions wrote in Greek, Arabic, and Latin, and he also recalls the fine work of Indians and Persians. The section on arithmetic includes the first known public announcement of the existence of the work of Diophantus in the West since antiquity (the first parenthetical remark is his):

> Although, through his skill with numbers, Pythagoras attained immortality among future generations, both because he submitted himself to wandering Egyptian and Arab teachers who were greatly skilled in that study, then because he tried to probe all the secrets of nature by the certain connection of numbers, nevertheless, Euclid made a much more worthy foundation of numbers in three of his books, the seventh, eighth and ninth, whence Jordanus gathered the ten books of elements of numbers and from this produced his three most beautiful books on given numbers.[218] Diophantus, however, produced thirteen most subtle books (which no one

215 Zinner (1990, 69) dates the discovery "between December, 1463 and February, 1464". Perhaps he proposed the December date presuming that Regiomontanus would have mentioned it to Bianchini in his second letter. Both Reich (2003, 80) and Allard (1980, 283) also assert that the manuscript is the one that belonged to Bessarion.
216 His full name was Abū al-ʿAbbās Aḥmad ibn Muḥammad ibn Kathīr al-Farghānī.
217 For the date, see Zinner (1990, 69–70).
218 Jordanus de Nemore wrote his *De Numeris Datis* in the early thirteenth century.

has yet translated from Greek into Latin), in which lie the very flower of all arithmetic, namely the art of *rei & census*, which today is called algebra after its Arabic name. Indeed, Latin authors treat many fragments of that most beautiful art, but apart from Giovanni Bianchini, an excellent man, I find a scarcity of greatly learned men in our own time. In our time, the *Quadripartitum Numerorum* is certainly thought to be quite distinguished, likewise the *Algorithmus Demonstratus* and the *Arithmetic* of Boethius, the introduction (of which) was taken from the Greek Nicomachus. Finally, Barlaam the Greek wrote his collection of theorems on computation in six books, which have not yet been carried over to Latin mathematics.[219]

The lack of any mention of Arabic writers, most obviously al-Khwārazmī, whose *Book on algebra* Regiomontanus owned in Latin translation, is no indication of an anti-Arabic or anti-medieval bias. In the section on astronomy Regiomontaus names Albategnius (al-Battānī, d. 929), Geber (Jābir ibn Aflaḥ, fl. 12th c.), and of course Alfraganus (al-Farghānī), the subject of the lectures. His treatment of arithmeticians may have been too brief to mention every worthy figure. Diophantus and Bianchini are the only names mentioned with regard to algebra, "the very flower of all arithmetic". Works and authors mentioned after Bianchini fall back to arithmetic. (Interpreting the reference to the *Quadripartitum Numerorum* as specifically an algebraic work would imply that Boethius also wrote on algebra, which he did not.)

Regiomontanus never translated Diophantus. He was unable to even begin many of the tasks he had set for himself, for in 1476 he died at the age of 40 in Rome, which had become stricken with a plague. For more than seven decades after the Padua oration, the only hints of interest in Diophantus that have been preserved are the production of copies of his works. It is true that the codices containing the *Arithmetica* and *On Polygonal Numbers*, Vat. gr. 191, Vat. gr. 304, and Marc. gr. 308, were loaned out several times in the first half of the century, but the borrowers were either interested in other works that were bound with Diophantus, or they were concerned solely with making copies to bulk up their own libraries.[220]

219 (Regiomontanus 1537, 4), translated in Byrne (2006, 54–55), slightly adjusted and completed.
220 On February 4, 1518, Pietro Bembo borrowed Vat. gr. 304 specifically for Ptolemy's *Geography* (Bertòla 1942, 55). On February 14, 1522, Cardinal Giovanni Piccolomini borrowed three codices containing mathematical texts, including Vat. gr. 304, all returned June 22. Piccolomini is not known to have had any particular interest in mathematics. These books were probably borrowed for copying (Bertòla 1942, 78–79). Pietro Aretino borrowed Vat. gr. 191 on July 7, 1522, almost certainly not for Diophantus. This codex contains 39 works by 19 different authors, a small portion of which contains the works of Diophantus (Bertòla 1942, 50; Vitrac 2018, 147). Cardinal Niccolò Ridolfi borrowed Vat. gr. 191 on September 24, 1531. At this time Ridolfi was borrowing many codices for the purpose of copying them for his personal library (Bertòla 1942, 109). Marc. gr. 308 was borrowed on February 28, 1546, by Diego Hurtado de Mendoza, again along

3.5.2 Diophantus among sixteenth-century mathematicians and lexicographers

Starting around the middle of the sixteenth century mathematicians writing on algebra often sought to give a brief account of the origin of the art, even if it might consist in merely naming some of the people who were instrumental in its development. Three "inventors" of algebra are repeated across several books: Diophantus, al-Khwārazmī, and Geber. "Geber" as a founder of algebra stems from a misreading. On the first page of his ca. 1460 abacus text *La Reghola de Algebra Amuchabale* Benedetto da Firenze wrote the phrases "reghola del geber" and "reghola d'algebra", both of which were intended to mean "rule of algebra". Later, Francesco Ghaligai misread the first phrase as meaning "Rule of Geber" in his printed 1521 *Summa de Arithmetica*, and he identified this Geber as "an Arabic name of great intelligence, and which some say is one whose name was Geber".[221] Ghaligai certainly had in mind the Andalusian astronomer Jābir ibn Aflaḥ, (fl. early 12th c.), whose name was Latinized as "Geber" in the late twelfth century by his translator Gerard of Cremona. It was then Michael Stifel, in his 1544 *Arithmetica Integra*, who made the identification explicit by calling this "inventor" (*autore*) of algebra "Gebro astronomo".[222] Al-Khwārazmī is first named in Girolamo Cardano's famous 1545 *Artis Magnae*, where the Arabic mathematician is referred to as "Mahomete, Mosis Arabis filio" ("Muḥammad son of Moses the Arab", i.e., Muḥammad ibn Mūsā). Cardano cites Fibonacci as his source, but no manuscript of Fibonacci gives this name.[223] Cardano most likely got it from Robert of Chester's or Gerard of Cremona's Latin translation of al-Khwārazmī's *Algebra*. References to Diophantus as an inventor of the art by sixteenth-century mathematicians, at least those published before Bombelli's 1572 *L'Algebra*, can all be traced back to the 1537 publication of Regiomontanus's Padua oration (no manuscript of the speech survives). Diophantus is first mentioned in Johannes Scheubel's 1550 Latin version of Euclid's *Elements*, which begins with a long "book" covering algebra. There, on page 1, he cites Regiomontanus's "praefatione Alphragani" as saying that "the discovery of these rules is ascribed to Diophantus, a Greek author". By reading more into Regiomontanus's words that is warranted, Scheubel thus becomes the first European to claim that algebra originated with Diophantus.

---

with many other books in those months, for copying for his personal library (Omont 1887, 658). (Because the new year began on March 1 [il capodano Veneto], the year is recorded as 1545.)

221 Translating Benedetto with the misreading: "here begins the text of Arabic Algebra in the rule of Geber which we call algebra. This rule of algebra, according to the translator Guglielmo de Lunis" ("chosì chomincia el testo de l'Aghabar arabico nella reghola del geber la quale noi diciamo algebra. La quale reghola d'algebra, secondo Guglielmo de Lunis translatore") (Benedetto da Firenze 1982, 1). Ghaligai wrote: "uno nome Arabo di grande intelligentia: & che alcuni dicono essere stato uno el quale nome era Geber" (Ghaligai 1521, f. 70v bis).

222 (Stifel 1544, f. 226v).

223 One manuscript of the *Liber Abaci* shows "Maumeht" in the margin of one page, but that would not give the name as Cardano cites it.

All three names are later cited by Jacques Peletier (1554), Caspar Peucer (1556), Juan Pérez de Moya (1573), and Guillaume Gosselin (1577). Antic Roca (1564) and Pedro Nuñez (1567) name both Diophantus and Geber, and Petrus Ramus mentions the six books of Diophantus in his brief rundown of Greek mathematicians (1569, 37). Niccolò Tartaglia (1560) and Simon Stevin (1585) name only al-Khwārazmī as the source of algebra. Most authors understood that algebra has a long history and that more than one person was involved with its development. Jacques Peletier (1517–82), for example, wrote in his *L'Algebre*:

> The first inventor of this art, according to some, was Geber the Arab. . . . According to others it was a Muḥammad son of Mūsā the Arab, as Gerolamo Cardano of Milan said, following Leonardo of Pisa. . . . I have also seen the book of Johann Scheubel, Mathematician at Tübingen, who attributes the invention of this art to a Greek, Diophantus, who left thirteen books. . . . I do not think that this Art, nor most of the others, was invented by a single author.
>
> (1554, 1–3)

Caspar Peucer (1525–1602), a German humanist mathematician and theologian, even briefly situated the three names in their historical context. His book, titled *Logistice Astronomica Hexacontadon et Scrupulorum Sexagesimorum, quam Algorythmum minutiarum & Physicalium vocant, Regulis explicata & demonstrationibus*, comprises two parts. The first part is a detailed introduction to sexagesimal calculation, relying heavily on Theon of Alexandria. The second part is on algebra, and it is there, in the beginning, that he writes about "Gebro Arabi", "Mahometen Moysis Arabis filium", and "Diophanto Pythagoraeo" (Peucer 1556, Lr). But Diophantus figures in the first part of the book, too. At one point Peucer cites Theon citing Diophantus's rule for multiplying species: "and Theon recited the Diophantine rule" ("Et recitat ex Diophanto regulam Theon"), which is followed by a quotation of Theon in Greek seven lines long that includes this quotation from the *Arithmetica*: "Because the unit, being immutable and always constant, any species multiplied by it will remain the same species".[224] Later, on folio Diiiv, we find "the rule of minutes in Diophantus" ("Estque Regula apud Diophantum de scrupulis") followed by another quotation from Theon, again in Greek: "When multiplying the parts of the unit, the species become different" (Diophantus 1893–95, II, 35.19–20). It only stands to reason that other readers would have also encountered Diophantus through Theon's commentary.

It was just about this time, in 1556, that Peucer was a dinner guest together with the philologist Joachim Camerarius (1500–74) at the home of Johann Ulrich Zasius (1521–70). At one point the conversation turned to logistic. Here is the description given by Camerarius in a letter he wrote to Zasius on November 27, 1556:

> It came to my mind what was said about this and the other liberal arts at that dinner (*convivio*) in your house, in which your pleasant invitation wished

---

224 (Peucer 1556, f. Cviiir). This Greek text is a variation of the text in Diophantus (1893–95, II, 35.2–11), which overlaps the first passage we translated in Section 3.2.1.

me and our friend Peucer to take part. When the conversation came to the authors of Logistic, and I mentioned the name of the Greek Diophantus, who is preserved in the Vatican Library, it seemed that there was a certain hope that a copy could become available to us. Then, inflamed by the desire to see it, perhaps too boldly, but not in an unfortunate way, I assigned to you, with your own accord, to manage the execution of this task, since of course you accepted voluntarily what was assigned to you, and the promise was given with the usual festivity with your elegant cup filled with excellent wine.[225]

The letter serves as a preface to Camerarius's 1557 book *De Graecis Latinisque Numerorum Notis*,[226] abbreviated as *De Logistica* after the prefatory letter. As befits a philologist, the book is not an introduction on how to perform calculations, but rather it reviews the names of numbers and the different notations from various cultures, mainly Greek, and it describes the arithmetical operations on them. He wrote in the introductory letter that his source of information is a Greek booklet or manuscript (*libellus*) dealing with logistic that he had obtained from Johann Baptist Haintzel (1524–81), a patrician in Augsburg who had traveled in Italy in the 1540s.[227] Camerarius cites a whole host of Classical authors in *De Logistica*, the large majority not mathematicians. Diophantus is not among them.[228]

In the chapter on division Camerarius gives Greek names for the powers that ultimately must have come from a manuscript of the *Arithmetica*. He introduces them like this (parenthetical clarifications are ours):

Now, the ranks (*incrementa*) of numbers are revealed in the following manner. In the beginning is a foundation and root of all of procreation. And from there arises a τετράγωνος, & this the logisticians call a δύναμιν, whose root is spoken as πλευρά. Δύναμις means, however, that which comes from the multiplication of an existing number (by itself). This is itself a number δυνάμει ("in power/square"), not in length or μήκει ("in length"), like ἡ διάμετρος δυνάμει διπλασίων ἐστὶ τῆς πλευρᾶς ("the diameter is duplicate in power of the side"). . . .

The common term for δύναμις is *Zensum*, then they increase to a cube, which is the third rank. In fact, a multitude of numbers arises abundantly

225 (Camerarius 1557, A 2v, 1569, C 1v – C 2r). We thank Dr. Michalis Filippou, who helped in the translation of this passage.

226 Full title: *De Graecis Latinisque Numerorum Notis. & praetetea Sarracenicis seu Indicis, cum indicatione elementorum eius, quam Logisticen graeci nominant (quae est methodus conficiendarum Rationum) & vocabulorum artis interpretatione, & aliis quibuidam ad hanc pertinentibus. Additae etiam sunt γνῶμαι graecae serie literarum expositae, ad usum puerilis institutionis. Studio Ioachimi Camerarii Pabenpergensis.*

227 (Camerarius 1557, A 2v, 1569, C 1v). Camerarius even devotes a section to the meaning of the term δύναμις in Plato's *Theaetetus* (Camerarius 1557, F 5v.10–F 7v; 1569, I 2v.3–I 5r.8).

228 He cites Aristotle, Plutarch, Euclid, Euripides, Plato, Lucian, Paul, Xenophon, Isocrates, Demosthenes, Herodotus, Thucydides, Aeschinus, Ammonius, Diogenes Laertius, Terence, Pindar, Cicero, Aulus Gellius, Eutocius, Nicomedes, Quintilian, and Plautus.

in this manner. For the multiplication of a square by a square is called a δυναμοδύναμις, (or commonly as) a *Zensdecenso*. For example, a four four times are 16, & a nine nine times (are) 81. Also, a side by a square number, which is multiplied ἐπὶ τῇ δυνάμει (by a power), produces a κύβος. And this is what is spoken by arithmeticians as ἴσον ἰσάκις ἴσον ("equal by equal by equal": Euclid, *Elements* Def. VII.19). For example, twice two are four, & twice four (are) 8, which is a cube. Likewise, thrice three are 9, & thrice 9 are 27, which is a cube of an odd root. But multiplying a cube by a square, which is ἐπὶ τῇ δυνάμει, this number presently is spoken as δυναμόκυβος, commonly as *sursolidum*. For example, twice two are four, & eight four times are 32, & thrice three are nine, & 27 nine times are 243. Further, for a δυναμοδύναμις ἐπὶ τῇ δυνάμει, which is the multiplication of a square of a square that multiplies a square τῆς δυνάμεως, in the first place (*id est primo*), makes the name δυναμοδυναμοδύναμις (6), commonly a *Zensicubus*. For example, 16 four times are 64, & 81 nine times are 729. This is a cubic and a square number. That which they call a doubled *bissursolidum* (7), & this comes from an even duplication of a root, (and) a *Zenzensdecenso* (8) comes from an odd triplication. The former of these (i.e., the *bissursolidum*) is the multiplication of a square by a square, produced (*ducta*) by a cube, (and) is a κυβοδυναμοδύναμις (7). For example, 16 eight times are 128, (and) 81 nine times are 729.[229] The latter (i.e., the *Zenzensdecenso*) is the multiplication τοῦ δυναμοκύβου by a cube. For example, 32 eight times are 256, (and) 243 twenty-seven times are 6561.[230]

The way Camerarius presents these names is reminiscent of the listing of names without regard to their application by Michael Psellus and the list by the pseudo-Psellus, translated in Section 3.4.1. The treatment is purely arithmetical, without any indication that the Greek or the Latinized German names could serve as unknowns in algebra. In fact, the first-degree term is not the algebraic ἀριθμός/*radix*/*res*, but the arithmetical πλευρά.

After a few more observations, Camerarius gives a table of the powers, using abbreviations, from the πλευρά (1st) up to the κυβοκυβυκυβόκυβος (18th), and on the following page he writes the names out in full, as shown in Figure 3.11.

The scheme is additive up to the eighth power, as in Diophantus, but with different names for the first and sixth powers. Above the eighth power, the names exhibit a hybrid additive/multiplicative scheme. The ninth power is the multiplicative κυβόκυβος, which is why the sixth power had to be expressed as a δυναμοδυναμοδύναμις. Karin Reich (2003, 82) accounts for the names of the higher powers beginning with the ninth, as shown in Table 3.9.

The names appear to be the result of a clash between the additive system of Diophantus and the multiplicative system of sixteenth-century algebraists. The

---

229 He should have written "81 twenty-seven times are 2,187".
230 (Camerarius 1557, C 4v.18, C 5r.18; Camerarius 1569, E 7v.2, E 8r.11).

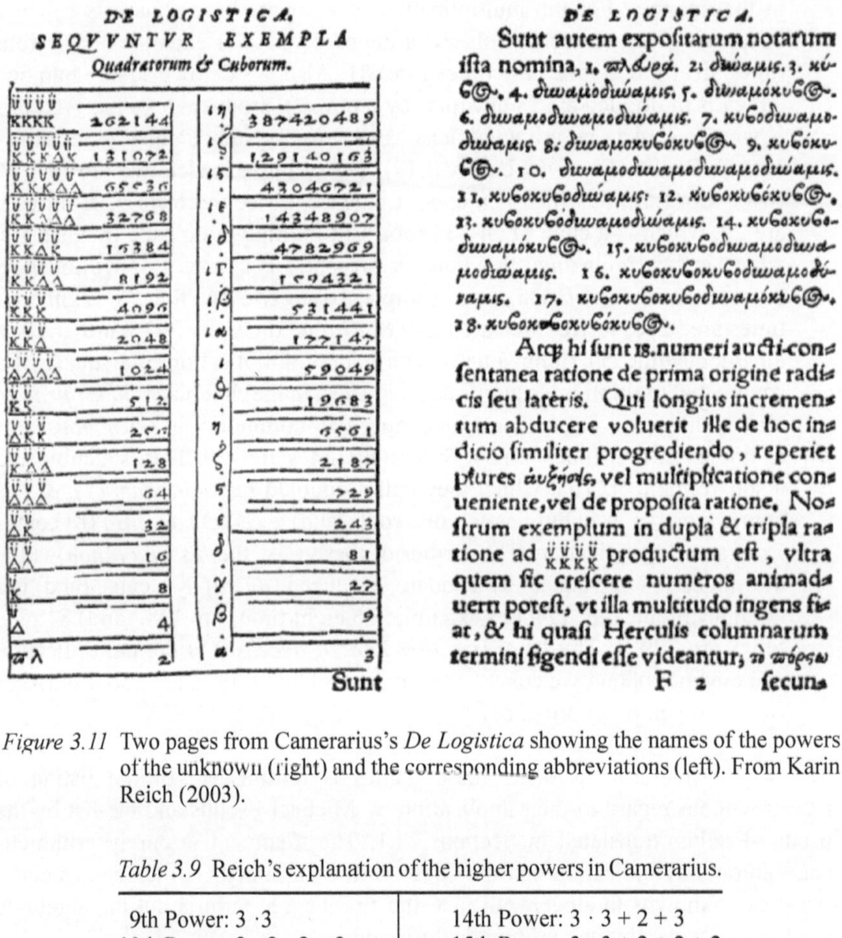

DE LOGISTICA.
SEQVVNTVR EXEMPLA
Quadratorum & Cuborum.

262144	387420489
131072	129140163
65536	43046721
32768	14348907
16384	4782969
8192	1594323
4096	531441
2048	177147
1024	59049
512	19683
256	6561
128	2187
64	729
32	243
16	81
8	27
4	9
2	3

Sunt

DE LOGISTICA.
Sunt autem expositarum notarum ista nomina, 1. πλευρά. 2. δύναμις. 3. κύβος. 4. δυναμοδύναμις. 5. δυναμόκυβος. 6. δυναμοδυναμοδύναμις. 7. κυβοδυναμοδύναμις. 8. δυναμοκυβόκυβος. 9. κυβόκυβος. 10. δυναμοδυναμοδυναμοδύναμις. 11. κυβοκυβοδύναμις. 12. κυβοκυβόκυβος. 13. κυβοκυβοδυναμοδύναμις. 14. κυβοκυβοδυναμόκυβος. 15. κυβοκυβοδυναμοδυναμοδύναμις. 16. κυβοκυβοκυβοδυναμοδύναμις. 17. κυβοκυβοκυβοδυναμόκυβος. 18. κυβοκυβοκυβόκυβος.

Atq; hi sunt 18. numeri aucti consentanea ratione de prima origine radicis seu lateris. Qui longius incrementum abducere voluerit ille de hoc indicio similiter progrediendo, reperiet plures αὐξήσεις, vel multiplicatione conveniente, vel de proposita ratione. Nostrum exemplum in dupla & tripla ratione ad ÜÜÜÜ KKKK productum est, vltra quem sic crescere numeros animaduerti potest, vt illa multitudo ingens fiat, & hi quasi Herculis columnarum termini figendi esse videantur, τὸ πόρσω

F 2 secun-

Figure 3.11 Two pages from Camerarius's *De Logistica* showing the names of the powers of the unknown (right) and the corresponding abbreviations (left). From Karin Reich (2003).

Table 3.9 Reich's explanation of the higher powers in Camerarius.

9th Power: 3 · 3	14th Power: 3 · 3 + 2 + 3
10th Power: 2 · 2 · 2 + 2	15th Power: 3 · 3 + 2 + 2 + 2
11th Power: 3 · 3 + 2	16th Power: 3 · 3 + 3 + 2 + 2
12th Power: 3 · 3 + 3	17th Power: 3 · 3 + 3 + 2 + 3
13th Power: 3 · 3 + 2 + 2	18th Power: 3 · 3 + 3 · 3

notation for the powers is curious, too. It may be that someone misread or was unable to typeset the superscript so that, for example, what we read in the manuscripts as "K^Y" ended up in the printed text as a K with a Ü above it. Or perhaps the superscript was simply translated into the German letter Ü. A few pages later the abbreviations are mentioned again, but now with two more terms added: "We designate ἄλογον or ἄρρητον ἀριθμόν thus: ς, (and) λεῖψιν, which, if anything should be lacking, is ⋎".[231] Diophantus had introduced the algebraic ἀριθμός as

231 The signs are shown as in Camerarius (1557). In Camerarius (1569) they are shown as ς and ϛ, respectively.

"ἄλογος ἀριθμός" ("unspoken Number"), and the sign in Camerarius resembles what is found in some manuscripts. The sign for λεῖψις (lacking, less) in Camerarius does not appear to be drawn well. These signs must have originally been copied from some manuscript of the *Arithmetica*. It is doubtful that it was Camerarius himself who saw it, for if he had, he would have known and cited the author's name. More likely, he took the names of the powers and their abbreviations from Haintzel's notebook, which did not cite the source.

Camerarius had already published his *De Logistica* in 1554 in Augsburg, Haintzel's hometown. The title page from that printing is reported to be the variation *De Logistica et Graecorum, Latinorum, ac Indorum, numerorum notis.*[232] We have not located any copy of this printing.

Camerarius continued asking his contacts for a manuscript of Diophantus for himself and for Peucer, but without any luck.[233] But he did finally see one, in 1572. That story is linked to Xylander, whom we cover later. Indeed, the search for Diophantus by humanists north of the Alps would pick up dramatically in the 1570s. In an August 14, 1571, letter, the Swiss humanist Conrad Dasypodius (1531–1601), professor of mathematics in Strasburg, asked his former colleague Hugo Blotius (1533–1608), who was then in Italy, to send him the works of Pappus and Diophantus (Oestermann 2020, 208). He would not have to wait as long as Camerarius, for he wrote in a letter to Heinrich Rantzau on April 5, 1573:

> Among the other old manuscripts I have acquired from the island of Patmos and surrounding regions the works of Hero of Alexandria, the most important mathematician, and likewise from the Peloponnese other works by him; furthermore a large handwritten codex in ancient Greek writing which had belonged to Cardinal Bessarion, to which he himself attests in his own hand, as well as six arithmetic books of Diophantus, which are particularly beautifully written, not to mention others, as it would be too much to list them. I will only mention the codex by Pappus from Alexandria which I had brought to me from Rome a few days ago and which I bought for 21 French crowns.
>
> (Translated in Oestermann 2020, 209)

It was in this same year that Dasypodius published his mathematical lexicon, titled *ΛΕΞΙΚΟΝ seu Dictionarium Mathematicum, in quo Definitiones, & Divisiones continentur scientiarum Mathematicarum*. This is a bilingual book, with the first part in Latin and the second part conveying nearly the same text in Greek. In the chapter "Vocabula Arithmetica"/"Ὀνόματα τῆς Ἀριθμητικῆς" Dasypolius copied several definitions from Euclid and Diophantus word for word, or sometimes nearly so, without naming his sources.[234] The names of the powers up to the

---

232 (Scheibel 1775–81, II, 357; Watt 1824, I, 187i–j). See also Smith (1908, 263).

233 For Peucer's interest, see Camerarius (1583, 113–14).

234 Dasypodius also copied Euclid's definitions 2 ("number"), 15 ("multiply"), 16 ("planar number"), and 17 ("solid number") from the same book. Although Dasypodius mentions Euclid once in the chapter, nowhere in the book is Diophantus named.

sixth are among the terms covered in this chapter. The Greek terms are given in the second part of the book and are translated into Latin in the first part, where contemporary Latin names are also given. We have collected them in Table 3.10.

Dasypodius makes no distinction between the Greek words τετράγωνος and δύναμις, both of which he translates as *quadratus*. He gives the definitions of the terms from τετράγωνος to κυβόκυβος in order.[235] Those for τετράγωνος and κύβος are copied from Euclid's definitions 18 and 19 in Book VII of the *Elements*, and those for the higher powers are all copied from Diophantus. Dasypodius characterizes these as being extensions of the arithmetical terms τετράγωνος and κύβος. For example, a δυναμόκυβος is the fifth power of a number, and his example is 243. Note that he imposes the Greek meanings on the Latin terms for the fifth and sixth powers that were commonly given a multiplicative meaning.

On the following page, Dasypodius copied Diophantus's explanations of the algebraic ἀριθμός and μονάς.[236] For the former, we have translated the passage in Diophantus as: "While, the (number) which has none of these characteristics, but merely has in itself a multitude of units, is called an unspoken Number". The algebraic ἀριθμός is "unspoken", or "unsayable" (ἄλογος), because it is unknown. Dasypodius translates the phrase ἄλογος ἀριθμός as "numerus irrationalis"

Table 3.10 Names of powers in Dasypodius's *ΛΕΞΙΚΟΝ*.

Greek	Dasypodius's translation	Contemporary Latin
ἀριθμός	numerus	radix
τετράγωνος/δύναμις	quadratus	census
κύβος	cubus	cubus
δυναμοδύναμις	quadrate quadratus	census census
δυναμόκυβος	quadrate cubus	censi cubus
κυβόκυβος	cubice cubus	cubus de cubus

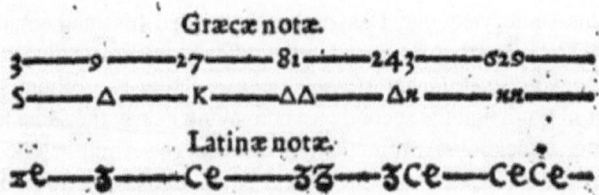

Figure 3.12 Notation for the powers in (Dasypodius 1573, 2v). The 629 should be 729.

235 (Dasypodius 1573, Latin f. 1v, l. 28ff, Greek f. 1v.15ff).
236 (Dasypodius 1573, Latin f. 3r, l. 11–18; Greek f. 2v, l. 7–11).

("irrational number"). Although this was a common meaning of the term, it makes no sense in the context of the definition, or with respect to the *Arithmetica* overall.

Dasypodius also shows the notation for the powers from the *Arithmetica* and compares it with the cossic signs of his own time (the latter typeset oddly; Figure 3.12). It might seem on first glance that Dasypodius was not familiar with algebra. In fact, he knew the algebra of his day well, but he practiced it in an antiquarian manner. In his 1570 book *Mathematicum, Complectens Praecepta: Mathematica, Astronomia, Logistica, Una cum Typis et Tabulis, ad explicationem eorundem necessariis*, he solves two problems by algebra (Dasypodius 1570, 382–84, 386–88). In the first problem, he obtains the simplified equation "19940 aureorum, esse census 60" ($19940 = 60x^2$ in modern notation), which he solves by writing "we will find the size of the *census*, which is one square, by proportions, which we find is 324".[237] He then works out the problem again, this time obtaining the simplified equation "censum unum & res 8 efficere summam 468 aureorum" ($x^2 + 8x = 468$). This he solves not by applying the arithmetical cossic rule that begins by halving the 8, but by drawing a geometric diagram and appealing directly to Euclid's *Elements*, Proposition II.6. In the other problem, the equation simplifies to $x^2 + 816 = 58x$, which he solves again by geometry, this time appealing to *Elements* II.5. In neither case does he name Euclid or cite the *Elements*. This way of solving simplified equations testifies to a desire on the part of Dasypodius to situate algebra as much as possible in the setting of Greek mathematics. With this in mind, his designation of the Latin *censi cubus* and *cubus de cubus* as the fifth and sixth powers instead of the sixth and ninth powers can be seen as another way to bend contemporary algebra toward a Greek foundation. His identification of Euclid's arithmetical terms τετράγωνος and κύβος with the Latin algebraic powers *radix* and *census* might also be a step in the same direction, and with that step the Diophantine higher powers are swept into an arithmetical role as well. We must keep in mind, too, that the lexicon is quite brief, so even if Dasypodius may have been aware of the distinction between arithmetical and algebraic meanings of the terms, it would not have suited him to diverge from the short and simple expanations he found in Euclid and Diophantus.

The translation error "numerus irrationalis" suggests that Dasypodius had not invested much time into studying Diophantus before writing his lexicon. How far into the *Arithmetica* he got is unknown. All we can say based on his lexicon is that he extracted some definitions from the beginning of Book I. It is unfortunate that he did not write more because he was a competent mathematician with original ideas about how contemporary algebra could be united with the Greek tradition.

3.5.3 *Bombelli's L'Algebra*

The few sources we have regarding the life of Rafael Bombelli (1526–1572/3) show that he spent his career as an engineer and architect. The native of Bologna

---

237 "investigabimus per rationem proportionum magnitudinem census, hoc est, quadrati unius, quam inveniemus esse 324" (Dasypodius 1570, 383.24).

engaged in various jobs in central Italy, the most notable being the reclamation of the marshes of the Val di Chiana south of Arezzo. It was during a pause in this work, probably between 1557 and 1560, that Bombelli composed the first draft of his *L'Algebra* in five Books.[238] This was designed as a comprehensive guide to the art, covering everything from basic principles to original investigations into the solutions of equations and the workings of complex numbers. Book I treats the arithmetic of square and cube roots, modeled on Book X of Euclid's *Elements*. Book II gives the vocabulary, notation, and rules of algebraic problem-solving, including his well-known and innovative treatment of cubic and quartic equations. Book III consists of a collection of worked-out problems, Book IV with algebra from the perspective of geometry, and Book V transforms many of the problems from Book III into problems in geometry.

The work was pubished only in 1572, by Giovanni Rossi in Bologna. The printed version, now with the fuller title *L'Algebra Parte Maggiore dell'Arimetica Divisa in Tre Libri*, lacks Books IV and V. We are fortunate, then, that drafts from the 1550s are extant. Two manuscripts were discovered in 1923 by Ettore Bortolotti: Codice B 1569 in the Biblioteca dell'Archiginnasio in Bologna contains the five books complete, while the other manuscript, designated MS 595 miscellanea O.12 of the Biblioteca Universitaria di Bologna, contains only Book III and part of Book IV (Bombelli 1929, 2017). Comparing the manuscripts with the 1572 printing, we see only minor modifications to Books I and II, but Book III underwent drastic revision. These changes were prompted mainly by Bombelli's discovery of a manuscript of Diophantus in the Vatican library. As he relates in the introduction to the published book (the parentheses are his),

> in these recent years, having found a Greek work of this discipline in the library of Our Lord in the Vatican, composed by a certain Diophantus, an Alexandrian Greek author who flourished in the time of Antoninus Pius, & having been shown to me by Mr. Antonio Maria Pazzi of Reggio, professor of mathematics in Rome, and with him judging the author to be very intelligent in numbers (although he does not deal with irrational numbers, but only in him one sees a perfect order of operation) he & I, to enrich the world with such a work, gave ourselves to translate it, and five books (of the seven that there are) we have translated; we were not able to finish the remainder due to the troubles that befell each of us, and in that work we found that many times he cites Indian authors, with which he made me know that this discipline was first applied by the Indians rather than the Arabs.[239]

---

238 Bombelli was baptized January 20, 1526 (Jayawardene 1963, 393), so he was probably born earlier that month. He died some time between June 22, 1572, when he wrote the dedicatory letter to his *L'Algebra*, and May 5, 1573, when a deed was drawn up that makes reference to his heirs (Jayawardene 1963, 395). On the date of the composition of the manuscript and on his engineering projects, see Jayawardene (1965, 304).

239 "ma questi anni passati, essendosi ritrovato una opera greca di questa disciplina nella libraria di Nostro Signore in Vaticano, composta da un certo Diofante Alessandrino Autor Greco, il quale

Antonio Maria Pazzi (d. 1585) is recorded as *lector ad mathematicam* at the University of Rome in the years 1559, 1566, 1567, and 1569,[240] so we cannot determine with any precision just when "these recent years" were. The manuscript they read must have been Vat. gr. 200, since its Book IV is divided into two books, making the total number of books seven.[241] That manuscript begins with the *Arithmetica*, and is followed by the commentary of Maximus Planudes on the first two books, Diophantus's work *On Polygonal Numbers*, and finally Planudes's treatise on Indian calculation. Bombelli may have mistaken the last work as part of the *Arithmetica*, which would account for his claim that Diophantus cites Indian authors.

Bombelli offers no explanation for placing Diophantus in the reign of the emperor Antoninus Pius (138–161 CE). Tannery proposed a possible line of reasoning based in a remark of Petrus Ramus in the 1569 book *Mathematicarum Libri Unus et Triginta*. After describing the six books of the *Arithmetica* of Diophantus, Ramus added "Scripserat & Diophantus harmonica" (Ramus 1569, 37.6). To account for this work in harmonics, Tannery suggested that the name "Leophantus", an author known to have written on the theory of rhythm, was misread by Ramus or Ramus's source as "Diophantus". Heath regarded the details of this theory as "conjectures of the wildest kind" (1885, 14) and devotes three pages to it (pp. 13–16). There is another, perhaps more plausible explanation, though we lack any direct evidence to support it: that someone, reading the reference to Diophantus in Theon's commentary on Ptolemy's *Almagest*, supposed that Diophantus and Ptolemy were contemporaries, and since Ptolemy is known to have been active in the time of Antoninus Pius, so was Diophantus.

In the manuscript of the 1550s Bombelli makes it clear that he was not happy with the traditional Italian names for the first two powers. He wrote that the first degree term

> has been called *cosa*, although in my opinion it is more appropriate if one says *quantità*, and having been an Arabic invention, it could be that in that

---

fù à tempo di Antonin Pio, & havendomela fatta vedere Messer Antonio Maria Pazzi Reggiano publico lettore delle Matematiche in Roma, e giudicatolo con lui Autore assai intelligente de numeri (ancorche non tratti de numeri irrationali, ma solo in lui si vede un perfetto ordine di operare) egli, & io, per arricchire il mondo di cosi fatta opera, ci dessimo à tradurlo, e cinque libri (delli sette, che sono) tradutti ne habbiamo; lo restante non havendo potuto finire per gli travagli avenuti all'uno, e all'altro, e in detta opera habbiamo ritrovato, ch'egli assai volte cita gli Autori Indiani, col che mi hà fatto conoscere, che questa disciplina appo gl'Indiani prima fù, che à gli Arabi" (Bombelli 1572, fourth page of letter to the reader). The spelling "Diophante" implies that he read the name as "Diophantēs" (Διοφάντης). We translate it as "Diophantus".

240 (Jayawardene 1963, 392 n. 8; 1973, 521 n. 24). Bortolotti wrote that Pazzi was professor for the years 1567–75, but he gives no source (Bombelli 1929, 14 n. 1).

241 (Allard 1980, 77, 1982–83, 70; Reich 2003, 84). Allard writes: "Les Arithmétiques de Diophante, en sept livres par division du livre IV en deux parties". They translated not just the "five Books", which would be Books I–III and IV^G, but also the first 20 out of the 30 problems of Book V^G, since they are included in Bombelli's revised Book III.

language *cosa* means *quantità*. . . . And the square of this said *quantità* has been called *Censo*, and I do not know where this name comes from. It is better, according to me, to say *quadrato di quantità*.[242]

Bombelli's reservation about the names was shared by others in his time. Some prominent algebraists writing in the years prior had in fact renamed the Italian or cossic terms for the powers and devised a corresponding notation for them. Cardano (1545) replaced *census*, the tradtional Latin name that he had used in his earlier book of 1539, with *quadratum*. Johann Scheubel (1550) left the first-degree term as the Latin *Radix*, but he renamed its powers as *Prima* (the second power), *Secunda*, *Tertia*, *Quarta*, etc., and he used their abbreviations ra., pri., se., ter., quar., etc. in his notation. Jean Borrel (1559) adopted instead geometric terms and notation. His first three – and only – powers are *lineam*, *quadratum*, and *cubum*, and in notation they are ρ, ◊, and ⌺ (the ρ was a traditional Italian sign for the first power). Also in 1560, Petrus Ramus used the geometric terms *latus*, *quadratum*, and *cubum* for the first three powers, with the abbreviations l, q, and c.

Despite his reservations, Bombelli retained the terms *cosa* and *censo* in his draft. His notation, though, was entirely new. In place of the co, ce, cu, ce ce, etc. of Luca Pacioli and other Italian algebraists, he adopted a numerical notation. The powers are shown as ⌣1⌣, ⌣2⌣, ⌣3⌣, etc., and in Book II of the manuscript the constant term was routinely designated with ⌣0⌣. He wrote the sign for the power above its multitude, so, for example, the polynomial that we would write as

$196 - 14x - \frac{3}{4}x^2$ is shown as $196 \; m \; \overset{\text{①}}{14} \; m \; \overset{\text{②}}{\frac{3}{4}}$ (Bombelli 2017, 60). In the printed edition the power was placed to the right of the multitude, though when shown in figures, the vertical arrangement was still used. Here is an example with the powers after the multitudes and which also shows his printed notation for square root:

$$\text{R.q.L} \; 18.\text{m}.\tfrac{1}{4} \; \overset{2}{} \; \text{I} \; \text{p}.\tfrac{1}{2} \; \overset{1}{} \; \rfloor \quad \text{(Bombelli 1572, 447). In modern nota-}$$

tion it would be written as $\sqrt{18 - \frac{1}{4}x^2} + \frac{1}{2}x$. Bombelli's notation is reminiscent of what we find in Nicolas Chuquet's manuscript *Triparty en la Science des Nombres* of 1484. Chuquet made his powers superscripts, so that the equation we write as $100\frac{1}{2}x^4 = 6x^6 + 72x^2$ is written as ".100.$^4\frac{1}{2}$. egaulx a .6.^{6} p. 72.^{2}".[243] Bombelli,

---

242 "La positione è sempre ponere che la valuta della cosa addimandata sia una quantità, la quale dagli inventori di questa arte è stata chiamata cosa, benché, quanto al mio giuditio, meglio se la confaceva dire quantità et essendo stata inventione arabica potrebbe essere che in quella lingua essa cosa significhi quantità, benché non lo affermo, perché tanto è a dire una quantità, quanto una cosa di numero. Et il quadrato di detta quantità è stato chiamato Censo, che manco non so dove sia derivato tal nome, et meglio quanto a me era dire quadrato di quantità" (Bombelli 2017, 19–20).

243 (Chuquet 1881, 225.14). Chuquet also wrote the exponent "0" for numbers, like ".12.^{0}" for 12, and he even indicated reciprocal powers by placing a "m" after the power. So ".7.$^{3.m.}$" is what we

who certainly was unaware of Chuquet's book, would have written these terms as 100 ½ $\overset{4}{\smile}$, 6 $\overset{6}{\smile}$, and 72 $\overset{2}{\smile}$.

The draft of Book III solves 156 problems using the new notation. These take the form of problems common in Italian abacus manuscripts of the previous century and in printed books of his time, like those in Luca Pacioli's *Summa de Arithmetica Geometria Proportioni & Proportionalita* (1494), Franceco Ghaligai's *Summa de Arithmetica* (1521), Francesco Feliciano's *Libro di Arithmetica & Geometria* (1526), and even Girolamo Cardano's *Practica Arithmetice, & Mensurandi Singularis* (Latin, 1539). Such collections follow a tradition in instruction stretching back through earlier medieval Italian and Latin works, across the Pyrenees and the Mediterranean to the earliest extant Arabic arithmetic and algebra books of the ninth century. As with the earlier texts in Arabic, Latin, and Italian, many of Bombelli's problems are posed in terms of unspecified units, like "Make for me from $\sqrt{60}$ two parts such that subtracting their squares one from the other leaves 6" or "Find for me 4 continuous and proportional quantities such that the second is 2 more than the first and the first multiplied by 8 makes the fourth".[244] Others are framed in some mercantile or otherwise practical context. The large majority of these are really abstract, since the situations they pose would never come up in real life. Problem 34: "Four people are to divide some money, which is 200 scudi. The first must have a third of the other three, the second a fourth of the other three, the third a fifth of the other three, and the fourth the rest. I ask, how much each one gets".[245] Other problems deal with two people who want to buy a horse (Problem 19), the different weights of the head, tail, and body of a fish (Problem 38), the purchase of three kinds of wool (Problem 145), two people enter into a partnership (Problem 46), etc. Some of Bombelli's problems are new, and a few, like 126, 135, and 145, give rise to irreducible cubic equations. Problem 126 asks: "Make for me from 16 two parts such that their squares added together, and multiplied by the difference of the greater square over the smaller square, make 1024".[246] The parts are named "8 $\overset{1}{m}$ 1" and "16 eguale a $\overset{3}{1}$ p $\overset{1}{64}$", and the equation simplifies to "16 eguale a $\overset{3}{1}$ p $\overset{1}{64}$" (our $16 - x^3 + 64x$). Bombelli then works it out according to the rule he gave in Book II. Recall that it is not through cubic equations directly, but through problems like these, sometimes posed in a mercantile

---

write as 7^{-3} (Chuquet 1881, 157.20–21). At least in his "exponents" Chuquet comes dangerously close to working with negative numbers.

244 "Fami di *Rq* 60 due parti che li lor quadrati tratti l'uno dell'altro resti 6", from Problem 31 (Bombelli 2017, 31); "Trovami 4 quantità continoe et proportionali, che la seconda sia 2 più che la prima et che la prima moltiplicata per 8 faccia la quarta", from Problem 115 (Bombelli 2017, 84).

245 "Quattro hanno a partire danari, cio è scudi 200, il primo ne dee havere il terzo de gl' altri tre, il secondo il quarto de gl'altri tre, il terzo il quinto de gl'altri tre, il quarto il reso. Dimando quanto ne tocca per uno" (Bombelli 2017, 33).

246 "Fammi di 16 due parti che li lor quadrati aggiunti insieme, et moltiplicati via la differentia che è dal quadrato maggiore al quadrato minore, faccia 1024" (Bombelli 2017, 89).

setting, that the famous challenges had been framed between Antonio Maria Fior and Tartaglia in 1535 and between Lodovico Ferrari and Tartaglia in 1548.

When composing his draft in the 1550s Bombelli was already inclined toward making his book a purely scientific work, and the *Arithmetica* gave him a justification for introducing changes in that direction. He announced in the beginning of the printed Book II that instead of *cosa* and *censo*, he will call the first two powers *tanto* and *potenza*. Regarding *tanto*, he writes: "I find that the Greek author Diophantus names it like this. This is no small argument, this being his own and true voice, he being such an ancient writer and of so much authority".[247] For the second power, too, "I am resolved to follow Diophantus . . . and call it *potenza*".[248] The Italian word *potenza* indeed means "power" and is a good translation of the Greek δύναμις. But *tanto* means "so much" or "a lot", so rather than being a translation of the Greek ἀριθμός, it is in fact Bombelli's own choice. Because he already used the word *numero* to mean the units in an expression, a literal translation of the Greek was out of the question. Italian algebraists already agreed with Diophantus on the term for the third power, *cubo* ("cube"), and Bombelli's higher powers follow the Italian multiplicative scheme: *potenza di potenza* (4th power), *primo relato* (5th), *potenza cuba* or *cubo di potenza* (6th), *secondo relato* (7th), *potenza di potenza di potenza* (8th), *cubo di cubo* (9th), *potenza del primo relato* (10th), *terzo relato* (11th), and *cubo di potenza di potenza* (12th) (Bombelli 1572, 204). Because he writes his expressions in notation, the awkwardness of the nomenclature does not impinge on his solutions.

The discovery of the Diophantus manuscript not only gave Bombelli a reason to abandon the traditional problems of Book III framed in terms of material units, but it also gave him the arithmetical problems to replace them. In the preface to Book III of the printed work he writes that he has "almost totally deviated from the use of writers of this discipline",[249] who for the most part framed their questions in terms of "selling, buying, bartering, (currency) exchange, interest, defalcation, alloys of coins and metals, weights, partnerships, and with loss and profit, games, and other numerous activities and human actions".[250] Instead, he writes,

> I have set myself in the spirit of truly teaching the discipline of the greater part of Arithmetic (called Algebra), imitating the ancient writers and some of the moderns; because those who have maintained the method mentioned above of [using] similar examples of human activities, more often have been

247 "io trovo, che Diofante Auttor Greco cosi la noma, il ch'è di non picciolo argomento, questa essere la sua propria, e vera voce, essendo egli Scrittore cosi antico, e di tanto valore" (Bombelli 1572, 201–02).

248 "mi son risoluto di seguitare Diofante (come hò fatto nel restante,) e chiamarlo potenza" (Bombelli 1572, 203).

249 "quasi totalmente habbia deviato dall'uso de scrittori di questa disciplina" (Bombelli 1572, 414).

250 "vendite, compere, restitutioni permute; cambij; interessi; deffalcationi; leghe di monete, di metalli; pesi; compagnie, e con perdita, e guadagno, giochi, e simili altre infinite azzioni, e operationi humane" (Bombelli 1572, 414), translation adapted from (Jayawardene 1973, 511 n. 7).

[concerned with] the practical rather than the scientific; yet clearly in every discipline one still sees today the teaching of the theoretical rather than the practical, thinking that the capacity of the human intellect must then be such that one can (possessing the theoretical) come to the use of the practical by himself.[251]

The revised Book III contains 271 problems. Of these, 55 are purely arithmetical problems carried over from the draft, 147 are problems translated from the *Arithmetica*, and 69 are new arithmetical problems.[252] Table 3.11, adjusted from that in (Reich 1968), lists problems taken from Diophantus.

To see how Bombelli presented the problems of Diophantus, we translate Problem 158, which he had translated from *Arithmetica* IV^G.16 (the parentheses are Bombelli's own) (Bombelli 1572, 539):

Problem CLVIII.

Divide a square number into three parts such that the square of the first added with the second, the square of the second added with the third, & the square of the third added with the first, their sum is a square number.

Posit that the second part is 4 $\underset{\smile}{1}$, and the first the side of a square such that adding 4 $\underset{\smile}{1}$[253] makes a square, which will be 1 $\underset{\smile}{1}$ m. 1, and satisfies the first condition. Now the square of the second (which is 16 $\underset{\smile}{2}$) if 8 $\underset{\smile}{1}$ p. 1 is added to it, will make a square, so posit the third to be 8 $\underset{\smile}{1}$ p. 1. It remains to be seen if all three together make a square. But all three together are 13 $\underset{\smile}{1}$, and are equal to a square, which is 169, so that the *Tanto* will be worth 13. Thus, the 4 *Tanti* of the second will be 52, the 8 *Tanti* of the third will be 104, & the *Tanto* of the first 13. Now, returning to the beginning, posit that the first is 13 $\underset{\smile}{2}$ m. 1, the second 52 $\underset{\smile}{2}$, & the third 104 $\underset{\smile}{2}$ p. 1, in order for all three together to make a square, and the square of the first with the second makes a square, the square of the second with the third makes a square, it remains for the square of the third with the first to make a square. The square of this third is 10816 $\underset{\smile}{4}$ p. 208 $\underset{\smile}{2}$ p. 1, which added with 13 $\underset{\smile}{2}$ m. 1, which is the first, makes 10816 $\underset{\smile}{4}$ p. 221 $\underset{\smile}{2}$, and this is equal to a square, the side of which is 104 $\underset{\smile}{2}$ p. however many *Tanti* you want, as long as its square is less than 221,

---

251 "io mi son posto nell'animo di veramente insegnare la disciplina della parte maggiore della Ari-metica (detta Algebra) immitando gli antichi scrittori, e qualche uno de Moderni; perche gli altri, che hanno tenuto quel modo detto disopra, di simili essempij di attioni humane, piú tosto hanno havuto del pratico; che del scientifico; e chiaramente in ogni disciplina si vede tutt'hora inseg-narsi la Teorica, e non la pratica, pensandosi, che la capacità dello intelletto humano debbia poi essere tale; ch'egli per se debbia (posedendo la Teorica) venire all'uso della pratica" (Bombelli 1572, 414).

252 The last problem is numbered 272, but there are two problems numbered as 49, and there is no 256 or 257. See Table 3 in Bombelli (2017, 152) for the 55 problems taken from the draft.

253 Text has 4 $\underset{\smile}{2}$.

Table 3.11 Problems of Bombelli (B) translated from Diophantus (D).

B	D	B	D	B	D	B	D	B	D	B	D
2	I.1	50	I.26	86	II.25	121	III.15	161	IV^G.19	202	V^G.1
8	I.2	51	I.29	88	II.26	122	III.16	162	IV^G.20	203	V^G.2
10	I.4	53	I.30	89	II.27	123	III.17	164	IV^G.21	209	V^G.3
11	I.5	54	I.31	90	II.28	124	III.18	167	IV^G.22	210	V^G.4
13	I.6	55	I.32	91	II.29	126	III.19	168	IV^G.23	211	V^G.5
14	I.7	56	I.33	92	II.30	127	III.20	169	IV^G.24	212	V^G.6
15	I.8	57	I.34	93	II.31	128	III.21	170	IV^G.25	213	lem1 V^G.7
16	I.9	58	I.35	96	II.32	129	IV^G.1	171	IV^G.26	215	lem2 V^G.7
18	I.10	59	I.37	97	II.33	133	IV^G.2	172	IV^G.27	216	V^G.7
19	I.11	60	I.39	98	II.34	134	IV^G.3	173	IV^G.28	217	lem V^G.8
21	I.12	61	II.8	99	II.35	136	IV^G.4	179	IV^G.29	218	V^G.8
26	I.13	62	II.9	100	III.1	137	IV^G.5	180	IV^G.30	219	V^G.9
27	I.15	63	II.10	101	III.2	138	IV^G.6	181	IV^G.31	220	V^G.10
28	I.16	66	II.11	102	III.3	140	IV^G.7	182	IV^G.32	221	V^G.11
30	I.17	67	II.13	103	III.4	141	IV^G.8	188	IV^G.33	222	V^G.12
31	I.18	69	II.14	104	III.5	142	IV^G.9	189	lem IV^G.34	225	V^G.13
35	I.19	70	II.15	105	III.6	148	IV^G.10	190	IV^G.34	226	V^G.14
36	I.20	72	II.16	106	III.7	149	IV^G.11	191	lem IV^G.35	232	V^G.15
37	I.21	73	II.17	110	III.8	150	IV^G.12	192	IV^G.35	233	V^G.16
41	I.22	74	II.19	111	III.9	151	IV^G.13	194	lem IV^G.36	234	V^G.17
42	I.23	78	II.20	113	III.10	152	IV^G.14	195	IV^G.36	235	V^G.18
44	I.25	81	II.21	114	III.11	153	IV^G.15	196	IV^G.37	236	V^G.19
49	I.27	83	II.22	116	III.12	158	IV^G.16	197	IV^G.38	237	V^G.20
49'	I.28	84	II.23	117	III.13	159	IV^G.17	200	IV^G.39		
		85	II.24	120	III.14	160	IV^G.18	201	IV^G.40		

and it is 104 ② p. 1 ①, whose square will be 10,816 ④ p. 208 ③ p. 1 ②. Removing like from like and reducing, we will have 208 ① equal to 220, so that equalized, the *Tanto* will be worth $1\frac{3}{52}$. Then the first number, or the first part, which was 13 ② m. 1,[254] will be $\frac{36,621}{2704}$; the second, which was 52 ②, will be $\frac{157,300}{2704}$; and the third, which was 104 ② p. 1, will be $\frac{317,304}{2704}$; & the divided square number will be $\frac{511,225}{2704}$, and they make what was proposed.

In this problem Bombelli makes the same assignments as Diophantus, so he arrives at the same solution. But in many problems he posits different given numbers, and he makes different assignments. Much like al-Karajī five and a half centuries earlier, he adapted the Diophantine problems to fit the style of his own work on algebra.

It is difficult to say what impact the Diophantine problems of *L'Algebra* had in the first half of the 1570s. Copies of the original printing are rare, and perhaps Bombelli's book only became better known with the second printing of 1579, four years after the appearance of Xylander's translation of the *Arithmetica*. In any case, people would not have known which problems came from Diophantus because Bombelli did not differentiate between his problems and those translated from the *Arithmetica*. We do know that his book was helpful to Bachet, whose 1621 edition and Latin translation became the standard for the succeeding two and a half centuries. Heath (1910, 21–22) noted that "Bachet admits his obligations to him, remarking that in many cases he found Bombelli's translation better than Xylander's and consequently very useful for the purpose of amending the latter".

3.5.4 Xylander's 1575 Latin translation

By the early 1560s, a century after Regiomontanus had announced his discovery of the *Arithmetica*, there were at least 14 manuscripts of Diophantus's work in European collections. Yet during that long span of time we know of no instance of a mathematician other than Regiomontanus reading it. One could argue that Diophantus remained largely unknown until the 1537 publication of the Padua oration, but all we find from mathematicians of the succeeding three decades are mentions of his name, in Scheubel (1550), Peletier (1554), Peucer (1556), and Roca (1564). What apparently prevented informed readings of the *Arithmetica* by mathematicians were the language barrier and limited access to manuscripts. Regiomontanus had the rare combination in the 1460s of being competent in both mathematics and Greek and having Bessarion's manuscript at hand. That combination would not occur again until ca. 1570 with Bombelli and Pazzi, and soon after with Wilhelm Holtzman, who published the first Latin translation of the work under his Hellenicized name Xylander in 1575.

---

254 Text has 53 ② m. 1.

How Xylander came into possession of a manuscript was not as straightforward as it was with Regiomontanus or Bombelli. His story begins with Gian Vincenzo Pinelli (1535–1601), an Italian humanist, scholar, and collector who amassed perhaps the largest and most important manuscript library in private hands at the time. Pinelli's parents were Genovese nobility who had relocated to Naples, and as a young man Pinelli traveled to Padua to study law. He never took a degree, but because the intellectual climate of Padua suited him, he settled there and remained for the rest of his life. Pinelli's house became a center for humanist interactions, and he maintained contact with many of Europe's most enlightened minds. Among his visitors and correspondents were Daniele Barbaro (1514–70, Joachim Camerarius the elder (whom we have already met), Paolo Manuzio (1512–74), and later, Galileo Galilei (1564–1642). Relevant for Diophantus is the mathematician and philologist Matteo Macigni (ca. 1510–82), professor of the Collegio dei Filosofi e dei Medici in Padua. Macigni read Greek and he, too, collected manuscripts. Among the authors in his library were Pappus, Ptolemy, Apollonius, Nicomachus, (pseudo-)Hero, Euclid (an epitome), and, as we are about to see, Diophantus as well.[255]

Around the middle of the 1500s Pinelli acquired the manuscript Mediolanensis Ambrosianus 91 Sup., which had been copied in the previous century from Bessarion's Marc. gr. 308. Pinelli's manuscript contains the *Arithmetica* with the commentary of Maximus Planudes on Books I and II written in the margins, Diophantus's *On Polygonal Numbers*, the fragment of the treatise on Indian numerals of Maximus Planudes, as well as a fragment of the *Pneumatica* of Hero (Allard 1982–83, 59). In the first half of the 1560s, in what appears to have been a reciprocal agreement between Pinelli and Macigni, Pinelli's preferred scribe Camillo Zanetti (d. 1587) copied for Macigni the Diophantine portions of the Vatican manuscript Vat. gr. 304 on 86 folios: the *Arithmetica*, the scholia of Maximus Planudes, and *On Polygonal Numbers*. Pinelli then loaned his Ambr. 91 Sup. to Macigni, who improved his own manuscript by making many marginal corrections to the text of the *Arithmetica* and by adding, on folios 86v–89, the treatise on Indian Numerals of Planudes that was lacking in the Vatican manuscript. Macigni's manuscript thus became the first to combine the two main families: it descends from the non-Planudean codex Vat. gr. 304, and it contains corrections that descend from the Planudean codex Marc. gr. 308. This made Macigni's copy superior to any other manuscript at the time, though it was still riddled with errors and omissions that stem back at least to Byzantine copyists. Pinelli then hired Zanetti, not later than 1565, to transcribe Macigni's manuscript so he, too, would have a copy of the improved text. Macigni's manuscript of Diophantus now resides in the Herzog-August Bibliothek in Wolfenbüttel as Guelferbytanus Gudianus gr. 1. Pinelli's copy is lost. After his death his library was scattered, and the manuscript may have been in one of the several chests of books tossed

---

255 On Pinelli and his library, see Grendler (1980). On Macigni's manuscripts, see Köhler and Milch-
    sack (1913, 1ff).

overboard in 1604 by Turkish pirates who had captured one of the three ships transporting his belongings to Naples.[256]

Although Pinelli had a serious interest in mathematics and science, he was not the kind of person who would have undertaken a study or translation of the *Arithmetica*. Macigni at least wrote a commentary in Latin on the first book, titled *Annotationes in librum primum Diophanti de arithmetica*, which occupies folios 90–94 of his manuscript (Köhler and Milchsack 1913, 1). This commentary remains unpublished, and we know of no further investigations on Macigni's part.

We turn now to another scholar with Paduan ties, the Hungarian diplomat Andreas Dudith (1533–89).[257] Dudith was partly educated in Italy, beginning in his late teens in Verona and Padua. Then, after some years traveling to Brussels, Paris, and London in the service of Cardinal Reginald Pole, he was back in Italy by early 1558 for a two-year stint in Padua to study law. There he continued cultivating erudite friendships with people such as Marc Antoine Muret (1526–85), Johannes Sambucus (1531–84), and the Venetian printer Paolo Manuzio. Dudith devoted most of his time to Greek and Latin rather than law, and it was Manuzio who published his Latin translation of *De Thucydidis Historia Iudicium* of Dionysius of Halicarnassus in 1560. We do not know when Dudith met Pinelli, but given their common interests and contacts they must have known each other at least as early as Dudith's stay in Padua. Another contact of Dudith's was Nicaise van Ellebode (1535–77), the Flemish philologist who stayed in Padua for nearly a decade, beginning late 1561 or early 1562 until the spring of 1571.[258] Ellebode resided in Pinelli's house, and in addition to his study of Greek literature, he took a degree in medicine. He and Dudith knew each other by 1561, when Dudith recommended Ellebode to Paolo Manuzio, though we do not know how they first met (Costil 1935, 93).

In 1560 Dudith was recalled to Hungary, where he was appointed bishop and embarked on a busy diplomatic career in the service of the Hapsburg emperor Ferdinand I, and later Maximilian II. While working in Krakow he fell in love with a Polish noblewoman, and in the spring of 1567 they were married. When news of the scandalous union became public later that year, Dudith found himself alienated from his Italian friends, including Pinelli and Manuzio, and he was soon excommunicated from the Catholic Church. Partly because of the fallout from his marriage, Dudith's career then entered a lull, during which time he turned to the study of astrology and mathematics. He needed a teacher, and it was Camerarius who recommended the mathematician Johannes Praetorius (1537–1617). From

---

256 On the manuscripts, see Köhler and Milchsack (1913, 1) and Allard (1982–83, 71, 114, 1984, 320, 1985, 299, 311).

257 For Dudith's biography, see Costil (1935) and Almási (2009). Details are taken from these sources.

258 Ellebode had been teaching in Trnava near Bratislava until 1561, and he was already in Padua on April 1 of the following year. After leaving Padua he arrived in Vienna on May 13, 1571 (Klaniczay 1973, 322, 324; Orbán 2021).

June[259] 1569 to January 1572 Praetorius lived in Dudith's house in Krakow, and it is during this time that Dudith took steps to acquire a manuscript of Diophantus. He knew that Pinelli had one, but he could not ask him directly for a copy. But Ellebode, one of the few contacts in Italy who had remained Dudith's friend, agreed to help. Pretending that he wanted the copy for himself, Ellebode was granted Pinelli's permission to have Zanetti and another scribe, Manuel Moro, copy the copy that Zanetti had recently made of Macigni's manuscript.[260] This work was accomplished in the first part of 1570. As of April 4 Dudith had not yet received his manuscript, though he knew that it consisted of 88 folios.[261] Dudith's manuscript has been identified as Vat. Reg. gr. 128 (Costil 1935, 297–303; Allard 1985).

Dudith was open to sharing his manuscript even before it reached him. On February 8, 1570, he wrote to Camerarius: "In Italy they are transcribing for me in Greek the books of Diophantus, whom Regiomontanus praises, 'on the thing and the *census*', as he himself calls it. If you would like, I'll see that a copy is sent to you, too".[262] Camerarius had been seeking a manuscript since at least 1556, so this must have been welcome news! In all likelihood Dudith was in possession of his manuscript by the middle of 1570, and in the second half of 1571 it was on loan. There are two passages that tell us something of the manuscript in this period, one from a letter of Camerarius and the other from Xylander's introduction to his published translation. On September 5, 1571, Camerarius wrote to Dudith:

> As far as the book of Diophantus is concerned, it has been sent to me and I thanked him, and I have indicated honestly that I would guard it. And me,

---

259 (Allard 1985, 315 n. 39).

260 On this manuscript, see Allard (1982–83, 71, 114, 1984, 320, 1985, 299, 311). Ellebode wrote to Adrian van der Mylen on April 20, 1571: "Several months ago Dudith requested that I arrange to have the *Arithmetica* of Diophantus transcribed. . . . I obtained a Diophantus from Pinelli and had it copied for Dudith, however pretending to transcribe it for myself, since as you know, Dudith's name is hateful, because he took a wife when he was bishop" ("Duditius abhinc aliquot menses me rogarat ut curarem sibi describenda Diophanti ἀριθμητικά. . . . Sumsi a Pinello Diophantum et Duditio exscribendum curavi, simulans tamen mihi describi, quoniam Duditii nomen, ut scis, odiosum est : duxit enim uxorem cum esset episcopus"). (quoted in Costil 1935, 296). The ellipses are Costil's.

261 Dudith wrote to Camerarius on April 4, 1570: "I have not yet received my Diophantus. There are eighty-eight sheets in folio, as we say, of large format paper. I am writing this to you so that you will understand that it is not nothing, although Regiomontanus, if I remember correctly, claims to have seen twelve books 'on the thing and the *census*'. So I wonder if this book is not a fragment of some of these books seen by Regiomontanus". ("Diophantem nondum accepi. Folia sunt in folio, ut loquimur, octoginta octo maioris papyri. Quod ideo scribo, ut intellegas non nihil esse. Quamvis Regiomontanus vester, si recte memini, duodecim libros se vidisse De re et censu affirmat, ut dubitem ne hic liber fragmentum aliquod illorum sit") (Dudith 1995, 128). See also Allard (1985, 306).

262 "Describuntur mihi Graece in Italia Diophanti libri, quos Regiomontanus dilaudat, De re et censu, ut ipse vocat. Si eius desiderio teneris, faciam ut ad te quoque perferatur" (Dudith 1995, 124). See also Allard (1985, 306).

after one period of life then another, because of this, I am not the one who will be able to accomplish the translation of this work. Therefore, I will do whatever your magnificence commands. I await your decision about this.[263]

The manuscript that Camerarius acquired would have been a copy of Dudith's. As we shall see shortly, it was made without Dudith's approval, though we can imagine that Camerarius felt that it had already been promised to him. No other related letters are extant.

In the introduction to his Latin translation Xylander recounts that in October 1571, while visiting Wittenberg, the mathematicians Sebastien Theodoric and Wolfgang Schuler showed him Dudith's manuscript of Diophantus. Xylander copied one problem from it, and on leaving Wittenberg he worked it out. He next met with the philosopher and physician Simone Simoni (1532–1602) in Leipzig. He showed Simoni the problem and expressed his desire to edit and translate the work. Simoni then relayed the request to Dudith, who agreed to loan his manuscript to Xylander (Diophantus 1575).

At this time Praetorius (1537–1616) accepted a position as professor of mathematics in Wittenberg, so in January 1572 he took the Diophantus manuscript with him and passed it off to Simoni in Leipzig on the way. Simoni then took it to Xylander in Heidelberg. This much can be derived from the February 1, 1572, letter that Dudith wrote to Praetorius:

> I have no doubt that you have already given Simonio the Diophantus intended for Xylander. If you have not yet done it, please do so without delay. Take great care, however, that what was copied from my (manuscript) does not appear before Xylander publishes mine with its translation. Whoever copied it acted treacherously, and Camerarius did not fulfill his duty as a friend. Also, I do not know if it is the act of a good man for allowing a copy be made without my permission.[264]

Although Camerarius had declared that he would not translate Diophantus in his letter to Dudith, he said nothing about publishing or commenting on the Greek text, and this may be what Dudith wanted to prevent. It is to Xylander that Dudith gave that privilege. Praetorius was now in Wittenberg, not far from Leipzig, where both Camerarius and Simoni lived. Supposing that Camerarius had intended to

---

263 "Quod ad Diophanti librum attinet, et eum ad me missum esse egi gratias, et significavi fideliter apud me custodiri, et me cum hac aetate tum alias ob causas eum non esse qui interpretando illo operam navare possim. Itaque facturum me, quidquid magnificentia tua iussisset. De quo indicium sententiae tuae exspecto" (Dudith 1995, 283).

264 "Diophantum te iam Simonio ad Xylandrum transferendum tradidisse non dubito. Id si nondum fecisti, oro te ut sine ulla mora facias. Da autem diligentem operam, ut non exeat is, qui ex meo descriptus est, antequam Xylander meum cum interpretatione sua in lucem emittat. Infideliter egit, qui descripsit, nec Camerarius amici officio functus est et haud scio an boni etiam viri, qui meo iniussu cuiquam describendi copiam fecerit" (Dudith 1995, 328).

publish the text and that the pressure from Dudith's friends did not dissuade him, then his death in 1574 would have put an end to the project.

Xylander thus would have received Dudith's manuscript in the early part of 1572. He completed the project of translation and commentary in only three years – he wrote the dedicatory letter on August 14, 1574, and the book was published in 1575 in Basel, with the title *Diophanti Alexandrini Rerum Arithmeticarum Libri sex, quorum primi duo adiecta habent Scholia, Maximi (ut coniectura est) Planudis. Item Liber de numeris Polygonis seu Multiangulis.* Xylander included the commentary of Planudes after each problem of Books I and II, and he gave his own commentary after most problems through about the middle of Book V. The fragment of *On Polygonal Numbers* occupies the last five pages, again with Xylander's commentary.

For his translation Xylander devised a notation that mimics the Greek, but with sixteenth-century modifications. He translated ἀριθμός, δύναμις, and κύβος as *numerus, quadratus,* and *cubus,* respectively, and he adopted their initials N, Q, and C in the notation. Following sixteenth-century practice, he put the multitude before the abbreviation, he has no abbreviation for units, and he wrote + and − for "and" and "lacking". Recall that the manuscripts are not uniform in showing the notation. Very often the name of the power is spelled out, and sometimes the notation for ἀριθμός or κύβος is present where the word takes its non-algebraic meaning. Like Bombelli before him and Clavius, Bachet, and Tannery after him,[265] Xylander made the editorial decision to write all algebraic expressions in notation, regardless of whether the manuscript showed it or not. Here are two examples of his notation from Book IV^G, the first from Problem 18 (Xylander's 19) and the second from Problem 28 (Xylander's 29). In both cases his manuscript showed the powers in words:[266]

$$\text{1 C C} \dagger \text{1 N} \dagger 64 \text{———} 16 \text{ C}$$

$$2\tfrac{1}{4} \text{ Q Q} \dagger 1\tfrac{1}{2} \text{ Q} \dagger \tfrac{1}{4} \text{———} \text{1 C ———} 3 \text{ N}$$

The task of translating the *Arithmetica* was not an easy one due to the number of textual corruptions that had crept into the manuscripts before the Renaissance copyists took over. It was common for numbers to have been miscopied or be missing altogether, words were sometimes left out, and many passages had become garbled. Most errors were easy to detect and fix, but for many the correct reading was elusive. Xylander did what he could to make sense of problematic passages, placing

---

265 Stevin and Viète wrote algebraic expressions in notation, too, but their versions of the problems entail wholesale rewordings of the solutions. Bachet kept the words in the Greek text and Xylander's notation in his translation.

266 (Diophantus 1575). In modern notation these are $x^6 + x + 64 - 16x^3$ and $2\tfrac{1}{4}x^4 + 1\tfrac{1}{2}x^2 + \tfrac{1}{4} - x^3 - 3x$, respectively. We have not seen Dudith's manuscript, but in both Marc. gr. 308 and Vat. gr. 304, the progenitors of Vat. Reg. gr. 128, the names of the powers in these examples are all spelled out (Marc. gr. 308, ff. 193r, col. a, l. 13–16, 204v, col. b, l. 14–17; Vat. gr. 304, ff. 97v, l. 3–4, 100v, l. 4–5).

LIBER II.   57.

vetur.) *Rurfus quadrati latera pofito* 1 N † 3, *quadratus fit* 1 Q † 6 N † 9. *Si pro altera parte ftatuamus* 6 N † 9. *pro reliqua* 4 N † 8, *aut* 2 N † 5, *perinde eft utrum ponas. quod multo adhuc magis uariabitur, fi* 1 N † 4 *quadrati latus ponas. Multo magis fi diuidendus numerus maior fit, & plures lateris pofitiones admittat. Quæ ego indicanda duxi, non etiam ftylo perfe-quenda, quod exercitationis anfam dediffe in tanta fecunditate fatis haberem.*

XVII.   *Inueniantur duo numeri quorum fit quæ præcipitur inter fe ratio: & uter-que cum quadrato qui proponitur coniunctus, quadratum numerum conficiat. Efto maior triplus minoris, & uterque adiecto nouem fiat quadratus. Heic à quo-cunque quadrato, cuius latus fit* 1 N † *aliquot unitates, 9 aufero; is numerus alter quæfitorum erit. Efto minor* 1 Q † 6 N, *erit maior* 3 Q † 18 N. *reftat ut ad hunc quo-que adiecto* 9, *fiat quadratus. At fit* 3 Q † 18 N † 9. *hoc ergo æquale eft quadrato. Fin-go quadrati latus* 2 N——3. *erit* 1 N, 30. *Ergo minor numerus eft* 1080. *maior* 3240. *quorum uterq, fi addas ei* 9, *eft quadratus.*

SCHOLION.

Quòd quadratum effingit à latere, in quo eft defectus, hanc habet rationem. 3 Q † 18 N † 9 *non nafcitur ab uno Numero: alioqui enim effet* 1 Q *duntaxat, cùm fint heic tr.a: ideo latus ponit* 2 N *cum defectu tali, ut Quadrati fuperent, Numeri autem deficiant, & unitates æquentur quantitate. fic enim unitates uniuerfæ hinc fub-mote, uniuerfas illinc fubmouebunt: & Quadrati de Quadratis auferentur, ut Quadrati fuperfit certæ Nume-rorum multitudini æqualis. Ergo addito utrobiq, defectu, & æqualibus amputatis, deminuentur notæ numerorum, fietq,* 1 N, 30, 1 Q *autem* 900. *Ergo minor, cum fit* 1 Q & 6 N, *erit* 1080. *maior* 3 Q † 18 N, 3240. *Et* 9 *ad* 1080 *adiectis, quadratum faciunt* 1089, *cuius latus* 33. *ad* 3140, *quadratum* 3249, *cuius latus* 57. *Et* 33 *funt* 1 N † 3, *latus quadrati* 1 Q † 6 N † 9. 57 *autem funt* 2 N——3, *latus quadrati* 4 Q † 9——12 N *Sunt au-tem* 4 Q † 9, 3609. *unde fi auferas* 12 N, *hoc eft* 360. *relinquentur* 3249.

XYLANDRI.

In pofitione hoc obferuabis facile, inueniri poffe numerum qui cum 9 *fit quadratus: idq, qua-dratu habere pro latere* 1 N, & *tot unitates, quot unitatibus radix quadrati propofiti conftat. heic* 3, *cum quadratum fit* 9. *Ergo* 1 N † 3 *in fe ductum facit* 1 Q † 6 N † 9. *Ergo alter nume-rorum eft* 1 Q † 6 N. *ad quem* 9 *fi addas, utique quadratum habebis. Hunc minorem ef-fe, placuit autori. nam fi maiorem ftatuiffet, minor* ⅓ Q † 2 N *fuiffet (superfedere autem minu-tijs interdum licet, cùm fcilicet non plus compendij in ijs eft quam in integris.) Nunc maior eft* 3 Q † 18 N. & 3 Q † 18 N † 9, *quadratus. Heic cum uideas tres fpecies,* Q, N, & *abfolutum numerum* 9, *elegantius eft induftria ita inftituere ratiocinationem, ut in æquatione harum una* [*Fictio lateris*] *prorfus elifa, reliqua comparentur. quo facit*——3, *qui* 9 *procreat, quæ* 9 *ab altera parte abole-* [*pro quadrato*] *ant. & quia defuturæ funt radices in nouo quadrato , quod multiplicationis doctrina patefit: ideo ponuntur radices in latere tot, ut in quadrato recens efficto fint plures* Q *quam* 3 Q, *qui e-rant in priore. Quia concife hæc funt ab autore dicta, æquationem fic fubijciamus in gratiam difcentium.*

2 N——3 *latus effingendi quadrati.*

2 N——3

——6 N † 9

4 Q ——6 N

4 Q ——12 N † 9 *quadratum, æquale quadrato* 3 Q † 18 N † 9.   *Primùm ab utraq, parte abijcio* 9. *deinde* 12 N *utriq, parti addo, & utriq,* 1 Q *adimo . fiunt* 30 N *æquales* 1 Q, *hoc eft, nominum facta, ut monuit interpres, deminutione,* 1 N *eft* 30. *Cætera habes in fcholio. Sed obiter memineris, heic quoq, uarias folutiones dari. nam quotquot ultra* 2 *pofueris radices*——3. *res aliter atq, aliter fuccedet . quod unico exemplo docere fatis eft. Sit latus effingendi quadrati* 3 N——3. *erit quadratus* 9 Q——18 N † 9. *æqualis* 3 Q † 18 N † 9. *æquatione côpofita,* 1 N *eft* 6. *numeri* 72 & 216. *utriq, additus* 9 *facit* 81 & 225. *quos quadratos effe nemo ignorat. Ita fi quæ-ras duos dupla proportionis numeros, quorû utriq,* 25 *additus quadratos faciat: cû alios inueni-es, tum* 600 & 1200, *latere pofito* 2 N——5, &c. *Per duplicationem æquationem has non nifi per-plexitate magna fe obijciente fieri poffe, experiendo fenties.*

XIIX.

greater importance on the mathematical meaning than on philological accuracy. If he could not fix a nonsensical passage, he translated it into nonsensical Latin and addressed the mathematical meaning as best he could in his commentary.[267]

Figure 3.13 shows page 57 of Xylander's translation. Problem II.16 (XVII in his numbering) begins on the sixth line, followed by the commentary of Planudes ("SCHOLION") and Xylander's commentary ("XYLANDRI"). One textual corruption is evident in the fourth line of the translation. Our translation from Tannery's Greek edition reads "From whatever square (formed) from a multitude of Numbers and ⟨3⟩ units", where the "3" was restored by Bachet in his 1621 edition. The 3 is necessary because the units in the square must be 9. Not aware that the "3" had been dropped, Xylander translated this passage as "heic a quocunque quadrato, cuius latus fit 1N + aliquot unitates" ("Now for whatever square, whose side makes 1N + some units"), making the multitude of Numbers 1, and transferring the word "multitude" (πλήθους) to the units as the word *aliquot*. He explains in his commentary that the units must be 3, but with his correction he lost the generality of what Diophantus intended by writing "1N". This adjustment did not cause any problem later in the solution, but in many other cases his reading made no sense.

3.5.5 *Gosselin, Stevin, and Clavius*

With the publication of Xylander's translation, mathematicians finally had complete access to the extant Diophantus, at least what was available in Europe. No longer did they need to rely solely on Regiomontanus's oration for knowledge about the existence of Greek algebra. Soon after publication the problems and techniques from the *Arithmetica* found their way into the works of other algebraists. In this section we review Guillaume Gosselin (1577), Simon Stevin (1585), and Christoph Clavius (1608), three premodern algebraists writing on Diophantus in the decades after Xylander's translation.

The earliest author we know to have benefited from the publication of Diophantus is Guillaume Gosselin (ca. 1552–ca. 1590), mathematician of the Collège de Cambrai of the University of Paris. So little is known of Gosselin's life that even the years of his birth and death must be approximated. He is described as being young when his translation of Tartaglia's arithmetic appeared in 1578, and according to Bachet, he died in a plague around 1590.[268] In the mid-1570s Gosselin was occupied with the writing of two books. The first to appear in print was his Latin algebra book *De Arte Magna Libri Quattuor*, published in Paris by Gilles Beys in 1577. The following year his French translation with commentary of the first two parts of Niccolò Tartaglia's 1560 *General Trattato di Numeri, et Misure*, titled *L'Arithmetique de Nicolas Tartaglia Brescian, Grand Mathematicien, et Prince des Praticlens*, was published by the same press. Later, in 1583,

---

267 See Morse (1981, 195ff) for a description of problems with the manuscripts and Xylander's approach to the text. On pp. 200ff. she examines a particularly confused example.
268 (Cifoletti 1992; Gosselin 2016; Diophantus 1621).

Gosselin's short guide to the teaching of mathematics, *De Ratione Discendae Docendaeque Mathematices Repetita Praelectio* (henceforth *Praelectio*) was printed.[269] These are Gosselin's only known writings.

There is no indication in his works that Gosselin had read Bombelli's *L'Algebra*. His acquaintance with Diophantus came from Xylander's translation and also through a manuscript that was later loaned to him by Davy du Perron (Cifoletti 1992, 64). Like Bombelli, Gosselin saw in the *Arithmetica* a classical, theoretical work on algebra that justified the turn away from the practical guides common in the 1570s. But contrary to Bombelli, his interest lay not with the *Arithmetica* as a source of solved problems, but as a source of new algebraic techniques. The two main methods of solving forthcoming equations, *al-istiqrā'* and the double equality, occupy a prominent place in Gosselin's treatment of algebra. These techniques were new to Europe. Despite the fact that several Arabic algebraists had given instructions and provided examples for solving indeterminate problems by these methods, none of their works had been translated into Latin. Gosselin became the first European to draw attention to them outside the context of the solutions to individual problems, which he did in all three of his books.[270] Together with rules for operating on expressions, rules for solving simplified equations, rules for working with independent unknowns, as well as other rules, the Diophantine techniques were part of the foundation of an algebra that had earned its place as a speculative art (Cifoletti 1992, 122ff).

In keeping with a theoretical presentation, *De Arte Magna Libri Quattuor* gives only a few solved problems. These are taken from several sources, including Pedro Nuñez, Jean Borrel, Girolamo Cardano, Luca Pacioli, and Diophantus (Gosselin 2016, 390ff). Although Gosselin mentions many problems from different books of the *Arithmetica*, he solves only four problems of Diophantus, all from Book I: 6, 18, 19, and 20.[271] He worked out I.6 and I.20 differently from Diophantus, using multiple independent unknowns. Perhaps the lack of problems would have been made up with the edition and commentary of the Greek text that he had been commissioned to write. Gosselin mentions this project in several places in his works, and he was even loaned a manuscript, but the plan was never carried out (Cifoletti 1992, 62–64).

We now move still farther north, to Leiden, home of the Flemish polymath Simon Stevin (1548–1620). During his career Stevin took on a variety of jobs in finance and engineering, and for some time he was also quartermaster general for the army. His many books range from the purely practical, including topics like bookkeeping and navigation, to the purely theoretical, including music theory and dialectics. He is well known in mathematics for his promotion of decimal

---

269 The *Praelectio* is known through a single copy in the Bibliothèque Nationale in Paris. It has been edited twice, first in Cifoletti (1992) accompanied with an English translation, and again in Gosselin (2016) with a French translation.

270 In *De Arte Magna*: (Gosselin 1577, 73rff, 2016, 374ff). In the translation of Tartaglia: (Gosselin 1578, II, ff. 88vff). In *Praelectio*: (Cifoletti 1992, 180, 217; Gosselin 2016, 467).

271 (Gosselin 1577, ff. 80r, 80v, 84v, 86r, 2016, 390, 392, 398, 400).

fractions in his 1585 booklet *De Thiende*, but for us it is his *L'Arithmetique*, written in French and published in Leiden that same year, that is of interest. The long subtitle describes its contents:

> Contenant les computationes des nombres Arithmetiques ou vulgaires : Aussi l'Algebre, avec les equations de cinc quantitez. Ensemble les quatre premiers livres d'Algebre de Diophante d'Alexandrie, maintenant premierement traduicts en François. Encore un livre particulier de la Pratique d'Arithmetique, contenant entre autres, Les Tables d'interest, La Disme; Et un traicté des Incommensurables grandeurs : Avec l'Explication du Dixiesme Livre d'Euclide.
>
> (Stevin 1585, title page)

With all that, it is not a surprise that the book takes up 890 pages. Stevin knew both Bombelli's *L'Algebra* and Xylander's translation of Diophantus. From Bombelli he adopted his algebraic notation, with small adjustments. Instead of designating the powers in semicircles, as $\lrcorner$, $\llcorner$, $\lrcorner$, etc., he put the numbers in full circles, as ①, ②, ③, etc. (Stevin 1585, 28). And instead of Bombelli's p and m, he wrote the signs "+" and "−". Like most authors writing on algebra, he included worked-out problems, which are, as was typical, given at the end of his exposition of the rules of algebra. He first gives 27 purely arithmetical problems that involve numbers and/or techniques not found in Diophantus, including irrational numbers, negative numbers, irreducible simplified equations of the third or fourth degree, and solutions involving multiple independent unknowns. One of these problems, XXIIII, is the same as Problem I.17 in the *Arithmetica*, and Problem XXIII is *Arithmetica* I.30. The solutions to both take advantage of two or more unknowns, and are thus different from the solutions in Diophantus. These 27 problems are followed by the translation of the first four books of the *Arithmetica*, as promised in the subtitle. Stevin wrote in the introduction to his translation that he did not manage to get to the last two books "pour empeschement d'autres occupations plus necessaires" (Stevin 1585, 433). Bombelli and Pazzi did not finish the fifth book for much the same reason, and perhaps it was also for lack of time that Xylander had only given detailed commentaries through the middle of Book V.

Stevin's translation was not intended to be literal. Like al-Karajī, Bombelli, and later Viète, Stevin wrote out the solutions of Diophantus in his own language, with his own format. A short example illustrating his approach is shown next. This is Problem II.24, numbered as 25 by Xylander. It consists of a nice melding of the rhetorical text of Diophantus with a tabular solution that includes the final values of the different expressions in the right-hand column. Stevin adopted this format for all problems, including his initial 27 problems that do not come from the *Arithmetica*. Later, in 1625, Albert Girard published a corrected and augmented edition of *L'Arithmetique* in which he provided translations of Books V^G and VI^G.[272]

---

272 By "arithmetical number" Stevin means a rational number. The first line after "Construction" reads "Soit la somme des quarrez" ("Let the sum of the squares"), which is an error. It should be "Soit le quarré de la somme" ("Let the square of the sum"). The "78th problem" is the 78th

QUESTION XXV.

Let us find two Arithmetical numbers such that the square of their sum, added to each number: The sums are squares with commensurable roots.

CONSTRUCTION

Let the sum of the squares	1②	$\frac{1}{121}$
And the smaller required number be such that added to the sum of the squares, the sum is a square according to the requirement, let	3②	$\frac{3}{121}$
And the greater number be also such that added to the sum of the squares (which is the first in order) the sum is a square according to the order, let	8②	$\frac{8}{121}$
The sum of the two numbers is 11②		
its square	121④	$\frac{1}{121}$
Equal to the first in order	1②	$\frac{1}{121}$

Which reduced, 121② will be equal to 1, & by the 78th problem, 1① will be worth $\frac{1}{11}$.

I say that $\frac{3}{121}$ & $\frac{8}{121}$ are the two required numbers.

Demonstration. The sum of the two numbers $\frac{3}{121}$ & $\frac{8}{121}$ is $\frac{11}{121}$, its square $\frac{121}{14641}$, adding to this the smaller number, $\frac{3}{121}$, makes the square $\frac{484}{14641}$, its root $\frac{22}{121}$. Likewise the said square $\frac{121}{14641}$, adding the greater number, $\frac{8}{121}$, makes the square $\frac{1089}{14641}$, its root $\frac{33}{121}$, these are therefore squares with commensurable roots according to the requirement; this is what was to be shown.

Stevin could not resolve the textual problems with Xylander's translation for Problems II.29, 30, III.12, 13, IV^G.3, 4, 7, 26, and 35 (Xylander's numbering). Stevin remarked at the beginning of IV^G.7: "Note. We describe this 7th question word-for-word because of errors in the text".[273] The translations of these problems are indeed literal, even keeping Xylander's notation.

The last premodern algebraist we will cover is Christoph Clavius (1538–1612),[274] whose Latin textbook *Algebra* was published in Rome in 1608. Clavius was born 70 years earlier in Bamberg, a little north of Nuremburg, and after studying for some time in Coimbra in Portugal, he spent his career in Rome as a Jesuit priest and scientist. He is best known for his work on the Gregorian calendar and for his defense of Ptolemaic astronomy, while in mathematics it is his commentary on Euclid's *Elements* (1574) that gained him fame. He intended the *Algebra* to be a clear, introductory textbook teaching the basics of the method. Although it contains a few minor innovations, the book reads like a cossic textbook from many decades earlier, and in fact he had apparently planned on writing the book as early as the 1580s. The book that most influenced Clavius was the *Arithmetica Integra* (1544) of Michael Stifel, from which Clavius adopted the cossic algebraic notation. Clavius also drew from the *Libro de Algebra en Arithmetica y Geometria* (1567) of Pedro Nuñez, and it is likely that the two mathematicians had met during Clavius's stay in Coimbra.

Clavius devotes 159 pages to the rules of algebra in Chapters 1–28, then 221 more pages to 321 worked-out problems in Chapters 29–32. He begins Chapter 1 by naming the same two "inventors" of algebra as Nuñez, "Gebrum Arabum Astronomum" and Diophantus.[275] Like Nuñez and other authors of the period 1550–74, Clavius brings up Diophantus in connection with Regiomontanus's Padua oration and does not mention Xylander's translation. Chapter 29 comprises 174 problems solved by algebra whose equations reduce to two terms. Without naming Diophantus, he reproduces 49 problems from Books I and II of the *Arithmetica* in this chapter. We list these in Table 3.12. Chapter 30 gives 47 problems whose simplified equations have three terms, Chapter 31 gives 70 mercantile and other practical problems, and Chapter 32 has 30 geometry problems solved by algebra. Back in Chapter 12, Clavius states nine impossible problems, and Problem 4 from this group is Diophantus's Problem I.14, but with a choice for the given number that violates the determination.

---

algebraic rule given earlier in the book on p. 376, for solving equations of the type $ax^2 = b$ (Stevin 1585, 502–03).

273 "Nota. Nous descriprons ceste 7e question de mot à mot, à cause des erreurs du texte" (Stevin 1585, 563).

274 See Rommevaux (2012) for details regarding Clavius's *Algebra*. The only person we know to have pointed out that Clavius solves problems taken from Diophantus is Chikara Sasaki (2003, 76–77).

275 (Clavius 1608, 4; Nuñez 1567, a ij v). While several sixteenth-century authors name Geber, Diophantus, and al-Khwārazmī, the only author we found who names only Geber and Diophantus is Nuñez.

Table 3.12 Problems from Clavius's chapter 29 taken from Diophantus.

Clav	Dioph						
1,2	I.1	18	I.13	71	I.26	87	I.33/II.5
7	I.8	19,21	I.14	72,73	I.29	88	II.6
8	I.9	42	I.31	74	I.32	89	II.7
9,31	I.2	57	I.15	75	I.34	90	II.8
10	I.3	58	I.16	76	I.34.1	91	II.9
11	I.4	59	I.17	77	I.34.2	92,93	II.10
12	I.5	60	I.18	78	I.35	98	II.11
13	I.6	61	I.19	80	I.37	99	II.12
14	I.7	62	I.21	81	I.38	100	II.13
15	I.10	63	I.22	82	I.38.1	101,139	II.14
16	I.11	64	I.23	84	I.38.3	102,140	II.15
17	I.12	67	I.24	85	I.38.4		
		68	I.25	86	I.39		

Unlike in al-Karajī, Bombelli, and Stevin, Clavius followed the format of problems in the *Arithmetica* by declaring given numbers only after stating the general enunciation. For example, Problem 59 starts out this way:

> 59 *To find four numbers such that each trio makes four proposed numbers, in such a way that a third part of the sum of the proposed numbers is greater than each individual proposed number.*
> The first, second, and third should be 20. The second, third, & fourth 22. The third, fourth, and first 24. And finally, the fourth, first, & second 27.[276]

In this problem Clavius adopts the same given values and solves the problem the same way as Diophantus. But in other problems he diverges from Diophantus. For example, in modern notation Problem I.17 of the *Arithmetica* asks for two numbers x and y such that $\begin{cases} x+a=m(y-a) \\ y+b=n(x-b) \end{cases}$ for given numbers a and b and given ratios m and n. Diophantus chooses $a = 30$, $b = 50$, $m = 2$, and $n = 3$, while in Clavius the numbers are $a = 40$, $b = 30$, $m = 4$, and $n = 9$. And where Diophantus begins by assigning y to be 1N 30u, Clavius instead assigns x to be 1N (1²℈ in his notation) and proceeds to set up and solve a different equation. Like al-Karajī, Bombelli, and Stevin, Clavius appropriated and adjusted the problems of Diophantus for his own use.

---

276 "59 Quatuor numeros invenire, ita ut terni quique efficiant quatuor numeros propositos; dummodo tertia pars summae propositorum numerorum maior sit singulis propositis numeris. Debeat primus, secundus, ac tertius facere 20. Secundis, tertius, & quartus 22. Tertius, quartus, ac primus 24. Quadratus denique, primus, & secundus 27" (Clavius 1608, 195).

What was Clavius's source for the problems taken from Diophantus? We know that he did not borrow them from Xylander because the wording of the enunciations is different in every problem. For example, Xylander's enunciation to Problem I.37 (his I.40) reads: "Postulantur duo numeri certam habentes rationem, ut minoris quadratus ad summam numerorus, datam rationem habeat" (Diophantus 1575, 41), while Clavius's Problem 80 reads: "Duos numeros in data proportione invenire, ita ut quadratus minoris ad eorum summam habeat proportionem datam quamlibet" (Clavius 1608, 213). Nor did Clavius borrow the problems from Bombelli, Stevin, or Viète. The most obvious evidence that neither Bombelli nor Viète was his source is that the 1608 *Algebra* contains some problems that are absent in both of those authors.[277] Stevin is not his source, either, because of the many ways Clavius often remained close to the Greek text where Stevin worked the problems his own way. For just one example, Clavius gives both of Diophantus's solutions to Problem II.8 (his 90), while Stevin gives only the first solution. We know that Clavius could read Greek (Lattis 1994, 22), so the easiest explanation for his source is that he read one of the manuscripts in the Vatican library.

3.5.6 *François Viète and the beginning of modern algebra*[278]

If there can be said to be revolutions in mathematics, then one was initiated in algebra with the work of François Viète. Not that Viète sought to overthrow traditional numerical algebra when he created a corresponding algebra for geometric magnitudes. He specifically intended his new specious logistic (*logistice speciosa*), as he called it, to be used for deriving geometrical theorems for the purpose of more efficiently generating astronomical tables. He promoted his creation as a restoration of a lost technique of the Greeks that could be applied in conjunction with numerical algebra, which he called "numerical logistic" (*logistice numerosa*). The unknowns and knowns in Viète's specious logistic are no longer numbers but are the non-arithmetized "sizes" of geometric magnitudes that are manipulated independently of any diagram, and with this geometric foundation came an entirely new understanding of polynomials and equations.

Diophantus figures largely in the work of Viète, both as a supposed ancient practitioner of Viète's specious logistic, and as a source for many of the problems that Viète included in his 1593 book *Zeteticorum*. In order to properly assess the ways that Diophantus was appropriated and reinterpreted by Viète, we will have to spend a few pages describing Viète's specious logistic, and how it differs from the premodern algebra that had been practiced thus far.

Viète introduced *logistice speciosa* in 1591 in his brief publication *In Artem Analyticem Isagoge*, which takes up only 11 pages after the dedicatory letter. On

---

277 Clavius mentioned Bombelli and Viète in his brief discussion of irreducible cubic equations, but he does not seem to have taken note of their worked-out problems (1608, 49).

278 This section is mainly condensed from Oaks (2018c), which gives an account of the nature of Viète's algebra.

the reverse of the title page, under the heading "Opus restitutæ Mathematicæ Analyseos Seu Algebrâ novâ", Viète lists several titles he proposed to write on his new algebra, most of which were eventually published (Viète 1591, 1v). The claim that his algebra is both a restoration of analysis and also new is made clearer in the dedication:

> Behold, the art which I present is new, but in truth so old, so spoiled and defiled by the barbarians, that I consider it necessary, in order to introduce an entirely new form into it, to think out and publish a new vocabulary, having gotten rid of all its pseudo-technical terms lest it should retain its filth and continue to stink in the old way.[279]

What was once a pristine algebra practiced by the Greeks, so Viète contended, had become "spoiled and defiled" over the centuries. The vocabulary and notation of the numerical algebra (*logistice numerosa*) of Diophantus had degenerated into the vocabulary and notation of cossic algebra, while the geometrical algebra (*logistice speciosa*) that Diophantus had practiced covertly, and whose principal stage Viète called "zetetics", had been forgotten altogether:

> Diophantus used zetetics most subtly of all in those books that have been collected in the *Arithmetica*. There he assuredly exhibits this method in numbers but not in species, for which it is nevertheless used. Because of this his ingenuity and quickness of mind are the most to be admired, for things that appear to be very subtle and abstruse in numerical logistic are quite familiar and even easy in specious logistic.[280]

Such is the kind of explanation we would expect from one determined to remain faithful to the norms of Greek mathematics and to uphold the Greek origin of algebra. Like his restoration of the lost *Tangencies* of Apollonius that he published in 1600 under the title *Apollonius Gallus*, his attribution of specious logistic to Diophantus gave his project the cultural legitimacy that many in his time valued.

The motivation for creating a geometrical algebra came from astronomy. Ptolemy had adopted the approach of Hipparchus, in which propositions in classical Greek geometry form the basis for calculating trigonometric and other

---

279 Translated in Klein (1968, 318). "Ecce ars quam profero nova est, aut demum ita vetusta, & a barbaris defaedata & conspurcata, ut novam omnino formam ei inducere, & albegatis omnibus suis pseudo-categorematis, ne quid suae spurcitiei retineret, & veternum redoleret" (Viète 1591, f. 2v).

280 Translation adjusted from Viète (1983, 27). "Zeteticem autem subtilissime omnium exercuit Diophantus in iis libris qui de re Arithmetica conscripti sunt. Eam vero tanquam per numeros, non etiam per species, quibus tamen usus est, institutam exhibuit, quo sua esset magis admirationi subtilitas & solertia, quando quae Logistae numeroso subtiliora adparent, & abstrusiora, ea utique specioso familiaria sunt & statim obvia" (Viète 1591, f. 8r; 1646, 10). The punctuation is different in the 1646 printing.

astronomical tables. The propositions are proven in the style of Euclid, where magnitudes without numerical measure are compared via ratio and proportion. Afterward, numerical values are assigned to the lines and calculations are performed. Geometric magnitudes are thus subject to two ways of measurement: they can be compared without any arithmetization as they are in Euclid, and numbers can be assigned to them and the calculations can shift to the realm of arithmetic. Inspired at least in part by the 1588 publication of the *Collection* of Pappus, whose Book VII covers geometrical analysis, Viète sought to create an algebra for deriving these geometrical theorems that could then serve as a basis for numerical calculation.

For his numerical logistic, Viète adopted the vocabulary and notation from Xylander's translation of Diophantus. Recall that the Greek names of the powers of the unknown are translated as *numerus, quadratum, cubus, quadrato-quadratum*, etc., and are abbreviated as N, Q, C, QQ, etc. In Viète this algebra remains premodern. Monomials still consist of a species and its multitude, and polynomials are still aggregations. For his new geometrical algebra, Viète chose a geometrical notation. Greek geometers had used single letters to label magnitudes whose sizes but not spatial relationships were relevant to the proposition at hand, like the lines A, B, and Γ in Proposition I.22 and region A in Proposition II.14 of Euclid's *Elements*. Viète similarly denoted the known and unknown magnitudes in his geometrical algebra with single letters A, B, G, D, Z, etc., more or less following the order of the transliterated Greek alphabet. To distinguish knowns from unknowns, he reserved consonants for the former and vowels for the latter. He calls these letters, or signs, the "species" of his specious logistic. These species and their powers are identified as being "magnitudes" (*magnitudines*) throughout his published works, where magnitudes are specifically geometric magnitudes of dimension 1, 2, 3, and as we shall see shortly, also 4, 5, etc.

Because there is no unit line in Euclidean geometry, dimension must be respected. A one-dimensional magnitude is called a *latus* (side), and is written in notation as a single letter: A, B, G, etc. The measure of a square with side of measure A is called "A quadratum" ("A square"), the measure of its cube is "A cubus" ("A cube"), followed by "A quadrato-quadratum" ("A square-square"), etc. Similarly for the other letters: "B quadratum" is the measure of a square with side B, etc. Magnitudes introduced in a problem with dimension greater than one are given other labels. "A planum" ("A plane"), "B planum", etc., designate the measures of two-dimensional magnitudes; "A solidum" ("A solid"), "B solidum", etc. designate three-dimensional magnitudes; and the labels continue with "plano-planum", "plano-solidum", following the pattern of the names of the powers.

In addition to solving problems with his new algebra, Viète sometimes gave algebraic versions of metrical equalities that he derived via Euclidean-style propositions in geometry. Proposition XVI of Viète's 1593 *Supplementum Geometriae* is a nice example in which he translates a geometrical equality based on a diagram into an equation in species, and from there he assigns a numerical value to the known species to form a numerical equation (Viète 1593b, f. 17v; 1646, 249).

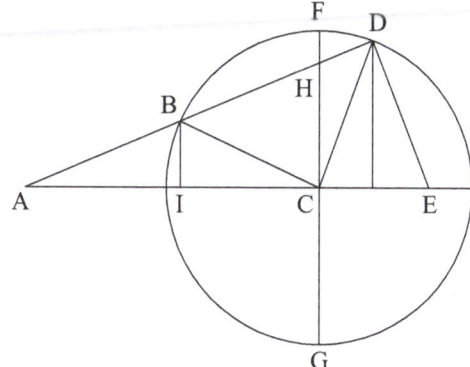

Figure 3.14 The diagram from Viète's *Supplementum Geometriae*, Proposition XVI.

The geometrical equality naturally makes reference to the letters labeling points in the diagram: "the cube of AC less triple the solid under AC & the square of AB is equal to the solid under CE & the square of AB".[281] True to the wording of similar equalities in Greek geometry, no operations are intended in this statement. The "cube of AC" is the geometric cube whose side is line AC. The "solid under AC & the square of AB" is the rectangular prism whose base is the square with side AB and whose altitude is line AC, and the "triple" means that there are three of them. What is more, the magnitudes in this statement are not drawn in position in the diagram. The rectangular prism cannot be drawn at all because AB and AC are not at right angles to each other, and further, if we are supposed to consider three of them, where would the other two be located? The equality is about geometrical *sizes*, without reference to spatial relationships. This equality is in line with the similar equalities of Greek authors beginning with Euclid, who routinely state metrical relationships like this one. There, too, the magnitudes often cannot exist in position, such as when Apollonius writes in *Conics* Proposition I.37 that the rectangle contained by ΛZ, ZE is equal to the sum of the rectangle contained by ΔE, EZ, and the square on ZE, even if no such rectangles can be drawn because the points Δ, Z, and E are colinear.[282]

Today we might reinterpret Viète's equality using operations borrowed from arithmetic, as $AC^3 - 3AC \cdot AB^2 = CE \cdot AB^2$. The idea of applying arithmetical operations to magnitudes was already fairly common in Viète's time, and he took it a step further with his new algebra by removing any reference to the diagram

281 "Cubus ex AC minus triplo solido sub AC & quadrato ex AB, aequalis est solido sub CE & quadrato ex AB". We translate *sub* as "under", in the sense of "subject to", or "determined by". This proposition is important for the trisection of an angle.
282 In Commandino's translation it is written as "rectangulum dfe rectangulo def una cum quadrato fe est aequale" (Apollonius 1566, f. 27r, l. 23).

in his equations. For this equality the unknown line AC is named A, and lines CE and AB are taken to be equal in length and are called Z. This new A is not the point of the same name in the diagram but is the algebraic unknown that represents the size of the line AC. Similarly, Z is a known species and not a point in the diagram: "A cubus minus Z quadrato ter in A, aequetur Z cubo". We translate this as "A cube minus thrice Z square by A, equals Z cube", and it is modernized in Witmer's 1983 translation as "$A^3 - 3Z^2 A = Z^3$" (Viète 1983, 404). Here the "A cubus" does not require a multitude to become a value, like the "1" that the numerical version "1C" must possess. The "A cubus" is already a value, and not a kind like the premodern "C". Moreover, the "Z quadrato", of which there are three copies, is explicitly multiplied by the "A" by virtue of the preposition *in* ("by"). Now monomials are built from the multiplication of values and not as a multitude–species pair, though a multitude is present if needed, like the "ter" in the second term. Throughout Viète's publications we find expressions like "A + B", "B in D", "$\dfrac{\overline{A\ cubus\ in\ E - E\ cubo\ in\ A}}{A - E}$", "$\overline{B + D}$ in A2", and "$\overline{G - B}$ cubus" (the overbar plays the role of parentheses).[283] For the first time, algebraic expressions can contain multiplications, the divisions that had been a patch to a system based in aggregations become legitimate, and a term can consist of a polynomial raised to a power.[284] We should also now reinterpret the words *plus* and *minus* and their signs + and − as operations, too.

On the other hand, there are certain kinds of expressions that are acceptable, and even common, in numerical algebra but make no sense in geometrical algebra. These are all consequences of the fact that the dimension of every term in specious logistic must be a positive integer. In numerical logistic it is fine to divide 3Q by 1Q to get 3u, but Viète can only divide a higher dimensional term by a lower dimensional term. He cannot divide B square by A square, for example. Similarly, he cannot take the reciprocal of any term. And where one can take square, cube, or other roots of any numerical amount, one cannot take, say, the square root of an odd-dimensional term in specious logistic because the dimension of the result would no longer be a whole number. This restriction does not seem to have limited the kinds of problems Viète solved, nor did it limit his options for working them out, even if his goal was numerical calculation.

Another major difference between specious and numerical logistic is that where coefficients in the latter must be specific, determinate numbers, those in the former must be indeterminate. One can name particular numbers, such as 12 or $\frac{14}{3}$, by virtue of the fact that numbers are by their nature all different from each other. The

283 Witmer translates these last three as $\dfrac{A^3E - E^3A}{A - E}$, (B + D)2A, and (G − B)^3, respectively (Viète 1615b, 44.10, 1646, 106.46, 1983, 211), (Viète 1593d, f. 18r, l. 17; 1646, 70.29, 1983, 138), and (Viète 1615b, 52.20, 1646, 110.42, 1983, 219). The overbar was added in the 1646 edition.

284 In numerical algebra polynomials were raised to powers too, but like multiplication, this was an operation that was performed in time, with an outcome. Multiplications and powers were not included in the static expressions admitted to equations.

number 12, for example, is even, composite, the double of a perfect number, etc., while $\frac{14}{3}$ is rational, lies between the integers 4 and 5, etc. In geometry without a unit line one cannot name a specific geometric size. Unlike numbers, lines are inherently all identical and only have a size in relation to other lines. For this reason the known magnitudes in an equation in species must all be designated with letters. But this obligatory generality gave specious logistic a powerful advantage that Viète exploited and which has been rightly touted as a critical aspect of his algebra. Never before had there been an algebra that accomodated undetermined knowns, and which thus permitted one to see in the solution to an equation how it arose from the given values.[285]

Viète next transfers his equation to numerical logistic, which means choosing a number for the consonant. He makes Z 1 and the unknown A 1N, to get: "1C − 3N, aequatur 1". Now he has returned to premodern algebra, where the 3N is a multitude–species pair, and the "−" is to be read as meaning "less" and not the modern "minus".

Certain problems needed to be overcome to make this geometrical algebra work. What was certainly the biggest hurdle was to somehow account for magnitudes of dimension greater than three. Because Viète was chiefly interested in numerical calulation for producing tables, it did not matter how such higher dimensional magnitudes were conceived. Instead, it was the *usefulness* of the propositions he derived while investigating angular sections that justified them. In modern notation, the main proposition shows how cos $n\theta$ and sin $n\theta$ can be found from cos θ and sin θ.[286] Viète worked out these relations geometrically, without using any algebra. In the process he found that the relations for angle $n\theta$ call for magnitudes of dimension n, and he did not stop with $n = 3$. For $n = 4$, for instance, he states one proportion as:

> In quadruple ratio, the second hypotenuse is similar to the square-square of the first hypotenuse; the base, to the square-square of the first base, *minus* sextuple the plane-plane under the square of the first perpendicular & the square of the base of the same, *plus* the square-square of the perpendicular.[287]

Viète then states these relations in algebraic terms, calling the base, altitude, and hypotenuse of the right triangle with acute angle θ as D, B, and Z, respectively. The ratio of the hypotenuse to the base in the triangle with acute angle 4θ is, in Witmer's modernized version, as Z^4 is to $D^4 − 6D^2B^2 + B^4$. It did not matter

---

285 See Montucla (1758, 488–89) and Oaks (2018c, 289).

286 Viète published his results on angular sections in two books: 1593c, f. 15r and 1595, f. 8, reprinted in Viète (1646, 314ff, 371ff). The propositions were later gathered together with further results by Alexander Anderson in Viète (1615a).

287 "In ratione quadruplâ, Hypotenusa secundi fit similis Quadrato-quadrato-Hypotenusae primi. Basis, Quadrato-quadrato basis primi, minus plano-plano sexies sub quadrato perpendiculi primi & quadrato basis eiusdem, plus quadrato-quadrato perpendiculi" (Viète 1593c, f. 15v; 1595, f. 9r; 1615b, 12, 1646, 289, 316, 372). The punctuation is slightly different in the 1646 printing.

that magnitudes like "the square-square of the first hypotenuse" are meaningless because when numbers are applied to the sides of the triangle with acute angle θ, the derived proportions give correct values for the sides of the triangle with acute angle $n\theta$.

Another problem to overcome was how to form equations if magnitudes have no absolute size. The objects of geometry are related through ratio and proportion, so Viète simply postulated that any proportion $a : b :: c : d$ is equivalent to the equation $ad = bc$, regardless of the dimensions of the terms. The third problem was how to add and subtract terms of different dimension. This he solved by multiplying lower-degree terms by known magnitudes so that all terms in a polynomial have the same dimension. We saw this earlier in the equation from *Supplementum Geometriae*, where each term has dimension three. For another example, in one proposition from his *De Recognitione Aequationum* (1615) Viète states the numerical equation "124992N − 1CC, aequari 249920". Because this is a sixth-degree equation, in any version in specious logistic the side (the first-degree term) must be multiplied by a five-dimensional coefficient, and the units must be six-dimensional. Viète writes it as "B plano-solidum in A − A cubo-cubo, aequetur Z solido-solido" ("B plane-solid by A − A cube-cube equals Z solid-solid"), which Witmer modernizes as "$B^{ps}A − A^6 = Z^{ss}$".[288]

In Viète all numbers are the measures of magnitudes, so all problems that he considers are geometry problems. There is no mention of an arithmetic independent of geometry in any of his works. And because his algebra was based in geometry, it was not lost on him that his new algebra could be applied to geometry problems outside of astronomy. He noted that for the last step, once a simplified equation or proportion has been found, one can either assign numerical values to the knowns and calculate the unknown, or one can take the final equation or proportion as instructions for the construction of a geometric solution. This does not mean that the species are abstract objects that transcend arithmetic and geometry, as Jacob Klein argued.[289] It is instead an acknowledgment of the two ways of measuring magnitudes, either through non-arithmetized ratio and proportion or through numerical calculation.

Viète gives a collection of solved problems by specious logistic in his *Zeteticorum* (1593/1600).[290] Eighty-two problems are presented, divided into five books. Athough these problems are stated in geometric terms, the only diagrams he shows are right triangles drawn in 11 problems in Books IV and V, and they serve only to illustrate the Pythagorean theorem. The kinds of problems solved are consistent with his other works in that they show his interest in metrical rather than spatial relationships of magnitudes. In fact, Viète gives only arithmetical

---

288 (Viète 1615b, 54, 1646, 111, 1983, 221), from Chapter XVIII, Proposition V in *De Recognitione Aequationum*. The equations are written as they appear in the 1646 printing.

289 (Klein 1968). Klein's thesis is addressed in Oaks (2018c, 290–93).

290 The first 16 folios were printed in Tours in 1593, and the remainder in Paris in 1600. Folio 16 in the 1593 printing ends in the middle of Zetetic IV.6 (Van Egmond 1985, 362).

solutions to his problems. Later, Jean-Louis Vaulezard added many geometrical solutions "En lignes" in his French translation of the *Zeteticorum* (Viète 1630), and other mathematicians interested in geometrical construction did the same in their solutions found through Viète's algebra, starting with Marino Ghetaldi in his *De Resolutione & Compositione Mathematica Libri Quinque* (Ghetaldi 1630).

Thirty-four of the problems in Viète's *Zeteticorum* are taken from Diophantus, and are listed in Table 3.13. Viète even mentions Diophantus by name in Problems IV.1, 2, 3, 7, and V.14.

Next we translate two problems of Diophantus as reworked by Viète.[291] The first is Problem I.2, Viète's version of *Arithmetica* I.4. This simple determinate problem shows the resolution of a proportion into an equation. Also, note that the ratio he posits is not numerical, but is given as the ratio of two sides R and S.

Zetetic II[292]

Given the difference of two sides, & their ratio, to find the sides.

Let B be the given difference of the two sides, and the given ratio of the smaller side to the greater be as R is to S. The sides are to be found.

The smaller side is A. Thus, the greater will be A + B. Therefore, A is to A + B as R is to S. Resolving this proportion, S by A will equal R by A, + R by B. And by translation with the contrary sign of affection, S by A, − R by A will

Table 3.13 Problems of Viète taken from Diophantus.[292]

Viète	Dioph	II.4	I.27	IV.8	II.12	V.7	III.10
I.1	I.1	II.6	I.28	IV.9	II.13	V.8	III.11
I.2	I.4	II.8	I.29	IV.10	lem1 $V^G.7$	V.9	$VI^G.3$
I.3	I.2	II.17	$IV^G.2$	IV.11	lem2 $V^G.7$	V.10	$VI^G.4$
I.4	I.7	II.18	$IV^G.1$	V.1	III.6	V.11	$VI^G.5$
I.5	I.9	IV.1	II.8	V.2	II.19	V.12	$V^G.5$
I.7	I.5	IV.2-3	II.9	V.3	III.7	V.13	$IV^G.31$
I.8	I.6	IV.6	II.10	V.4	III.8	V.14	$V^G.30$
II.3	I.30	IV.7	II.11	V.5	III.9		

291 We silently correct whatever typographical errors there are in one printing or the other. No error occurs in both printings. Also, in a few instances one printing has *plus* in place of +, or *minus* in place of −. We have used the symbols in each case.

292 (Viète 1593d, f. 1r, 1646, 42).

293 Assembled with the help of Reich (1968) and Viète (1983). For Viète's Problem V.2, the given ratio from Diophantus's II.19 is taken to be 1:1; in V.9 Viète also requires the given number to be the sum of two squares; for Viète's V.12, the three conditions that the sum of the squares added to the product of the squares is a square is omitted; and Viète's V.14, whose enunciation is already stated in algebraic terms, corresponds to the sub-problem solved during the solution to Diophantus's $V^G.30$.

equal R by B, & dividing everything by S − R, $\frac{R\,by\,B}{S-R}$ will equal A. Hence, as S − R is to R, so B is to A.

Or, the greater side is E. Thus, the smaller will be E − B. Therefore, E is to E − B as S is to R. Resolving this proportion, R by E will equal S by E, − S by B. And by proper transposition, S by E, − R by E will equal S by B. Whence, as S − R is to S, so B is to E.

Given, then, the difference of two sides, & their ratio, the sides are found. Indeed,

The difference between two similar sides is to the similar greater or smaller side as the difference of the true sides is to the true greater or smaller side.[294]

Let B be 12, R 2, S 3. This makes A 24, E 36.

Viète solves this problem twice, first making his unknown the smaller side, and then the greater side. In this problem he prefers a simplified proportion to a simplified equation, but the two are equivalent in his eyes. Also, in this and the next problem translated, at the end numbers are assigned to the given magnitudes and the unknown is found numerically. In some other problems, after assigning numbers to the givens Viète translates the equation in species into the corresponding numerical equation.

Next we translate an indeterminate problem, IV.7, which is adapted from *Arithmetica* Problem II.10. Diophantus solves it twice, first by the double equality and then by *al-istiqrā'*. Viète, too, gives both solutions. Here he must ensure that the answer is rational when solved in numbers, so he asks to find "a plane in number" (*numero planum*), similar to Stevin asking for "arithmetical numbers" in his versions of Diophantus's problems.

Zetetic VII[295]

To find a plane in number, which added to either of two given planes, produces a square.

Let the two given planes be B plane, D plane. It is necessary to find another plane which, added to either B plane or D plane, makes a square in number.

That added plane is A plane. Thus, B plane + A plane equals a square, and again D plane + A plane equals a square. This, then, establishes a double equation, as Diophantus says. Moreover, let B plane be greater than D plane. Therefore, the difference of the squares to be formed is B plane − D plane. Now, the square of the sum of two sides exceeds the square of their difference by quadruple the rectangle under the sides. Thus, B plane − D plane will be understood to be quadruple the rectangle under the sides. Whence let B plane + A plane be the square of the sum of the sides, (and) D plane + A plane

---

294 The "true sides" (*laterum verorum*) are A and A + B, while the "similar sides" (*similium laterum*) are R and S. A and A + B are "similar" to R and S by virtue of the proportion A : A − B :: R : S.

295 (Viète 1593d, f. 17r, 1646, 68).

the square of the difference. Further, A plane (is) the square of the sum of the sides, with B plane removed,[296] or a square of the difference of the sides, with D plane removed.

The problem becomes that of resolving $\frac{\text{B plane}-\text{D plane}}{4}$, which is the rectangle under the sides, into the two sides under which it is made. One is G, which is greater than the difference of L B plane & L D plane,[297] or less than their sum. The other is $\frac{\text{B plane}-\text{D plane}}{\text{G}4}$. Therefore, the side of the greater of the squares will be $\frac{\text{B plane}-\text{D plane}+\text{G square }4}{\text{G}4}$, of the smaller $\frac{\text{B plane}-\text{D plane}=^{298}\text{G square }4}{\text{G}4}$.

Let B plane be 192, D plane 128. The difference is 64, quadruple the rect-
angle under the two sides. Thus, the simple (rectangle) is 16, made from the
sides 1 & 16, their sum 17, difference 15, & the sum of the squares (is) 289,
removing 192, leaves 97. Thus, 192 + 197 makes the square of the sum of the
sides, which is 289; & consequently 128 + 97 makes the square of the differ-
ence, which is 225. So the problem is satisfied.

On the other hand, one could also work it out this way. Since adding the same plane to either B plane or D plane produces a square, (suppose that) that plane is A square − B plane. Thus, adding B plane to it will make a square, namely A square. Therefore, it remains for D plane + A square − B plane to equal a square. It is formed by F − A. Thus, A square + F square − F by A 2 will equal D plane, + A square − B plane. Arranging this equality well, $\frac{\text{F square}+\text{B plane}-\text{D plane}}{\text{F}2}$ will equal A.

Let B plane be 18, D plane 9, F 9. (This) makes A 5, the added plane 7,
which added to 18 makes 25, to 9 makes 16, the squares of 5 & 4.

Viète works the double equality differently than Diophantus. He first finds the difference of the squares B plane + A plane and D plane + A plane to be B plane − D plane. Now, if we write the difference of the two squares as $a^2 - b^2$, instead of factoring this into $(a - b)(a + b)$ and then finding a and b in terms of these factors as Diophantus does, Viète considers $a^2 - b^2$ as being $(s + t)^2 - (s - t)^2 = 4st$ for two sides s and t. So in this problem $st = \frac{\text{B plane}-\text{D plane}}{4}$, and if t is named G, then s will be $\frac{\text{B plane}-\text{D plane}}{\text{G}4}$. The solution would not have been any more difficult if he had followed Diophantus's method of finding a and b. Viète then finds the sides of the two squares, but he does not write out the solution. If he had, the value

296 The translation "with B plane removed" comes from "multatum B plano", where the word *multatum* indicates that A plane is fined B plane, as if it were a financial penalty. Similarly for the
"D plane removed" just after.

297 The "L" in "L B plane & L D plane" in the 1593 printing is an abbreviation for "Latus" ("side"),
the geometrical term for square root. The 1646 printing shows "$\sqrt{\ }$", with no overbar, in place of
"L". We would write the roots as "$\sqrt{\text{B plane}}$" and "$\sqrt{\text{D plane}}$".

298 Viète used "=" to mean the difference between two amounts if he did not know which was greater.
He had no sign for "equals". The notation does not indicate what constitutes the left term in this
difference. Using the 1646 overbar, it would have been written as "$\overline{\text{B plane}-\text{D plane}}=\text{G square }4$".

of A plane from one of the equations could have been written as something like "A plane equals $\frac{\text{B plane} - \text{D plane} + \text{G square 4 square}}{\text{G square 16}}$ − B plane", or, using Witmer's notation,
$$A^p = \frac{\left(B^p - D^p + 4G^2\right)^2}{16G^2} - B^p.$$

Many solutions to problems in Diophantus are made simpler in species because numerical values can be assigned at the end, after the unknown A or E has been found in terms of the other species. In many problems Diophantus finds himself in a dead end because the multitudes in his equation do not lead to a positive rational solution, so he has to retrace his steps to see how the unacceptable numbers arose. This happens in Problem III.10, for example, where the forthcoming equation 52P 12u = □″ cannot be solved by *al-istiqrā'* because neither 52 nor 12 is a square. Diophantus must then look to see where the 52 came from, so he can make different choices so that he gets a square instead of 52. Viète solves this same problem as Zetetic V.7, and in his version he finds in place of 52 the expression " $\frac{\text{B quad.} - \text{Z plano}}{A}$ in $\frac{\text{D quad.} - \text{Z plano}}{Z}$ " (Viète 1593d, f. 22r, l. 37; 1646, 78.20). No backtracking of his steps is necessary since he knows that it is enough to ensure that B quad.− Z plano and B quad.− Z plano are both squares when he assigns numbers to the knowns at the end.

In some other problems Diophantus must find the value of 1N in an equation for which some multitude is not yet determined. In Problem $V^G.30$, for example, he wishes to solve the forthcoming equation 1P ℓ 60u = □ by assigning $\sqrt{\square}$ to be 1N less a multitude of units. Without specifying these units he finds the value of 1N to be "a certain number multiplied by itself and having received in addition 60 units, which has been divided by its double". We do not know how Diophantus found this solution, whether he had some notation for it that he wrote on the wax tablet, or if he performed it mentally. In species, though, it is easy to work out, since multitudes (coefficients) are by their nature undetermined. In Viète's version, his Problem V.14, this multitude is called F, and the equation is solved just like any other. Appropriate numerical values are assigned only at the end, after the inequalities that delimit the choice of F have been worked out. It is true, as Viète wrote, that "things that appear to be very subtle and abstruse in numerical logistic are quite familiar and even easy in specious logistic".

3.5.7 *Diophantus in the twilight of premodern algebra*

Xylander's premature death in 1576 put an end to his project of publishing the Greek text of Diophantus's works. People would have to wait for close to a half-century more for the appearance of Claude-Gaspard Bachet's (1581–1638) *editio princeps* of 1621. In the intervening years there were at least two people who had a plan to publish an edition of the Greek text. In addition to Gosselin, whom we have already mentioned, there was the Neapolitan mathematician Joseph Auria. Auria actually produced a corrected edition of the Greek text with Latin translation at the end of the sixteenth or the beginning of the seventeenth century, but it never made it to print (Allard 1983a; Meskens 2010, 161–62).

Bachet's *Diophanti Alexandrini Arithmeticorum Libri Sex, et De Numeris Multiangulis*[299] contains the Greek text of the *Arithmetica* and the fragment on polygonal numbers, accompanied by a Latin translation and a comprehensive commentary. He also included his reconstruction of the lost *Porisms* and an appendix consisting of 53 propositions relating to polygonal numbers. Bachet worked mainly from codex Par. gr. 2379, which had been copied for Cardinal Ridolfi in 1545, and which descends from Vat. gr. 200 and Vat. gr. 191 (Allard 1982–83, 124ff). Bachet devoted enormous effort to correcting the innumerable flaws in the text, including those that had eluded Xylander. Contrary to Xylander, he aimed to produce an edition as close as possible to what Diophantus himself had written, so instead of printing corrupt passages as they stood and then trying to deduce the underlying mathematics in his commentary, Bachet worked at making philologically plausible corrections so the text itself makes sense (Morse 1981, 215ff). Then in his commentary he typically investigated the general method that he saw behind Diophantus's particular solutions (Morse 1981, 234ff). The result was for the time excellent both philologically and mathematically, and his work remained the standard edition until the 1890s, when Tannery published his edition based on all manuscripts known at that time.

For his Latin translation Bachet began with Xylander's, which he adjusted to agree with his emended text. For his edition he wrote the powers as they appear in the Paris manuscript, whether in words or notation. He kept Xylander's notation in the translation, and like Xylander he used it for all algebraic expressions and equations. To make sense of the problems he read Xylander closely, of course, but he also benefited from Bombelli and Viète, each of whom is cited several times in his commentary. He even consulted Clavius's *Algebra* for at least one problem (Diophantus 1621, 98), but there is no mention of Stevin.

It was not long after Bachet's edition was published that people began to modify Viète's algebra. Thomas Harriot (1631), William Oughtred (1631), and James Hume (1636) all adapted the algebra of species to arithmetic so that their letters stand for the values of numbers. Like Vaulezard's translation of Viète, these authors also dropped the preposition *in* so that multiplication is indicated by juxtaposition. Because there is no need to conserve dimension, their equations look much more modern. Harriot, for example, states equations like "$aaa + 95,400.a = 1,819,459$" for what we would write as $a^3 + 95,400a = 1,819,459$ (Harriot 1631, 136), and Oughtred can multiply $B + 1$ by A to get $BA + A$ (Oughtred 1631, 8). Soon after, Descartes posited a unit line and wrote geometrical equations like "$y^4 + 16y^3 + 71yy - 4y - 420 = 0$" (Descartes 1637, 376). Even if practical textbooks teaching the old cossic algebra continued to be printed into the eighteenth century, advances in algebra after Descartes were all conducted in species. Viète's creation of specious logistic thus unintentionally ensured the imminent obsolescence of

---

299 On Bachet, see Collet and Itard (1947). For details on Bachet's approach and his sources, see Diophantus (1621, letter to the reader), Heath (1910, 27–28), and Morse (1981, 215ff.).

Diophantus as an algebraist, and from the perspective of algebra, Bachet's edition was a historical work almost from the start.

But that was not the end of Diophantus's relevance to mathematics. It was Bachet's edition, with the editor's particular interest in generalizing problems, that inspired Pierre de Fermat, around 1640, to initiate the modern study of number theory. That begins a new chapter in the legacy of Diophantus, the one for which he is best known today.

4 Structure and language of the *Arithmetica*

Because the *Arithmetica* is the only algebraic work in Greek that has come down to us, it should not be surprising that the structure of the core part of its solutions is unlike the structure of the solutions to problems or theorems in other ancient Greek books. Even so, the overall organization of Diophantine problems is generally in agreement with what we read in Euclid and other authors. We begin this chapter with an investigation into this structure, after which we examine the language Diophantus employs in different parts of his solutions. This language is equally idiosyncratic when compared with other Greek works but agrees generally with what we read in medieval and Renaissance algebraic solutions in other languages.

4.1 The structure of Diophantine problems

4.1.1 Proclus on Euclid

It was over seven centuries after Euclid's time that the philosopher Proclus composed his *Commentary on the First Book of Euclid's Elements*. In it he famously listed six formal divisions of a proposition: πρότασις (*protasis*, enunciation), ἔκθεσις (*ekthesis*, setting-out), διορισμός (*diorismos*, definition of goal), κατασκευή (*kataskeuē*, construction), ἀπόδειξις (*apodeixis*, proof), and συμπέρασμα (*sumperasma*, conclusion).[1] As a brief example, we can break down *Elements* Proposition I.6 according to this scheme:

Enunciation (stated in general terms)

If in a triangle two angles be equal to one another, the sides which subtend the equal angles will also be equal to one another.

---

1 (Proclus 1873, 203.1–207.25; 1970, 159). The beginning of this passage is also found in the pseudo-Heronean *Definitions* (Hero 1912, 120.21–122.19). The translation we employ for the Greek terms is taken from Heath (1926, I, 129), as modified in (Netz 1999a, 10–11). Alternative translations have been proposed, for example, by Vitrac (2005b), Acerbi in Diophantus (2011), and Sidoli (2018b).

DOI: 10.4324/9781315171470-5

Setting-out (instantiation)

Let *ABC* be a triangle having the angle *ABC* equal to the angle *ACB*;

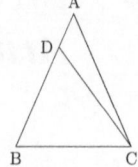

Definition of goal (the instantiated enunciation)

I say that the side *AB* is also equal to the side *AC*

Construction

For, if *AB* is unequal to *AC*, one of them is greater. Let *AB* be greater; and from *AB* the greater let *DB* be cut off equal to *AC* the less; let *DC* be joined.

Proof

Then, since *DB* is equal to *AC*, and *BC* is common, the two sides *DB*, *BC* are equal to the two sides *AC*, *CB* respectively; and the angle *DBC* is equal to the angle *ACB*; therefore the base *DC* is equal to the base *AB*, and the triangle *DBC* will be equal to the triangle *ACB*, the less to the greater: which is absurd. Therefore *AB* is not unequal to *AC*; it is therefore equal to it.

Conclusion

Thus, if a triangle has two angles equal to one another then the sides subtending the equal angles will also be equal to one another. (Which is) the very thing it was required to show.[2]

The idea of classifying the different parts of a proposition likely originated with a commentator, perhaps Proclus himself, since none of our extant mathematical texts break down the different parts of their propositions (Netz 1999b). Even though authors may not have called attention to the structure of their theorems and problems, it will be profitable for us to consider Proclus's divisions as a starting point for understanding the structure of the problems in Diophantus. We can already note that there is at least some agreement for the first parts and the last for some problems of the *Arithmetica*. For example, Problem IV.39 begins and ends:

Enunciation

We want to find a cubic number such that if we multiply a square of its side by two given numbers and we subtract the cube from each of them, leaves from each of them a square number.

Setting-out

Let the two given numbers be three and seven.

2 Translated in Heath (1926 I, 255), except the conclusion, which is translated in Fitzpatrick (2008).

Definition of goal

We want to find a cubic number such that if we multiply a square of its side by three and by seven, and we subtract the cube from each of the products, then the remainder from each of them is a square number.

. . .

Conclusion

We have found a cubic number according to the stipulation that we stipulated, and it is twenty-seven eighths of an eighth. And that is what we wanted to find.

Very often one or more of these four parts is omitted or combined in the *Arithmetica*. Many problems have no given numbers, so there is no need for the setting-out. In the Greek books there are no conclusions, and in Problems I.5–11, where there is what can be considered a definition of goal, it is combined with the setting-out. Similarly, Proclus himself noted that not all parts are present in every problem of the *Elements*.[3] For example, in some propositions the construction is absent because the setting-out is sufficient (Proclus 1873, 204.3–5; 1970, 159).

As Netz (1999b, 293) has observed, even if Proclus claimed that this scheme applies to "[e]very problem and every theorem" (Proclus 1873, 203.1–2, 1970, 159), he appears to have classified these six parts specifically with Euclid's theorems from Book I in mind. Recall that theorems assert the truth of some statement, while problems ask for something to be constructed. Problems are a little different from theorems in their organization in that they have no need for a conclusion, and they give a second definition of goal after the construction.

4.1.2 Analysis (ἀνάλυσις) and synthesis (σύνθεσις)

Variations like the ones just described will remain fairly superficial as long as the main argument, designated by Proclus as the "construction" and "proof", is synthetic. That is, the line of argument begins with given hypotheses and proceeds forward through their implications until the truth of the theorem is verified or the desired construction is achieved. Arguments by analysis take a different form. There one begins by presuming that what is sought is true or has been achieved, and through a series of steps one arrives at something independently known to be true or constructible. Four extant ancient mathematics texts give definitions of analysis and synthesis: Hero, as reported in al-Nayrīzī's ninth-century commentary on Euclid's *Elements* (Lo Bello 2009, 22–23); an anonymous scholium to the

---

3 Proclus stated that the "most essential" parts are the enunciation, the proof, and the conclusion, though in fact the conclusion is missing in many of Euclid's propositions (Proclus 1873, 203.17–18, 1970, 159).

first five propositions of Book XIII of the *Elements* (Euclid 1973 IV, 198.12–16); another anonymous scholium, appended to the previous one in the margin of Vat. gr. 190, a codex of the first half of the eleventh century (f. 233v; Euclid 1977 V, 2, 309.3–8); and most famously the introduction to Book VII of the *Collection* of Pappus of Alexandria (Pappus 1986 I, 82–85). The first anonymous scholiast defined the terms like this:

> What is analysis, and what is synthesis? Analysis is the assumption of what one is seeking, as if it were established, by way of its consequences, to something that is established to be true. Synthesis is the assumption of what is established, by way of its consequences, to something that one is seeking.[4]

Pappus distinguished between theorematic and problematic analysis. After describing the former he writes:

> In the case of the problematic kind, we assume the proposition as something we know, then, proceeding through its consequences, as if true, to something established, if the established thing is possible and obtainable, which is what mathematicians call "given", the required thing will also be possible, and again the proof will be the reverse of the analysis; but should we meet with something established to be impossible, then the problem too will be impossible.
>
> (tr. Jones [Pappus 1986, 84])

When both are present in a proposition in Greek geometry, the analysis is given first, followed by the synthesis. And while a synthesis consists of a "construction" and then a "proof", as we saw earlier in Euclid, an analysis in geometrical problem solving consists of a "transformation" and a "resolution".[5] The transformation begins by positing the desired construction as having been achieved and continues with further constructions and logical inferences, terminating with the finding of something independently constructible or given. To verify that the synthesis will indeed be successful, the resolution begins with this given, and through a chain of givens shows that the sought-after figure is also given. In theorematic analysis there is no resolution because nothing is being constructed or calculated.

4.1.3 Determinations

There is another part of a proposition that is common in solutions to problems in Greek geometry, and which is often present in problems in the *Arithmetica* as well. In some cases the choice of givens affects the solvability of the problem,

---

4 Translated by Jones in Pappus (1986, vol II, 381). The last phrase "that one is seeking" is Jones's correction. The text has "that is established to be true", which is clearly an error.
5 These terms were first coined by Hankel (1874, 137–50), and have become standard. See also Hintikka and Remes (1974, 22ff) and Berggren and Van Brummelen (2000, 5–16).

so a separate analytical investigation is conducted to determine the relationship between the givens and the possible solutions. This investigation is called a διορισμός (*diorismos*, determination) by geometers, the same word Proclus used for the setting-out, but here with a completely different meaning. Just after the passage describing problematic analysis and synthesis translated above, Pappus defined the διορισμός as "the preliminary distinction of when, how, and in how many ways the problem will be possible" (Pappus 1986, 84). For example, Proposition II.7 of Archimedes's *On the Sphere and the Cylinder* asks: "To cut a section from a given sphere with a plane, such that the section has a given ratio to the cone that has the same base and height as the section".[6] The determination examines the givens via the diagram from the transformation and concludes that "it is necessary that the ratio given for the synthesis be greater than that which three has to two".[7]

Saito and Sidoli (2010, 588) explain that when present, the determination is naturally situated after the analysis and before the synthesis. Thus, a complete analyzed problem consists of three main parts: (i) an analysis, consisting of a transformation and a resolution; (ii) a determination; and (iii) a synthesis, consisting of a construction and proof. In some published propositions in which the analysis and determination are omitted and only the conclusion of the determination is stated, that conclusion can be found attached to the enunciation. We find this in Euclid's *Elements* Propositions I.22, VI.28, and XI.23, and in Apollonius's *Conics* Propositions VI.29 and VI.32 (Toomer in Apollonius 1990, lxxxv). For example, the enunciation and the conclusion of the determination of *Elements* I.22 read: "Out of three straight lines, which are equal to three given straight lines, to construct a triangle: thus, it is necessary that two of the straight lines taken together in any manner should be greater than the remaining one" (Heath 1926, I, 292). In Diophantus, too, the conclusions of determinations are situated after the enunciation, as seen here in Proposition I.9:

> To subtract the same number from two given numbers and make the remainders have to one another a given ratio. It is certainly necessary for the given ratio to be greater than the ratio which the greater of the givens has to the smaller.

The word διορισμός does not appear in the *Arithmetica*, but the cognate noun προσδιορισμός appears twice (36.6 and 340.9), as does the related verb διορίζεσθαι (424.14 and 428.21), both in connection with solvability.[8] In some problems the

---

6 Translated in Saito and Sidoli (2010, 589).
7 Translated in Saito and Sidoli (2010, 590).
8 See "Index graecitatis apud Diophantum" (Diophantus 1893–95, II, 281), *s.v.* προσδιορισμός, and Tannery's explanation "conditio, limitatio datorum ita ut problema possibile sit". Acerbi raised the possibility that the word προσδιορισμός with the meaning of a condition of solvability originates with Diophantus himself (in Diophantus 2011, 16). The word also appears in the pseudo-Heronian *Definitions* (Hero 1912, 156.7–8, 11).

determination can be found by simply examining the enunciation, but in others, such as I.21, IV.21, and IV.22, Diophantus clearly derived it from the particular calculations that he made during the course of solving the problem. In Problem I.21 the determination even depends on the particular algebraic assignments he is about to make. When solving a problem, then, Diophantus often found the determination as a consequence of his algebraic solution, and he then placed its conclusion after the enunciation when writing his book. Other problems for which the determination was evidently found after the algebraic solution are I.9, 14, 27, 28, 30, IV.17–20, V.7–12, and IV^G.34, 35.

4.1.4 Algebra

We now turn our attention from geometrical to arithmetical problem solving. Recall that the solution of a problem in premodern algebra can be broken down into four stages (see Section 2.4.2):

1   One or more unknowns are named, operations are performed on these names according to the conditions of the enunciation, and an equation is set up.
2   The equation is simplified to a standard form.
3   The simplified equation is solved.
4   The values of the sought-after unknowns are calculated from the value of the Number/Thing.

In Diophantus, we can also add a stage 5 in which the answers found in stage 4 are applied to the enunciation to prove that the solution works. This "proof" is consistently given in the Arabic books, but in the Greek books it is explained only in Problems I.4–11, and there only briefly. In Problem I.4, for example, it is expressed as: "And it is fulfilled, since the greater is the quintuple of the smaller, and the difference is 20 units". In other problems in the Greek books we only get a short phrase like "and they fulfill what was proposed", "and they fulfill the problem", or, in 19 cases, mainly in Book IV^G, "and the proof (ἀπόδειξις) is obvious".

Among the stages of solving a problem by algebra, stage 1 is the most difficult. This is especially true for Diophantus, who insisted that his simplified equations consist of two terms of consecutive powers, or, when that is not possible, that a rational solution is guaranteed by some other means. There are three kinds of actions that take place in stage 1: (i) the naming of unknowns, (ii) operations on these names, and (iii) the setting-up of partial (forthcoming) or complete equations. (These correspond to the three columns we show in the tabular summaries in our commentary.) In many complex problems these three actions become intermingled and are performed toward the goal of achieving a positive rational solution. Stages 2 and 3, on the other hand, are short, mechanical processes.

Stages 1–3 together fit the definitions of analysis given in the ancient texts since by assigning names to unknowns and then operating on them one is presuming that they are something we know, and stage 3 ends with the finding of the value of 1 Number, which is a truly given value. In fact, algebra was later recognized as belonging to analysis by al-Samaw'al (11th c.) and especially by François Viète (1591) and his successors.[9] Al-Samaw'al, who also authored a lost commentary on the *Arithmetica*, included a short section titled "That the art of algebra (*al-jabr*) is a part of the art of analysis (*al-taḥlīl*)" in his *Dazzling Book on the Science of Calculation*. In it he explains (Rashed 2021, 63.7):

> And whoever wants to find a sought-after (amount) or prove a certain proposition, let him make what is sought-after assumed,[10] and see what necessarily follows from that, and what necessarily follows from its necessities until one arrives at a known simple element. If it is true, then synthesize its decomposed[11] parts and start from where the analysis ended. If these simple elements are false, then we have learned the impossibility of what is sought and we abandon it.
>
> For example, we want to find two numbers such that one of them is three times the other and their surface[12] is 75 units. We suppose that we have found the two sought-after numbers, and we call the smaller by a name which expresses its form without restricting its amount, which is the Thing. The other will then necessarily be 3 Things, since we supposed that one of them is three times the other. We multiply one of them by the other, so the result of the multiplication is three *Māl*s, and that should equal[13] seventy-five. If we divide them both by the greatest number that measures them, which is 3, then it results that the *Māl* is 25 units, which is equal to a square of the Thing, so the Thing is five. We supposed that the smaller is a Thing, so it is five, and we supposed that the other is three Things, so it is fifteen. We have found the two numbers that we wanted, and they are five and 15. This process is the one required in the art of algebra, and it is the same that the art of analysis requires.

Al-Samaw'al follows this with a non-analytical, and thus non-algebraic, solution to the same problem that invokes Proposition VII.18 of Euclid's *Elements*, and he concludes the short section with an attempted algebraic solution of an impossible problem.

Only two other Arabic texts are known to associate algebra with analysis. One of them is Qusṭā's translation of the *Arithmetica*, specifically in remarks made

---

9 In Europe the one person before Viète who is known to have identified algebra as belonging to analysis is Lazarus Schöner, in his 1586 edition of *Arithmetices libri duo, et Algebrae* of Petrus Ramus (Ramus 1586, 322; Sasaki 2003, 256).

10 The word behind "assumed", as in "posited", is *mafrūḍ*, which has the same root as the verb *faraḍa*, which we translate in al-Samaw'al's example as "suppose". In the Arabic Diophantus we translate this verb as "assign". Here we write "we suppose", as in "we posit", since "We assign that we have found the two" would be awkward.

11 As in Greek, the Arabic word for "analysis" also means "decomposition". We could have translated this as "its analyzed parts".

12 The word *musaṭṭaḥ* ("surface") is a geometric term adopted in Arabic arithmetic to mean "product".

13 Only in this short section does al-Samaw'al use the word *sawiya* rather than *'adala* for "equal". We do not know why.

in Problems IV.37, 42a, and 43. There the assignment of unknowns is said to belong to the analysis, while the calculation of the unknowns in stage 4 begins the synthesis. Later, in the late thirteenth century, al-Fārisī seemed to have been reacting against the accepted opinion when he asserted that "Algebra is the science of finding unknown quantities by the properties of proportional numbers (i.e., the powers of the unknown) successively from one by synthesis (*tarkīb*), and not analysis (*taḥlīl*)" (al-Fārisī 1994, 461.19; Oaks 2019, 111–12). We disagree with each other as to whether Diophantus himself recognized his method of solution as being analytic. One of us, Christianidis, holds that these are Arabic interpolations and do not originate with Diophantus. He makes his arguments in (Christianidis 2021b). The other of us, Oaks, holds that what we read in the Arabic translation is perfectly consistent with what Diophantus himself could have written. He expresses his view in a note appended at the end of the commentary to Problem IV.42 beginning on page 615.

4.1.5 The parts of a Diophantine problem

We can now identify the general structure of a Diophantine problem. Next are two simple problems, one from the Greek books and one from the Arabic books, that contain as many parts as can be expected in their respective languages.

Greek Book I, Problem 9	Arabic Book IV, Problem 18

1 πρότασις (enunciation)

To subtract the same number from two given numbers and make the remainders have to one another a given ratio.	We want to find two cubic numbers whose sides are in a given ratio, and if each of them is multiplied by a given number, what is gathered from one of them is a square, and from the other is a side of that square.

2 προσδιορισμός (determination)

It is certainly necessary for the given ratio to be greater than the ratio which the greater of the givens has to the smaller.	It is necessary that the given number be a cube.

3 ἔκθεσις (setting-out, instantiation)

Now, let it be proposed to subtract from 20 and 100 the same number, and make the greater the sextuple of the smaller.	Let the given ratio be the ratio of three times, and the given number be eight units.

4 διορισμός (definition of goal, or instantiated enunciation)

	We want to find two cubic numbers such that a side of one of them is with respect to a side of the other in the ratio of three times, and the multiplication of the greater by the eight units is a square number, and the multiplication of the smaller by the eight units is a side of that square.

5 Stage 1: Naming unknowns, performing operations, establishing an equation

Let what is subtracted from each number be assigned to be 1 Number. And when it is subtracted from the 100, there remain 100 units lacking 1 Number; while, when (it is subtracted) from the 20, there remain 20 units lacking 1 Number. And the greater should be the sextuple of the smaller. Six times the smaller, therefore, are equal to the greater. But six times the smaller makes 120 units lacking 6 Numbers. These are equal to 100 units lacking 1 Number.

We assign the smaller cube to be from a side of one Thing, thus, it is one Cube, and a side of the greater cube is three Things, so it is twenty-seven Cubes. When we multiply the twenty-seven Cubes by eight units, it gives two hundred sixteen Cubes, and when we multiply the one Cube by eight units, it gives eight Cubes. Since the two hundred sixteen Cubes is a square from a side of the eight Cubes, which is sixty-four Cube Cubes, the two hundred sixteen Cubes Equal sixty-four Cube Cubes.

6 Stages 2–3: Simplifying and solving the equation

Let the lacking be added in common and let likes be subtracted from likes. There remain 5 Numbers equal to 20 units, and the Number becomes 4 units.

When we divide them by the one Cube, which is the lesser in degree of the two sides, it gives two hundred sixteen units Equal sixty-four Cubes. Therefore, the one Cube is three units and three eighths of a unit, and the one Cube is a cube from a side of one Thing, and the three units and three eighths of a unit is a cube from a side of one and a half. Thus, the one Thing Equals one and a half.

7 Stage 4: Calculating the desired numbers

To the hypostases: I assigned what is subtracted from each number to be 1 Number; it will be 4 units.

Therefore, the smaller cube is three units and three eighths of a unit, and the greater cube, whose side is four units and a half, is ninety-one and an eighth.

8 ἀπόδειξις (proof)

And if it is subtracted from the 100, there remain 96 units; while if (it is subtracted) from the 20, there remain 16 units. And it is fulfilled that the greater is the sextuple of the smaller.

And when this greater cube is multiplied by the eight units, it amounts to seven hundred twenty-nine, which is a square from a side of twenty-seven units, which is what is gathered from multiplying the smaller cube, which was shown to be three and three eighths, by the given number, which is eight units.

9 συμπέρασμα (conclusion)

We have found two numbers according to the stipulation that we stipulated. And that is what we wanted to find.

We have shown in Christianidis and Oaks (2013) that there is a direct correspondence between Diophantus's arrangement of the sections of the introduction and the successive stages that a Diophantine solution displays. The introduction was organized apparently to reflect these stages. To illustrate this

Table 4.1 The structural correspondence between the introduction and worked-out problems.

Introduction		Problems I.9 and IV.18		
Content	Reference	Stages	I.9	IV.18
Prolegomena	2.3–13			
Enunciation of the problems: the language of the enunciations	2.14–4.11	Generic enunciation	1	1
		Determination	2	2
		Instantiation	3	3
		Instantiated enunciation		4
The language of the solutions: the technical terms and their abbreviations	4.12–6.21	Assignment of names	5	5
Operations with technical terms	6.22–14.10	Operations on the names	5	5
Statement of the equation	14.11–12	Statement of the equation	5	5
Simplification of the equation	14.12–24	Simplification and solution of the equation	6	6
		Answer to the problem and test proof	7 8	7 8
		Conclusion		9
Concluding paragraph	14.25–16.7			

concordance, we put side by side in Table 4.1 the order in which the contents of the introduction are unfolded and discussed, and the successive parts the solutions of the two problems just shown.

4.2 The language of Diophantine problems

With no other works on Greek algebra for comparison, it is to be expected that many of the terms and phrases common in Diophantus occur very rarely or not at all, or are applied with different meanings, in other texts of the ancient Greek mathematical literature. Algebraic solutions unfold in a very particular way so that the different parts of a solved problem are charged with different goals and are expressed with different vocabulary. We review this vocabulary now, following roughly the progression of the parts of the solution outlined earlier.

4.2.1 The names of the unknowns and their notation

The technical terms expressing the unknown and its powers in both the Greek and Arabic books of the *Arithmetica* are covered in Section 2.4.1, while the corresponding notational signs are discussed in Section 2.8. Here, we merely expand some issues which either are briefly discussed there or are not addressed at all.

The technical terms for the unknown and its powers are shown in Table 4.2. Even if our modern notation carries a different meaning from the premodern names, we put their equivalent in the right column.

The extant Greek books do not contain terms corresponding to powers above the sixth. However, we saw in Section 3.4.1 that three codices of the fourteenth and fifteenth centuries, which carry texts of Michael Psellus, contain a fragment that preserves names of most of the powers up to the 13th, while Psellus himself preserves names for the 7th, 8th, and 9th powers drawn from unknown sources. The flexibility that these texts exhibit regarding names of the higher powers, and the same flexibility we see in Arabic, Latin, Italian, and German algebra described in Section 3.1.2, suggest that Diophantus, too, might have written his powers above the fourth in various ways in his solutions, such as writing either δυναμόκυβος or κυβοδύναμις for the fifth power.

Table 4.2 Technical terms for the powers of the unknown in the Greek and the Arabic text of the *Arithmetica*.

Greek term	Translation	Arabic term			Translation	
μονάς	unit	واحد		*wāḥid*	unit	1
ἀριθμός	Number	شيء	*shay'*	جذر *jidhr*	Thing/Root	x
δύναμις	Power	مال	*māl*		Māl	x^2
κύβος	Cube	كعب		*ka'b*	Cube	x^3
δυναμοδύναμις	Power-Power	*māl māl*			Māl Māl	x^4
δυναμόκυβος	Power-Cube	*māl ka'b* or *ka'b māl*			Māl Cube or Cube Māl	x^5
κυβόκυβος	Cube-Cube	*ka'b ka'b*			Cube Cube	x^6
						x^7
		ka'b ka'b māl or *māl māl māl*			Cube Cube Māl or Māl Māl Māl Māl	x^8
		ka'b ka'b ka'b			Cube Cube Cube	x^9
ἀριθμοστόν	Inverse Number	جزء من شيء		*juz' min shay'*	part of a Thing	$\frac{1}{x}$
δυναμοστόν	Inverse Power	جزء من مال		*juz' min māl*	part of a Māl	$\frac{1}{x^2}$
κυβοστόν	Inverse Cube					$\frac{1}{x^3}$
δυναμοδυναμοστόν	Inverse Power-Power					$\frac{1}{x^4}$
δυναμοκυβοστόν	Inverse Power-Cube					$\frac{1}{x^5}$
κυβοκυβοστόν	Inverse Cube-Cube					$\frac{1}{x^6}$

The names of the powers of the unknown together with their "signs" (σημεῖα) are introduced by Diophantus in the introduction of the first book (4.14–6.8). Thus, the "Power" is written as $Δ^Y$, from the first two letters of the corresponding Greek term ΔΥΝΑΜΙΣ, the "Cube" is written as K^Y, from the first two letters of the Greek word ΚΥΒΟΣ, and similarly for the higher powers: $Δ^YΔ$ for ΔΥΝΑΜΟΔΥΝΑΜΙΣ, $ΔK^Y$ for ΔΥΝΑΜΟΚΥΒΟΣ, and K^YK for ΚΥΒΟΚΥΒΟΣ. The technical term "Number", corresponding to the first power of the unknown, is printed in Tannery's edition as S, while he shows the sign $\overset{.}{M}$ for the word "unit" (ΜΟΝΑΣ).[14] The ς-shaped sign for "Number" adopted by Tannery looks like the sign which appears in some of the oldest manuscripts of the *Arithmetica*. Specifically, the sign appears in Matrit. 4678, Marc. gr. 308, and the two Vatican manuscripts Vat. gr., 191 and 304, in a form which roughly looks like a final ς ($\mathcal{S}$, $\mathcal{S}$, S), although in the manuscript of Madrid, as well as in several later manuscripts, different forms, like $\mathbf{Y}$, $\mathbf{C_{o}}$, and $\mathbf{\zeta}$, are also found (the latter being used in the manuscript of Madrid in additions made by a second hand).[15] These signs are often duplicated to express the plural. The ς-shaped sign is also used in the late thirteenth-century manuscript Vat. gr. 1087, which includes a Diophantine solution by Theon of a mensuration problem discussed by Ptolemy in *Almagest* XIII.3 (see Section 3.2.1), as well as in the famous ninth-century manuscript Vat. gr. 1594, which carries the same problem.

The manuscripts containing the Greek text of the *Arithmetica* are not consistent in using this notation, and manuscripts are not even consistent with each other, testimony to the liberties taken by copyists. In many places full words are used instead of abbreviations, while the ς-shaped sign for the unknown quantity is sometimes used for the word ἀριθμός with the nontechnical meaning of a common noun. In Figure 4.1 we see a portion of the first solution to Problem III.6 as it appears in the sixteenth-century codex Par. gr. 2379. The word ἀριθμούς ("numbers") in the enunciation ("To find three numbers") is written by a duplicated ς-shaped sign, just as the technical expression "2 Numbers" (S $\bar{\varsigma}$) is written in the second line.

Tannery put an end to this inconsistency in his edition. He uses the ς-shaped sign only for the word ἀριθμός, with the technical meaning of the algebraic unknown, and he writes the word in full when it has the meaning of a common noun. Thus, Tannery never uses the ς-shaped sign in the enunciations. Similarly, abbreviations for powers and inverse powers of the unknown are systematically used only in the

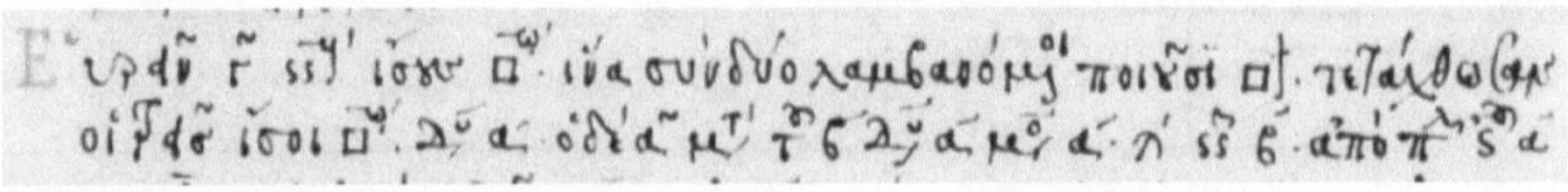

Figure 4.1 The beginning of Problem III.6 in Par. gr. 2379.

14 Cf. Diophantus (1893–95, II, xxxiv–xlii) and Heath (1910, 32ff).
15 Cf. Heath (1910, 35–36) and Allard (1980, 34–36).

body of the solutions. The inconsistency between the printed modern edition of the *Arithmetica* and the manuscripts as regards the use of abbreviations has led scholars to make misleading inferences, which went so far as to question the very distinction between technical and untechnical terms in Diophantus's text.[16] Yet, Tannery was fully aware of the fact that there are in the *Arithmetica* two different kinds of terms. Fully aware were also other editors and translators, like Xylander, C.-G. Bachet de Méziriac, P. Ver Eecke, A. Czwalina, and E. S. Stamatis, to mention only a few. Others have questioned the use of abbreviations by Diophantus. For example, G. J. Toomer writes in his entry on Diophantus in the third edition of *The Oxford Classical Dictionary* (1996) – a statement repeated by Toomer and Netz in the fourth edition of the same *Dictionary* (2012) – that:

> In the Greek (but not the Arabic) MSS the words for the unknown (ἀριθμός) and its powers up to the sixth degree are represented by symbols, as is the operation for minus, so that the equations appear in a primitive algebraical notation, but it is possible that this was introduced in Byzantine times rather than by the author.

Now, even without notation one can easily recognize the difference between technical and nontechnical terms simply from the context, even in cases in which a manuscript has, for example, "I assigned the sought-after number (ἀριθμόν) to be 1 number (ἀριθμοῦ ἑνός)", as in Problem I.7. Only the second "number" is the technical "Number", i.e., the algebraic unknown. Recall also that the few books on Arabic algebra that show notation do so only in figures to show students what should be put down on the dust-board so that the powers are always written in words in the running text. There, too, the distinction between common nouns and technical terms is always clear. We do not know what Diophantus wrote in his autograph. He may have made the distinction explicit by writing the notation for the names of the powers and writing the word out in full where it is a common noun. But he may have written out the names of the powers in full throughout his solutions to problems, or he may even have intermittently and inconsistently shown the notation, as we see in our manuscripts. Whichever he did is not important. Diophantus introduces the notation as a tool for those who intend to solve problems of their own invention. Like the similar notations in Arabic, the signs Diophantus describes in his introduction are primarily for use on a temporary medium, such as the wax tablet, for working through the solution for the first time (see Section 2.8). The *Arithmetica* is, after all, a pedagogical work. The inconsistencies in our manuscripts show that it remained up to the copyist whether to write the power in notation or use the word.

---

16 Cf. the following statement: "Tannery introduced a distinction between some terms which he systematically rendered as symbols and other terms which he always wrote down in full, thereby establishing
two different kinds of terms, in contrast to the manuscripts which use abbreviations for both kinds in
comparable ways" (Chemla 2012, 36).

Diophantus certainly practiced what he taught. He would have used this notation when first solving the problems that he shows in his book, and these notational solutions would have been handy in preparation for producing his published work, the ἔκδοσις (edition). Tiziano Dorandi has studied the preparatory work that precedes "edited" ancient Greek and Latin books, specifically books on history, medicine, zoology, etc. (but not on mathematics). He writes:

> Une pratique répandue parmi les auteurs antiques, surtout ceux d'œuvres scientifiques ou techniques, et qui précédait la rédaction d'un texte, était la lecture des sources ou de la littérature 'secondaire' se rapportant au sujet que l'écrivain avait l'intention de traiter. On formait souvent des recueils à l'aide de ces notes (ὑπομνήματα), et l'auteur y puisait les matériaux nécessaires à la composition d'un traité sur un sujet déterminé.
>
> (Dorandi 2000, 27)

Preparatory work for a book like the *Arithmetica* would include not only notes for the preparation of the "edition" (i.e., the edited manuscript) but first and foremost the very solution of the problems, in all likelihood in a terse way, and the preservation of the brief solutions to a material support, in view of the future publication, in which the solutions will be presented in a detailed form. Seen under such a perspective, the "signs" are elements of the solving activity that precedes the "edition" and are not necessarily components of the final product, that is, of the "edited manuscript", which is addressed to a large audience. They belong to the machinery the mathematician has at his disposal when he tries to solve the problems.[17]

4.2.2 *Assigning values to given numbers*

In the instantiation of Problem I.2 we read "Ἐπιτετάχθω δὴ τὸν $\overline{\xi}$ διελεῖν"[18] ("Now, let it be proposed to divide 60"). Ἐπιτετάχθω is third person, singular, perfect tense, imperative mood, middle/passive voice of the verb ἐπιτάσσειν. This is the most common verb used in the Greek books for assigning a value to a given number. The meaning of ἐπιτάσσω is "to propose", "to give", "to prescribe". Thus, ὁ ἐπιταχθείς

---

17 Cf. the following statement from Christianidis (2021a): "The context in which Diophantus introduces the abbreviations in the preface to the *Arithmetica* suggests that they were intended to be used by the actor of problem-solving, the problem-solver. Therefore, they were not intended to be used in the published work but rather to serve the stage of its preparation. Indeed, the abbreviations were perfectly suited to serve the great deal of work which had to be done before the final reduction (the 'edition') of the treatise and which included the solving activity and the preservation of provisional versions of the solutions to a material support, in view of future publication."

18 It seems that the pronoun ἡμῖν (for us, to us) is implied, so the full sentence should have been ἐπιτετάχθω δὴ (ἡμῖν) τὸν διελεῖν, "now, let it be $\overline{\xi}$ proposed (for us) to divide the (number) 60". Cf. the following sentence from Iamblichus's *In Nicomachi Arithmeticam Introductionem*: "προστετάχθω γὰρ ἡμῖν λόγου χάριν ἀριθμοὺς ἐκθέσθαι τέσσαρας" (Iamblichus 1975, 63.20–21). For the sake of simplicity, we translate phrases like the above by "now, let it be proposed to divide".

ἀριθμός is "the proposed number", "the given number", "the prescribed number". Throughout this book we adopted the first form, "the proposed number", in accordance with Tannery, who writes, "dans Diophante, ἐπιταχθεὶς ἀριθμός signifie toujours 'nombre proposé'" (Tannery 1912j, 435), a translation he endorsed later also in his Latin translation of the *Arithmetica* ("propositum numerum").

It must be stressed that the basic form of the verb ἐπιτάσσειν, i.e., the verb τάσσειν, is never used by Diophantus in the enunciations of problems, either generic or instantiated. The verb τάσσειν is employed mainly in the core part of the solutions, specifically in the phrases by which the unknowns are named (see Section 4.2.3). There the verb has the meaning "to assign a name".

The other verb frequently employed in the Greek text for assigning a value to a given number is the verb "to be". Thus, in Problem II.14 we read "Ἔστω τὸν κ̅ διελεῖν εἰς δύο ἀριθμούς" ("Let it be to divide 20 into two numbers"). Ἔστω is third person, singular, present tense, imperative mood, active voice of the verb εἶναι (to be).

In the Arabic books the construction "Let (the given number/multiple/ratio) be" occurs 42 times in the setting-out. Problem IV.17 has "Let the given ratio be the ratio of twenty times", while in Problem V.6 it is "Let the given numbers be seven and four". This construction is also used to assign numbers other than the sought-after numbers in 16 cases during the course of the solution, such as this example from Problem IV.26: "Let one of the two parts be four hundred and the other be two hundred twenty-five". Sometimes the verb فرض (*faraḍa*) "assign" is used to declare given numbers in the setting-out (10 times), while other times جعل (*ja'ala*) "make" is used (2 times). For example, Problem IV.14: "We assign the two numbers to be five and ten".

4.2.3 Assigning algebraic names to sought-after numbers

In the Greek books

Among the verbs that are used in the Greek books for naming the unknowns, by far the most common is the verb τάσσειν, "to assign". Other words and expressions are sometimes used, and all are listed here.

- Assignment with the verb τάσσειν, "to assign"

Problem I.1 asks to divide 100 units into two numbers whose difference is 40 units. The phrase by which Diophantus names one of the two sought-after numbers is – the abbreviation in Tannery's edition being resolved – "τετάχθω ὁ ἐλάσσων ἀριθμοῦ α̅", or even "τετάχθω ὁ ἐλάσσων ἀριθμοῦ ἑνός". Τετάχθω is third person, singular, perfect tense, imperative mood, middle/passive voice of the verb τάσσειν, which we translate by "to assign". Thus, the literal translation of the aforementioned phrase is "Let the smaller be assigned of one Number". Yet, even with the abbreviation resolved, the phrase is elliptical. Clearly the noun ἀριθμός (in nominative case) is implied after the adjective ἐλάσσων and before the same noun ἀριθμοῦ (now in genitive case) that occurs in the phrase. Thus, expressed more completely, the phrase should be "τετάχθω ὁ ἐλάσσων (ἀριθμὸς) ἀριθμοῦ ἑνός", lit. "let the smaller (number) be assigned of one Number". Now, there is one more word implied, the verb εἶναι (to be), before ἀριθμοῦ ἑνός. Thus, the full sentence should be as follows: "τετάχθω ὁ ἐλάσσων (ἀριθμὸς) (εἶναι) ἀριθμοῦ ἑνός", or "τετάχθω ὁ ἐλάσσων (ἀριθμὸς) ἀριθμοῦ (εἶναι) ἑνός". Accordingly, for the previous phrase we propose the following translation: "Let the smaller (number) be assigned to be 1 Number", or merely "Let the smaller be assigned to be 1 Number". The translation "Let the smaller be assigned to be of 1 Number", in which the genitive ἀριθμοῦ ἑνός is respected, would be, of course, more accurate, though we adopted the former for simplicity.

• Assignment with the verb εἶναι, "to be"

Problem IV^G.19 asks to find three numbers indeterminately such that the product of any two, together with one unit, makes a square. The solution contains the following passage:

> I form a square from however many Numbers and 1 unit (I want); *let it be 1 Number, 1 unit*. Therefore, the square itself will be 1 Power, 2 Numbers, 1 unit. If I subtract the 1 unit, there remain 1 Power, 2 Numbers; it will be the product of the first and second. *Let the second be 1 Number*. Therefore, the first will be 1 Number, 2 units.

The two phrases written in italics in this passage are shown in Tannery's edition (the abbreviations being resolved) as "ἔστω ἀριθμὸς ᾱ μονὰς ᾱ" and "ἔστω ὁ δεύτερος ἀριθμοῦ ᾱ".[19] As said before, the same verb is used for assigning values to given numbers. An example in which both uses occur in the same phrase is "let the added number be 1 Number and let it be (added) to 2 and 3", in Problem II.11.

• Assignment with the verb πλάσσειν, "to form"

The act of making a non-square (or cubic) polynomial equal to a square (or cube) by an appropriate assignment for the side of the square/cube was called *al-istiqrā'* by Arabic algebraists (see Section 2.7.2). In the Greek books of the *Arithmetica* the

---

19 Matr. 4678 (f. 96r), Vat. gr. 304 (f. 97v), Marc. gr. 308 (f. 194r), and Vat. gr. 191 (f. 375v) have all ἀριθμῶν Δ̄ (genitive, plural: "of 4 Numbers"), the word ἀριθμῶν being written in all cases in full. Given the similarity of Δ (4) with A (1) one may suspect that the oldest archetype of the *Arithmetica* had the sign for ἀριθμός (ϟ), without a case ending, followed by the number A (1); it seems that during the textual transmission the latter was miscopied as Δ̄ (4), and the sign was resolved in the plural form of ἀριθμός, i.e., ἀριθμῶν, instead of the singular Δ̄ form ἀριθμοῦ, Ā which would be the correct writing. If this suggestion is correct, then we may infer that abbreviations were amply used in the manuscript of Diophantus.

verb πλάσσειν is reserved almost exclusively for making these *al-istiqrā'* assignments. For example, in the process of solving Problem II.13 Diophantus arrives at the forthcoming (i.e., half-completed) equation "1 Power lacking 1 unit is equal to a square". At this point he writes: "I form the square from 1 Number lacking 2 units", thus assigning a name to a side of the unnamed square. The Greek text is "πλάσσω τὸν τετράγωνον ἀπὸ ἀριθμοῦ ᾱ λείψει μονάδος ᾱ". Again this phrase is elliptical, the noun πλευρᾶς ("side") being implied after the preposition ἀπό and before the noun ἀριθμοῦ. Therefore, the complete phrase should be "πλάσσω τὸν τετράγωνον ἀπὸ (πλευρᾶς) ἀριθμοῦ ᾱ λείψει μονάδος ᾱ", whose translation is "I form the square from (a side) of 1 Number lacking 1 unit". For simplicity, in our translation of the elliptical form we do not employ the construction with the genitive.

- Assignment with the formulas ἐν ἀριθμοῖς, "in terms of Numbers", or ἐν δυνάμει, "in terms of a Power"

It is through these phrases that determinate numbers referring to units or parts of a unit are set to be multitudes (in modern terms, coefficients) of algebraic terms. An example of the former is found in Problem II.34:

> I assign the three to be 5 2' units (for) the first, 2 units (for) the second, and a half of a unit (for) the third. And it is obvious that the square of each one of them, if it receives in addition the 12, makes (each one) a square: one 12 4', another 16, and the other 42 4'. So, I assign them to be in (terms of) Numbers, the first to be 5 2' Numbers, the second 2 Numbers, and the third a half of a Number.

Problem II.25 contains an example in which the numbers are made multitudes of Powers.

> I first take a certain square from which, if I remove two numbers whatever, I leave a square. Let it be 16. Indeed, if this lacks 12 units, it becomes a square, and again if (it lacks) 7 units, it becomes a square. So, I assign them once again to be in (terms of a) Power – so that the one is 12 Powers, the other 7 Powers, and the square of the sum 16 Powers – and it is established that the square of the sum lacking either makes a square.

The constructions just given are by far the most commonly used in the Greek books for making algebraic assignments. Occasionally, Diophantus assigns algebraic names with other constructions. Thus, in the first solution to Problem III.6, we find an assignment with the verb ποιεῖν, "to make". The text is: "Again, since we require that the second together with the third make a square, let them make 1 Power, 1 unit lacking 2 Numbers, (which is a square) from a side of 1 Number lacking 1 unit". A rare construction is with the verb ἐκκεῖσθαι, "to set out". One occurrence is in Problem V^G.15 at a point in the solution in which Diophantus is led to the problem of finding three numbers such that any one of them, together with 1 unit, makes a cube, and moreover the sum of the three is a square. Then he writes:

"ἐκκείσθω ἡ μὲν τοῦ πρώτου πλευρὰ ἀριθμοῦ ā μονάδος ā, ἡ δὲ τοῦ δευτέρου μονάδων β̄ λείψει ἀριθμοῦ ā, ἡ δὲ τοῦ τρίτου μονάδων β̄" (the abbreviations in Tannery's printed text being resolved). The syntax is again elliptical. A more complete version would be "ἐκκείσθω ἡ μὲν τοῦ πρώτου (κύβου) πλευρὰ ἀριθμοῦ ā μονάδος ā, ἡ δὲ τοῦ δευτέρου (κύβου) (πλευρὰ) μονάδων β̄ λείψει ἀριθμοῦ ā, ἡ δὲ τοῦ τρίτου (κύβου) (πλευρὰ) μονάδων β̄". Also, the genitive clause of the names assigned makes us suggest that the verb εἶναι is implied before the names. Thus, the full sentence should be "ἐκκείσθω ἡ μὲν τοῦ πρώτου (κύβου) πλευρὰ (εἶναι) ἀριθμοῦ ā μονάδος ā, ἡ δὲ τοῦ δευτέρου (κύβου) (πλευρὰ) (εἶναι) μονάδων β̄ λείψει ἀριθμοῦ ā, ἡ δὲ τοῦ τρίτου (κύβου) (πλευρὰ) (εἶναι) μονάδων β̄". Ἐκκείσθω is third person, singular, present tense, imperative mood, middle/passive voice of the verb ἐκκεῖσθαι, the meaning of which is "to set out". Thus, we can translate the passage as "Let the side of the first (cube) be set out (to be) of 1 Number, 1 unit, the (side) of the second (cube), of 2 units lacking 1 Number, and the (side) of the third (cube), of 2 units". Again, for simplicity we have adopted the translation "Let the side of the first (cube) be set out to be 1 Number, 1 unit, the (side) of the second, 2 units lacking 1 Number, and the (side) of the third, 2 units". Another rare construction is with the verb ὑποκεῖσθαι, which we translate as "to be posited". One occurrence is in Problem 8 of Book VI[G], where we read "καὶ πάλιν ὑποκείσθω ⟨μία⟩ τῶν ὀρθῶν ἀριθμοῦ ā" ("Once again let ⟨one⟩ of the perpendiculars be posited to be 1 Number"). Other ways of making assignments are scarce. One example is found in Problem II.14. The text is: "Set out (ἔκθου) two numbers so that (the sum of) their squares is smaller than 20 units. Let them be 2 and 3".[20]

Before moving to the Arabic books two more cases should be mentioned, which occur quite often in the Greek text of the *Arithmetica*. Sometimes the verb in the assignment is implied. This occurs in Problem V[G].15, where we read: "9 Powers, 14 units lacking 9 Numbers (become) equal to a square, (say) the one from a side of 3 Numbers lacking 4 units". The second case is when the assignment depends on previous assignments and involves operations, that is, according to the classification we proposed in Section 2.7.2, when it is a "derivative assignment". Indeed, this is a very common case. It occurs in almost every problem of the *Arithmetica* in which more than one assignment is made. We find it in the very first problem, which asks to divide the number 100 into two numbers differing by 40 units. The text is "Τετάχθω ὁ ἐλάσσων ἀριθμοῦ ā· ὁ ἄρα μείζων ἔσται ἀριθμοῦ ā μονάδων μ̄". Translation: "Let the smaller be assigned to be 1 Number. Therefore, the greater will be 1 Number, 40 units".

In the Arabic books

•   Assigning algebraic names with the verb فرض (*faraḍa*), "to assign"

---

20 In Planudes's commentary the assignment of algebraic names is labeled ἔκθεσις (setting-out). Clearly the word ἔκθεσις is employed by him with the nontechnical meaning of "setting-out of terms in a series". See LSJ (= Liddell, Scott, and Jones 1996), *s.v.* ἔκθεσις, A.VI.

The verb *faraḍa*, translated from the Greek τάσσειν, is by far the most common verb for assigning algebraic names. For example, in Problem IV.19: "We assign the sought-after number to be one Thing", or in Problem IV.13: "We assign the cube to be from a side of a Thing, so that it is one Cube". The same wording is often given for *al-istiqrā'* assignments, too. One example from Problem IV.2: "Let us assign a side of the square to be seven Things, so that it is forty-nine *Māl*s."

- Assignments with the verb جعل (ja'ala), "to make"

The verb *ja'ala* is used in place of *faraḍa* for the initial assignments of algebraic names 12 times, such as this from Problem VI.1: "Let us make a side of the square a Thing, so the square is a *Māl*". *Ja'ala* is used to make assignments in the stage of *al-istiqrā'* 17 times. From Problem IV.34: "We make the greater square Equated to the Cube and the four *Māl*s be twenty *Māl*s and a fourth of a *Māl*". *Ja'ala* is also used to make an assignment that satisfies a second condition four times. After assigning a square to be a *Māl* in Problem VI.15, we find "And we make its excess over the smaller square be two Things and one, so the smaller number is a *Māl* less two Things and (less) one".

In Arabic books that contain worked-out problems solved by algebra, by far the most common verb for making assignments before the mid-thirteenth century is *ja'ala*. The earliest author we know who also used *faraḍa* is Ibn al-Khawwām in the later thirteenth century. For the few instances of *ja'ala* in our manuscript we do not know if the translator of the *Arithmetica* slipped back to the standard way of assigning names in Arabic or if *ja'ala* is translated from the Greek ποιεῖν (*poiein*). *Ja'ala* is used for other purposes, too, like to make an equation from two expressions (twice), to assign given numbers (twice), to make a number the multitude of a particular species (four times), and to make a lacking quantity common to the two sides of an equation (the stage of *al-jabr*, once).

- Assignment with the verb عمل ('amila), "to form"

This is the Arabic version of the Greek πλάσσειν, and as in the Greek, it is used as a kind of assignment only to "form" a square or a cube from its side. In this example from Problem IV.14 it is used in conjunction with *faraḍa*: "Let us form the square from a side of however many Things we wish. Let us assign it to be from a side of two Things, so that it is four *Māl*s". In other Arabic algebra books the word *ja'ala*, and not *'amila*, is used for this purpose.

- Making a determinate number a multitude of a power of the unknown, with the verb جعل (ja'ala), "to make"

In five instances a number (or numbers) is made a power. In Problem VI.2: "it is forty units and twenty-four parts of twenty-five parts of the unit. Let us make it

Cube Cubes, so it is forty Cube Cubes and twenty-four parts of twenty-five parts of a Cube Cube".

- Assignment with the implied verb "to be", in the construction "let . . . be"

There is only one example of an algebraic naming in this manner, in the context of *al-istiqra*. This occurs in the second solution to Problem VI.22: "Let the cubic number be eight Cubes". (In the first solution "assign" is employed in this stage.) In one instance in Problem VII.10 the sought-after number is declared to be a specific number using this construction: "Let the square number be sixty-four". Compare this with the previous problem, where it is given as "Let us assign the square number to be sixty-four". In three other instances the construction "Let . . . to be" is used to name unknowns after one of the other verbs (assign or form) has announced the plan. For example, in Problem VII.6: "Let us assign the number composed of the sum of the two numbers be a *Māl*, and we divide a *Māl* into two square parts. Let one of them be sixteen parts of twenty-five parts of a *Māl* and the other part be nine parts of twenty-five parts of a *Māl*".

4.2.4 The language of the operations

In the introduction of the *Arithmetica*, after having explained all multiplications with algebraic terms, having given the rule for multiplying extant and lacking terms, and having announced that the divisions are manifest, Diophantus continues with a pedagogical advice to his addressee Dionysius:

> Then, it is good that you who are beginning this study should have acquired practice in the addition, subtraction, and multiplications concerning the species; and how to add extant and lacking species, not of the same multitude, to other species, themselves either extant or likewise extant and lacking, and how from extant and other lacking species to subtract other (species), either extant or likewise extant and lacking.
>
> (14.3–10)

In this paragraph, Diophantus describes the operations he will be practicing in problem solving, and indeed the operations we encounter in the *Arithmetica* on the species of the "arithmetical theory" are, after all, the same four operations stemming from arithmetic, that is, addition, subtraction, multiplication, and division, and in few cases the extraction of square or higher roots. Hence, it is not surprising that the wording of an operation is the same whether the involved terms are concrete numbers, algebraic expressions, or both.

In the Greek books

As already mentioned in Section 2.2.2, a typical operation requires two verbal phrases, like this one by which Diophantus opens the long list of multiplications mentioned earlier: "A Number multiplied by a Number makes a Power". The first verb, πολυπλασιάζειν or πολλαπλασιάζειν, here in participle, "multiplied", expresses the operation to be performed, and the second verb, ποιεῖν, "to make", announces the outcome. The outcome verb varies. In the present case the verb ποιεῖν is used, but other verbs such as γίνεσθαι, whose range of meanings includes "to become", "to result"; εἶναι "to be"; καταλείπειν "to leave", when the operation is a subtraction, etc., are used as well. All verbs and expressions which are used in the Greek text of the *Arithmetica* to announce the outcome of an operation are surveyed in the glossary. It is important to note that verbal clauses comprising the adjective "equal", like "is equal", "becomes equal", etc., which are used to state equations, are never used by Diophantus to introduce the result of an operation. Next we give a potpourri of the most common expressions used in the Greek books for stating operations.

The most common verb for addition is προστιθέναι, usually translated as "to add".[21] Thus, in Problem I.8 it is asked to add (προσθεῖναι) the same required number to the given numbers 100 and 20 and make the greater of the resulting numbers triple the smaller. Diophantus sets the sought-after number as "1 Number", and he says: "And when it is added (προστεθῇ) to the 100, it will be (ἔσται) 1 Number, 100 units; while, when (it is added) to the 20, it becomes (γίνεται) 1 Number, 20 units".

Apart from the verb προστιθέναι, other verbs, and even an adjective, are also used for addition. Thus, in Problem I.2 the addition of "1 Number" and "3 Numbers" is expressed by the formula, "the two added together (συντεθέντες) are 4 Numbers". The verb expressing the operation here is συντιθέναι, the first meaning of which according to LSJ *Greek-English Lexicon* is "to place/put together".[22] In Problem I.1 Diophantus assigns the two sought-after numbers to be "1 Number" and "1 Number, 40 units", respectively, then he says: "The two together (συναμφότεροι), therefore, become (γίνονται) 2 Numbers, 40 units". Here the adjective συναμφότεροι, meaning "the two together", is used to express the addition.[23] And even if the word is an adjective, it is used as if it were a noun

---

21 Literally προσ-τιθέναι means "to put (to set, to place) to (besides, in addition)", therefore "to bring together", "to add". Thus in Plato's *Cratylus* (418a) we read in H. N. Fowler's translation (Plato 1926): "See, Hermogenes, how true my words are when I say that by adding (προστιθέντες) and taking away letters people alter the sense of words so that even by very slight changes they sometimes make them mean the opposite of what they meant before".

22 See LSJ (= Liddell, Scott, and Jones 1996), *s.v.* συντίθημι.

23 The literal meaning of the word συν-αμφότερος – a quite typical adjective of the Greek mathematical language (Acerbi 2012) – is to bring two things together as one. In this sense it is used, for example, in Plato's *Alcibiades I* (130a): "Socrates: Now, here is a remark from which no one, I think, can dissent. Alcibiades: What is it? Socrates: That man must be one of three things. Alcibiades: What things? Socrates: Soul, body, or both together (συναμφότερον) as one whole. Alcibiades: Very well" (Plato 1955). Also illuminating is the use of the term by Euclid in Proposition 45 of the *Data*: "If a triangle has one angle given, and both together the sides about the given angle as one have a given ratio to the remaining side, the triangle is given in form". What we translate as "both together as one" is Euclid's expression συναμφότεραι ὡς μία (Euclid 1896, 82.11–12). It

meaning "the two together". In such expressions there is no verb for the operation, though one is implied, since the idea of the operation taking place in time can
be understood from the verb announcing the outcome, which is either the future
tense of "to be" or the verb "become". We understand "the two together" as somehow implying the verb "taken", as "the two taken together". The same adjective
συναμφότεροι is used also for the sum, along with σύνθεμα and σύνθεσις.

In Problem II.17 we encounter another verb expressing addition, the verb
προσλαμβάνειν, "to receive in addition": "But, when 4 Numbers lacking 6 units
receive in addition (προσλάβωσιν) 2 Numbers, 5 units, they become 6 Numbers
lacking 1 unit". The verb λαμβάνειν, "to receive", is also used in the same problem
to express the addition: "It remains, also, for this, if it receives (λαβόντα) from
the middle the sixth and 7 units and if it gives (δόντα) the seventh and 8 units, to
become 6 Numbers lacking 1 unit".

The word δόντα, participle aorist of the verb διδόναι, "to give", is used in
the previous passage from Problem II.17 to state the operation of subtraction.
However, the most common verb for subtraction is ἀφαιρεῖν or ἀφελεῖν, usually
translated as "to subtract", "to remove". Thus, in Problem I.7, we read: "Let the
sought-after number be assigned to be 1 Number. And if from this I subtract
(ἀφέλω) the 100, there remains 1 Number lacking 100 units, while if (I subtract)
the 20, there remains 1 Number lacking 20 units".

Sometimes the verb λείπειν, "to lack", "to be decreased", is used to denote
subtraction. An example is given in Problem IV^G.24 in which Diophantus writes,
"indeed, thrice the 2 Numbers, if they lack (λείψωσιν) 2 Numbers, make (ποιοῦσι)
twice the 2 Numbers". The verb ποιοῦσι announcing the result shows that we have
here a subtraction described by the verb λείψωσιν. Another example is given in
Problem IV^G.22: "Let the sought-after number be 1 Number. If this is doubled and
decreased (λείψας) by a dyad, it makes 2 Numbers lacking 2 units".[24] The verb
λείπειν is related to the noun λεῖψις, meaning "lacking", in algebraic expressions.

---

seems that the extension ὡς εἷς/μία/ἕν, "as one", is implied when the adjective συναμφότεροι/
συναμφότεραι/συναμφότερα is used alone.

24 The use of the verb λείπειν with a meaning close to "to subtract" allows us to interpret the corresponding noun λεῖψις as taking, in a few cases, a meaning close to subtraction. One such case is
in Problem II.12, which asks to subtract the same (required) number from two given numbers (9
and 21) so as to make each of the remainders a square. Diophantus names the required number "9
units lacking 1 Power", and he finds the Number to be 4 8ths. In calculating the numerical value
of the required number (the one which was posited as 9u ℓ 1P) he writes: "The 9 units amount to
72 eighths, that is, 576 64ths; and the lacking (λεῖψις) of 1 Power removes (ἀφαιρεῖ) from them 16
64ths, and fulfills what was proposed", meaning that because in the term 9u ℓ 1P what is lacking
is 1P, we have to subtract from 9. We must say, how $\frac{16}{64}$ ever, that the use of λεῖψις with a meaning
close to subtraction is rather an exception. The standard word for subtraction is ἀφαίρεσις, and the
corresponding verb is ἀφαιρεῖν or ἀφελεῖν; cf. the phrase of the introduction "it is good that you
who are beginning this study should have acquired practice in addition (συνθέσει), subtraction
(ἀφαιρέσει), and the multiplications (πολλαπλασιασμοῖς) concerning the species". On the other
hand, the common meaning of λεῖψις throughout *Arithmetica* is "lacking", that is, the lacking part
that is missing from a term so as to be completed.

Occasionally the verb αἴρειν, "to remove", is also used. Thus, in the lemma preceding Problem IV^G.34, "1 Number lacking 1 unit" is removed from 8 units. The text reads: "Let the second be 1 Number lacking 1 unit. I remove (αἴρω) these from 8 units. There remain 9 units lacking 1 Number".

The outcome of a subtraction is often announced by the adjective λοιπός, lit. "remaining", "remainder", and the verb "to be", the last being in most cases implied. We translate such clauses either literally as "the remaining/remainder is" or by equivalent forms, such as "there remains/remain" or "is/are left".

The common verb for multiplying two quantities is πολλαπλασιάζειν or πολυπλασιάζειν. An example was given in the beginning of this section. To give another example we quote from Problem I.26, where, after the sought-after number has been named "1 Number", Diophantus writes: "And when it is multiplied (πολλαπλασιασθῇ) by the 200 units it makes 200 Numbers, while, when (it is multiplied) by the 5 units it makes 5 Numbers".

Often, the preposition ἐπί, "by", is used either in accompanying the verb denoting the multiplication or alone, when the corresponding verb is implied. This is the case of the famous phrase of the introduction, "A lacking (species) multiplied by (ἐπί) a lacking (species) makes an extant (species); a lacking (species) by an extant (species) makes a lacking (species)". Regarding the preposition ἐπί it is important to notice that in some cases it appears as part of the wording καὶ πάντα ἐπί, "and everything by", the implied operation being, of course, multiplication. This construction most often appears when Diophantus is simplifying an equation that contains inverse powers or, more generally, fractions, and a multiplication of the terms by the denominator is needed.

We should note that quite often the product of two numbers is expressed by the elliptic expression "ὁ ὑπ' αὐτῶν", lit. "the by them", that is, "the (number contained) by them". Throughout our translation we write it as "their product". The expression "ὁ ἐκ τοῦ πολλαπλασιασμοῦ", "the from the multiplication", which is a shortened form of the phrase "ὁ ἐκ τοῦ πολλαπλασιασμοῦ γενόμενος ἀριθμός", "the number produced from the multiplication", is rarely employed, as in Problem I.14.

Squaring, τετραγωνίζειν, "to square", is a special case of multiplication. The verb appears rarely. Problem I.26 provides an example: "if I square (τετραγωνίσω) the 5 Numbers, 25 Powers become equal to 200 Numbers".

Μερίζειν, "to divide", is the verb used for division. Division itself is called μερισμός, as we see in the sentence, already cited, of the introduction, "And the multiplications having been clarified for you, the divisions of the species just described are obvious". Within the *Arithmetica* the verb μερίζειν is accompanied either by the preposition εἰς, or by the preposition παρά, both constructions meaning "divide by".[25] Thus, in the lemma preceding Problem IV^G.34, having found the expression "9 units lacking 1 Number", Diophantus writes: "I divide these by (ταῦτα μερίζω εἰς) the (second number) increased by 1 unit, that is, by 1 Number". An example of the syntax with παρά is given in Problem IV^G.34:

---

25 See LSJ (= Liddell, Scott, and Jones 1996), *s.v.* μερίζω, A.1.

Let the second be assigned to be 1 Number lacking 1 unit. If I remove this from 8 units, and I divide (the remainder) by (μερίσω παρά) the number which is 1 unit greater than the second, the first will be 9 inverse Numbers lacking 1 unit.[26]

Another verb used for division is παραβάλλειν, which is known in Greek geometry with the meaning "to apply". It is accompanied, as the verb μερίζειν, either by εἰς or by παρά. Thus, in Problem V^G.10 we read: "Divide by (παράβαλε παρά) the multitude of the Powers; the Number becomes not greater than 67 17ths". In the same problem we encounter also the syntax παραβάλλειν εἰς: "Therefore, we are reduced to finding a certain number which, if made six times, and being divided by (παραβληθεὶς εἰς) its square increased by 1 unit, makes the quotient (παραβολή) greater than 17 12ths but smaller than 19 12ths".

Occasionally, the preposition παρά is used alone to denote a division, the accompanying verb, either μερίζειν or παραβάλλειν, being omitted. Thus, in the lemma preceding Problem V^G.7 we read: "Now, (divide) again the square (number) 4 by (παρά) 6 4ths, they become ⟨16 6ths⟩". As we said with regard to the preposition ἐπί and multiplication, the preposition παρά, also, is used sometimes alone, as part of the expression καὶ πάντα παρά, "and everything by", the implied operation being now the division. This construction appears most often when in a quasi-simplified equation a division should be made so as to obtain an even simpler equation.

Diophantus employs the verb διαιρεῖν or διελεῖν to "partition" or "divide" a number into two or more parts. The corresponding noun is διαίρεσις.[27] Thus, the very first problem of the *Arithmetica* is, "To divide (διελεῖν) a proposed number into two numbers having a given difference", while the famous Problem II.8 is, "To divide (διελεῖν) a proposed square (number) into two squares".

An unusual expression for division is through the verb μετρεῖν, "to measure", accompanied by κατά + noun in accusative case. Thus, an expression of the form, "ὁ {A} μετρεῖ τὸν {B} κατὰ τὸν {G}" means "{A} measures {B} according to {G}", that is, B can be factored as A · G. The corresponding noun μέτρησις, "measurement", is used to denote the factorization. Thus, a sentence like, "ἡ μέτρησις: ὁ {A} μετρεῖ τὸν {B} κατὰ τὸν {G}" means "the measurement: {A} measures {B} according to {G}". Problem V^G.1 provides a clear-cut example of this operation: "Their difference is 1 Power lacking 6 2' Numbers. The

---

26 An interesting example of division is the following, lifted from the lemma preceding Problem IV^G.36: "If, therefore, we assign the second to be whatever (multitude) of Numbers, and we multiply it by the ratio, it makes 3 Numbers. And if (this) is divided by (μερισθῇ εἰς) the excess by which the second number exceeds the ratio, that is by 1 Number lacking 3 units, the first becomes 3 Numbers in a part of 1 Number lacking 3 units". The division performed here is the division of "3 Numbers" by the expression "1 Number lacking 3 units", and the result, announced with the verb γίνεται (becomes), is "3 Numbers in a part of 1 Number lacking 3 units". We have here the result of a division to be expressed as a common fraction. Cf. Section 2.6.2; more on the wording of fractions in Section 4.2.7.

27 See "Index graecitatis apud Diophantum", *s.v.* διαίρεσις (Diophantus 1893–95, II, 266).

measurement: 1 Number measures it according to 1 Number lacking 6 2′ units". In modern language, x measures the polynomial $x^2 - 6\frac{1}{2}x$ according to $x - 6\frac{1}{2}$.

In the Arabic books

Operations in the Arabic books follow the same pattern as in the Greek books, which is also the pattern of Arabic books on arithmetic and algebra generally. Operations are expressed as an act performed in time which produces a particular outcome. Thus, two verbs are present, one for the operation and one for the outcome. As elsewhere in Arabic arithmetic and algebra, no verb or adjective meaning "equal", including *mithl*, *sawiya*, *ka-*, and *'adala* (the latter used to state algebraic equations), is ever employed to announce the outcome of an operation.

The verbs for addition in our book are زاد (*zāda*; add, increase), جمع (*jama'a*; add up, add together), and ضاف (*ḍāfa*; add, subjoin). The first of these is the most common, and its associated preposition is على (*'alā*, to). An example with this verb is: "when we add the two Things to the four Things, that gives six Things", from Problem IV.34. Perhaps it would be more literal to translate phrases of the form "*zāda A 'alā B*" as "increase *A* by *B*", but we write instead "add A to B". As we find it applied in our text, *jama'a* does not require a preposition. From Problem VI.17: "We want to find three square numbers such that if added together, their sum is a square". The preposition for *ḍāfa* is الى (*ilā*, to), and one example is: "We add it to the Cube. What is gathered from that is five hundred seventy-six", from Problem IV.10.

The outcome verbs for addition are most often either كان (*kāna*; give, get, be) or اجتمع من (*ijtama'a min*; amount to, gather from), and rarely we find صار (*ṣāra*, become). A common form of كان is فيكون (*fayakūn*, so [it] gives), where the prefix *fa-* indicates the beginning of a new verbal phrase. Sometimes the "sum" (مجتمع, *mujtama'*) is named, and the outcome verb is the implied "to be", as in the example from Problem VI.17. In four instances from the enunciations and conclusions of Problems IV.23, 25, and 29 (lines 623, 677, 706, and 796), an addition is indicated using the adjective مجموع (*majmū'*, added together), instead of the related verb *jama'a*. The enunciation to IV.23 reads: "We want to find two square numbers such that their squares added together are a cube". The lack of a verb for the operation is not important here because the operation is not actually being performed. The phrase "their squares added together" is to be read as meaning the result of the addition.

In other Arabic arithmetic and algebra books the word حمل (*ḥamala*, "to take on", as in extra passengers or more debt) is occasionally used to mean "add", but it is absent in our text. Its preposition is *'alā*. Also, the verb *jama'a* is used by other authors with the preposition *ilā*.

For subtraction the text shows two verbs, both of which are common: ناقص (*naqaṣa*; subtract, diminish) and لقي (*laqiya*; cast away). The preposition for a subtraction for both verbs is من (*min*, from), though rarely we find مما (*mimmā*; lit. "from what"). The outcome verb for a subtraction is بقي (*baqiya*; remain, leave). Here is an example using *naqaṣa* from Problem V.16: "The three *Māl*s and one less three Things, if subtracted from the cube, leave a cube, which is, as we said, a Cube and three Things less three *Māl*s and less one". This next example, with *laqiya*, is the subtraction of a common amount from the two sides of an equation: "Thus, the Cube and ten *Māl*s Equal sixteen *Māl*s. Let us cast away the common ten *Māl*s, leaving six *Māl*s Equal a Cube". We find *laqiya* employed three times as often to subtract a common amount from the two sides of an equation than for other subtractions, and *naqaṣa* is far more common outside the context of simplifying equations. This is also true in Arabic algebra generally. The choice of verb, then, lies more with Arabic use than with the Greek verbs that they translate.

For the verb *naqaṣa* the translator switched to a variation on the wording of the outcome from Books IV and V to Books VI and VII. Instead of "there remains" with *baqiya*, we find in Books VI and VII "the remainder is" with the related noun باقى (*bāqiya*; remainder) in conjunction with the verb *kāna*. From Problem VI.11: "if we subtract a Cube Cube from a square of the three Cubes, then the remainder is eight Cube Cubes". There are occasionally other variations on these standard forms, such as this one from Problem IV.42 (line 1406) with the outcome verb *kāna*: "if we subtract a given number from it, which is sixteen, it gives a square", or this one from IV.34: "We take the difference between these two squares, which is eight *Māl*s", using اخذ (*akhadha*; take). *Akhadha* is used for other operations, too, like "take nine times the square that comes from its side" and "if we take half of the sum of the two numbers". Several other verbs for subtraction are found in other Arabic books, like *saqaṭa* (drop), *ṭaraḥa* (throw away), *'azala* (remove), and *dhahaba* (eliminate).

The most common verb for multiplication is ضرب (*ḍaraba*, multiply), but sometimes we find ضعف (*ḍa'afa*, duplicate) instead. The preposition for *ḍaraba* is فى (*fī*, by), and for *ḍa'afa* it is ب (*bi-*, with). One multiplies a number *by* another number, but one duplicates a number *with* a number, which means to duplicate it that many times. The outcome verbs for multiplication are the same as those for addition. Here are two examples: "if you multiply the four Things by its side it amounts to ten Things" (Problem IV.15) and "Then we duplicate it with the third number, which is a *Māl*, to get nine parts of sixteen parts of a *Māl*." (Problem VI.18).

Division is expressed with one verb, قسم (*qasama*, divide), and its preposition is على (*'alā*, by). The outcome verb for division in our manuscript is usually *kāna*, though quite often it is خرج (*kharaja*; result in). The latter is the more common outcome verb in other Arabic books on arithmetic and algebra. Here are examples of each: "We divide thirty-six *Māl*s by a Thing to get thirty-six Things" (Problem VI.8) and "we divide nine *Māl*s by a *Māl*, resulting in nine units" (Problem VII.6). As with other operations, sometimes the noun for the operation, here "quotient" (قسم, *qism*), is used in conjunction with *kana*: "if one of them is divided by the other, then the quotient is a square" (Problem IV.22). There are other ways that the outcome is phrased, like "We also divide nine units by sixteen parts of twenty-five parts of a *Māl*, so the result of the division is fourteen parts and half an eighth of a part of a *Māl*" (Problem VI.23). Often, as here, "result" is still a verb so that the phrase would more literally be translated as something like "what results from the division", but we write "the result of the division" to make the reading easier.

The Arabic books do not perform square or higher roots of numbers or expressions.

4.2.5 The statement of the equation

This issue has been covered in Section 2.5.3, so here we merely provide a few additions for the Greek books. The standard expression in stating equations in Greek is by the adjective ἴσος, "equal", and the verb "to be". Thus, the equation 30N = 5P in Problem II.1 is stated: ἀριθμοὶ ἄρα $\overline{λ}$ ἴσοι εἰσὶ δυνάμεσι $\overline{ε}$, "Thus, 30 Numbers are equal to 5 Powers". In a fully written equation the word "equal" is part of the clause "is/are equal", as in the example just quoted. In most cases, however, only the adjective "equal" is written, the verb being implied, as in the clause ταῦτα ἴσα, "these (are) equal", or as in phrases like ἀριθμοὶ ἄρα $\overline{δ}$ ἴσοι μονάσι $\overline{ξ}$, "thus, 4 Numbers (are) equal to 60 units" from Problem I.2. Sometimes, when two different algebraic expressions for a single amount are found, the equality is stated only with the implied verb "to be", as in this example from Problem I.20: "Thus, the three (together) are 5 Numbers, but, also, 100 units". Other instances are in Problems I.21 and III.4. And last, often the result of an operation becomes the first part of an equation, as in the following example from the lemma to Problem IV^G.35: "Their product, when it lacks their sum, makes 2 Numbers lacking 3 units, (which are) equal to 8 units". Running the result of an operation into an equation like this is common in Arabic algebra, too.

The verb ἰσοῦν, "to equate", is used rarely, and never with the meaning of stating a final equation. It rather signifies either the task of stating an equation for which the two parts have already been established or stating a half-constructed, forthcoming equation of which only one part has been established. In both cases the equation is, so to speak, in-becoming, and this is manifest by the presence of verbs like "it remains (for us)", "then we must", etc. Therefore, the verb ἰσοῦν is always used with reference to a task that remains for us to fulfill in order to have the equation. For example, in Problem IV^G.28, the text has: "It remains (for us) to equate their product, that is, 196 units lacking 1 Power, to 36 units". Problem V^G.6 provides an example of an *al-istiqrā'* case: "It remains (for us) to equate 4 Powers, 4 Numbers, 4 units to a square". Another example, from Problem IV^G.15, is expressed with conditional syntax: "And if I equate 20 units, 300 inverse Powers with 32 units, the Number becomes 5 units". In some cases, the verb ἰσοῦν is employed together with the adjective ἴσος. In such cases we translate the former either by "to set up the equation" or by "to solve the equation". An example of the former is Problem IV^G.31: "I return to the initial (problem), to set up the equation: 3 Numbers, 18 units lacking 1 Power equal to a square". Problem IV^G.39 provides an example of the latter: "when we solve such an equation" (cf.

footnote 105 in page 466). In Problem II.11 the passive voice of the verb ἰσοῦν is employed with the meaning "to solve". Thus, referring to the treatment of the "double equality", the text has "it is solved in this manner".

4.2.6 Simplifying the equation

All complete equations in both the Greek and the Arabic books are polynomial equations. Recall that equations state the equivalence of two aggregations of the powers of the unknown. Consequently, there are five possible steps that might be required to simplify an equation: (i) add a deleted amount to the other side, (ii) subtract like amounts from the two sides, (iii) divide each term by a power (or, in one case, a polynomial), (iv) multiply each term by a power or a number, or (v) take the square or higher root of both sides. The first of these steps is associated with the word *al-jabr* (restoration) in Arabic, and the second with *al-muqābala* (confrontation).

In the Greek books

The standard clause in Greek by which steps (i) and (ii) are performed in the simplification of an equation, i.e., the "restoration" and the "confrontation", is "κοινὴ προσκείσθω ἡ λεῖψις, καὶ ἀφηρήσθω ἀπὸ ὁμοίων ὅμοια", "let the lacking be added in common, and let likes be subtracted from likes" (Problem I.9). The second part is often condensed to "καὶ ἀπὸ ὁμοίων ὅμοια", "and likes from likes", as in Problem II.8. Sometimes the two steps are stated separately, each step being followed by its own calculation, as in Problem I.7: "Let the lacking be added in common. It becomes 3 Numbers equal to 1 Number and 280 units. And let likes be subtracted from likes. Then, 2 Numbers are equal to 280 units". When only one of these two steps is required, only that part of the clause is stated.

The expression "πάντα παρὰ ἀριθμόν", "everything by a Number", is used to denote the division of all terms of an equation by a Number, and in other examples in the Greek books Diophantus divides everything by a Power or by a Cube. In one problem, VI^G.17, Diophantus seems to implicitly divide both sides by a polynomial: "1 Power, 2 Numbers, 3 units, is equal to 1 Cube, ⟨3⟩ Numbers ⟨lacking 3 Powers⟩, 1 ⟨unit⟩, from which the Number is found to be 4 units." The equation simplifies to 4P 4u = 1C 1N, which, dividing both sides by 1P 1u, reduces to 1N = 4u. Moreover, for the purpose of eliminating inverse powers or fractions, the expression "πάντα ἐπί", "everything by", is used to denote multiplication. In Problem IV^G.38: "But the three are 18 inverse Powers, (which are) equal to 1 Power. And everything by 1 Power. It becomes 1 Power-Power equal to 18 units". Similarly, in the second solution of Problem I.21 the fractions are eliminated in the equation $1N \, \ell \, \frac{1}{9}N = 11\frac{1}{9}u$: "Everything 9 times. Therefore, 8 Numbers are equal to 100 units". Similarly, in Problem II.29 the forthcoming equation that in modern form is $\frac{25}{16}x^2 + \frac{25}{16} = \square$ is simplified by taking "everything 16 times", where the 16 is necessarily a square. An example of taking the square roots of both sides of the equation, from Problem II.24, is: "Thus, 121 Power-Powers are equal to 1 Power. Therefore, the side will be also equal to the side. Thus, 1 Number is equal to 11 Powers".

In the Arabic books

The meanings of the term *al-jabr* (restoration) for step (i) and of the phrase *al-jabr wa-l-muqābala* (restoration and confrontation) for steps (i) and/or (ii) in Arabic algebra are presented in Section 2.6.3. Both the term and the phrase appear in Qusṭā's translation of the *Arithmetica*. Step (i) is worded with *al-jabr* three times. In Problem IV.11: "Thus, the Cube less six *Māl*s Equals four *Māl*s. We restore the Cube by the six *Māl*s, and we add it to the four *Māl*s. So one Cube Equals ten *Māl*s". In the other two instances, in Problems VI.13 and 14, the step is less specific. In VI.14: "Restore each side of the two sides by what is lacking from it, and add its same to the other side". The phrase *al-jabr wa-l-muqābala* appears six times where only step (i) is required. In Problem VI.3: "thirty-six Cube Cubes, and they Equal a *Māl Māl* less sixty-four Cube Cubes. We restore and confront, to get a *Māl Māl* Equals one hundred Cube Cubes". In these nine instances total, the wording of Diophantus was reformulated into wording common in Arabic algebra. But in 12 other instances we find something closer to the Diophantine way of adding a common amount to both sides. In Problem IV.26 the text reads "a Cube Cube less sixteen *Māl Māl*s Equals nine *Māl Māl*s. We add the sixteen *Māl Māl*s in common to the two sides to get a Cube Cube Equals twenty-five *Māl Māl*s". The amount that is added is specified each time, while in Greek we find the more generic "let the lacking be added in common" in nearly every instance. So even here we evidently have a rewording.

For step (ii), the confrontation of like amounts, our text expresses it as was already common in other books on algebra in Arabic, beginning with al-Khwārazmī and Abū Kāmil. For example, in Problem IV.29, faced with the equation 1CC 16CCM = 36CC, the text has: "Let us cast away the sixteen Cubes Cube *Māl*s in common from both sides, leaving a Cube Cube Cube (which) Equals twenty Cube Cube *Māl*s". By far the most common verb for this step in Arabic algebra is *laqiya* (cast away), and indeed it is used in 27 out of the 28 times that this step is spelled out in Books IV–VII. In 22 of the 28 instances the amount that is subtracted from both sides is stated, again making the wording more specific than the Greek phrase "let likes be subtracted from likes". The phrase *al-jabr wa-l-muqābala* is not used when only step (ii) is needed.

When both (i) and (ii) are required, our text applies the phrase *al-jabr wa-l-muqābala* six times. A typical example is this from Problem IV.13: "a part of eight parts of a Cube, and that Equals a Cube less seven *Māl*s. We restore and confront, leaving seven eighths of a Cube Equal seven *Māl*s". In other instances the two steps are stated individually, as described earlier.

Step (iii) is performed on at least 74 equations across the 4 Arabic books. The operation is explained 17 times using the word *aq'ad*, which takes the meaning of "lesser" in degree. The Greek books have no such word. From Problem IV.18: "the two hundred sixteen Cubes Equal sixty-four Cube Cubes. When we divide them by the one Cube, which is the lesser in degree of the two sides, it gives two hundred sixteen units Equal sixty-four Cubes". Three times as often it is worded without *aq'ad*, as in this example from Problem VI.3: "a *Māl Māl* Equals one hundred Cube Cubes. We divide each of these by a *Māl Māl* to get one Equals one hundred *Māl*s". Steps (iv) and (v) are not performed in the Arabic books.

4.2.7 The language of fractions

In the Greek books

Fractions are written in Tannery's edition in different ways. When the fraction is a submultiple, that is, when it is the reciprocal of an integer, only the denominator, say m, is written, either with the case ending –ov, m^{ov} (e.g. γ^{ov} for 3rd, ε^{ov} for 5th), or with a χ-like sign as a superscript, $m\times$ (e.g., $\gamma\times$, $\varepsilon\times$). We translate such unitary fractions either by "an mth" (e.g., a 3rd, a 21st) or by "m'" (like 2′, 4′); especially when a unitary fraction is part of a mixed number we use always the notation with a prime. In Tannery's text the form "m^{th}" is also used for ordinal numbers, as in Problem I.15, where the ordinals τὸν πρῶτον, "the first" and τὸν δεύτερον, "the second", are written τὸν α^{ov} ("the 1st") and τὸν β^{ov} ("the 2nd") (or τοῦ β^{ov} in genitive), respectively. In our translation such ordinal numbers are written in full, in conformity with writings we find in the oldest manuscripts. For example, the whole sentence of the instantiation of Problem I.15 appears in Matrit. 4678 in a way whose literal translation is: "Now, let it be proposed that the first, if it receives 30 units from the second, becomes its double, while the second, if it receives 50 units from the first, becomes its triple" (f. 64r, lines 1–2 from bottom, and f. 64v, lines 1–2 from top).

In Tannery's text special symbols are used for $\frac{1}{2}$ (namely, ∠′, ἥμισυ, i.e., "half"), and for $\frac{2}{3}$ (namely, ω̄). In our translation the former is written either in full, "half", or as 2′, while we write the latter, which occurs only four times in the Greek books, as 3″.

When the numerator is not unity three forms appear in Tannery's text:

1   Numerator and denominator are written as modern fractions but with the denominator being placed above the numerator, the two being separated by a horizontal line. As Heath points out (1910, 45), this writing is used also for cases in which numerator and denominator are written in the manuscripts in full. Occasionally Tannery writes the denominator as a superscript, as in Problem I.39, where we find the fraction $\frac{15}{4}$ written as $\overline{\iota\varepsilon}^{\delta}$ (78.26). In Vat. gr. 304 (f. 84v, line 11) this fraction is written "$\overline{\Gamma}$ καὶ τριῶν $\delta^{\omega v}$", "3 and three 4ths".

2   Numbers expressing numerator and denominator are written one after the
other in the same line, the numerator first, with a horizontal line above it, and
then the denominator, often with the respective case ending.

The two methods are seen in the following sentence from Problem II.12:

$$\text{αἱ μὲν } \bar{\vartheta} \ \dot{M} \text{ συνάγουσιν } \overline{oβ} \ η^a, \text{ τουτέστι } \overline{\varphi o\varsigma}^{\xi\delta}$$

We see here two fractions, $\frac{72}{8}$ and $\frac{576}{64}$, the first written as $\overline{72}$ 8ths, the second as
$\frac{64}{576}$. The word τουτέστι, connecting the two fractions, meaning "that is to say",
shows that the two forms are used interchangeably. Also, they were uttered in
the same manner, and it seems that the latter is in fact an evolution of the former.
The intermediary stages in this evolution were, very likely, forms in which the
denominator (with or without the case ending) is written as a superscript. The
stages of the evolution, in the example of the last fraction, can be illustrated in the
following diagram:

$$\overline{\varphi o\varsigma}\,\xi\delta^a \to \overline{\varphi o\varsigma}\,\varsigma^{\xi\delta^a} \to \overline{\varphi o\varsigma}\,^{\xi\delta} \to \frac{\xi\delta}{\varphi o\varsigma}$$

i.e.,

$$\overline{576}\,64^{ths} \to \overline{576}\,^{64^{ths}} \to \overline{576}\,^{64} \to \frac{64}{576}.$$

The manuscript tradition of the *Arithmetica* provides ample evidence for this
evolution. For example, in Matrit. 4678 we find both forms, $\overline{oβ}^{\eta a}$ and $\frac{\xi\delta}{\varphi o\varsigma}$, in f.
76r, line 11. Because of this, we express all of these forms in one way, as "m n^{ths}".

3   A third way of expressing fractions is by putting between the numerator and
denominator the word μορίου or ἐν μορίῳ – genitive and dative case, respectively, of μόριον (meaning primarily "part") – followed by number word(s),
often written as numeral(s), in genitive or dative, respectively. In both cases
the species to which the numbers refer (unit, Number, Power, etc.) are in
genitive. For example, we find in Problem I.25 the fraction we write as $\frac{47}{90}$
written as μονάδων μζ ἐν μορίῳ μονάδος ϟ°. A literal translation of this
expression would be "of 47 units in a 90th part of a unit", meaning "47 units
with 90 units for a denominator". In our translation we adopt the nearly literal
version, "47 units in a 90th part of a unit". Similarly, in Problem IV^G.36 the
fraction $\frac{3x}{1x-3}$ in Tannery's edition is written as $s\bar{\gamma}$ ἐν μορίῳ $s\,\bar{a}\,\wedge\,\dot{M}\bar{\gamma}$. This is
written in Matrit. 4678 as

(f. 105r, last line), literally translated as "of 3 Numbers in a part of 1 Number
lacking 3 units", meaning "3 Numbers with 1 Number lacking 3 units for a

denominator". Again, in our translation we adopt the literal form without the genitive in the beginning. On the other hand, in Problem IV^G.25 we find the fraction $\frac{8}{1x^2+1x}$ written in Matrit. 4678 (f. 99r) as

which we translate as "8 units of a part of 1 Power 1 Number", meaning "8 units with denominator 1 Power 1 Number".

In short, the two variants of this third way in which the manuscripts of Diophantus express fractions, with the word μόριον ("part") either in dative, μορίῳ, or in genitive, μορίου, are translated as "*m* in an nth part of" and "*m* of a part of *n*", respectively, by which it is meant "*m* with *n* for a denominator" and "*m* with denominator *n*".

Finally, mixed numbers, that is, whole numbers with a fractional part, are common in the Greek books. In every instance the fraction is a unit fraction. Tannery's text shows these in the form $\bar{m}$ nx, and we write them in our translation as *m n*'. Examples: 5 4' 12' corresponds to $5\frac{1}{4}\frac{1}{12}$ (our $5\frac{1}{3}$), 2 2' corresponds to $2\frac{1}{2}$, the latter written in Tannery's text as $\bar{\beta}$ ∠' (which in Problem II.13 is written out as μονάδες δύο καὶ μονάδος ἥμισυ, "2 units and a half of a unit").

Table 4.3 Fractions designated by "*m* of a part of *n*".

	Reference	Numerator	Denominator
1	186.5	65u	12,768u
2	186.7	12,675,000u	(163,021,824u)
3	186.8	15,615,600u	(163,021,824u)
4	186.8–9	8,517,600u	(163,021,824u)
5	246.13–14	8u	1P 1N
6	246.19	1P 1N 8u	1P 1N
7	256.17	1,507,984u	262,144u
8	286.23	7P ℓ 24N	1P 12u ℓ 7N
9	288.7–8	7P ℓ 24N	1P 12u ℓ 7N
10	288.9–10	12P	1P 12u ℓ 7N
11	288.11	12P	1P 12u ℓ 7N
12	288.12–13	35P ℓ 120N	1P 12u ℓ 7N
13	332.3	18,421,264N*	484,996u
14	332.4	42,954,916N*	(484,996u)
15	332.4–5	69,923,044N*	(484,996u)
16	332.5–6	131,299,224N	484,996u
17	332.8	131,299,224u	1,629,586,560u
18	332.11–12	781,543u	9,699,920u

Table 4.4 Fractions designated by "*m* in an *n*th part of".

	Reference	Numerator	Denominator
1	60.6	47u	90u
2	286.8–9	3N	1N ℓ 3u
3	286.18	3N	1N ℓ 3u
4	286.18–19	4N	1N ℓ 3u
5	286.22–23	12P	1P 12u ℓ 7N
6.	340.10	6N	1P 1u
7	370.15	25,600CC	1,221,025
8	370.17	25,600PP	1,221,025
9	370.19	25,600PP	1,221,025
10	374.4	14,400P	28,561u
11	416.12	12u	1P ℓ 30u
12	416.13	144u	1PP 900u ℓ 36P
13	416.14–15	60P 2520u	1PP 900u ℓ 36P
14	418.19–20	4u	1P ℓ 6u
15	418.21	16u	1PP 36u ℓ 12P
16	418.22–23	24P 24u	1PP 36u ℓ 12P
17	420.16	4u	"the excess"
18	420.20	96u	1PP 36u ℓ 12P
19	420.21	12u	6u ℓ 1P
20	420.22	72u ℓ 12P	1PP 36u ℓ 12P
21	422.1–2	96u	1PP 36u ℓ 12P
22	422.2	12P 24u	1PP 36u ℓ 12P
23	424.4	3u	6u ℓ 1P
24	424.5–6	54u	1PP 36u ℓ 12P
25	424.6–7	54u	1PP 36u ℓ 12P
26	424.8	90u ℓ 15P	1PP 36u ℓ 12P
27	424.9–10	15P ℓ 36u	1PP 36u ℓ 12P
28	438.14	2C 3P 1N	1P 2N 1u
29	438.15	2N 1u	1N 1u
30	442.2–3	8u	cube of 1P ℓ 2u
31	442.3–4	8u	square of 1P ℓ 2u
32	442.4–5	2u	1P ℓ 2u
33	442.6	2PP	cube of 1P ℓ 2u

Tables 4.3 and 4.4 summarize all fractions that appear in the Greek text of the *Arithmetica* and which are designated by the twin formulas "*m* of a part of *n*" and "*m* in a nth part of".

In the Arabic books

The Arabic fractions associated with finger reckoning, that is, the method of mental calculation in which intermediate results were "stored" by positioning the fingers in particular ways, was practiced before the advent of Islam across the Graeco-Roman world and the Middle East. One set of fractions traditionally associated with this system was based in the first nine unit fractions, which were called in Arabic the "heads": "a half", "a third", up to "a tenth". Other fractions were expressed as sums of these unit fractions or were approximated with them. For example, $\frac{3}{4}$ was spoken as "a half and a fourth", and $\frac{1}{16}$ as "half an eighth".

By the early ninth century, when al-Khwārazmī wrote his treatise on algebra, the range of fractions had expanded. Multiples of unit fractions were now allowed, and the language of "parts" had been adopted for general fractions. The former now included such fractions as al-Khwārazmī's "three fourths", "four ninths", and for $\frac{9}{25}$, "a fifth and four fifths of a fifth". Generally, the formulation in terms of parts was reserved for fractions inexpressible in term of the heads, like this one from Abū Kāmil (2012, 557.12): "twenty-eight parts of one hundred forty-three parts of a unit". Here the unit is partitioned into 143 parts, and the fraction is 28 of those parts. Al-Khwārazmī also gives $\frac{9}{25}$ in terms of parts, as "nine parts of twenty-five parts". Sometimes this formulation is condensed even further, like "five parts of eighty" in Problem IV.14 of our text. When the fraction is of a power of the unknown, it is the power that is partitioned. For example, in Problem IV.14 we find "a part of ten parts of a Cube". Here one Cube is partitioned into ten parts, and the fraction is one of those parts. This wording reinforces the idea that the fraction (coefficient) and the power are not two separate amounts brought together to form the term but are two aspects of a single amount.

One unit-fraction representation that remained in common use after the ninth century was "a half and a fourth" for $\frac{3}{4}$. This formulation is found in later Arabic authors like al-Uqlīdisī (10th c.) and al-Khayyām (11th c.). It occurs in Problem IV.22 of Qusṭā's translation, though it is also consistent with Greek unit fractions as well. The representation "three fourths" is given in many other problems. Also, the "a half and an eighth" in Problem IV.15 remained the usual way of expressing $\frac{5}{8}$, and was not replaced with "a half and a fourth of a half".

Fractions in either system could accompany a whole part, like these examples in al-Khwārazmī (2009, 271.5, 269.8): "a *māl* and a sixth of an eighth of a *māl*", where we would write "$1\frac{1}{48}$ of a *māl*" (The word *māl* here is an amount of money to be distributed in an inheritance), and "eight shares and one hundred twenty parts of one hundred fifty three parts of a share" for $8\frac{120}{153}$ shares of an inheritance. Improper fractions were avoided in Arabic, though they do appear occasionally.

Many fractions in the Arabic books can thus be identified as being either characteristically Greek or characteristically Arabic. Improper fractions, like "six fifths" in Problem VII.18, and "eighty-one parts of seventeen parts of one" in Problem IV.23 reflect Diophantus's way. Mixed fractions in which the fractional part has a numerator greater than 1, like "sixty-nine and four ninths" for $69\frac{4}{9}$ in Problem IV.24 and "three hundred seven units and thirty-five parts of sixty-four parts" for $307\frac{35}{64}$ in Problem IV.22, are the result of a reworking in Arabic. Some fractions appear to be a hybrid of the two, like "eighty-one eighths of an eighth" for $\frac{81}{64}$. The "eighty-one eighths" is a Greek improper fraction, while the "eighths of an eighth" is the result of an Arabic rewriting. Diophantus would have written this as $\frac{81}{64}$, while it would ordinarily have been expressed in Arabic as "one and two eighths and an eighth of an eighth".

4.2.8 Other technical terms in the Greek algebra of Diophantus

Some other terms employed by Diophantus are either rare or are used in a different way in other Greek mathematics texts.

Diophantus uses the term εἶδος (*eidos*; species), in a way different from other Greek authors. We discussed this difference at the end of Section 2.5.1, and we explained how the species of algebraic number in Diophantus relates to species of numbers in Greek arithmetic in Section 2.2.1.

The word πλασματικόν (*plasmatikon*) appears at the end of the determinations of Problems I.27–28, I.30, and in Arabic translation as *al-muhayya'at* in Problems IV.17, IV.19, and V.7. Our interpretation of the term, which underlies the translation we adopted for the clause ἔστι τοῦτο πλασματικόν, conforms with the explanation of the meaning of the clause provided by an anonymous comment transmitted by some Greek manuscripts, such as Vat. gr. 304 (f. 82v) and Vat. gr. 191 (f. 365r), and published by Tannery among the "Testimonia Vetera in Diophantum": "*plasmatikon*: that is, arising not by a certain fabrication, but which is displayed together with the formation itself" (Diophantus 1893–95, II, 260.16–18). There have been many suggestions for the meaning of word πλασματικόν. For a survey we refer the reader to the study (Acerbi 2009) and the bibliography provided there.

The term ὑπόστασις (*hypostasis*) and the corresponding expression Ἐπὶ τὰς ὑποστάσεις, which we translate as "To the hypostases", is scarcely used in Greek mathematical texts other than Diophantus's *Arithmetica*. The principal meaning of the term is the numerical value of an unknown quantity, either named or unnamed, as we infer from the sentence "the hypostasis of the Number is not manifest" from Problem I.39. Most often the term is used in the clause "To the hypostases", meaning something like "let us now proceed to the calculation of the numerical values of the sought-after numbers". The fact that this clause, with "hypostases" in plural, is used even when there is only one sought-after number in a problem, suggesting that the expression "To the hypostases" was regarded by Diophantus as a subtitle for the last part of the textual unit of a worked-out proposition, in which numerical values of the sought-after numbers are calculated (Christianidis 2015).

Another term is παρισότης (*parisotēs*) and its cognate πάρισος (*parisos*), which are employed four times in the *Arithmetica*, in Problems V^G.11 and V^G.14. The meaning is "almost equality" and "almost equal", respectively. It is this word that lies behind Fermat's term "*adequality*". The Greek word comes from ancient rhetoric. It belongs to the "figures of style" and denotes a succession of a number of coordinate clauses which have approximately the same length, based on their number of words or syllables. The famous phrase "Veni vidi vici" (I came; I saw; I conquered), attributed to Julius Caesar, is an example of παρισότης (cf. footnote 25, p. 10).

And last, the first nongrammatical word in the opening line of the *Arithmetica*, εὕρεσις (*heuresis*), is used only once in the treatise, and which we translate as "invention": "Knowing that you are anxious, my most honourable Dionysius, to learn the 'invention' of problems in numbers". We explain our choice of "invention" in the corresponding footnote in the translation. One of us has argued that by using this word Diophantus intends the most important part of a problem solving by algebra, which is the establishment of the equation, and that this interpretation makes Diophantus's use of εὕρεσις comparable to its rhetorical use, i.e., the working out of arguments to answer a question (Christianidis 2007).

5   The didactic aspect of the *Arithmetica*

The didactic dimension of the *Arithmetica* is manifested by the fact that Diophantus addresses the book to Dionysius, who, aiming to develop his own capacity for "inventing" problems in numbers, had asked Diophantus to teach him how to solve problems using the method that Diophantus himself was practicing. In addition to this, the didactic character is reflected throughout the introduction to the book. For example, after defining the technical terms of the "arithmetical theory" and how to operate with them, Diophantus gives Dionysius the following advice, as a teacher to his student:

> And the multiplications having been clarified for you, the divisions of the species just described are obvious. Then, it is good that you who are beginning this study should have acquired practice in addition, subtraction, and the multiplications concerning the species; and how to add extant and lacking species, not of the same multitude, to other species, themselves either extant or likewise extant and lacking, and how from extant and other lacking species to subtract other (species), either extant or likewise extant and lacking.
>
> (14.1–10)

One could object that the very character of the introduction, written in epistolary form, favors such comments and remarks. Yet the didactic tone can also be recognized in the main part of the treatise, in the worked-out problems. For example, Diophantus makes the following comment in Problem III.19, which asks to find four numbers such that if any one of them is added to or subtracted from the square of their sum, gives a square:

> Since in any right-angled triangle the square on the hypotenuse, if it receives in addition twice the (product of the sides) about the right angle, or if it lacks (it), makes a square, I search first for four right-angled triangles having equal hypotenuses; this is the same as dividing a square into two squares (in four ways); *but we learned how to divide a given square into two squares in an unlimited number of ways.*
>
> (182.22–184.4, emphasis added)

DOI: 10.4324/9781315171470-6

The concluding sentence of this passage refers of course to Problem II.8, "To divide a proposed square into two squares", that was treated earlier, hence the use of the verb "we learned" (i.e., how to solve it).

Throughout the *Arithmetica* there is a plethora of similar indications of a didactic concern of Diophantus. The most striking of all are the dozens of explanations, dispersed in almost all problems, by which Diophantus explains resolutory steps. For example, in Problem II.26, which asks to find two numbers such that their product, if it receives in addition either, makes a square, while the sides of the squares, if added together, make a proposed number, Diophantus begins the solution by explaining the way he found the names he assigns to the two sought-after numbers:

> Since, *if there are two numbers the greater of which is the quadruple of the smaller except one unit, then their product makes a square if it receives in addition the smaller*, I assign the smaller to be 1 Number, and the greater 4 Numbers less 1 unit. And it comes out, similarly, that their product makes a square if it receives in addition the smaller.
>
> (122.9–14, emphasis added)

The problem requires the finding of two numbers satisfying certain conditions. Diophantus makes the lesser sought-after number to be "1 Number" (1N), the greater to be "4 Numbers lacking 1 unit" (4N ℓ 1u), and he explains his choices by evoking an algorithm we write as $n \times (4n - 1) + n \rightarrow \square$, which can be read as representing one of the conditions of the problem. In Table 5.1 we reproduce in abbreviated symbolic manner the treatment of Problem II.26 by presenting in separate columns steps of the solution and explanations:

The analysis in Table 5.1 shows that the steps of the solution are interspersed with a non-negligible amount of material serving didactic purposes, a fact that was recognized by the Byzantine scholiast Maximus Planudes, who leaves out all such material in the abbreviated presentation of the solution to that problem (Diophantus 1893–1895, II, 243.1–9), which we reproduce in Table 5.2.

Heuristic explanations and reminders of mileposts of the procedure like the ones we saw in the previous example, are provided quite often throughout the *Arithmetica*, bearing witness of Diophantus's care for the didactic aim of his book.

But the didactic setting of the *Arithmetica* is by no means confined to such auxiliary comments on the part of Diophantus. Problem 27 of the Greek book IV displays a specific didactic technique that Diophantus uses occasionally in his solutions, which can be described as "constructive dead-end". In this problem we are seeking two numbers such that their product, if it lacks either one, makes a cube. In other words, we are seeking two numbers having the property to produce cubes when combined in the following two ways:

> (1st sought-after number) × (2nd sought-after number) – (1st sought-after number), and
> (1st sought-after number) × (2nd sought-after number) – (2nd sought-after number).

Table 5.1 Problem II.26. The left column has the resolutory steps, the right column the heuristic explanation.

To find x, y such that
$x \cdot y + y = \square$; $x \cdot y + x = \square'$; $\sqrt{\square} + \sqrt{\square'} = 6$

Step	Explanation
	Since $n \times (4n - 1) + n \rightarrow \square$,
I assign $y := 1$N (the smaller) $x := 4$N ℓ 1u (the greater)	
	Indeed, their product makes a square if it receives in addition the smaller. If it receives the greater (4N ℓ 1u) it must also make a square, from a side of 6u ℓ 2N, so that the sides of the two (squares), when added together, make 6.
From the second condition, Mult. 1N by 4N ℓ 1u, then add 4N ℓ 1u → 4P 3N ℓ 1u Square 6u ℓ 2N →4P 36u ℓ 24N Thus, 4P 3N ℓ 1u = 4P 36u ℓ 24N, so 1N = $\frac{37}{27}$ To the hypostases: $y = \frac{37}{27}$, $y = \frac{121}{27}$	

Table 5.2 Planudes's tabular setting of the solution of Problem II.26.

setting-out	4N ℓ 1		1N
	4P 3N ℓ 1u		
	6u ℓ 2N		
multiplication	4P 36u ℓ 24N	eq.	4P 3N ℓ 1u
addition	4P 37u	eq.	4P 27N
subtraction	37u		27N
division	$\frac{37}{27}$		1N
hypostases	$\frac{121}{27}$		$\frac{37}{27}$

Diophantus starts by assigning the following names to the numbers sought: he sets the first sought-after number "8 Numbers" (8N) and the second sought-after number "1 Power, 1 unit" (1P 1u). Then, the first condition gives, after the operations have been performed, 8C ("eight Cubes"), which is indeed a cube. The second condition, however, leads to the expression 8C 8N ℓ 1P ℓ 1u, which must be a cube, and this, says Diophantus, is impossible (250.15).[1] This means that the path

---

1 Heath points out (1910, 186 n. 1) that the expression $8x^3 + 8x - 1x^2 - 1$ can be made a cube by positing the side of the cube as being either $2x - \frac{1}{12}$ or $\frac{8}{3}x - 1$.

he has followed leads to a dead end, so his assignments must be modified. He returns to the beginning and renames the two sought-after numbers:

> 1st sought-after number ≔ 8N 1u (8 Numbers, 1 unit),
> 2nd sought-after number ≔ 1P (1 Power).

With these names the second condition is fulfilled, while the first condition leads to the expression 8C 1P ℓ 8N ℓ 1u, which must be a cube. This is done by making it the side of the cube 2N ℓ 1u, and from the resulting equation he finds the value of 1N, and from it, the values of the two sought-after numbers.

The obvious question arising from this example, given that the first attempt to solve the problem was sterile, is why did Diophantus include it in his text? If his purpose was simply to solve the problem, to write down and present the solution, then would it not have been sufficient to present only the second attempt, the one that actually solves the problem? The answer to this question is that Diophantus incorporated the unsuccessful attempt in his text because he had his reader in mind. He was acting as the teacher toward his student. The inclusion of the sterile attempt in his book can be interpreted as a didactic technique, which aims to attract the interest of the reader, to elicit his participation, his involvement, to draw his attention to a choice of names for the sought-after numbers that once made would have not led to the solution.

Our last comment regarding the didactic aspect of the *Arithmetica* is that Diophantus's concern for his pupil is displayed not solely in one or another particular solution, as seen in the previous example, but in the large-scale organization of the book as a whole. To this end, the last paragraph of the introduction is particularly revealing:

> From now on, we shall follow the way of the propositions, having at our disposal much collected material on the species themselves. Since (the things involved) are many in number and large in size, and for this reason they are mastered slowly by those who acquire them, and (since) they include things that are difficult to memorize, *I thought it worthwhile to divide those that are amenable (to being divided), and moreover to arrange those at the beginning, as elements, from the simpler to the more complex, as it seems convenient to do. In this way, indeed, they will be easier to follow for the beginners and their working out will be memorized.*
>
> (14.25–16.6, emphasis added)

The arrangement and the progression of the material, from the simplest to the most complex, are two key features that demonstrate what was described earlier as the didactic character in the large scale of Diophantus's treatise. This issue has been thoroughly discussed with respect to the first three books of the *Arithmetica* by Bernard and Christianidis (2012). The two authors argue that the "collected material on the species" and the indeterminate "things" of which Diophantus speaks, refer to a multitude of techniques – designated in the paper "methods of invention", yet the designation "methods of assignment" would be more appropriate – a kind of "toolbox" to be used for assigning names to the unknowns of a problem. The reason these things are not explained in the introduction – where, instead, they are described either by the allusive expression "collected material on the species" or by a number of participles and pronouns which are all indeterminate – is because *they shall be discovered along the way*, just as the beginning of the passage explicitly says. That is, *within the treatments of the problems* there is something very important to be learned regarding the method of solution taught by Diophantus and its implementation in practice. This *mathesis* unfolds gradually, as the study of the treatise goes on. In other words, the treatise itself is organized in such a way as to conform to this progressivity, and this is what we called the didactic aspect in the large scale of the *Arithmetica.*

Part II

Translation

Notes on the translation

Three styles of parentheses/brackets are used in the translation. Passages in angular brackets ⟨. . .⟩ are additions that a copyist evidently omitted by mistake and were restored to the text by an editor. For the Greek books the editor is indicated in footnotes. For the Arabic text, they are Sesiano's additions unless noted otherwise. Passages in square brackets [. . .] are identified as probably being interpolations by Tannery for the Greek and by Sesiano for the Arabic. In the Arabic manuscript Sesiano remarks that these are passages "which complete or clarify the text in some way, or which, simply, do not render its comprehension difficult" (p. 29). In our translation we have in some cases transferred such passages in footnotes. Words in parentheses (. . .) are inserted by us to make the reading easier.

Capital letters are used for translations of the names of the powers, such as "Number"/"Thing" and "Cube" in Greek and in Arabic, and also for the algebraic "Equals" in Arabic.

Line numbers from Sesiano's edition are given at the start of each paragraph of Books IV–VII. For the Greek books page numbers are given referring to the start of each page in Tannery's edition.

In footnotes, observations and notes made by Sesiano are indicated by /s/, those made by Rashed are indicated by /r/, and those made by both are indicated by /sr/.

DOI: 10.4324/9781315171470-7

Book I

Knowing that you are anxious, my most honorable Dionysius, to learn the "invention"[1] of problems in numbers, I have tried (to expose how) to hypostasize the nature and power (subsisting) in numbers, beginning with the foundations on which the subject is built.

Well, perhaps the subject appears rather difficult, in as much as it is not yet familiar – and the souls of beginners have little hope for successful accomplishment – but you, with your eagerness and my teaching, will find it easy to grasp; for keenness supported by teaching is a swift road to knowledge.

But besides these things, since you know that all numbers are made up of a certain multitude of units, it is clear that their formation has no limit. Therefore, among them are

- squares (τετραγώνων), which are formed when a certain number is multiplied by itself, and this number is called a side (πλευρά) of the square;
- cubes (κύβων), which are formed when squares are multiplied by their sides;

---

1 The word εὕρεσιν (2.3), which we translate as "invention", should be understood as meaning a distinct part in the process of solving a problem and not the process as a whole. Tannery's Latin translation of εὕρεσις is "solutio". Heath translates the passage by "Knowing, my most esteemed friend Dionysius, that you are anxious to learn how to investigate problems in numbers" (1910, 129); however, in the introduction of his book he presents another interpretation of the passage, proposing two possible translations of εὕρεσις: "Knowing you, O Dionysius, to be anxious to learn the solution (or, perhaps, 'discovery,' εὕρεσιν) of problems in numbers" (1910, 9). None of the previous translations is fully satisfactory. The translation of εὕρεσις we adopted is "invention", in accordance with Hultsch's suggestion in interpreting a passage from the fourth Book of Pappus's *Collection*, where Pappus discusses the classification of geometric problems, which reads as follows: "But those problems which are solved when there is assumed toward their *heuresis* one or several of the sections of the cone are called solid". See Hultsch (1876–78, I, 270.8–11); the same passage occurs also at p. 54.12–14; see also the "Index graecitatis" (1876–78, III), *s.v.* εὕρεσις. The same translation is also adopted by Heike Sefrin-Weis in her translation of the fourth Book of Pappus's *Collection*; see Pappus (2010, 144).

DOI: 10.4324/9781315171470-8

4
- others,[2] which are formed when squares are multiplied by themselves;
- others,[3] which are formed when squares are multiplied by the cubes from the same side;
- others,[4] which are formed when cubes are multiplied by themselves.

And it is from the addition, subtraction, and multiplication of these (numbers), and from the ratio which they bear to one another, or, even, each one to its own side, that most arithmetical problems are formed;[5] they can be solved if you follow the way that I will show.

Now, it is an established fact that each of these numbers, once it has been given an abbreviated denomination, is an element of the arithmetical theory. So,

- the square is called a Power, and its sign is Δ marked with the sign Y, i.e. Δ^Y, δύναμις;
- the (cube is called) a Cube, and its sign is K marked with the sign Y, i.e. K^Y, κύβος;
- the square multiplied by itself (is called) a Power-Power, and its sign is two deltas marked with the sign Y, i.e. $\Delta^Y\Delta$, δυναμοδύναμις;
- the square multiplied by the cube from the same side (is called) a Power-Cube, and its sign is ΔK marked with the sign Y, i.e. ΔK^Y, δυναμόκυβος;

6
- the cube multiplied by itself (is called) a Cube | -Cube, and its sign is two kappas marked with the sign Y, i.e. $K^Y K$, κυβόκυβος.
- While, the (number) which has none of these characteristics, but merely has in itself a multitude of units, is called an unspoken[6] Number (ἀριθμός), and its sign is **Ϛ**.[7]

---

2 Tannery's text, and the manuscripts, have here the word δυναμοδυνάμεων.
3 Again Tannery's text, and the manuscripts, have here the word δυναμοκύβων.
4 Similarly, Tannery's text, and the manuscripts, have here the word κυβοκύβων.
5 Lit. "are woven", πλέκεσθαι.
6 ἄλογος. The range of meanings of this term comprises, among others, unspeakable, unutterable, impossible to be pronounced. Here it simply means that the value is unknown, so it cannot be spoken. It does not mean "irrational" in this instance.
7 The translation of this sentence refers to the text as it appears in the manuscripts and in Allard's edition (1980, 375.11–13), not as it is published by Tannery. The text published by Allard reads as follows: "ὁ δὲ μηδὲν τούτων τῶν ἰδιωμάτων κτησάμενος, ἔχων δὲ ἐν ἑαυτῷ πλῆθος μονάδων, ἄλογος ἀριθμὸς καλεῖται, καὶ ἔστιν αὐτοῦ σημεῖον τὸ **Ϛ**". Tannery "amended" the text, on the basis of a letter of Michael Psellus that he published in the second volume of the edition of Diophantus (1893–95, II, 37–42), by substituting the word ἄλογος (unspoken), which is between the words μονάδων and ἀριθμός, with ἀόριστον (undetermined), and shifting the comma after μονάδων, thus making the word between μονάδων and ἀριθμός to be syntactically connected with the former (ἔχων δὲ ἐν ἑαυτῷ πλῆθος μονάδων ἀόριστον, "having in itself an *undetermined* multitude of units") and not with the latter (ἄλογος ἀριθμὸς καλεῖται, "is called an *unspoken* Number"). For Tannery's view, see (1912e, 429–30). The most recent discussion of the whole issue is Acerbi (2015). [N.B.: The sign for ἄλογος ἀριθμός is

• But there is also another sign, the immutable (element) of the determinate (numbers), the unit, and its sign is M marked with the sign O, i.e., 𝔐.[8]

And just as the homonymous parts of the numbers are called after the numbers – of three is the third, and of four the fourth – so the homonymous parts of the just named numbers, also, will be called after the numbers:

(the homonymous part)	of the Number,	the inverse Number,[9]
	of the Power,	the inverse Power,
	of the Cube,	the inverse Cube,
	of the Power-Power,	the inverse Power-Power,
	of the Power-Cube,	the inverse Power-Cube,
	of the Cube-Cube,	the inverse Cube-Cube;

and each of them will have on the sign of the homonymous number the mark ×, distinguishing the species.

Then, having exposed to you the denomination of each of the numbers, I will proceed to their multiplications; you will find them obvious, because they are pretty well expressed in the name itself.

A Number multiplied	by a Number makes	a Power,
	by a Power,	a Cube,
	by a Cube,	a Power-Power,
	by a Power-Power,	a Power-Cube,
	by a Power-Cube,	a Cube-Cube.
A Power (multiplied)	by a Power,	a Power-Power,
	by a Cube,	a Power-Cube,
	by a Power-Power,	a Cube-Cube.
A Cube (multiplied)	by a Cube,	a Cube-Cube.

8

While every number multiplied by its own homonymous part makes a unit. Now, because the unit is immutable and always constant, any species multiplied by it will remain the same species. While the homonymous parts, when multiplied by themselves, will make parts homonymous with the numbers.

E.g., an inverse Number (multiplied)

	by an inverse Number,	makes	an inverse Power,
	by an inverse Power,		an inverse Cube,

---

reproduced from the codex Marc. gr. 308 (**B**). For the sigla of the Greek manuscripts, see Section 1.3, Table 1.1.]

8 The sign for μονάς is also reproduced from **B**.

9 ἀριθμοστόν, lit. "Number-th", such as when we make, for example, "seventh" from "seven", as Diophantus does himself in several problems. The same applies to the other terms: δυναμοστόν is literally "Power-th", κυβοστόν is "Cube-th", etc.

by an inverse Cube,	an inverse Power-Power,
by an inverse Power-Power,	an inverse Power-Cube,
by an inverse Power-Cube,	an inverse Cube-Cube,

and the same will arise homonymously.[10]

10   An inverse Number

by a Power,	a Number,
by a Cube,	a Power,
by a Power-Power,	a Cube,
by a Power-Cube,	a Power-Power,
by a Cube-Cube,	a Power-Cube.

An inverse Power

by a Number,	an inverse Number,
by a Cube,	a Number,
by a Power-Power,	a Power,
by a Power-Cube,	a Cube,
by a Cube-Cube,	a Power-Power.

An inverse Cube

by a Number,	an inverse Power,
by a Power,	an inverse Number,
by a Power-Power,	a Number,
by a Power-Cube,	a Power,
by a Cube-Cube,	a Cube.

12   An inverse Power-Power

by a Number,	an inverse Cube,
by a Power,	an inverse Power,
by a Cube,	an inverse Number,
by a Power-Cube,	a Number,
by a Cube-Cube,	a Power.

An inverse Power-Cube

by a Number,	an inverse Power-Power,
by a Power,	an inverse Cube,
by a Cube,	an inverse Power,
by a Power-Power,	an inverse Number,
by a Cube-Cube,	a Number.

An inverse Cube-Cube

by a Number,	an inverse Power-Cube,
by a Power,	an inverse Power-Power,
by a Cube,	an inverse Cube,
by a Power-Power,	an inverse Power,
by a Power-Cube,	an inverse Number.

10 Probably with this phrase Diophantus intends species of power higher that the sixth.
   Such species occur in the Arabic books.

A lacking (amount) multiplied by a lacking (amount) makes an extant (amount); a lacking (amount) by an extant (amount) makes a lacking (amount); and the sign of the lacking is a truncated Ψ turned upside down, i.e., ⋏.[11]

And the multiplications having been clarified for you, the divisions of the species just described are obvious. Then, it is good that you who are beginning this study should have acquired practice in the addition, subtraction, and multiplications concerning the species; and how to add extant and lacking species, not of the same multitude, to other species, themselves either extant or likewise extant and lacking, and how from extant and other lacking species to subtract other (species), either extant or likewise extant and lacking.

14

After that, if there result from a problem certain species equal to the same species, but not of the same multitude, it will be necessary to subtract likes from likes from either side, until it results in one species equal to one species. But if by chance there be on either (side) or on both lacking species, it will be necessary to add the lacking species on both sides, until the species on each side become extant, and again to subtract likes from likes until one species is left on each side. All this should be worked out with subtlety within the hypostases[12] of the propositions, if possible until one species be left equal to one species; and later I will also show you how, when two species are left equal to one, such a case is solved.

From now on, we shall follow the way of the propositions,[13] having at our disposal much collected material on the species themselves. Since (the things involved) are many in number and large in size, and for this reason they are mastered slowly by those | who acquire them, and (since) they include things that are difficult to memorize, I thought it worthwhile to divide those that are amenable (to being divided), and moreover to arrange those at the beginning, as elements, from the simpler to the more complex, as it seems convenient to do. In this way, indeed, they will be easier to follow for the beginners and their working out[14] will be memorized. As for their treatment, it has been done in thirteen books.

16

1. To divide a proposed number into two numbers having a given difference.

Now, let the given number be 100, and the difference be 40 units. To find the numbers.

Let the smaller be assigned to be 1 Number. Therefore, the greater will be 1 Number, 40 units. The two together, therefore, become 2 Numbers, 40 units. But they were given to be 100 units. Thus, 100 units are equal to 2 Numbers, 40 units.

11 A more literal translation of this passage would be as follows: "A deficit multiplied by a deficit makes an existence; a deficit by an existence makes a deficit; and the sign of the deficit is a truncated Ψ turned upside down, i.e., ⋏". [N.B.: The sign is reproduced from **A**; Tannery's sign looks like ⋏.]

12 The meaning of this expression should be something like "in the context of the hypostases".

13 In other words, we shall follow the examples of the problems contained in the rest of the work.

14 See LSJ (= Liddell, Scott, and Jones 1996), *s.v.* ἀγωγή, II.6.

And likes from likes.[15] From the 100 (units) I subtract 40 units, and from the 2 Numbers and the 40 units, similarly, 40 units.[16] The remaining 2 Numbers are equal to 60 units; therefore, each one becomes 30 units.

To the hypostases: the smaller will be 30 units, the greater 70 units, and the proof is obvious.

2. It is required to divide a proposed number into two numbers having a given ratio.

Now, let it be proposed to divide 60 into two numbers having a triple ratio.

Let the smaller be assigned to be 1 Number. Therefore, the greater will be 3 Numbers, so the greater is the triple of the smaller. Then the two must be equal to 60 units. But the two, when added together, are 4 Numbers. Thus, 4 Numbers are equal to 60 units; therefore, the Number is 15 units.

Thus, the smaller will be 15 units, and the greater 45 units.

3. To divide a proposed number into two numbers having a given ratio and difference.[17]

Now, let it be proposed to divide 80 into two numbers so that the greater is the triple of the smaller and, moreover, exceeds (it) by (an additional) 4 units.

Let the smaller be assigned to be 1 Number. Therefore, the greater will be 3 Numbers and 4 units; so the greater, being the triple of the smaller, exceeds moreover by (an additional) four units. Then, I want the two to be equal to 80 units. But the two, when added together, are 4 Numbers and 4 units. Thus, 4 Numbers and 4 units are equal to 80 units.

And I subtract likes from likes. Then, there remain 76 units equal to 4 Numbers, and the Number becomes 19 units.

---

15 This is a shortened form of the sentence καὶ ἀφηρήσθω ἀπὸ ὁμοίων ὅμοια, "and let likes be subtracted from likes", that Diophantus writes out fully in other instances, such as in Problem I.9.

16 The phrase "and from . . . 40 units", appears in **B**, but is omitted in **AVT**. Tannery believes it is interpolated.

17 The Greek text is ἐν λόγῳ καὶ ὑπεροχῇ τῇ δοθείσῃ, lit. "in a given ratio and excess". As Heath points out (1910, 132 n. 1), this phrase has the same meaning as Euclid's phrase δοθέντι μείζων ἢ ἐν λόγῳ, defined in *Data*, def. 11, which reads "a magnitude is greater in ratio than a magnitude by a given (magnitude) if, when the given magnitude be subtracted, the remainder has a given ratio to the same". In modern language, a magnitude m is by a given magnitude k greater in a ratio λ than a magnitude n if $m - k = \lambda n$ or $m = \lambda n + k$. The same relation reappears in Diophantus in Problem II.7, this time framed as in Euclid, δοθέντι ἀριθμῷ μείζων ᾖ ἢ ἐν λόγῳ. Problem II.7 stipulates to find two numbers such that the difference of their squares is by a given number greater in ratio than their difference.

To the hypostases: thus, the smaller number will be 19 units, and the greater 61 units.[18]

4. To find two numbers having a given ratio, such that their difference is also given.

Now, let it be proposed that the greater is the quintuple of the smaller, and their difference makes 20 units.

Let the smaller be assigned to be 1 Number. Therefore, the greater will be 5 Numbers. Then, I want 5 Numbers to exceed 1 Number by 20 units. But their difference is 4 Numbers. These are equal to 20 units.

The smaller number will be 5 units, and the greater 25 units. And it is fulfilled, since the greater is the quintuple of the smaller, and the difference is 20 units.

5. To divide a proposed number into two numbers such that given parts, not the same, of each of the two from the division, if added together, make a given number. It is certainly necessary[19] for the given number to be given in such a way that it lies in the interval between the two numbers produced when the given parts, not the same, are taken from the (number) proposed in the beginning.

Now, let it be proposed to divide 100 into two numbers such that the third of the first and the fifth of the second, if added together, make 30 units.

I assign[20] the fifth of the second to be 1 Number; (the second) itself, therefore, will be 5 Numbers. Then, the third of the first will be 30 units lacking 1 Number; (the first) itself, therefore, will be 90 units lacking 3 Numbers. Then, I want the two, if added together to make 100 units. But the two, when added together, make 2 Numbers and 90 units. These are equal to 100 units.

And likes from likes. Then, there remain 10 units equal to 2 Numbers; therefore, the Number will be 5 units.[21]

To the hypostases: I assigned the fifth of the second to be 1 Number; it will be 5 units; (the second) itself, therefore, will be 25 units. And (I assigned) the third of the

---

18 After "61 units" Tannery's text has in square brackets, continuing the same sentence: "the 4 units that I subtracted from the 80 units having been added. Indeed, I subtracted (them) to find of how many units each number will be, and then, after knowing of how many each one is, I add the 4 units to the greater number". Tannery considers this passage to be interpolated. Most likely Tannery is right. It must be noticed, however, that the passage appears in the four manuscripts **ABVT**.

19 Δεῖ δή. This is the clause by which Diophantus introduces the determinations (conditions of solubility). In this context, the particle δή signifies that the reader is expected to agree as strongly with the statement of the determination as the author (cf. van der Pas 2014). Throughout the work this clause, when used to introduce a determination, is translated "It is certainly necessary".

20 Lit. "I assigned". This syntax in the past tense, very rare in the Greek text, is employed in the Arabic books.

21 The phrase is omitted in **A**; **V** and **T** have ὁ εἷς ἔσται ἀριθμὸς μονάδων ε̄, "the one Number will be 5 units".

22 first, to be 30 units lacking 1 Number; it will be | 25 units; (the first) itself, therefore, will be 75 units. And it is fulfilled, since the third of the first and the fifth of the second is 30 units, and these, when added together, make the proposed number.[22]

6. To divide a proposed number into two numbers such that a given part of the first exceeds a given part of the second by a given number. It is certainly necessary for the given number to be smaller than the number produced when, from the (number) proposed in the beginning, that given part is taken which exceeds (the other part).

Now, let it be proposed to divide 100 into two numbers such that the fourth of the first exceeds the sixth of the second by 20 units.

I assign[23] the sixth of the second to be 1 Number; (the second) itself, therefore, will be 6 Numbers. Then, the fourth of the first will be 1 Number and 20 units; (the first) itself, therefore, will be 4 Numbers and 80 units. Then, I want the two, if added together, to make 100 units. But the two, when added together, make 10 Numbers and 80 units. These are equal to 100 units.

Likes from likes. Then, 10 Numbers are equal to 20 units, and the number becomes 2 units.

To the hypostases: I assigned the sixth of the second to be 1 Number; it will be 2 units; (the second) itself, therefore, will be 12 units. And (I assigned) the fourth of the first, to be 1 Number and 20 units; it will be 22 units; (the first) itself, therefore, will be 88 units. And it is fulfilled, since the fourth of the first exceeds the sixth of the second by 20 units, (and) these, when added together, make the proposed number.[24]

24 **7.** From the same number to subtract two given numbers and make the remainders have to one another a given ratio.

Now, let it be proposed to subtract from the same number 100 and 20, and make the greater the triple of the smaller.

Let the sought-after (number) be assigned to be 1 Number. And if from this I subtract the 100, there remains 1 Number lacking 100 units, while if (I subtract) the 20, there remains 1 Number lacking 20 units. And the greater should be the triple of the smaller. Thrice the smaller, therefore, is equal to the greater. But thrice the smaller is 3 Numbers lacking 300 units. These are equal to 1 Number lacking 20 units.

---

22 Tannery considers the passage "and these, when . . . number" to be interpolated, but it is present in the main mss. (**ABVT**).

23 Lit. "I assigned".

24 Again the concluding phrase ("these, when . . . number") is considered by Tannery to be interpolated, but it is present in the manuscripts.

Let the lacking be added in common. It becomes 3 Numbers equal to 1 Number and 280 units. And let likes be subtracted from likes. Then, 2 Numbers are equal to 280 units, and the Number becomes 140 units.

To the hypostases:[25] I assigned the sought-after number to be 1 Number; therefore, it will be 140 units. And if I subtract the 100 from it, there remain 40 units, while if (I subtract) the 20, there remain 120 units. And it is fulfilled that the greater is the triple of the smaller.

8. To add the same number to two given numbers and make the produced (numbers) have to one another a given ratio. It is certainly necessary for the given ratio to be smaller than the ratio which the greater of the givens has to the smaller.

Now, let it be proposed to add the same number to 100 and 20, and to make the greater the triple of the smaller.

Let what is added to each number be assigned to be 1 Number. And when it is added to the 100, it will be 1 Number, 100 units; while, when (it is added) to the 20, it becomes 1 Number, 20 units. And the greater should be the triple of the smaller. Thrice the smaller, therefore, are equal to the greater. But thrice the smaller becomes 3 Numbers, 60 units. These are equal to 1 Number, 100 units.

Likes from likes. There remain 2 Numbers equal to 40 units, and the Number becomes 20 units.

To the hypostases: I assigned what is added to each number to be 1 Number; it will be 20 units. And if it is added to the 100, they become 120 units; while if (it is added) to the 20, they become 40 units. And it is fulfilled that the greater is the triple of the smaller.

9. To subtract the same number from two given numbers and make the remainders have to one another a given ratio. It is certainly necessary for the given ratio to be greater than the ratio which the greater of the givens has to the smaller.

Now, let it be proposed to subtract from 20 and 100 the same number, and make the greater the sextuple of the smaller.

Let what is subtracted from each number be assigned to be 1 Number. And when it is subtracted from the 100, there remain 100 units lacking 1 Number; while, when (it is subtracted) from the 20, there remain 20 units lacking 1 Number. And the greater should be the sextuple of the smaller. Six times the smaller, therefore, are equal to the greater. But six times the smaller makes 120 units lacking 6 Numbers. These are equal to 100 units lacking 1 Number.

26

---

25 Note the plural number of "hypostases" despite the fact that the sought-after number in this problem is only one. It seems that the expression "To the hypostases" is like a subtitle for the last part of the textual unit of a worked-out proposition, in which numerical value(s) of the sought-after number(s) is(are) calculated. See Christianidis (2015).

Let the lacking be added in common, and let likes be subtracted from likes. There remain 5 Numbers equal to 20 units, and the Number becomes 4 units.

28   To the hypostases: I assigned what is subtracted from each number to be 1 Number; it will be 4 units. And if it is subtracted from the 100, there remain 96 units; while if (it is subtracted) from the 20, there remain 16 units. And it is fulfilled that the greater is the sextuple of the smaller.

10. Given two numbers, to add to the smaller of them and to subtract from the greater the same number, and make the produced (number from the addition) have to the remainder a given ratio.

Let it be proposed to add to 20 and to subtract from 100 the same number, and make the greater the quadruple of the smaller.

Let what is added to and subtracted from each number be assigned to be 1 Number. And if it is added to the 20, it becomes 1 Number, 20 units; while, if it is subtracted from the 100, it becomes 100 units lacking 1 Number. And the greater should be the quadruple of the smaller. Four times the smaller, therefore, are equal to the greater. But four times the smaller makes 400 units lacking 4 Numbers. These are equal to 1 Number, 20 units.

Let the lacking be added in common, and let likes be subtracted from likes. There remain 5 Numbers equal to 380 units, and the Number becomes 76 units.

To the hypostases: I assigned what is added to and subtracted from each number to be 1 Number; it will be 76 units. And if 76 units are added to the 20, they become 96 units; while if they are subtracted from the 100, there remain 24 units. And it is fulfilled that the greater is the quadruple of the smaller.

30   **11**. Given two numbers, to add the first to, and subtract the second from the same number, and make the produced (numbers) have to one another a given ratio.

Let it be proposed to add 20 to, and subtract 100 from the same number, and make the greater the triple of the smaller.

Let the sought-after (number) be 1 Number. And if we add to it 20 units, it becomes 1 Number, 20 units; while, if 100 units are subtracted from it, there remains 1 Number lacking 100 units. And the greater should be the triple of the smaller. Thrice the smaller, therefore, are equal to the greater. But thrice the smaller makes 3 Numbers lacking 300 units. Thus, 3 Numbers lacking 300 units are equal to 1 Number, 20 units.

Let the lacking be added in common, and let likes be subtracted from likes. Then, 320 units are equal to 2 Numbers, and the Number becomes 160 units.

To the hypostases: thus, the greater will be 180 units, and the smaller 60 units. And it is fulfilled that the greater is the triple of the smaller.

12. To divide a proposed number into two numbers twice, such that one of the (numbers) from the first division has to one of the (numbers) from the second division a given ratio, and the other of the (numbers) from the second division has to the other of the (numbers) from the first division a given ratio.

Now, let it be proposed to divide 100 into two numbers | twice, such that the greater of the (numbers) from the first division is the double of the smaller of the (numbers) from the second division, and the greater of the (numbers) from the second division is the triple of the smaller of the (numbers) from the first division.

Let the smaller (number) from the second division be assigned to be 1 Number. Therefore, the greater of the (numbers) from the first division will be 2 Numbers. Then, the smaller of the (numbers) from the first division will be 100 units lacking 2 Numbers. And since the greater of the (numbers) from the second division is the triple of that, it will be 300 units lacking 6 Numbers. It remains, also, for the (numbers) of the second division, if added together, to make 100 units. But when added together they make 300 units lacking 5 Numbers. These are equal to 100 units, and the Number becomes 40 units.

To the hypostases: I assigned the greater of the (numbers) from the first division to be 2 Numbers; it will be 80 units; and the smaller ⟨of the numbers from⟩[26] the same division to be 100 units lacking 2 Numbers; it will be 20 units; and the greater from the second division to be 300 units lacking 6 Numbers; it will be 60 units; and the smaller from the second division to be 1 Number; it will be 40 units. And the proof is obvious.

13. To divide a proposed number into two numbers three times, such that one of the (numbers) from the first division has to one of the (numbers) from the second division a given ratio; and the other (number) from the second division has to one of the (numbers) from the third division a given ratio; and, moreover, the other (number) from the third division has to the other of the (numbers) from the first division a given ratio.

Now, let it be proposed to divide 100 into two numbers three times, such that the greater (number) from[27] the first division is the triple of the smaller (number) from the second; and the greater (number) from the second division is the double of the smaller (number) from the third; and, moreover, the greater (number) from the third division is the quadruple of the smaller (number) from the first.

Let the smaller (number) from the third division be assigned to be 1 Number. Therefore, the greater (number) from the second division will be 2 Numbers. And since the partitioned (number) as a whole[28] is 100 units, therefore, the smaller (number) from the second division will be 100 units lacking 2 Numbers. And since the greater (number) from the first division is the triple of that, it will

32

34

---

26 Supplemented by Tannery.
27 We write "greater (number) from" where the text says literally "the greater of the (numbers) from". We made this change throughout the problem for "greater (number) from" and "smaller (number) from".
28 ἡ ὅλη διαίρεσις, lit. "the whole partition".

be 300 units lacking 6 Numbers. Therefore, the smaller (number) from the first division will be 6 Numbers lacking 200 units. And since the greater (number) from the third division is the quadruple of that, it will be 24 Numbers lacking 800 units. It remains, also, for the (numbers) of the third division, if added together, to make 100 units. But when added together they make 25 Numbers lacking 800 units. These are equal to 100 units, and the Number becomes 36 units.

To the hypostases: the smaller (number) from the third division will be 36 units, and the greater 64. The smaller (number) from the first division will be 16 units, and the greater 84. And the smaller (number) from the second division will be 28 units, and the greater 72. And it is manifest that they fulfill the problem.

14. To find two numbers such that the (number) produced from the multiplication has to the (number) produced from the addition a given ratio. It is certainly necessary for the supposed multitude of units of one | of the numbers to be greater than the (number) homonymous with the given ratio.[29]

36

Now, let it be proposed that the (number) produced from the multiplication has to the (number) produced from the addition a triple ratio.

Let one of them be assigned to be 1 Number, and the other, according to the determination, greater than 3 units. Let it be 12 units. And so, their product is 12 Numbers and their sum is 1 Number, 12 units. It remains for 12 Numbers to be the triple of 1 Number, 12 units. Thus, three times the smaller are equal to the greater, and the Number becomes 4 units.

One of them will be 4 units, the other 12 units, and they fulfill the problem.

15. To find two numbers such that each one, if it receives from the other a proposed number, has to the remainder a proposed ratio.

Now, let it be proposed that the first, if it receives 30 units from the second, becomes its double, while the second, if it receives 50 units from the first, becomes its triple.

Let the second be assigned to be 1 Number and the 30 units that it gives. The first, therefore, will be 2 Numbers lacking 30 units, so that, after receiving 30 units from the second, it will become its double. It remains, also, for the second, if it receives 50 units from the first, to become its triple. But having given 50 units, the first is left with 2 Numbers lacking 80 units. On the other hand the second, having received the 50 units, becomes 1 Number, 80 units. It remains for 1 Number, 80 units to be the triple of 2 Numbers lacking 80 units. Thus, three times the smaller are equal to the greater, and the Number becomes 64 units.

---

29 Actually, both numbers must be greater than the numerical value of the ratio. Diophantus writes the requirement for only "one of the numbers" because one of them will be given a determinate value in the beginning of the solution. Also, Diophantus did not regard ratios as being numbers, so he must write "the number homonymous" with it, such as in the present problem, where the number homonymous with "triple" is "three".

So, the first will be 98 units, the second 94 units, and they fulfill the problem.

16. To find three numbers such that, taken by twos, make proposed numbers. It is certainly necessary for half of the three proposed (numbers) to be greater than each one of them.

Now, let it be proposed that the first with the second, if added together, make 20 units, and the second with the third make 30 units, while the third with the first make 40 units.

Let the three be assigned to be 1 Number. And since the first and the second make 20 units, if I subtract 20 units from 1 Number I shall have that the third is 1 Number lacking 20 units. By the same (reasoning), the first will be 1 Number lacking 30 units, and the second 1 Number lacking 40 units. It remains for the three numbers, if added together, to be equal to 1 Number. But the three, when added together, make 3 Numbers lacking 90 units. These are equal to 1 Number, and the Number becomes 45 units.

To the hypostases: the first will be 15 units, the second 5 units, and the third 25 units. And the proof is obvious.

17. To find four numbers such that, added by threes, make proposed numbers. It is certainly necessary that one-third of the four (together) be greater than each one of them.

Now, let it be proposed that the three following in order from the first, if added together, make 20 units, the three following the second make 22 units, the three following the third make 24 units, and the three following the fourth make 27 units.

Let the four (together) be assigned to be 1 Number. Therefore, if from 1 Number I subtract the first three, that is 20 units, I shall have as a remainder the | fourth, (namely) 1 Number lacking 20 units. By the same (reasoning), the first will be 1 Number lacking 22 units, the second, 1 Number lacking 24 units, and the third, 1 Number lacking 27 units. It remains for the four numbers, if added together, to be equal to 1 Number. But the four, when added together, make 4 Numbers lacking 93 units. These are equal to 1 Number, and the Number becomes 31 units.

To the hypostases: the first will be 9 units, the second 7 units, the third 4 units, and the fourth 11 units. And they fulfill the problem.

18. To find three numbers such that, taken by twos, exceed the remaining one by a proposed number.

Now, let it be proposed that the first and the second exceed the third by 20 units, the second and the third exceed the first by 30 units, and the third and the first exceed the second by 40 units.

Let the three be assigned to be 2 Numbers. And since the first and the second exceed the third by 20 units, if the third is added in common, (then) the three are twice the third, and the excess of 20 units. If, therefore, from the three, that is 2

Numbers, I subtract 20 units, I shall have that twice the third is 2 Numbers lacking 20 units; the third alone,[30] therefore, will be 1 Number lacking 10 units. Now, by the same (reasoning), the first will be 1 Number lacking 15 units, and the second, 1 Number lacking 20 units. It remains for the three to be equal to 2 Numbers. But the three, when added together, make 3 Numbers lacking 45 units. These are equal to 2 Numbers, and the Number becomes 45 units.

To the hypostases: the first will be 30 units, the second 25 units, and the third 35 units. And they fulfill what was proposed.[31]

42 **[Otherwise]**[32]

Since the first and the second exceed the third by 20 units, let the third be 1 Number. Therefore, the sum of both the first and the second will be 1 Number, 20 units. Again, since the second and the third exceed the first by 30 units, I assign the second to be (made up) of so many units as half of the 20 and 30 is, that is 25 units. And since the first and the second are 1 Number, 20 units, of which the second is 25 units; therefore, the remaining first will be 1 Number lacking 5 units. Further, it is also required that the third together with the first exceed the second by 40 units. But the first together with the third is 2 Numbers lacking 5 units, (which,) therefore, are equal to 65 units.

Let the lacking be added in common. Thus, 2 Numbers are equal to 70 units, and the Number becomes 35 units.

To the hypostases: I assigned the first to be 1 Number lacking 5 units: it will be 30 units; the second 25 units; and the third 1 Number: it will be 35 units.

19. To find four numbers such that, taken by threes, exceed the remaining one by a proposed number. It is certainly necessary for half of the four differences to be greater than any one of them.

Now, let it be proposed that the three following in order from the first, if added together, exceed the fourth by 20 units; the three following the second exceed the first by 30 units; the three following the third, similarly, exceed the second by 40 units; and, moreover, the three following the fourth exceed the third by 50 units.

Let the four be assigned to be 2 Numbers. And since the three following in 44 order from the first exceed the fourth by 20 units; and | the first three exceed the

---

30 Lit. "once the third".

31 We translate in this manner the phrase ποιοῦσι τὰ τῆς προτάσεως. What is implied in this elliptic phrase seems to be the word ἐπιτάγματα. Thus, the complete passage should have been ποιοῦσι τὰ ἐπιτάγματα τῆς προτάσεως, "they fulfill the proposed tasks of the proposition". Cf. "Index graecitatis apud Diophantum" (Diophantus 1893–95, II, 270–71), *s.v.* ἐπίταγμα.

32 This alternative solution, as well as the alternative solutions of problems 19 and 21, do not appear in **T**, but they appear in **ABV**. Tannery believes that they must be attributed to an ancient scholiast.

fourth by as much as the four exceed twice the fourth; and the four are 2 Numbers; therefore, 2 Numbers exceed twice the fourth by 20 units. The double of the fourth, therefore, will be 2 Numbers lacking 20 units; so, this will be 1 Number lacking 10 units. Now, by the same (reasoning), the first will be 1 Number lacking 15 units, the second 1 Number lacking 20 units, and, moreover, the third (will be) 1 Number lacking 25 units. It remains for the four to be equal to 2 Numbers. But the four are 4 Numbers lacking 70 units. These are equal to 2 Numbers, and the Number becomes 35 units.

To the hypostases: the first will be 20 units, the second 15 units, the third 10 units, and the fourth 25 units. And they fulfill the problem.

[Otherwise]

Since the three following in order from the first exceed the fourth by 20 units, let the fourth be assigned to be 1 Number. Therefore, the three will be 1 Number, 20 units. Again, since the three following in order from the second exceed the first by 30 units, let the sum of both the second and the third be assigned to be as many units as half of the two excesses – I mean of course the 20 and the 30 – that is, 25 units. And since the three following in order from the first are 1 Number, 20 units, of which the second and the third are 25 units, the remaining first, therefore, will be 1 Number lacking 5 units. And since the three following the second exceed the first by 30 units, and the three following the third exceed the second by 40 units; therefore, the sum of both the third and the fourth will be 35 units. The remaining third, therefore, will be 35 units lacking 1 Number. But the second and the third are 25 units, of which the third is 35 units lacking 1 Number. The remaining second, therefore, will be 1 Number lacking 10 units. It remains for the three following in order from the fourth to exceed the third | by 50 units. But the three, when added together, make 3 Numbers lacking 15 units, while the third is 35 units lacking 1 Number. Now, it is further required that 3 Numbers lacking 15 units exceed 35 units lacking 1 Number by 50 units. Thus, 85 units lacking 1 Number are equal to 3 Numbers lacking 15 units, and the Number becomes 25 units.

To the hypostases: I assigned the first to be 1 Number lacking 5 units; it will be 20 units; the second, likewise, will be 15 units, the third 10 units, and the fourth 25 units.

20. To divide a proposed number into three numbers such that each of the extremes, if it receives in addition the mean, has to the other of the extremes a given ratio.

Now, let it be proposed to divide 100 into three numbers such that the first and the second are the triple of the third, and the second and third are the quadruple of the first.

Let the third be assigned to be 1 Number. And since the first and the second are the triple of the third, let the two be assigned to be 3 Numbers. Therefore, the three (together) are 4 Numbers. These are equal to 100 units, and the Number becomes 25 units.

To the hypostases: I assigned the third to be 1 Number: it will be 25 units; and the first and second to be 3 Numbers: they will be 75 units.

46

Again, since the second and the third are the quadruple of the first, let the first be assigned to be 1 Number.[33] Therefore, the second and the third will be 4 Numbers. Thus, the three (together) are 5 Numbers, but, also, 100 units; and the Number becomes 20 units.

Thus, the first will be 20 units, while the second and the third, 80 units, of which 25 units is the third. The remaining second, therefore, will be 55 units. And they fulfill what was proposed.

21. To find three numbers such that the greatest exceeds the middle by a given part of the least, the middle | exceeds the least by a given part of the greatest, and the least exceeds a given part of the middle by a given number. It is certainly necessary for the middle to exceed the least by such a part of the greatest that the (number) homonymous with that part, if multiplied by the difference of the middle and the least, makes the multitude of Numbers in it greater than (that) in the middle.

Now, let it be proposed that the greatest exceeds the middle by a third part of the least, the middle exceeds the least by a third part of the greatest, and the least exceeds the third part of the middle by 10 units.

Well, let the smaller be assigned to be 1 Number and the 10 units by which it exceeds the middle. Therefore, the middle will be 3 Numbers, so that the least will have a third part of the middle and 10 units. Or even as follows: let the middle be assigned to be 3 Numbers. And since I want the least to exceed a third part of the middle itself by 10 units, it will be 1 Number and 10 units. It remains, also, for the middle to exceed the least by a third part of the first. But the middle exceeds the least by 2 Numbers lacking 10 units. These, therefore, are a third part of the greatest. Therefore, the greatest itself will be 6 Numbers lacking 30 units. Then, it is also required for the greatest to exceed the middle by a third part of the least. But the greatest exceeds the middle by 3 Numbers lacking 30 units. These, therefore, are a third part of the least. Therefore, the least will be 9 Numbers lacking 90 units. But it was found, also, to be 1 Number, 10 units; and so the Number becomes 12 2′ units.

Thus, the third will be 22 2′ units, the middle 37 2′ units, the greatest 45 units, and they fulfill what was proposed.

[50] [Otherwise]

To find three numbers such that the greatest exceeds the middle by a given part of the least, the middle exceeds the least by a given part of the greatest, and the

---

33 Heath (1910, 137 n. 1) notes: "As already remarked, Diophantus does not use a second symbol for the second unknown, but makes ἀριθμός do duty for the second as well as for the first". But there would have been no need for a new symbol to name the second unknown, since in this case Diophantus splits up the problem into two separate solutions, which are solved in sequence, and two independed equations are set up and solved.

least exceeds a given part of the middle by a given number.[34] It is certainly necessary for the given part of the greatest to be given of such a size that, if added to the least, it makes the Numbers in it smaller than the ones of the middle that were assumed in the beginning.

Let, again, the smaller be assigned to be 1 Number and the 10 units by which it exceeds a third part of the middle. Therefore, the middle will be 3 Numbers, so that the least will exceed a third part of the middle by 10 units. Again, since I want the greatest to exceed the middle by a third part of the least, if I add to the middle a third part of the least, the greatest will be 3 3′ Numbers, 3 3′ units. Further, it is also required the middle to be equal to the least and a third part of the greatest. But the least together with a third part of the greatest are 2 9′ Numbers and 11 9′ units. These are equal to the 3 Numbers of the middle.

Likes from likes. Thus, 1 Number lacking 9′ is equal to 11 9′. Everything 9 times. Therefore, 8 Numbers are equal to 100 units, and the Number becomes 12 2′ units. And the proof is the same as above.

22. To find three numbers such that, if each gives a proposed part of itself to the next one, givers and receivers become equal.

Now, let it be proposed that the first gives to the second its third, the second to the third a fourth,[35] and, moreover, the third to the first a fifth, and after the exchange[36] they become equal.

Let the first be assigned to be certain Numbers having a third part, since it gives a third; so, let it be 3 Numbers; and the second, certain units having a fourth part, since it gives a fourth; so, let it be 4 units. And so, having given and having received, the second becomes 1 Number, 3 units. It remains, also, for the first, after giving and receiving, to become 1 Number, 3 units. But having given its third, (namely) 1 Number, and having received 3 units lacking 1 Number, (the first) becomes 1 Number, 3 units. Therefore, 3 units lacking 1 Number are a fifth part of the third. The latter, therefore, is 15 units lacking 5 Numbers. It is further required for the third, if it gives its fifth, and if it receives from the second its[37] fourth, (namely) 1 unit, to become 1 Number, 3 units. But when it gives its fifth, (namely) 3 units lacking 1 Number, it is left with 12 units lacking 4 Numbers; while, when it receives from the

52

---

34 The restatement of the enunciation is omitted in Tannery's edition, which has only Εὑρεῖν κ. τ. ἑ., "To find etc." However the enunciation is restated in the four main witnesses of the manuscript tradition of the *Arithmetica* (see **A** [f. 67r–v], **B** [f. 97v], **V** [f. 364r], and **T** [f. 81v]), as well as in Allard's edition (1980, 395.23–26). Tannery believes that this solution was added by an ancient scholiast.

35 Here, and in similar situations, "its fourth" is the intended meaning.

36 The Greek word is ἀντίδοσις, a term which is borrowed from the language of law. It refers to a specific legal procedure which was in effect in Athens in the fourth century BCE providing the possibility of integral or partial exchange of properties between parties. See more in Christianidis and Sialaros (2022).

37 Lit. "the fourth". We have made this adjustment in similar situations. In some cases the reflexive possessive "ἑαυτοῦ" for "its" is present.

second a fourth, (namely) 1 unit, it becomes 13 units lacking 4 Numbers. These are equal to 1 Number, 3 units, and the Number becomes 2 units.

To the hypostases: the first will be 6 units, the second 4 units, and the third 5 units. And what was proposed is obvious.

54

23. To find four numbers such that, if each gives a proposed part of itself to the next one, givers and receivers become equal.

Let it be proposed that the first gives to the second a third, the second to the third a fourth, the third to the fourth a fifth, and, moreover, the fourth to the first a sixth, and after the exchange they become equal.

Let the first be assigned to be certain Numbers having a third part, since it gives a third; let it be 3 Numbers; and the second, certain units having a fourth part, since it gives a fourth; let it be 4 units. Therefore, having given its fourth, (namely) 1 unit, and having received from the first the third, (namely) 1 Number, the second becomes 1 Number, 3 units. Then it is also required that the first, if it gives its third, (namely) 1 Number, and if it receives from the fourth its sixth, becomes 1 Number, 3 units. But when it gives 1 Number, it is left with 2 Numbers. Therefore, after receiving a sixth of the fourth, the latter must become 1 Number, 3 units. Thus, 3 units lacking 1 Number are a sixth part of the fourth. The fourth itself, therefore, will be 18 units lacking 6 Numbers. It remains, also, for the fourth, if it gives its sixth, and if it receives from the third its fifth, to become 1 Number, 3 units. But when it gives its sixth, (namely) 3 units lacking 1 Number, the remainder is 15 units lacking 5 Numbers. Therefore, after receiving a fifth of the third, the latter must become 1 Number, 3 units. But when it receives 6 Numbers lacking 12 units it becomes 1 Number, 3 units. Therefore, 6 Numbers lacking 12 units are a fifth part of the third. The

56 latter, therefore, will be 30 Numbers lacking 60 units. | Then, it is also required that the third, if it gives its fifth, and if it receives from the second its fourth, becomes 1 Number, 3 units. But when it gives its fifth, (namely) 6 Numbers lacking 12 units, it is left with 24 Numbers lacking 48 units; while, when it receives from the second its fourth, it becomes 24 Numbers lacking 47 units. These are equal to 1 Number, 3 units, and the Number becomes 50 23rds.

To the hypostases: the first will be 150 (23rds), the second 92 (23rds), the third 120 (23rds), and the fourth 114 (23rds). Let the part be eliminated. It is clear that the first will be 150 units, the second 92, the third 120, and the fourth 114. And they fulfill what was proposed.

24. To find three numbers such that, if each one receives from the other two, as if they were one,[38] a proposed part, they become equal.

Now, let it be proposed that the first receives from the other two, as if they were one, a third, the second receives from the other two, as if they were one, a fourth,

---

38 That is, from the sum of the other two.

and the third receives from the other two, as if they were one, a fifth, and they become equal.

Let the first be assigned to be 1 Number, while the other two certain units having, for the sake of convenience, a third part, since they give a third; let them be 3 units. The three, therefore, will be 1 Number, 3 units, and the first, having received from the other two a third, is left with 1 Number, 1 unit. Then, it is also required that the second, if it receives from the ⟨other⟩[39] two, as if they were one, a fourth, becomes 1 Number, 1 unit. Everything four times. | Therefore, four times   58
the second, having received in addition the two, becomes three times the second, having received in addition the three. Therefore, three times the second, having received in addition the three, becomes 4 Numbers, 4 units. If, therefore, I subtract the three from them, the remaining 3 Numbers, 1 unit are thrice the second. The second itself, therefore, will be 1 Number, 3' of a unit. Then, it is also required that the third, if it receives from other two, as if they were one, a fifth, becomes 1 Number, 1 unit. Similarly, everything five times. And by the same (reasoning), the third is found to be 1 Number, a half of a unit. It remains for the three, if added together, to become equal to 1 Number, 3 units; and the Number becomes 13 12ths.

And by removing the part, the first will be 13 units, the second 17 units, and the third 19 units. And they fulfill what was proposed.

25. To find four numbers such that, if each one receives from the other three, as if they were one, a proposed part, they become equal.

Now, let it be proposed that the first receives from the other three, as if they were one, a third, the second receives from the other three, as if they were one, a fourth, the third similarly a fifth, and the fourth a sixth, and they become equal.

Let the first be assigned to be 1 Number, and the other three certain units having a third part, since they give a third; let them be 3 units. The first, therefore, having received from the other three, as if they were one, a third, becomes 1 Number, 1 unit. | Then, it is also required that the second, if it receives from the other three, as   60
if they were one, a fourth, becomes 1 Number, 1 unit. Again, similarly, everything four times. And by the same (reasoning), the second is found to be 1 Number, 3' of a unit, the third 1 Number, 2' of a unit, and the fourth 1 Number, 3 5ths of a unit. It remains for the four, if added together, to be equal to 1 Number, 3 units; and the Number is found to be 47 units in a 90th part of a unit.

The first will be 47 units, the second 77 units, the third 92 units, and the fourth 101 units. And they fulfill what was proposed.

26. Given two numbers to find some other number which, if multiplied by either, makes one of them a square, and the other a side of that square.

---

39 Supplemented by Tannery.

Let the two given numbers be 200 and 5, and let the sought-after (number) be 1 Number. And when it is multiplied by the 200 units it makes 200 Numbers, while, when (it is multiplied) by the 5 units it makes 5 Numbers. Now, one of them must be a square, and the other its side. Well then, if I square the 5 Numbers, 25 Powers become equal to 200 Numbers. Everything by a Number. Then, 25 Numbers are equal to 200 units, and the Number becomes 8 units, and it fulfills what was proposed.

27. To find two numbers such that their addition and multiplication make given numbers. It is certainly necessary that, of the (numbers) found, the square on the half of | the sum of both exceeds their product by a square. And this arises with the formation (of the problem).[40]

Now, let it be proposed that their addition makes 20 units, while the multiplication makes 96 units.

Let their difference be assigned to be 2 Numbers. And since their sum is 20 units, if I bisect it, each of the two from the partition will be half the sum, (namely) 10 units. And if, on the one hand, I add half of the difference, which is 1 Number, to one of the (two) from the partition, and, on the other hand, I subtract (it) from the other, the sum remains, again, 20 units, and the difference 2 Numbers. Accordingly, let the greater be assigned to be 1 Number and the 10 units of half of the sum. The smaller, therefore, will be 10 units lacking 1 Number. And the sum remains 20 units, and the difference 2 Numbers. It remains, also, for their product to make 96 units. But their product is 100 units lacking 1 Power. These are equal to 96 units, and the Number becomes 2 units.

Thus, the greater will be 12 units, the smaller 8 units, and they fulfill what was proposed.

28. To find two numbers such that their addition and the addition of their squares make given numbers. It is certainly necessary that twice the sum of their squares exceeds the square of their sum by a square. And this also arises with the formation (of the problem).

Now, let it be proposed that their addition makes 20 units, while the addition of their squares makes 208 units.

Well, let their difference be assigned to be 2 Numbers. And let the greater be 1 Number and 10 units, again the half of the sum, and the smaller, 10 units lacking 1 Number. And again their sum remains 20 units, and the difference 2 Numbers. It remains, also, for the sum of their squares to make 208 units. But the sum of their squares makes 2 Powers, 200 units. These are equal to 208 units, and the Number becomes 2 units.

To the hypostases: the greater will be 12 units, the smaller 8 units, and they fulfill what was proposed.

---

40 We translate the clause ἔστι δὲ τοῦτο πλασματικόν as "arises with the formation (of the problem)". We discussed the term πλασματικόν in Section 4.2.8 of the Introduction.

29. To find two numbers such that their addition and the difference of their squares make given numbers.

Now, let it be proposed that their addition makes 20 units, while the difference of their squares makes 80 units.

Let their difference be assigned to be 2 Numbers. Likewise, the greater will be 1 Number, 10 units, and the smaller, 10 units lacking 1 Number. And again their sum remains 20 units, and the difference 2 Numbers. It remains, also, for the difference of their squares to make 80 units. But the difference of their squares is 40 Numbers. These are equal to 80 units.

And again the greater is found to be 12 units, the smaller 8 units, and again they fulfill the problem.

30. To find two numbers such that their difference and multiplication make given   66
numbers. It is certainly necessary that four times their product, together with the square of their difference, makes a square. And this also arises with the formation (of the problem).

Now, let it be proposed that their difference is 4 units, and the multiplication 96 units.

Let their sum be assigned to be 2 Numbers. But we have, also, that the difference is 4 units. Similarly, the greater will be 1 Number, 2 units, the smaller 1 Number lacking 2 units, and their sum remains 2 Numbers, and the difference 4 units. It remains, also, for their multiplication to make 96 units. But their multiplication is 1 Power lacking 4 units. These are equal to 96 units.

And again the greater is 12 units, the smaller 8 units, and they fulfill the problem.

31. To find two numbers having to one another a given ratio, and such that the sum of their squares has to their sum a given ratio.

Now, let it be proposed that the greater is the triple of the smaller, and the sum of their squares is the quintuple of their sum.

Let the smaller be assigned to be 1 Number. Therefore, the greater will be 3 Numbers. It remains for the sum of their squares ⟨to be the quintuple of (their) sum. But the sum of their squares⟩[41] makes 10 Powers, while their sum, 4 Numbers. Therefore, 10 Powers are the quintuple of 4 Numbers. | Thus, 20 Numbers are   68
equal to 10 Powers, and the Number becomes 2 units.

The smaller will be 2 units, the greater 6 units, and they fulfill what was proposed.

32. To find two numbers in a given ratio, such that the sum of their squares has to their difference a given ratio.

Now, let it be proposed that the greater is the triple of the smaller, and the sum of their squares is the decuple of their difference.

---

41 Supplemented by Bachet.

Let the smaller be assigned to be 1 Number. Therefore, the greater will be 3 Numbers. Then, I want the sum of their squares to be the decuple of their difference. But the sum of their squares makes 10 Powers, while their difference, 2 Numbers. Therefore, 10 Powers are the decuple of 2 Numbers. And everything by a Number. Thus, 10 Numbers are equal to 20 units, and the Number becomes 2 units.

And, again, the smaller will be 2 units, the greater 6 units, and they fulfill what was proposed.

33. To find two numbers in a given ratio, and such that the difference of their squares has to the sum of the two (numbers) a given ratio.

70   Now, let it be proposed that the greater is the triple of the smaller, | and the difference of their squares is the sextuple of the sum of the two (numbers).

Let the smaller be assigned to be 1 Number. Therefore, the greater will be 3 Numbers. It remains, also, for the difference of their squares to be the sextuple of the sum of the two (numbers). But the difference of their squares is 8 Powers, while the sum of the two, 4 Numbers. Therefore, 8 Powers are the sextuple of 4 Numbers. Thus, 24 Numbers are equal to 8 Powers, and the Number becomes 3 units.

⟨And so, the smaller will be 3 units, the greater 9 units⟩,[42] and they fulfill the problem.

34. To find two numbers in a given ratio, and such that the difference of their squares has to their difference a given ratio.

Now, let it be proposed that the greater is the triple of the smaller, and the difference of their squares is the duodecuple of their difference.

Again, let the smaller be assigned to be 1 Number. The greater, therefore, will be 3 Numbers. It remains, also, for the difference of their squares to be the duodecuple of their difference. But the difference of their squares is 8 Powers. These, therefore, are the duodecuple of 2 Numbers. Thus, 24 Numbers are equal to 8 Powers, and the Number becomes, again, 3 units.

And the proof is obvious.

(Corollaries:) And likewise, by the same (reasoning), can also be found:

72   (34.1) two numbers having to one another a given | ratio, such that their product has to their sum a given ratio,

(34.2) and, again, two numbers having to one another a given ratio, such that their product has to their difference a given ratio.

35. To find two numbers in a given ratio, such that the square of the smaller has to the greater a given ratio.

42 Supplemented by Tannery.

Now, let it be proposed that the greater is the triple of the smaller, and the square of the smaller is the sextuple of the greater.

Let, again, the smaller be assigned to be 1 Number. Therefore, the greater will be 3 Numbers. It remains, also, for the square of the smaller to be the sextuple of the greater. But the square of the smaller is 1 Power. Therefore, 1 Power is the sextuple of 3 Numbers. Thus, 18 Numbers are equal to 1 Power, and the Number becomes 18 units.

The smaller will be 18 units, the greater 54 units, and they fulfill the problem.

36. To find two numbers in a given ratio, such that the square of the smaller has to the smaller itself a given ratio.

Now, let it be proposed that the greater is the triple of the smaller, and the square of the smaller is the sextuple of the smaller itself.

Similarly, the greater will be 3 Numbers, the smaller 1 Number, and it is estab-lished that the greater is the triple of the smaller. It remains, also, for the square of the smaller to be the sextuple of the smaller itself. Therefore, 1 Power is the sextuple of 1 Number. Thus, 6 Numbers are equal to 1 Power, and the Number becomes 6 units.

74

The smaller will be 6 units, the greater 18 units, and they fulfill the problem.

37. To find two numbers in a given ratio, such that the square of the smaller has to their sum a given ratio.

Let it be proposed that the greater is the triple of the smaller, and the square of the smaller is the double of the sum of the two.

Again, similarly, the greater will be 3 Numbers, and the smaller 1 Number. It remains, also, for the square of the smaller to be the double of the sum of the two. But the square of the smaller is 1 Power, and the sum of the two, 4 Numbers. Therefore, 1 Power is the double of 4 Numbers. Thus, 8 Numbers are equal to 1 Power, ⟨and⟩[43] the Number becomes 8 units.

The smaller will be 8 units, the greater 24 units, and they fulfill what was proposed.

38. To find two numbers in a given ratio, such that the square of the smaller has to their difference a given ratio.

Now, let it be proposed that the greater is the triple of the smaller, and the square of the smaller is the sextuple of their difference.

76

Again, similarly, the greater will be 3 Numbers, and the smaller 1 Number. It remains, also, for the square of the smaller to be the sextuple of their difference.

---

43 Supplemented by Bachet.

Therefore, 1 Power is the sextuple of 2 Numbers. Thus, 12 Numbers are equal to 1 Power, so the Number will be 12 units.

Therefore, the smaller will be 12 units, the greater 36 units, and they fulfill what was proposed.

(Corollaries:) And likewise, by the same (reasoning), one can find:

(38.1) two numbers in a given ratio, such that the square of the greater has to the smaller a given ratio.

(38.2) And, again, two numbers in a given ratio, such that the square of the greater has to the greater itself a given ratio.

(38.3) And, similarly, two numbers in a given ratio, such that the square of the greater has to their sum a given ratio.

(38.4) And, moreover, two numbers in a given ratio, such that the square of the greater has to their difference a given ratio.

39. Given two numbers, to find another number such that, the three being set out, if added together by twos and | multiplied by the remaining one, produce three numbers in equal difference.

Let the two given numbers be 3 and 5, and let it be required to find another number such that, if added together by twos, and multiplied by the remaining one, produce three numbers in equal excess.

Let the required number be 1 Number. And when added together with 5 units, it becomes 1 Number, 5 units; and when (this) is multiplied by the other, that is the 3, they become 3 Numbers, 15 units. Again, when 1 Number is added together with 3 units, it becomes 1 Number, 3 units; and when (this) is multiplied by 5 units, it becomes 5 Numbers, 15 units. And moreover, when 5 units are added together with 3 units, and the resulting 8 units are multiplied by 1 Number, they become 8 Numbers. It is obvious that the (value) of 3 Numbers, 15 units[44] will never be the greatest. Indeed the (value) of 5 Numbers, 15 units is greater than it. Therefore, the (value) of 3 Numbers, 15 units is either the middle or the least, while the (value) of 5 Numbers, 15 units is either the greatest or the middle. As for the (value) of 8 Numbers, it can either be the greatest or the middle or the least, since the numerical value of the Number is unknown.[45] So, first, let the greatest be assigned to be the (value) of 5 Numbers and 15 units, the least the (value) of

---

44 ὁ τῶν ἀριθμῶν γ̄ μονάδων ῑε, lit. "the of the 3 Numbers 15 units". This expression is found for the first time in this problem, in which it occurs 13 times. Judging by the first definite article ὁ, the phrase should be translated as "the (number) of the 3 Numbers 15 units", by which is meant the numerical value of the expression 3 Numbers 15 units after the value of "1 Number" has been found. Thus, we write "the (value) of". The similar expression ταῦτα διπλάσιά εἰσι τῶν τοῦ μέσου, lit. "these are the double of the of the middle", is also found in the same problem.

45 ἄδηλον εἶναι τὴν τοῦ ἀριθμοῦ ὑπόστασιν, lit. "the hypostasis of the Number is not manifest".

3 Numbers, 15 units, and the middle, of course, the (value) of 8 Numbers. But if there are three numbers in equal excess, the greatest and the least, if added together, are twice the middle. And the greatest and the least are 8 Numbers, 30 units. These are equal to 16 Numbers, and the Number becomes 15 4ths. This much will be the number sought for and fulfilling what was proposed.

But now, let greatest be the (number) of 5 Numbers, 15 units, middle the (number) of 3 Numbers, 15 units, and least the (number) 8 Numbers. And if there are three numbers in equal excess, as much as the greatest exceeds the middle, so much the middle exceeds the least. But the greatest exceeds the middle by 2 Numbers, while the middle (exceeds) the least by 15 units lacking 5 Numbers. Thus, 15 units lacking 5 Numbers are equal to 2 Numbers, and the Number becomes 15 7ths. This much will be the number sought for and doing the problem.

80

And now let greatest be the (number) of 8 Numbers, middle the (number) of 5 Numbers, 15 units, and least the (number) of 3 Numbers, 15 units. Then since, again, the greatest and the least are twice the middle, and the greatest and the least are 11 Numbers, 15 units, the latter is twice the middle. But the middle is 5 Numbers, 15 units. Thus, 10 Numbers, 30 units are equal to 11 Numbers, 15 units. The required number, therefore, will be 15 units, and it fulfills what was proposed.

Book II

1. To find two numbers such that their sum has to the sum of their squares a given ratio.[1]

Now, let it be proposed that their sum is a tenth part of the sum of their squares.

Let the smaller be assigned to be 1 Number, and the greater 2 Numbers. Their sum becomes 3 Numbers, while the sum of their squares, 5 Powers. Therefore, 3 Numbers should be a tenth part of 5 Powers. Thus, 30 Numbers are equal to 5 Powers, and the Number becomes 6 units.

Thus, the smaller will be 6 units, the greater 12 units, and they fulfill the problem.

2. To find two numbers such that their difference has to the difference of their squares a given ratio.[2]

Now, let it be proposed that their difference is a sixth part of the difference of their squares.

Let the smaller be assigned to be 1 Number, and the greater 2 Numbers. Their difference becomes 1 Number, while the (difference) of their squares, 3 Powers. Therefore, 1 Number should be a sixth part of 3 Powers. Thus, 6 Numbers are equal to 3 Powers, and the Number becomes 2 units.

The smaller will be 2 units, the greater 4 units, and they fulfill the problem.

3. To find two numbers such that the (number) produced from the multiplication has to their sum, or to the difference, a given ratio.[3]

Now, let it be proposed, first, that the (number) produced from the multiplication is the sextuple of their sum.

1 Cf. Problem I.31.
2 Cf. Problem I.34.
3 Cf. Problem I.34, corollaries 1 and 2.

DOI: 10.4324/9781315171470-9

Let the sought-after (numbers) be assigned to be 1 Number and 2 Numbers; indeed, they can also be proposed in a given ratio.[4] Therefore, the (number) produced from their multiplication will be 2 Powers, while their sum, 3 Numbers. Therefore, 2 Powers should be the sextuple of 3 Numbers. Thus, 18 Numbers are equal to 2 Powers. Everything by a Number.[5] Thus, 18 units are equal to 2 Numbers, and the Number becomes 9 units.

The first will be 9 units, the second 18 units, and they fulfill the problem.

If it is proposed that the (number) produced from the multiplication be the sextuple of the difference, the (number) produced from the multiplication will be, again, 2 Powers, and the difference, 1 Number. Again, 6 Numbers are equal to 2 Powers, and the Number becomes 3 units.

The first will be 3 units, the second 6 units, and again they fulfill the problem.

86

4. To find two numbers such that the (number) composed of their squares has to their difference a given ratio.[6]

Now, let it be proposed that the (number) composed of their squares is the decuple of their difference.

Let, again, one be assigned to be 1 Number, and the other 2 Numbers. Therefore, the (number) composed of their squares will be 5 Powers, and their difference, 1 Number. Therefore, 5 Powers should be the decuple of 1 Number. Thus, 5 Powers are equal to 10 Numbers, and the Number becomes 2 units.

The first will be 2 units, the second 4 units, and they fulfill the problem.

5. To find two numbers such that the difference of their squares has to the sum of the two (numbers) a given ratio.[7]

Now, let it be proposed that the difference of their squares is the sextuple of the sum of the two (numbers).

Let, again, one of the sought-after (numbers) be 1 Number, and the other 2 Numbers. And the difference of their squares becomes 3 Powers, while the sum of the two, 3 Numbers. Therefore, 3 Powers should be the sextuple of 3 Numbers.[8] Thus, 3 Powers are equal to 18 Numbers, and the Number becomes 6 units.

And the proof is obvious.

---

4 This is precisely the case of Problem I.34 in which the two sought-after numbers are requested to be in a given ratio.

5 πάντα παρὰ ἀριθμόν. The phrase appears in the margin of **A**, and is absent in **B**. The phrase appears in **T**, without article before ἀριθμόν, and in **V** with article before ἀριθμόν (πάντα παρὰ τὸν ἀριθμόν).

6 Cf. Problem I.32.

7 Cf. Problem I.33.

8 This sentence is present in **B** but missing from **ATV**.

88

6. To find two numbers with a given difference such that the difference of their squares exceeds their difference by a given number. It is certainly necessary that the square of their difference be smaller than the sum of the said difference and the given (excess) of the squares over their difference.

Now, let it be proposed that their difference is 2 units, and that the difference of their squares exceeds their difference by 20 units.

Well, let the smaller be assigned to be 1 Number. Therefore, the greater will be 1 Number, 2 units. And it is established that their difference is 2 units, while the difference of their squares is 4 Numbers, 4 units. Therefore, 4 Numbers, 4 units should exceed 2 units by 20 units. Thus, 4 Numbers, 4 units are equal to 22 units, and the Number becomes 4 2′ units.

The smaller will be 4 2′ units, the greater 6 2′ units, and they fulfill what was proposed.

7. To find two numbers such that the difference of their squares is greater in ratio[9] than their difference by a given number.

Let it be proposed that the difference of their squares is the triple of their difference and moreover exceeds (it) by 10 units. It is certainly necessary that the square of their difference be smaller than the sum of the triple of the difference and the given 10 units.

90

Let their excess be assigned to be 2 units, and the smaller (number), 1 Number. The greater, therefore, will be 1 Number, 2 units. Therefore, 4 Numbers, 4 units should be the triple of 2 units and moreover exceed (it) by 10 units. Thus, three times 2 units, together with 10 units, are equal to 4 Numbers, 4 units. But three times 2 units, together with 10 units, become 16 units. These are equal to 4 Numbers, 4 units, and the Number becomes 3 units.

The smaller will be 3 units, the greater 5 units, and they fulfill the problem.

8. To divide a proposed square (number) into two squares.

Now, let it be proposed to divide 16 into two squares.

And let the first be assigned to be 1 Power. Therefore, the other will be 16 units lacking 1 Power. Thus, 16 units lacking 1 Power should be equal to a square. I form the square from (a side of) whatever (multitude) of Numbers lacking as many units as there are in the side of the 16 units. Let it be 2 Numbers lacking 4 units. Therefore, the square itself will be 4 Powers, 16 units lacking 16 Numbers. These are equal to 16 units lacking 1 Power.

Let the lacking be added in common and likes from likes. Thus, 5 Powers are equal to 16 Numbers, and the Number becomes 16 5ths.

The one will be 256 25ths, the other 144 25ths, and the two added together make 400 25ths, that is 16 units, and each of them is a square.

---

9 See the explanation provided for this expression in Problem I.3.

[92] Otherwise

Now, let (it be proposed) again to divide the square (number) 16 into two squares.

Let, again, the side of the first be assigned to be 1 Number, and (the side) of the other whatever (multitude) of Numbers lacking as many units as there are in the side of the divided (square). Well, let it be 2 Numbers lacking 4 units. Therefore, one of the squares will be 1 Power, and the other, 4 Powers, 16 units lacking 16 Numbers. Well then, I want the two, when added together, to be equal to 16 units. Thus, 5 Powers, 16 units lacking 16 Numbers are equal to 16 units, and the Number becomes 16 5ths.

The side of the first will be 16 5ths, therefore (the first) itself will be 256 25ths; and (the side) of the second 12 5ths, therefore (the second) itself will be 144 25ths. And the proof is obvious.

9. To divide a given number which is composed of two squares, again, into two other squares.

Let (it be proposed) to divide 13, which is composed of the squares 4 and 9, again, into two other squares.

Let the sides, 2 units and 3 units, of the aforesaid squares be taken, and let one of the sides of the sought-after squares be assigned to be 1 Number, 2 units, and the other whatever (multitude) of Numbers lacking as many units as the side of the other has. Let it be 2 Numbers lacking 3 units. So, one of the squares becomes 1 Power, 4 Numbers, 4 units, and the other, 4 Powers, 9 units lacking 12 Numbers. | 94
It remains for the two, if added together, to make 13 units. But the two, when added together, make 5 Powers, 13 units lacking 8 Numbers. These are equal to 13 units, and the Number becomes 8 5ths.

To the hypostases: I assigned the side of the first to be 1 Number, 2 units; it will be 18 5ths; and the side of the other, 2 Numbers lacking 3 units; it will be one (5th). And the squares themselves will be 324 25ths the one, one (25th) the other. And the two, when added together, make 325 25ths, which amount to the proposed 13 units.

10. To find two square numbers with a given difference.

Now, let it be proposed that their difference is 60 units.

Let the side of the one be assigned to be 1 Number, and the side of the other 1 Number and whatever (multitude) of units you wish, provided that the square of the units neither exceed nor equal[10] the given excess; indeed in this manner the problem will hold good,[11] since one species will be left equal to one species.[12] Let

---

10 The clause "nor equal" (μήτε μὴν ἴσος ᾖ) appears in **T**, while in **A** and in **V** it is added between the lines. The clause is absent in **B**.

11 συσταθήσεται τὸ πρόβλημα. According to the *Patristic Greek lexicon* of G. W. H. Lampe, *s.v.* συνίστημι, the range of meanings of which includes "stand", "hold good", "remain valid", "be consistent".

12 This last word "species" appears only in **B**.

it be 1 Number, 3 units. The squares themselves, therefore, will be 1 Power, and 1 Power, 6 Numbers, 9 units, and their difference, 6 Numbers, 9 units. These are equal to 60 units, and the Number becomes 8 2' units.

96   The side of the first will be 8 2' units, and (the side) of the other 11 2' units; and one of the squares themselves will be 72 4' units, and the other 132 4' units; and what was proposed is obvious.

11. To add the same number to two given numbers and make each of them a square.

Now, let the added (number) be 1 Number and let it be (added) to 2 and 3. Therefore, the one will be 1 Number, 2 units, the other, 1 Number, 3 units, and they are equal to squares. This kind is called a "double-equality" and it is solved in this manner. Considering the difference, look for two numbers such that their product makes the difference: 4 units and a fourth of a unit are (such numbers). Of these, either half the difference (multiplied) by itself is equal to the smaller, or half the sum (multiplied) by itself is equal to the greater. But the half of the difference (multiplied) by itself is 225 64ths. These are equal to 1 Number, 2 units, and the Number becomes 97 64ths. While the half of the sum (multiplied) by itself is 289 64ths. These are equal to the greater, that is 1 Number, 3 units, and the Number again becomes 97 64ths.

Thus, the added (number) will be 97 64ths, and what was proposed is obvious.

98   Now, in order not to fall into a double equality, it must be shown as follows:

To find a number which, if added to each of the (numbers) 2 and 3, makes a square.

First, I look for a certain number which, if it receives in addition 2 units makes a square, or, alternatively, a number which makes a square if it receives in addition 3 units. Then, from whatever square I subtract the units, this will be the sought-after (number). Well, let it be on the 2 units, and let them be subtracted from 1 Power. 1 Power lacking 2 units will remain, and it is manifest that, if it receives in addition 2 units it makes a square. It remains, also, that if it receives in addition 3 units, it makes a square. But if it receives in addition 3 units, it becomes 1 Power, 1 unit. These are equal to a square. I form the square from 1 Number lacking so many units so that the numerical value[13] of the Power exceeds the lacking units set out before, namely, in the present case, the 2 units; indeed, in this manner again, in either part one species will be left equal to one (species). So, let it be (formed) from 1 Number lacking 4 units. Therefore, the square itself will be 1 Power, 16 units lacking 8 Numbers. These are equal to 1 Power, 1 unit.

Let the lacking be added in common, and let likes be subtracted from likes. 8 Numbers remain equal to 15 units, and the Number becomes 15 8ths.

To the hypostases: the added (number) will be 97 64ths.

_____

13 ὑπόστασις.

12. To subtract the same number from two given numbers and make each of the remainders a square.

Now, let it be proposed to subtract from 9 and 21 the same number and make each   100
of the remainders a square.

Whatever the square may be that I subtract from each of them, I assign (the sought-after number) to be the remainder; indeed, when this is subtracted, it leaves the square. So, let the square which is subtracted from the 9 units be 1 Power. There remains 9 units lacking 1 Power. Then, it is also required to subtract 9 units lacking 1 Power from 21 units and make a square. But, when I subtract 9 units lacking 1 Power from 21 units, there remains 1 Power, 12 units. These are equal to a square. I form the square from 1 Number lacking so many units so that their square is greater than the 12 units; indeed, in this manner, in either part one species will again be left equal to one (species). So, let it be 4 units. Therefore, the square itself will be 1 Power, 16 units lacking 8 Numbers. These are equal to 1 Power, 12 units.

Likes from likes. There remain 8 Numbers equal to 4 units, and the Number becomes 4 8ths.

The 9 units amount to 72 eighths, that is, 576 64ths; and the lacking of 1 Power removes from them 16 64ths, and fulfills what was proposed.

13. From the same number to subtract two given numbers and make each of the remainders a square.

⟨Now, let it be proposed to subtract from the same number 6 and 7, and make each of the remainders a square⟩.[14]

Let the sought-after (number) be assigned to be 1 Number. And if I subtract 6   102
units from it, there remains 1 Number lacking 6 units equal to a square; while, if I subtract 7 units, there remains 1 Number lacking 7 units equal to a square. And again, in this (problem), similarly, a double-equality appears. Since the difference, being 1 unit, is contained by 2 units and a half of a unit, the Number is found to be 121 16ths, and it makes the problem.

In order not to arrive at a double equation, it must be investigated in this manner:

I first search for a number from which, if I subtract 6 units, makes a square; it is clear that when I add the 6 units to this square, that (number) will be the sought-after (number). So, let (the square) be 1 Power. Therefore, the sought-after (number) will be 1 Power, 6 units. And it is manifest that if I subtract 6 units from it, the remainder will be a square. Then, it is also required to subtract 7 units from 1 Power, 6 units, and make a square. Thus, 1 Power lacking 1 unit is equal to a square. I form the square from 1 Number lacking 2 units. Therefore, the square

---

14 This sentence is lacking in **ABVT** but is added in **A** by a second hand (Chortasmenos's?) between
    the lines (f. 76r).

itself will be 1 Power, 4 units lacking 4 Numbers. These are equal to 1 Power lacking 1 unit, and the Number becomes 5 4ths.

The sought-after (number) will be 121 16ths, and it fulfills the problem.

14. To divide a given number into two numbers and to find a square which, if it receives in addition each of the two from the division, makes a square.

104 Let it be to divide 20 into two numbers.

Set out two numbers so that (the sum of) their squares is smaller than 20 units. Let them be 2 and 3. And if 1 Number is added to each of them, one of the squares will be 1 Power, 4 Numbers, 4 units, and the other 1 Power, 6 Numbers, 9 units. If, therefore, I remove from each of them the Power, that is the square, I shall have the sought-after (numbers), which clearly make a square if they receive in addition a square. But when I subtract 1 Power, the remainders will be 4 Numbers, 4 units, and 6 Numbers, 9 units. Therefore, their sum, that is 10 Numbers, 13 units, should be equal to 20 units, and the Number becomes 7 10ths.

The one will be 68 10ths, the other 132 10ths, and they fulfill what was proposed.

15. To divide a given number into two numbers and to find in addition a square which, if it lacks either, makes a square.

Let it be proposed, again, to divide 20 into two numbers.

And let the sought-after square be assigned to be (formed) from a side of 1 Number and so many units so that its square does not exceed the 20. Let it be 1 Number, 2 units. Therefore, the square will be 1 Power, 4 Numbers, 4 units. And it is clear that if it lacks 4 Numbers, 4 units, it leaves a square; and, similarly, if it lacks 2 Numbers, 3 units, it leaves a square: 1 Power, 2 Numbers, 1 unit. Accordingly, I assign the first to be 4 Numbers, 4 units, the second 2 Numbers, 3 units,
106 and the sought-after (square) 1 Power, 4 Numbers, 4 units; | and if this lacks either one, it makes a square. It remains for the two (together) to be equal to the divided (number). But the two make 6 Numbers, 7 units. These are equal to 20 units.

Likes from likes, and the Number becomes 13 6ths.

The one will be 76 6ths, the other 44 6ths, the square 625 36ths, and they fulfill what was proposed.

16. To find two numbers in a given ratio such that each of them, together with a proposed square, makes a square.

Now, let it be proposed that the greater is the triple of the smaller, and each of them, together with 9 units, makes a square.

From whatever square (formed) from a multitude of Numbers and (3)[15] units, if I subtract 9 units, this will be one of the sought-after (numbers). So, let the smaller be 1 Power, 6 Numbers. Therefore, the greater will be 3 Powers, 18 Numbers.

---

15 Supplemented by Bachet.

Then this, if it receives in addition 9 units, should make a square. But when it receives in addition 9 units, they become 3 Powers, 18 Numbers, 9 units. These are equal to a square. I form the square from 2 Numbers lacking 3 units, and the Number becomes 30 units.

The smaller will be 1080 units, the greater 3240 units, and together with 9 units they fulfill what was proposed.

17. To find three numbers so that if each gives to the next a proposed part of itself, in order, and a given number besides, givers and receivers become equal.

Now, let it be proposed that the first gives to the second a fifth and 6 units besides; the second (gives) to the third a sixth and 7 units; and the third (gives) to the first a seventh and 8 units.

Let the first be assigned to be 5 Numbers, and the second, similarly, 6 Numbers. And when the second receives from the first 1 Number, 6 units, it becomes 7 Numbers, 6 units; while, when it gives to the third a sixth, (namely) 1 Number, and 7 units, it becomes 6 Numbers lacking 1 unit. But, when the first gives its fifth and 6 units besides, it becomes 4 Numbers lacking 6 units. This, therefore, if it receives from the third the seventh and 8 units, should become 6 Numbers lacking 1 unit. But, when 4 Numbers lacking 6 units receive in addition 2 Numbers, 5 units, they become 6 Numbers lacking 1 unit. Therefore, 2 Numbers and 5 units are a seventh of the third and 8 units besides. If, therefore, from 2 Numbers, 5 units I subtract 8 units, the remaining 2 Numbers lacking 3 units are a seventh part of the third. The latter, therefore, will be 14 Numbers lacking 21 units. It remains, also, for this, if it receives from the middle the sixth and 7 units, and if it gives the seventh and 8 units, to become 6 Numbers lacking 1 unit. But when it gives the seventh and 8 units, | the remainder is 12 Numbers lacking 26 units; while, when it receives from the middle the sixth[16] and 7 units, it becomes 13 Numbers lacking 19 units. These are equal to 6 Numbers lacking 1 unit, and the Number becomes 18 7ths.

The first will be 90 7ths, the second 108 7ths, and the third 105 7ths. And these numbers fulfill what was proposed.

18. To divide a given number into three numbers so that if each of the (numbers resulting) from the division gives to the next a proposed part of itself, in order, and a given number besides, (then) givers and receivers become equal.

Now, let it be proposed to divide 80 into three numbers so that the first gives to the second a fifth and 6 units besides, the second to the third a sixth and 7 units, and the third to the first a seventh and 8 units, so that, after the exchange, they become equal.

---

16 Tannery's text has ζ^{ον} (7th).

⟨Otherwise the 17th⟩[17]

Let the first be assigned to be 5 Numbers, and the second 12 units. And having received from the first a fifth, (namely) 1 Number, and 6 units, the second becomes 1 Number, 18 units; while, having given to the third a sixth and 7 units besides, it becomes 1 Number, 9 units. It remains, also, for the other two, after giving and receiving, to become 1 Number, 9 units. | But, when the first gives a fifth of itself and 6 units, the remainder is 4 Numbers lacking 6 units. This, therefore, when it receives a seventh of the third and 8 units, should become 1 Number, 9 units. But, when it receives 15 units lacking 3 Numbers, it becomes 1 Number, 9 units. Therefore, 15 units lacking 3 Numbers are a seventh of the third and 8 units besides. If, therefore, we subtract 8 units from 15 units lacking 3 Numbers, we shall have the seventh of the third, (namely) 7 units lacking 3 Numbers. The (third) itself, therefore, will be 49 units lacking 21 Numbers. It remains, also, for the latter, if it receives from the middle a sixth and 7 units, and if it gives to the first a seventh and 8 units, to become 1 Number and 9 units. But after giving and receiving, it becomes 43 units lacking 18 Numbers. These are equal to 1 Number, 9 units, and the Number becomes 34 19ths.

The first will be 170 19ths, the second 228 19ths, and the third 217 19ths.

19. To find three square (numbers) such that the difference of the greatest and the middle has to the difference of the middle and the least a given ratio.

Now, let it be proposed that the difference is the triple of the difference.

Let the smaller be assigned to be 1 Power, and the middle 1 Power, 2 Numbers, 1 unit, that is, from a side of 1 Number, 1 unit. Therefore, the greatest will be 1 Power, 8 Numbers, 4 units. And so, 1 Power, 8 Numbers, 4 units should be equal to a square. I form the square from ⟨1⟩[18] Number, in order to have the Power, and | so many units besides, so that the other species produced in the square, (namely) the Numbers and the units, do not exceed, in multitude, the 8 Numbers and the 4 units respectively, but one falls short and the other exceeds. Now, let it be 3 units. Therefore, the square itself will be 1 Power, 6 Numbers, 9 units. These are equal to 1 Power, 8 Numbers, 4 units, and the Number becomes 2 2' units.

To the hypostases: the greatest will be 30 4' units, the least 6 4' units, and the middle 12 4' units, and they fulfill the problem.

20. To find two numbers such that the square of each, if it receives in addition the other, makes a square.

Let the first be assigned to be 1 Number, and the second 1 unit, 2 Numbers, so that the square of the first, if it receives in addition the second, makes a square. It remains, also, for the square of the second, if it receives in addition the first, to make a square. But the square of the second, when it receives in addition the first, makes 4 Powers, 5 Numbers, 1 unit. These are equal to a square. I form the square from 2 Numbers lacking 2 units. Therefore, this will be 4 Powers, 4 units lacking 8 Numbers, and the Number becomes 3 13ths.

The first will be 3 13ths, the second 19 13ths, and they fulfill the problem.

21. To find two numbers such that the square of each, if it lacks the other, makes a square.

Let the smaller be assigned to be 1 Number and whatever (multitude) of units; well, let it be 1 unit; and the greater, 1 Power less than the square of the smaller, so that the square of the smaller lacking the greater produces a square. And since the square of the smaller is 1 Power, 2 Numbers, 1 unit, the greater, therefore, will be the (terms of the expression) after the Power, (that is) 2 Numbers, 1 unit. And it is established that the square of the smaller lacking the greater makes a square. It is further required that the square of the greater, (namely) 4 Powers, 4 Numbers, 1 unit, lacking the smaller, makes a square. But the square of the greater lacking the smaller makes 4 Powers, 3 Numbers. These are equal to a square. I form the square from 3 Numbers, and the Number becomes 3 5ths.

The smaller will be 8 5ths, the greater 11 5ths, and they fulfill what was proposed.

22. To find two numbers such that the square of each of them, if it receives in addition the sum of the two, makes a square.

Let the smaller be assigned to be 1 Number, and the greater 1 Number, 1 unit, so that the square of the smaller, that is 1 Power, if it receives in addition the sum of the two, that is 2 Numbers, 1 unit, makes a square. It remains, also, for the square of the greater, if it receives in addition the sum of the two, to make a square. But the square of the greater, when it receives in addition the sum of the two, becomes 1 Power, 4 Numbers, 2 units. These are equal to a square. | I form the square from 1 Number lacking 2 units. Therefore, the square itself will be 1 Power, 4 units lacking 4 Numbers, and the Number becomes 2 8ths.

The smaller will be 2 8ths, the greater 10 8ths, and they fulfill the problem.

23. To find two numbers such that the square of each of them lacking the sum of the two makes a square.

Let the smaller be assigned to be 1 Number, and the greater 1 Number, 1 unit, so that, similarly, the square of the greater lacking the sum of the two makes a square. Then, the square of the smaller lacking the sum of the two should also make a square. So, it will be 1 Power lacking 2 Numbers, 1 unit. These are equal to a square. I form the square from a side of 1 Number lacking 3 units. Thus, 1 Power, 9 units lacking 6 Numbers are equal to 1 Power lacking 2 Numbers, 1 unit, and the Number becomes 2 2′ units.

The smaller will be 2 2′ units, the greater 3 2′ units, and they fulfill the problem.

24. To find two numbers such that the square of the sum, if it receives in addition either, makes a square.

116

118

Since 1 Power, if it receives in addition 3 Powers or 8 Powers, makes a square, I assign one of the sought-after numbers to be 3 Powers, the other 8 Powers, and the square of the sum, 1 Power. And it is established that the square of the sum, if it receives in addition | either, makes a square. And since the sum of the two is 11 Powers, the square of the sum of the two, therefore, will be 121 Power-Powers. But it is also 1 Power. Thus, 121 Power-Powers are equal to 1 Power. Therefore, the side will be also equal to the side. Thus, 1 Number is equal to 11 Powers. And everything by a Number. Therefore, 11 Numbers are equal to 1 unit, and the Number becomes an 11th of a unit.

120

To the hypostases: the one will be 3 121sts, the other 8 (121sts), while the square of the sum, 121 14,641sts, and they fulfill the problem.

25. To find two numbers such that the square of the sum lacking either makes a square.

I first take a certain square from which, if I remove two numbers whatever, I leave a square. Let it be 16. Indeed, if this lacks 12 units, it becomes a square, and again if (it lacks) 7 units, it becomes a square. So, I assign them once again[19] to be in (terms of a) Power – so that the one is 12 Powers, the other 7 Powers, and the square of the sum 16 Powers – and it is established that the square of the sum lacking either makes a square. Then, the square of the sum should be made equal to 16 Powers, so the side, also, to the side, that is 19 Powers, (should be made) equal to 4 Numbers, and the Number becomes 4 19ths.

122

The first will be 192 361sts, the second 112 361sts, and they fulfill the problem.

26. To find two numbers such that their product, if it receives in addition either, makes a square, while the sides of the squares, if added together, make a proposed number.

Now, let it be proposed to make 6.

Since, if there are two numbers the greater of which is the quadruple of the smaller except one unit, then their product makes a square if it receives in addition the smaller, I assign the smaller to be 1 Number, and the greater, 4 Numbers lacking 1 unit. And it comes out, similarly, that their product makes a square if it receives in addition the smaller. It remains, also, for their product, if it receives the greater, that is 4 Numbers lacking 1 unit, to make a square – the side of which is 6 units lacking the 2 Numbers of the side of the smaller (square) – so that, according to the problem, the sides of the two (squares), if added together, make 6 units. But their product, when it receives in addition the greater, makes 4 Powers, 3 Numbers lacking 1 unit, while the square of 6 units lacking 2 Numbers is

---

19 As in the previous problem, two concrete numbers are set in terms of a Power, hence the adverb "again".

4 Powers, 36 units lacking 24 Numbers. These are equal to each other, and the Number becomes 37 27ths.

To the hypostases: I assigned the smaller to be 1 Number, it will be 37 (27ths); the greater, 4 Numbers lacking 1 unit, it will be 121 (27ths); and what was proposed is fulfilled.

27. To find two numbers such that their product lacking either makes a square, while the sides of the squares, if added together, make a given number.

Now, let it be proposed (to make) 5.

Since, if there are two numbers the greater of which is the quadruple of the smaller and one unit, (then) their product lacking the smaller makes a square, I assign the greater to be 4 Numbers, 1 unit, and the smaller 1 Number, and so their product lacking the smaller makes a square. It remains, also, for their product lacking the greater to make a square; (and for) the sides of these (squares) to yield the proposed 5 units. But their product lacking the greater becomes 4 Powers lacking 3 Numbers, 1 unit. These are equal to a square, the one from a side of 5 units lacking 2 Numbers, and the Number becomes 26 17ths.

The smaller will be 26 (17ths), the greater 121 (17ths), and they fulfill what was proposed.

28. To find two square numbers such that their product, if it receives in addition either, makes a square.

If I assign one of the squares to be 1 Power, and the other square to be a unit, then their product will be $\langle 1 \rangle^{20}$ Power, (which is) a square. The latter, therefore, if it receives in addition either, should make a square. Thus, we are reduced to seeking which square makes a square if it receives in addition a unit.

Let the square which I wish to be (made) from them be assigned to be 1 Power. Then, if this receives in addition 1 unit, becomes 1 Power, 1 unit. This must be equal to a square. I form the square from a side of 1 Number lacking 2 units. This will be equal to 1 Power, 1 unit, and the Number becomes 3 4ths. The one will be 9 16ths, and the other 16 (16ths). And it comes out that their product, if it receives in addition a unit, makes a square.

Then, it is also required that their product, if it receives in addition the second, makes a square. And since their product is 9 16ths, let it now be posited in (terms of a) Power, that is, (having taken) everything 16 times, (let it be) 9 Powers, 9 units. Thus, 9 Powers, 9 units are equal to a square. I form the square from a side of 3 Numbers lacking 4 units. Thus, the square itself will be 9 Powers, 16 units lacking 24 Numbers, and the Number becomes 7 24ths.

The first will be 324 576ths, the second 49 576ths, and they fulfill the problem.

---

20 Supplemented by Bachet.

29. To find two square numbers such that their product lacking either makes a square.

If I assign the first to be 1 Power, and the other 1 unit, their product will be 1 Power. Then, the latter lacking 1 unit should make a square. But the Power is a square. Therefore, we are reduced to seeking which square lacking 1 unit makes a square. | 25 16ths is (such) a square. Indeed, this lacking the 16 16ths of a unit[21] makes the square 9 16ths. So, I assign one to be 1 Power, the other 25 16ths, and so their product lacking 1 Power makes a square. Therefore, their product lacking 25 16ths units should be equal to a square. But their product lacking 25 16ths units becomes 25 16ths Powers lacking 25 16ths units. These are equal to a square. Everything sixteen times ⟨and the 25th⟩.[22] I form the square from 1 Number lacking 4 units. Thus, the square itself will be 1 Power, 16 units lacking 8 Numbers, equal to 1 Power lacking 1 unit, and the Number becomes 17 8ths.

The first will be 289 64ths, the second 100 64ths, and they fulfill the problem.

30. To find two numbers such that their product, if it receives in addition or if it lacks the sum of both, makes a square.

Since for any two numbers, their squares, being added together, if they receive or lack twice their product, make a square, we set out two numbers, 2 and 3. And it is manifest that the sum of their squares, together with twice their product, which yields 25 units, makes a square, and again if one subtracts from the sum of their squares twice their product it becomes a square, (namely) a unit. Accordingly, I assign their product to be 13 Powers. | So, let the one be assigned to be 1 Number, and the other 13 Numbers, and so their product becomes 13 Powers. Therefore, 13 Powers, if they receive or lack 12 Powers, make a square. Thus, 12 Powers should be equal to the sum of the two. But the sum of the two is 14 Numbers. Thus, 12 Powers are equal to 14 Numbers, and the Number becomes 14 12ths, that is 7 6ths.

So, the one is 1 Number, it will be 7 6ths; the other, 13 Numbers, it will be 91 6ths; and they fulfill the problem.

31. To find two numbers equal to a square, such that their product, if it receives in addition, or if it lacks, the sum of both, makes a square.

Since, if there are two numbers of which one is the double of the other, their squares, being added together, if they lack or receive in addition twice their product, make a square, we set out the (numbers) 4 and 2. Then, let them be assigned

---

21 Tannery's text has τῆς Μ° $\overline{\iota\varsigma}^{\iota\varsigma}$, that is "the 16 16ths of a unit". The four mss **ABTV** have all τῶν τῆς μονάδος ἑκκαίδεκα $\overline{\iota\varsigma}^{ων}$, "the 11 16ths of a unit".

22 The phrase καὶ τὸ κε°ⁿ, "and the 25th", added by Tannery, is not found in any manuscript. Bachet had already added a more detailed phrase: Πάντα ἑκκαιδεκάκις, καὶ παρὰ τὸν κε. γίνεται Δ^{Y} $\overline{\alpha}$ λείψει μονάδος $\overline{\alpha}$ ἴση τετραγώνῳ, "Everything sixteen times, and (we take) the 25th. It yields 1 Power lacking 1 unit equal to a square" (Diophantus 1621, 127).

to be in (terms of a) Power. Their product is 20 Powers, while the sum of the two, 16 Powers. Let the one be 2 Numbers, the other 10 Numbers, the sum of both 12 Numbers, but, also, 16 Powers. Thus, 16 Powers are equal to 12 Numbers, ⟨and the Number becomes 12 16ths⟩,[23] that is 3 4ths.

The one will be 6 (4ths), the other 30 (4ths), and they fulfill the problem.   132

32. To find three numbers such that the square of any one of them, if it receives in addition the next one, makes a square.

Let the first be assigned to be 1 Number. And since, if a number is one unit greater than the double of a number, the square of the smaller, if it receives the greater, makes a square, let the second be assigned to be one unit greater than the double of the first; so, it is clear that it will be 2 Numbers, 1 unit; and moreover, the third, one unit greater than the double of that; it will be 4 Numbers, 3 units. And it comes out that the square of the first, if it receives in addition the second, becomes a square, (namely) 1 Power, 2 Numbers, 1 unit; and similarly, the square of the second, if it receives the third, makes a square, (namely) 4 Powers, 8 Numbers, 4 units. Then, also, the square of the third, if it receives in addition the first, should make a square. But the square of the third, when it receives in addition the first, makes 16 Powers, 25 Numbers, 9 units. These are equal to a square. I form the square from a side of 4 Numbers lacking 4 units. Therefore, this will be 16 Powers, 16 units lacking 32 Numbers, and the Number becomes 7 57ths.

The first will be 7 (57ths), the second 71 (57ths), the third 199 (57ths), and they fulfill the problem.

33. To find three numbers such that the square of any one of them, if it lacks the next one, makes a square.

Since, if a number is the double of a number except one unit, the square of the smaller, lacking the | greater, makes a square, I assign the first to be 1 Number, 1   134
unit, the second, similarly, 2 Numbers, 1 unit, the third 4 Numbers, 1 unit, and it comes out that the square of the first, lacking the second, makes a square, and also the square of the second, lacking the third, makes a square. It remains, also, for the square of the third, lacking the first, to make a square. But the square of the third, lacking the first, makes 16 Powers, 7 Numbers. These are equal to a square. I form the square from 5 Numbers. Thus, 25 Powers are equal to 16 Powers, 7 Numbers, and the Number becomes 7 9ths.

The first will be 16 (9ths), the second 23 (9ths), the third 37 (9ths), and what was proposed is fulfilled.

34. To find three numbers such that the square of any one of them, if it receives in addition the (number) composed of all three, makes a square.

---

23 Supplemented by Bachet.

Since, if a number is measured by a certain number, and we take the one according to which it is measured, and (between) the one that measures and the one according to which it measures we subtract the smaller from the greater, the square of the half of the remainder, if it receives in addition the initial (number), makes a square, I assign the (number) composed of all three to be of a certain (multitude of) Powers having three measuring numbers. So, let it be 12. Indeed, 1 unit measures it according to 12, 2 units according to 6, and 3 units according to 4. And if I subtract the measuring number from the one according to which it measures, and I take the halves of the remainders, I assign the three to be 5 2′ units (for) the first, 2 units (for) the second, and a half of a unit (for) the third. | And it is obvious that the square of each one of them, if it receives in addition the 12, makes (each one) a square: one 12 4′, another 16, and the other 42 4′. So, I assign them to be in (terms of) Numbers, the first to be 5 2′ Numbers, the second 2 Numbers, and the third a half of a Number. And the (number) composed of all three must be equal to 12 Powers. But the (number) composed of all three is 8 Numbers. Thus, 8 Numbers are equal to 12 Powers, and the Number becomes 4 6ths.

The first will be 22 (6ths), the second 8 (6ths), the third 2 (6ths), and what was proposed is fulfilled.

35. To find three numbers such that the square of any one of them, if it lacks the (number) composed of all three, makes a square.

Similarly, I assign a certain number which has three measuring numbers. Let it again be the 12. And adding the measuring number to the one according to which it measures, and taking the half, I assign one of the three numbers to be 6 2′ Numbers, another 4 Numbers, and the other 3 2′ Numbers. And it comes out that the square of each one, if it lacks the 12, makes a square. Then, the three (together) must be equal to 12 Powers. But the three, when added together, make 14 Numbers. Thus, 14 Numbers are equal to 12 Powers, and the Number becomes 7 6ths.

The first will be 45 and a half (6ths), the second 28 (6ths), the third 24 and a half (6ths), and they fulfill what was proposed.

Book III

1. To find three numbers such that the square of any one of them, if subtracted from the (number) composed of all three, makes a square.[1]

Set out two squares, the one from 1 Number, the other from 2 Numbers. Their squares (together) become 5 Powers. I assign the (number) composed of all three to be 5 Powers, one of the sought-after numbers to be 1 Number, the other 2 Numbers, and so two of the proposed tasks are fulfilled. And since we have the 5 divided into two squares, the unit and the tetrad, let us divide it again, as has been shown before,[2] into two other squares, (namely) into 4 25ths and 121 25ths. I assign now the third to be (made) of a side of one of them. Let it be 2 fifths of a Number. And it is established, again, that its square, when subtracted from the sum of the two, makes a square, (namely) 121 25ths. Then, all three should | be equal to 5 Powers. But the three are 3 Numbers and 2 fifths (of a Number), and the Number becomes 85 125ths.[3]

The first will be 85 (125ths), the second 170 (125ths), the third 34 (125ths), and they fulfill what was proposed.

2. To find three numbers such that the square of the (number) composed of all three, if it receives in addition any one of them, makes a square.

---

1 The fact that problems 1–4 of Book III are very similar to problems 34–35 of Book II makes Tannery suspect that they crept into the text of the *Arithmetica* from some ancient commentary.

2 See Problem II.9.

3 Diophantus solves here the equation $5x^2 = 3\frac{2}{5}x$, regarding $3\frac{2}{5}$ as aggregation of 3 with 2 5ths, to get the solution 85 125ths. This number is found by dividing each of the 3 and the 2 5ths by 5, to get 3 5ths and 2 25ths, and then by performing the cross-multiplication to get the numerator $3 \cdot 25 + 2 \cdot 5 = 85$ and the denominator $5 \cdot 25 = 125$. He did not first convert the $3 + 2$ 5ths to 17 5ths since that would give the answer 17 25ths. But while the latter operation can be conceived within the conceptual framework of unit fractions, the former cannot. The cross-multiplication to get the numerator and the multiplication of the denominators to get the denominator shows that Diophantus conceived of expressions like 3 5ths, 17 25ths, or 85 125ths as common fractions and he made operations on them. The denominator 125 is not written in the manuscripts. In **A** (f. 82r) it is added by a second hand. The same scholiast added that the result of the addition 3 and 2 5ths is 17 5ths, and that the number 85 125ths is the same as 17 25ths. The scholiast apparently did not recognize the addition of common fractions in Diophantus's calculation.

DOI: 10.4324/9781315171470-10

Let the square of the (number) composed of all three be assigned to be 1 Power. I assign the first to be 3 Powers, the second to be 8 Powers, and the third to be 15 Powers, so that the square of the (number) composed of all three, that is the 1 Power, if it receives in addition any one of them, makes a square, one 4 Powers, ⟨another 9 Powers⟩,[4] and the other 16 Powers. And the three, if added together, should become equal to the side of the square (of the sum) of the three, that is to 1 Number. But the three, when added together, make 26 Powers, and the Number becomes one ⟨26th⟩.[5]

Thus, the first will be 3 676ths, the second 8 676ths, the third 15 676ths, and they fulfill the problem.

3. To find three numbers such that the square of the (number) composed of all three, if it lacks any one of them, makes a square.

142   Let the (number) composed of all three be assigned to be 4 Numbers, and | its square 16 Powers, which, if it lacks 7 Powers or 12 Powers or 15 Powers, makes a square. I assign, therefore, the first to be 7 Powers, the second to be 12 Powers, and the third to be 15 Powers. It remains for the (number) composed of all three to be equal to the three (squares). But the (number) composed of all three has been supposed to be 4 Numbers, while all three are 34 Powers; and the Number becomes 2 17ths, while the Power, 4 289ths.

The first will be 28 (289ths), the second 48 (289ths), the third 60 (289ths), and they fulfill the problem.

4. To find three numbers such that the square of the (number) composed of all three, if subtracted from any one of them, makes a square.

Let the (number) composed of all three be assigned to be 1 Number, and its square 1 Power, and let one the three be 2 Powers, another 5 Powers, and the other 10 Powers. And it is established that each one of them, if it lacks the square of the (number) composed of all three, that is the 1 Power, makes a square. Since the square of the (number) composed of all three has, of course, the (number) composed of all three as a side, therefore, the sum of the three is 1 Number; but, also, 17 Powers; and the Number becomes one ⟨17th⟩,[6] while the Power one ⟨289th⟩.[7]

144   The first will be 2 (289ths), the second 5 (289ths), the third 10 (289ths), and they fulfill what was proposed.

5. To find three numbers equal to a square, such that, taken by twos, exceed the remaining one by a square.

Let the three be assigned to be equal to a square from 1 Number, 1 unit, that is, to 1 Power, 2 Numbers, 1 unit, and among them let the first and the second exceed the

---

4 The phrase is missing in **ABVT** but is added in **A** between the lines by a second hand.
5 Supplemented by Bachet.
6 Supplemented by Bachet.
7 Supplemented by Bachet.

third by 1 unit. The third, therefore, will be 2′ of a Power, 1 Number, so that the first and the second will exceed the third by one unit. Again, the second and the third exceed the first by a square. Let them exceed by 1 Power. Similarly, the first will be 1 Number, 2′ of a unit, and so we shall have as the second the remainder, (namely) 2′ of a Power, 2′ of a unit. Then it is required that the first and the third exceed the second by a square. But the first and the third exceed the middle by 2 Numbers. These are equal to a square, that is, (say,) to 16 units, and the Number becomes 8 units.

The first will be 8 2′ units, the second 32 2′ units, the third 40 units, and they fulfill what was proposed.

[146] Otherwise[8]

First I look for three numbers[9] equal to a square. But when I add two numbers,[10] for example the (number) 4 and the (number) 9, and ask what square, if it receives in addition 13, makes a square, I will find 36; and the three squares will be equal to one square. Then, we are reduced to seeking to find three numbers which, (taken) by twos, exceed the remaining one by a given number: the first together with the second (exceed) the third by 4 units; the second together with the third (exceed) the first by 9 units; while the third together with the first (exceed) the second by 36 units. But this has been shown before,[11] and the first will be 20 units, the second 6 2′ units, and the third 22 2′ units, and they fulfill what was proposed.

6. To find three numbers equal to a square, so that, taken by twos, they make a square.

Let the three be assigned to be equal to a square, (namely) to 1 Power, 2 Numbers, 1 unit; and the first together with the second, 1 Power; then the remaining third will be 2 Numbers, 1 unit. Again, since we require that the second together with the third make a square, let them make 1 Power, 1 unit lacking 2 Numbers, (which is a square) from a side of 1 Number lacking 1 unit. But all three are | 1 Power, 2 Numbers, 1 unit. Therefore, the remaining first will be 4 Numbers. But also, together with the second, it was assigned to be 1 Power. The second, therefore, will be 1 Power lacking 4 Numbers. Further, it is required to equate the first together with the third – yielding 6 Numbers, 1 unit – to a square; let it be equal to 121 units, and the Number becomes 20 units.

148

The first will be 80 units, the second 320 units, the third 41 units, and they fulfill the proposed task.

---

8 Heath's comment (1910, 157): "We should naturally suppose that this alternative solution, like others, was interpolated. But we are reluctant to think so because the solution is so elegant that it can hardly be attributed to a scholiast. If the solution is not genuine, we have here an illustration of the truth that, however ingenious they are, Diophantus' solutions are not always the best imaginable (Loria, *Le scienze esatte nell'antica Grecia*, Libro v, pp. 138–9). In this case the more elegant solution is the alternative one".

9 Bachet added τετραγώνους, "squares".

10 Again Bachet has added the implied word τετραγώνους.

11 See Problem I.18.

Otherwise[12]

Let all three be assigned to be 1 Power, 2 Numbers, 1 unit; and let the first and the second be 1 Power; then the remaining third will be 2 Numbers, 1 unit. And, also, let the second together with the third be 1 Power, 1 unit lacking 2 Numbers, of which the third is 2 Numbers, 1 unit. Therefore, the remaining second will be 1 Power lacking 4 units. But, also, the first together with the second are 1 Power, of which the second is 1 Power lacking 4 units. Therefore, the remaining first will be 4 Numbers. And the three added together make the assigned square, 1 Power, 2 Numbers, 1 unit, while the first together with the second, and the second together with the third, make a square. | Further, it is required to equate the third together with the first – yielding 6 Numbers, 1 unit – to a square; let it be to 36 units, and the Number becomes 35 6ths.

The first will be 140 6ths, i.e., 840 36ths, the second 385 36ths, the third 456 36ths, and they fulfill the problem.

7. To find three numbers in equal excess[13] such that, taken by twos, make a square.

First, I look for three ⟨square⟩[14] numbers so that they are in equal excess, and half the sum of the three is greater than any one of them. Well, let the first be assigned to be 1 Power, and the second 1 Power, 2 Numbers, 1 unit; so, their excess is 2 Numbers, 1 unit. And if I add the 2 Numbers, 1 unit to the second, the third becomes 1 Power, 4 Numbers, 2 units. These are equal to a square, (say) the one from a side of 1 Number lacking 8 units. The square becomes 1 Power, 64 units lacking 16 Numbers; (this is) equal to 1 Power, 4 Numbers, 2 units, and the Number becomes 62 20ths, that is 31 10ths.

The first will be 961, the second 1681, and the third 2401, and they fulfill what was required in the problem, that is, (to find) three squares in equal difference, and (such that) half the (sum of the) three is greater than any one of them.

Now I return to what was proposed, that is, | to find three numbers in equal excess, such that, taken by twos, make a square. I first look for three squares in equal excess. But this was shown before, and the squares are 961 the first, 1681 the second, and 2401 the third. Now, it is required for the first and the second to make 961 units, the second and the third, 2401 ⟨units⟩[15] – indeed they can be taken interchangeably because of the difference[16] – while the third and the first, 1681 units. Let all three be assigned to be 1 Number. And since the three are 1 Number, if, therefore, I subtract the 961 units of the first and second, I shall have the third, (namely) 1 Number lacking 961 units. Again, since from 1 Number I subtract the 2401 units of the second and the third, I shall have the first, (namely) 1 Number

150

152

---

12 This alternative solution is essentially identical with the earlier one, the only difference between the two being that in the former the polynomial "6 Numbers, 1 unit" is equated with the square number 121, while, in the latter, with 36. It is probably interpolated.

13 That is, three numbers forming an arithmetical progression.

14 Supplemented by Xylander and Bachet.

15 Supplemented by Tannery.

16 Bachet inserted here the word ἴσην, that is, "because of the equal difference".

⟨lacking 2401 units. And again, if from 1 Number I subtract the 1681 units of the third and the first, I shall have the second, (namely) 1 Number⟩[17] lacking 1681 units. It remains for the three, if added together, to be equal to 1 Number, and the Number becomes 2521 2′.

And so, the first will be 120 2′ units, the second 840 2′ units, and the third 1560 2′ units, and the proposed task is fulfilled.

8. Given a certain number, to find three other (numbers) such that the (number) composed of any two, if it receives in addition the given (number), makes a square, and, moreover, the three added together, if they receive in addition the given (number), make a square.   154

Let the given (number) be 3 units; the (number) composed of the first two be 1 Power, 4 Numbers, 1 unit, so that together with the 3 units it makes a square; the next two be 1 Power, 6 Numbers, 6 units; and all three be 1 Power, 8 Numbers, 13 units, so that together with 3 units they also make a square. And since the three are 1 Power, 8 Numbers, 13 units, of which the first two are 1 Power, 4 Numbers, 1 unit, therefore, the remaining third is 4 Numbers, 12 units. Again, since the three are 1 Power, 8 Numbers, 13 units, of which the second and third are 1 Power, 6 Numbers, 6 units, therefore, the remaining first is 2 Numbers, 7 units. But, also, the first and the second are 1 Power, 4 Numbers, 1 unit. Therefore, the remaining second will be 1 Power, 2 Numbers lacking 6 units. It remains, also, for the first together with the third, if it receives in addition 3 units, to make a square. But when the first together with the third receives in addition 3 units, it becomes 6 Numbers, 22 units. These are equal to a square; let it be to 100, and the Number becomes 13 units.

The first will be 33 units, the second 189 units, and the third 64 units, and they fulfill the problem.

9. Given a certain number to find three other (numbers) such that the (number) composed of any two, if it lacks the given (number), makes a square, and moreover the three added together, if they lack the given (number), make a square.   156

Let, again, the given (number) be 3 units; the (number) composed of the first two be 1 Power, 3 units, so that, if it lacks the 3 units it makes a square; the next two be 1 Power, 2 Numbers, 4 units; and all three be 1 Power, 4 Numbers, 7 units, so that, if they lack the three units they also make a square. And since all three are 1 Power, 4 Numbers, 7 units, of which the first and the second are 1 Power, 3 units, therefore, the remaining third is 4 Numbers, 4 units. Again, since the second and the third are 1 Power, 2 Numbers, 4 units, of which the third is 4 Numbers, 4 units, therefore, the remaining second will be 1 Power lacking 2 Numbers. But, the first and the second are 1 Power, 3 units, of which the second is 1 Power lacking 2 Numbers, therefore, the remaining first will be 2 Numbers,

---

17 Text supplemented by Bachet.

3 units. Then, It remains, also, for the third together with the first, lacking 3 units, to make a square. But the third together with the first, lacking 3 units, is 6 Numbers, 4 units. These are equal to a square; let it be to 64, and the Number becomes 10 units.

To the hypostases: the first will be 23 units, the second 80 units, the third 44 units, and they fulfill what was proposed.

10. To find three numbers such that the product of any two, if it receives in addition a given number, makes a square.

Now, let it be proposed (that it receives) 12.

Since we seek for the product of the first and second, if it receives in addition 12, to make a square, if, therefore, I subtract 12 from a certain square I shall have the product of the first and second. So, let the square be 25 units. If, therefore, I subtract 12 from this, I shall have as the remainder the product of the first and second, (namely) 13 units. So, let the first be 13 units, the second 1 unit, and let them be assigned to be in (terms of) Numbers, so that their product make 13 units. So, let the first be 13 Numbers, and the second ⟨1⟩[18] inverse Number. If, moreover, I subtract 12 from another square, I shall have the product of the second and third. Let it be from 16. Therefore, the remaining product of the second and third will be 4 units. Let them be assigned, again, to be in (terms of) Numbers, so that their product makes 4 units, of which the second is 1 inverse Number. Therefore, the remaining third will be 4 Numbers. Then it is required, also, that the product of the first and third, together with 12 units, makes a square. But the product of the first and third is 52 Powers. Therefore, 52 Powers together with 12 units must make a square. And if the multitude of the 13 units of the first number were a square, the setting up of the equation would be easy. But since it is not, I am reduced to finding two numbers such that their product is a square and, moreover, each of them, together with 12 units, makes a square. And if, instead of numbers, I find squares, then their product will be a square.

So, it resulted (in the problem) of finding two squares each of which together with 12 units makes a square. But this is easy[19] and, | as we have said, makes setting up the equation easy. The one is 4, and the other 4′. Indeed each of them, together with 12 units, makes a square.

Having found these numbers I return to the initial (problem), and I assign the first number to be 4 Numbers, the second 1 inverse Number, and the third a 4th of a Number. It remains, also, for the product of the first and third, together with 12 units, to make a square. But the product of the first and third is 1 Power. Thus, 1 Power together with 12 units is equal to a square. I form the square from a side of 1 Number, 3 units. The square itself, therefore, will be 1 Power, 6 Numbers, 9 units, and the Number becomes 2′, and the proposed task is fulfilled.

---

18 Added by a second hand in **A**.
19 The working out is indicated in Problem II.34.

11. To find three numbers such that the product of any two, if it lacks a given (number), makes a square.

Now, let it be proposed (that it lacks) 10.

Since we seek for the product of the first and second, lacking 10 units, to make a square, if, therefore, I add 10 units to a certain square, I shall have their product. Let it be (added) to 4. Therefore, the product of the first and second will be 14 units. Let the first be 14 units; then the second will be 1 unit. And let them, again, be assigned to be in (terms of) Numbers, so that their product make 14 units; and let the first be 14 Numbers, and the second, an inverse Number. | Again, if I add 10 units to another square, I shall have the product of the second and third. Let it be (added) to 9. Therefore, the product of the second and third will be 19 units, of which the second is 1 inverse Number. The remaining third, therefore, will be 19 Numbers. Then it is required, also, that the product of the third and first, lacking 10 units, ⟨makes a square⟩.[20] ⟨But the product of the third and first, lacking 10 units⟩,[21] becomes 266 Powers lacking 10 units. These are equal to a square. And because of what was said in the previous (problem), I am reduced to finding two squares each of which, lacking 10 units, makes a square. But this is easy.

Indeed, you will find it by seeking which square lacking 10 units makes a square. And, since, if a unit is added to a certain number, and we take the square of the half of the obtained (number), and from the produced square we subtract the initial (number), the remainder will again be a square, (so) I add 1 unit to 10 units, and if I take the square of the half the obtained (number), that is 5 2', then if I subtract the 10 units from the 30 4' units, I shall have 20 4' units, a square from a side of 4 2'. So, I assign the first to be 30 4', and the third 1 Power. | Then, if 10 units are subtracted from 1 Power, the remainder should be made a square. Therefore, 1 Power lacking 10 units is equal to a square. I form the square from a side of 1 Number lacking 2 units. (The square) itself, therefore, will be 1 Power, 4 units lacking 4 Numbers, and the Number becomes 3 2' units. Since I assigned the third to be 1 Power, it will be 12 4'; and the first is 30 4'; which, if they lack 10 units make squares.

I return to what was initially required, and I assign the first to be 30 4' Numbers, the second an inverse Number, the third 12 4' Numbers, so, then, the product of the first and third becomes 370 2' 16' Powers. Thus, the latter lacking 10 units is equal to a square. And in order (for the multitude of the) Powers to be whole (number), I produce them 16 times. Thus, 5929 Powers lacking 160 units are equal to a square, (say) the one from a side of 77 Numbers lacking 2 units, that is to 5929 Powers, 4 units lacking 308 Numbers; and the Number becomes 41 77ths.

I assigned the first to be 30 4' Numbers, it will be 1240 4' 77ths; the second an inverse Number, it will be 77 41sts; and the third 12 4' Numbers, it will be 502 4' 77ths; and what was proposed is fulfilled.

162

164

---

20 Supplemented by Bachet.
21 Supplemented by Tannery.

12. To find three numbers such that the product of any two, if it receives in addition the other one, makes a square.

166 Since we seek for the product of the first and second, if it receives in addition | the remaining one, to make a square, if, therefore, having set out a certain square we assign a certain part of it to be the third, and the remainder to be the product of the first and second, we will have fulfilled one of the proposed tasks. Let the square be formed from 1 Number, 3 units. This, therefore, will be 1 Power, 6 Numbers, 9 units. Let the third be assigned to be 9 units. Therefore, the remaining 1 Power, 6 Numbers will be the product of the first and second. Let the first be assigned to be 1 Number. The remaining second, therefore, will be ⟨1 Number, 6 units. Then it is required, also, for the product of the second and third, if it receives in addition the first – thus becoming⟩[22] 10 Numbers, 54 units – to be equal to a square; and, moreover, for the product of the third and first, if it receives in addition the second – thus becoming 10 Numbers, 6 units – again to be equal to a square. So, a double equality arises, and their difference is 48 units. Therefore, two squares should be found with a difference of 48 units. But this is easy[23] and can be done in an unlimited number of ways: the smaller is 16 units, the greater 64 units, and with respect to either one if I produce the equality, I shall find the numerical value[24] of the Number. Indeed, if we say that the 64 units of the greater are equal to 10 Numbers, 54 units, the Number is found to be 1 unit; again, if we say that the 16 units of the smaller are equal to 10 Numbers, 54 units, the Number is found to be 1 unit.

To the hypostases: the first will be 1 unit, the second 7 units, and the third, 9 units, and they fulfill the proposed task.

13. To find three numbers such that the product of any two, if it lacks the other one, makes a square.

168 Let the first be assigned to be 1 Number, and the second, 1 Number, 4 units. Their product, therefore, will be 1 Power, 4 Numbers. This, then, if it lacks the third, should make a square. So, if I assign the third to be 4 Numbers, ⟨one of the proposed tasks will be fulfilled. Then, it is further required that the product of the second and third, if it lacks the first, makes a square⟩,[25] and, also, that the product of the third and first, if it lacks the second, makes a square. But the product of the second and third, if it lacks the first, is 4 Powers, 15 Numbers, (and this) is equal to a square; while the product of the third and first, if it lacks the second, is 4 Powers lacking 1 Number, 4 units, (and this) is equal to a square. And, again, a double equation arises. Since their difference happens to be 16 Numbers, 4 units, I look for two numbers whose product is 16 Numbers, 4 units. 4 units and 4 Numbers, 1 unit are (such numbers). So, again, either half their sum (multiplied) by itself is equal to the greater, or half the difference (multiplied) by itself is equal to the smaller, and the Number is found to be 25 20ths.

---

22 Text restored initially by Bachet, and then by Tannery.
23 See Problem II.10.
24 ὑπόστασιν.
25 Text restored by Auria, Bachet, and finally by Tannery.

The first will be 25 (20ths), the second 105 (20ths), the third 100 (20ths), and what was proposed is fulfilled.

14. To find three numbers such that the product of any two, if it receives in addition the square of the other one, makes a square.

Let the first be assigned to be 1 Number, the second 4 Numbers, 4 units, and the third 1 unit, so that two of the proposed tasks are fulfilled. It remains, also, for the product of the third and first, if it receives in addition the square of the second, to make a square. But the product of the third and first, when it receives in addition the square of the second, makes 16 Powers, 33 Numbers, 16 units. These are equal to a square, (say) the one from a side of 4 Numbers lacking 5 units, that is, 16 Powers, 25 units lacking 40 Numbers, and the Number becomes 9 73rds.

The first will be 9 (73rds), the second 328 (73rds), the third 73 (73rds), and they fulfill the problem.

170

15. To find three numbers such that the product of any two, if it receives in addition their sum, makes a square.

Now, the product of any two consecutive squares, if it receives in addition their sum, makes a square. Accordingly, let the first be assigned to be 4 units, and the second 9 units, so that the square produced by them, (namely) 36 units, if it receives in addition their sum, makes a square. It remains, also, for the product of the second and third, if it receives in addition their sum, (to make a square), and moreover, for the product of the third and first, if it receives in addition their sum, to make a square. Let the third be assigned to be 1 Number. The product of the second and third, when it receives in addition their sum, becomes 10 Numbers, 9 units, (and this is) equal to a square. And moreover, the product of the third and first, when it receives in addition their sum, (becomes) 5 Numbers, 4 units, (and this is) equal to a square. And here again a double equation arises, and the difference is 5 Numbers, 5 units. So, I again look for two numbers whose product is 5 Numbers, 5 units. And one of (the numbers) | whose product makes that difference is 1 Number, 1 unit, and the other 5 units. And, likewise,[26] either half their sum (multiplied) by itself is equal to the greater, or half the difference (multiplied) ⟨by itself⟩[27] is equal to the smaller, and the Number becomes 28 units.

The first is 4 units, the second 9 units, and the third 28 units. And they fulfill what was proposed.

172

Otherwise

To find three numbers such that the product of any two, if it receives in addition their sum, makes a square.

---

26 The manuscripts have here the phrase τὸ ἐν τῷ δευτέρῳ, "the in the second", referring to the second problem before it. Tannery thinks it is interpolated.

27 Supplemented by Bachet.

Let the first be assigned to be 1 Number, the second 3 units, and so their product together with their sum becomes 4 Numbers, 3 units. These are equal to a square; let it be to 25 units, and the Number becomes 5 2′ units. The first will be 5 2′ units, the second 3 units, and one of the proposed tasks is fulfilled; indeed, their product together with their sum makes the square 25. Then it is also required that the product of the second and third, and moreover the product of the third and first, if they receive in addition their sum, make squares.[28] Let the third be assigned to be 1 Number. The product of the second and third, when it receives in addition their sum becomes, again, 4 Numbers, 3 units, while the | product of the third and first, ⟨if it receives their sum⟩,[29] 6 2′ Numbers, 5 2′. Each of them is equal to a square. But since the multitude of the Numbers and the units in the one (expression) is greater (than that in the other), nor do they have a ratio that a square has to a square, the assignment[30] made is unsuccessful.

174

So, we are reduced ⟨to⟩[31] finding two numbers such that their product, if it receives in addition their sum, makes a square, and moreover, ⟨the numbers one unit greater than them⟩[32] have to each other a ratio that a square has to a square. Since if a number is the quadruple of (another) number and by 3 units greater, the (numbers) which are one unit greater than them have to each other the ratio that a square number has to a square number, I assign one to be 1 Number, and the other 4 Numbers, 3 units. Then, their product together with their sum, must be equal to a square. But their product together with their sum is 4 Powers, 8 Numbers, 3 units. These are equal to a square. I form the square from 2 Numbers lacking 3 units, and the square becomes 4 Powers, 9 units lacking 12 Numbers, so the Number becomes 6 20ths, that is 3 10ths. The first will be 3 10ths, the second 42 10ths, that is 4 5′ units, and one of the proposed tasks is fulfilled.

It remains for the product of the second and third, together with their sum, to make a square. I assign the third to be 1 Number, while the second is 4 5′ units; their product, together with their sum, becomes 5 5′ Numbers, 4 5′ units. These are equal to a square. | Again, since the third is 1 Number, and the first is 3 10ths, their product, together with their sum, will be 13 10ths Numbers, 3 10ths units. These are equal to a square. I multiply the 5 5′ Numbers, 4 5′ units by 25. They result in 130 Numbers, 105 units equal to a square. And likewise, the 13 10ths Numbers, 3 10ths units by 100. They result in 130 Numbers, 30 units equal, again, to a square. Their difference is 75 units and, again, it results in a double equality, and the Number is found to be 7 10ths.

176

The third will be 7 10ths, while the first was 3 10ths, and the second, 42 10ths, and they fulfill the proposed task.

28 The Greek text is in singular ("a square") and applies to both expressions.
29 Supplemented by Bachet.
30 ὑπόστασις. In this context the word means the expression assigned to an unknown, from which the numerical value of the unknown will be determined.
31 Supplemented by Bachet.
32 Supplemented by Bachet.

16. To find three numbers such that the product of any two, if it lacks their sum, makes a square.

Similarly to the previous problem,[33] let the first be assigned to be 1 Number, the second whatever units (we wish), and I will likewise come to a dead end. So, in order to have the multitude of Numbers having to the multitude of Numbers the ratio that a square number has to a square number, we are reduced to looking for two numbers such that their product, if it lacks their sum, makes a square, ⟨and, moreover, the numbers one unit smaller than them have to each other the ratio that a square number has to a square number⟩.[34] And now, since, if a number is the quadruple of a number lacking 3 units, the (numbers) one unit smaller than them have to each other | the ratio that a square number (has) to a square number.[35]   178
Therefore, I assign the first to be 1 Number, 1 unit, and the other 4 Numbers, 1 unit. And it remains for their product lacking the sum of both, which is 4 Powers lacking 1 unit, to be equal to a square; (say) to the one from a side of 2 Numbers lacking 2 units, that is to 4 Powers, 4 units lacking 8 Numbers, and the Number becomes 5 8ths. The first will be 13 8ths, the second 28 8ths, and one of the proposed tasks has been fulfilled.

And since the first is 13 8ths and the second 3 2′ units, I assign the third to be 1 Number. So, it remains for the product of the second and third, which comes out to be 3 2′ Numbers, lacking the sum of the two, (namely) 1 Number, 3 2′ units – (this sum) being 2 2′ Numbers lacking 3 2′ units – to be equal to a square. ⟨These four times. They become 10 Numbers lacking 14 units⟩.[36] And the product of the third and first becomes 13 8ths Numbers, lacking their sum becomes 5 8ths Numbers lacking 13 8ths units, (which is) equal to a square. These sixteen times. They become 10 Numbers lacking 26 units. And their excess is 12 units, which are the product of 2 units | and 6 units. Half their sum multiplied by itself becomes 16   180
units, (which is) equal to the greater, that is to 10 Numbers lacking 14 units, and the Number becomes 3 units.

The third will be 3 units, that is 24 8ths; but we have, also, the first, 13 8ths, and the second, 3 2′ units, that is 28 8ths, and they fulfill the problem.

17. To find two numbers such that their product, if it receives in addition their sum, or either one, makes a square.

Let the one be assigned to be 1 Number, and the other 4 Numbers lacking 1 unit, since if a number is the quadruple of a number lacking one unit, their product, if it receives in addition the smaller, makes a square. Next, we must also establish

---

33 Lit. "to the previous of it".
34 Text restored by Bachet, by analogy with the corresponding phrase of the previous problem.
35 There is at this point the following passage, which is clearly an interpolation: "because the 1 unit being subtracted from each of them, the lessening becomes 4 units and 1, and it is manifest that if a quadruple (ratio) is subtracted from a quadruple ratio, the (ratio which is) left will be also quadruple, that is (a ratio) of a square to a square".
36 Supplemented by Auria.

the other two proposed tasks, so that their product, if it receives in addition ⟨the second, will make a square, and moreover their product, if it receives in addition⟩[37] their sum, will make a square. But their product, when it receives in addition the second, becomes 4 Powers, 3 Numbers lacking 1 unit, (which is) equal to a square; while, when their product receives in addition their sum, becomes 4 Powers, 4 Numbers lacking 1 unit, (which is) equal to a square. And so, a double equality arises, and their difference is 1 Number, which is contained by a 4th of a unit and 4 Numbers; and the Number is found to be 65 224ths.

182   The one will be 65 (224ths), the other 36 (224ths), and they fulfill the problem.

18. To find two numbers such that their product, if it lacks either of them, or their sum, makes a square.

Let the one be assigned to be 1 Number, 1 unit, and the other 4 Numbers, since if a number is the quadruple of a number lacking 4 units, their product, if it lacks the greater, makes a square. Then, their product, if it lacks the smaller, must make a square, and moreover their product, if it lacks their sum, must make a square. But when their product lacks the smaller it becomes 4 Powers, 3 Numbers lacking 1 unit, while when their product lacks their sum (it becomes) 4 Powers lacking 1 Number, 1 unit. (The two are) equal to a square. Their difference is 4 Numbers. I assign one to be 4 Numbers, the other 1 unit, and the Number becomes 1 4′ units.

The one will be 2 4′ units, the other 5 units, and the proof is obvious.

19. To find four numbers such that the square of the (number) composed of all four, if it receives in addition any one (of them), or if it lacks (any one of them), makes a square.

Since in any right-angled triangle the square on the hypotenuse, if it receives in addition twice the (product of the sides) about the right angle, or if it lacks (it), makes a square, I search first for four right-angled triangles | having equal hypotenuses; this is the same as dividing a square into two squares ⟨in four ways⟩;[38] but we learned how to divide a given square into two squares in an unlimited number of ways.[39] So, now, let us set out two right-angled triangles (comprised) by the smallest numbers,[40] namely (the triangles) 3, 4, 5 and 5, 12, 13. Multiply each of the set out (triangles) by the hypotenuse of the other, and so the first triangle will be 39, 52, 65, and the other, 25, 60, 65. And they are right-angled, having equal hypotenuses. Further, the 65 is divided naturally into (two) squares in two ways: into 16 and 49, but, also, 64 and a unit. This is due to the fact that the number 65 is contained by the (numbers) 13 and 5, each of which is divided into two squares.[41] I take now the sides of the set out (squares) 49 and 16; they are 7 and

184

37 Text restored successively by Auria and Bachet.
38 τετραχῶς, added by Tannery; Bachet's edition has instead τετράκις, "four times".
39 See Problem II.8.
40 What is meant is that the numbers are the smallest for their respective proportions. That is, they are relatively prime.
41 See Section 3.3.4, where we discuss the propositions of al-Khāzin inspired by this problem.

4; and I form the right-angled triangle by means of the two numbers, 7 and 4, and it is (the triangle) 33, 56, 65. And likewise, (I take) the sides 8 and 1 of the (squares) 64 and a unit, and I form by means of them a right-angled triangle, the sides of which are 16, 63, 65. And so we have four right-angled triangles having equal hypotenuses.

So, coming back to the initial problem, I assign the (number) composed of the four (numbers) to be 65 Numbers, while each of these four to be of as many Powers as | the quadruple of the area, (namely,) the first ⟨4056 Powers, the second 3000 Powers, the third⟩[42] 3696 Powers, and, moreover, the fourth 2016 Powers. And so all four are 12,768 Powers, (which are) equal to 65 Numbers, and the Number becomes 65 of a part of 12,768 units.

To the hypostases: the first will be 17,136,600 (of a part), ⟨the second 12,675,000⟩[43] of the same part, the third 15,615,600 of the same part, and the fourth 8,517,600 of the same part, the part being 163,021,824.

20. To divide a given number into two numbers and to find a square which, if it lacks either of the (numbers) from the division, makes a square.[44]

Now, let the given (number) be 10 units.

Let the square to be found be assigned to be 1 Power, 2 Numbers, 1 unit. If it lacks 2 Numbers, 1 unit a square is left, and if (it lacks) 4 Numbers again a square is left. So, I assign the first to be 2 Numbers, 1 unit, and the other 4 Numbers. | These, if added together, must make the given (number). But when added together they become 6 Numbers, 1 unit. These are equal to 10 units, and the Number becomes 1 2′ units.

To the hypostases: the first will be 4 units, the second 6 units, and the square, 6 4′ units.

21. To divide a given number into two numbers and to find for them a square which, if it receives in addition each of the (numbers) from the division, makes a square.

Let the given (number) be 20 units.

And now let the square be assigned to be 1 Power, 2 Numbers, 1 unit. If I add to this 2 Numbers, 3 units, it becomes a square; but, also, if I add 4 Numbers, 8 units. Their sum, therefore, is 6 Numbers, 11 units. ⟨These are equal to 20 units, and the Number becomes 1 2′ units⟩.[45]

The first of the (numbers) from the division will be 6 units, the second 14 units, and the square 6 4′ units, and the proof is obvious.

186

188

---

42 Supplemented by Bachet.

43 Supplemented by Tannery.

44 Heath (1910, 167) notes: "This problem and the next are the same as II.15, 14 respectively. It may therefore be doubted whether the solutions here given are genuine, especially as interpolations from ancient commentaries occur most at the beginning and end of Books".

45 Sentence supplemented by Bachet.

Book IV (Arabic)

1   The fourth Book of the work of Diophantus of Alexandria, on squares and cubes,[1] translated from the Greek language to the Arabic language by Qusṭā ibn Lūqā of Baalbek, from the hand of Muḥammad ibn Abī Bakr ibn Ḥakīr[2] the astrologer[3] and written in the year five hundred ninety-five of the hegira.[4]

6   In the name of God, the Merciful, the Compassionate. The fourth Book of the work of Diophantus, on squares and cubes.

8   I have presented in the preceding parts of the book on arithmetical problems many problems in which we reached, after restoration and confrontation, one species Equals one species, which gave specifically the linear and the planar number, and also gave a pair of them.[5]

11   And I did this in stages through which the beginner can learn them and acquire knowledge of their meanings. I also find, lest you miss something which you could learn pertaining to this art, that I should also write for you in what follows many problems of the kind that deal with the species of number which is called "solid" and also what one obtains by composing it with the first two species.[6] And I will follow the same path, advancing you from one step to the next, and from

---

1 The phrase "on squares and cubes" applies to Book IV only, and not to the *Arithmetica* as a whole.

2 Sesiano (1982, 22 n. 6) notes: "Gulchīn-I Ma ʿānī, in his description, reads *Jāgīr* (with the Persian *gāf*) meaning the land obtained as a reward for services (its possessor being the *Jāgīr-dār*)". Rashed adopts the Persian reading.

3 Sesiano translates *al-munajjim* as "astronomer", while Rashed has "astrologue". Rashed provides more than three pages explaining this choice (Diophantus 1984, 3, 99–102). Overall, we agree with his evidence, though there can be no definitive translation for such a slippery term.

4 Lines 1–5 occupy the title page of the manuscript. The first line, in black ink, is shown in the manuscript in large, bold *kufic* letters, and the rest of the passage is written in red ink in *naskhī* script immediately beneath (Sesiano 1982, 24; Rashed in Diophantus 1984, 3, Photo 1 and p. LXIV). The treatise begins on the next page, with line 6. The text on the title page together with the explicit at line 3588 shows that the manuscript, covering only Books IV to VII, is not missing any leaves.

5 This covers the three simple equations. The equation $ax = b$ contains the linear number ax, $ax^2 = b$ contains the planar number ax^2, and $ax^2 = bx$ contains both of them.

6 The "solid" species is the Cube, and composing it with the first two species, Things and *Māl*s, give the higher powers *Māl Māl*s, *Māl* Cubes, etc.

DOI: 10.4324/9781315171470-11

one kind to the next, for the benefit of experience and practice. Once you have come to know what I have described, you will be in a position to find the answer to many problems which are not described here, since I have described for you the way to invent[7] many problems, and I have presented you with an example of each of their species.

I say that every *Māl*[8] multiplied by its side gives a Cube. So whenever you divide the Cube by the *Māl*, it results in a side of the Cube, and when divided by a Thing [which is a root of that *Māl*] it results in a *Māl*. When you multiply the Cube[9] by a Thing, the result is the same as what results from multiplying the *Māl* by itself, and that is called a *Māl Māl*. So when a *Māl* of the *Māl* is divided by a Cube, it results in a Thing [which is a root of the *Māl*] and when divided by a *Māl*, it results in a *Māl*, and when divided by a Thing [which is a root of the *Māl*] it results in a Cube. When a *Māl* of the *Māl* is multiplied by a Thing [which is a root of the *Māl*] the result is the same as what comes from multiplying the Cube by the *Māl*, and it is called a *Māl* Cube. When a *Māl* Cube is divided by a Thing [which is a root of the *Māl*] it results in a *Māl Māl*, and when divided by a *Māl*, it results in a Cube, and when divided by a Cube, it results in a *Māl*, and when divided by a *Māl Māl*, it results in a Thing [which is a root of the *Māl*]. And when a *Māl* Cube is multiplied by a Thing, the result is the same as what comes from multiplying the Cube by itself, and from multiplying the *Māl* by a *Māl Māl*, and it is called a Cube Cube. And when a Cube of the Cube is divided by a Thing [which is a root of the *Māl*], it results in a *Māl* Cube, and when divided by a *Māl*, results in a *Māl Māl*, and when divided by a Cube, results in a Cube, and when divided by a *Māl Māl*, results in a *Māl*, and when divided by a *Māl* Cube, results in a Thing [which is a root of the *Māl*].[10]

If, after the restoration and confrontation [– I mean by restoration the addition of what is lacking to both sides and by confrontation the casting away of what is equal from both sides] we end up with one species of these species – for which I have (already) described the duplication[11] of one with the other and the division of one by

20

35

---

7 We translate *fī wujūd* as "to invent". Lane (1893, 2985) gives this definition of *wujūd*, which seems to fit the present meaning best: "the finding, &c., by the intellect, or by means of the intellect". By writing "invent" we suggest that the Arabic word translates the Greek εὕρεσιν found also in the beginning of Book I. See page 276, footnote 1 on its meaning.

8 The text has "square" (*murabbaʿ*) in error here and also on lines 130, 467, and 816. /s/

9 The text has "cube" (*mukaʿʿab*). Out of 1834 instances of either "cube" or "Cube", the manuscript shows the wrong word 31 times (about 1.7%), 15 of them in the first 351 lines.

10 This entire paragraph, and perhaps also the following one at line 35, appears to have been composed specifically for a manuscript that was intended to start with Book IV since the rules for multiplying the powers were already given by Diophantus in the introduction to the work.
    If the phrases "which is a root of the *Māl*" are, as Sesiano proposes, interpolations, then they are interpolations to an interpolated passage.

11 The "duplication" (*taḍāʿīf*) of a number with an integer is the copying of that number as many times the integer. In Arabic mathematics this was conceived as a different operation from "multiplication" (*maḍrūb*). One cannot duplicate, say, a number with an irrational root, but one can multiply them. In the present context the two words can be interchanged.

the other – Equals another species, we should divide everything by one of the lesser in degree of the two sides so that it results in one species Equals a number.

41   **1.** We want to find two cubic numbers whose sum is a square number.

We assign a side of the smaller cube to be one Thing, so that its cube is one Cube. And we assign a side of the greater cube to be however many Things we wish, so we assign it to be two Things. Therefore, the greater cube is eight Cubes, and their sum is nine Cubes. We require that to Equal a square. We form the square from a side of however many Things we wish, so we form it from a side of six Things, so that it is thirty-six *Māl*s. Therefore, the nine Cubes Equal thirty-six *Māl*s. Since the side that has the *Māl*s is lesser in degree than the other side, we divide everything by one *Māl*. When you divide the nine Cubes by one *Māl*, it gives nine Things [which are nine roots of the *Māl*]. And for the thirty-six *Māl*s, when you divide ⟨them⟩ by one *Māl*, it results in a number, which is thirty-six units. Therefore, the nine Things [which are Roots] Equal thirty-six units, so the one Thing is equal to four units.

52   And since we assigned the ⟨smaller⟩ cube to be from a side of a Thing, its side is four, so the smaller cube is sixty-four. And since we assigned the greater cube to be from a side of two Things, ⟨its side⟩ is eight units. Therefore, the greater cube is five hundred twelve, and the sum of the cubes is five hundred seventy-six, which is a square from a side of twenty-four.

57   We have found two cubic numbers whose sum is a square. The smaller is sixty-four and the greater is five hundred twelve. And that is what we wanted to show.[12]

59   **2.** We want to find two cubic numbers whose difference is a square number.

60   We assign the smaller cube to be from a side of one Thing, so it is one Cube. And we assign a side of the greater to be however many Things we want, so let us assign it to be from a side of two Things, so that the greater cube is eight Cubes, and their difference is seven Cubes, and they Equal a square number. Let us assign a side of the square to be seven Things, so that it is forty-nine *Māl*s. Thus, the seven Cubes Equal forty-nine *Māl*s. The side that has the *Māl*s is the lesser in degree of the two sides, so we divide everything by one *Māl* which results in seven Things Equal forty-nine units. The one Thing then Equals seven units.

67   And because we assigned the smaller cube to be from a side of one Thing, it is three hundred forty-three. And a side of the greater, because it comes from two Things, is fourteen, and the greater cube is two thousand seven hundred forty-four. Their difference is two thousand four hundred one, which is a square whose side is forty-nine.

---

12 This is an error for "find". The same error occurs in the corresponding places in problems IV.2 through IV.6. See also line 3330 for a similar error. /s/

We have found two cubic numbers whose difference is a square number. And that is what we wanted to show.

71

3. We want to find two square numbers whose sum is a cubic number.

73

We assign the smaller square to be a *Māl*, and we assign the greater square to be four *Māl*s. Thus, the sum of the squares is five *Māl*s, and we require that to Equal a cubic number. Let us form it from a side of however many ⟨Things⟩ we wish. Let us again assign it to be from ⟨a side of⟩ one Thing, so that it is one Cube. Thus, the five *Māl*s Equal one Cube. Since the side that has the *Māl*s is the lesser in degree of the two sides, we divide everything by one *Māl* to get one Thing Equals five units.

74

Since we assigned the smaller square to be a *Māl*, and the *Māl*, since it comes from multiplying the Thing by itself, and the Thing was found to be five units, the *Māl* is twenty-five units. And since we made the greater square four *Māl*s, it is one hundred. The sum of the two squares is one hundred twenty-five, which is a cubic number, and its side is five units.

79

We have found two square numbers whose sum is a cubic number, and they are one hundred and twenty-five.[13] And that is what we wanted to show.

84

4. We want to find two square numbers whose difference is a cubic number.

86

We assign a side of the smaller square to be a Thing, and a side of the greater to be however many Things we wish. Let a side of this be five Things, so that the greater square is twenty-five *Māl*s and the smaller is one *Māl*. Their difference is twenty-four *Māl*s, and that Equals a cube. Let us assign the cube to be from a side of however many Things we wish. Let us assign it to be from a side of two Things. Then twenty-four *Māl*s Equal eight Cubes, since the cube that comes from the two Things is eight Cubes. We again divide everything by one *Māl*, to get eight Things Equal twenty-four units. Thus, the one Thing is three units.

87

Since we assigned the smaller square to be from a side of one Thing, and we assigned a side of the greater square to be five Things, a side of the smaller is three and a side of the greater is fifteen, and the smaller square is nine, and the greater square is two hundred twenty-five. Their difference is two hundred sixteen, which is a cubic number from a side of six units.

94

We have found two square numbers whose difference is a cubic number, and they are two hundred twenty-five and nine. And that is what we wanted to show.

98

5. We want to find two square numbers containing a cubic number.

100

We assign the smaller to be a *Māl* and the greater to be from a side of however many Things we wish. We assign it to be from a side of two Things. The greater

13 The text reads "which is one hundred twenty-five". /s/

square is four *Māl*s, and they contain four *Māl Māl*s, which Equal a cubic number. We assign the cube to be from a side of two Things, so that it is eight Cubes. Thus, four *Māl Māl*s Equal eight Cubes. We divide everything by a Cube to get eight units Equal four Things, since when you divide eight Cubes by a Cube it results in eight units. [And because the one (multiplied) by a Cube is a Cube, if one divides the Cube by the Cube it results in one.] And when you divide four *Māl Māl*s by a Cube it results in four Things. Thus, four Things Equal eight units, so the one Thing Equals two.

109   Since we made the smaller square a *Māl*, it is four units, since the *Māl* comes from multiplying the Thing by itself. And since we made the greater square four *Māl*s, it is sixteen. And the number contained by these two squares is sixty-four, which is a cube whose side is four units.

114   We have found two square numbers containing a cubic number, and they are the four and the sixteen. And that is what we wanted to show.

116   **6.** We want to find two numbers, a square and a cube, containing a square number.

117   We assign a side of the square to be whatever we wish of Things. We assign it to be one Thing, so that the square is one *Māl*. And again, we assign a side of the cube to be whatever we want of Things. We assign it to be two Things, so that ⟨the cube⟩ is eight Cubes. And what they contain – namely, the *Māl* and the eight Cubes – are eight *Māl* Cubes, and that Equals a square. If we were to assign a side of the square to be Things, it would amount to ⟨*Māl*s, giving *Māl* Cubes Equal⟩ *Māl*s. We would then need to divide the two sides by the *Māl*s, which would then give Cubes Equal units [since, as I mentioned, whenever you divide Cubes of *Māl*s by the *Māl*s, it results in Cubes].[14] So we assign a side of the square to consist of *Māl*s, however many we wish. So we assign it to consist of four *Māl*s, so that the square is sixteen *Māl Māl*s. Thus, eight Cube *Māl*s Equal sixteen *Māl Māl*s.

126   We divide everything by a *Māl Māl*, since it is the lesser in degree of the two sides. When you divide sixteen *Māl Māl*s by a *Māl Māl*, it results in sixteen units. And for eight Cube *Māl*s, when you divide that by a *Māl Māl*, it results in eight Things. So eight Things Equal sixteen units; thus, the Thing is two.

130   Since we assigned a side of the square[15] to be a Thing, the square is four units. And the cube, since we assigned it to be from a side of two Things, is sixty-four. The number they contain – that is, the square which is four and the cube which is sixty-four – is two hundred fifty-six, which is a square whose side is sixteen units.

134   We have found two numbers, one of them a square and the other a cube, which contain a square number, and they are the four and the sixty-four. And that is what we wanted to show.

---

14 We believe this passage to be an interpolation.
15 The text has "the *Māl*".

7. We now want to find two numbers, one of them a square and the other a cube,   136
which contain a cubic number.

We assign a side of the square to be a Thing, so the square is a *Māl*. And we   138
assign a side of the cube to be whatever we wish of Things. We assign it to be
four Things, so that the cube is sixty-four Cubes. They contain sixty-four Cube
*Māl*s, and they Equal a cubic number. If we were to assign a side of the cube to
consist of Things, then the cube would consist of Cubes, and if we were to
Equate it with Cube *Māl*s, we would need to divide everything by a Cube, which
would then give us *Māl*s Equal units. We would then need the units ⟨that Equal
the one *Māl*⟩ to be a square. But if we were to assign it to consist of *Māl*s, then
the cube would consist of Cube Cubes. If we were to Equate that with Cube
*Māl*s, then we would need to divide ⟨the two sides⟩ by a Cube *Māl*, which would
then give us Things Equal units. So we assign a side of the cube to consist of
two *Māl*s, so the cube is eight Cube Cubes. Thus, eight Cube Cubes Equal sixty-
four Cube *Māl*s.

We divide everything by a Cube *Māl*, since it is the lesser in degree of the two   148
sides. Our result from dividing eight Cubes of the Cube by a Cube *Māl* is eight
Things, and from dividing the sixty-four *Māl* Cubes by a *Māl* Cube is sixty-
four units. Thus, eight Things Equal sixty-four units, so the one Thing is eight
units.

Since we assigned a side of the square to be a Thing, it is sixty-four. And the cube,   152
since we assigned its side to be four Things, its side is thirty-two, and the cube is
thirty-two thousand seven hundred sixty-eight. If we multiply that by the square,
which is sixty-four, that amounts to a cubic number, since each of them is a cube.

We have found two numbers according to the stipulation that we stipulated. And   157
that is what we wanted to find.

8. We want to find two cubic numbers containing a square number.   159

If, in this problem, we were to again assign a side of the smaller cube to be one   160
Thing, then the smaller cube would be one Cube. ⟨And if⟩ we were to assign the
greater to be from a side of however many ⟨Things⟩ we wish, for instance if we
were to assign it to be from a side of two Things, then the greater cube would be
eight Cubes. These contain eight Cube Cubes, and we would require that to be
equal to a square. It is not correct for us to assign a side of this square to be Things,
since a square of Things is *Māl*s, and if you were to Equate (them) with Cube
Cubes, and you were to divide by the lesser in degree of the two sides, which is
*Māl*s, it would result in a *Māl Māl* Equals units. ⟨But if we were to assign a side
of the square to be *Māl*s, then the square would be *Māl Māl*s. Then if we were to
Equate that with Cube Cubes, we would need to divide the two sides by a *Māl*
Māl, to then get *Māl*s Equal units⟩. We thus require that the units Equated to the
one *Māl* be a square. Because of that, we are led to look for a square and a cubic
number containing a square number [which will be easier to work out]. In the

same way as mentioned above,[16] we find one of the two numbers, which is the square, to be four, and the other, which is the cube, to be sixty-four. These two numbers contain two hundred fifty-six, which is a square whose side is sixteen units. And that is what we wanted to find.

176   **9.** We want to find two cubic numbers containing a square.

We assign a side of the greater cube to be four Things, and a side of the smaller cube to be one Thing. So the greater cube is sixty-four Cubes and the smaller cube is one Cube. These contain sixty-four Cube Cubes, and we require that to be Equal to a square number. We assign a side of the square to consist of *Māl*s, so its number[17] is equal to a side of the square that comes from multiplying the sixty-four by the four, which is two hundred fifty-six. ⟨(This has) a side of sixteen units. Thus, we assign a side of the square to be⟩ sixteen *Māl*s, so that its square is two hundred fifty-six *Māl Māl*s. Thus, sixty-four Cube Cubes Equal two hundred fifty-six *Māl Māl*s.

185   We divide everything by a *Māl Māl*, since it is the lesser in degree of the two sides. When you divide the sixty-four Cube Cubes by a *Māl Māl*, it results in sixty-four *Māl*s. And when we divide the two hundred fifty-six *Māl*s of the *Māl* by a *Māl Māl*, it results in two hundred fifty-six units. Thus, the sixty-four *Māl*s Equal two hundred fifty-six units, so the one *Māl* Equals four units. The *Māl* is a square and the four is a square, so their sides are equal. And a side of the *Māl* is a Thing, and a side of the four is two, so the Thing is two.

191   Since we assigned a side of the smaller cube to be one Thing, the ⟨smaller⟩ cube is eight units. And since we assigned a side of the greater cube to be four Things, which is eight units, the greater cube is five hundred twelve. When we multiply it by the smaller cube, it amounts to the number they contain, which is four thousand ninety-six, which is a square, and its side is sixty-four.

197   We have found two cubic numbers containing a square number, and they are the eight and the five hundred twelve. And that is what we wanted to find.

199   If we want to find a cubic number such that if we divide it by a cube it results in a square number, we search for a square number such that if we multiply it by the other cubic number that we also search for, then what is gathered from the multiplication is a cubic number. If we find that, then what is gathered from multiplying one of them by the other is the cubic number that we want.

203   Similarly, if we want to find a square number such that if we divide it by a square it results in a cube, we work it out in reverse order from the preceding. And anything we search for by means of division of the kind just mentioned is done similarly, because it is the same, since division is nothing but the inverse of multiplication.

---

16 In Problem IV.6. /sr/
17 I.e., its multitude, or what we call its coefficient.

10. We want to find a cubic number such that if we add to it the same as[18] the   207
square that comes from its side however many times we wish, amounts to a square
number.

We assign the cube to be from a side of one Thing, so it is one Cube. And we   209
assign the number of times to be ten, and we add to a Cube ten times ⟨a square⟩
of a side of the cube, which is a *Māl*. This gives a Cube and ten *Māl*s, and that
Equals a square. We assign that square to be from a side of Things, so its square
is more than ten *Māl*s, in order to make the confrontation possible.[19] We assign
that to be from a side of four Things, so the square is sixteen *Māl*s. Thus, the
Cube and ten *Māl*s Equal sixteen *Māl*s. Let us cast away the common ten *Māl*s,
leaving six *Māl*s Equal a Cube. We divide that by a *Māl*, resulting in one Thing
Equals six units.

Thus, the Cube is two hundred sixteen, and a square of the side is thirty-six,   216
and ten times that is three hundred sixty. We add it to the Cube. What is gathered from that is five hundred seventy-six, which is a square from a side of
twenty-four.

We have found a cubic number such that if we add to it ten times the square that   220
comes from its side, becomes, after the addition, a square number. It is two hundred sixteen, and its side is six. And that is what we wanted to find.

11. We want to find a cubic number such that if we subtract from it the same as   223
the square that comes from its side however many times we wish, leaves a square
number.

We assign the cube to be from a side of one Thing, so it is one Cube. And we   225
assign the number of times to be six. We want the remainder from the Cube,[20]
after subtracting the six *Māl*s, to be a square. We assign the square to be from a
side (consisting) of Things, however many we wish. We assign it to be from a
side of two Things, so that its square is four *Māl*s. Thus, the Cube less six *Māl*s
Equals four *Māl*s.

We restore the Cube by the six *Māl*s, and we add them to the four *Māl*s. So one   229
Cube Equals ten *Māl*s. We divide everything by a *Māl*, resulting for us in one
Thing Equals ten units.

---

18 For the modern reader, this could be simplified by omitting "the same as" (*mithl*). Arabic authors
    often included *mithl* when the number in question is calculated from another number already in play.
    In another example from Problem VII.8 we read: "And every square number, if we add the same as
    two of its roots and one". Here the "two of its roots" comes from the "square number", but because
    the "square number" must remain intact, another new number *the same as* (or, equal to) two of its
    roots must be declared.

19 I.e., in order that confronting the two sides to set up, simplify, and solve the equation results in a
    (positive) solution.

20 We follow Sesiano, who corrects "we suppose" (*nafriḍ*) to "remainder from" (*yabqā min*), and
    "cube" to "Cube". Rashed leaves the text as it is and translates it as: "Nous voulons supposer que
    le cube" (Diophantus 1984, 3, 14.8).

231   Since we assigned a side of the cube to be one Thing, it is one thousand, while a square of the side is one hundred. Six times the hundred is six hundred, which leaves from the thousand, after ⟨subtracting⟩ six hundred, four hundred, which is a square number whose side is twenty.

235   We have found a cubic number such that if we subtract from it the same as a square of its side six times, leaves a square number. It is one thousand, and its side is ten.

237   **12**. We want to find a cubic number such that if we add to it the same as the square that comes from its side however many of times we wish, then what is gathered from it is a cubic number.

239   We assign a side of the cube to be one Thing, so that it is one Cube. And we add to it the number of times (of the *Māl*) we want, which is what we assigned before,[21] to get a Cube and ten *Māl*s, which Equal a cube. We form the cube from a side of two Things, so that eight Cubes Equal one Cube and ten *Māl*s.

242   We cast away the common Cube, leaving ten *Māl*s Equal seven Cubes. We divide that by a *Māl* to get seven Things Equal ten units. So the one Thing is ten sevenths, and the Cube is one thousand of the quantity, which is a seventh of a seventh of a seventh. When we add to it ten times the square, which is one hundred sevenths of a seventh, that is, seven thousand sevenths of a seventh of a seventh, that amounts to eight thousand sevenths of a seventh of a seventh, which is a cube from a side of twenty sevenths.

249   We have found a cube that exhibits the stipulation that we stipulated, and it is one thousand sevenths of a seventh of a seventh, with side of ten sevenths. And that is what we wanted to find.

251   **13**. We want to find a cubic number such that if we subtract from it the same as the square that comes from its side however many times we wish, leaves a cubic number.

253   We assign the cube to be from a side of a Thing, so that it is one Cube. And we assign the number of times to be seven, so that what remains is a Cube less seven *Māl*s, and that Equals a cubic number. We assign a side of the cube to be some portion[22] of a Thing, so we assign it to be from a side of half a Thing, so that the cube is a part of eight parts of a Cube, and that Equals a Cube less seven *Māl*s.

257   We restore and confront, leaving seven eighths of a Cube Equal seven *Māl*s. We divide everything by a *Māl*, resulting in seven units Equal seven eighths of a Thing.

---

21 In Problem IV.10. /s/

22 The text uses the word *baʿḍ* and not the usual word for "part", which is *juzʾ*. Had he written "some part of a Thing" it would be interpreted not as a fraction of a Thing but as some multitude of the reciprocal of a Thing.

Thus, the one Thing is eight units, and the cube is five hundred twelve. When we subtract seven times the sixty-four from it, it leaves sixty-four, which is a cube.

And we can work that out in another way. We make a side of the first cube be however many Things we wish. We make it two Things, so that the cube is eight Cubes. This leaves [the difference between the Cube and the eight Cubes as][23] seven Cubes Equal seven times the square that comes from a side of the greater cube, and a side of the greater cube is two Things, and its square is four *Māl*s, and seven times that is twenty-eight *Māl*s. So twenty-eight *Māl*s Equal seven Cubes. We divide everything by a *Māl* to get twenty-eight units Equal seven Things, and the one Thing Equals four units.

Therefore, the smaller cube is sixty-four [and its side was assigned to be one Thing]. For the greater cube, since its side was assigned to be two Things, its side is eight units, so the cube is five hundred twelve. It is clear that the other cube, which is the greater, exceeds the smaller cube by seven times the square from a side of the greater cube, and that is the stipulation that we stipulated in this problem. And that is what we wanted to find.

14. We want to find a number such that if we multiply it by two given numbers, then one of them is a cube and the other is a square.

We assign the two numbers to be five and ten, and we want to find a number such that if we multiply it by the ten, it gives a cube, and if we multiply it by the five, it gives a square.

We assign the sought-after number to be a Thing, and we multiply it by the five to get five Things. Then we multiply it by the ten to get ten Things. And we want to Equate the ten Things with a cubic number, and the five Things with a square number. We assign the square Equated to five Things to be whatever part we wish or whatever parts we wish[24] of a square of a side of the cube Equated to the ten Things; furthermore, that a side of the part be commensurable to a side of the whole, which is to say that the part is a square. Or, we assign a square of a side of the cube to be whatever part or whatever parts of the square Equated to the five Things, and furthermore, that it be a square.

Let us assign a square of a side of the cube to be a fourth of the square Equated to the five Things. Then a square of a side of the cube Equated to the ten Things is one Thing and a fourth of a Thing. So this square, which is the one Thing and a fourth of a Thing, if we multiply it by its side, gives ten Things. If we then divide the ten Things by the Thing and a fourth of a Thing, then what results is a side of the cube Equated

262

269

275

277

278

286

---

23 Sesiano notes (1982, 30): "This addition by an Arabic reader was occasioned by a lacuna in the text. Observe that the whole of the second part of problem IV.13 is confused; this confusion may quite possibly antedate, at least in part, the translation into Arabic".

24 For "part" and "parts" see Euclid's *Elements*, definitions VII.3 and VII.4, and also Diophantus's Problem IV^G.33. /s/

to the ten Things. And what results from dividing the ten Things by the Thing and a fourth ⟨of a Thing⟩ is eight units [since the Things, if you multiply (them) by the units, give Things]. So the eight units are a side of the cube Equated to the ten Things, which is a side of the square Equated to the one Thing and a fourth of a Thing. And the cube which comes ⟨from a side⟩ of eight units is five hundred twelve, and that Equals ten Things. So the one Thing is fifty-one and a fifth. Likewise, a square of the eight units is sixty-four, and that Equals one Thing and a fourth of a Thing. Thus, the one Thing is four fifths of the sixty-four, which is fifty-one and a fifth. When we multiply the fifty-one and a fifth by the ten, it amounts to five hundred twelve, which is a cubic number. And when we multiply it by the five, it gives two hundred fifty-six, which is a square whose side is sixteen.

302   We have found a number such that if we multiply it by two given numbers, which are the ten and the five, then it gives a cubic number from its multiplication by the ten, and it gives a square number from its multiplication by the five, which is what we wanted to find.

305   Suppose we want what is gathered from multiplying the Thing by the five to be the cube, and from its multiplication by the ten to be the square number. Again, we Equate five Things with a cubic number, and we Equate ten Things with a square number.[25] We make the square number that is from a side of the cube Equated to the five Things again be a fourth ⟨of the square⟩ Equated to the ten Things, so that a square of a side of the cube Equated to the five Things is two Things and half a Thing.[26] If we divide the five Things by it, what results from that is a side of the cube Equated to the five Things. But when we divide the five Things by the two Things and a half ⟨of a Thing⟩, then what results is two. And if a side of the cube Equated to the five Things is two, then the cube that is Equated to the five Things is eight units, and the one Thing is eight fifths of a unit. When we multiply it by the five it amounts to forty fifths, that is, eight units, which is a cubic number. And when we multiply it by the ten it amounts to eighty fifths, that is, sixteen units, which is a square whose side is four.[27]

318   Let us assign in the first problem that[28] the square that is Equated to the five Things be in the ratio of a fourth with respect to the square that comes from a side of the cube Equated to the ten ⟨Things⟩, so that a square of a side of the cube Equated to the ten Things is twenty Things. Then when you divide the ten Things

---

25 Lit. "five Things we Equate them with a cubic number, and ten Things we Equate them with a square number".

26 The text has at this point what appears to be the last line copied again, a little sloppily, by mistake by a copyist: "Thus ⟨a square⟩ of a side of the cube Equated to the five ⟨Things⟩ is two Things and half a Thing". The two restorations are due to Rashed, who keeps the passage in his edition. /s/

27 The text has "(a) square", which Sesiano corrects to "four" (line 317). Rashed keeps "(a) square".

28 The text reads here: "Let us assign in the first problem that [a square of a side of the cube of] the square". Sesiano regards what we write in brackets as an error, and he deletes it. Rashed deletes only "the cube of", and he notes: "Il est probable que l'espression 'carré du côté du carré' sert ici à conserver la symétrie avec 'carré du côté du cube', et pour cette raison nous l'avons gardée".

by the twenty Things, the quotient is a half, and that is a side of the cube Equated to the ten Things.

And the cube that comes from the half is an eighth of a unit. Thus, the ten Things Equal an eighth of a unit, so the one Thing is a part of eighty parts. When we multiply it by the five, it amounts to five parts of eighty, that is, one part of sixteen, which is a square whose side is a fourth of a unit. And when multiplied by the ten it gives ten parts of eighty, that is, an eighth of a unit, which is a cube whose side is half of a unit.

If we assign in the reversed problem[29] that the square that is Equated to the ten Things be ⟨in the ratio of a fourth⟩ of a square of a side of the cube Equated to the five Things, ⟨then a square of a side of the cube Equated to the five Things⟩ is forty Things. When we divide the five Things by it, it results in a part of eight parts of a unit. And a side of the cube Equated to the five Things is an eighth of a unit, and the cube is one part of five hundred twelve, and the five Things Equal one part of five hundred twelve. And the one Thing equals one part of two thousand five hundred sixty. So when we multiply it by the ten it gives ten parts of two thousand five hundred sixty, that is, one part of two hundred fifty-six, which is a square from a side of one part of sixteen. And when we multiply it by the five, it gives five parts of two thousand five hundred sixty, that is, one part of five hundred twelve, which is a cube from a side of an eighth of a unit.

We have found a number such that if we multiply it by each of the ten and the five, it gives a square number and a cubic number.

And we can work it out again another way. We assign the cube that is gathered from multiplying the sought-after number by the ten to be from a side of however many Things we wish. Let us assign it to be from a side of a Thing, so that it is one Cube and the sought-after number is a part of ten parts of a Cube. We require this to be the part which, if multiplied by the five, amounts to a square number. But when we multiply one part of ten parts of a Cube by five units, it amounts to five parts ⟨of ten parts of a Cube⟩, that is, half a Cube, and that is Equal to a square number. Let us form the square from a side of however many Things we wish. Let us assign it to be from a side of two Things, so that it is four *Māls*. Then half of the Cube Equals four *Māls*. We divide everything by a *Māl*, to get half a thing Equals four units, so the one Thing is eight units.

And since we assigned the cube that is gathered from multiplying the sought-after number by the ten to be from a side of one Thing, its side is eight units and the cube is five hundred twelve. We divide the five hundred twelve by ten. The result of this is the sought-after number, which is fifty-one and a fifth.

And if we wish, we can assign a side of the square that is gathered from multiplying the sought-after number by the five to be however many Things we wish. Let

323

328

341

343

353

358

---

29 I.e., with the square and the cube reversed.

us assign it to be from a side of one Thing, so that it is one *Māl*. So the sought-after number is one part of five parts of a *Māl*. When we multiply it by the ten units, it amounts to ten parts of five parts of a *Māl*, that is, two *Māl*s, and that Equals a cubic number. We assign the cube to be from a side of however many Things we wish. Let us assign it to be from ⟨a side⟩ of one Thing, so that it is one Cube. Thus, two *Māl*s Equal one Cube. We divide everything by a *Māl* to get one Thing Equals two units.

365   And since we assigned the square to be from a side of one Thing, its side is two units and it is four units. So the sought-after number, if we multiply it by the five units, amounts to four units. Thus, the sought-after number is four-fifths. When we multiply it by the five, it amounts to twenty fifths, ⟨that is⟩, four units, which is a square. And when we multiply it by the ten, it amounts to forty fifths, that is, eight units, which is a cube.

371   We have found a number such that if we multiply it by the ten and by the five, amounts to a square number and a cubic number.[30]

373   **15.** We want to find a number such that if we multiply it by two given numbers, gives from its multiplication by one of them a cubic number, and from the other the square that comes from a side of that cube.

376   Let one of the given numbers be four and the other ten. We want to find a number such that if we multiply it by the ten, it gives a cubic number, and if we multiply it by (the) four, it amounts to the square that comes from a side of the cube, or vice versa: the procedure is the same.

379   The rule from the previous problem[31] is that we assign the ⟨sought-after⟩ number to be a Thing, so the cube is ten Things, and the square that comes from its side is four Things; and that a side of the cube, if multiplied by itself, amounts to four Things, and the whole cube is ten Things. Since, if you multiply the four Things by its side it amounts to ten Things, we divide the ten Things by the four Things, so a side of the cube is two and a half, the square of which is six units and a fourth. Thus, the four Things are Equal[32] to six units and a fourth. We multiply all of that by four because of the part, which is the fourth, to get sixteen Things Equal twenty-five units; thus, the one Thing is twenty-five parts of sixteen parts.

388   According to the method in the second procedure,[33] we make the cube that comes from multiplying the sought-after number ⟨by the ten to be from a side of however many Things we wish. We assign it to be from a side of one Thing, making it one Cube. Thus, the sought-after number is⟩ a tenth of the Cube,[34] and what is

---

30 The text has "a cubic number and a square number". /s/
31 Lit. "And its rule in the preceding", which is the rule at line 277. /s/
32 Text has "equal" (*musāwiya*).
33 The rule in Problem IV.14 starting at line 318. /s/
34 Rashed's reconstruction (Diophantus 1984, 3, 24.3) of this passage translates to "According to the second procedure, we make ⟨the sought-after number from⟩ the cube which comes from multiplying the sought-after number ⟨by the ten⟩, a tenth of the Cube".

gathered from its multiplication by the four is four tenths of a Cube. Thus, four tenths of a Cube Equal a square of a side of the cube, which is a Thing, and its square is one *Māl*.

We multiply everything with it[35] by ten because of the part, which is the tenth, to get four Cubes Equal ten *Māl*s. We then divide everything by a *Māl* to get four Things Equal ten units, so the Thing Equals two units and a half, and[36] a square of a side of the cube is six units and a fourth. Therefore, the sought-after number is twenty-five parts of sixteen parts. It is clear that this number, when multiplied by the four, amounts to one hundred parts of sixteen parts, which is a square, and when multiplied by the ten, amounts to two hundred fifty parts of sixteen. And it is clear that the two hundred fifty ⟨parts⟩[37] of the sixteen are fifteen units and a half and an eighth, which is a cube, and its side is two and a half, and its square is six units and a fourth. Similarly, when you multiply the twenty-five parts of sixteen by four, it gives one hundred parts of sixteen, which is six units and a fourth, which is a square whose side is two and a half.

393

We have found a number such that if we multiply it by two given numbers, then the result ⟨from multiplying it by⟩[38] one of them is a cubic number, and from multiplying it by the other is the square that comes from a side of that cube.

406

And if we want to find two numbers such that one of them is with respect to the other in a given ratio, and one of them is a cubic number and the other number is a square, we assign the ratio to be the ratio of three to one. We first assign two numbers so that one of them is three times the other. Then, following the same procedure as before, we look for a number such that if we multiply it by each of the two given numbers amounts to a square number and a cubic number.

409

Thus, we have found two numbers in the ratio of three times: one of them is a cube and the other is a square, since for any number multiplied by two numbers, what is gathered from the multiplications are in (the same) ratio as the original numbers.[39]

414

16. We want to find two numbers such that if we multiply them by a given number, what is gathered from one of them by it is a cubic number, and from the other is a side of that cube.

417

---

35 The phrase "everything with it" is a standard phrase in Arabic algebra meaning all the other terms in the equation.

36 The text has "side" at this point, which Rashed regards as an error and omits from his translation. Sesiano sees "side" as the beginning of a lost phrase. Our translation of his reconstruction is "and ⟨a side of the cube is also two units and a half. And four times the sought-after number, which Equals⟩ a square of a side of the cube, is six and a fourth".

37 Rashed's restoration.

38 Restoration by both Sesiano and Rashed, though Sesiano places it in the apparatus.

39 Euclid's *Elements*, Proposition VII.17. /s/

419   We assign the number to be ten, and we want to find two numbers such that if we multiply them by ten, then what is gathered from multiplying the ten by one of them is a cubic number, and if we multiply the ten by the other, then what is gathered is a side of that cube.

421   Let us assign the first number to be however many Things we wish, so let us assign it to be one Thing, and we multiply it by ten to get ten Things, which is a side of the cube. Therefore, the cube that is gathered from multiplying the second number by the ten is one thousand Cubes. And we assign the second number to be however many *Māl*s we wish, so let us assign it to be three hundred *Māl*s, and we multiply it by the ten to get three thousand *Māl*s. Thus, one thousand Cubes Equal three thousand *Māl*s. We divide everything by one *Māl* to get: one thousand Things Equal three thousand units. Therefore, the one Thing is three units.

428   Since we assigned the first number to be one Thing, it is three units. And since we assigned the second number to be three hundred *Māl*s, and the *Māl* is nine, it is two thousand seven hundred. When we multiply the second number by the ten it gives twenty-seven thousand, and when we multiply the first number by the ten, it amounts to thirty. And the thirty is a side of the cube which is twenty-seven thousand.

433   We have found two numbers according to the stipulation that we stipulated, and they are the three and the two thousand seven hundred. And that is what we wanted to find.

435   **17.** We want to find two square numbers whose sides are in a given ratio, and if each of them is multiplied by a given number, then what is gathered from one of them is a cube, and from the other is a side of that cube. It is necessary that the number that is for the given ratio together with the given number contain a square number commensurable with the second, which is called *plasmatikon*.[40]

440   Let the given ratio be the ratio of twenty times, and the given number be five units. We want to find two square numbers such that a side of one of them is with respect to a side of the other in the ratio of twenty times, and if the greater square is multiplied by five units, it amounts to a cubic number, and if the smaller square is multiplied by the five, then it amounts to a side of that cube.

---

40 For the phrase "commensurable with the second, which is called *plasmatikon*" we follow Rashed's edition. The reading is problematic since we cannot find an explanation for the phrase "commensurable with the second". See Diophantus (1984, 3, 133–38) for Rashed's discussion. The Arabic word *al-muhayya'at* translates the Greek πλασματικόν, which takes the meaning "arises with the formation (of the problem)". Rashed translates the word as "convenablement déterminé". In the two other problems in which *al-muhayya'at* occurs, IV.19 and V.7, we translate *hādhihī min* al-*al-muhayya'at* as "this (kind of restriction) arises with the formation of the problems". Sesiano proposes his own reconstruction of the passage (1982, 99), which he translates as: "These (problems), from their feasibility, are those called the 'constructible' ones", and he explains his interpretation of *al-muhayya'at/plasmatikon* in (1982, 99 n. 48; 192). See note 40 to Book I on the use of the word in the Greek books.

We assign a side of the smaller square to be one Thing, so that it is one *Māl*. Thus,   444
a side of the greater square is twenty Things, and the greater square is four hundred *Māl*s. We multiply the four hundred *Māl*s by five to get two thousand *Māl*s,
and we multiply the one *Māl* by five units to get five *Māl*s. And since a stipulation
of this problem is that the two thousand *Māl*s is a cube of a side of the five *Māl*s,
we multiply the five *Māl*s by itself, then by itself, to get one hundred twenty-five
Cube Cubes. Then the one hundred twenty-five Cube Cubes Equal two thousand
*Māl*s.

We divide everything by one of the lesser in degree of the two sides, namely, the   451
Māl. Thus, one hundred twenty-five *Māl Māl*s Equal two thousand units. Therefore, the one *Māl Māl* Equals sixteen units. A *Māl* of a *Māl* is a square with square
side, and similarly the sixteen units is a square number with square side. They are
equal, so the sides of their sides are also equal, and a side of a side of the *Mal Māl*
is one Thing, and a side of a side of the sixteen is two units. Therefore, the one
Thing Equals two units.

Since we formed the smaller square from a side of one Thing, it is four units. And   456
since we formed the greater square from (a side of) twenty Things, its side is forty
units, and it is one thousand six hundred. And when we multiply the one thousand
six hundred by the given number, which is five units, it gives eight thousand,
which is a cube from a side of twenty units. This is what is gathered from multiplying the smaller square, which has been shown to be four units, by the given
number, which is five units.

We have found two numbers according to the stipulation that we stipulated, and   463
they are the four and the one thousand six hundred. And that is what we wanted
to find.

18. We want to find two cubic numbers whose sides are in a given ratio, and if   465
each of them is multiplied by a given number, what is gathered from one of them
is a square, and from the other is a side of that square. It is necessary that the given
number be a cube.

Let the given ratio be the ratio of three times, and the given number be eight units.   469
We want to find two cubic numbers such that a side of one of them is with respect
to a side of the other in the ratio of three times, and the multiplication of the
greater by the eight units is a square number, and the multiplication of the smaller
by the eight units is a side of that square.

We assign the smaller cube to be from a side of one Thing; thus, it is one Cube,   473
and a side of the greater cube is three Things, so it is twenty-seven Cubes. When
we multiply the twenty-seven Cubes by eight units, it gives two hundred sixteen
Cubes, and when we multiply the one Cube by eight units, it gives eight Cubes.
Since the two hundred sixteen Cubes is a square from a side of the eight Cubes,
which is sixty-four Cube Cubes, the two hundred sixteen Cubes Equal sixty-four
Cube Cubes.

479   When we divide them by the one Cube, which is the lesser in degree of the two sides, it gives two hundred sixteen units Equal sixty-four Cubes. Therefore, the one Cube is three units and three eighths of a unit, and the one Cube is a cube from a side of one Thing, and the three units and three eighths of a unit is a cube from a side of one and a half. Thus, the one Thing Equals one and a half.

484   Therefore, the smaller cube is three units and three eighths of a unit, and the greater cube, whose side is four units and a half, is ninety-one and an eighth. And when this greater cube is multiplied by the eight units, it amounts to seven hundred twenty-nine, which is a square from a side of twenty-seven units, which is what is gathered from multiplying the smaller cube, which was shown to be three and three eighths, by the given number, which is eight units.

491   We have found two numbers according to the stipulation that we stipulated. And that is what we wanted to find.

493   **19.** We want to find a number such that if we multiply it by two given numbers, gives from its multiplication by one of them a cube, and from its multiplication by the other a side of that cube. It is necessary that the two given numbers contain a square number. And this (kind of restriction) also arises with the formation of the problems.

497   Let one of the given numbers be five units and the other twenty units. We assign the sought-after number to be one Thing. We multiply it by the five units to get five Things, and we likewise multiply it by the twenty to get twenty Things. Since the twenty Things is a cube from a side of five Things, and a side of any cube, if multiplied by its square, gives the cube, and the cube is twenty Things, then when you divide it by its side, which is five Things, it gives a square of a side of twenty Things. But when you divide the twenty Things by the five Things, it gives four units. So the four units is a square from a side of five Things. So a side of the four, which is two, Equals five Things. Thus, the one Thing is two fifths of a unit.

506   When we multiply it by the twenty it gives eight units, which is a cube from a side of two units, which is what is gathered from multiplying ⟨the sought-after number⟩, which was shown to be two fifths of a unit, by the other given number, which is five units.

510   We have found a number such that if we multiply it by the two given numbers, one of which is five units and the other twenty units, gives from its multiplication by the twenty a cube, and from its multiplication by the five a side of that cube, and it is two-fifths of a unit. And that is what we wanted to find.

514   **20.** We want to find a cubic number such that if we multiply it by two given numbers, gives from its multiplication by one of them a square, and from its multiplication by the other a side of that square. It is necessary that a square of one of the given numbers count the other number by a cubic number.

Let one of the given numbers be five units and the other number be two hundred   518
units. We want to find a cubic number such that if we multiply it by the two hundred, it amounts to a square, and if we multiply it by the five units, it amounts to
a side of that square.

We assign the sought-after cube to be from a side of one Thing, so that it is one   521
Cube. And we multiply it by the two hundred to get two hundred Cubes. And ⟨we
also multiply it by the five to get⟩ five Cubes. Since two hundred Cubes is a square
from a side of five Cubes, and every square divided by its side gives a result equal
to its side, and when you divide two hundred Cubes by five Cubes it gives forty
units; thus, the five Cubes Equal forty units. Therefore, the one Cube Equals eight
units, and the one Cube is a cube from a side of one Thing, and the eight is a cube
from a side of two units. So the Thing we assigned as a side of the sought-after
cube is two units, and the cube is eight units.

When we multiply it by the two hundred, it amounts to one thousand six hundred,   529
and when we multiply it by the five, it amounts to forty, which is a side ⟨of the
square⟩, which is one thousand six hundred.

We have found a cubic number such that if we multiply it by the two given num-   532
bers, which are the two hundred and the five units, gives from its multiplication
by the two hundred a square, and from its multiplication by the five a side of that
square, which is eight units. And that is what we wanted to find.

21. We want to find a square number such that if we multiply it by two given   536
numbers, gives from its multiplication by one of them a cube, and from its multiplication by the other a side of that cube. It is necessary that the given numbers
contain a square number whose side is a ⟨square⟩.

Let one of the given numbers be two and the other number be forty and a half. It   540
is clear that the planar number that is contained by these two numbers, which is
eighty-one, is a square with a square side. We want to find a square number such
that if we multiply it by the forty and a half and by the two, gives from its multiplication by the forty and a half a cube, and from its multiplication by the two a
side of that cube.

We assign the square to be one *Māl*, and we multiply it by each of the given num-   544
bers so that one of the products is forty *Māl*s and half a *Māl*, and the other product
is two *Māl*s. Since the forty *Māl*s and half a *Māl* is a cube from a side of two
*Māl*s, and every ⟨cube⟩ divided by its side gives a square of that side, and when
you divide the forty *Māl*s and half a *Māl* by two *Māl*s it gives the quotient twenty
and a fourth, then a square of the two *Māl*s Equals twenty and a fourth. A side of
the twenty and a fourth is four units and a half, so the one *Māl* Equals two units
and a fourth, which is a square from a side one and a half. And when this square,
which is two and a fourth, is multiplied by one of the given numbers, which is
forty and a half, it gives ninety-one and an eighth, which is a cube from a side of
four units and a half. This is what is gathered from multiplying the sought-after

square number, which was shown to be two and a fourth, by the other given number, which is two.

557 We have found a square number according to the stipulation that we stipulated, and it is two and a fourth. And that is what we wanted to find.

559 It was necessary for us that the two given numbers satisfy the stipulation we described. For I say that if we assign the sought-after square to be one *Māl*, then we multiply it by each of the given numbers, then each of the products is *Māl*s. One of these products is a cube from a side of *Māl*s, which is the other product. And if the one that is the cube is divided by the one that is the side, the result from the division is a number (that is) Equated to a square of the *Māl*s that is the side. Therefore, the number that results from the division should be a square, in order that its side be a number Equated to the *Māl*s that is the side. Therefore, it is necessary that the two given numbers, if one of them is divided by the other, give a square. Because of this, the two numbers, one of them multiplied by the other, is also a square. And because the number that is a side of the square number – (the square being) the quotient that resulted from dividing one of the two numbers by the other – is Equal to the *Māl*s whose number is equal to the divisor of the two multiplied numbers, it is also necessary that if this number is divided by the (multitude of the) *Māl*s Equated to it, then the quotient will be a square, so that the one *Māl* Equals a square number. Therefore, it is necessary that if one of the given numbers is divided by the other, then the quotient will be a square, and also that if its side is divided by the number of the divisor, then the quotient will be a square, or said another way, also that the product of this side by the given number of the divisor will be a square. And for two numbers, if one of them divided by the other gives a square, and if its side divided by the number of the divisor gives a square, then these two numbers, if one of them is multiplied by the other, gives a square with a square side. And that is what was to be shown.

582 **22.** We want to find a cubic number such that if we multiply it by two given numbers, it gives a cube and a side of that cube.

584 It is necessary for us to first find the characteristic of the two given numbers. We say that if we assign the sought-after cube to be one Cube, and we multiply it by each of the given numbers, then both of the products will be Cubes. One of these products is a cube from a side of ⟨the other⟩ product. And of these two products, if you divide the Cubes that are the cube by the Cubes that are its side, the result from the division will be a number Equated to a square of the Cubes that are the side; and therefore, the number that results from the division should be a square in order that its side be Equated to the Cubes that are the side. Because of this, we assign the two given numbers so that if one of them is divided by the other, then the quotient is a square. And also, the number that is a side of the square number that resulted from the division is Equated to the Cubes that are the side, and its number is equal to the number of the divisor of the two given numbers. If this number is divided by the Cubes Equated ⟨to it⟩, the quotient should be a cube,

in order that the one Cube Equals a cubic number. So the characteristic of these two numbers, being now complete, is that if one of them is divided by the other, then the quotient will be a square, and if a side of this square is divided by the number of the divisor, then the quotient will be a cube.

We should (now) find these two numbers. We assign one of them to be two units, and we want to find the other number. Because one of these numbers divided by the other gives a square which, if its side is divided by the number of the divisor then the quotient is a cube, we should look for a number that, if we divide it by the two, gives a cube, and this is six units and a half and a fourth. And the six units and a half and a fourth is a side of the square that results from dividing one of the two numbers by the other, and the square that comes from the six and a half and a fourth is forty-five and a half and half an eighth. And the number that you get from dividing it[41] by the two[, this number we mentioned,] is ninety-one and an eighth. Thus, the other number we were looking for is ninety-one and an eighth. And you can learn the characteristics mentioned for the given numbers in the previous problems in the same way and find them.

600

Since one of the given numbers is two and the other number is ninety-one and an eighth, we want to find a cubic number such that if we multiply it by the ninety-one and an eighth, it gives a cube, and if we multiply it by the two, it gives a side of that cube.

611

We assign the cube to be one Cube, and we work it out like we did in the previous problems. We know that the sought-after cube is three units and three eighths, which, when multiplied by the ninety-one and an eighth gives a cube, which is three hundred seven units and thirty-five parts of sixty-four parts, and when multiplied by the two gives six units and a half and a fourth of a unit, which is a side of the cube that is three hundred seven units and thirty-five parts of sixty-four parts.

614

We have found a cubic number according to the stipulation that we stipulated. And that is what we wanted to find.

621

23. We want to find two square numbers such that their squares added together are a cube.

623

We assign one of the squares to be one *Māl* and the other square to be from a side of however many Things we wish. Let us assign it to be from a side of two Things, so that it is four *Māl*s. And the squares of these squares: as for the smaller, it is a *Māl Māl*, and as for the greater, it is sixteen *Māl Māl*s. They add up to seventeen *Māl Māl*s, and it Equals a cubic number. We form the cube from a side of three Things, so that it is twenty-seven Cubes. Thus, the seventeen *Māl Māl*s Equal twenty-seven Cubes. Therefore, seventeen Things Equal twenty-seven units. Thus, the one Thing is twenty-seven parts of seventeen parts of a unit.

624

---

41 The "it" is the "forty-five and a half and half an eighth". See Sesiano's footnote 49 (1982, 103), where he reconstructs how the text here became awkward at the hands of copyists.

631   And since we assigned the smaller square to be from a side of one Thing, its side is twenty-seven parts of seventeen parts, and the smaller square is seven hundred twenty-nine parts of two hundred eighty-nine parts of a unit. And since we assigned the greater square to be from a side of two Things, a side of the greater square is fifty-four parts of seventeen parts, and the greater square is two thousand nine hundred sixteen parts of two hundred eighty-nine parts of a unit. And because of that, a square of the smaller square is also five hundred ⟨thirty-one thousand four hundred forty-one parts of eighty-three thousand five hundred⟩ twenty-one parts of a unit, and for a square of the greater square, it is eight million five hundred three thousand fifty-six parts of eighty-three thousand five hundred twenty-one parts of a unit. The sum of these squares is nine million thirty-four thousand four hundred ninety-seven parts of eighty-three thousand five hundred twenty-one parts of a unit, which is also five hundred thirty-one thousand four hundred forty-one parts of four thousand nine hundred thirteen parts of a unit, which is a cube whose side is eighty-one parts of seventeen parts of a unit.

648   We have found two square numbers according to the stipulation that we stipulated, and they are the seven hundred twenty-nine parts and the two thousand nine hundred sixteen parts of two hundred eighty-nine parts. And that is what we wanted to find.

651   **24.** We want to find two square numbers such that the difference of their squares is a cubic number.

652   We assign the smaller square to be from a side of one Thing and the greater square to be from a side of two Things, so that the smaller is one *Māl*, and the greater is four *Māl*s, and the difference of their squares is fifteen *Māl Māl*s, which is a cube. Let us assign it to be from a side of five Things. And for every cube divided by its side, what results from the division is equal to a square of its side. And when you divide the fifteen *Māl Māl*s, which is a cube whose side is five Things, by its side, which is five Things, then the quotient is three Cubes. So the three Cubes are a square whose side is five Things. And the square that comes from the five Things is twenty-five *Māl*s. Thus, the three Cubes Equal twenty-five *Māl*s. And when we divide them by the *Māl*, which is the lesser in degree of the two sides, it gives three Things Equal twenty-five units. Thus, the one Thing Equals eight units and a third.

662   And since we assigned the smaller square to be from a side of one Thing, and the greater square from a side of two Things, a side of the smaller square is eight units and a third, and a side of the greater square is sixteen units and two thirds of a unit, and the smaller square is sixty-nine and four ninths, and the greater square is two hundred seventy-seven and seven ninths, and a square of the smaller square is four thousand eight hundred twenty-two and forty-three parts of eighty-one parts of a unit, and a square of the greater square is seventy-seven thousand one hundred sixty and forty parts of eighty-one parts of a unit. And the difference of these two squares is seventy-two thousand three hundred thirty-seven and seventy-eight

parts of eighty-one parts of a unit, which is twenty-six parts of twenty-seven parts of a unit, which is a cube whose side is forty-one and two thirds of a unit.

We have found two ⟨square⟩ numbers according to the stipulation that we stipulated, and they are the sixty-nine and four ninths and the two hundred seventy-seven and seven ninths. And that is what we wanted to find. 674

25. We want to find two numbers, a square and a cube, such that their squares added together are a square. 677

We assign the cube to be from a side of one Thing, so that it is one Cube. And we assign the square to be from a side of however many Things we wish. Let us assign it to be from a side of two Things, so that it is four *Māl*s. And their squares: as for a square of the cube, it is one Cube Cube, and as for a square of the square, it is sixteen *Māl Māl*s. Their sum is one Cube Cube and sixteen *Māl Māl*s, and that Equals a square number. We should find the number that is a side of this square, so we say that we assign this side to be *Māl*s, so the square Equated to the one Cube Cube and sixteen *Māl*s of the *Māl*s is *Māl Māl*s. Then, if we subtract the sixteen *Māl Māl*s in common from both[42] sides, it leaves *Māl Māl*s Equal a Cube Cube. And if we divide them by the one *Māl Māl*, which is the lesser in degree of the two sides, it gives a *Māl* Equals a number. This number should be a square, since it Equals one *Māl*. But this number is an excess in *Māl*s of the *Māl*s, which is a square number over the sixteen. Therefore, the number of *Māl*s of the *Māl*s should be a square number exceeding the sixteen by a square number. Because of that we come to look for two square numbers whose difference is sixteen. We find the greater square to be twenty-five and the smaller square to be nine units. So we make the square Equated to the Cube Cube and the sixteen *Māl Māl*s to be twenty-five *Māl Māl*s, which is from a side of five *Māl*s. 679

We cast away the sixteen *Māl Māl*s in common from both sides, leaving a Cube Cube Equals nine *Māl Māl*s. Therefore, one *Māl* Equals nine units, and the *Māl* is a square ⟨with⟩ a side of one Thing, and the nine units is a square whose side is three units. So the one Thing is three units. 696

And since we assigned the cube to be from a side of one Thing, its side is three units, so it is twenty-seven. And since we assigned the square to be from a side of two Things, its side is six units, and it is thirty-six. And their squares: as for a square of the twenty-seven, it is seven hundred twenty-nine, and as for a square of the thirty-six, it is one thousand two hundred ninety-six. And their sum is two thousand twenty-five, which is a square whose side is forty-five. 699

We have found two numbers, a cube and a square, such that their squares added together are a square, and they are the twenty-seven and the thirty-six. And that is what we wanted to find. 706

---

42 Sesiano corrects "to" ('*alā*) to "from both" (*min kalā*). Rashed follows the manuscript and translates "communs aux deux membres" (Diophantus 1984, 3, 41.9).

708   **26.** We want to find two numbers, a cube and a square, such that the difference of their squares is a square number.

710   We assign the cube to be one Cube and the square to be four *Māl*s. So a square of the cube is a Cube Cube, and a square of the square is sixteen *Māl Māl*s. And we want their difference to be a square number. Let us first require that ⟨a square⟩ of the cube exceed ⟨a square⟩ of the square by a square number. So we say, a Cube Cube less sixteen *Māl Māl*s Equals a square number. Following the example that we saw in the problem preceding this problem, we look for (how many) *Māl*s should be assigned as a side of this square. We find it to be three *Māl*s, and the square that comes from it to be nine *Māl Māl*s. So a Cube Cube less sixteen *Māl Māl*s Equals nine *Māl Māl*s.

717   We add the sixteen *Māl Māl*s in common to the two sides to get a Cube Cube Equals twenty-five *Māl Māl*s. Therefore, the one *Māl* Equals twenty-five units, and the *Māl* is a square whose side is one Thing, and the twenty-five is a square whose side is five units, and the one Thing Equals five units.

721   And since we assigned the Cube to be from a side of one Thing, its side is five units, and it is one hundred twenty-five. And since we assigned the square to be from a side of two Things, its side is ten units, and it is one hundred. And their squares: as for a square of the one hundred twenty-five, it is fifteen thousand six hundred twenty-five, and as for a square of the one hundred, it is ten thousand. Their difference is five thousand six hundred twenty-five, which is a square whose side is seventy-five.

728   We have found two numbers, a cube and a square, such that the excess ⟨of a square⟩ of the cube over ⟨a square⟩ of the square is a square number, and they are the one hundred and the one hundred twenty-five.

730   Similarly, let us require that the excess of a square of the square over a square of the cube be a square. We assign the cube to be one Cube and the square to be from a side of five Things, so that we get six hundred twenty-five *Māl Māl*s less one Cube Cube Equal a square number. Let us look for a side of this square. We say that if we assign ⟨it⟩ to be *Māl*s, then its square Equated to the six hundred twenty-five *Māl Māl*s less a Cube Cube is *Māl Māl*s. If we add a Cube of the Cube in common to both sides, it becomes six hundred twenty-five *Māl*s of the *Māl*s Equal a Cube Cube and *Māl Māl*s.[43] Then after ⟨the confrontation and⟩ the division there will remain a *Māl* Equals a number. This number should be a square. But this number is the excess of the product of the twenty-five ⟨by itself⟩ over a number of *Māl Māl*s that is a square of the sought-after side. So we should divide the six hundred twenty-five, which is a square number, into

---

43 At this point Rashed introduces this restoration: "⟨Then, if we subtract the *Māl Māl*s common to the two sides, they leave six hundred twenty-five *Māl Māl*s less *Māl Māl*s Equal a Cube Cube.⟩" He thus does not need Sesiano's next restoration, "⟨the confrontation and⟩".

two square numbers, according to what we explained in the second Book.[44] Let one of the two parts be four hundred and the other be two hundred twenty-five. And we assign the square Equated to the six hundred twenty-five *Māl Māl*s less a Cube Cube to be (a number of) *Māl Māl*s equal to one of these two parts. Let us assign it to be two hundred twenty-five *Māl Māl*s. After the restoration and confrontation and the division, there remains a *Māl* Equals four hundred. Therefore, the Thing, which we assigned to be a side of the cube, is twenty, and it is eight thousand.

And ⟨a side⟩ of the square, since we assigned it to be five Things, is one hundred, and it is ten thousand. And their squares: as for a square of the cube, which is eight thousand, it is sixty-four million, and as for a square of the square, which is ten thousand, it is one hundred million. Their difference is thirty-six million, which is a square whose side is six thousand.

747

We have found two numbers, a cube and a square, such that a square of the square exceeds a square of the cube by a square number, and they are the ten thousand and the eight thousand. And that is what we wanted to find.

752

27. We want to find two numbers, a cube and a square, such that a square of the cube together with a given multiple of the square number is a square number.

755

We assign the cube to be one ⟨Cube⟩, and we multiply it by itself to get one Cube Cube. And we assign the square to be from a side of *Māl*s, however many we wish. Let us assign it to be from a side of two *Māl*s, so it is four *Māl Māl*s. And let the given multiple be five, and we multiply the four *Māl Māl*s by five to get twenty *Māl Māl*s. And we add it to the one Cube Cube to get a Cube Cube and twenty *Māl Māl*s, and that is a square. Let us look for two squares whose difference is twenty,[45] and they are thirty six and sixteen. We make the square, which is a Cube Cube and twenty *Māl Māl*s, Equal thirty-six *Māl Māl*s.

757

We subtract the twenty *Māl Māl*s in common from both sides, leaving sixteen *Māl Māl*s Equal a Cube Cube. And let us divide all of that by a *Māl Māl*, giving sixteen Equals a *Māl*; and the sixteen is a square whose side is four units. So four Equals a side of the *Māl*, which is one Thing.

764

And since we assigned the cube to be one Cube, its side is four units, and it is sixty-four. And since we assigned a side of the square to be two *Māl*s, its side is thirty-two, and it is one thousand twenty-four. And five times that is five thousand one hundred twenty. We add that to the cubic number[46] to get five thousand one hundred eighty-four, and that is a square number whose side is seventy-two.

767

---

44 Problem II.8. /s/

45 Problem II.10. /s/

46 This should be added to *a square of* the cubic number, or 4096, to get 9216, which is a square number with a side of 96. /s/ Rashed (Diophantus 1984, 3, 46) offers this restoration in a footnote

773   We have found two numbers, a square and a cube, and a square of the cube together with five times the square number is a square number, and they are the sixty-four and the one thousand twenty-four. And that is what we wanted to find.

776   **28**. We want to find two numbers, a cube and a square, such that a square of the square together with a given multiple of the cubic number is a square number.

778   Let the given multiple be ten times, and we make the cube one Cube, and we multiply it by the ten to get ten Cubes. And we assign a side of the square to be two Things, so it is four *Māl*s, and its square is sixteen *Māl Māl*s. And we add that to the ten Cubes, to get sixteen *Māl Māl*s and ten Cubes, and that Equals a square number. We assign the square to be from a side of six *Māl*s. And for every square divided by its side, the result of the division is equal[47] to its side. Thus, we divide the sixteen *Māl Māl*s and the ten Cubes by six *Māl*s, to get two *Māl*s and two thirds of a *Māl* and a Thing and two thirds of a Thing, and that Equals six *Māl*s.

785   We subtract the two *Māl*s and two thirds of a *Māl* in common from both sides, leaving three *Māl*s and a third Equal a Thing and two thirds of a Thing. Therefore, three Things and a third Equal one and two thirds of a unit, and the one Thing Equals a half.

788   Since we assigned the cube to be from a side of one Thing, its side is a half, the cube is an eighth, and ten times it is one and a fourth. And since we assigned the square to be from a side of two Things, its side is one and it is also one. And when we add that[48] to the one and a fourth which is ten times the cube, it gives a square number, which is two and a fourth, and its side is one and a half.

793   We have found two numbers, a cube and a square, such that a square of the square together with ten times the cubic number is a square number, and they are the one and the eighth of a unit. And that is what we wanted to find.

796   **29**. We want to find two numbers, a cube and a square, such that a cube of the cube and a square of the square added together are a square number.

798   We assign the cube to be one Cube, ⟨so that⟩ its cube is a Cube Cube by a Cube, and that is called a Cube Cube Cube. And we assign ⟨a side⟩ of the square to be whatever we wish of *Māl*s. Let us assign it to be two *Māl*s, so that the square is four *Māl Māl*s, and its square is sixteen *Māl Māl*s ⟨by a *Māl*⟩ *Māl*,[49] and that Equals sixteen Cube

---

to his translation: "au carré du cube, on a neuf mille deux cent seize, nombre carré don't le côté est quatre-vingt-seize".

47   The text has "Equal" (*mu'ādilān*), which is an error for "equal" with a different root, such as *musāwiya*.

48   Rather, we should add *a square of* that since the enunciation specifies "a square of the square". /s/

49   Rashed restores this and the next instance as "*Māl Māl Māl Māl*s".

Cubes multiplied by a *Māl*, one of which is called a Cube Cube *Māl*. So a Cube Cube Cube together with the sixteen Cube Cube *Māl*s Equal a square number. Let us also assign its side to be whatever we wish of *Māl Māl*s. We assign it to be from a side of six *Māl Māl*s, and we multiply it by itself to get thirty-six *Māl Māl*s ⟨by a *Māl*⟩ *Māl*, that is, thirty-six Cube Cube *Māl*s. So the Cube Cube Cube together with the sixteen Cube Cube *Māl*s Equal thirty-six Cube Cube *Māl*s.

Let us cast away the sixteen Cubes Cube *Māl*s in common from both sides, leaving a Cube Cube Cube Equals twenty Cube Cube *Māl*s. And we divide each of them by one of the lesser in degree of the two sides, which is a Cube Cube *Māl*. Thus, when you divide the twenty Cube Cube *Māl*s by a Cube Cube *Māl*, it gives you twenty units. And the Cube Cube Cube is nothing but what comes from multiplying a Cube ⟨Cube⟩ by a Cube, which is also what comes from multiplying a Cube Cube *Māl* by a Thing. So when a Cube Cube Cube is divided by a Cube Cube *Māl* the quotient is one Thing, and the one Thing Equals twenty units.   807

Since we assigned a side of the cube to be one Thing, a side of the cube is twenty units, and the cube is eight thousand. And since we assigned the square to be from a side of two *Māl*s, and the *Māl* is four hundred, a side ⟨of the square⟩ is eight hundred, and the square is six hundred forty thousand, and a cube of the cube is five hundred twelve billion, and a square of the square is four hundred nine billion six hundred million. And their sum is nine hundred twenty-one billion six hundred million, and that is a square number whose side is nine hundred sixty thousand.   814

We have found two numbers, a cube and a square, such that a cube of the cube together with a square of the square add up to a square number, and they are eight ⟨thousand⟩ and six hundred forty thousand. And that is what we wanted to find.   823

30. We want to find two numbers, a cube and a square, such that the excess of a cube of the cube over a square of the square is a square number.   826

We assign the cube to be one Cube, so that its cube is a Cube Cube by a Cube, that is, ⟨what⟩ is called a Cube Cube Cube. And we assign the square to be from a side of two *Māl*s, so that the square is four *Māl Māl*s, and its square is sixteen *Māl Māl*s by a *Māl Māl*, that is, sixteen Cube Cube *Māl*s. So a Cube Cube Cube less sixteen Cube Cube *Māl*s Equals a square number. Let us assign its side to be two *Māl Māl*s so that its square is four *Māl Māl*s by a *Māl Māl*, that is, four Cube Cube *Māl*s. Thus, a Cube Cube Cube less sixteen Cube Cube *Māl*s Equals four Cube ⟨Cube⟩ *Māl*s.   828

Let us add the sixteen Cube Cube *Māl*s in common to the two sides, to get a Cube Cube Cube Equals twenty Cube Cube *Māl*s. And let us divide everything by a Cube Cube *Māl*, which is one of the lesser in degree of the two sides, to get, after the division, one Thing Equals twenty units.   835

Since we assigned a side of the cube to be one Thing, its side is twenty and it is eight thousand. And since we assigned a side of the square to be two *Māl*s, and the *Māl* is four hundred, a side of the square is eight hundred, and the square is six hundred   838 forty thousand. As for a cube of the cube, it is clearly five hundred twelve billion. And as for a square of the square, it is four hundred nine billion six hundred million. And their difference, that is, the excess of a cube of the cube over a square of the square, is one hundred two billion four hundred million, which is a square number whose side is three hundred twenty thousand. And it was shown in the previous problem that the sum of these two numbers is also a square number.

848   We have found two numbers, a cube and a square, such that the excess of a cube of the cube over a square of the square is a square number, and they are the eight thousand and the six hundred forty thousand. And that is what we wanted to find.

851   Thus, it is clear that we have also found two numbers, a cube and a square, such that a cube of the cube, if a square of the square is added to it, amounts to a square number, and if a square of the square is subtracted from a cube of the cube, then the remainder is a square number, and these are also the two numbers.

855   **31**. We want to find two numbers, a square and a cube, such that the excess of a square of the square over a cube of the cube is a square number.

857   We assign the cube to be one ⟨Cube⟩, so its cube is a Cube Cube by a Cube, which is called a Cube Cube Cube. And we assign a side of the square to be two *Māl*s, so the square is four *Māl Māl*s, and its square is sixteen *Māl Māl*s by a *Māl* ⟨*Māl*⟩, which is called a Cube Cube *Māl*. So the sixteen Cube Cube *Māl*s, which is a square of the square number, exceeds a Cube Cube ⟨Cube⟩ by a square number. Let us assign a side of that square to be two *Māl Māl*s. And every square divided by its side, what results from the division is equal to its side. Therefore, the sixteen Cube Cube *Māl*s less the Cube Cube Cube, when we divide them by two *Māl Māl*s, give a result from the division that Equals two *Māl Māl*s. But when the sixteen Cube Cube *Māl*s less a Cube Cube Cube is divided by two *Māl Māl*s,[50] then ⟨for⟩ the sixteen Cube Cube *Māl*s, since they come from multiplying sixteen *Māl Māl*s by a *Māl Māl*, when you divide them by two *Māl Māl*s, the quotient is eight *Māl Māl*s. And for the Cube Cube Cube, since it comes from multiplying a Cube Cube by a Cube, and the Cube Cube comes from multiplying a *Māl Māl* by a *Māl*, the Cube Cube Cube comes from multiplying a *Māl Māl* by a *Māl* Cube. So when the Cube Cube Cube is divided by two *Māl Māl*s, the quotient is half a *Māl* Cube. Then what results from the division is eight *Māl Māl*s less half a *Māl* Cube, and that Equals two *Māl Māl*s.

873   We make half a *Māl* ⟨Cube⟩ common, adding it to both sides, giving eight *Māl Māl*s Equal two *Māl Māl*s and half a *Māl* Cube. Let us cast away the two *Māl Māl*s in common from both sides, so there remains half a *Māl* Cube Equals six *Māl* ⟨*Māl*s⟩. And after the division there remains half a Thing Equals six units, so the one Thing Equals twelve units.

---

50 The passage "But . . . *Māl Māl*s" is not translated by Sesiano.

And since we assigned a side of the cube to be one Thing, its side is twelve, and the cube is one thousand seven hundred twenty-eight. And since we assigned a side of the square to be two *Māl*s, and the *Māl* is one hundred forty-four, since the Thing is twelve, a side of the square is two hundred eighty-eight, and the square is eighty-two thousand nine hundred forty-four. As for a cube of the cube, it is five billion one hundred fifty-nine million seven hundred eighty thousand three hundred fifty-two. And as for a square of the square, it is six billion eight hundred seventy-nine million seven hundred seven thousand one hundred thirty-six. And the excess of this number over a cube of the cube is one billion seven hundred nineteen million nine hundred twenty-six thousand seven hundred eighty-four, which is a square number whose side is forty-one thousand four hundred seventy-two.

878

We have found two numbers according to the stipulation that we wanted, and they are the one thousand seven hundred twenty-eight and the eighty-two thousand nine hundred forty-four. And that is what we wanted to find.

890

32. We want to find two numbers, a cube and a square, such that a cube of the cube together with a given multiple of what comes from multiplying the square by the cube is a square number.

893

Let the given multiple be five. We assign the cube to be one Cube, so its cube is a Cube Cube Cube, and we assign the square to be ⟨from⟩ a side of two Cubes, so the square is four Cube Cubes. And we multiply that by the cubic number that we assigned to be one Cube, so it amounts to four Cube Cube Cubes, and five times that is twenty Cube Cube Cubes. We add it to a cube of the cube, so it amounts to twenty-one Cube Cube Cubes, and that Equals a square number. Let us assign its side to be seven *Māl Māl*s. The square is forty-nine Cube Cubes by a *Māl*, and that Equals twenty-one Cube Cube Cubes. Let us divide each of them by a Cube Cube *Māl*, to get twenty-one Things Equal forty-nine units. Thus, the one Thing Equals two and a third.

895

And since we assigned the cube to be from a side of one Thing, a side of the cube is two and a third of a unit. Therefore, the cube, because its side is seven thirds, is three hundred forty three parts of twenty-seven parts. And since we assigned a side of the square to be two Cubes, a side of the square is six hundred eighty-six parts of twenty-seven parts of a unit, and the square is four hundred seventy thousand five hundred ninety-six parts of seven hundred twenty-nine parts of a unit. As for a cube of the cube, it is forty million three hundred fifty-three thousand six hundred seven parts of nineteen thousand six hundred eighty-three parts of a unit. And as for what comes from multiplying the square number by the cubic number, it is one hundred sixty-one million four hundred fourteen thousand four hundred twenty-eight parts of nineteen thousand six hundred eighty-three parts of a unit, and five times that is eight hundred seven million seventy-two thousand one hundred forty parts. And when we add that to a cube of the cube, their sum is eight hundred forty-seven million four hundred twenty-five thousand seven hundred forty-seven parts of nineteen thousand six hundred eighty-three parts of a unit.

904

And that is also two hundred eighty-two million four hundred seventy-five thousand two hundred forty-nine parts of six thousand five hundred sixty-one parts of a unit, and that is a square number whose side is sixteen thousand eight hundred seven parts of eighty-one parts of a unit.

925   We have found two numbers according to the stipulation that we stipulated, and they are the three hundred forty-three parts of twenty-seven parts of a unit and the four hundred seventy thousand five hundred ninety-six parts of seven hundred twenty-nine parts of a unit. And that is what we wanted to find.

929   **33**. We want to find two numbers, a cube and a square, such that a cube of the cube exceeds a given multiple of what comes from multiplying the square number by the cubic number by a square number.

932   Let the given multiple be three times. We assign the cube to be one Cube, so its cube is a Cube Cube Cube. And we assign the square to be from a side of half a Cube, so the square is a fourth of a Cube Cube. We multiply that by the cubic number, which we assigned to be one Cube, resulting in a fourth of a Cube Cube Cube, and three times that is three fourths of a Cube Cube Cube. We subtract it from a cube of the cube, leaving a fourth of a Cube Cube Cube Cube Equals a square number. Let us assign its side to be whatever we wish of *Māl Māl*s, so let us assign it to be one *Māl Māl*. Thus, a Cube Cube *Māl* Equals a fourth of a Cube Cube Cube, and after the division it gives a fourth of a Thing Equals one, so the whole Thing Equals four units.

940   Since we assigned a side of the cube to be one Thing, its side is four units, and the cube is sixty-four. And since we assigned the square to be from a side of half a Cube, a side of the square is thirty-two, so the square is one thousand twenty-four. As for a cube of the cube, it is two hundred sixty-two thousand one hundred forty-four. And as for what comes from multiplying the square number by the cubic number, it is sixty-five thousand five hundred thirty-six. And three times that is one hundred ninety-six thousand six hundred eight, and when we subtract that from a cube of the cube, it leaves sixty-five thousand five hundred thirty-six, which is a square from a side of two hundred fifty-six.

949   We have found two numbers according to the stipulation that we stipulated, and they are the sixty-four and the one thousand twenty-four. And that is what we wanted to find.

951   In addition to the example we have just described, we find that the remaining variations of this kind of problem are: to find two numbers, a cube and a square, such that a square of the square together with a given multiple of what comes from multiplying the square number by the cubic number is a square number, and also a cube of the square together with a given multiple of what comes from multiplying the square number by the cubic number is a square number, and also vice versa, and whatever else is similar.

34. We want to find two numbers, a cube and a square, such that the cube, if the square is added to it, amounts to a square number, and similarly, if the square is subtracted from it, leaves a square number.

956

We assign the cube to be one Cube, and we assign the square to be four *Māl*s, so a Cube and four *Māl*s Equal a square number, and a Cube less four *Māl*s also Equals a square number. We solve this by working out the double equality. We take the difference between these two squares, which is eight *Māl*s, and we look for two numbers such that the product of one of them by the other is eight *Māl*s, and they are two Things and four Things. Their difference is two Things, and half of the two Things is one ⟨Thing⟩, and its square is one *Māl*, and that Equals a Cube less four *Māl*s.

958

When we add the four *Māl*s in common to both sides, it gives the one Cube Equals five *Māl*s. Likewise, when we add the two Things to the four Things, that gives six Things, and its half is three Things, and a square of the three Things is nine *Māl*s, and that Equals a Cube and four *Māl*s. We cast away the four *Māl*s in common from both sides, leaving one Cube Equals five *Māl*s. So both sides of the Equation are equal, each of them turning out to be a Cube Equals five *Māl*s. Let us then divide all of that by a *Māl*, to get one Thing Equals five units.

964

Therefore, a side of the cube is five units, and the cube is one hundred twenty-five. And a side of the square is ten units, and the square is one hundred, which, if added to the cubic number, amounts to two hundred twenty-five, which is a square number whose side is fifteen, while if subtracted from the cubic number, leaves twenty-five, which is a square number whose side is five units.

971

We can also work this out without the ⟨double⟩ equality. We say: because the Cube and the four *Māl*s Equal a square number, If we assign its side to be Things, then the square is *Māl*s (which) Equal a Cube and four *Māl*s. And if we subtract the four *Māl*s in common from both sides, it leaves a Cube Equals *Māl*s. And if we divide them by a *Māl*, as for the Cube it gives a Thing, and as for the *Māl*s it gives a number [equal to the number of *Māl*s]. Therefore, the number that was assigned in the problem to be one Thing is equal to the number of the remaining *Māl*s. Likewise, because the Cube less four *Māl*s Equals a square number, if we also assign its side to be Things, then the square is *Māl*s, and if we add the four *Māl*s in common to both sides, it amounts to a Cube Equals *Māl*s. Therefore, the number that was assigned in the problem to be a Thing is equal to the number of the *Māl*s added together. Therefore, the number of *Māl*s remaining in the first Equation should be equal to the number of *Māl*s gathered in the second Equation. But the *Māl*s remaining in the first Equation is what remained of a square number after subtracting four units, and the *Māl*s gathered in the second Equation is the number gathered from a square number and four units.

977

Because of this, we look for two square numbers such that if we subtract four units from the greater, and we add four units to the smaller, they will be equal.

991

Then we should look for two square numbers whose difference is eight units,[51] and they are twelve and a fourth and twenty and a fourth. We make the greater square Equated to the Cube and the four *Māl*s be twenty *Māl*s and a fourth of a *Māl*, and the smaller square Equated to the Cube less four *Māl*s be twelve *Māl*s and a fourth of a *Māl*. Thus, each of the Equations turns out to be one Cube Equals sixteen *Māl*s and a fourth of a *Māl*. Therefore, the one Thing Equals sixteen units and a fourth of a unit.

998   Since we assigned a side of the cube to be one Thing, a side of the cube is sixteen and a fourth, and the cube is four thousand two hundred ninety-one ⟨and⟩ a part of sixty-four parts of a unit. And since we assigned a side of the square to be two Things, a side of the square is thirty-two and a half, and the square is one thousand fifty-six and a fourth. When we add it to the cubic number, it amounts to five thousand three hundred forty-seven and seventeen parts of sixty-four, which is a square number whose side is seventy-three and an eighth of a unit. And when we subtract ⟨it⟩ from the cubic number, it leaves three thousand two hundred thirty-four and forty-nine parts of sixty-four parts of a unit, which is a square whose side is fifty-six and seven eighths.

1008   We have found two numbers, a cube and a square, such that the cubic number, if the square number is added to it, amounts to a square number, and if the square number is subtracted from it, also leaves a square number.

1011   **35**. We want to find two numbers, a cube and a square, such that the square number, if the cubic number is added to it, amounts to a square number, and if the cubic number is subtracted from it, leaves a square number.

1014   We assign the cube to be one Cube, and the square to be four *Māl*s, so four *Māl*s and a Cube Equal a square number, and four *Māl*s less a Cube Equal a square number. If we assign a side of the square Equated to the four *Māl*s and the Cube to be Things, then the square is *Māl*s (which) Equal four *Māl*s and a Cube. And if we subtract the four *Māl*s in common from both sides, it leaves a Cube Equals *Māl*s, and the number that was assigned in the problem as ⟨a Thing⟩ is equal to the number of remaining *Māl*s. Likewise, if we assign a side of the square Equated to the four *Māl*s less a Cube to be Things, then the square is *Māl*s (which) Equal four *Māl*s less a Cube. If we add the Cube in common to both sides it becomes *Māl*s and a Cube Equal four *Māl*s. And if we subtract the *Māl*s in common from both sides, it leaves a Cube Equals *Māl*s. And the number assigned in the problem to be a Thing is likewise equal to the number of the remaining *Māl*s.

1025   So the (number of) *Māl*s remaining in the first Equation should be equal to the number of *Māl*s remaining in the second Equation. But the *Māl*s remaining in the first Equation is a square number (with) four subtracted from it, and the *Māl*s remaining in the second Equation is a square number subtracted from four. Thus, we say: a square less four units Equals four units less another square. And if we add the four units excluded from the first square in common to both sides, it amounts to[52] a square Equals eight units less a square. And if we add the other square in common to both sides it amounts to two squares, and they Equal eight units.

But the eight is made up of two equal square numbers, so we should divide the eight into two other square numbers, as we explained in the second Book.[53] Let one of the numbers[54] be four parts of twenty-five of a unit, and the other be seven and twenty-one parts of twenty-five parts of a unit. We make the square that is Equated to the four *Māl*s and a Cube be seven *Māl*s and twenty-one parts of twenty-five parts of a *Māl*, and the square Equated to the four *Māl*s less a Cube be four parts of twenty-five parts of a *Māl*. Each of the Equations turns out to be, after the restoration and confrontation, three *Māl*s and twenty-one parts of twenty-five parts of a *Māl* Equal a Cube. And when we divide them by a *Māl*, it gives three and four fifths and a fifth of a fifth of a unit Equal one Thing.

1033

Since we assigned the cube to be from a side of one Thing, a side of the cube is ninety-six parts of twenty-five parts, and the cube is eight hundred eighty-four thousand seven hundred thirty-six parts of fifteen thousand six hundred twenty-five parts of a unit. And since we assigned a side of the square to be two Things, a side of the square is one hundred ninety-two parts of twenty-five parts of a unit, and the square is thirty-six thousand eight hundred sixty-four parts of six hundred twenty-five parts of a unit, which is also nine hundred twenty-one thousand six hundred parts of fifteen thousand six hundred twenty-five parts of a unit. And when we add that to the cubic number, it amounts to one million eight hundred six thousand three hundred thirty-six parts ⟨of fifteen thousand six hundred twenty-five parts⟩, and that is a square number whose side is one thousand three hundred forty-four parts of one hundred twenty-five parts of a unit. And when we subtract the cubic number from it, it leaves thirty-six thousand eight hundred sixty-four parts of fifteen thousand six hundred twenty-five parts of a unit, ⟨and that is a square number whose side is one hundred ninety-two parts of one hundred twenty-five parts of a unit⟩.

1043

We have found two numbers, a cube and a square, such that the square, if the cube is added to it,[55] gives a square number, and if the cube is subtracted from it, leaves a square number. They are the eight hundred eighty-four thousand seven hundred thirty-six parts and nine hundred twenty-one thousand six hundred parts

1059

---

52 The text has "remains" (*baqiya*), that Sesiano restores to "amounts to" (line 1031). Rashed makes no correction here (Diophantus 1984, 3, 63.12), and also at lines 1215 and 1263 (his 74.12 and 77.12), claiming that the word is not an error but takes the meaning of "il demeure". See his discussion in (1984, III, 151). We have never seen *baqiya* used in this context or with this meaning in any other Arabic mathematics text, so we adopt Sesiano's correction.

53 Problem II.9. /sr/

54 Lit. "parts".

55 The text has "cube" and "square" switched. Sesiano makes the correction (line 1059), while Rashed does not (Diophantus 1984, 3, 65.10).

of fifteen thousand six hundred twenty-five parts of a unit. And that is what we wanted to find.

1065   **36**. We want to find a cubic number such that if we add to it a given multiple of the square that comes from its side, amounts to a square number, and if we subtract from it another given multiple of the square that comes from its side, leaves a square number.

1068   Let the appended multiple be four times, and the lacking multiple be five times. We want to find a cubic number such that if we add to it four times the square that comes from its side amounts to a square number, and if we subtract from it five times the square that comes from its side leaves a square number.

1071   We assign the cube to be one Cube, so that the square that comes from its side is one *Māl*. And we look for two square numbers such that if we subtract from the greater four units and we add to the smaller five units, they are Equal. That is, we look for two square numbers whose difference is nine units.[56] We find that one of the squares is sixteen and the other square is twenty-five. And we add to the cube four times the square that comes from multiplying its side by itself to get one Cube and four *Māl*s, which is a square number. Let us make the square Equated to it be (a number of) *Māl*s equal to the greater square of the two squares whose difference is nine units, which is twenty-five *Māl*s. We cast away the four *Māl*s in common from both sides, leaving one Cube Equals twenty-one *Māl*s. Likewise, we subtract from the cube five times the square that comes from its side, which is five *Māl*s, leaving a Cube less five *Māl*s, and that Equals a square number. Let us make the square Equated to it be (a number of) *Māl*s equal to the smaller square of the two squares whose difference is nine units, which is sixteen *Māl*s. And we add the five *Māl*s lacking from the Cube in common to both sides to get one Cube Equals twenty-one *Māl*s. Thus, both Equations turn out to be one Cube Equals twenty-one *Māl*s. Let us divide each of them by a *Māl* to get one Thing Equals twenty-one.

1088   Since we assigned a side of the cube to be one Thing, a side of the cube is twenty-one, and the cube is nine thousand two hundred sixty-one, and the square that comes from multiplying its side by itself is four hundred forty-one, and four times that is ⟨one thousand⟩ seven hundred sixty-four. When we add it to the cubic number, it amounts to eleven thousand twenty-five, which is a square whose side is one hundred five units. And five times a square of a side of the cube is two thousand two hundred five units. When we subtract it from the cubic number, it leaves seven thousand fifty-six, which is a square number whose side is eighty-four.

1096   We have found a cubic number such that if we add to it four times the square that comes from its side amounts to a square number, and if we subtract ⟨from it⟩ five

---

56 Problem II.10. /s/

times the square that comes from its side leaves a square number, and it is nine thousand two hundred sixty-one. And that is what we wanted to find.

We also note that if we want the appended multiple to be five and the lacking multiple to be four, a side of the cube would be twenty and the cube eight thousand. If we add to it five times the square that comes from its side, which is two thousand, it amounts to ten thousand, which is a square number whose side is one hundred. And if we subtracted from it four times the square that comes from its side, which is one thousand six hundred, it leaves six thousand four hundred, which is a square number whose side is eighty.

1100

37. We want to find a cubic number such that if we multiply the square that comes from its side by two given numbers, and we add what is gathered from each of them to the cubic number, amounts to a square number.

1106

Let one of the numbers be five and the other ten. We want to find a cubic number such that if we multiply a square of its side by five and by ten, and we add what is gathered from their multiplications by each of them to the cubic number, amounts to a square number.

1109

We assign the cube to be one Cube, and we multiply a square of its side, which is one *Māl*, by the five and by the ten, so they give five *Māl*s and ten *Māl*s. We add each of them to the Cube to get a Cube and five *Māl*s Equal a square number, and a Cube and ten *Māl*s Equal a square number. If we make a side of the square that is a Cube and five *Māl*s be Things, then its square is *Māl*s. Then, if we subtract the five *Māl*s in common from both sides, it leaves a Cube Equals *Māl*s. Thus, it is clear that the number assigned in this problem to be a Thing is equal to the number of remaining *Māl*s. And also, if we make a side of the square that is a Cube and ten *Māl*s be Things, then its square is *Māl*s. And if we subtract the ten *Māl*s in common from both sides, it leaves a Cube Equals *Māl*s. Therefore, the number assigned to be a Thing in this analysis is equal to the number of remaining *Māl*s.

1111

So the remaining *Māl*s in the first Equation should be equal to the remaining *Māl*s in the second Equation. But the remaining *Māl*s in the first Equation is a square number less five units, and the remaining *Māl*s in the second Equation is a square less ten units, and so we should find two square numbers such that if we subtract ten units from the greater and five units from the smaller, (they are) equal. Thus, we say, a square less five units Equals the other square less ten units. And we add the ten units in common to both sides, so it amounts to a square and five units Equal a square.

1122

So we should look for two squares whose difference is five units, and so that the smaller is more than five units.[57] Let the smaller square be fifty-three and seven ninths, and its side is seven units and a third of a unit, and the greater square be

1129

57 Problem II.10. /s/

fifty-eight and seven ninths, and its side is seven units and two thirds of a unit. We make the square Equated to the Cube and five *Māl*s be fifty-three *Māl*s and seven ninths of a *Māl*, and we make the square Equated to the Cube and ten *Māl*s be fifty-eight *Māl*s and seven ninths of a *Māl*. Each of the Equations turns out to be one Cube Equals forty-eight *Māl*s and seven ninths of a *Māl*. When we divide them by one *Māl*, it gives one Thing Equals forty-eight units and seven ninths of a unit.

1138   Since we assigned a side of the cube to be one Thing, its side is four hundred thirty-nine ninths, and the cube is eighty-four million six hundred four thousand five hundred nineteen ninths of a ninth of a ninth, which is also seven hundred sixty-one million four hundred forty thousand six hundred seventy-one ninths of a ninth of a ninth of a ninth, and a square of a side of the cube is one hundred ninety-two thousand seven hundred twenty-one ninths of a ninth, which is also fifteen million six hundred ten thousand four hundred ninths of a ninth of a ninth of a ninth ⟨and⟩ a ninth of a ninth of a ninth of a ninth of a unit. And when we multiply that by five units, it amounts to seventy-eight million fifty-two thousand five ninths of a ninth of a ninth of a ninth. And when we add that to the cubic number, it amounts to eight hundred thirty-nine million four hundred ninety-two thousand six hundred seventy-six ninths of a ninth of a ninth of a ninth ⟨which is a square whose side is twenty-eight thousand nine hundred seventy-four ninths of a ninth⟩. Also, when we multiply ⟨a square⟩ of a side of the cube by ten units, it amounts to one hundred fifty-six million one hundred four thousand ten ninths of a ninth of a ninth of a ninth. And when we add it to the cubic number, it amounts to nine hundred seventeen million five hundred forty-four thousand six hundred eighty-one ninths of a ninth of a ninth of a ninth, which is a square whose side is thirty thousand two hundred ninety-one ninths of a ninth.

1157   We have found a cubic number according to the stipulation that we stipulated and these are the two numbers that we mentioned.[58]

1159   **38**. We now want to find a cubic number such that if we multiply the square that comes from its side by two given numbers, and we subtract either of them from the cubic number, it leaves a square number.

1162   Let one of the numbers be five units and the other ten units. We want to find a cubic number such that if we multiply a square of its side by five and by ten, and we subtract what is gathered from either of them from the cubic number, it leaves a square number.

1164   Like before, we make the cube one Cube, and we multiply a square of its side, which is one *Māl*, by five and by ten to get five *Māl*s and ten *Māl*s. We subtract each of

---

58 Sesiano writes, "Not only the sense but also the unusual wording of this final statement suggest that we have here an interpolation made by some reader or copyist; did he have in mind the two forms given for x^3?" (1982, 115 n. 72).

them from the cubic number, so there remain a Cube less five *Māl*s and a Cube less ten *Māl*s, and each of them Equals a square number. As for the Things that are a side of the square Equated to the Cube lacking five *Māl*s, if five *Māl*s are added to its square, then the sum of that is *Māl*s, whose number is the number assigned to be a Thing in the problem. While, as for the Things that are a side of the square Equated to the Cube lacking ten *Māl*s, if ten *Māl*s are added to its square, then the sum of that is *Māl*s, whose number is the number assigned to be a Thing in the problem.

Therefore, we should take up ⟨the search for⟩[59] two square numbers such that if we add five units to the greater and we add ten units to the smaller, they are equal. So we say, a greater square and five units Equal a small[60] square and ten units. We subtract the five units in common from both sides, leaving a small square and five units Equal a great square, and the difference of the two squares is five units. Let us look for two square numbers whose difference is five units, whatever two numbers they may be. Let the smaller be four units and the greater be nine units. And we make the square Equated to the Cube lacking five *Māl*s be nine *Māl*s, and we make the square Equated to the Cube lacking ten *Māl*s be four *Māl*s. Each of the Equations turns out to be a Cube Equals fourteen *Māl*s, and the one Thing Equals fourteen units.   1173

And since we assigned a side of the cube to be one Thing, its side is fourteen, and the cube is two thousand seven hundred forty-four. The square that comes from its side is one hundred ninety-six. When we multiply it by five units, it results in nine hundred eighty, and when we subtract it from the cubic number, it leaves one thousand seven hundred sixty-four, which is a square whose side is forty-two. Likewise, when we multiply a square of a side of the cube by ten units, it amounts to one thousand nine hundred sixty, and when we subtract it from the cubic number, it leaves seven hundred eighty-four, which is a square whose side is twenty-eight.   1183

We have found a cubic number according to the stipulation that we stipulated, and it is the two thousand seven hundred forty-four. And that is what we wanted to find.   1193

39. We want to find a cubic number such that if we multiply a square of its side by two given numbers and we subtract the cube from each of them, leaves from each of them a square number.   1194

Let the two given numbers be three and seven. We want to find a cubic number such that if we multiply a square of its side by three and by seven, and we subtract the cube from each of the products, it leaves from each of them a square number.   1197

---

59 Rashed's restoration.

60 Sesiano notes (1982, 115 n. 75): "The positive form is used when no strict comparison is involved. The 'larger' [our 'greater'] (*a'ẓam*) written before may be a scribal error". The same words are found in problems IV.39, IV.40, IV.41, and IV.43.

1199   Let us assign the cube to be one Cube, and we multiply a square of its side, which is one *Māl*, by three and by seven. We subtract the Cube from each of them, leaving three *Māl*s less a Cube Equal a square, and seven *Māl*s less a Cube Equal a square. We make a side of the square Equated to the three *Māl*s lacking a Cube be Things. We multiply it by itself, so it becomes: *Māl*s Equal three *Māl*s except a Cube. We add the Cube in common to both sides, so it becomes: *Māl*s and a Cube Equal three *Māl*s. And if we subtract the common *Māl*s from the three *Māl*s, it leaves a Cube Equals *Māl*s. Thus, the Thing is equal to the number of remaining *Māl*s. Likewise, if we make a side of the square Equated to the seven *Māl*s lacking a Cube to be Things, and we multiply it by itself and we restore and confront, it also leaves one Cube Equals the remainder of the seven ⟨*Māl*s⟩. And the Thing is likewise equal ⟨to the remainder⟩ of the seven. Therefore, the remaining *Māl*s from the three *Māl*s must be equal to the remaining *Māl*s from the seven *Māl*s. But the remainder from the three *Māl*s is three less a square number, and the remainder from the seven *Māl*s is seven less a square number. Therefore, three less a square number Equal seven less a square number. We add each of the two squares in common to both sides, so it amounts to seven and a small square Equal three and a great square. We cast away the common three, leaving a great square Equals a small square and four units.

1216   Therefore, we should look for two square numbers whose difference is four units and so that the smaller is less than three units.[61] They are two and a fourth, and six units and a fourth. We make the square Equated to the three *Māl*s lacking a Cube be two *Māl*s and a fourth of a *Māl*, and we make the square Equated to the seven *Māl*s lacking a Cube be six *Māl*s and a fourth of a *Māl*. Thus, each of the Equations turns out to be a Cube Equals three fourths of a *Māl*. Therefore, the Thing is three fourths of a unit.

1222   The Cube is twenty-seven eighths of an eighth, and a square of a side of the Cube is thirty-six eighths of an eighth. And when we multiply it by three, it amounts to one hundred eight eighths of an eighth. And when we subtract the cubic number from it, it leaves eighty-one eighths of an eighth, which is a square whose side is nine eighths. Similarly, when we multiply a square of a side of the cube, which is thirty-six eighths of an eighth, by seven units, it amounts to two hundred fifty-two eighths of an eighth. When the cubic number is subtracted from it, it leaves two hundred twenty-five eighths of an eighth, which is a square number whose side is fifteen eighths.

1230   We have found a cubic number according to the stipulation that we stipulated, and it is twenty-seven eighths of an eighth. And that is what we wanted to find.

1232   **40.** We want to find two numbers, a square and a cube, such that a square of the square, if the cube is added to it, amounts to a square number, and if the cube is subtracted from it, it leaves a square number.

_____

61 Problem II.10. /s/

Let us assign the square to be from a side of two Things, so the square is four
Māls, and its square is sixteen *Māl Māls*. And let us assign the cube to be from a
side of however many Things we wish. Let us assign it to be from a side of four
Things, so the cube is sixty-four Cubes. We add this cube to sixteen *Māl Māls*,
and we subtract it from it, to get sixteen *Māl Māls* and sixty-four Cubes Equal a
square number, and sixteen *Māl Māls* less sixty-four Cubes Equal a square number. Then we take up the search for a Thing that makes the two sides of the Equation equal, as we did before.[62] So we say, if we assign a side of the square Equated
to the sixteen *Māl Māls* and sixty-four Cubes to be *Māls*, then its square is *Māl
Māls* (which) Equal sixteen *Māl Māls* and sixty-four Cubes. If we subtract the
sixteen *Māl Māls* in common from both sides, it leaves sixty-four Cubes Equal
Māl Māls. And if we divide all of that by a Cube, it gives Things Equal sixty-four
units. Therefore, the number assigned in the problem ⟨to be a Thing⟩ is equal to
what results from dividing the number of Cubes, which is sixty-four, by the number of remaining *Māl Māls*. Similarly, if we assign a side of the square Equated
to the sixteen *Māl Māls* less sixty-four Cubes to be *Māls*, then its square is *Māl
Māls* (which) Equal sixteen *Māl Māls* less sixty-four Cubes. If we add the lacking Cubes in common to both sides, it becomes *Māl Māls* and sixty-four Cubes
Equal sixteen *Māl Māls*. And if we subtract the common *Māl Māls*, it leaves you
with sixty-four Cubes Equal *Māl Māls*. If we divide all of that by a Cube, it gives
sixty-four units Equal Things. Therefore, the Thing is the number that results
from dividing the sixty-four by the number of the remaining ⟨*Māl*⟩ *Māls*.

Therefore, the number of ⟨*Māl*⟩ *Māls* remaining in the first Equation is equal to
the number of ⟨*Māl*⟩ *Māls* remaining in the second Equation. But the number of
⟨*Māl*⟩ *Māls* remaining in the first Equation is a square number less sixteen, and
the ⟨number of *Māl*⟩[63] *Māls* remaining in the second Equation is sixteen less a
square number. Thus, a great square number less sixteen Equals sixteen less a
small square number. If we add the small square in common and we also add the
sixteen lacking from the great square in common to both sides, it amounts to a
great square and a small square Equal thirty-two units. But the thirty-two is made
up of two equal squares, so it can be divided into two different square numbers.[64]
Let us divide (it), and let one of the squares be sixteen fifths of a fifth and the other
square be thirty-one and nine fifths of a fifth. And we make the square Equated
to the sixteen *Māl Māls* and sixty-four Cubes be thirty-one *Māl Māls* and nine
fifths of a fifth of a *Māl Māl*, and the square Equated to the sixteen *Māl Māls* less
sixty-four Cubes be sixteen fifths of a fifth of a *Māl Māl*. Each of the Equations
turns out to be sixty-four Cubes Equal fifteen *Māl Māls* and nine fifths of a fifth
of a *Māl* ⟨*Māl*⟩.

---

62 Solutions by "equating the corresponding sides" had been given in the second solution to problem
   IV.34, and in IV.35–39. /s/
63 Rashed's restoration. Sesiano only restores "*Māl*".
64 Problem II.9. /s/

1271   We divide all of that by a Cube to get fifteen Things and nine fifths of a fifth of a Thing Equal sixty-four units. Therefore, the one Thing is what results from dividing one thousand six hundred by three hundred eighty-four, which is four units and a sixth of a unit.

1274   Since we assigned a side of the square to be two Things, a side of the square is eight units and a third of a unit, and the square is sixty-nine units and four ninths of a unit, and a square of the square is four thousand eight hundred twenty-two and four ninths and seven ninths of a ninth. And since we assigned a side of the cube to be four Things, a side of the cube is sixteen units and two thirds of a unit, and the cube is four thousand six hundred twenty-nine and five ninths and two thirds of a ninth, which, when added to the number gathered from multiplying the square number by itself, amounts to nine thousand four hundred fifty-two and a ninth and four ninths of a ninth, which is a square number whose side is ninety-seven units and two ninths of a unit. And when this number is subtracted from a square of the square number, it leaves one hundred ninety-two and eight ninths and a ninth of a ninth, which is a square number whose side is thirteen units and eight ninths of a unit.

1286   We have found two numbers according to the stipulation that we stipulated, and they are the two numbers that we determined.

1288   **41.** We want to find two other[65] numbers, a cube and a square, such that the cubic number, if a square of the square is added to it, it amounts to a square number, and (if) a square of the square is subtracted from it, it leaves a square number.

1291   In the way we described, we say: sixty-four Cubes and sixteen *Māl Māl*s Equal a square number. If we make its side *Māl*s, then the square is *Māl Māl*s (which) Equal sixty-four Cubes and sixteen *Māl Māl*s. If we subtract the sixteen *Māl Māl*s in common from both sides, it leaves sixty-four Cubes Equal *Māl Māl*s. If we divide them by a Cube, it gives sixty-four units Equal Things, so the Thing is what results from dividing the sixty-four by the number of remaining *Māl Māl*s. Likewise, sixty-four Cubes less sixteen *Māl Māl*s Equal a square number. If we make its side *Māl*s, then its square is *Māl Māl*s (which) Equal sixty-four Cubes less sixteen *Māl Māl*s. We add the sixteen *Māl Māl*s in common to both sides, so it amounts to *Māl Māl*s Equal sixty-four Cubes. And if we divide them by a Cube, it gives Things Equal sixty-four units. Again, the Thing is what results from dividing the sixty-four by the number of gathered ⟨*Māl*⟩ *Māl*s.

1303   Thus, the number of ⟨*Māl*⟩ *Māl*s remaining in the first Equation, which is a square number less sixteen, should be equal to the number of ⟨*Māl*⟩ *Māl*s gathered in the

---

65 Sesiano notes (1982, 118 n. 81): "The presence of this 'other' and the abrupt beginning of the problem may indicate that nos. 40 and 41 were once considered to be a single problem with two subdivisions".

second Equation, which is a square number and sixteen. Thus, a square number less sixteen Equals another square and sixteen. We add the lacking sixteen in common to both sides. It amounts to a (small) square and thirty-two units Equal a great square. Therefore, we look for two square numbers whose difference is thirty-two units.[66] Thus, the greater square is more than sixteen. Let the smaller square be four units and the greater square be thirty-six. We assign ⟨the square⟩ Equated to the sixty-four Cubes and sixteen *Māl Māl*s to be thirty-six *Māl Māl*s, and the square Equated to the sixty-four Cubes less sixteen *Māl Māl*s to be four *Māl Māl*s. Each of the Equations turns out to be sixty-four Cubes Equal twenty *Māl Māl*s. Let us divide each of them by a Cube, to get twenty Things Equal sixty-four units. So the one Thing is three units and a fifth of a unit.

Since we assigned the square to be from a side of two Things, ⟨its side⟩ is six   1317
units and two fifths of a unit, and the square is forty units and four fifths of a unit and four fifths of a fifth of a unit, and a square of the square is one thousand six hundred seventy-seven and four hundred fifty-one parts of six hundred twenty-five parts of a unit. And since we assigned a side of the cube to be four Things, a side of the cube is twelve and four fifths, and the cube is two thousand ninety-seven and ninety-five parts of six hundred twenty-five parts of a unit. And when a square of the square is added to it, it amounts to three thousand seven hundred seventy-four and five hundred forty-six parts of six hundred twenty-five parts of a unit, which is a square whose side is sixty-one and eleven parts of twenty-five parts of a unit. And when a square of the square number is subtracted from this cube, it leaves four hundred nineteen units and two hundred sixty-nine parts of six hundred twenty-five parts of a unit, which is a square number whose side is twenty units and twelve parts of twenty-five parts of a unit.

We have found two numbers according to the stipulation that we stipulated, and   1331
they are the two numbers that we determined. And that is what we wanted to find.

42. We want to find two numbers, a cube and a square, such that the sum of a cube   1333
of the cube and a square of the square is a square number, and their difference is a square number.

We assign a side of the cube to be whatever we wish of Things. Let us assign it to   1335
be two Things, so ⟨the cube⟩ is eight Cubes, and a cube of that cube is five hundred twelve Cube Cube Cubes. Similarly, we assign the square to be from a side of however many we wish of *Māl*s. Let us assign it to be from a side of four *Māl*s, so that the square is sixteen *Māl Māl*s, and a square of the square is two hundred fifty-six *Māl Māl*s by a *Māl Māl*, which is called one Cube Cube *Māl*.

Let us first require that a cube of the cube, if a square of the square is added to it,   1341
gives a square, and (if) a square of the square is subtracted from it, leaves a square. We previously found two numbers with this characteristic by chance, without

---

66 Problem II.10. /s/

intending to find them.[67] We want to relate here the way to find them. We say: five hundred twelve Cube Cube Cubes and two hundred fifty-six Cube Cube *Māl*s Equal a square number. Similarly, five hundred twelve Cube Cube Cubes less two hundred fifty-six Cube Cube *Māl*s Equal a square.

1348   If we wish, we can work this out by the method of the double equality, which is that we take the difference between these two squares, which is five hundred twelve *Māl Māl*s by a *Māl Māl*, and we look for two numbers of *Māl Māl*s such that, if we multiply one of them by the other, then what is gathered from the multiplication is five hundred twelve *Māl Māl*s by a *Māl Māl*. Then, if we take half of the sum of the two numbers and we multiply what is gathered by itself and we confront it with the greater square, which is five hundred twelve Cube Cube Cubes and two hundred fifty-six Cube Cube *Māl*s, then we take ⟨half of⟩ the difference between the two numbers, and we multiply the outcome by itself and we confront it with the smaller square, which is five hundred twelve Cube Cube Cubes less two hundred fifty-six *Māl Māl*s by a *Māl Māl*, then each of the two confrontations[68] results in five hundred twelve Cube Cube Cubes Equal the same number of Cube Cube *Māl*s. And if we divide each of them by one of the lesser in degree of them, which is a Cube Cube *Māl*, it gives five hundred twelve Things Equal a number, and from that we can find the Thing. And once the Thing is found, we can return to the hypostases we established.[69] After knowing the Thing, we can synthesize everything in the problem.

1363   And if we wish, we can work this out by seeking to make the two sides of the Equation equal, as we explained in the preceding problems,[70] which is that we say: if we make a side of the greater square *Māl Māl*s, then its square is Cube Cube *Māl*s (which) Equal (the expression for) the greater square. And if we subtract the two hundred fifty-six Cube Cube *Māl*s in common from both sides, it leaves five hundred twelve Cube Cube Cubes Equal Cube Cube *Māl*s. If we then divide them by a Cube Cube *Māl*, it gives five hundred twelve Things Equal a number. Therefore[71] the number is equal to the number of remaining Cube Cube *Māl*s which, if divided by five hundred twelve, gives as a result the number assigned to be a Thing in the problem.

67 Diophantus found the numbers in Problems IV.29 and IV.30. /s/
68 The confrontation of two expressions results in an equation, and in this case "the two confrontations" (*al-mutaqābilatayn*, at line 1357) both "result in" the same equation. Sesiano regards *al-mutaqābilatayn* to be an error and substitutes it with *al-muʿādalatayn* ("the two Equations"). Rashed does not correct the text, but he nevertheless translates *al-mutaqābilatayn* as "des deux équations". He gives his explanation of the word in (Diophantus 1984, 3, 155–56).
69 I.e., return to the initial assignments. /s/
70 Solutions by "equating the corresponding sides" had been given in the second solution to Problem IV.34, and in Problems IV.35–41. /s/
71 Sesiano corrects "likewise" (*ka-dhālika*) to "therefore" (*li-dhālika*) (line 1369). Rashed does not make the correction (Diophantus 1984, 3, 83.16).

And likewise, if we assign a side of the smaller square to be *Māl Māl*s, then its square is Cube Cube *Māl*s (which) Equal (the expression for) the smaller square. And if we add the two hundred fifty-six Cube Cube *Māl*s in common to both sides, it amounts to Cube Cube *Māl*s Equal five hundred twelve Cube Cube Cubes. And if this is divided by one of the lesser in degree of the two sides, which is a Cube Cube *Māl*, it gives five hundred twelve Things Equal a number. Therefore, if that number is divided by five hundred twelve, then the number that results from the division is the assigned ⟨number⟩ of Things in the problem.

1371

Thus, the number of Cube Cube *Māl*s remaining in the first Equation should be equal to the number of Cube Cube *Māl*s gathered in the second Equation. But the number of Cube Cube *Māl*s remaining in the ⟨first⟩ Equation ⟨is a square number less two hundred fifty-six, and the number of Cube Cube *Māl*s gathered in the⟩ second ⟨Equation⟩ is a square number and two hundred fifty-six. Therefore, we should look for two ⟨square⟩[72] numbers whose difference is double the two hundred fifty-six, that is, five hundred twelve.[73] And once we have found them, we make the greater (square) Cube Cube *Māl*s, and we Equate it with (the expression for) the greater square. And we make the smaller (square) Cube Cube *Māl*s and we Equate it with (the expression for) the smaller square. Each of the Equations then turns out to be, after that, five hundred twelve Things Equal the same number, and from that we can find the Thing whose value we are looking for. Then we return to perform the synthesis of the problem.

1379

If we wish, we can say: five hundred twelve Cube Cube Cubes and two hundred fifty-six Cube Cube *Māl*s Equal a square, and five hundred twelve Cube Cube Cubes less two hundred fifty-six Cube Cube *Māl*s Equal a square. And for any square divided by a square, the result of the division is a square. Thus, we divide the five hundred twelve Cube Cube Cubes and the two hundred fifty-six Cube Cube *Māl*s by a square, which can be a Cube Cube *Māl* or four Cube Cube *Māl*s or nine Cube Cube *Māl*s or sixteen Cube Cube *Māl*s or by whatever we wish among the square numbers after we make any one of them Cube Cube *Māl*s. As for Cube Cube *Māl*s, the result of its division by Cube Cube *Māl*s is a number, and for Cube Cube Cubes, the result is Things. Let us suppose[74] that we divide them by sixteen Cube Cube *Māl*s. Then the result of the division is thirty-two Things and sixteen units. And by the same amount that we divide this square, let us divide the other square, which is five hundred twelve Cube Cube Cubes less two hundred fifty-six Cube Cube *Māl*s, to get thirty-two Things less sixteen units. Thus, thirty-two Things and sixteen units are a square, and thirty-two Things less

1391

---

72 Rashed's restoration. Sesiano adds "(square)" to his translation.

73 Problem II.10. /s/

74 Rashed's word here is *faraḍa*, the same word we translate as "assign", and which he translates with "Supposons" (Diophantus 1984, 3, 85.12). The word in Sesiano's edition is *nazala*, "to put down" in this context. He translates it as "Suppose", following other translators (1982, 457, line 1400).

sixteen units are a square. Let us look for a number such that, if we add a given number to it, which is sixteen, it gives a square, and if we subtract a given number from it, which is sixteen, it gives a square. Once we have found that number, we divide it by thirty-two, so what results from the division is the Thing. Once we know it, we return to synthesize the problem according to the way we set it up in the analysis. Most problems of "the self-same sides" (like the ones) presented above can be worked out by the method we (just) explained.

1412   Now let us look for a square of the square such that, if a cube of the cube is added to it, gives a square, and if a cube of the cube is subtracted from it, leaves a square number. Again, let us say as before: two hundred fifty-six Cube Cube *Māl*s and five hundred twelve Cube Cube Cubes Equal a square, and two hundred fifty-six Cube Cube *Māl*s less five hundred twelve Cube Cube Cubes Equal a square. We work this out by looking for equality in both sides of the Equation, as we explained above for this kind of problem.[75] In the end, we divide double the two hundred fifty-six, which is a square number, and its double is five hundred twelve, into two different square numbers.[76] Let the smaller of the two square numbers be ten units and six parts of twenty-five parts of a unit, whose side is three units and a fifth of a unit, and the greater square number be five hundred one and nineteen parts of twenty-five parts of a unit, whose side is twenty-two and two fifths of a unit. We make the smaller of these two squares Equal to (the expression for) the smaller of the first of the two squares, and the greater of them Equal to (the expression for) the greater of them.[77] And each of the two Equations turns out to be five hundred twelve Cube Cube Cubes Equal two hundred forty-five Cube Cube *Māl*s and nineteen parts of twenty-five parts of a Cube Cube *Māl*. Let us divide each of them by a Cube Cube *Māl*, to get five hundred twelve Things Equal two hundred forty-five units and nineteen parts of twenty-five parts of a unit. Therefore, the one Thing is twelve parts of twenty-five.

1432   And since we assigned a side of the cube to be two Things, a side of the cube is twenty-four parts of twenty-five parts of a unit, and the cube is thirteen thousand eight hundred twenty-four parts of a cube of the twenty-five, and a cube of this cube is two trillion six hundred forty-one billion eight hundred seven million five hundred forty thousand two hundred twenty-four parts of a cube ⟨of a cube⟩ of the twenty-five, which is also one hundred five billion six hundred seventy-two million three hundred one thousand six hundred eight ⟨parts⟩ and twenty-four parts ⟨of twenty-five parts⟩ of one part of a square of a square of the six hundred twenty-five.

1441   And the square: since we assigned its side to be four *Māl*s, and the *Māl* is one hundred forty-four parts of six hundred twenty-five – and that is because the Thing is

75 Solutions by "equating the corresponding sides" had been given in the second solution to Problem IV.34, in Problems IV.35–41, and in the second solution to IV.42(a). /s/
76 Problem II.9. /s/
77 He should have mentioned that we make both squares Cube Cube *Māl*s first. /s/

twelve parts of twenty-five – a side of the square is five hundred seventy-six parts ⟨of six hundred twenty-five parts⟩ and the square is three hundred thirty-one thousand seven hundred seventy-six parts of a square of the six hundred twenty-five, and a square of this square is one hundred ten billion seventy-five million three hundred fourteen thousand one hundred seventy-six parts of a square of a square of the six hundred twenty-five, which, if a cube of the cubic number is added to it, amounts to two hundred fifteen billion seven hundred forty-seven million six hundred fifteen thousand seven hundred eighty-four parts and twenty-four parts of twenty five of a part of a square of a square of the six hundred twenty-five, which is a square whose side is four hundred sixty-four thousand four hundred eighty-six parts and two fifths of a part of a square of the six hundred twenty-five. And if a cube of the cubic number is subtracted from a square of this square, it leaves four billion four hundred three million twelve thousand five hundred sixty-seven parts and a part of twenty-five parts of a part of a square of a square of the six hundred twenty-five, which is a square whose side is sixty-six thousand three hundred fifty-five parts and a fifth of a part of a square of the six hundred twenty-⟨five⟩.

We have found two numbers according to what we wanted, and they are the two numbers that we determined. And that is what we wanted to find.   1461

43. We want to find two numbers, a cube and a square, such that a cube of the cube, if we add a given multiple of a square of the square to it, amounts to a square number, and if we subtract a given multiple of a square of the square from it, leaves a square number.   1463

Let us assign the cube to be one Cube, so its cube is a Cube Cube Cube. And we assign a side of the square to be whatever we wish of *Māls*. Let us assign it to be from a side of two *Māls*, so it is four *Māl Māls*, and a square of the square is sixteen Cube Cube *Māls*. And let the added given multiple be one time and a fourth of a time, and the subtracted given multiple be a half and a fourth of a time. We add to a cube of the cube one and a fourth of ⟨a square⟩ of the square, which is twenty Cube Cube *Māls*, which gives a Cube Cube Cube and twenty Cube Cube *Māls*, which Equal a square number. And let us subtract from a cube of the cube a half and a fourth of a square of the square, which is twelve Cube Cube *Māls*, which gives a Cube Cube Cube less twelve Cube Cube *Māls* Equals a square number.   1466

So if we assign a side of the square Equated to the Cube Cube Cube and the twenty Cube Cube *Māls* to be *Māl Māls*, then its square is *Māl Māls* by *Māl Māls*, of which one of them is called a Cube Cube *Māl*. If we Equate it with ⟨the Cube⟩ Cube Cube and the twenty Cube Cube *Māls*, then we subtract the twenty ⟨Cube Cube *Māls*⟩ in common, it leaves a Cube Cube Cube Equals Cube Cube *Māls*, its number being equal to a square less twenty, and that is the number assigned to be a Thing in this solution. Likewise, if we assign a side of the square Equated to the Cube Cube Cube less twelve Cube Cube *Māls* to be *Māl Māls*, then its square is Cube Cube *Māls*. And if we add to it the twelve Cube Cube *Māls* lacking from the   1474

Cube Cube Cube and we make them added in common to both sides, it amounts to a Cube Cube Cube Equals Cube Cube *Māl*s, its number being equal to a square and twelve, and that is (also) the number assigned to be a Thing in the problem. So a ⟨great⟩[78] square less twenty Equals a small square and twelve. And we add the twenty in common to the two sides together, giving a small square and thirty-two Equal a great square. The small square is four units, and when thirty-two is added to it, it amounts to thirty-six, which is the great square. So we make the square Equated to the Cube Cube Cube and twenty Cube Cube *Māl*s be thirty-six Cube Cube *Māl*s, and the square Equated to the other square be four Cube Cube *Māl*s. Both Equations turn out to be, after the restoration and confrontation and division, a Thing Equals sixteen units.

1493   We will now perform the synthesis of the problem in the way we did the analysis. A side of the cube: we made it a Thing, so it is sixteen units, and the cube is four thousand ninety-six. And a side of the square: we assigned it to be two *Māl*s, and the *Māl* is two hundred fifty-six, so a side of the square is five hundred twelve and the square is two hundred sixty-two thousand one hundred forty-four. As for a cube of the cube, it is sixty-eight billion seven hundred nineteen million four hundred seventy-six thousand seven hundred thirty-six. And for a square of the square, it is also the same as this number. So a cube of the cube is a square equal to what is gathered from multiplying the square number by itself. And if a square of the square and its fourth are added to it, then what is gathered from that is twice a square of the square and its fourth, and that is a square number whose side is one and a half of the square number. Also, if three fourths of a square of the square are subtracted from it, the remainder is a fourth of a square of the square number, and that is a square whose side is half of the square number.

1507   We have found two numbers according to the description that we described, and they are the two numbers we determined. And that is what we wanted to find.

1509   **44.** We want to find two numbers, a cube and a square, such that a square of the square number, if we multiply it by two ⟨given⟩ numbers and we add a cube of the cube to either of them, gives from its addition to each of them a square number, or if we subtract either of them from a cube of the cube then what remains is a square number, or if we subtract from either of them ⟨a cube⟩ of the cube, then what remains from each of them is a square number.

1515   Let one of the given numbers be three and the other eight. We want to find two numbers, a cube and a square, such that a square of the square, if we multiply it by three and by eight, and we add what is gathered from either of them to a cube of the cube, gives from the addition of each of them a square number, or if we subtract what is gathered from either of them from a cube of the cube, then what remains from a cube of the cube, after subtracting either of them, is a square

---

78 Rashed's restoration.

number, or if we subtract ⟨a cube⟩ of the cube from either of them, it leaves from each of them a square number.

44[a]. So let us investigate the first of the three. We assign a side of the cube to be one Thing, so ⟨the cube is a Cube and⟩ its cube is a Cube Cube Cube. And we assign the square to be from a side of two *Māl*s, so the square is four *Māl Māl*s and a square of the square is sixteen Cube Cube *Māl*s. When we multiply it by three and by eight, they amount to forty-eight Cube Cube *Māl*s and one hundred twenty-eight Cube Cube *Māl*s. And when we add each of them to a cube of the cube, they give a Cube Cube Cube and forty-eight Cube Cube *Māl*s, and a Cube Cube Cube and one hundred twenty-eight Cube Cube *Māl*s.

1522

Each of them is a square, and any square divided by a square gives a square, so let us divide each of them by a square, and let that square be a Cube Cube *Māl*. One of the quotients is a Thing and forty-eight units, which Equals a square number, since it is the result of dividing a square by a square. And the other quotient is a Thing and one hundred twenty-⟨eight⟩ units, which Equals a square, since it is the result of dividing a square by a square. Then the Thing, if forty-eight units is added to it, amounts to a square, and if one hundred twenty-eight is added to it, also becomes a square. Let us look for a number that, if we add it to these two numbers, then each of them becomes a square,[79] and it is sixteen units. The Thing, then, is sixteen.

1529

And since we assigned a side of the cube to be a Thing, its side is sixteen, and the cube is the cubic number that we found in the previous problem, and its cube is likewise the number that is its cube in the previous problem. Similarly, a square of the square is also equal to a cube of the cube, and if multiplied by three units and increased by a cube of the cube, it amounts to four times a square of the square, which is a square whose side is twice the square number. Similarly, if that is multiplied by eight units and increased by a cube of the cube, it amounts to nine times a square of the square, which is a square whose side is three times the square number.

1537

We have found two numbers, a cube and a square, such that if we multiply a square of the square by three and by eight, then we add either of them to a cube of the cube, what is gathered from either of them is a square number. They are: for the cube, four thousand ninety-six, and for the square, two hundred sixty-two thousand one hundred forty-four.

1545

[44b.] Similarly, let us investigate the second of the three. If we again make the cube one Cube and the square four *Māl Māl*s, we get two squares, one of them a Cube Cube Cube less forty-eight Cube Cube *Māl*s, and the other a Cube Cube Cube less one hundred twenty-eight Cube Cube *Māl*s. And for every square

1549

---

79 Problem II.11. /s/ In his commentary Sesiano refers to Problem II.10. In fact, both procedures can be used.

divided by a square, the quotient is also a square. And let the square by which we divide a Cube Cube Cube less ⟨forty-eight Cube Cube *Māl*s and a Cube Cube Cube less one hundred twenty-eight Cube Cube *Māl*s⟩ be a Cube Cube *Māl*, which is what comes from multiplying a *Māl Māl* by itself. One of the quotients is a Thing less forty-eight and the other is a Thing less one hundred twenty-eight, and each of them is a square. Let us look for a number such that if we subtract forty-eight from it, leaves a square number, and if we subtract one hundred twenty-eight from it, also leaves a square number.[80] That number is taken to be a Thing in the treatment of the problem, and it is one hundred ninety-two.

1560   Since a side of the cube that we found in the previous problem is sixteen, and a side of the cube (in this problem) is one hundred ninety-two, a side of this cube is with respect to[81] a side of that cube in the ratio of twelve times. So this cube is with respect to that cube in the ratio of a cube of twelve to one. And since we assigned a side of the square to be two *Māl*s, and this Thing is with respect to that Thing from the previous problem in the ratio of twelve to one, this *Māl* is with respect to that *Māl* in the ratio of a square of twelve to one. Likewise (for) a side of the square with respect to a side of that square. As for the square with respect to the square, (they) are in the ratio of ⟨a square⟩ of one hundred forty-four to one. And as for a cube of this cube with respect to a cube of that cube, (they) are in the ratio of a cube of the cube that comes from (the ratio of) twelve to one. And as for a square of the square with respect to a square of that square, (they) are in the ratio of a square of a square of one hundred forty-four to one, and a square of that square is equal to a cube of that cube. Therefore, a cube of that cube is a square, ⟨and⟩ one is also a cube (and a) square.[82]

1573   Therefore, a side of this cube is twelve and the cube is one thousand seven hundred twenty-eight, and a side of the square is one hundred forty-four and the square is twenty thousand seven hundred thirty-six, and a cube of this cube is five billion one hundred fifty-nine million seven hundred eighty thousand three hundred fifty-two, and a square of this square is four hundred twenty-nine million nine hundred eighty-one thousand six hundred ninety-six. And three times a square of the square is one billion two hundred eighty-nine million nine hundred forty-⟨five⟩ thousand eighty-eight. And if we subtract it from a cube of the cube, it leaves three billion eight hundred sixty-nine million eight hundred thirty-five thousand two hundred sixty-four, which is a square whose side is sixty-two thousand two hundred eight. And eight times a square of the square is three billion four hundred thirty-nine million eight hundred fifty-three thousand five hundred sixty-eight which, if subtracted from a cube of the cube, leaves one billion seven hundred nineteen million

---

80 Problem II.13. /s/ In his commentary Sesiano refers to Problem II.10. In fact, both procedures can be used.

81 In this passage every instance of "with respect to" is the word *'inda*.

82 I.e. a perfect fifth power. Sesiano translates this as "a square cube", and Rashed translates it as "un cube et un carré" (Diophantus 1984, 3, 96.7).

nine hundred twenty-six thousand seven hundred eighty-four, which is a square whose side is forty-one thousand four hundred seventy-two.

We have found two numbers, a cube and a square, such that if we multiply a square of the square by three and by eight, and we subtract either of them from a cube of the cubic number, leaves a square number. And they are the two numbers we found.

1591

[**44c.**] Let us now investigate the remaining variation of the three that we set out. We say, forty-eight Cube Cube *Māl*s less a Cube Cube Cube Equal a square, and one hundred twenty-eight Cube Cube *Māl*s less a Cube Cube Cube Equal a square. Let us divide them by a Cube Cube *Māl* to get that one of the quotients is forty-eight units less a Thing, and the other is one hundred twenty-eight units less a Thing, and each of them is a square. Let us look for a number which, if we subtract it from the forty-eight and from the one hundred twenty-eight, leaves from either of them a square.[83] Let it be forty-seven, and that is the number assigned to be a Thing in the solution of this problem.

1594

And since we assigned a side of the cube to be a Thing, its side is forty-seven, so the cube is one hundred three thousand eight hundred twenty-three. And since we assigned a side of the square to be two *Māl*s, and the *Māl* is two thousand two hundred nine units, a side of the square is four thousand four hundred eighteen, and the square is nineteen million five hundred eighteen thousand seven hundred twenty-four. And a cube of the cube, if subtracted from three times a square of this square, gives a square remainder whose side is four million eight hundred seventy-nine thousand six hundred eighty-one, and if subtracted from eight times a square of the square, it leaves a square whose side is forty-three million nine hundred seventeen thousand one hundred twenty-nine.

1601

We have found two numbers, a cube and a square, such that a square of the square, it multiplied by three and by eight, and a cube of the cube is subtracted from ⟨each⟩ of them, leaves from each of them a square number, and they are the two numbers we determined. And that is what we wanted to find.

1611

End of the fourth Book[84] of the work of Diophantus on the squares and cubes, which consists of forty-four problems.

1615

---

83 Problem II.12. /s/. In his commentary Sesiano refers to Problem II.10. In fact, both procedures can be used.
84 The text has *qawl*, a word meaning something closer to "section".

Book V (Arabic)

1617   In the name of God, the Merciful, the Compassionate. The fifth Book of the work of Diophantus of Alexandria on arithmetical problems.

1619   **1.** We want to find two numbers, a square and a cube, such that a square of the square, if we add to it a given multiple of the cubic number, amounts to a square number, and if we subtract from it another given multiple of the cubic number, leaves a square number.

1622   Let the appended multiple be four and the lacking (multiple) be three. We want to find two numbers according to what we described.

1623   We make a side of the square one Thing, so that the square is one *Māl* and a square of the square is one *Māl Māl*, which, together with four times a certain cube Equals a square, and less three times that cube also Equals a square. Thus, the cube Equals a certain quantity that has to the *Māl Māl* a given[1] ratio such that if four times it is added to the *Māl Māl*, it gives a square, and if three times it is subtracted from the *Māl Māl*, it leaves a square. We look for three square numbers such that the ratio of the excess of the greater over the middle to the excess of the middle over the smaller is equal to the ratio of the four to the three.[2] Let these numbers be eighty-one and forty-nine and twenty-five. If we make the *Māl Māl* forty-nine parts, then the quantity in the given ratio to it – which, if four times it, which is thirty-two parts of forty-nine parts, is added to the *Māl* ⟨*Māl*⟩ gives a square; and if three times it, which is twenty-four parts of forty-nine parts, is subtracted from the *Māl Māl* leaves a square – is eight parts of forty-nine parts of the *Māl Māl*. So the sought-after cube Equals eight parts of forty-nine parts of a *Māl Māl*. Let us assign the cube to be from a side of however many Things we wish. Let us assign it to be from a side of two Things, so the cube is eight Cubes. Thus,

---

1 In this problem, the text has "a given (*mafrūḍa*) ratio", while in the corresponding passages in problems V.2 and V.3 we find "a known (*ma'lūm*) ratio". The two words were used interchangeably for this purpose in Arabic mathematics (Sidoli and Isahaya 2018, 210). Also, the ratio will truly be "given" (i.e., numerically determinable) only after Diophantus chooses his three squares in the next step.
2 Problem II.19. /s/

DOI: 10.4324/9781315171470-12

the eight Cubes Equal eight parts of forty-nine parts of a *Māl Māl*. Let us divide each of them by a Cube, so eight parts of forty-nine parts of a Thing Equal eight units, and the one Thing Equals forty-nine units.

Therefore, a side of the square is forty-nine, and the square is two thousand four hundred one. And since we assigned a side of the cube to be two Things, a side of the cube is ninety-eight, and the cube is nine hundred forty-one thousand one hundred ninety-two. And as for a square of the square, it is five million seven hundred sixty-four thousand eight hundred one. And when four times the cubic number, which is three million seven hundred sixty-four thousand seven hundred sixty-eight, is added to it, it amounts to nine million five hundred twenty-nine thousand five hundred sixty-nine, which is a square whose side is three thousand eighty-seven. And when three times the cubic number, which is two million eight hundred twenty-three thousand five hundred seventy-six, is subtracted from it, it leaves two million nine hundred forty-one thousand two hundred twenty-five, which is a square number whose side is one thousand seven hundred fifteen. 1642

We have found two numbers according to the stipulation that we wanted. And that is what we wanted to find. 1655

2. We want to find two numbers, a square and a cube, such that the cubic number, if we multiply it by two given numbers and we add each of them to a square of the square, then each of them amounts to a square. 1657

We make the two given numbers twelve units and five units and we assign a side of the square to be one Thing. Thus, the square is one *Māl* and its square is one *Māl Māl*, which, together with twelve cubes, Equal a square, while together with five times that cube also Equal a square. Then let us look for the quantity in the known ratio with respect to the *Māl Māl* for which, if twelve times it is added to it, gives a square, and if five times it is added to it, also gives a square. Therefore, we are led to look for three square numbers such that the ratio of the excess of the greater over the middle is to the excess of the middle over the smaller as the ratio of the excess of the twelve over the five to the five, that is, the ratio of one time and two fifths of a time.[3] Let these numbers be sixteen and nine and four. If we make the *Māl Māl* (consist of) four parts, it is clear that the quantity in a given ratio with respect to it which, if five times it, that is five parts, is added to it, gives a square, and if twelve times it, that is twelve parts, is added to it, gives a square, is a fourth of a *Māl Māl*. So a fourth of a *Māl Māl* Equals a cubic number. Let us assign the cube to be from a side of two Things. Therefore, the cube is eight Cubes, which Equal a fourth of a *Māl Māl*. Let us divide each of them by a Cube, to get a fourth of a Thing Equals eight units. Therefore, the Thing Equals thirty-two. 1660

3 Problem II.19. /s/

1676   Thus, a side of the square is thirty-two, so the square is one thousand twenty-four, and a square of the square is one million forty-eight thousand five hundred seventy-six. And since we assigned a side of the cube to be two Things, a side of the cube is sixty-four, so the cube is ⟨two hundred⟩ sixty-two thousand one hundred forty-four which, when multiplied by twelve, amounts to three million one hundred forty-five thousand seven hundred twenty-eight. And when that is added to a square of the square, it amounts to four million one hundred ninety-four thousand three hundred four, which is a square whose side is two thousand forty-eight. And also, when the cubic number is multiplied by five units, it amounts to one million three hundred ten thousand seven hundred twenty, and if that is added to a square of the square, it amounts to two million three hundred fifty-nine thousand two hundred ninety-six, which is a square whose side is one thousand five hundred thirty-six.

1688   We have found two numbers according to the stipulation that we wanted, and they are the two numbers we found.

1690   **3.** We want to find two other numbers, a cube and a square, such that the cube, if we multiply it by two given numbers and we subtract either of them from a square of the square, leaves a square.

1693   Let one of the given numbers be twelve and the other seven units.

1694   We again assign the square to be one *Māl*, so that a square of the square is again one *Māl Māl*. Thus, a *Māl Māl* less twelve cubes Equals a square, while less seven cubes also Equals a square. Let us look for the quantity in a known ratio with respect to the *Māl Māl* such that if twelve times it is subtracted from the *Māl* ⟨*Māl*⟩, it leaves a square, and if seven times it is subtracted from it, it also leaves a square. Thus, we look for three square numbers such that the ratio of the excess of the greater over the middle is to the excess of the middle over the smaller as the ratio of seven to its deficiency from the twelve,[4] which are the numbers we mentioned: sixteen and nine and four. Therefore, the quantity in a known ratio with respect to the *Māl Māl* that we established is a part of sixteen parts of a *Māl Māl*. Therefore, the cube Equals a part of sixteen parts of a *Māl Māl*. We assign a side of the cube to be half a Thing, so the cube is an eighth of a Cube. Thus, an eighth of a Cube Equals a part of sixteen of a *Māl Māl*. Therefore, half an eighth of a Thing Equals an eighth of a unit, so the one Thing Equals two units.

1707   Therefore, the square is four units, and a square of the square is sixteen units. And since we assigned a side of the cube to be half a Thing, a side of the cube is one, and the cube is also one, and when (it is) multiplied by twelve and by seven, and each of them is subtracted from a square of the square, it leaves a square.

1711   **4.** We want to find two numbers, a square and a cube, such that a square of the square, if we add to it a given multiple of a cube of the cube, amounts to a square

----

4 The deficiency from the 12 is the difference between 7 and 12, which is 5.

number, and if we subtract from it another given multiple of a cube of the cube, also leaves a square number.

Let the appended multiple be five and the lacking multiple be three.   1714

And let us assign the cube to be from a side of one Thing, so it is one Cube, and its   1714
cube is one Cube Cube Cube. And we assign the square to be from a side of two
*Māl*s, so the square is four *Māl Māl*s, and its square is sixteen Cube Cube *Māl*s.
Thus, sixteen Cube Cube *Māl*s and five Cube Cube Cubes Equal a square, and
sixteen Cube Cube *Māl*s less three Cube Cube Cubes Equal a square. And (for)
any square divided by a square, the quotient is a square. So let us divide each
of these two squares by the square which is one Cube Cube *Māl*. One of the
quotients is sixteen units and five Things, and the other is sixteen units less three
Things, and each of them is a square. And any square number increased by five
times its fourth is a square, and decreased by three times its fourth also gives a
square. Thus, the one Thing is a fourth of the sixteen, which is four units.

Because we assigned a side of the cube to be one Thing, a side of the cube is four   1725
units and the cube is sixty-four. And since we assigned a side of the square to
be two *Māl*s, and the *Māl* is sixteen units, a side of the square is thirty-two and
the square is one thousand twenty-four. And as for a square of the square, it is
one million forty-eight thousand five hundred seventy-six. And as for a cube of
the cube, it is two hundred sixty-two thousand one hundred forty-four, which,
when five times it is added to a square of the square, amounts to two million three
hundred fifty-nine thousand two hundred ninety-six, which is a square whose side
is one thousand five hundred thirty-six. And when three times it is subtracted from
a square of the square, it leaves two hundred sixty-two thousand one hundred
forty-four, which is a square whose side is five hundred twelve units.

We have found two numbers according to the stipulation that we stipulated, and   1737
they are the two numbers we found.

5. We want to find two numbers, a cube and a square, such that a cube of the cube,   1739
if we multiply it by two given numbers and we add either one of them to a square
of the square, amounts to a square number.

We make one of the given numbers twelve and the other five units. We want to   1742
find two numbers according to what we described.

We assign a side of the cube to be one Thing, so the cube is one Cube and its   1743
cube is one Cube Cube Cube. And we assign a side of the square to be two *Māl*s,
so the square is four *Māl Māl*s and a square of the square is sixteen Cube Cube
*Māl*s. Therefore, sixteen Cube Cube *Māl*s and twelve Cube Cube Cubes Equal a
square, and sixteen Cube Cube *Māl*s and five Cube Cube Cubes Equal ⟨a square⟩.
And (for) any square divided by a square, the quotient is a square. So let us divide
each of them by the square which is a Cube Cube *Māl*. Because of that, each of
the sixteen units and twelve Things, and sixteen units and five Things, is a square.
But any square increased by five times its fourth gives a square, and if increased

by twelve times its fourth also gives a square. Thus, the one Thing is a fourth of sixteen, that is, four units.

1754   The cube is then sixty-four and the square is one thousand twenty-four. And it is clear that a square of this square, when twelve times a cube of this cube, which is three million one hundred forty-five thousand seven hundred twenty-eight, is added to it, becomes four million one hundred ninety-four thousand three hundred four units, which is a square whose side is two thousand forty-eight. As was shown in the preceding problem, if five times a cube of the cube is added to it, it also becomes a square.

1761   **6.** We want to find two numbers, a cube and a square, such that a cube of the cube, if we multiply it by two given numbers and we subtract either one of them from a square of the square, leaves a square.

1764   Let the given numbers be seven and four.

1764   We form the cube from a side of one Thing, so the cube is one Cube, and its cube is one Cube Cube Cube. And we form the square from a side of three *Māl*s, so the square is nine *Māl Māl*s, and a square of the square is eighty-one Cube Cube *Māl*s. Therefore, eighty-one Cube Cube *Māl*s less seven Cube Cube Cubes Equal a square, and less four Cube Cube Cubes also Equal a square. Let us divide each of them by the square which is one Cube Cube *Māl*. Then eighty-one less seven Things Equals a square, and eighty-one less four Things also Equals a square. Let us look for the determined quantity in any square such that if seven times it is subtracted from the square, it leaves a square, and if four times it is also subtracted from the square, it leaves a square. One looks for that the same way as before.[5] Let that quantity be eight ninths of a ninth. Thus, the eighty-one, if seven times eight ninths of its ninth, that is, fifty-six, are subtracted from it, leaves a square, which is twenty-five. And if four times eight ninths of its ninth, that is, thirty-two, is subtracted from it, it leaves a square, which is forty-nine. Therefore, the one Thing is eight ninths of a ninth of the eighty-one, which is eight units.

1779   And since we assigned a side of the cube to be one Thing,[6] the cube is five hundred twelve. And since we assigned a side of the square to be three *Māl*s, and the *Māl* is sixty-four, a side of the square is one hundred ninety-two, and the square is thirty-six thousand eight hundred sixty-four, and a square of the square is one billion three hundred fifty-eight million nine hundred fifty-four thousand four hundred ninety-six. And as for a cube of the cube, it is one hundred thirty-four million two hundred seventeen thousand seven hundred twenty-eight, which, when seven times it is subtracted from a square of the square, leaves four hundred nineteen

---

5 Problem II.19. /s/. Rashed and Houzel say instead Problem II.16 (2013, 305).

6 Sesiano corrects *wa li-hādhā* ("and for this") to *wāḥid* ("one") (line 1780). Rashed leaves the text as is, so the passage reads "et pour cela le cube est" (Diophantus 1984, 4, 10.13).

million four hundred thirty thousand four hundred, which is a square whose side is twenty thousand four hundred eighty. And when four times it is subtracted from a square of the square, it leaves eight hundred twenty-two million eighty-three thousand five hundred eighty-four, which is a square whose side is twenty-eight thousand six hundred seventy-two.

We have found two numbers according to the stipulation that we stipulated, and they are, for the cube, five hundred twelve, and for the square, thirty-six thousand eight hundred sixty-four. And that is what we wanted to find. 1792

7. We want to find two numbers such that their sum and the sum of their cubes are equal to two given numbers. It is necessary that four times the given number for the sum of the cubes of the two numbers exceed a cube of the given number for the sum of the two numbers by a number which, if divided by three times the given number for the sum of the two numbers, then the quotient is a square, and if multiplied by three fourths of the given number for the sum of the two numbers, gives a square. This (kind of restriction) arises with the formation of the problems. 1795

Let the given number for the sum of the two numbers be twenty units, and the ⟨given⟩ number for the sum of the cubes of the two numbers be two thousand two hundred forty. We want to find two numbers such that their sum is twenty units and the sum of their cubes is two thousand two hundred forty units. 1802

We make the difference of the two numbers two Things, so one of them is ten units and a Thing and the other is ten units except a Thing. And we form from each of them a cube. Whenever we want to form a cube from a side composed of two different species we should avoid errors due to the (presence of) a multitude of species. We take the cubes of each of the two different species, and we add to them three times what is gathered from multiplying a square of each of them by the other species, so what is gathered [from the multiplication] is composed of four species, which is the cube that comes from the sum of the two different species. And if one of the species is excluded from the other, we take a cube of the greater and we add to it three times what is gathered from multiplying a square of the smaller species by the greater species, and we cast away from it a cube of the smaller species and three times what is gathered from multiplying ⟨a square⟩ of the greater species by the smaller species. What remains is the cube that comes from the difference between the two different species. 1805

Therefore, it happens that the cube that comes from a side of ten units and a Thing is what is gathered from a cube of the ten, which is one thousand, and from a cube of the Thing, which is a Cube, and three times what is gathered from multiplying the ten by a square of the Thing, which is thirty *Māl*s, and also three times what is gathered from multiplying the Thing by a square of the ten, which is three hundred Things. Thus, the cube that comes from ten and a Thing is one thousand units and one Cube and three hundred Things and thirty *Māl*s. Likewise, the cube that comes from a side of ten units except a Thing is 1816 also equal to a cube of the ten, which is one thousand, and to three times what is gathered from multiplying the ten by a square of the Thing, which is a *Māl*, and that is thirty *Māl*s, less what is gathered from a cube of the Thing, which is one Cube, and less three times what is gathered from multiplying the Thing by a square of the ten, and that is three hundred Things. Thus, the cube that comes from ten units except a Thing is one thousand (units) and thirty *Māl*s less a Cube and three hundred Things. The sum of these two cubes is two thousand (units) and sixty *Māl*s, and this is because the Cube and the three hundred Things lacking in one of them absorb[7] the Cube and the three hundred Things present in the cube of the other. So the two thousand and the sixty *Māl*s Equal two thousand two hundred forty units.

1831   Let us cast away the two thousand that is on one of the two sides from the number that is on the other side, leaving sixty *Māl*s Equal two hundred forty units.[8] Therefore, the one *Māl* is four units. Each of them is a square, so their sides are also equal. But a side of the *Māl* is one Thing, and a side of the four units is two units, so the one Thing is two units.

1835   Since we assigned the greater number of the two sought-after numbers to be ten units and a Thing, this number is twelve units. And since we assigned the smaller number to be ten units except a Thing, it is eight units. A cube of the greater number is one thousand seven hundred twenty-eight, and a cube of the smaller number is five hundred twelve units, and their sum is two thousand two hundred forty units.

1841   We have found two numbers such that their sum is twenty units and the sum of their cubes is two thousand two hundred forty units, and they are twelve and eight units. And that is what we wanted to find.

1844   **8.** We want to find two numbers such that their difference and the difference of their cubes are equal to two given numbers. It is necessary that four times the given number for the difference of the two cubes exceed a cube of the given number for the difference of the two numbers by a number which, if divided by three times the given number ⟨for the difference of the two numbers⟩ gives a square, and if multiplied by three fourths of the number that is for the difference of the two numbers gives a square.

---

7 Rashed's text has ينتهبها, which he translates as "absorbé" (Diophantus 1984, 4, 13.18). Sesiano has تذهبها, which he translates as "is cancelled". He notes that the text shows سهمها (line 1829). We follow Rashed's reading, though Sesiano's fits well, too. See also Rashed's discussion on this word (Diophantus 1984, 4, 121–22).

8 Rashed corrects and restores this sentence differently but with the same mathematical meaning: "Let us cast away the two thousand that is on one of the two sides ⟨and we cast it away⟩ from the number that is on the other side, leaving {what is on one of the sides} sixty *Māl*s ⟨and that⟩ Equals two hundred forty units" (in {brackets} is text that Sesiano omitted).

Let the given number for the difference of the two numbers be ten units, and the given number for the difference of the two cubes be two thousand one hundred seventy units. We want to find two numbers such that their difference is ten units and the difference of their cubes is two thousand one hundred seventy units. 1850

We assign the sum of the two numbers to be two Things, so one of them is a Thing and five units, and the other is a Thing less five units, so that their difference is ten units. We form a cube from each of them. The cube whose side is a Thing and five units, as we have explained,[9] is equal to a cube of the Thing, which is a Cube, and a cube of the five, which is one hundred twenty-five, and three times what is gathered from multiplying a square of the Thing by the five, which is fifteen *Māl*s, and three times what is gathered from multiplying a square of the five by the Thing, which is seventy-five Things. Thus, the cube that comes from a side of a Thing and five units is a Cube and one hundred twenty-five and fifteen *Māl*s and seventy-five Things. And the cube whose side is a Thing less five units is equal to a cube of the Thing, which is a Cube, and three times what is gathered from multiplying a square of the five, which is lacking from the Thing, by the Thing, which is seventy-five Things, less a cube of the five, which is one hundred twenty-five, and less three times what is gathered from multiplying a square of the Thing by the five units, which is fifteen *Māl*s. Thus, the cube that comes from a side of a Thing less five units is a Cube and seventy-five Things less fifteen *Māl*s, and less one hundred twenty-five units. Let us cast away this cube from the first cube, to get the remainder two hundred fifty (units) and thirty *Māl*s. This is because the lacking fifteen *Māl*s and one hundred twenty-five in this cube, because they are lacking, are present ⟨and are added to the fifteen *Māl*s and the one hundred twenty-five present⟩ in the other cube, and the Cube and the seventy-five Things vanish in each of them. Therefore, the two hundred fifty and the thirty *Māl*s Equal two thousand one hundred seventy units. 1853

Let us cast away the two hundred fifty in common from both sides, leaving one thousand nine hundred twenty units Equal thirty *Māl*s. Therefore, the *Māl* is sixty-four. Each of them is a square, so their sides are equal, and a side of the *Māl* is one Thing, and a side of the sixty-four is eight units. Therefore, the one Thing is eight units. 1874

And since we assigned the greater number to be a Thing and five units, it is thirteen units, and as we assigned the smaller number to be a Thing less five units, the smaller is three units. And their cubes: as for a cube of the greater, it is two thousand one hundred ninety-seven, and as for a cube of the smaller, it is twenty-seven, and their difference is two thousand one hundred seventy. 1878

We have found two numbers such that their difference is ten units and the difference of their cubes is ⟨two thousand⟩ one hundred seventy, and they are thirteen and three units. And that is what we wanted to find. 1883

---

9 In the previous problem. /s/

1885   9. We want to divide a given number into two parts such that the sum of their cubes is a given multiple of a square of their difference. It is necessary that the given multiple be greater than three fourths of the given number by a number containing, with a cube of the given number, a square number.

1889   Let the given number be twenty, and the multiple be one hundred forty times. We want to divide the twenty into two parts such that the sum of their cubes is one hundred forty times a square of the difference between them.

1891   Let us again assign the difference of the two parts to be two Things, so that one of the parts is ten units and a Thing and the other is ten units less a Thing. The sum of their cubes, according to the way we explained above,[10] is two thousand units and sixty *Māl*s, and a square of the difference of the two numbers is four *Māl*s. Therefore, two thousand units and sixty *Māl*s Equal one hundred forty times the four *Māl*s, that is, five hundred sixty *Māl*s.

1896   We cast away the sixty *Māl*s in common from both sides, leaving two thousand units Equal five hundred *Māl*s. Therefore, the one *Māl* Equals four units, and a side of the *Māl* is one Thing, and a side of the four units is two units. Thus, the one Thing Equals two units.

1899   Since we assigned one of the parts to be ten units and a Thing, it is twelve units. And since we assigned the other part to be ten units less a Thing, it is eight units. And a cube of the twelve, when a cube of the eight is added to it, gives two thousand two hundred forty units, and the difference of the parts is four units, and its square is sixteen units. And the two thousand two hundred forty units are one hundred forty times the sixteen units, which is a square of the difference between the two parts that we found.

1906   We have divided the twenty into two parts according to what we wanted. The greater part is twelve units and the smaller is eight units. And that is what we wanted to do.

1908   10. We want to find two numbers such that their difference is a given number, and the difference of their cubes is with respect to a square of their sum in a given ratio. It is necessary that the number that is for the given ratio be greater than three fourths of the given number for the difference of the two numbers ⟨by a number containing, with a cube of the given number for the difference of the two numbers, a square number⟩.[11]

1913   Let the given number for the difference of the two sought-after numbers be ten units, and the number that is for the given ratio be eight times and an eighth of a time. We want to find two numbers such that their difference is ten units, and the

---

10 Problem V.7. /s/
11 Rashed does not make this restoration, but he notes in his commentary the inadequacy of what appears in the manuscript (Diophantus 1984, 4, XXXI).

ratio of the difference of their cubes is to a square of their sum in the ratio of the eight and an eighth to the one.

We assign their sum to be two Things, and we make one of the two numbers a Thing and five units and the other a Thing less five units, so that their difference is ten units. We take the difference between their cubes, to get two hundred fifty and thirty *Māl*s, and a square of the sum of the two numbers is four *Māl*s. Then the two hundred fifty and the thirty *Māl*s Equal eight times and an eighth of a time the four *Māl*s, and that is thirty-two *Māl*s and half a *Māl*. 1916

Let us cast away the thirty *Māl*s in common from both sides, leaving two hundred fifty units Equal two *Māl*s and half a *Māl*. Therefore, the one *Māl* Equals one hundred, and therefore, the Thing is ten units. 1921

Since we assigned one of the two numbers to be a Thing and five units, it is fifteen units. And since we assigned the other number to be a Thing less five units, it is five units. A cube of the fifteen is three thousand three hundred seventy-five, and a cube of the five is one hundred twenty-five, and their difference is three thousand two hundred fifty, and a square of the sum of the two numbers is four hundred, and the ratio of the three thousand two hundred fifty to the four hundred is the ratio of eight times and an eighth of a time. 1924

We have found two numbers such that their difference is ten units and the difference of their cubes is eight times and an eighth of a time a square of their sum, and they are the fifteen units and the five units. And that is what we wanted to find. 1930

11. We want to find two numbers such that their difference is a given number and the sum of their cubes is with respect to their sum in a given ratio It is necessary that the number that is for the given ratio exceed three fourths of a square of the given number for the difference of the two numbers by a square number. 1933

Let the difference of the two numbers be four units, and the number that is for the given ratio be twenty-eight times. We want to find two numbers such that their difference is four units, and the sum of their cubes is with respect to their sum in the ratio of twenty-eight times. 1937

We assign the sum of the two numbers to be two Things, so one of them is a Thing and two units, and the other is a Thing less two units. A cube of the greater is a Cube and eight units and six *Māl*s and twelve Things, and a cube of the smaller is a Cube and twelve Things less six *Māl*s and less eight units. Their sum is two Cubes and twenty-four Things, and this is because the six *Māl*s and eight units lacking in a cube of the smaller number unite with[12] the eight units and six 1939

---

12 We translate as "unite with" what Sesiano reads as تحيّزها (line 1944). Rashed reads instead تجبرها ("you restore it") (Diophantus 1984, 4, 20.15). He explains: "Le mot « compenser » traduit ici le verbe « najbur ». Cette traduction peut d'ailleurs être proposée pour remplacer « restaurer »".

*Māl*s present in a cube of the greater number. Thus, the two Cubes and twenty-four Things Equal twenty-eight times the sum of the two numbers, which is two Things, and that is fifty-six Things.

1947   We cast away the twenty-four Things in common from both sides, leaving two Cubes Equal thirty-two Things. We divide each of them by one Thing to get two *Māl*s Equal thirty-two units. Thus, the *Māl* Equals sixteen units. And the *Māl* is a square whose side is one Thing, and the sixteen is a square whose side is four units; thus, the one Thing Equals four units.

1952   Since we assigned the greater number to be one Thing and two, the greater number is six units. And since we assigned the smaller number to be a Thing less two units, the smaller number is two units. A cube of the greater is two hundred sixteen units, and a cube of the smaller is eight units. The sum of these two cubes is two hundred twenty-four units, which is twenty-eight times the sum of the two numbers, which is eight units.

1958   We have found two numbers such that their difference is four units, and the sum of their cubes is twenty-eight times their sum, and they are six units and two units. And that is what we wanted to find.

1961   **12.** We want to divide a given number into two parts such that the difference of their cubes is a given multiple of their difference. It is necessary that the number that is for the given ratio again exceed three fourths of a square of the given number by a square number.

1965   Let the given number be eight units, and the multiple that is for the given ratio be the ratio of fifty-two times. We want to divide the eight into two numbers such that the difference of their cubes is fifty-two times their difference.

1967   We assign the difference of the two numbers to be two Things, so the greater part is four units and a Thing, and the smaller part is four units less a Thing. And a cube of the greater part is sixty-four units and a Cube and forty-eight Things and twelve *Māl*s, and a cube of the smaller part is sixty-four units and twelve *Māl*s less a Cube and less forty-eight Things. And their difference is two Cubes and ninety-six Things. The two Cubes and ninety-six Things Equal fifty-two times the difference of the two numbers, which is two Things, and that is one hundred Things and four Things.

1974   We cast away the ninety-six Things in common from both sides, leaving two Cubes Equal eight Things. We divide each of them by a Thing to get two *Māl*s

---

Un tel usage de ce terme est bien propre à l'époque. Il suffit en effet de parcourir l'algèbre d'Abū Kāmil pour rencontrer le même emploi". He then explains that Abū Kāmil used the word to simplify an equation of the form $x^2 - ax = g(x)$ to one of the form $x^2 = g(x) + ax$ (Diophantus 1984, 4, 122). But this is the kind of "restoration" common to Arabic algebra, while the verb in the present situation is applied in a completely different way.

Equal eight units. Therefore, the one *Māl* Equals four units, and the one Thing Equals two units.

And since we assigned the greater part to be four units and a Thing, the greater part is six units. And since we assigned the smaller part to be four units less a Thing, the smaller part is two units. And a cube of the greater part is two hundred sixteen units, and a cube of the smaller part is eight units, and their difference is two hundred eight units, which is fifty-two times the difference of the two parts, which is four units.

1977

We have divided the eight into two parts such that the difference of their cubes is fifty-two times their difference, and they are six and two. And that is what we wanted to do.

1983

13. We want to find a cubic number such that if we add a given number to a given multiple of a square of its side, then that is equal to the sum of two numbers, each of which, if added to the cube, gives a cube.

1985

Let the given number be thirty units and the given multiple be nine times. We want to find a cubic number such that if we add nine times the square that comes from its side to thirty units, then that is equal to two numbers, each of which, if added to the cubic number, becomes a cube.

1988

Let us assign a side of the cube to be one Thing, so the cube is one Cube. And let us take nine times the square that comes from its side, which is nine *Māl*s. We add that to the thirty to get nine *Māl*s and thirty units. Since these nine *Māl*s and thirty units are equal to two numbers, each of which, if added to the cube, which is one Cube, becomes a cube, if we form two cubes from two sides, each of them a Thing and some units, and we take the excess of each of them over the cube and we put the two excesses in place of the two numbers and we add them up and we Equate them with the nine *Māl*s and thirty units, we will arrive at what we are looking for. But these two excesses are composed of *Māl*s and Things and number. It is therefore necessary that these *Māl*s that are in the two excesses add up to nine *Māl*s, and the number that is with them be less than thirty, in order that we end up with a number Equals a Thing.

1991

We want to form ⟨the two cubes⟩ from two sides such that each of them is a Thing and a number, in such a way that the sum of the *Māl*s that are in each of the cubes is nine *Māl*s and the units are less than thirty, which is the given number. But the *Māl*s present in each of the cubes is three times each of the two numbers added to the Thing in the sides of the cubes, and the sum of the units that are in the two cubes is the sum of their cubes. Thus, the sum of the two numbers added to the Thing should be three units, in order that three times it be the number of *Māl*s, which is nine. And we should divide the three into two parts such that the sum of their cubes is less than thirty, and they are two and one.

2002

We form one of the cubes from a side of a Thing and two units, giving a Cube and six *Māl*s and twelve Things and eight units. And we form the other cube from a

2011

side of a Thing and one, giving a Cube and three *Māl*s and three Things and one. Since the six *Māl*s and twelve Things and eight units, if added to a Cube, give a cube, and likewise for the three *Māl*s and three Things and one, we make their sum, which is nine *Māl*s and fifteen Things and nine units, as we mentioned, and they are Equal to the nine *Māl*s and thirty units.

2018   We cast away the nine *Māl*s in common from both sides, leaving fifteen Things and nine units Equal thirty units. Then we cast away the nine units in common from both sides, leaving fifteen Things Equal twenty-one units. Therefore, the one Thing is one and two fifths of a unit.

2021   Since we assigned a side of the sought-after cube to be one Thing, it is seven fifths of a unit, and the cube is two units and ninety-three parts of one hundred twenty-five parts of a unit. As for a square of a side of the cube, it is one and twenty-four parts of twenty-five of a unit. And nine times that is seventeen units and sixteen parts of twenty-five parts, that is, eighty parts of one hundred twenty-five parts. And when we add that to thirty, it gives forty-seven units and eighty parts of one hundred twenty-five parts. And as we assigned one part of these summed numbers to be six *Māl*s and twelve Things and eight units, as for the six *Māl*s, it is eleven units and ninety-five parts of one hundred twenty-five parts of a unit, and the twelve Things are sixteen units and one hundred parts of one hundred twenty-five parts of a unit. So the entire first number is thirty-six units and seventy parts of one hundred twenty five parts of a unit. And the other number is what is left from the forty-seven units and eighty parts of one hundred twenty-five, which is eleven units and ten parts of one hundred twenty-five. And when we add the first number of these two numbers to the cubic number, which is two units and ninety-three parts of one hundred twenty-five, the sum of that is thirty-nine units and thirty-eight parts of one hundred twenty-five parts, which is a cubic number whose side is three units and two fifths of a unit. And when we add the second number to the cubic number, the sum of that is thirteen units and one hundred parts and three parts of one hundred twenty-five parts of a unit. And that is a cubic number whose side is two units and two fifths of a unit.

2043   We have found a cubic number such that if we add nine times a square of its side to thirty units, then that is equal to two numbers, each of them, if added to the cubic number, becomes a cube, which is the cube that we found. And that is what we wanted to find.

2047   One should know that this problem can be worked out this way whenever a cube of a third of the number of times is smaller than four times the given number.

2049   **14.** We want to find a cubic number such that if we subtract a given number from a given multiple of a square of its side, then that is equal to two numbers which, if either of them is subtracted from the cube, leaves a cube.

2052   Let the given number be twenty-six and the given multiple be nine times. We want to find a cubic number according to what we described.

We assign a side of the cube to be one Thing, so the cube is one Cube. And we take   2053
nine times a square of a side of the cube, which is a *Māl*, and that is nine *Māl*s.
We cast away the given number from it, leaving nine *Māl*s less twenty-six units,
which is equal to two numbers, either of them, if subtracted from the cube, leaves
a cube. As we explained in the previous problem, let us form two cubes in such a
way that each of them is from a side of a Thing less a number, so that the sum of
the lacking *Māl*s in them are nine *Māl*s. We have no need in this problem for the
sum of the two numbers that are in them to be less than the given units. Rather,
they may be any (values). Let us form one of the two cubes from a side of a Thing
less two units, so that the cube is a Cube and twelve Things less six *Māl*s and less
eight units. And let us form the other from a side of a Thing less one, so the cube is
a Cube and three Things less three *Māl*s and less one. Now the six *Māl*s and eight
units and less twelve Things, if we subtract them from the cube, become a cube.
And likewise, the three *Māl*s and one less three Things, if we subtract them from
the cube, also leave a cube. Let us make the sum of these two numbers Equal to
the nine *Māl*s less twenty-six units. But their sum is nine *Māl*s and nine units less
fifteen Things, so they Equal nine *Māl*s less twenty-six units.

Let us add the twenty-six units to both sides, and similarly the fifteen Things. We   2071
cast away the nine *Māl*s in common from both sides. There remains, after the restoration and confrontation, fifteen Things Equal thirty-five units. Therefore, the
Thing is two units and a third of a unit.

Since we assigned a side of the cube to be one Thing, a side of the cube is two   2074
units and a third of a unit, and the cube is twelve units and nineteen parts of
twenty-seven parts of a unit. And a square of a side of the cube is five units and
twelve parts of twenty-seven parts, and nine times that is forty-nine units. Let us
cast away from it the given twenty-six, so there remain twenty-three units. We
assigned one part of these twenty-three to be six *Māl*s and eight units less twelve
Things. But the six *Māl*s are thirty-two and eighteen parts of twenty-seven, and
the twelve Things are twenty-eight units; thus, the greater number of the two
numbers is twelve units and eighteen parts of twenty-seven of a unit. And thus,
the smaller number is ten units and nine parts of twenty-seven parts of a unit.
But the greater of these two numbers, when subtracted from the cube, which, as
we explained, is twelve units and nineteen parts of twenty-seven parts of a unit,
leaves one part of twenty-seven parts of a unit, which is a cube whose side is a
third of a unit. And when we subtract the smaller number from the cubic number,
it leaves two units and ten parts of twenty-seven parts of a unit, which is a cube
whose side is one and a third of a unit.

We have found a cubic number according to the restriction that we restricted, and   2092
that is what we wanted to find.

15. We want to find a cubic number such that if we subtract a given number from   2094
a given multiple of a square of its side, then that is equal to two numbers which, if
one of them is added to the cube, gives a cube, and if the other is subtracted from
the cube, also leaves a cube.

2097 Let the given multiple be nine times, and the given number be eighteen units. We want to find a cubic number such that if eighteen units are subtracted from nine times a square of its side, then the remainder from that is two numbers such that one of them, ⟨if added⟩ to the cube amounts to a cube, and if the other, if subtracted from the cube leaves a cube.

2101 We assign a side of the cube to be one Thing, so the cube is one Cube. We take nine times a square of its side, which is nine *Māl*s. We cast away eighteen units from it. Then we form two cubes in such a way that a side of one of them is a Thing and a number, and a side of the other is a Thing less a number, so that the *Māl*s present in one of the cubes with the *Māl*s lacking in the other cube are nine *Māl*s. Let the first cube be from a side of ⟨a Thing⟩ less two units, so it is a Cube and twelve Things less eight units and less six *Māl*s. And let the other cube be from a side of a Thing and one, so that it is a Cube and three *Māl*s and three Things and one. Thus, the six *Māl*s and eight units less twelve Things, if we subtract ⟨them⟩ from the sought-after cube, which is one Cube, then the remainder becomes a cube. And the three *Māl*s and one and three Things, if added to the sought-after cube, amount to a cube. Let us make the sum of these two numbers Equal to the nine *Māl*s less eighteen units. But their sum is nine *Māl*s and nine units less nine Things; thus, they Equal nine *Māl*s less eighteen units.

2114 Let us restore that and confront them, leaving, after the restoration and confrontation, twenty-seven units Equal nine Things. Thus, the one Thing Equals three units.

2116 And since we assigned a side of the cube to be one Thing, a side of the cube is three units, and the cube is twenty-seven. And a square of a side of the cube is nine units, and nine times that is eighty-one. Let us cast away from it the given number, which is eighteen units, leaving sixty-three. And as we assigned one of the two numbers to be six *Māl*s and eight units less twelve Things, it is, therefore, twenty-six units. And the other number is the remainder from sixty-three, which is thirty-seven units. And the twenty-six, when subtracted from the cube, which is twenty-seven, leaves one, which is a cube. And the thirty-seven, when added to the cube, which is twenty-seven, amounts to sixty-four, which is a cube whose side is four units.

2126 We have found a cube according to the stipulation that we stipulated. And that is what we wanted to find.

2128 **16**. We want to find a cubic number such that if we subtract a given number from a given multiple of a square of its side, then the remainder from that is equal to two numbers which, if one of them is subtracted from the cube it leaves a cube, and if the cube is subtracted from the other number, it leaves a cube.

2132 Again, let the given multiple be nine times, and the given number be sixteen. We want to find a cubic number such that if we subtract sixteen units from nine times a square of its side, then what remains from that is equal to two numbers which, if one of them is subtracted from the cube it leaves a cube, and if the cube is subtracted from the other number it leaves a cube.

We again make the cube one Cube, and we subtract sixteen units from nine times a square of its side, and we form two cubes in such a way that a side of one of them is a Thing less a number, and a side of the other is a number less a Thing. And let the *Māl*s that are in them[13] be nine *Māl*s. We form one of the cubes from a side of a Thing less one, so it is a Cube and three Things less three *Māl*s and less one. And we form the other cube from a side of two units less a Thing, so it is eight units and six *Māl*s less a Cube and less twelve Things. The three *Māl*s and one less three Things, if subtracted from the cube, leave a cube, which is, as we said, a Cube and three Things less three *Māl*s and less one. And the six *Māl*s and eight units less twelve Things, if the cube, which is one Cube, is subtracted from it, leave a cube, which is, as we also said, eight units and six *Māl*s less twelve Things and less a Cube. Let their sum be Equal to the nine *Māl*s less sixteen. But their sum is nine *Māl*s and nine units less fifteen Things, which thus Equal nine *Māl*s less sixteen units.

2136

Let us restore that and confront them. So it turns out, after the restoration and confrontation, to be fifteen Things Equal twenty-five units. Therefore, the one Thing is one and two thirds of a unit, which is a side of the cube. Therefore, the cube is four units and seventeen parts of twenty-seven parts of a unit. As for a square of a side of the cube, it is two units and twenty-one parts of twenty-seven parts, and nine times that is twenty-five. Let us cast away the sixteen from it, leaving nine units. And as we assigned the number subtracted from the cube, (which) consists of two numbers whose sum is nine units, to be three *Māl*s and one less three Things, and the three *Māl*s are eight units and a third, and the three Things are five units, then this number that we mentioned is four units and a third of a unit. And the other number remaining from the nine units is four and two thirds of a unit. And the four units and a third of a unit, when subtracted from the cube, which is four units and seventeen parts of twenty-seven parts, leaves eight parts of twenty-seven parts of a unit, which is a cube whose side is two thirds of a unit. And for the other number, which is four units and two thirds of a unit, when the cube is subtracted from it, leaves one part of twenty-seven parts of a unit, which is a cube whose side is a third of a unit.

2149

We have found a cubic number according to the stipulation that we stipulated. And that is what we wanted to find.

2166

End of the fifth Book of the work of Diophantus on the arithmetical problems, which consists of sixteen problems.

2168

---

13 Lit. "which are located in them".

Book VI (Arabic)

2170   In the name of God, the Merciful, the Compassionate. The sixth Book of the work of Diophantus.

2172   **1.** We want to find two numbers, one of them a cube and the other a square, such that their sides are in a given ratio, ⟨and⟩[1] if their squares are added then it gives a square number.

2174   Let the given ratio be the ratio of two times. We want to find two numbers, one of them a cube and the other a square, such that a side of the cube is twice a side of the square, ⟨and⟩[2] if ⟨their squares⟩ are added, then it gives a square number.

2176   Let us make a side of the square a Thing, so the square is a *Māl*, and a side of the cube two Things, so the cube is eight Cubes. The sum of a square of the cube and ⟨a square⟩ of the square is sixty-four Cube Cubes and a *Māl Māl*, and we require that to be a square. Let us look for a square number such that if sixty-four units are subtracted from it, then the remainder is a square. This is easily found by what we explained above in our work,[3] and that is one hundred units. Let us make it ⟨Cube Cubes, so it is⟩ one hundred Cube Cubes, and we confront them with sixty-four Cube Cubes and a *Māl Māl*.

2183   We cast away what is common, leaving thirty-six Cube Cubes Equal a *Māl Māl*. We divide the two sides by the lesser in degree of them, which is a *Māl Māl*, to get one Equals thirty-six *Māl*s. Thus, the *Māl* is a part of thirty-six parts of a unit, and the Thing is a part of six parts of a unit, which is a side of the square number, and a side of the cubic number is twice that, which is two parts of six parts of the unit, and the cube is eight parts of two hundred sixteen parts of the unit. So when its square, which is sixty-four parts of forty-six thousand six hundred fifty-six parts of the unit, is added to a square of the square number, which is thirty-

---

1 Rashed's restoration.
2 Rashed's restoration.
3 Problem II.10. /s/

DOI: 10.4324/9781315171470-13

six parts of forty-six thousand six hundred fifty-six parts, it gives one hundred parts of forty-six thousand six hundred fifty-six, which is a square number whose side is ten parts of two hundred sixteen parts of the unit.

We have found two numbers according to the restriction that we restricted, and they are eight parts of two hundred sixteen parts of the unit, and six parts of two hundred sixteen parts of the unit. And that is what we wanted to find.

2194

2. We want to find two numbers, one of them a cube and the other a square, such that their sides are in a given ratio, ⟨and⟩[4] if a square of the square is subtracted from ⟨a square⟩ of the cube, then the remainder is a square.

2197

Let the given ratio be the ratio of two times. We want to find two numbers, one of them a cube and the other a square, such that a side of the cube is twice a side of the square, and if a square of the square is subtracted from a square of the cube, then the remainder is a square.

2200

Let us make a side of the square number a Thing, and a side of the cubic number two Things, so the square number is a *Māl* and its square is a *Māl Māl*, and the cubic number is eight Cubes and its square is sixty-four Cube Cubes. If we subtract a *Māl Māl* from sixty-four Cube ⟨Cubes⟩, then the remainder is sixty-four Cube Cubes less a *Māl Māl*, and we require that to be a square number. Let us look for a square number such that if we subtract it from sixty-four, then the remainder is a square number. This is easily found according to what we explained above:[5] it is forty units and twenty-four parts of twenty-five parts of the unit. Let us make it Cube Cubes, so it is forty Cube Cubes and twenty-four parts of twenty-five parts of a Cube Cube. So they Equal sixty-four Cube Cubes less a *Māl Māl*.

2202

We confront[6] them and we cast away what is common, leaving twenty-three Cube Cubes and a part of twenty-five parts of a Cube Cube Equal a *Māl Māl*. We divide the two sides by a *Māl Māl* to get one Equals twenty-three *Māl*s and a part of twenty-five of a *Māl*. So the *Māl* is twenty-five parts of five hundred seventy-six parts of the unit, and the Thing is five parts of twenty-four parts of a unit.

2212

As we made a side of the cubic number two Things, and that is five parts of twelve parts of the unit, the cube is one hundred twenty-five parts of one thousand seven hundred twenty-eight parts of the unit, and its square is fifteen thousand six hundred twenty-five parts of two million nine hundred eighty-five thousand nine hundred eighty-four parts. And when we subtract from it a square of the square number, which is five thousand six hundred twenty-five parts of two million nine hundred eighty-five thousand nine hundred eighty-four, the remainder is ten

2217

---

thousand parts of two million nine hundred eighty-five thousand nine hundred eighty-four parts. And that is a square number, and its side is one hundred parts of one thousand seven hundred twenty-eight parts.

2228   We have found two numbers according to the restriction that we restricted, and they are one hundred twenty-five parts of one thousand seven hundred twenty-eight parts of the unit, and seventy-five parts of one thousand seven hundred twenty-eight parts of the unit. And that is what we wanted to find.

2232   3. We want to find two numbers, one of them a cube and the other a square, such that their sides are in a given ratio, ⟨and⟩[7] if we subtract a square of the cube from a square of the square number, then the remainder is a square.

2235   Let the given ratio be the ratio of two times. We want to find two numbers, one of them a cube and the other a square, such that a side of the cube is with respect to a side of the square in the ratio of two times, and if a square of the cubic number is subtracted from a square of the square number, then the remainder is a square.

2238   We make a side of the square number a Thing, so a side of the cube is two Things and the cube is eight Cubes, and its square is sixty-four Cube Cubes. And as we made a side of the square number a Thing, the square is a *Māl* and its square is a *Māl Māl*. We cast away from it a square of the cube, which is sixty-four Cube Cubes, leaving a *Māl Māl* less sixty-four Cube Cubes. We require that to be a square. We want to look for a square number such that if sixty-four is added to it, amounts to a square.[8] And that is thirty-six, and its side is six. So we make a side of the *Māl Māl* less sixty-four Cube Cubes be six Cubes, and we multiply it by itself to get thirty-six Cube Cubes, and they Equal a *Māl Māl* less sixty-four Cube Cubes.

2246   We restore and confront, to get a *Māl Māl* Equals one hundred Cube Cubes. We divide each of these by a *Māl Māl* to get one Equals one hundred *Māl*s; thus, the *Māl* is a part of one hundred, which is a tenth of a tenth of a unit, and the Thing is a part of ten, which is a tenth of a unit.

2249   As we made the cube ⟨from a side of⟩ two Things, its side is two parts of ten, and the cube is eight parts of one thousand, and its square is sixty-four parts of one million. And when we subtract it from a square of the square number, which is one hundred parts of one million, the remainder is thirty-six parts of one million, which is a square number, and its side is six parts of one thousand parts of the unit.

2255   We have found two numbers according to the restriction that we restricted, and they are eight parts of one thousand parts and ten parts of one thousand parts. And that is what we wanted to find.

---

7 Rashed's restoration.
8 Problem II.10. /s/

4. We want to find two numbers, one of them a cube and the other a square, such that a side of the cube is with respect to a side of the square in a given ratio, ⟨and⟩[9] if a square of the cube is added to the number they contain, then what is gathered is a square.

2257

Let the ratio be the ratio of five times. We want to find two numbers, one of them a cube and the other a square, such that a side of the cube is five times a side of the square, and if a square of the cubic number is added to the number they contain, then what is gathered from that is a square.

2260

Let us assign a side of the square to be a Thing, so the square is a *Māl*, and a side of the cube to be five Things, so the cube is one hundred twenty-five Cubes. The number they contain is one hundred twenty-five Cubes multiplied by a *Māl*. We add to that a square of the cube [and a square of the cube is fifteen thousand six hundred twenty-five Cubes Cubes] to get that the sum of that is fifteen thousand six hundred twenty-five Cube Cubes and one hundred twenty-five Cubes multiplied by a *Māl*. We require that to be a square. Let us look for a square number such that if we subtract fifteen thousand six hundred twenty-five from it, then the remainder is a small number. It is not necessary for us that the remainder be a square number. And that number is fifteen thousand eight hundred seventy-six, and its side is one hundred twenty-six. Let us make[10] a side of fifteen thousand six hundred twenty-five Cube Cubes and a *Māl* multiplied by one hundred twenty-five Cubes be one hundred twenty-six Cubes. We multiply it by itself to get fifteen thousand eight hundred seventy-six Cube Cubes. They Equal fifteen thousand six hundred twenty-five Cube Cubes and one hundred twenty-five Cubes multiplied by a *Māl*.

2263

We cast away the fifteen thousand six hundred twenty-five Cube Cubes in common from the two sides, leaving two hundred fifty-one Cube Cubes Equal one hundred twenty-five Cubes multiplied by a *Māl*. Divide the two sides by a Cube multiplied by a *Māl* to get one hundred twenty-five units Equal two hundred fifty-one Things. Thus, the Thing is one hundred twenty-five parts of two hundred fifty-one, which is a side of the square, and the square is fifteen thousand six hundred twenty-five parts of a square of two hundred fifty-one, and that is sixty-three thousand one.

2277

A side of the cubic number is five times a side of the square number, and that is six hundred twenty-five parts of two hundred fifty-one, so the cube is two hundred

2284

---

9 Rashed's restoration.

10 Sesiano reads the start of this verbal phrase as "فلنجعله ذلك", while Rashed reads "فلنجعل ضلع". Sesiano translates: "Let us make it – [that (is, the side of) $15,625x^6 + 125x^5 - (x^3$'s, so it becomes)] $126x^3$"; (1982, 141). (We have added the [brackets] that Sesiano includes in his Arabic text to indicate what he believes to be an interpolation.) Rashed passes no judgment on possible interpolation and translates the passage as "Posons le côté de quinze mille six cent vingt-cinq cubo-cubes plus un carré mulitplié par cent vingt-cinq *cubes*, cent vingt-cinq *cubes*" (Diophantus 1984, 4, 41). We translate Rashed's text because it requires fewer restorations.

forty-four million one hundred forty thousand six hundred twenty-five parts of two million five hundred sixty-three thousand one.[11] We are satisfied that the validity of the solution to this problem is sufficient (for the solution to apply) to others like it.

2290   We have found two numbers according to the restriction that we restricted, and they are fifteen thousand six hundred twenty-five parts of sixty-three thousand one, and two hundred forty-four million one hundred forty thousand six hundred twenty-five parts of two million five hundred sixty-three thousand one.[12] And that is what we wanted to find.

2295   **5**. We want to find two numbers, one of them a cube and the other a square, such that a side of the cube is equal to a side of the square, ⟨and⟩[13] if the same as a square of the square number is added to the number they contain, then the sum of that is a square.

2298   Let us make a side of the square a Thing, so the square is a *Māl*. A side of the cube is also a Thing, so the cube is a Cube. The number they contain is a Cube multiplied by a *Māl*. We add to it a square of the square number, which is a *Māl Māl*, to get a Cube multiplied by a *Māl* and a *Māl Māl*, which Equals a square number. Let us assign its side to be two *Māl*s, so this gives four *Māl Māl*s Equal a *Māl Māl* and a *Māl* multiplied by a Cube.

2302   We cast away the common *Māl Māl*, leaving a *Māl* multiplied by a Cube Equals three *Māl Māl*s. We divide everything we have by a *Māl Māl* to get a Thing Equals three units, which is a side of the square, and the square is nine units. A side of the cube is also equal to a side of the square, so it is three units, and the cubic number is twenty-seven units, and the number they contain is what comes from multiplying nine by twenty-seven, and that is two hundred forty-three. And when a square of the square number, which is eighty-one, is added to it, then the sum of that is three hundred twenty-four, and that is a square number whose side is eighteen.

2311   We have found two numbers according to the restriction that we restricted. And that is what we wanted to find.

2313   **6**. We want to find two numbers, one of them a square and the other a cube, such that a side of the cube is equal to a side of the square, and if we subtract a square of the cubic number from the number they contain, then the remainder is a square number.

11 The number should be 15,813,251. Sesiano traces the error to a corruption in the Greek text, where "251 by 63001" was misread as "2563001" (1982, 246–47). Rashed proposes essentially the same explanation for the error and writes 15,813,251 in his text in quotation marks (Diophantus 1984, 4, 122–23).

12 Again, the text has 2,563,001 in error. See the previous footnote.

13 Rashed's restoration.

Let us assign a side of the square number to be a Thing, so the square number is   2316
a *Māl*. A side of the cubic number is also a Thing, so the cubic number is a Cube.
We are required to subtract a square of a Cube from the number that is contained
by a Cube and a *Māl*, but the number that is contained by a Cube and a *Māl* is a
Cube multiplied by a *Māl*. When we subtract from it a square of the cubic number, which is a Cube Cube, then the remainder is a Cube multiplied by a *Māl* less
a Cube Cube. We require that to be a square. Let us make its side one Cube. We
multiply it by itself to get a Cube Cube. It Equals a Cube multiplied by a *Māl* less
a Cube Cube.

Let us add a Cube Cube to the two sides and divide the two sides by the lesser in   2323
degree of the two sides, and that is a Cube multiplied by a *Māl*, to get two Things
Equal one. Thus, the one Thing Equals a half.

As we made a side of the square a Thing, the square is a part of four, which is a   2325
fourth of a unit. A side of the cube is also half of a unit, so the cube is an eighth of
a unit. And since the number they contain is a part of thirty-two, when a square of
the cube, which is a part of sixty-four parts, is cast away from it, the remainder is
a part of sixty-four, which is a square whose side is a part of eight.

We have found two numbers according to the restriction that we restricted, and they   2331
are a fourth of a unit and an eighth of a unit. And that is what we wanted to find.

7. We want to find two numbers, one of them a cube and the other a square, such   2333
that their sides are equal, ⟨and⟩[14] if a square of the square number is subtracted
from the number they contain, then the remainder is a square.

Let us assign a side of the square number to be a Thing, so the square number is a   2336
Māl. And since a side of the cube is equal to a side of the square number, the cubic
number must be a Cube. Thus, the number they contain is a Cube multiplied by a
Māl. And if we cast away from it a square of the square number, which is a *Māl*
Māl, then the remainder is a Cube multiplied by a *Māl* ⟨with a *Māl Māl* removed
from it⟩.[15] We require that be a square. Let us assign to it a side of a *Māl*, and we
multiply it by itself to get a *Māl Māl* (which) Equals a Cube multiplied by a *Māl*,
with a *Māl Māl* removed from it.[16] We then restore and confront to get that the
Thing is two.

As we assigned a side of the square to be a Thing, it is two, and the square is four   2342
units. A side of the cube is also two, and the cube is eight. And since the square is

---

14 Rashed's restoration.
15 Our restoration. Sesiano restored it as "⟨less a *Māl Māl*⟩". See the next footnote for an explanation.
16 The unusual word choice, with "reduced (*manqūṣ*) by a *Māl Māl*" instead of "less a *Māl Māl*", is
a consequence of writing "a Cube multiplied by a *Māl*". Had he written "a Cube multiplied by a
Māl less a *Māl Māl*", it could have been interpreted as the product of "a Cube" by "a *Māl* less
a *Māl Māl*". Abū Kāmil, too, adjusts his wording for the sake of clarity in similar instances in his
Book on Algebra.

four and the cube is eight, the number they contain is ⟨thirty-⟩two units. And when we subtract the same as a square of the square number from it, the remainder is sixteen units, and that is a square number whose side is four.

2348   We have found two numbers according to the restriction that we restricted, and they are eight units, (and)[17] four units. And that is what we wanted to find.

2350   **8.** We want to find two numbers, one of them a cube and the other a square, such that if the number they contain is increased by the same as its side, then the sum of that is a square.

2352   We assign the cubic number to be sixty-four units, and the square number to be a *Māl*, so the number they contain is sixty-four *Māl*s. But if we add to it the same as its side, which is eight Things, then the sum is sixty-four *Māl*s and eight Things. We require[18] that to be a square. Let us assign its side to be whatever we wish of Things, as long as it is more than eight ⟨Things⟩. We make it ten Things, and we multiply it by itself, which gives one hundred *Māl*s. Thus, they Equal sixty-four *Māl*s and eight Things.

2358   We cast away sixty-four *Māl*s from the two sides, leaving thirty-six *Māl*s Equal eight Things. We divide thirty-six *Māl*s by a Thing to get thirty-six Things, and we divide eight Things by a Thing to get eight units. Thus, the eight units Equal the thirty-six Things, and the Thing is two parts of nine.

2361   And as we assigned a side of the square to be a Thing, the square is ⟨four⟩ parts of eighty-one parts of the unit, which is the square number. And the cubic number is sixty-four units, and the number they contain is two hundred fifty-six parts of eighty-one parts of the unit. When we add to it its side, which is sixteen parts of nine, that is, one hundred forty-four parts of eighty-one, the sum of that is four hundred parts of eighty-one, which is a square number, and its side is twenty parts of nine.

2369   We have found two numbers according to the restriction that we restricted, and they are sixty-four units, (and) four parts of eighty-one parts of a unit. And that is what we wanted to find.

2372   **9.** We want to find two numbers, one of them a cube and the other a square, that contain a number such that if its side is subtracted from it, then the remainder is a square.

---

17 Rashed restores the "⟨and⟩" to the text, while Sesiano adds "(and)" to his translation. Sesiano writes "Towards the end of the manuscript, the majority of the problems' conclusions states the results obtained without any particle of coordination to connect them (cf. note 703), as if in some earlier copy they had either been separated by red dots or put on different lines" (1982, 37). From here on we follow Sesiano and write the added word as "(and)".

18 Sesiano corrects "we make" (نجعل) to "we require" (نحتاج) (line 2355). Rashed instead translates the passage as "Faisons en sorte que ce soit un carré" (Diophantus 1984, 4, 8.1).

We assign the cubic number to be sixty-four units, and the square number to be a *Māl*, so the number they contain is sixty-four *Māl*s. And if we subtract its side from it, it leaves sixty-four *Māl*s less eight Things. We require that to be a square. Let us assign its side to be whatever we wish of Things, as long as it is less than eight Things. We assign it to be seven Things, and we multiply it by itself, to get forty-nine *Māl*s. These Equal sixty-four *Māl*s less eight Things.

We restore and confront, to get fifteen *Māl*s Equal eight Things. We divide that ⟨by⟩ a Thing to get fifteen Things Equal eight units. Thus, the Thing is eight parts of fifteen parts of the unit.

2374

As we assigned a side of the square number to be a Thing, the square number is sixty-four parts of two hundred twenty-five parts of the unit. And as the ⟨cubic⟩ number ⟨is sixty-four units, the number⟩ they contain is four thousand ninety-six parts of two hundred twenty-five parts. And when we subtract from it its side, which is sixty-four parts of fifteen, that is, nine hundred sixty parts of two hundred twenty-five, then the remainder is three thousand one hundred thirty-six parts of two hundred twenty-five, which is a square number, and its side is fifty-six parts of fifteen.

2382

We have found two numbers according to the restriction that we restricted, and they are sixty-four units and sixty-four parts of two hundred twenty-five.[19] And that is what we wanted to find.

2390

10. We want to find two numbers, one of them a cube and the other a square, such that if the number they contain is subtracted from its side, then the remainder is a square number.

2393

Let us assign the cubic number to be sixty-four units, and the square number to be a *Māl*, so the number they contain is sixty-four *Māl*s. If we subtract sixty-four *Māl*s from a side of sixty-four *Māl*s, which is eight Things, then the remainder is eight Things less sixty-four *Māl*s, and we require that to be a square. We make its side however many Things we wish, so we make it four Things. Thus, sixteen *Māl*s Equal eight Things less sixty-four *Māl*s. We restore and confront, to get eight Things Equal eighty *Māl*s. We divide the two sides by a Thing to get eight units Equal eighty Things, and the Thing is a part of ten.

2395

As we assigned a side of the square to be a Thing, the square is a part of one hundred parts of the unit. And the cubic number is sixty-four units, so the number that they contain is sixty-four parts of one hundred parts, and when we subtract it from its side, which is eight parts of ten, that is, eighty parts of one hundred, then the

2402

---

19 The second number is written in the manuscript as "eight parts of fifteen parts of the unit, that is, sixty-four parts of two hundred twenty-five". Rashed amends this to read "⟨a square of⟩ eight parts of fifteen parts" (Diophantus 1984, 4, 48.6), while Sesiano believes that the phrase "that is, sixty-four parts of two hundred twenty-five" is a marginal correction that crept into the text (line 2391). We have shortened the text to conform with the style of the answers to other problems.

remainder is sixteen parts of one hundred, which is a square number whose side is four parts of ten.

2408   We have found two numbers according to the restriction that we restricted, and they are sixty-four units and a part of one hundred parts of the unit. And that is what we wanted to find.

2410   **11.** We want to find a cubic number such that if we add it to its square, then what is gathered is a square number.

2412   We assign the cubic number to be from a side of a Thing, so the cubic number is a Cube. And if we add it to its square, which is a Cube Cube, then the sum is a Cube Cube and a Cube. We require that to be a square. Let us assign its side to be from a number of ⟨Cubes such that if we subtract one Cube Cube from its square, then the remainder is a cube, such as⟩[20] three Cubes. So if we subtract a Cube Cube from a square of the three Cubes, then the remainder is eight Cube Cubes, which is a cubic number. If we confront it with a cubic number, the problem can be solved and working it out is not impossible. So let us multiply the three Cubes by itself to get nine Cube Cubes. They Equal a Cube Cube and a Cube.

2419   We cast away the common Cube Cube, leaving eight Cube Cubes Equal a Cube. We divide the two sides by a Cube, resulting in eight Cubes Equal one. Thus, the Cube is an eighth of a unit, which is a part of eight. When we add to it its square, which is a part of sixty-four parts of the unit, then the sum is nine parts of sixty-four parts of the unit, which is a square number, and its side is three parts of eight.

2425   We have found a number according to the restriction that we restricted, and it is a part of eight parts of the unit. And that is what we wanted to find.

2427   **12.** We want to find two square numbers such that if the quotient of the greater (divided) by the smaller is added to the greater, then the sum is a square, and also, if added to the smaller, then ⟨the sum⟩ is a square.

2430   Let us assign the smaller number to be a *Māl*, and we make the quotient of the greater (divided) by the smaller be half a *Māl* and half an eighth of a *Māl*. If we add it to a *Māl* then the sum is a square. And the greater number is half a *Māl Māl* and half an eighth of a *Māl Māl*. So if we add half a *Māl* and half an eighth of a *Māl* to it, it gives half a *Māl Māl* and half an eighth of a *Māl Māl* and half a *Māl* and half an eighth of a *Māl*. We require that to be a square number.

2434   Let us look for a square number such that if we subtract a half and half an eighth from it, then the remainder is a square number, and let us ensure that the square that remains is less than ⟨eighty-⟩one ⟨parts of two hundred fifty-six parts of the

---

20   Rashed makes no restoration here. He translates the sentence as "Supposons son côté un nombre qui soit trois *cubes*" (Diophantus 1984, 4, 49).

unit).[21] Finding that is easy, given what we explained in the second Book,[22] and that is one hundred sixty-nine parts of two hundred fifty-six parts of the unit, and its side is thirteen parts of sixteen parts of the unit. And we know that when we subtract half of a unit and half an eighth of a unit, that is, one hundred forty-four parts of two hundred fifty-six parts, from one hundred sixty-nine parts of two hundred fifty-six parts of the unit, then the remainder is twenty-five parts of two hundred fifty-six part of the unit, and that is a square number whose side is five parts of sixteen parts. So let us assign a side of half a *Māl Māl* and half an eighth of a *Māl Māl* and half a *Māl* and half an eighth of a *Māl* to be thirteen parts of sixteen parts of a *Māl*. We multiply it by itself to get one hundred sixty-nine parts of two hundred fifty-six parts of a *Māl Māl*. It Equals half a *Māl Māl* and half an eighth of a *Māl Māl* and half a *Māl* and half an eighth of a *Māl*.

Let us cast away the common half a *Māl Māl* and half an eighth of a *Māl Māl*, leaving twenty-five parts of two hundred fifty-six parts of a *Māl Māl* Equal half a *Māl* and half an eighth of a *Māl*. And let us multiply everything we have by ten and six parts of twenty-five, to get: a *Māl Māl* Equals five *Māl*s and nineteen parts of twenty-five parts of a *Māl*. We divide the two sides by a *Māl*, which results in: a *Māl* Equals five units and nineteen parts of twenty-five parts of the unit.　　2448

As we assigned the smaller number to be a *Māl*, it is five units and nineteen parts of twenty-five parts of the unit. Let us multiply that by twenty-five to get one hundred forty-four parts of twenty-five parts. And as we assigned the greater number to be half a *Māl Māl* and half an eighth of a *Māl Māl*, we know it is eleven thousand parts and six hundred sixty-four parts of six hundred twenty-five parts of the unit. So let us make the one hundred forty-four parts of twenty-five parts, which is the smaller square, parts of six hundred twenty-five, and that is that we multiply it by twenty-five, to get that the smaller square is three thousand six hundred parts of six hundred twenty-five. And the quotient of the greater square (divided) by the smaller square is three units and six parts of twenty-five of the unit. Let us make that parts of six hundred twenty-five, to get two thousand parts and twenty-five parts of six hundred twenty-five. When we add it to the greater square, which is eleven thousand six hundred sixty-four parts of six hundred twenty-five, it gives thirteen thousand six hundred eighty-nine parts of six hundred twenty-five of the unit. And that is a square number, and its side is one hundred seventeen parts of twenty-five parts. Also, let us add the two thousand twenty-five parts of six hundred twenty-five ⟨to the smaller square, which is three thousand six hundred parts of six hundred twenty-five, so the sum is five thousand six hundred twenty-five　　2454

---

21 This requirement ensures that it is the greater that is divided by the smaller. Rashed does not make the restoration, and translates the text as "faisons en sorte que le carré qui reste soit inférieur à un". He does calculate in his commentary that this remainder must be less than $\left(\frac{9}{16}\right)^2$, but he only notes that this number is less than 1, as the uncorrected text states (Diophantus 1984, 4, LXIII, 51.1).

22 Problem II.10. /s/

parts of six hundred twenty-five⟩, and that is a square number whose side is seventy-five parts of twenty-five.

2474   We have found two numbers according to the restriction that we restricted, and they are eleven thousand parts and six hundred parts and sixty-four parts of six hundred twenty-five parts of the unit, ⟨and⟩ three thousand parts and six hundred parts of six hundred twenty-five parts of the unit. And that is what we wanted to find.

2478   **13**. We want to find two square numbers such that if the result of dividing the greater by the smaller is subtracted from either of them, then the remainder is a square.

2480   Let us assign a side of the smaller square to be a Thing, so the smaller square is a *Māl*. And we make what results from dividing the greater square by the smaller square, which is a *Māl*, something which, if we subtract ⟨it⟩ from a *Māl*, then the remainder is a square. And we also require that which we subtract from the *Māl* to be a square. Let us divide the *Māl* into two square parts,[23] and they[24] are sixteen parts of twenty-five parts of a *Māl*, and nine parts of twenty-five parts of a *Māl*. Let us make the nine parts of twenty-five parts of a *Māl* the result of dividing the greater square by a *Māl*. We multiply the nine parts of twenty-five parts of a *Māl* by a *Māl* to get nine parts of twenty-five parts of a *Māl Māl*, and that is the greater number. It is clear that if we subtract what results from dividing the greater number, which is nine parts of twenty-five parts of a *Māl Māl*, by the smaller number, which is a *Māl* – the result of this division being nine parts of twenty-five parts of a *Māl* – if we subtract it from the smaller number, namely, a *Māl*, then the remainder is sixteen parts of twenty-five parts of a *Māl*, which is a square whose side is four fifths of a Thing. We want to subtract nine parts of twenty-five parts of a *Māl* from the greater number, which is nine parts of twenty-five parts of a *Māl Māl*, so it leaves a square number. But when we subtract nine parts of twenty-five parts ⟨of a *Māl* from nine parts of twenty-five parts⟩ of a *Māl Māl*, then the remainder is nine parts of twenty-five parts of a *Māl Māl* less nine parts of twenty-five parts of a *Māl*, ⟨and that⟩ Equals a square number.

2499   Let us look for a square number such that if we subtract it from nine parts of twenty-five, then the remainder is a square number,[25] and that is eighty-one parts of six hundred twenty-five parts of the unit. ⟨When we subtract it from nine parts of twenty-five, that is, two hundred twenty-five parts of six hundred twenty-five parts of the unit⟩, the remainder is one hundred forty-four parts of six hundred twenty-five, which is a square number whose side is twelve parts of twenty-five parts of the unit. Now that we have arrived at what we were

---

23 Problem II.8. /s/
24 The text has *dhalika* ("that").
25 Problem II.8. /s/

looking for, let us make a root of the nine parts of twenty-five parts of a *Māl Māl* less nine parts of twenty-five parts of a *Māl* be ⟨nine parts of twenty-five parts of a *Māl*⟩. We multiply it by itself to get eighty-one parts of six hundred twenty-five parts of a *Māl Māl*. This Equals nine parts of twenty-five parts of a *Māl Māl* less nine parts of twenty-five parts of a *Māl*, that is, two hundred twenty-five parts of six hundred twenty-five parts of a *Māl Māl* less two hundred twenty-five parts of six hundred twenty-five parts of a *Māl*.

We restore and confront and cast away the common (amounts), leaving: one hundred forty-four parts of six hundred twenty-five parts of a *Māl Māl* Equal two hundred twenty-five parts ⟨of six hundred twenty-five parts⟩ of a *Māl*. Divide the two sides by a *Māl* to get one hundred forty-four parts of six hundred twenty-five parts of a *Māl* Equal two hundred twenty-five parts of six hundred twenty-five parts of a unit. And the *Māl* Equals one and eighty-one parts of one hundred forty-four parts of the unit, that is, one and nine parts of sixteen parts.   2513

As we assigned the smaller square to be a *Māl*, it is twenty-five parts of sixteen parts of the unit. And the greater number is nine parts of twenty-five parts of a square of the smaller number, and that is two hundred twenty-five parts of two hundred fifty-six parts of the unit. And the division of the greater number, which is two hundred twenty-five parts of two hundred fifty-six parts of a unit, by the smaller number, which is twenty-five parts of sixteen, that is, four hundred parts of two hundred fifty-six parts of a unit, results in a half of a unit and half an eighth of a unit, that is, one hundred forty-four parts of two hundred fifty-six parts. And when we subtract it from one of the two squares, that is, from the four hundred parts of two hundred fifty-six parts of the unit, the remainder is two hundred fifty-six parts of two hundred fifty-six parts, that is, one, which is a square whose side is one. And also, if we subtract the result of the division, which is one hundred forty-four parts of two hundred fifty-six parts of the unit, from the (other) square, which is two hundred twenty-five parts of two hundred fifty-six, then the remainder is eighty-one parts of two hundred fifty-six, which is a square, and its side is nine parts of sixteen.   2520

We have found two numbers according to the restriction that we restricted, and they are four hundred parts of two hundred fifty-six parts of a unit, (and) two hundred twenty-five parts of two hundred fifty-six parts of the unit. And that is what we wanted to find.   2536

Our goal in this problem was for the number of the dividend to be the larger number, but the way we worked it out the larger number is the divisor. Our procedure was correct (and) without doubt we have established it. So we will work this problem out a second time according to what we originally intended, with the division of the greater square by the smaller square, and let this be easier than the preceding solution.   2539

We assign the smaller square to be from a side of one and two thirds of a unit, so the square is two units and seven ninths of a unit. And we assign the greater   2544 square to be from a side of a Thing, so the greater square is a *Māl*. We divide the greater square, which is a *Māl*, by the smaller square, which is two units and seven ninths of a unit. The result of the division is nine parts of twenty-five parts of a *Māl*. If we subtract it from the greater square, which is a *Māl*, then the remainder is sixteen parts of twenty-five parts of a *Māl*, which is a square number whose side is four fifths of a Thing. Also, we subtract the result of the division, which is nine parts of twenty-five parts of a *Māl*, from the smaller square, which is two units and seven ninths of a unit, leaving two units and seven ninths of a unit less nine parts of twenty-five parts of a *Māl*. We require that to be a square. We assign its side to be one and two thirds of a unit less a Thing and a fifth of a Thing, and we multiply it by itself to get two units and seven ninths of a unit and a *Māl* and eleven parts of twenty-five parts of a *Māl* less four Things. This Equals two units and seven ninths of a unit less nine parts of twenty-five parts of a *Māl*.

2557 Restore each side of the two sides by what is subtracted from it, and add the same to the other side, and cast away the common like terms, leaving a *Māl* and four fifths of a *Māl* Equal four Things. Divide both sides by a Thing to get a Thing and four fifths of a Thing Equal four units. Thus, the one Thing is two units and two ninths, and a side of the greater square is a Thing, so its side is two units and two ninths, and the greater square is four hundred parts of eighty-one parts of a unit. And if we divide it by the smaller square, which is two units and seven ninths of a unit, that is, two hundred twenty-five parts of eighty-one parts of the unit, then that which results from the division is one and seven ninths, that is, one hundred forty-four parts of eighty-one parts. And if we subtract it from the greater square, which is four hundred parts of eighty-one parts, then the remainder is two hundred fifty-six parts ⟨of eighty-one parts⟩, which is a square number whose side is sixteen parts of nine. And if we subtract it from the smaller square, which is two hundred twenty-five parts of eighty-one parts, then the remainder is eighty-one parts of eighty-one, that is, one, which is a square whose side is one.

2573 We have found two numbers according to the restriction that we restricted, and they are four hundred parts of eighty-one parts of the unit, (and) two hundred twenty-five parts of eighty-one parts of the unit. And that is what we wanted to find.

2576 **14.** We want to find two square numbers such that if the greater is divided by the smaller, the result of the division is something which, if we subtract the greater square from it, then the remainder is a square, and if we also subtract the smaller square from it, then the remainder is a square.

2579 Let us assign a side of the greater square to be a Thing, so the greater square is a *Māl*. And we also assign a side of the smaller square to be four fifths of a unit, so the smaller square is sixteen parts of twenty-five parts of a unit. It is clear that when we divide the greater square, which is a *Māl*, by the smaller square, which is sixteen parts of twenty-five parts of a unit, the result of the division is a *Māl* and nine parts of sixteen parts of a *Māl*. And when we subtract from it the greater square, which is a *Māl*, the remainder is nine parts of sixteen parts of a *Māl*, which is a square whose side is three-fourths of a Thing. We cast away from it the smaller square, which is sixteen parts of twenty-five parts of the unit, giving the remainder twenty-five parts of sixteen parts of a *Māl* less sixteen parts of twenty-five parts of a unit. We require that to be a square. Let us assign to it a side of a Thing and a fourth of a Thing less two units. We multiply it by itself to get twenty-five parts of sixteen parts of a *Māl* and four units less five Things, and that Equals twenty-five parts of sixteen parts of a *Māl* less sixteen parts of twenty-five parts of a unit.

We restore each of the two sides by how much it was reduced and we add its same to the other side, and we cast away the common (amounts), leaving five Things Equal four units and sixteen parts of twenty-five parts of a unit. Thus, the one Thing is a fifth of four units and sixteen (parts) of[26] twenty-five parts of a unit, and that is one hundred sixteen parts ⟨of⟩ one hundred twenty-five parts of the unit.   2592

And as we assigned a side of the greater square to be a Thing, its side is one hundred sixteen parts of one hundred twenty-five parts of the unit, so the square, consequently, is thirteen thousand four hundred fifty-six parts of fifteen thousand six hundred twenty-five. When we divide it by the smaller square, which is sixteen parts of twenty-five, that is, ten thousand parts of fifteen thousand six hundred twenty-five, then it is one and three thousand parts and four hundred fifty-six parts of ten thousand, that is, twenty-one thousand twenty-five parts of fifteen thousand six hundred twenty-five.   2597

When we subtract from it the greater square, which is thirteen thousand four hundred fifty-six parts, the remainder is seven thousand five hundred sixty-nine parts of fifteen thousand six hundred twenty-five, and that is a square number whose side is eighty-seven parts of one hundred twenty-five parts. Likewise, when we subtract from it the smaller number, which is ten thousand parts, then the remainder is eleven thousand parts and twenty-five parts of fifteen thousand six hundred twenty-five, and that is a square number whose side is one hundred parts and five parts of one hundred twenty-five parts of the unit.   2604

We have found two numbers according to the restriction that we restricted, and they are thirteen thousand four hundred fifty-six parts of fifteen thousand six hundred twenty-five parts of the unit, (and) ten thousand parts of fifteen thousand six hundred twenty-five parts of the unit. And that is what we wanted to find.   2612

15. We want to find two square numbers such that if the excess of the greater over the smaller is added to either of them, then the sum is a square.   2616

---

26 Rashed (Diophantus 1984, 4, 60.11) is in error here by adding ⟨a hundred and⟩ to the text. Instead of $\frac{1}{5}$ of (4 and $\frac{16}{25}$), which is intended, he reads it as ($\frac{1}{5}$ of 4) and $\frac{16}{125}$.

2618   Let us assign a side of the greater square to be a Thing, so the greater square is a *Māl*. And we make its excess over the smaller square be two Things and one, so the smaller number is a *Māl* less two Things and one. And it is clear that when we add the excess of the greater of the two numbers ⟨over the smaller⟩, and that is two Things and one, to the smaller, which is a *Māl* less two Things and one, then the sum of that is a *Māl*, which is the greater square number, which is a square. And when we add the excess of the greater square over the smaller square, which is two Things and one, to the greater square, which is a *Māl*, it gives a *Māl* and two Thing and one, and that is a square number, and its side is a Thing and one. We require that the smaller number, which is a *Māl* less two Things and one, be a square. Let us assign its side to be a Thing less two units. We multiply it by itself to get a *Māl* and four units less four Things, and that Equals a *Māl* less two Things and one.

2628   We add ⟨four Things⟩ to a *Māl* and four units less four Things to get a *Māl* and four units. Likewise, we add four Things to a *Māl* less two Things and one to get a *Māl* and two Things less one. We add one to the two sides together, and we cast away the common (amounts), so that it leaves one species Equals ⟨one⟩ species. There remains five units Equal two Things, so the one Thing is two units and a half.

2633   As we assigned a side of the greater square to be a Thing, its side is two and a half, and the greater square is six units and a fourth. If we cast away from it two Things and one, and that is, six units, then the remainder is a fourth of a unit, which is the smaller square. And it is clear that the excess of the greater square, which is six units and a fourth of a unit, over the smaller square, which is a fourth of a unit, is six units. And when added to the greater square, the sum of that is twelve and a fourth, and that is a square number whose side is three and a half. And likewise, if added to the smaller square, the sum of that is six and a fourth, and that is a square number and its side is two and a half.

2641   We have found two numbers according to the restriction that we restricted, and they are six units and a fourth of a unit, (and) a fourth of a unit. And that is what we wanted to find.

2643   **16.** We want to find two square numbers such that if the excess of the greater over the smaller is subtracted from the greater, then the remainder is a square number, and also, if it is subtracted from the smaller, then the remainder is a square number.

2646   Let us assign a side of the greater square to be a Thing, so the greater square is a *Māl*. We make its excess over the smaller square be two Things less one, so the smaller square is a *Māl* and one less two Things. It is clear that when we subtract the excess of the greater square over the smaller square, which is two Things less one, from the greater square, which is a *Māl*, the remainder[27] is a *Māl* and one

---

27 Rashed (Diophantus 1984, 4, 63.19) follows the text at this point, which translates to "which is the smaller square, ⟨which⟩ is a *Māl* and one less two Things". Sesiano reverses the two phrases and adds the "⟨square⟩" at the end (line 2650).

less two Things, which is the smaller square, which is ⟨a square⟩. And when we subtract the excess of the greater square over the smaller square, which is two Things less one, from the smaller square, which is a *Māl* and one less two Things, then the remainder is a *Māl* and two units less four Things. We require that to be a square. Let us assign its side to be a Thing less four units. We multiply it by itself to get a *Māl* and sixteen units less eight Things. It Equals a *Māl* and two units less four Things. We add eight Things to the two sides together, and we cast away the common *Māl* and two, leaving four Things Equal fourteen units, and the one Thing is three units and a half.

As we assigned a side of the greater square to be a Thing, its side is three units   2657
and a half, and its square is twelve units and a fourth of a unit. We cast away from it the same as two Roots[28] less one, and that is six units, leaving six units and a fourth of a unit, and that number is the smaller square. And the excess of the greater square over the smaller square is six units. So when we subtract six units from the greater square, then the remainder is six and a fourth, and that is a square number whose side is two units and a half. And when we also subtract the six units from the smaller square, then the remainder is a fourth of a unit, and that is a square number and its side is a half of a unit.

We have found two numbers according to the restriction that we restricted, and   2665
they are twelve units and a fourth of a unit, (and) six units and a fourth of a unit. And that is what we wanted to find.

17. We want to find three square numbers such that if added together, their sum   2667
is a square, and the first of these numbers is equal to a side of the second, and the second is equal to a side of the third.

Let us assign the first to be a *Māl*, so the second is a *Māl Māl*, since it is a square of   2670
the *Māl* [and the *Māl* is equal to a side of the second].[29] And the third is a *Māl Māl Māl Māl*, which is equal to a square of the second [and the second is its side]. The sum of the three numbers is a *Māl Māl Māl Māl* and a *Māl Māl* and a *Māl*, and we require that to be a square number. Let us assign to it a side of a *Māl Māl* and half of a unit. We multiply it by itself to get a *Māl Māl Māl Māl* and a *Māl Māl* and a fourth of a unit. This Equals a *Māl Māl Māl Māl* and a *Māl Māl* and a *Māl*. We cast away the common like terms, leaving a *Māl* Equals a fourth of a unit.

As we assigned the first number of the three numbers to be a *Māl*, it is a fourth   2676
of a unit, which is equal to a side of the second, and the second is half an eighth of a unit. And the second is likewise equal to a side of the third, and the third is a part of two hundred fifty-six parts of a unit. Adding together these three numbers gives eighty-one parts of two hundred fifty-six parts of the unit, which is a square number, and its side is nine parts of sixteen.

---

28 The text has "Root" in place of "Thing". This is not incorrect but is unusual for this text.
29 Sesiano notes that the word *al-thānī* ("the second") could be read as *al-māl* ("the *Māl*") (1982, 393 n. 771). Rashed's text shows "*al-māl* ⟨*māl*⟩" (Diophantus 1984, 4, 65.6).

2682   We have found three numbers according to the restriction that we restricted, and they are a fourth of a unit, (and) half an eighth of a unit, (and) a part of two hundred fifty-six parts of a unit. And that is what we wanted to find.

2685   **18**. We want to find three square numbers such that if we duplicate the first number with the second number, then what is gathered with the third number, and we add to the outcome the number composed of the sum of the three numbers, then the sum of that is a square number.

2688   Let us assign the first number to be one, the second to be half of a unit and half an eighth of a unit, and the third to be a *Māl*. Then we duplicate the first, which is one, with the second, which is nine parts of sixteen, to get nine parts of sixteen parts of the unit. Then we duplicate it with the third number, which is a *Māl*, to get nine parts of sixteen parts of a *Māl*. We add to it the number composed of the sum of the three numbers, and that is a *Māl* and twenty-five parts of sixteen parts of a unit. The sum of that is twenty-five parts of sixteen parts of a *Māl* and twenty-five parts of sixteen parts of the unit. We require that to be a square number. Let us assign to it a side of a Thing and a fourth of a Thing and a fourth of a unit. We multiply that by itself to get twenty-five parts of sixteen parts of ⟨a *Māl* and ten parts of sixteen parts of⟩ a Thing and a part of sixteen parts of a unit. This Equals twenty-five parts of sixteen parts of a *Māl* and twenty-five parts of sixteen parts of a unit.

2700   We cast away the common like terms, leaving twenty-four parts of sixteen parts of a unit Equal ten parts of sixteen parts of a Thing. Thus, the whole Thing Equals two units and two fifths of a unit.

2702   As we assigned a side of the third number to be a Thing, its side is two and two fifths of a unit, and the third number is one hundred forty-four parts of twenty-five parts of the unit; the first number is one, as we assigned it; and the second number is nine parts of sixteen parts of a unit, as we assigned it. We duplicate the first number with the second number, then what is gathered with the third number, to get eighty-one parts of twenty-five parts of the unit, that is, one thousand two hundred ninety-six parts of four hundred parts of the unit. We add to it the number composed of the three numbers, which is one hundred forty-four parts of twenty-five parts, and one, and nine parts of sixteen parts of the unit, that is, two thousand nine hundred twenty-nine parts of four hundred, so the sum is four thousand parts and two hundred parts and twenty-five parts of four hundred parts of the unit. That is a square number, and its side is sixty-five parts of twenty parts of the unit.

2715   We have found three numbers according to the restriction that we restricted, and they are one hundred forty-four parts of twenty-five parts of the unit, and one, and nine parts of sixteen parts of the unit. And that is what we wanted to find.

2718   **19**. We want to find three square numbers such that if one duplicates the first with the second, and what is gathered with the third, and one subtracts from the outcome the number composed of the sum of the three numbers, then the remainder of that is a square.

Let us assign the first number to be one and the second to be one and nine parts   2721
of sixteen parts, and the third to be a *Māl*. We duplicate the first with the second,
and what is gathered with the third, to get a *Māl* and nine parts of sixteen parts
of a *Māl*. We subtract from it the number composed of the sum of the three numbers, and that is a *Māl* and two units and nine parts of sixteen parts of a unit. The
remainder, then, is nine parts of sixteen parts of a *Māl* less two units and nine parts
of sixteen parts of a unit. We want that to be a square. We assign it to be from a
side of three fourths of a Thing less a fourth of a unit, and we multiply it by itself
to get nine parts of sixteen parts of a *Māl* and a part of sixteen parts of a unit less
six parts of sixteen parts of a Thing. That Equals nine parts of sixteen parts of a
Māl less two units and nine parts of sixteen parts of a unit.

We add to the two sides the sum of two units and nine parts of sixteen parts of a   2731
unit and six parts of sixteen parts of a Thing, to get, after the addition, nine parts
of sixteen parts of a *Māl* and six parts of sixteen parts of a Thing Equal nine parts of
sixteen parts of a *Māl* and two units and ten parts of sixteen parts of a unit. We cast
away the nine parts ⟨of sixteen parts of⟩ the *Māl* in common from the two sides,
leaving forty-two parts of sixteen parts of a unit Equal six parts of sixteen parts of
a Thing, and the Thing Equals seven units.

As we assigned a side of the third square to be a Thing, it is seven units, and the   2738
third square is forty-nine units. And the first square is one, according to what we
assigned, and the second square is one and nine parts of sixteen parts of a unit, as
we assigned. We duplicate the first square with the second square, then what is
gathered with the third square. The outcome of that is seventy-six units and nine
parts of sixteen parts of a unit. When we subtract from it the number composed
of the sum of the three numbers, and that is fifty-one and nine parts of sixteen
parts of a unit, the remainder is twenty-five units, which is a square number and
its side is five.

We have found three numbers according to the restriction that we restricted, and   2747
they are forty-nine units, (and) one, (and) one and nine parts of sixteen parts of a
unit. And that is what we wanted to find.

20. We want to find three square numbers such that if one duplicates the first with   2750
the second, and what is gathered with the third, then we subtract the outcome of
that from the number composed of the sum of the three numbers, then that which
remains is a square.

Let us assign the first square to be four units, and the second square to be four   2753
parts of twenty-five parts of a unit, and the third square to be a *Māl*. Then we
duplicate the first square with the second square, then what is gathered with the
third square, which gives sixteen parts of twenty-five parts of a *Māl*. Let us subtract it from the number composed of the sum of the three numbers, which is a

Māl and four units and four parts of twenty-five parts of a unit. The remainder is nine parts of twenty-five ⟨parts⟩[30] of a *Māl* and four units and four parts of twenty-five parts of a unit. We require that to be a square. Let us assign its side to be three fifths of a Thing and one, and we multiply it by itself to get nine parts of twenty-five parts of a *Māl* and a Thing and a fifth of a Thing and one. This Equals nine parts of twenty-five parts of a *Māl* and one hundred parts and four parts of twenty-five parts of a unit.

2763   We cast away the nine parts of twenty-five parts of a *Māl* and the one in common, so that it leaves one species Equals one species. Thus, it leaves thirty parts of twenty-five parts of a Thing Equal seventy-nine parts of twenty-five parts of a unit. Therefore, the one Thing Equals seventy-nine parts of thirty parts of the unit.

2768   As we assigned ⟨the third⟩ square number to be a *Māl*, its side is seventy-nine parts of thirty parts of the unit, so the square is six thousand two hundred forty-one parts of nine hundred parts of the unit [so it is the third number]. And the first number is four units, as we assigned, and the second number is four parts of twenty-five parts of the unit, according to what we assigned. So if we duplicate the first number, namely, the four units, with the second number, namely, the four parts of twenty-five parts of the unit, then we duplicate what is gathered with the third number, namely, six thousand parts and two hundred parts and forty-one parts of nine hundred parts of a unit, it gives ninety-nine thousand eight hundred fifty-six parts of twenty-two thousand five hundred. If we (then) subtract it from the number composed of the three numbers, and that is four units and four parts of twenty-five parts of a unit and six thousand parts and two hundred ⟨parts⟩ and forty-one parts of nine hundred parts of a unit, that is, two hundred ⟨forty-nine⟩ thousand six hundred twenty-five parts of twenty-two thousand five hundred, then the remainder is one hundred thousand parts and forty-nine thousand parts and seven hundred sixty-nine parts ⟨of twenty-two thousand five hundred⟩, which is a square number, and its side is three hundred parts and eighty-seven parts of a unit hundred fifty parts.

2785   We have found three numbers according to the restriction that we restricted, and they are four units, (and) four parts of twenty-five parts of the unit, (and) six thousand parts and two hundred forty-one parts of nine hundred parts of the unit. And that is what we wanted to find.

2789   **21.** We want to find two square numbers such that if the number composed of their sum is added to a square of either of them, then the sum is a square.

2791   For any square number, adding to it the same as its side and a fourth of a unit gives a square. Accordingly, we should assign one of the two numbers to be a *Māl*, so its square is a *Māl Māl*. And if one adds to a *Māl Māl* the same as its

---

30 Rashed's restoration. Sesiano puts "(parts)" in his translation.

side and a fourth of a unit, then that gives a *Māl Māl* and a *Māl* and a fourth of a unit, which is a square number whose side is a *Māl* and a half of a unit. It is clear that the number composed of the sum of the two numbers is a *Māl* and a fourth of a unit, and as we assigned one of the two numbers to be a *Māl*, the other number is a fourth of a unit. And when we add to its square, which is half an eighth of a unit, the number composed of the sum of the two numbers, and that is a *Māl* and a fourth of a unit, then the sum of that is a *Māl* and five parts of sixteen parts of a unit. We require that to be a square. We assign its side to be a Thing and a half of a unit, and we multiply it by itself to get a *Māl* ⟨and a Thing and a fourth of a unit. That Equals a *Māl*⟩ and five parts of sixteen parts of a unit.

We cast away a *Māl* and a fourth of a unit from the two sides together, leaving a part of sixteen parts of a unit Equals a Thing. So the Thing is a part of sixteen parts of a unit.   2801

As we assigned one of the two squares to be a *Māl*, its side is a part of sixteen parts of a unit, and the square is a part of two hundred fifty-six of a unit, and the other number is a fourth of a unit, as we assigned. The number composed of their sum is sixty-five parts of two hundred fifty-six parts of a unit. When added to a square of one of the numbers, which is sixteen parts of two hundred fifty-six of a unit, it gives eighty-one (parts) of two hundred fifty-six parts of a unit, which is a square number, and its side is nine parts of sixteen parts of a unit. Likewise, when we add it to a square of the other number, namely, a part of sixty-five thousand five hundred thirty-six, then the sum is sixteen thousand six hundred forty-one parts of sixty-five thousand five hundred thirty-six, which is a square number, and its side is one hundred twenty-nine parts of two hundred fifty-six.   2803

We have found two numbers according to the restriction that we restricted, and they are a fourth of a unit, (and) a part of two hundred fifty-six parts of a unit. And that is what we wanted to find.   2814

22. We want to find two square numbers such that if added together (they) give a square number, and if one of them is duplicated with the other (they) give a cubic number.   2816

Since the cubic number comes from the duplication of a number with itself, and again what is gathered with that same number, we assign one of the two square numbers to be a *Māl*, and we duplicate it with itself to get a *Māl Māl*; and let us assign the second number to be a *Māl Māl*. It is clear that if we duplicate one of the numbers, which is a *Māl*, with the other number, which is a *Māl Māl*, it gives a Cube Cube, which is a cubic number, since it comes from a number duplicated with itself and what is gathered with that same number. And if we add the two square numbers together then their sum is a *Māl Māl* and a *Māl*. We require that to be a square. Let us assign to it a side of a *Māl* and a fourth of a *Māl*, and we multiply it by itself to get a *Māl Māl* and nine parts of sixteen parts   2818 of a *Māl Māl*. This Equals a *Māl Māl* and a *Māl*. We cast away the *Māl Māl* in common from the two sides, leaving nine parts of sixteen parts of a *Māl Māl* Equal a *Māl*. Let us divide the two sides by a *Māl*, so it results in nine parts of sixteen parts of a *Māl* Equal one. Therefore, the whole *Māl* Equals sixteen parts of nine parts of a unit.

2830   As we assigned one of the two numbers to be a *Māl*, one of the two numbers is sixteen parts of nine parts of a unit. And the other number is two hundred parts and fifty-six parts of eighty-one parts of the unit. And when we duplicate sixteen parts of nine parts of a unit with two hundred parts and fifty-six parts of eighty-one parts of a unit, it amounts to four thousand parts and ninety-six parts of seven hundred twenty-nine parts of a unit, and that is a cubic number whose side is sixteen parts of nine parts of a unit. And also, when we add the two square numbers together, then the sum is four hundred parts of eighty-one, and that is a square number and its side is twenty parts of nine.

2839   We have found two numbers according to the restriction that we restricted, and they are sixteen parts of nine, (and) two hundred parts and fifty-six parts of eighty-one. And that is what we wanted to find.

2842   We want to work out this problem another way that is easier than the first way. We look for two square numbers such that if added together, then their sum is a square, and that is sixteen *Māl*s and nine *Māl*s. We duplicate one of them with the other, to get one hundred forty-four *Māl Māl*s, which Equals a cubic number. Let the cubic number be eight Cubes, so the one hundred forty-four *Māl Māl*s Equal eight Cubes. We divide both sides by a Cube to get one hundred forty-four Things Equal eight units. Thus, the one Thing is a part of eighteen parts of a unit.

2849   As we assigned one of the two square numbers to be nine *Māl*s, its side is three Things, and that is a part of six parts of a unit. We multiply it by itself to get a part of thirty-six parts of a unit, which is one of the two numbers. The other number was assigned to be sixteen *Māl*s, and its side is four Things, and that is two parts of nine parts of a unit. We multiply it by itself to get ⟨four⟩ parts of eighty-one parts of a unit, which is the other number. And it is clear that if we add the two square numbers together, then their sum is twenty-five parts of three hundred twenty-four, and that is a square number, and its side is five parts of eighteen. And if we duplicate one of the two numbers, namely, a part of thirty-six parts of a unit, with the other number, namely, four parts of eighty-one, it gives four parts of two thousand nine hundred sixteen, that is, one part of seven hundred twenty-nine, which is a cubic number, and its side is a part of nine parts of a unit.

2861   We have found two numbers according to the restriction that we restricted, and they are a part of thirty-six parts of a unit, (and) four parts of eighty-one parts of a unit. And that is what we wanted to find.

2864   **23.** We want to find two square numbers such that if a given square number is divided by either of them and the results of the two divisions are added together, then the sum is a square number; and if the three numbers, that is, the two sought-after numbers and the given number, are added together, then the sum is a square.

Let the given square number be nine units. We want to find two square numbers such that if nine units are divided by either of them and the results from the division are added together, then that gives a square number; and if the three numbers, that is, the two sought-after numbers and the given nine units, are added together, then the sum is a square number.

2868

And we know that whenever we divide a square number into two square parts, (and) we then divide a square number by each of the parts, then the sum of what results from the division(s) is a square number. Let us assign a square number and divide it into two square parts.[31] The number that we assign is a *Māl*, and we divide it into two ⟨square⟩ parts, one of them being nine parts of twenty-five parts of a *Māl* and the other part being sixteen parts of twenty-five parts of a *Māl*. Let these two parts be the sought-after numbers. We divide nine units by nine parts of twenty-five parts of a *Māl*, which results in twenty-five parts of a *Māl*. We also divide nine units by sixteen parts of twenty-five parts of a *Māl*, so the division results in fourteen parts and half an eighth of a part of a *Māl*. When the results of the two divisions are added together, the sum is thirty-nine parts and half an eighth of a part of a *Māl*, which is a square number whose side is six parts and a fourth of a part of a Thing. But when we add together the three numbers, that is, the two sought-after numbers and the given nine, then the sum is a *Māl* and nine units. We require that to be a square. Let us assign its side to be a Thing and one. We multiply it by itself to get a *Māl* and two Things and one. This Equals a *Māl* and nine units. Cast away a *Māl* and one from the two sides so that it leaves one species Equals one species. It leaves two Things Equal eight units, so the Thing is four units.

2872

One of the two sought-after numbers is sixteen parts of twenty-five parts of a *Māl*, and its side is four fifths of a Thing; therefore, its side is four fifths of four, and that is sixteen fifths. We multiply it by itself to get two hundred fifty-six parts of twenty-five, and that is one of the two ⟨sought-after⟩ numbers. Likewise, the other number is nine parts of twenty-five parts of a *Māl*, and its side is three fifths of a Thing, and the Thing is four units; therefore, its side is twelve fifths. We multiply it by itself to get one hundred forty-four parts of twenty-five parts of a unit, which is the other sought-after number. When we divide the given number, namely, the nine units, which is two hundred parts and twenty-five parts of twenty-five parts, by one of the two numbers, which is two hundred parts and fifty-six parts of twenty-five, then the result of the division is two hundred parts and twenty-five parts of two hundred parts and fifty-six parts. Also, when we divide the nine units, that is, the two hundred parts and twenty-five parts of twenty-five, by the other number, namely, the one hundred forty-four parts of twenty-five parts, then the result of the

2888

31 Problem II.8.

division is two hundred parts and twenty-five parts of one hundred forty-four parts, that is, four hundred parts of two hundred fifty-six parts. Then if we add that to the result of dividing the nine by the other number, which is two hundred parts and twenty-five parts of two hundred fifty-six, then the sum of that is six hundred parts and twenty-five parts of two hundred fifty-six, and that is a square number whose side is twenty-five parts of sixteen parts of a unit. So if we add the three numbers together, namely, the two hundred fifty-six parts of twenty-five parts of the unit, and the one hundred forty-four parts of twenty-five, and the nine units, which is two hundred parts and twenty-five parts of twenty-five, then the sum of that is six hundred twenty-five parts of twenty-five, which is twenty-five units, which is a square number, and its side is five units.

2914   We have found two numbers according to the restriction that we restricted, and they are two hundred parts and fifty-six parts of twenty-five parts of a unit, (and) one hundred forty-four parts of twenty-five parts of the unit. And that is what we wanted to find.

2918   End of the sixth Book of the work of Diophantus, and in this book there are twenty-three problems of the arithmetical problems.

Book VII (Arabic)

In the name of God, the Merciful, the compassionate. The seventh Book of the work of Diophantus. 2920

Our goal in this Book is to present many arithmetical problems without diverging from the previous type of problems of the fourth and fifth Books, (even if) they differ in species, to serve in the strengthening of skill and in the growth of experience and practice. 2922

1. We want to find three cubic numbers such that a side of the first is with respect to a side of the second in a given ratio, and a side of the second is with respect to a side of the third in (the same) given ratio, and if the first number is duplicated with the second number, and the outcome is duplicated with the third number, then that gives a square number. 2926

Let the given ratio be the ratio of two times. We want to find three cubic numbers such that a side of the first is twice a side of the second, and a side of the second is twice a side of the third, and if the first of the three numbers is duplicated with the second number, and the outcome with the third number, then that gives a square number. 2930

Let us assign a side of the third number to be a Thing, so the third number is a Cube. And we assign a side of the second number to be two Things, since it is two times a side of the third number, so the second number is eight Cubes. And we assign a side of the first number to be four Things, since it is two times a side of the second number, so the first number is sixty-four Cubes. If we duplicate the first number, which is sixty-four Cubes, with the second number, which is eight Cubes, and the outcome with the third number, which is a Cube, then that is five hundred twelve Cube Cube Cubes. We require that to be a square. Let us assign to it a side of thirty-two *Māl Māl*s. We multiply it by itself to get one thousand twenty-four *Māl Māl Māl Māl*s. This Equals five hundred twelve Cube Cube Cubes. 2933

We divide the five hundred twelve Cube Cube Cubes by a *Māl Māl Māl Māl*, resulting in five hundred twelve Things. And we divide one thousand twenty-four *Māl Māl Māl Māl*s by a *Māl Māl Māl Māl* to get one thousand twenty-four units, which Equal five hundred twelve Things. Thus, the one Thing is two units. 2942

DOI: 10.4324/9781315171470-14

2946   As we assigned a side of the third number to be a Thing, and the Thing is two, a side of the third number is two units, and the third number is eight units. And as we assigned a side of the second number to be two Things, since it is two times a side of the third number, and the two Things are four, the second number is sixty-four units. And as we assigned a side of the first number to be four Things, since it is two times a side of the second number, and the Thing is two units, a side of the first number is four times two units, and that is eight units. Thus, the first number is five hundred twelve units, and if we duplicate the first number, which is five hundred twelve units, with the second number, which is sixty-four units, the outcome of that is thirty-two thousand seven hundred sixty-eight units. We duplicate it with the third number, which is eight units, to get two hundred sixty-two thousand one hundred forty-four units, which is a square number whose side is five hundred twelve.

2959   We have found three numbers according to the restriction that we restricted, and they are five hundred twelve units, sixty-four units, and eight units. And that is what we wanted to find.

2961   **2.** We want to find three cubic numbers which are also squares, and if the first number is duplicated with the second number, and also what is gathered is duplicated with the third number, then that is a square with a square side.

2964   Let us assign the first number to be a part of sixty-four parts, which is a cubic number. Its side is a fourth of a unit, which is also a square number, and its side is half of a unit. And we assign the second number to be sixty-four units, which is a cubic number. Its side is four units, which is also a square number, and its side is two units. And we assign the third number to be a Cube Cube, which is a cubic number. Its side is a *Māl*, which is also a square number, and its side is a Cube.[1] And if the first number, namely, the part of sixty-four parts of the unit, is duplicated with the second number, namely, sixty-four units, the outcome of that is one. And if the one is duplicated with the third number, namely, a Cube Cube, the outcome of that is a Cube Cube. We require its side to be a square [and by 'its side' is meant here 'its root']. A side of a Cube Cube is a Cube, so we Equate a square number, namely, four *Māl*s, with a Cube, and we divide the two sides by a *Māl*, to get: a Thing Equals four units. Thus, it is the Thing, which is a side of the Cube, and the Cube is sixty-four.

2975   As we assigned the third number to be a Cube Cube, which comes from multiplying a Cube by itself, and the Cube is sixty-four units, we multiply sixty-four units by itself to get four thousand ninety-six units, and that is the third number. And if we duplicate the first number, which is a part of sixty-four parts of a unit, with the second number, which is sixty-four units, the outcome of that is one. And if

---

1 Note that to match the wording of the first two assignments, the text should have "a Thing" here. See Sesiano's note (1982, 157 n. 3).

we duplicate the third number, which is four thousand ninety-six, with the one, then that gives four thousand ninety-six, which is a square number, and its side is sixty-four, which is also a square number, whose side is eight units.

We have found three numbers according to the restriction that we restricted, and   2983
they are a part of sixty-four parts of a unit, sixty-four units, and four thousand ninety-six units. And that is what we wanted to find.

3. We want to find a square number with a square side such that if we divide it into   2986
three parts, then each part is a cube.[2]

Let us assign a side of this number to be a *Māl*, so the number is a *Māl Māl*. We   2988
want to divide a *Māl Māl* into three parts such that each part is a cube. Let us assign the first part to be a Cube, and the second part to be eight Cubes, and the third part to be sixty-four Cubes. It is clear that each part of these parts is a cube. But the sum of the three parts is seventy-three Cubes. This Equals the partitioned number, namely, the *Māl Māl*. We divide a *Māl Māl* by a Cube to get a Thing, and we divide seventy-three Cubes by a Cube to get seventy-three units, which Equal a Thing. Thus, the Thing is seventy-three units.

As we assigned a side of the partitioned number to be a *Māl*, it is a square of sev-   2995
enty-three units, and that is five thousand three hundred twenty-nine units. And the partitioned number is a square of five thousand three hundred twenty-nine, which is twenty-eight million three hundred ninety-eight thousand two hundred forty-one. And as we assigned one of the parts to be a Cube, and the Cube comes from multiplying seventy-three by seventy-three, and the outcome by seventy-three, and that is three hundred eighty-nine thousand seventeen units, it is the cubic number that is one of the three parts. And the next part is eight times it, since we assigned it to be eight Cubes, and that is three million one hundred twelve thousand one hundred thirty-six. And the other part is sixty-four Cubes, which is the same as the first part sixty-four times, and that is twenty-four million eight hundred ninety-seven thousand eighty-eight units. It is clear that if we add together these three parts for which each part is a cube, then the cubic number from their sum is twenty-eight million three hundred ninety-eight thousand two hundred forty-one, and that is the partitioned number, which is a square number with a square side.

We have found a number according to the restriction that we restricted, and it is   3009
twenty-eight million three hundred ninety-eight thousand two hundred forty-one. And that is what we wanted to find.

4. We want to divide a cubic number with a square side into three parts such that   3011
each part is a square.

2 The enunciation is misstated. It suggests that if the number is divided into any three parts what-soever, then each part will be a cube. It should say that the number *can* be divided in this way. /s/

3014   Let us assign a side of the cube to be a *Māl*, so the cube is a Cube Cube. We want to divide a Cube Cube into three parts, each part being a square. Let us look for three numbers which, if added together, then the sum is a square, and each of them is a square. Finding that is easy according to what (we did) before.[3] Thus, one of the numbers is one, the second is four units, and the third is four ninths of a unit. Let us make each number of these numbers *Māl Māl*s, so the first number is a *Māl Māl* and the second is four *Māl Māl*s and the third is four ninths of a *Māl Māl*. And because we want to divide the cubic number into three square parts, let us assign each part of the three parts to be a number of these three numbers, (so) their sum ⟨is that cubic number⟩. The number composed of their sum is forty-nine parts of nine parts of a *Māl Māl*, and that Equals the cubic number, namely, the Cube Cube. We divide each of these by a *Māl Māl* to get a *Māl* Equals forty-nine parts of nine parts of a unit.

3026   And because we assigned a side of the cubic number to be a *Māl*, and the *Māl* is forty-nine parts of nine parts of the unit, it is a side of the cube. And the cubic number comes from multiplying forty-nine parts by itself, and what is gathered by forty-nine parts, and that is one hundred seventeen thousand six hundred forty-nine parts of seven hundred twenty-nine parts of the unit. And because we assigned one of the parts to be a *Māl*,[4] it is forty-nine parts of nine parts ⟨of the unit. And⟩ because we assigned the second part to be four *Māl*s, it is one hundred ninety-six parts of nine, and the third part is four ninths of a *Māl*, and that is one hundred ninety-six parts of seven hundred twenty-nine[5] parts of the unit. And if we add together these three parts, we get the same as the cubic number.

3036   We have found a number according to the restriction that we restricted, and it is one hundred seventeen thousand six hundred forty-nine parts of seven hundred twenty-nine. And that is what we wanted to find.

3039   **5.** We want to find a cubic number with a cubic side which, if duplicated with two numbers, one of them a cube and the other a square, and these are added together, then the ⟨sum⟩ is a square number.

3041   Let us assign a side of the cubic number to be a cubic number. Let it be eight units, so the cubic number is five hundred twelve units. We want to find two numbers, one of them a cube and the other a square such that if each of them is duplicated

---

3 Problem III.5 (otherwise). /sr/

4 The part was assigned to be a *Māl Māl*, not a *Māl*. The calculations that follow should all be multiplied by another *Māl*, or $\frac{49}{9}$. Sesiano (1982, 265) attributes the error to the author of the major commentary. Rashed does not mention the error, though he gives correct values in his commentary. The correct values for the three numbers are, $\frac{86,436}{729}$, and $\frac{9604}{729}$. /s/

5 This fraction has the numerator of the incorrect value (see the previous footnote) but the denominator of the correct value.

with five hundred twelve units and these are added together, then the sum is a square. Let us assign the cubic number to be a Cube, and the square to be a *Māl*. We multiply a Cube and a *Māl* by five hundred twelve to get ⟨the sum⟩ five hundred twelve Cubes and five hundred twelve *Māl*s. We require that to be a square number. We assign to it a side of sixty-four Things, and we multiply it by itself to get four thousand *Māl*s and ninety-six *Māl*s. They Equal five hundred Cubes and twelve Cubes and five hundred *Māl*s and twelve *Māl*s.

We cast away five hundred twelve *Māl*s from the two sides, leaving five hundred   3050
twelve Cubes Equal three thousand five hundred eighty-four *Māl*s. We divide the two sides by a *Māl*, resulting in three thousand five hundred eighty-four units Equal five hundred Things and twelve Things, so the one Thing is seven units.

And because we assigned the square number to be a *Māl*, and its side is a Thing,   3054
and the Thing is seven units, and the *Māl* is forty-nine units, ⟨therefore, the square number is forty-nine units⟩. Likewise, because we assigned the cubic number to be a Cube, and the Cube comes from multiplying the *Māl* by the Thing, the cubic number is three hundred forty-three. If we duplicate the cubic number whose side we assigned to be a cube, namely, the five hundred twelve, with the cubic number, which is three hundred forty-three units, that gives one hundred seventy-five thousand six hundred sixteen. Likewise, if we duplicate the five hundred twelve with the square number, namely, forty-nine units, then that is twenty-five thousand eighty-eight. If the one hundred seventy-five thousand six hundred sixteen is added to it, then the sum is two hundred thousand seven hundred four, and that is a square number, and its side is four hundred forty-eight.

We have found a number according to the restriction that we restricted, and it is   3066
five hundred twelve units. And that is what we wanted to find.

6. We want to find two square numbers such that the number composed of their   3068
sum is a square, and if one of them is duplicated with the other, then what is gathered is with respect to the number composed of the sum of the two numbers in a given ratio. But ⟨the number for⟩[6] the given ratio is nothing but a square number, since for any two square numbers, ⟨the number for⟩ the ratio of the larger to the smaller is nothing but a square number, and also the size of the smaller with respect to[7] the larger is nothing but a square.

Let the given ratio be the ratio of nine times.   3074

---

6 This and the next restoration are ours. Sesiano also made this suggestion (1982, 159 n. 12).

7 The preposition in the text is *ilā* ("to"), which indicates that it should be accompanied with *nisba* ("ratio"). But the text has *miqdār* ("size") instead, so the preposition should be *min* or *ʿindā* ("with respect to").

3074   Let us assign the number composed of the sum of the two numbers to be a *Māl*, and we divide a *Māl* into two square parts.[8] Let one of them be sixteen parts of twenty-five parts of a *Māl* and the other part be nine parts of twenty-five parts of a *Māl*. And if we duplicate one of the parts with the other, then that is one hundred forty-four parts of six hundred twenty-five parts of a *Māl Māl*. We require that to Equal nine times the number composed of the two square numbers, namely, nine *Māl*s. We divide one hundred forty-four parts of six hundred twenty-five parts of a *Māl Māl* by a *Māl*, resulting in one hundred forty-four parts of six hundred twenty-five parts of a *Māl*. And we divide nine *Māl*s by a *Māl*, resulting in nine units. This is Equal to the one hundred forty-four parts of six hundred twenty-five parts of a *Māl*. The whole *Māl*, then, Equals thirty-nine units and a part of sixteen parts of a unit.

3086   One of the two numbers is sixteen parts of twenty-five parts of a *Māl*, and that is twenty-five units. And the other number is nine parts of twenty-five parts of a *Māl*, and that is fourteen units and a part of sixteen of a unit. And the sum of the two numbers is thirty-nine and half an eighth, which is a square number whose side is six and a fourth. And if we duplicate one of the two numbers with the other, and that is twenty-five by[9] fourteen and half an eighth, ⟨then that gives three hundred fifty-one units and a half and half an eighth⟩, which is nine times the sum of the two numbers, namely, the thirty-nine and half an eighth.

3094   We have found two numbers according to the restriction that we restricted, and they are twenty-five units, ⟨and⟩ fourteen units and half an eighth of a unit. And that is what we wanted to find.

3096   7. We want to divide a square number with a cubic side into three parts (so that) any two parts added together (make) their sum a square.

3098   Let us assign a side of the square number to be a Cube, so the square number is a Cube Cube. We want to divide a Cube Cube into three parts (so that) any two parts added together (make) their sum a square. ⟨Let us look for three numbers (so that) any two of the numbers added together is a square⟩, and the number composed of the three numbers is a square. Finding that is easy according to what we explained in the sixth problem of the third Book. Thus, the first number is eighty units, the second is three hundred twenty units, and the third is forty-one units. And the sum of the three numbers is four hundred forty-one units. Let us substitute units with *Māl Māl*s, so the sum of the three numbers is four hundred forty-one *Māl Māl*s. They Equal a Cube Cube.

3106   We divide the two sides by a *Māl Māl*. The result of dividing a Cube Cube by a *Māl Māl* is a *Māl*, and the result of dividing four hundred forty-one *Māl*

---

8 Problem II.8. /s/
9 Note the switch to the preposition for multiplication.

*Māl*s by a *Māl* ⟨*Māl*⟩ is four hundred forty-one units. They Equal a *Māl*, so the *Māl* is four hundred forty-one units. The *Māl Māl* is what is gathered from multiplying four hundred forty-one by itself, and that is one hundred ninety-four thousand four hundred eighty-one.

Because we assigned one of the three parts to be eighty *Māl Māl*s, (it) is fifteen million five hundred fifty-eight thousand four hundred eighty units. Likewise, because we assigned the second part to be three hundred twenty *Māl Māl*s, (it) is sixty-two million two hundred thirty-three thousand nine hundred twenty. And likewise, because we assigned the third part to be forty-one *Māl Māl*s, (it) is seven million nine hundred seventy-three thousand seven hundred twenty-one units. The number divided into these three parts is the number composed of their sum and is eighty-five million seven hundred sixty-six thousand one hundred twenty-one, and that is a square number and its side is nine thousand two hundred sixty-one. And the side of this is a cubic number, its side being twenty-one.   3112

Because the first part of the three parts is fifteen million five hundred fifty-eight thousand four hundred eighty units and the second part is sixty-two million two hundred thirty-three thousand nine hundred twenty units, the number composed of their sum is seventy-seven million seven hundred ninety-two thousand four hundred units, which is a square number, and its side is eight thousand eight hundred twenty units. Likewise, because the second part is sixty-two million two hundred thirty-three thousand nine hundred twenty units and the third part is seven million nine hundred seventy-three thousand seven hundred twenty-one units, the number composed of their sum is seventy million two hundred seven thousand six hundred forty-one, and that is a square number, and its side is eight thousand three hundred seventy-nine. Likewise, because the third part is seven million nine hundred seventy-three thousand seven hundred twenty-one units, and the first part is fifteen million five hundred fifty-eight thousand four hundred eighty units, the number composed of their sum is twenty-three million five hundred thirty-two thousand two hundred one, and that is a square number and its side is four thousand eight hundred fifty-one units.   3122

We have found a number according to the restriction that we restricted, and it is eighty-five million seven hundred sixty-six thousand one hundred twenty-one. And that is what we wanted to find.   3139

Now that we have reached the end in solving this problem, we want to work it out a second way that is easier than the first way. Let us begin with the question, before working it out: we want to find a square number with a cubic side which, if divided into three parts, then any two parts, ⟨if added together⟩, are a square number.[10]   3142

Let us assign the square number that we want to divide to be sixty-four units, which is a square number and its side is a cube. We want to divide sixty-four   3145

---

10 This is misstated in the same way as the enunciation to Problem VII.3. /s/

units into three parts, any two parts (of which) added together make their sum a square. Let us look for three numbers which, if added together, make their sum a square, and any two of them added together make their sum a square. We explained this in the sixth problem of the third Book, an explanation we dispense with (to avoid) repetition. One of these three sought-after numbers is three hundred twenty units, the second number ⟨is eighty units⟩, and the third number is forty-one units. And the number composed of the sum of these numbers is four hundred forty-one units, which is a square number.

3154   If this had been the number we wanted to divide, then we would have arrived at what we wanted. But the number we want to divide is sixty-four units. So let us take from each number of the three numbers whose sum is four hundred forty-one units ⟨a number⟩ whose size with respect to the number that we took it from is as the size of the sixty-four with respect to the four hundred forty-one. And that is that we multiply each number of the three numbers by sixty-four, then each product is parts of four hundred forty-one. Because the first number is three hundred twenty units, if we duplicate it with sixty-four units, (it becomes) twenty thousand four hundred eighty, and it is parts of four hundred forty-one. And because the second part is eighty units, if we duplicate it with sixty-four, (it becomes) five thousand one hundred twenty, and it is parts of four hundred forty-one parts of the unit. Likewise, because the third part is forty-one units, if we duplicate it with sixty-four, (it becomes) two thousand six hundred twenty-four, and it is parts of four hundred forty-one.

3166   So we have divided the sixty-four into three parts such that when the first and the second are added together, their sum is twenty-five thousand six hundred parts of four hundred forty-one, and that is a square number and its side is one hundred sixty parts of twenty-one. And if the second and the third are added together, then their sum is seven thousand seven hundred forty-four parts of four hundred forty-one, and that is a square number and its side is eighty-eight parts of twenty-one. And if the third and the first are added together, then their sum is twenty-three thousand one hundred ⟨four parts of⟩ four hundred forty-one, which is a square number and its side is one hundred fifty-two parts of twenty-one.

3175   We have found a number according to the restriction that we restricted, and it is sixty-four units, and we have divided it into three parts which are twenty thousand four hundred eighty parts of four hundred forty-one, (and) five thousand one hundred twenty parts of four hundred forty-one, (and) two thousand six hundred twenty-four parts of four hundred forty-one. And that is what we wanted to find.

3180   **8.** We want to find a square number with a cubic side such that if a certain number is added to it, it gives a square, and if twice that number is also added to it, it gives a square.

3182   Let us assign the square number to be sixty-four units, which is a square number with a cubic side. We want to find a number which, if (it) is added to sixty-four, then the sum is a square, and if twice it is added to sixty-four, then the sum is also a square. Let us look for that in a square number other than sixty-four. We search for a square number such that if we add a certain number to it gives a square, and if we add to it twice that number to it then the sum is a square. And every square number, if we add (to it) the same as two of its roots and one, then the sum is a square. We assign the square number to be a *Māl* and we add two of its roots and one to it, to get a *Māl* and two Things and one, and that is a square number and its side is a Thing and one. When we add to the *Māl* twice two Things and one, that is, four Things and two, then the sum is a *Māl* and four Things and two. We want that to be a square. Let us assign to it a side of a Thing less two. We multiply it by itself to get a *Māl* and four units less four Things. This Equals a *Māl* and four Things and two units.

We add four Things to everything we have,[11] and we cast away a *Māl* with[12] a 3193
Māl, leaving eight Things and two Equal four units. Then we cast away two units from the two sides, leaving eight Things Equal two. Thus, the one Thing is a fourth of a unit, and the *Māl* is half an eighth of a unit. And the number added to it is two Things and one, and that is one and a half. And the other number added to it is double the one and a half, and that is three.

Let us multiply ⟨each of⟩ these by sixteen, so that the *Māl* is a whole number. The 3198
Māl is one and the number added to it is twenty-four and the other number added to it is double the twenty-four, and that is forty-eight. And it is clear that if we add twenty-four units to the one, it gives twenty-five units, and that is a square number. Also, if we add double the twenty-four, that is, forty-eight, to the one, it gives forty-nine, and that is a square number.

And if the assigned square number had been one, then we would have arrived at 3205
what we wanted. But it is sixty-four. And because the sixty-four counts the one sixty-four times, we must duplicate each of the two added numbers, that is, the twenty-four and the forty-eight, with sixty-four. If we duplicate twenty-four with sixty-four, that gives one thousand five hundred thirty-six, which is the number added to the sixty-four. And if we duplicate forty-eight with sixty-four, then the outcome of that is three thousand seventy-two, and that is double the first number. And if we add one thousand five hundred thirty-six to the sixty-four, that gives one thousand six hundred, and that is a square number and its side is forty. And if we add double the one thousand five hundred thirty-six, that is, three thousand seventy-two, to the sixty-four, then the outcome of that is three thousand one hundred thirty-six, which is a square number and its side is fifty-six.

---

11 He means to add four Things to both sides of the equation, not "to everything we have", i.e., to each term in the equation. One can write "to everything we have" when it is a multiplication or division, but not when it is an addition or subtraction.

12 The preposition *bi* ("with") instead of *min* ("from") is used because he is subtracting a number from itself, rather than a smaller number from a greater number. This is how it was worded in medieval Arabic.

3217   We have found two numbers, one of them double the other, such that if either of them is added to a square number with a cubic side, then the sum is a square. And they are one thousand five hundred thirty-six units, (and) three thousand seventy-two units. And that is what we wanted to find.

3221   **9.** We want to find a square number with a cubic side such that if we subtract a certain number from it, then the remainder is a square, and if we also subtract twice that number from it, then the remainder is a square.

3224   Let us assign the square number to be sixty-four, which is a square number, and its side is a cube. We want to find a number such that if we subtract it from sixty-four, then the remainder is a square, and if we subtract twice it from sixty-four, then the remainder is a square. We look for that in a square number other than sixty-four. We search for a square number such that if we subtract a certain number from it, then the remainder is a square, and if we subtract the double of that number from it, then it gives a square. And every square number, (if) we subtract the same as two of its roots less one from it, then the remainder is a square. Let us assign the square to be a *Māl*, and we subtract the same as two of its roots less one from it, then the remainder is a square. When we subtract from it twice two of its roots less one, that is, four of its roots less two, then the remainder is a *Māl* and two units less four Things. We require that to be a square. Let us assign to it a side of a Thing less three units. We multiply it by itself to get a *Māl* and nine units less six Things. They Equal a *Māl* and two units less four Things.

3235   We cast away a *Māl* and two units less four Things from the two sides together,[13] leaving two Things Equal seven units. So the one Thing is three units and a half, and the *Māl* is twelve units and a fourth, and the two numbers subtracted from it are six and twelve. Then we multiply all of them by four so that the *Māl* is a whole number. Thus, the *Māl* is forty-nine and the two subtracted numbers are twenty-four and forty-eight. Now, if the square number had been forty-nine, then we would have arrived at what we wanted. But it is sixty-four, and the sixty-four counts the forty-nine one time and fifteen parts of forty-nine ⟨parts of a time⟩. For this reason, we should likewise add to the two subtracted ⟨numbers⟩, that is, the twenty-four and the forty-eight, fifteen parts of forty-nine parts of them.

---

13 Sesiano notes, "The performed operation is simply wrong in terms of ancient mathematics, and certainly goes back to a commentator; the restoration with 6*x* should have preceded the removal of $x^2 + 2$, in order to avoid arriving at an expression equal to zero" (1982, 163 n. 20). Sometimes Arabic algebraists stated and solved equations with "nothing" on one side, including Ibn al-Bannā', Ibn Badr, Ibn al-Qunfūdh, and al-Fārisī. The other unconventional aspect of this simplification is that Diophantus, and Arabic algebraists as well, would ordinarily not "cast away" diminished amounts but would restore and confront in separate steps. But Diophantus has just stated the rule about subtracting "the same as two of its roots less one" from a square. So, while the simplification here is unconventional, it may not be interpolated.

So let us multiply the twenty-four by sixty-four, and that is one thousand five 3245
hundred thirty-six, ⟨and that is⟩[14] parts of forty-nine parts of the unit, and that is
the number subtracted from the sixty-four. Likewise, we multiply forty-eight by
sixty-four, so that is three thousand seventy-two, and that is parts of forty-nine,
which is the other number subtracted from the sixty-four, and it is double the first
number. And because the first number is one thousand five hundred thirty-six
parts of forty-nine, when we subtract it from the sixty-four, that is, three thousand
one hundred thirty-six parts of forty-nine, then the remainder is one thousand
six hundred parts of forty-nine, and that is a square number, and its side is forty
sevenths. Likewise, because the second number is double the first number, that
is, three thousand seventy-two ⟨parts of forty-nine⟩, when we subtract it from the
sixty-four, that is, three thousand one hundred thirty-six parts of forty-nine, then
the remainder is sixty-four parts of forty-nine, and that is a square number, and its
side is eight parts of seven.

We have found two numbers, one of them double the other, such that if either of 3259
them is subtracted from a square number with a cubic side, then the remainder is
a square. And they are three thousand seventy-two parts of forty-nine, (and) one
thousand five hundred thirty-six parts of forty-nine. And that is what we wanted
to find.

10. We want to find a square number with a cubic side and a number such that 3263
if we add it to it, then the sum is a square, and if we subtract it from the square
number, then the remainder is a square.

Let the square number be sixty-four. We look for a square number other than 3265
sixty-four such that if a certain number is added to it, then the sum is a square, and
if we subtract that number from it, then the remainder is a square. Now, any square
number, (if) we subtract two of its roots less one from it, then the remainder is a
square. So let us assign the square number to be a *Māl*, and the number subtracted
from it to be two Things less one. If we add two Things less one to a *Māl*, then the
sum is a *Māl* and two Things less one. We require that to be a square. Let us assign
to it a side of a Thing less three units. We multiply it by itself to get a *Māl* and nine
units less six Things. This Equals a *Māl* and two Things less one. We cast away
a *Māl* less six Things and one from the two sides, leaving eight Things Equal ten
units,[15] and the one Thing Equals one and a fourth, and the *Māl* is twenty-five
parts of sixteen parts of the unit.

The subtracted number is two Things less one, and that is twenty-four parts of 3275
sixteen parts of the unit, and similarly for the number added to it. Let us multiply

everything by sixteen so the *Māl* is a whole number. Thus, the *Māl* is twenty-five and the number added to it is twenty-four and the subtracted number is twenty-four.

3280   If the number which we assigned were twenty-five, it would be what we were looking for. But the assigned number is sixty-four, and the sixty-four counts the twenty-five two times and fourteen parts of twenty-five parts of a time. Thus, we should multiply the added number, which is also the subtracted number, that is, the twenty-four, by sixty-four, and that gives one thousand five hundred thirty-six, which is parts of twenty-five, and that is the number which we add to sixty-four and subtract from sixty-four.

3286   It is clear that when we add the one thousand five hundred thirty-six parts of twenty-five ⟨to sixty-four, that is, one thousand six hundred parts of twenty-five⟩, then that gives three thousand one hundred thirty-six ⟨parts of twenty-five⟩, which is a square number, and its side is fifty-six parts of five. And when we subtract one thousand five hundred thirty-six parts of twenty-five parts from sixty-four, that is, one thousand six hundred parts of twenty-five, then the remainder is sixty-four parts of twenty-five, and that is a square number and its side is eight parts of five.

3294   We have found a number such that if we subtract it from a square number with a cubic side, then the remainder is a square, and if we add it to it, then the sum is a square, and it is one thousand five hundred thirty-six parts of twenty-five parts of the unit. And that is what we wanted to find.

3298   **11.** We want to divide a given square number into two parts such that if one of them is added to it, then the sum is a square, and if the other is subtracted from it, then the remainder is a square.

3300   Let the given number be twenty-five units. We want to divide twenty-five into two parts such that if we add one of them to the twenty-five, then the sum is a square number, and if we subtract the other part from the twenty-five, then the remainder is a square.

3303   Let us seek to find a certain square (that) we will divide into two parts such that if we add one of them to it and we subtract the other from it, then, after the addition and subtraction, (it) is a square. We know that when we add to a *Māl* two of its roots and one, then the sum is a *Māl* and two Things and one, and that is a square number. And when we subtract from the *Māl* two of its roots less one, then the remainder is a *Māl* and one less two Roots, and that is a square number. And since we want the added and subtracted numbers to add up to a *Māl*, and they in fact add up to four Things, the four Things Equal a *Māl*. We divide each of them by a Thing to get a Thing Equals four units.

3310   And because the Thing is a side of the *Māl*, the *Māl* is sixteen, and the number added to it is two Things and one, and that is nine units; and the number subtracted from it is two Things less one, and that is seven units. And the nine and the seven, if added together, give sixteen units, becoming what we were looking for.

But the assigned number is twenty-five units. Let us multiply the nine units by the twenty-five to get two hundred twenty-five. Then we divide it by the sixteen to get two hundred twenty-five parts of sixteen parts, and that is one of the parts of the twenty-five, that is, the added part. Likewise let us multiply the seven units by the twenty-five ⟨to get one hundred seventy-five⟩. And we divide that by sixteen to get one hundred seventy-five parts of sixteen, which is the other part, that is, the (part) subtracted from the twenty-five. 3314

And it is clear that when we add the two hundred twenty-five parts to the twenty-five units, that is, four hundred parts of sixteen, then the sum is six hundred parts and twenty-five parts of sixteen, and that is a square number whose side is twenty-five parts of four. Likewise, when we subtract the other part, that is, the one hundred seventy-five parts of sixteen, from the four hundred parts, the remainder is two hundred twenty-five parts of sixteen, and that is a square number, and its side is fifteen parts of four. And if the two parts are added together, it gives twenty-five units. 3321

We have divided the twenty-five into two parts according to the restriction that we restricted, and they are two hundred twenty-five parts of sixteen, (and) one hundred seventy-five parts of sixteen. And that is what we wanted to find.[16] 3328

And because one cannot find a square number such that if we divide it ⟨into two parts⟩ and we add either of them to it, gives a square, we want to do something we can. 3331

12. So we say: we want to divide a given square number into two parts such that if we subtract either of them from it, then the remainder is a square. 3333

Let the given number be twenty-five units. We want to divide twenty-five units into two parts such that if we subtract either of them from it, then the remainder is a square. 3335

Let us look for this condition in some square. And every square divided into two square parts, if we subtract either of the parts from the square, then the remainder, which is the other part, is a square. We have already covered this in our work.[17] One of the two parts is sixteen and the other part is nine. 3337

We have divided the twenty-five into two parts such that if we subtract either of them from twenty-five, then the remainder is a square, and they are nine units, (and) sixteen units. And that is what we wanted to do. 3341

---

16 The text should read "And that is what we wanted to do". See line 58 for a similar error. /s/
17 Problem II.8. /s/

3344   **13**. We want to divide a given (square) number into three parts such that if any of them is added to it, then the sum is a square.

3346   Let the given number be twenty-five units. We want to divide twenty-five units into three parts such that if any one of them is added to the twenty-five, then the sum is a square.

3348   And because, if we divide a square number into three parts and we add each part to the divided number then three numbers are gathered such that the number composed of their sum is the same as four times the divided number, because of that, if we divide the twenty-five into three parts and we add each part to the twenty-five, then the sum of the three numbers is one hundred units. So let us divide the one hundred into three square parts, and let each part be more than twenty-five. We have already covered in our treatise how to divide the square number into square parts,[18] so we dispense with repetition of the work. One of the parts is thirty-six, another part is thirty units and three hundred seventy parts of eight hundred forty-one parts of the unit, and the other part is thirty-three units and four hundred seventy-one parts of eight hundred forty-one parts of the unit.

3359   And as each of these three parts is composed of nothing but the twenty-five and one part of the parts of the twenty-five, if we subtract twenty-five from each of these three parts, then the remainder from each part is a part of the parts of the twenty-five. When we subtract twenty-five from thirty-six, the remainder is eleven, which is the first part of the parts of the twenty-five. Likewise let us subtract the twenty-five from the second part, which is thirty-three units and four hundred seventy-one parts of eight hundred forty-one, so the remainder is eight units and four hundred seventy-one parts of eight hundred forty-one, which is the second part of the parts of the twenty-five. Likewise, if we subtract the twenty-five from the third part, which is thirty units and three hundred seventy parts of eight hundred forty-one, then the remainder is five units and three hundred seventy parts of eight hundred forty-one, which is the third part of the parts of the twenty-five. Adding these three parts together gives twenty-five, and adding any one of them to the twenty-five makes the sum a square number.

3374   We have divided the twenty-five into three parts such that if any one of them is added to the twenty-five, then the sum is a square number, and the first of these parts is eleven units, (and) the second is eight units and four hundred seventy-one parts of eight hundred forty-one parts of the unit, (and) the third is five units and three hundred seventy parts of eight hundred forty-one parts. And that is what we wanted to do.

---

18 Problem II.8, iterated. /s/

14. We want to divide a given square number into three parts such that if any one of them is subtracted from it, then the remainder is a square.

3379

Let us assign the square number to be twenty-five units. We want to divide twenty-five units into three parts such that if we subtract any one of the parts from the twenty-five, then the remainder is a square.

3381

Because of this, if we divide the twenty-five into three parts and we subtract each of the parts from the twenty-five, we will have found three numbers such that the number composed of their sum is fifty. Let us divide the fifty into three square parts, and let each of them be less than twenty-five. We have previously shown in our book how to divide a number into square parts.[19] So as before, let us dispense with the repetition. One of the parts is sixteen units, another part is twenty-two units and three parts of one hundred sixty-nine parts of the unit, and the other part is eleven units and one hundred sixty-six parts of one hundred sixty-nine parts of the unit.

3383

And as each part of these parts is equal to the twenty-five if each of its parts is subtracted from it, we must cast away each part of these three parts from the twenty-five, so (each) remainder from the twenty-five is (a part of the) parts of the twenty-five. But when we cast away sixteen units from twenty-five, then the remainder is nine units, which is the first part of the parts of the twenty-five. Likewise, when we cast away twenty-two units and three parts of one hundred sixty-nine parts from twenty-five, then the remainder is two units and one hundred sixty-six parts of one hundred sixty-nine, which is another part. Likewise, when we cast away eleven units and one hundred sixty-six parts of one hundred sixty-nine from twenty-five, then the remainder is thirteen units and three parts of one hundred sixty-nine parts of the unit, ⟨which is the other part⟩.

3391

And the number composed of these three parts – that is, the nine and the two units and one hundred sixty-six parts of one hundred sixty-nine, and the thirteen and three parts of one hundred sixty-nine – is twenty-five units. And if any part of these three parts is subtracted from the twenty-five, then the remainder is a square number.

3402

We have divided the twenty-five into three parts such that if any one of them is subtracted from the twenty-five, then the remainder is a square number, and they are thirteen units and three parts of one hundred sixty-nine parts of a unit, (and) nine units, (and) eleven units[20] and one hundred sixty-six parts of one hundred sixty-nine. And that is what we wanted to do.

3406

15. We want to divide a given square number into four parts such that two parts of the four parts, if either of them is subtracted from the given square number, then the remainder is a square; and similarly, the two remaining parts of the four parts,

3411

---

19 Problem II.9. /s/
20 This is an error. It should be "two units". /s/

if either of them is added to the given square number, then the sum is a square number.

3416   Let the given square number be twenty-five units. We want to divide twenty-five into four parts such that two parts of the four parts, if either of them is subtracted from the twenty-five, then the remainder is a square; and (the other) two parts, if either of them is added to the twenty-five, then the sum is a square.

3420   Let us look for this condition in some square number. Because, if we add to a square number, namely, a *Māl*, the same as twice its side and one, then the sum is a square, we make one of the parts two Things and one. And likewise, if we add to a *Māl* the same as the quadruple of its side and four units, then that amounts to a square, so let us assign the other added part to be four Things and four units. Thus, the number composed of the two added parts is six Things and five units. Likewise, if we subtract from the *Māl* the same as twice its side less one, namely, two Things less one, then the remainder is a square, so we assign one of the subtracted parts to be two Things less one. Likewise, because, if we subtract from the square number, namely, the *Māl*, the same as the quadruple of its side less four units, then the remainder is a square number, we make the other subtracted part four Things less four units. And the number composed of the two ⟨subtracted⟩ parts is six Things less five units. And the number composed of the two added parts was already six Things and five units; therefore, the number composed of the four parts is twelve Things. They Equal a *Māl* since our goal is to divide the *Māl* into four parts. We divide a *Māl* by a Thing to get a Thing, and we divide twelve Things by a Thing to get twelve units, so the Thing Equals twelve units.

3436   The Thing is a side of the *Māl*, so the *Māl* is one hundred forty-four units. One of the two parts added to it is two Things and one, so it is twenty-five units. And the other added part is four Things and four units, so it is fifty-two units. Likewise, one of the two subtracted parts is two Things less one, so it is twenty-three; and the other part is four Things less four units, so it is forty-four units. We have reached what we wanted in what was required of this square number, but we have not (yet) reached what we want to conclude the problem.

3443   If the given number had been one hundred forty-four, then we would have arrived at what we wanted. But it is twenty-five units. And because of that, we should multiply each part of the parts of the one hundred forty-four by the twenty-five and divide what is gathered from that by one hundred forty-four. When we multiply the first of the parts, which is twenty-five, by the twenty-five, then that amounts to six hundred twenty-five. And when we divide it by one hundred forty-four, it gives six hundred twenty-five parts of one hundred forty-four, and that is one of the two parts added to the twenty-five. Likewise, because the other added part is fifty-two units, we multiply fifty-two units by twenty-five, so that is one thousand three hundred, ⟨and if we divide it by one hundred forty-four, it gives one thousand three hundred⟩ parts of one hundred forty-four, which is the other added part. Likewise, because one of the two subtracted parts is twenty-three units, we multiply twenty-five by twenty-three to get five hundred seventy-five, and we divide it by one hundred forty-four to get five hundred seventy-five parts of one hundred forty-four, which is the (first) number subtracted from the twenty-five. Likewise, because the other subtracted part is forty-four units, we multiply forty-four units by twenty-five to get one thousand one hundred, (and we divide it by one hundred forty-four to get one thousand one hundred) parts of one hundred forty-four, which is the other part subtracted from the twenty-five.

And it is clear that when we add these four parts together, they become twenty-five units. And if we add either of the two added parts to the twenty-five, then the sum is a square, and if we subtract either of the two subtracted parts from the twenty-five, then the remainder is a square.   3462

We have divided twenty-five into four parts according to the stipulation that we stipulated. The added (parts) are six hundred twenty-five parts of one hundred forty-four (and) one thousand three hundred parts of one hundred forty-four, (and) the subtracted (parts) are five hundred seventy-five parts of one hundred forty-four (and) one thousand one hundred parts of one hundred forty-four. And that is what we wanted to do.   3466

And the same way we worked this out, we can work out a problem asking: We want to divide a given square number into eight parts such that four (of those) parts, if we add any one of them to the given square, then the sum is a square, and the (remaining) four parts, if we subtract any one of them from the given square, then what remains is a square number.   3471

16. We want to find three square numbers which are also proportional, such that if the first is subtracted from the second then the remainder is a square, and if the second is subtracted from the third, then the remainder is a square.   3476

For three square numbers which are also proportional, if the first is subtracted from the second and the remainder is a square, then it is in the nature of these numbers that if the second is subtracted from the third, then the remainder is (also) a square. Let us assign the first number to be one, and the third number to be a *Māl Māl*. Because of this, the second number is a *Māl*. If we subtract the first number, which is one, from the second number, which is a *Māl*, the remainder is a *Māl* less one. We require that to be a square number. Let us assign to it a side of a Thing less two units. We multiply it by itself to get a *Māl* and four units less four Things, which Equals a *Māl* less one.   3479

We add four Things and one to the two sides, to get a *Māl* and four Things Equal a *Māl* and five units. We cast away the common *Māl*, leaving five units Equal four Things, so the one Thing is one and a fourth.   3486

Because we assigned the second number to be a *Māl*, and its side is a Thing, and the Thing is one and a fourth, namely, five parts of four, the *Māl* is twenty-five   3489 parts of sixteen parts of the unit. The third number was assigned to be a *Māl Māl*, which comes from multiplying the *Māl* by itself, and that is six hundred twenty-five parts ⟨of two hundred fifty-six. So the third number is six hundred twenty-five parts⟩ of two hundred fifty-six parts of the unit. And the first number is what we assigned it to be, namely, we assigned it to be one.

3495   When we subtract the first number, which is one, from the second number, which is twenty-five parts of sixteen, then the remainder is nine parts of sixteen parts of the unit, which is a square number whose side is three parts of four. Likewise, when we subtract the second number, which is twenty-five parts of sixteen parts, that is, four hundred parts of two hundred fifty-six parts from the third number, which is six hundred twenty-five parts of two hundred fifty-six, the remainder is two hundred twenty-five parts of two hundred fifty-six parts of the unit, which is a square number and its side is fifteen parts of sixteen.

3504   We have found three numbers according to the restriction that we restricted, and they are one, (and) four hundred parts of two hundred fifty-six, (and) six hundred twenty-five parts of two hundred fifty-six. And that is what we wanted to find.

3507   **17.** We want to find four square numbers which are also proportional, such that the number composed of their sum is a square.

3509   For four proportional numbers, the product of the first by the fourth is equal to the second by the third. We assign the first square number to be one, and the fourth to be sixteen *Māls*, and the second number to be *Māls* such that if we add it to sixteen *Māls*, then the sum is *Māls*, its number being a square. That is nine *Māls*, since if we add nine *Māls* to sixteen *Māls*, then ⟨the sum⟩ is twenty-five *Māls*, which is a square number, and its side is five Things. And the second number, if multiplied by the third, is the same as the first number if duplicated with the fourth. And the first number, if duplicated with the fourth, gives sixteen *Māls*. We divide sixteen *Māls* by nine *Māls* to get one and seven ninths, which is the third number. Because ⟨of that⟩, the number composed of the four numbers is twenty-five *Māls* and two units and seven ninths of a unit. We require that to be a square. Let us assign to it a side of five Things and a third of a unit. We multiply it by itself to get twenty-five *Māls* and three Things and a third of a Thing and a ninth of a unit. This Equals twenty-five *Māls* and two units and seven ninths of a unit.

3522   We cast away the common ⟨terms⟩ from the two sides, leaving three Things and a third ⟨of a Thing⟩ Equal two units and two thirds of a unit. Thus, the one Thing is eight parts of ten parts of the unit.

3524   Because a side of the second number is three Things, and the ⟨second⟩ number is nine *Māls*, its side is twenty-four parts of ten and the second number is five hundred seventy-six parts of one hundred. Likewise, because the fourth was assigned to be sixteen *Māls* and its side is four Things and the Thing is eight parts of ten, four Things are thirty-two parts of ten, which is a side of the fourth number, and the fourth number is one thousand twenty-four parts of one hundred of the unit. And because we assigned the first number to be one, it is one, as we assigned it. And as we assigned the third number to be one and seven ninths, it is, as we assigned it, one and seven ninths. And each number of these four numbers is a square, and the number composed of their sum is sixteen thousand nine hundred parts of nine hundred, which is a square number, and its side is one hundred thirty parts of thirty parts of the unit.

We have found four numbers according to the restriction that we restricted, and they are in order one, (and) five hundred seventy-six parts of one hundred, (and) one and seven ninths, (and) one thousand twenty-four parts of one hundred parts of a unit. And that is what we wanted to find.   3535

18. We want to find four square numbers which are also proportional, such that if   3539
the first is subtracted from the second, then the remainder is a square; and if the second is subtracted from the third, then the remainder is a square; and if the third is subtracted from the fourth, then the remainder is a square.

We have found that it is in the nature of number that (for) any four propor-   3542
tional numbers which are also squares, if the first number is subtracted from the second number and the remainder is a square, then also if the third number is subtracted from the fourth. number, then the remainder is a square. Because of this, we look for four square proportional numbers (such that if the first is subtracted from the second, then the remainder is a square, and if the second is subtracted from the third, then the remainder is a square). Let us assign the first number to be whatever we wish of units, as long as it is a square. Let us assign it to be nine units. And because if we subtract the first from the second then the remainder is a square, let us assign the second to be whatever we wish of square numbers such that if nine units are subtracted from it, then the remainder is a square. Let us assign it to be twenty-five units. And let us assign the fourth number to be whatever we want of *Māl*s, as long as it is a square. Let us assign it to be one *Māl*. Therefore, the product of the first, which is nine units, by the fourth number, which is a *Māl*, is nine *Māl*s. It is required that the product of the second number, which is twenty-five units, by the third number, also be nine *Māl*s. Therefore, the third number is nine parts of twenty-five parts of a *Māl*. If we subtract the second number, which is twenty-five units, from the third number, which is nine parts of twenty-five parts of a *Māl*, then the remainder is nine parts of twenty-five parts of a *Māl* less twenty-five units. We require that to be a square. Let us assign to it a side of three fifths of a Thing less one. We multiply it by itself to get nine parts of twenty-five parts of a *Māl* and one less six fifths of a Thing. That Equals nine parts of twenty-five parts of a *Māl* less twenty-five units.

Let us add to the two sides six fifths of a Thing and twenty-five units together,   3562
and we cast away the nine parts in common, leaving six fifths of a Thing Equal twenty-six units. Thus, the one Thing is one hundred thirty parts of six. .

Because the fourth number was assigned to be a *Māl* and its side is a Thing, and   3565
the Thing is one hundred thirty parts of six, the fourth number is sixteen thousand

nine hundred parts of thirty-six parts of the unit. Likewise, because the third number is nine parts of twenty-five parts of a *Māl*, it is six thousand eighty-four parts of thirty-six parts of the unit.

3570   When we subtract the first number, which is nine units, from the second number, which is twenty-five units, then the remainder is sixteen units, which is a square number, and its side is four units. And when we subtract the second number, which is twenty-five units, that is, nine hundred parts of thirty-six parts of the unit, from the third number, which is six thousand eighty-four parts of thirty-six parts of the unit, then the remainder is five thousand one hundred eighty-four parts of thirty-six parts of the unit, which is a square number, and its side is seventy-two parts of six parts of the unit. Similarly, when we subtract the third number, which is six thousand eighty-four parts of thirty-six from the fourth number, which is sixteen thousand nine hundred parts of thirty-six, then the remainder is ten thousand parts and eight hundred parts and sixteen parts of thirty-six parts of the unit, which is a square number whose side is one hundred four parts of six.

3582   We have found four numbers according to the restriction that we restricted, and they are nine units, and twenty-five units, and six thousand eighty-four parts of thirty-six, and sixteen thousand nine hundred parts of thirty-six parts of the unit. And that is what we wanted to find.

3586   End of the seventh Book of the work of Diophantus on algebra (*al-jabr wa'l-muqābala*), which consists of eighteen problems.

3588   The work is finished. Praise be to God, Lord of the Two Worlds.[21] The copying was completed on the date of Friday the third of Ṣafar in the year five hundred ninety-five.[22] Glory to the Exalted God and prayers for His prophet Muḥammad and his family.

---

21 The "Two Worlds" are the worlds of humans and *jinn* (Lane 1863–93, 2141).
22 This date corresponds to December 4, 1198 (Sesiano 1982, 22; Rashed in Diophantus 1984, 3, LXVIII).

Book IV^G

1. To divide a given number into two cubes, whose sides (together) are given.   190

Now, let it be (proposed) to divide the number 370 into two cubes whose sides (together) are 10 units.

Let the side of the first cube be assigned to be 1 Number, 5 units, that is, half of the sides. Then, the side of the other will be 5 units lacking 1 Number. Therefore, the cubes themselves (together) will be 30 Powers, 250 units. These are equal to 370 units, that is, to the given (number), and the Number becomes 2 units.

To the hypostases: the side of the first cube will be 7 units, the (side) of the second 3 units, while the cubes themselves will be 343 the first, 27 the second.

2. To find two numbers such that their difference makes a given (number), and, moreover, the difference of their cubes (makes a given number).

Now, let their difference make 6 units, and the difference of their cubes 504 units

Let, again, the side of the greater cube be assigned to be 1 Number, ⟨3 units, and   192
that of the smaller 1 Number⟩[1] lacking 3 units. And it is established that their difference is 6 units. Then, it is required that the difference of their cubes be 504 units. But the difference of their cubes is 18 Powers, 54 units. These are equal to 504 units, and the Number becomes 5 units.

To the hypostases: the side of the greater cube will be 8 units, the (side) of the smaller 2 units, while the cubes themselves will be 512 the one, 8 the other. And the proof is obvious.

3. To multiply the same number by a square and (by its) side, and make the side a cube, and the square the side of the cube.

Let the square be assigned to be 1 Power. Therefore, its side will be 1 Number. And let the number to be multiplied be however many cubic inverse Numbers; well, let it be 8 inverse Numbers. So, when we multiply (this) by 1 Power, we find

---

1 Supplemented by Bachet.

DOI: 10.4324/9781315171470-15

8 Numbers, while, when (we multiply it) by $\langle 1 \rangle^2$ Number, we find 8 units. And we want the 8 Numbers to be a cubic side of 8 units. Thus, 2 units are equal to 8 Numbers, and the Number becomes 2 8ths, while the multiplied number is 32 units. And if we wish not to apply parts, we shall find 8 Numbers equal to 2 units, and the Number becomes a 4th.[3]

194   To the hypostases: the square will be a 16th, the side a 4th, and the multiplied (number) 32; for, if the Number is a 4th, the inverse Number is 4 units.[4] And the proof is obvious.

4. To add the same number to a square and (to its) side and make the same.

Let the square be 1 Power. Therefore, the side will be 1 Number. And let the added (number) be of so many Powers so that, together with 1 Power, it makes a square. Let it be 3 Powers. These, added to $\langle 1 \rangle^5$ Power make a square, while (added) to 1 Number make 3 Powers, 1 Number. These are equal to the side of the square, (i.e.,) of the 4 Powers, that is to 2 Numbers, and the Number becomes one 3rd.

To the hypostases: the square will be one 9th, the side one 3rd, and the added number 3 9ths.

5. To add the same number to a square and (to its) side and make the opposite.

Let the square be 1 Power. Therefore, the side will be 1 Number. And in order to make the side a square, let the added (number) be however many square Powers lacking the Numbers of the side of the square (we want). Let it be 4 Powers lacking 1 Number.
196   ⟨These, added to 1 Number, make | a square, while, (added) to 1 Power, make 5 Powers lacking 1 Number⟩.[6] These are equal to 2 Numbers – the side of the square which is produced from the addition – and the Number becomes 3 5ths.

To the hypostases: the square will be 9 25ths, the side 3 5ths, and the added (number) 21 25ths.

6. To add the same square (number) to a cube and to a square and make the same.

Let the cube be 1 Cube, and the square however many square Powers; let it be 9 Powers. Since we want a certain square, together with 9 Powers, to make a square, I set out two numbers whose product is 9 units. So, let (them) be 1 unit and 9 units. If I subtract the unit from the 9, and I multiply half of the remainder

---

2 Following Bachet, Tannery's text has "⟨1⟩ Number".
3 Tannery points out that this remark refers to the fractional notation used, i.e., the 2 8ths is shown with denominator over the numerator $\left(\frac{\eta}{\beta} \right)$, while the 4th is shown as the delta with the superscript chi (δ^χ). Regarding the different ways fractions are expressed in the Diophantine text we refer the reader to our discussion in Section 4.2.7.
4 This phrase attests that ἀριθμός and ἀριθμοστόν are understood as inverse to each other, as we today understand m and $\frac{1}{m}$, from which the translation "inverse Number" we adopted for the latter.
5 Following Bachet, Tannery's text has "⟨1⟩ Power".
6 Supplemented by Tannery.

by itself, I shall have 16 units. This, if it receives in addition the 9, makes a square. Accordingly, I assign the added square to be 16 Powers. And if it is added to the 9 Powers, it becomes a square, while, if (it is added) to the 1 Cube, it becomes 1 Cube, 16 Powers. These are equal to a cube, say to 8 Cubes, and the Number becomes 16 7ths.

To the hypostases: the cube will be 4096 343rds, the square 2304 49ths, and the square added to them, 4096 49ths.

7. To add the same square (number) to a cube and to a square and make the opposite.   198

Let the cube be the first, the square be the second, and the square added to them be the third.[7]

Since I want the square, the third (number), which is added to the square, the second (number), to make a cube, let it make the cube which is the first (number). Therefore, the first exceeds the second by the third, that is by a square; for the third is a square. Now, if I set out two numbers whatever, their squares, if they receive in addition, or if they lack, twice their product, make a square. Accordingly, two numbers being set out, I must assign the first (number) to be the (sum) of their squares – since the first is equal to the two squares, the sought-after and the added, (i.e.,) the third and the second square – and the third to be twice their product. But the third is a square; therefore, twice their product is also a square. Let the one be assigned to be 1 Number, and the other 2 Numbers, so that twice their product is a square. Accordingly, taking their squares, I assign the first to be 5 Powers, and twice their product, (i.e.,) the third, to be 4 Powers. Then, it remains for the second to be 1 Power; for, together with the third, it is equal to the first. It remains for the first to make a cube. Thus, 5 Powers are equal to 1 Cube, and the Number becomes 5 ⟨units⟩.[8]

To the hypostases: the cube, the first, will be 125 units; the square, the second, 25 ⟨units⟩;[9] and the added square, the third, 100 units. And the proof is obvious.

Otherwise   200

Let the cube be the first, the square be the second, and the added square be the third.

So, since I want the added square, having been added to the second, that is to a square, to make a cube, let it make the first. Again, since (I want) the first, having been added together with the third, to make a square, I am reduced to finding two

---

7 This way of assigning the names "first number", "second number", and "third number" is for reference only. They do not function as algebraic names because they are not subject to operations that result in compound expressions.
8 Supplemented by Tannery.
9 Supplemented by Tannery.

squares whose sum with one of them makes a square.[10] Let the first of the two squares be assigned to be 1 Power, and the second 4 units. Their sum, together with one of them, becomes 2 Powers, 4 units, (which are) equal to a square, (say) the one from a side of 2 Numbers lacking 2 units. The square becomes 4 Powers, ⟨4 units⟩[11] lacking 8 Numbers, and the Number becomes 4 units. To the hypostases: the one will be 4, and the other 16.

Now, assign the square added to them to be 16 Powers, and the second, 4 Powers. Therefore, the first will be 20 Powers; for we want this to be equal to the sum of both. Then, 20 Powers must be equal to 1 Cube, and the Number becomes 20 units.

To the hypostases: the first will be 8000, the second 1600, and the added (number) 6400. And this can be shown in an unlimited number of ways.

8. To add the same number to a cube and to (its) side and make the same.

202   Let the added (number) be 1 Number, and the side of the cube however many Numbers; let it be 2 Numbers; the cube, therefore, is 8 Cubes. Therefore, if 1 Number is added to 2 Numbers, they become 3 Numbers, while if (it is added) to 8 Cubes, they become 8 Cubes, 1 Number. These are equal to 27 Cubes. Let the 8 Cubes be subtracted. So, there remain 19 Cubes equal to 1 Number. Everything by a Number. Therefore, 19 Powers are equal to 1 unit. And the one unit is a square; if 19, the multitude of the Powers,[12] were also a square, the equality would be solvable. But the 19 Powers result from the excess that 27 Cubes have (over) 8 Cubes, and the 27 Cubes are a cube from 3 Numbers, while the 8 Cubes are a cube from 2 Numbers. Accordingly, the 19 arose from the excess that the cube from 3 Numbers has (over) the cube from 2 Numbers. But 2, on the one hand, are the supposed Numbers, and 3, on the other, are always one unit greater than the arbitrary multitude of the Numbers of the side.

Thus, I am reduced to finding two numbers exceeding one another by 1 unit, so that the difference of their cubes makes a square. Let the one be 1 Number, and the other 1 Number, 1 unit. The difference of their cubes is 3 Powers, 3 Numbers, 1 unit. These are equal to a square, (say) to the one from a side of 1 unit lacking 2 Numbers, so the Number becomes 7 units. To the hypostases: the one will be 7, and the other 8.

So, I return to the initial (problem) and assign the added (number) to be 1 Number, and the side of the cube to be 7 Numbers. The cube, therefore, will be 343 Cubes, and when the (1) Number is added to each of them it makes one of them 8 Numbers and the other 343 Cubes, 1 Number. We want, therefore, these to be a

---

10   The text has at this point the following phrase, considered by Tannery to be interpolated: "surely for the reason that, because the two squares, the one which is added to the second and the second, make a cube, that is the first".

11   Supplemented by Bachet.

12   Lit. "if the multitude of the 19 Powers".

cube having 8 Numbers as its side. | Thus, 512 Cubes are equal to 343 Cubes, 1 Number, and the Number becomes one ⟨13th⟩.[13]

To the hypostases: the cube will be 343 2197ths, the side 7 13ths, and the added (number) one (13th).

9. To add the same number to a cube and to (its) side and make the opposite.

Let the cube be however many cubic Cubes; so, let it be 8 (Cubes); therefore, its side will be 2 Numbers. ⟨And in order to make the side a cube, let the added (number) be a cubic (multitude) of Cubes lacking 2 Numbers⟩,[14] namely those of the side of the cube, that is, 27 Cubes lacking 2 Numbers. And if they are added to the 2 Numbers, they make 27 Cubes, which is a cube from a side of 3 Numbers, while, if (they are added) to the 8 Cubes, they make 35 Cubes lacking 2 Numbers. Now, we want these to be a cubic side of the produced 27 Cubes, that is, 3 Numbers. Thus, 35 Cubes lacking 2 Numbers are equal to 3 Numbers. And so, 5 Numbers become equal to 35 Cubes. And everything by a Number. Thus, 35 Powers are equal to 5 units. And so, the Number turns out to be inexpressible, for the species does not have to the species the ratio that a square number has to a square number. But the 35 Powers, on the one hand, are the sum of two cubes, 27 and 8, while the 5 units, on the other, (arise) from the addition of their sides.

Thus, I am reduced to finding two cubes which, | added together, will have   206
to their sides, added together, the ratio that a square number has to a square number. Let their sides added together be however many units; well, let them be 2. And let the side of the first cube be assigned to be 1 Number. Therefore, the side of the other will be 2 units lacking 1 Number. And their cubes, when added together, make 6 Powers, 8 units lacking 12 Numbers. So, we want these to have to the sum of their sides, that is to 2 units, the ratio that a square number has to a ⟨square⟩[15] number. And 2 units are the double of a square. Therefore, 6 Powers, 8 units lacking 12 Numbers are also the double of a square. Thus, their half is equal to a square, that is, 3 Powers, 4 units lacking 6 Numbers become equal, (say) to the (square) from 2 units lacking 4 Numbers, and the Number becomes 10 13ths.

To the hypostases: the one will be 10 13ths, the other 16 13ths. I eliminate the 13ths, and (take) the half; therefore, the sides of these cubes are 5 and 8.

So, I return to the initial (problem) and I assign the side of the cube to be 5 Numbers; the cube, therefore, will be 125 Cubes; and (I assign) the added (number) to be from the cube of 8, that is, 512 Cubes lacking 5 Numbers. This, when added to 5 Numbers, makes a cube, while, when added to 125 Cubes, makes 637 Cubes lacking 5 Numbers. Thus, we want these to be a cubic side of 512 Cubes.

---

13 Supplemented by Tannery.
14 Supplemented by Tannery.
15 Supplemented by Bachet.

Therefore, 8 Numbers are equal to 637 Cubes lacking 5 Numbers, and the Number becomes one ⟨7th⟩.[16]

To the hypostases: the cube will be 125 343rds, the[17] side 5 7ths, and the added number 267 343rds.

208   **10**. To find two cubes equal to their own sides.[18]

So, let the sides of the cubes be in (terms of) Numbers, one of them 2 Numbers, the other 3 Numbers. Therefore, the cubes, when added together, will make 35 Cubes, (which are) equal to (the sum of) the sides, that is, to 5 Numbers. And everything by a Number. Thus, 35 Powers are equal to 5 units, and the Number turns out to be inexpressible.

But the 35 Powers are the sum of two cubes, 8 and 27, while the 5 units (arise from) their sides added together. Thus, I am reduced to finding two cubes which, having been added together, and having been divided by their sides added together, make the quotient a square. But this was shown before,[19] and the sides of the cubes are 8 Numbers, and 5 Numbers.

So, I return to the initial (problem) and I assign one of the sides of the cubes to be 8 Numbers, the other 5 Numbers, and so the cubes, when added together, become 637 Cubes. These are equal to (the sum of) the sides, that is, to 13 Numbers, and the Number becomes one ⟨7th⟩.[20]

To the hypostases: the side of the first cube will be 5 (7ths), of the other 8 (7ths), while the cubes themselves will be 125 343rds and 512 343rds.

11. To find two cubes whose difference is equal to the difference of their sides.

Let one of their sides be 2 Numbers, the other 3 Numbers. So, the difference of their cubes is 19 Cubes, while the difference of the sides, 1 Number. Thus, 1 Num-
210   ber is equal to 19 Cubes, | and the Number turns out to be inexpressible, for the species does not have to the species the ratio of a square to a square.

So, I am reduced to finding two cubes so that their difference has to the difference of their sides the ratio that a square ⟨number⟩[21] (has) to a square number. Let the sides of the cubes be, the one 1 Number, the other 1 Number, 1 unit, so that their difference is a square, namely 1 unit. And since the side of the one is 1 Number, and (the side) of the other is 1 unit and 1 Number, therefore, the difference of the sides will be 1 unit, ⟨while the difference of the cubes, 3 Powers, 3 Numbers, 1 unit⟩.[22] Accordingly, we want 3 Powers, 3 Numbers, 1 unit to have to 1 unit – the difference of the sides – the ratio that a square number (has) to a

---

16 Supplemented by Tannery.
17 Tannery's edition has here the masculine article ὁ, instead of the feminine ἡ (p. 206.24).
18 That is, the sum of the cubes is equal to the sum of the sides of the cubes.
19 In the previous problem.
20 Supplemented by Tannery.
21 Supplemented by Tannery.
22 Supplemented first by Auria, then, with a slight modification, by Bachet.

square number. Therefore, their product must be a square. But their product is 3 Powers, 3 Numbers, 1 unit. These are equal to a square, (say) to the one from a side of 1 unit lacking 2 Numbers, and the Number becomes 7 units. To the hypostases: the sides will be 7 and 8.

I return to the initial (problem) and I assign one of the sides of the cubes to be 7 Numbers, the other, 8 Numbers. So, their difference is 1 Number, while the difference of their cubes, 169 Cubes. Thus, 169 Cubes are equal to 1 Number, and the Number becomes one ⟨13th⟩.[23]

To the hypostases: the sides of the cubes will be 7 (13ths) and 8 (13ths).

12. To find two numbers such that the cube of the greater, if it receives in addition the smaller number, is equal to the cube of the smaller if it receives in addition the greater number.

Let the one be 2 Numbers, and the other 3 Numbers. And the cube of the greater number, when it receives in addition the smaller, makes 27 Cubes, 2 Numbers, while the cube of the smaller, when it receives in addition the greater, makes 8 Cubes, 3 Numbers. Thus, 8 Cubes, 3 Numbers are equal to 27 Cubes, 2 Numbers. And everything by a Number: 19 Powers become equal to 1 unit, and the Number is inexpressible.

But the 19 Powers are the difference of two cubes, while 1 unit is the difference of their sides. So, I am reduced to finding two cubes whose difference has to the difference of their sides the ratio that a square number has to a square number. But this was shown before,[24] and the sides of the cubes are 7 and 8.

So, I return to the initial (problem) and I assign one to be 7 Numbers, and the other 8 Numbers. And so, 343 Cubes, 8 Numbers become equal to 512 Cubes, 7 Numbers, and the Number becomes one ⟨13th⟩.[25]

To the hypostases: the one will be 7 (13ths), the other 8 (13ths), and the proof is obvious.

13. To find two numbers such that either one of them, or their sum, or their difference, together with one unit, makes a square.

So, if I subtract 1 unit from a certain square I shall have the first. | I form a certain square from however many Numbers and 1 unit; so, let it be 3 Numbers, 1 unit. This (square), therefore, will be 9 Powers, 6 Numbers, 1 unit; and if I remove the one unit, I assign the first to be 9 Powers, 6 Numbers. Again, since we want the first and the second together with 1 unit to make a square, and both the first and the second together with 1 unit are ⟨the second together with 1 unit⟩[26] and 9

212

214

23 Supplemented by Tannery.
24 See the previous problem.
25 Supplemented by Tannery.
26 Supplemented by Tannery.

Powers, 6 Numbers, and the second together with 1 unit is a square, I am led to looking for a square which, together with 9 Powers, 6 Numbers, makes a square.

I set out two numbers whose product is 9 Powers, 6 Numbers. ⟨9 Numbers, 6 units measure (them) according to 1 Number. And if I assign the side of the smaller to be half of their difference, it will be 4 Numbers, 3 units.⟩[27] These (multiplied) by themselves become 16 Powers, 24 Numbers, 9 units. I subtract 1 unit and I assign the second to be 16 Powers, 24 Numbers, 8 units. But the first was 9 Powers, 6 Numbers. And so, each together with 1 unit makes a square. It remains for their difference together with 1 unit, which is 7 Powers, 18 Numbers, 9 units, to be equal to a square, (say) to the one from a side of 3 units lacking 3 Numbers, and the Number becomes 18 units.

To the hypostases: the first will be 3024, the second 5624, and the proof is obvious.

216 **14**. To find three square numbers which, if added together, will be equal to their differences added together.

Since, on the one hand, the difference of the greatest and the middle, and the difference of the middle and the least, and the difference of the greatest and the least, are equal to the three (squares), and, on the other hand, the (sum of) the differences of the three (squares) is twice the difference of the greatest and the least, therefore, twice the difference of the greatest and the least is equal to all three. Let the least[28] square be assigned to be 1 unit, and the greatest, 1 Power, 2 Numbers, 1 unit. So, twice the difference of the greatest and the least is 2 Powers, 4 Numbers. But (these) are the three squares, two of which are 1 Power, 2 Numbers, 2 units. ⟨Therefore, the remaining middle (square) will be 1 Power, 2 Numbers lacking 2 units⟩.[29] These, therefore, must be equal to a square, say to the one from a side of 1 Number, 4 units, and the Number becomes 9 5ths.

To the hypostases: the greatest will be 196 25ths, the middle 121 25ths, and the least 1 unit. And everything 25 times. The greatest will be 196, the middle 121, and the least 25.

15. To find three numbers such that any two added together and multiplied by the remaining one make given numbers.

218 Now, let it be proposed that the sum of the first and the second, if multiplied by the third, makes 35 units; the sum of the second and the third, if multiplied by the first, makes 27 units; and, moreover, the sum of the first and the third, if multiplied by the second, makes 32 units.

Let the third be assigned to be 1 Number. Then, the first and the second are 35 inverse Numbers. Let the first be 10 inverse Numbers; the second will be 25

---

27 Supplemented by Tannery.
28 Lit. "smaller".
29 Supplemented by Bachet.

inverse Numbers. And so, two (of the) proposed tasks remain: that the sum of the second and the third, (if multiplied) by the first, make 27 units, ⟨and also that the sum of the first and third, (if multiplied) by the second, make 32 units⟩.[30] But the second and the third, (when multiplied) by the first, ⟨make⟩[31] 10 units, 250 inverse Powers. Thus, 10 units together with 250 inverse Powers are equal to 27 units. On the other hand, the third and the first, (when multiplied) by the second, make 25 units, 250 inverse Powers, (which are) equal to 32 units, while 10 units and 250 inverse Powers are equal to 27 units. And the units exceed the units by 5 units. So, if the 25 units, 250 inverse Powers also exceeded (the) 10 units, 250 inverse Powers by 5 units, the excesses[32] would have been equal. But (the) 25 units come from the second, while the 10 units come from the first. Thus, we want their excess to be 5 units. But these two, the first and the second, are not arbitrary, but, if added together, are 35 units. Thus, I am led to dividing the 35 | into two   220
numbers such that one exceeds the other by 5 units.[33] And so, the one is 15 and the other 20. I assign the first to be 15 inverse Numbers, and the second 20 inverse Numbers. And the sum of the second and the third, (when multiplied) by the first, makes 15 units, 300 inverse Powers, (which are) equal to 27 units; while the sum of the first and the third, (when multiplied) by the second, makes 20 units, 300 inverse Powers, (which are) equal to 32 units. And if I equate 20 units, 300 inverse Powers with 32 units, the Number becomes 5 units.

To the hypostases: the first will be 3 units, the second 4 units, and the third 5 units.

16. To find ⟨three⟩[34] numbers equal to a square, such that the square of each one of them, if it receives in addition the next one, makes a square.

Let the middle be assigned to be however many Numbers (we want). Let it be 4 Numbers. And since I want the square of the first, if it receives in addition the second, to make a square, I am reduced to finding a square which, if it receives in addition 4 Numbers, makes a square. First, look for two numbers whose product is 4 Numbers. 2 Numbers measure (4 Numbers) according to 2 units. And if I assign the first to be half their difference, it will be 1 Number lacking 1 unit, so, (the proposed task – that) the square of the first, if it receives in addition the second, makes a square – is fulfilled. The square of the middle, also, if it receives in addition the third, must make a square; that is, 16 Powers together with the third | (must) ⟨make⟩[35] a square. If, therefore, I subtract the 16 Powers from a   222
certain square, I shall have the third. I assign the square to be from the side of the 16 Powers, (i.e.,) 4 Numbers, 1 unit. This square, therefore, will be 16 Powers, 8

30 Supplemented by Tannery.
31 Supplemented by Bachet.
32 Lit. "excess".
33 This is Problem I.1.
34 Supplemented by Bachet.
35 Supplemented by Bachet.

Numbers, 1 unit. Therefore, if I subtract the 16 Powers, the remainder, 8 Numbers, 1 unit, will be the third. Again, since I want all three to be equal to a square, and all three are 13 Numbers, these are equal to a square; let them be (equal) to the square (multitude of) 169 Powers,[36] so the Number becomes 13 Powers.

To the hypostases: the first will be 13 Powers lacking 1 unit, the second 52 Powers, the third 104 Powers, 1 unit, and I have fulfilled three of the proposed tasks indeterminately.

It remains, also, for the square of the third, that is, 10,816 Power-Powers, 208 Powers, 1 unit, together with the first, that is, 13 Powers lacking 1 unit, to make a square. But it makes 10,816 Power-Powers, 221 Powers, (which are) equal to a square. Everything by a Power. Thus, 10,816 Powers, 221 units become equal to a square, (say) to the one from a side of 104 Numbers, 1 unit, and the Number becomes 55 52nds.

To the hypostases: the first will be 36,621 2704ths, the second 157,300 2704ths, and the third 317,304 2704ths.

17. To find three numbers equal to a square, such that the square of each one of them, if it lacks the next one, makes a square.

Let, again, the middle be assigned to be 4 Numbers. And since I want the square of the first, if it lacks the second, that is the 4 Numbers, | to make a square, I am reduced ⟨to⟩[37] finding a square which, if it lacks 4 Numbers, makes a square. So, I search first for two numbers whose product is 4 Numbers. But 2 units measure 4 Numbers according to 2 Numbers. Now, if I take half of their sum, I assign the first to be 1 Number, 1 unit, and one of the proposed tasks has been fulfilled. Again, since I want the square of the second, that is 16 Powers, if it lacks the third, to make a square, if, therefore, we remove from the 16 Powers a certain square, (let it be the one) from 4 Numbers lacking 1 unit, they result in 16 Powers, 1 unit lacking 8 Numbers. I subtract these from 16 Powers. There remain 8 Numbers lacking 1 unit. So, I assign the third to be 8 Numbers lacking 1 unit, and another proposed task has been fulfilled. Again, since I want the three, that is, 13 Numbers, to be equal to a square, let the equal (square) be 169 Powers, and the Number becomes 13 Powers.

To the hypostases: the first will be 13 Powers, 1 unit, the second 52 Powers, the third 104 Powers lacking 1 unit, and again I have fulfilled three of the proposed tasks indeterminately.

It remains, also, for the square of the third, if it lacks the first, to make a square. But the square of the third, when it lacks the first, makes 10,816 Power-Powers lacking 221 Powers. (These are) equal to a square. And everything by a Power: 10,816 Powers lacking 221 units become equal to a square, (say) to the one from a side of 104 Numbers lacking 1 unit, and the Number becomes 111 104ths.

---

36 This is a second unknown. See the commentary on this problem.
37 Supplemented by Bachet.

To the hypostases: the first will be 170,989 10,816ths, the second 640,692 10,816ths, and the third 1,270,568 10,816ths.

18. To find two numbers such that the cube of the first, if it receives in addition the second, makes a cube, and the square of the second, if it receives in addition the first, makes a square.   226

Let the first be assigned to be 1 Number. Therefore, the second will be a cubic (multitude of) units,[38] (let it be) 8, lacking 1 Cube. And so, the cube of the first, if it receives in addition the second, becomes a cube. It remains, also, for the square of the second, if it receives in addition the first, to make a square. But the square of the second, when it receives in addition the first, makes 1 Cube-Cube, 1 Number, 64 units lacking 16 Cubes. ⟨These are equal to a square, say to the one from a side of 1 Cube, 8 units, that is to 1 Cube-Cube, 16 Cubes, 64 units⟩.[39] And the lacking having been added in common, and the likes having been subtracted from likes, there remain 32 Cubes equal to 1 Number. And everything by a Number: 32 Powers are equal to 1 unit. The unit is a square; so, if the 32 Powers were also a square, then the setting-up of the equation would have been accomplished. But the 32 (of the) Powers come from twice (the) 16 (of the) Cubes, while the 16 (of the) Cubes come from (the product of) twice the 8 (of the) units and the 1 (of the) Cube, that is, twice the 8 (of the) units. Therefore, the 32 Powers (come) from four times the 8 units.

So, I am led to finding a cube which, taken four times, makes a square. Let the sought-after (cube) be 1 Cube. This, taken four times, makes 4 Cubes, (which are) equal to a square, let it be to 16 Powers, and the Number becomes 4 units. To the hypostases: the cube will be 64 units.

Thus, I assign the second to be 64 units lacking 1 Cube. So, it remains for the square of the second, if it receives in addition the first, to make a square. But the square of the second, when it receives in addition the first, makes 1 Cube-Cube, 4096 units, 1 Number lacking 128 Cubes. These are equal to a square, (say) to the one from a side of 1 Cube, | 64 units, and the square becomes 1 Cube-Cube,   228
4096 units, 128 Cubes. And so, there remain 256 Cubes equal to 1 Number, and the Number becomes one ⟨16th⟩.[40]

To the hypostases: the first will be one 16th, and the second 262,143 4096ths.

19. To find three numbers indeterminately such that the product of any two, together with one unit, makes a square.

Since I want the product of the first and second, together with 1 unit, to make a square, if, therefore, from a certain square I subtract the unit I shall have the product of the first and second. I form a square from however many Numbers and

---

38 Lit. "cubic units, 8", meaning "cubic units however many, let it be 8".
39 Supplemented by Bachet.
40 Supplemented by Tannery.

1 unit (I want); let it be 1 Number, 1 unit. Therefore, the square itself will be 1 Power, 2 Numbers, 1 unit. If I subtract the 1 unit, there remain 1 Power, 2 Numbers; it will be the product of the first and second. Let the second be 1 Number. Therefore, the first will be 1 Number, 2 units. Again, since I want the product of the second and third, with 1 unit, to make a square, if, similarly, I subtract from a certain square one unit, I shall have the product of the second and third. Let the square be formed from 3 Numbers, 1 unit. The square will be 9 Powers, 6 Numbers, 1 unit. If, therefore, I subtract one unit, they result in 9 Powers, 6 Numbers. Therefore, the product of the second and third must be 9 Powers, 6 Numbers, of which the second is 1 Number. The remaining third, therefore, will be 9 Numbers, 6 units. | Again, since I want the product of the first and third, together with one unit, to make a square, and the product of the first and third, together with one unit, is 9 Powers, 24 Numbers, 13 units, (this must be) equal to a square. And the (multitude of) Powers is a square.[41] ⟨If the (multitude of) units were also a square⟩,[42] and twice the product of the sides of the (multitudes of) Powers and units were equal to the (multitude of) Numbers, then all three proposed tasks would have been fulfilled indeterminately. But the 13 units come from the product of the 2 units and 6 units, together with 1 unit; but the 2 units, on the one hand, come from twice the product of 1 Number and 1 unit, while the 6 units, on the other hand, come again from twice the product of 3 Numbers and 1 unit. I want twice the (multitude of) Numbers (multiplied) by twice the (multitude of) Numbers, together with 1 unit, to make a square. But twice the (multitude of) Numbers by twice the (multitude of) Numbers is four times the product of the (multitudes of) Numbers. So, I want four times their product, together with 1 unit, to make a square. But, of course, four times the product of any two numbers, together with the square of their difference, makes a square. So, if we establish that the square of their difference be 1 unit, (then) four times their product, together with 1 unit, makes a square. So, if the square of their difference is 1 unit, their difference is also 1 unit. Accordingly, we must form (the squares) from consecutive Numbers, and 1 unit. (Let them be) from 1 Number and 1 unit, and from 2 Numbers, 1 unit. And the square of 1 Number, 1 unit will be 1 Power, 2 Numbers, 1 unit. If I subtract the unit, the remainder is 1 Power, 2 Numbers. The product of the first and second, therefore, must be 1 Power, 2 Numbers. Let the second be assigned to be 1 Number. Therefore, the remaining first will be 1 Number, 2 units. Again, since the square of 2 Numbers, 1 unit is 4 Powers, 4 Numbers, 1 unit, | if, similarly, I subtract the one unit, the remainder is 4 Powers, 4 Numbers. So, the product of the second and third must be 4 Powers, 4 Numbers, of which the second is 1 Number. Therefore, the remaining third will be 4 Numbers, 4 units.

And thus, it has been solved indeterminately, so that the product of any two (numbers), together with one unit, makes a square, whatever (value) one might

230

232

---

41 Lit. "I have the Powers (to be) square".
42 Supplemented by Bachet.

wish the Number to be. Indeed, to seek indeterminately means that the hyposta-sis[43] (of the unknown) be such that whatever one might wish the Number to be, when he brings (it) to the hypostases he will fulfill the proposed task.

20. To find four numbers such that the product of any two, if it receives in addition one unit, makes a square.

Since I want the product of the first and second, together with one unit, to make a square, if, therefore, from a certain square I remove one unit, I shall have the product of the first and second. I form a square from 1 Number, 1 unit, so this square becomes 1 Power, 2 Numbers, 1 unit. If I subtract the 1 unit, the remainder, (i.e.,) the product of the first and second, becomes 1 Power, 2 Numbers. Let the first be 1 Number. ⟨Therefore, the second will be 1 Number⟩,[44] 2 units. Again, since I want the product of the first and third, together with one unit, to make a square, I form a square from 2 Numbers, 1 unit, (i.e.,) the next in order[45] by virtue of what was shown before, and taking their product, I remove the one unit, and I assign the product of the first and third to be 4 Powers, 4 Numbers, of which the first is 1 Number. Therefore, the remaining third is 4 Numbers, 4 units. | Again, since I want the product of the first and fourth, together with one unit, to make a square, I form a square from the next in order,[46] (namely, from) 3 Numbers, 1 unit. And taking the product and subtracting one unit, I shall have the product of the first and fourth, (namely,) 9 Powers, 6 Numbers, of which the first is 1 Number. Therefore, the remaining fourth will be 9 Numbers, 6 units. And since it comes out that the product of the third and fourth, together with one unit, makes a square, but the product of the second and fourth, together with one unit, makes 9 Powers, 24 Numbers, 13 units, (this) is equal to a square. (Let it be) to the one from 3 Numbers lacking 4 units, and the Number becomes one ⟨16th⟩.[47]

234

To the hypostases: the first will be one (16th), the second 33 (16ths), the third 68 (16ths), and the fourth 105 (16ths).

21. To find three numbers in proportion, such that the difference of any two is a square.

Let the smaller be assigned to be 1 Number, the middle 1 Number, 4 units – in order that the difference be a square – and the third 1 Number, 13 units – in order that its difference with respect to the middle be also a square. If, moreover, the difference of the greatest and the least were a square, then (the requirement) that

---

43 In this context the word ὑπόστασις means the expression from which the numerical value of the unknown is determined.

44 Restored by Bachet. **ABT** have "ἔστω ὁ πρῶτος ἀριθμοῦ ἑνὸς μονάδος μιᾶς", i.e., "let the first be 1 Number, 1 unit". An alternative restoration could be "Let the first be ⟨1 Number. Therefore the second will be⟩ 1 Number, 2 units".

45 The "next in order" refers to the multitude of Numbers in the expression.

46 Similarly, "next in order" in respect of the multitude of Numbers in the expression.

47 Supplemented by Tannery.

the difference of any two is a square would have been accomplished indeterminately. But the greatest exceeds the least by 13 units, and the 13 units are the sum of the squares 4 and 9. Thus, I am led to finding two squares equal to one square.

236 | But this is easy by means of a right-angled triangle. Now, 9 and 16 are (such squares). So, I assign the least to be 1 Number, the middle 1 Number, 9 units, the third 1 Number, 25 units, and the difference of any two is a square. It remains for these to be in proportion. But if three numbers are in proportion, the product of the extremes is equal to the square of the middle. But the product of the greatest and least, that is, the product of the extremes, is 1 Power, 25 Numbers, while the square of the middle is 1 Power, 18 Numbers, 81 units. (This) is equal to 1 Power, 25 Numbers, and the Number becomes 81 7ths.

To the hypostases: the first will be 81 (7ths), the second 144 (7ths), and the third 256 (7ths).

22. To find three numbers such that the solid (number formed) from them, if it receives in addition each of them, makes a square.

Let the solid (number formed) from the three be assigned to be 1 Power, 2 Numbers, and the first, 1 unit, so that the solid from the three, together with the first, makes a square. Again, since I want the solid from the three, together with the second, to make a square, if, therefore, from a certain square I remove 1 Power, 2 Numbers, I shall have the second. I form a square from 1 Number, 3 units. Its square lacking 1 Power, 2 Numbers makes 4 Numbers, 9 units. So, I assign the

238 second to be 4 Numbers, 9 units. | But since the solid from the three is 1 Power, 2 Numbers, while the product of the first and second is 4 Numbers, 9 units, if, therefore, I divide 1 Power, 2 Numbers by 4 Numbers, 9 units, I shall have the third. But the division is not possible. In order that the division be possible, it is necessary that as 1 Power is to 4 Numbers, so 2 Numbers is to 9 units; and alternately, as 1 Power is to 2 Numbers, so 4 Numbers is to 9 units. But the 1 Power is in multitude half the 2 Numbers. So, if (the) 4 Numbers were also (in multitude) half the 9 units, the division would be (possible). Now, the 4 (Numbers) come from the excess that 6 Numbers exceed 2 Numbers. But the 6 Numbers come from twice the product of 3 units and 1 Number, that is, twice the 3 units; while the 9 units is the square of 3 units. Thus, I am reduced to finding a certain number – as (in this instance) the 3 units – which, if taken two times and decreased by a dyad, will be half of its square.

Let the sought-after (number) be 1 Number. If this is doubled and decreased by a dyad, it makes 2 Numbers lacking 2 units; while its square is 1 Power. Thus, we want 2 Numbers lacking 2 units to be half of 1 Power. Thus, 1 Power is equal to 4 Numbers lacking 4 units, and the Number becomes 2 units.

Now, I return to the initial (problem). I had the first number (as) 1 unit, and the solid (formed) from the three, 1 Power, 2 Numbers. But the solid from the three, if it receives in addition the second, must also make a square. If, therefore, from a certain square I subtract 1 Power, 2 Numbers I shall have the second. I form the square from 1 Number and so many units so that the (multitude of) units, if taken two times and decreased by a dyad, is half of their square. But this has been shown before, and | it is 2 units. I form the square from 1 Number, 2 units. Therefore, the square will be 1 Power, 4 Numbers, 4 units. If I remove the solid from the three, that is, 1 Power, 2 units, the remainder, (that is,) the second, will be 2 Numbers, 4 units. And the product of the first and second is ⟨2 Numbers, 4 units. If, therefore, I divide the solid from the three, that is 1 Power, 2 Numbers, by the product of the first and second⟩,[48] that is, by 2 Numbers, 4 units, I shall have the third. But the division gives 2′ of a Number. It remains for the solid from the three, together with the third, to make a square. But the solid from the three, together with the third, is 1 Power, 2 2′ Numbers. These are equal to a square, (say) to 4 Powers, and the Number becomes 5 6ths.

To the hypostases: the first will be 6 (6ths), the second 34 (6ths), and the third 2 2′ (6ths).

23. To find three numbers such that the solid (number formed) from them, if it lacks any one (of them), makes a square.

Let the first be assigned to be 1 Number, and the solid from the three, 1 Power, 1 Number; and so, if it lacks the first, it makes a square. And since the solid from the three is 1 Power, 1 Number, while the first is 1 Number, therefore, the product of the second and third will be 1 Number, 1 unit. Let the second be 1 unit. The remaining third, therefore, will be 1 Number, 1 unit. It remains for the solid from the three, if it lacks the second or the third, to make a square. But when it lacks the one, it makes 1 Power, 1 Number lacking 1 unit, (which are) equal to a square; while, when (it lacks) the other, it makes 1 Power lacking 1 unit, (which are) equal to a square. | And so, a double equality arises. I take the difference, which is 1 Number. I set out two numbers whose product is of such a size. Let this 1 Number be measured by 2′ of a unit according to 2 Numbers, that is, according to twice the side of the Power. And, as you know, the equation is made with them, and the Number becomes 17 8ths.

To the hypostases: the first will be 17 (8ths), the second 1 unit, and the third 25 8ths.

24. To divide a given number into two numbers and make their product a cube except (its) side.

Now, let the given (number) be 6.

Let the first be assigned to be 1 Number. Therefore, the remaining second will be 6 units lacking 1 Number. Then, their product must be a cube except (its) side. But their product will be 6 Numbers lacking 1 Power. These are equal to a cube except (its) side. I form a cube from however many Numbers lacking 1 unit (I want). So, let it be from 2 Numbers lacking 1 unit. And its cube, if it lacks itself, makes 8 Cubes, 4 Numbers lacking 12 Powers. These are equal to 6 Numbers lacking 1 Power. And

240

242

---

48 Supplemented by Tannery.

if the (multitudes of) Numbers in either part of the equation were equal, it would be reduced to equate Cubes equal to Powers, and the Number would be expressible. Now, the 4[49] Numbers come from the excess over 2 Numbers, that is, from thrice the 2 Numbers; indeed, thrice the 2 Numbers, if they lack 2 Numbers, | make twice the 2 Numbers. But the 6 was taken arbitrarily according to the supposition. Thus, I am reduced to finding a certain number – like the 2, (the multitude of)[50] Numbers – which taken two times makes 6, and it is 3. Accordingly, I seek (to make) 6 Numbers lacking 1 Power equal to a cube except (its) side. I now assign the side of a cube to be from 3 Numbers lacking 1 unit. And so its cube, if it lacks this, makes 27 Cubes, 6 Numbers lacking 27 Powers, (which are) equal to 6 Numbers lacking 1 Power, and the Number becomes 26 27ths.

To the hypostases: the first will be 26 (27ths), and the second 136 (27ths).

25. To divide a given number into three numbers such that the solid (number formed) from them makes a cube whose side is equal to their differences added together.

Let the given (number) be 4.

Since the solid from the three is a cube, let it be 8 Cubes, whose side is 2 Numbers. But the difference of the second and the first, and the difference of the third and second, and moreover (the difference) of the third and the first, is twice the difference of the third and the first – that is, if there are three unequal numbers, the (sum of the) differences of the three is twice the difference of the extremes. But we have in the hypostasis[51] of the side of the cube 2 Numbers, and the 2 Numbers must be the (sum of the) differences of the three (numbers). Therefore, the third exceeds the first by 1 Number. Let the first be 2 Numbers, or however many (Numbers we wish). Therefore, the third will be 3 Numbers. And since the solid from the three | is 8 Cubes, while the product of ⟨the⟩[52] first and third is 6 Powers, the remaining second, therefore, will be 1 3′ Numbers. And if the second were greater than the first and smaller than the third, what was required would have been accomplished. But the second resulted from the division of 8 by the product of (the multitudes of the) first and third. But (the multitudes of) the first and the third are not arbitrary, since they differ by a unit. Thus, I am reduced to finding two numbers, one of which exceeds the other by a unit, so that if 8 is divided by their product make a certain (number) which is greater than the smaller but smaller than the greater.

244

246

---

49 Tannery's text has "2".
50 The text literally reads "like the 2 Numbers", where this "2 Numbers" is understood as simply a 2, for which the species is irrelevant. Here, like in other places, we have tried to clarify this for the modern reader by adding "(the multitude of)", which would not be necessary for a Greek reader.
51 That is, in the expression, namely, "2 Numbers", assigned to the side of the cube.
52 Supplemented by Bachet.

Let the smaller be assigned to be 1 Number. The greater, therefore, will be 1 Number, 1 unit. And if I divide by their product, that is, by 1 Power, 1 Number, the middle will be found to be 8 units of a part of 1 Power, 1 Number. And we want this to be greater than 1 Number but smaller than 1 Number, 1 unit. And since their difference is 1 unit, therefore, the difference of the first[53] and the second[54] is smaller than 1 unit. Accordingly, the second together with 1 unit is greater than the first. But the second, having received in addition the unit, and being divided by[55] 1 Power, 1 Number, becomes 1 Power, 1 Number, 8 units of a part of 1 Power, 1 Number. So, these are greater than 1 Number, 1 unit. And everything by the part. 1 Power, 1 Number, 8 units are greater than 1 Cube, 2 Powers, 1 Number. And likes from likes. So, 8 units become greater than 1 Cube, 1 Power. I form a cube having (in its expression) 1 Cube, 1 Power. Therefore, the side of the cube will be 1 Number, 3' of a unit. But, since 8 units are greater than | 1 Cube, 1 Power, and the cube from 1 Number, 3' of a unit is also greater than 1 Cube, 1 Power, (then,) if I also equate the sides, that is, 2 units equal (to) 1 Number, 3' of a unit, the Number becomes 5 3rds.

248

To the hypostases: the first will be 8 3rds, the second 9 5ths, the third 5 3rds. And everything in 15ths.[56] The first will be 40, the second 27, the third 25 – the common part, 15, having been eliminated – and so three numbers have been found such that the solid (formed) from them is a cube having as a side their differences added together.

Accordingly, I assign the first to be 40 Numbers, the second ⟨27 Numbers, the third⟩[57] 25 Numbers. And so, the solid (number formed) from the three is a cube whose side is equal to their differences added together. Further, the (sum of the) three must be equated to the given units. But the given units were 4. Therefore, 92 Numbers are equal to 4 units, and the Number becomes 1 ⟨23rd⟩.[58]

To the hypostases: the first will be 40 (23rds), the second 27 (23rds), and the third 25 (23rds).

26. To find two numbers such that their product, if it receives in addition either one, makes a cube.

I assign the first to be from a cubic (multitude) of Numbers; so, let it be 8; the second, | 1 Power lacking 1 unit. And so I have satisfied one (of the) proposed tasks; indeed, their product, if it receives in addition the first, makes a cube. Further, their product, if it receives in addition the second, must make a cube. But their product, when it receives in addition the second, makes 8 Cubes, 1 Power lacking

250

---

53 That is, the greater of the two numbers, namely, the number named "1 Number, 1 unit".
54 That is, the middle number, namely, "8 units of a part of 1 Power, 1 Number".
55 The text has "ἀναλυθεὶς εἰς", lit. "resolved into".
56 That is, determine how many fifteenths each number is. 8 3rds, for example, is 40 fiteenths.
57 Supplemented by Bachet.
58 Supplemented by Tannery.

8 Numbers, 1 unit. These are equal to a cube. I form the cube from 2 Numbers lacking 1 unit, and the Number becomes 14 13ths.

To the hypostases: the first will be 112 13ths, and the second 27 169ths.

27. To find two numbers such that their product, if it lacks either one, makes a cube.

Similarly, let the first be assigned to be a cubic (multitude) of Numbers, (say) 8, the second 1 Power, 1 unit again, and so their product, if it lacks ⟨the first,⟩ becomes ⟨a cube. Again, their product, when it lacks⟩[59] the second, makes 8 Cubes, 8 Numbers lacking 1 Power, 1 unit. These are equal to a cube. But (this) is impossible.[60] So then I assign again the one to be a cubic (multitude) of Numbers, 1 unit; let it be 8 Numbers, 1 unit; and the other, 1 Power. And so, their product, if it lacks the second, becomes a cube. Again, their product, when it lacks the first, makes 8 Cubes, 1 Power lacking 8 Numbers, 1 unit. These are equal to a cube, (say) to the one from a side of 2 Numbers lacking 1 unit, and the Number becomes 14 13ths.

252 To the hypostases: the first will be 125 13ths, and the other 196 169ths.

28. To find two numbers such that their product, if it receives in addition, or if it lacks, the sum of the two, makes a cube.

So, since their product, together with the sum of the two, makes a cube, let it make 64 units. Again, since their product, if it lacks the sum of the two, makes ⟨a cube, let it make⟩[61] 8 units. Therefore, twice the sum of the two, which constitutes their difference, will be 56 units. Therefore, the sum of the two will be 28 units. But, also, their product, together with the sum of the two, makes 64 units. Then, their product will be 36 units. Thus, I am reduced to finding two numbers ⟨which, added together, make⟩[62] 28 units and whose product is 36 units.

Let the greater be assigned to be 1 Number, 14 units. Therefore, the smaller will be 14 units lacking 1 Number. It remains (for us) to equate their product, that is, 196 units lacking 1 Power, to 36 units, and it results in 1 Power equal to 160 units. Now, if (the) 160 units were a square, the proposed task would have been accomplished. But the 160 units is the excess by which 196 units exceed the 36. But the 196 units is the square of 14 units, while 14 is half of the 28. Accordingly, 196 is half of the 28 (multiplied) by itself. But the 28 is half of the 56. So, the 14 is a fourth of 56. But 56 | is the difference of two cubes, 64 and 8, while 36 is half of the sum of the two cubes. Thus, I am reduced to finding two cubes, a fourth of

254

---

59 Supplemented by Bachet.

60 For this claim, see the commentary to this problem.

61 Supplemented by Bachet.

62 Bachet suplemented the phrase "οἱ συντεθέντες ποιοῦσι", adopted in our translation. Tannery adopted in his edition the phrase "ὥστε συναμφότερον ποιεῖν", "so that the sum makes", which was proposed by Auria.

the difference of which, multiplied by itself, and lacking half of the sum of the two, makes a square.

Let the side of the greater cube be 1 Number, 1 unit, and of the smaller, 1 Number lacking 1 unit. And so, the greater cube becomes ⟨1 Cube⟩,[63] 3 Powers, 3 Numbers, 1 unit, while the smaller, 1 Cube, 3 Numbers lacking 3 Powers, 1 unit; and a fourth of their difference, 1 2′ Powers, 2′ of a unit. These, (multiplied) by themselves, become 2 ⟨4′⟩ [64] Power-Powers, 1 2′ Powers, 4′ of a unit. If these lack half of the sum of the two cubes, which is 1 Cube, 3 Numbers, there remain 2 4′ Power-Powers, 1 2′ Powers, 4′ of a unit lacking 1 Cube, 3 Numbers, (which are) equal to a square. And everything four times because of the part. It results in 9 Power-Powers, 6 Powers, 1 unit lacking 4 Cubes, 12 Numbers. These are equal to a square, (say) to the one from a side of 3 Powers, 1 unit lacking 6 Numbers. This (square), therefore, will be 9 Power-Powers, 42 Powers, 1 unit lacking 36 Cubes, 12 Numbers, (which are) equal to 9 Power-Powers, 6 Powers, 1 unit lacking 4 Cubes, 12 Numbers. Let the lacking be added in common and likes from likes. Then, 32 Cubes are equal to 36 Powers, and the Number becomes 9 8ths.

To the hypostases: I assigned one of the sides of the cubes to be 1 Number, 1 unit, and the other 1 Number lacking 1 unit; the one will be 17 (8ths), the other 1 (8th). The cubes themselves, therefore, will be 4913 512ths and 1 (512th).

So, I return to the initial (problem) and I seek for their product, together with the sum of the two, to make the cube 4913 (512ths), while their product, if it lacks the sum of the two, to make the cube 1 (512th). Since their product together with the sum of the two makes a cube, namely, 4913 512ths, and their product, when it lacks the sum of the two, makes a cube, namely, 1 512th, therefore, twice the sum of the two is their difference, namely, 4912 (512ths). Accordingly, the sum of the two will be 2456 (512ths). But their product together with the sum of the two is 4913 (512ths), of which the sum of the two is 2456 (512ths). Their product, therefore, will be 2457 512ths. But this demonstration has been shown before, in the first Book,[65] and it will be also shown now, for the sake of the (present) problem. Let the first be assigned to be 1 Number and (a multitude of) units which is half the sum of the two, that is 1228 512ths. The second will be 1228 512ths units lacking 1 Number. So, the sum of the two is 2456 512ths, while their product is 1,507,984 of a part of 262,144 lacking 1 Power. These are equal to 2457 512ths units. And everything by ⟨the⟩[66] part, that is by 262,144, and likes from likes. It results in 262,144 Powers, (which are) equal to 250,000, and the Number becomes 500 512ths of a unit.

To the hypostases: the first will be 1728 (512ths), the second 728 (512ths), and the proof is obvious.

256

258

---

63 Supplemented by Bachet.
64 Supplemented by Bachet.
65 Problem I.27.
66 Supplemented by Tannery.

Otherwise

To find two numbers such that their product, if it receives in addition, or if it lacks, the sum of the two, makes a cube.

In a (problem) such as this one, if any square number is partitioned into its side and the remainder, (then) it makes their product, together with the sum of the two, a cube. So, let the square be assigned to be 1 Power, and let it be partitioned into the side and the remainder. It will be 1 Number, and 1 Power lacking 1 Number. And their product, together with the sum of the two, is a cube. Further, their product, if it lacks the sum of the two, must make a cube. But their product, when it lacks the sum of the two, makes 1 Cube lacking 2 Powers. These are equal to a cube smaller than 1 Cube. I form an 8th of a Cube. And everything 8 times. 8 Cubes lacking 16 Powers become equal to 1 Cube, and the Number becomes 16 7ths.

To the hypostases: the first will be 16 7ths, and the second 144 49ths.

29. To find four ⟨square⟩[67] numbers which added together and receiving in addition their own sides added together, make a given number.

Now, let (them make) 12.

260   Since any square, if it receives in addition its own side and 4′ of a unit, makes a square – the side of which, lacking 2′ of a unit, makes a certain number which is a side of the initial square – therefore, when, on the one hand, the four numbers receive in addition their own sides, they make 12 units, and when, on the other hand, they receive in addition 4 fourths besides, they make four squares. But the 12 units together with 4 fourths, which is 1 unit, make 13 units. Therefore, I must divide the 13 units into four squares, and from the sides, if from each side I subtract 2′ of a unit, I shall have the sides of the four squares. But the 13 is divided into two squares, 4 and 9. Again, each of them is divided into two squares, (namely,) into 64 25ths and 36 25ths, and 144 25ths and 81 25ths. Well then, taking the side of each one, (namely,) 8 5ths, ⟨6 5ths, 12 5ths⟩,[68] 9 5ths, I remove 2′ of a unit from the side of each one, and the sides of the sought-after squares will be 11 10ths, 7 10ths, 19 10ths, 13 10ths. Thus, the squares themselves are 121 100ths, 49 100ths, 361 100ths, and 169 100ths.

30. To find four square (numbers) which added together and lacking their own sides added together, make a given number.

262   Now, let (them make) 4 units.

So, since the first, if it lacks its side, and the second, if it lacks its side, and the third, and the fourth, if they similarly lack (their sides), ⟨must⟩[69] make 4 units, but also, (since) any square, if it lacks its side and receives in addition 4′ of a unit,

---

67 Supplemented by Bachet.
68 Supplemented by Bachet.
69 Supplemented by Auria.

makes a square, whose side, if it receives in addition 2′ of a unit, makes the side of the initial square, then the four (squares), if they lack their own sides and receive in addition 4 fourths of a unit, that is 1 unit, will make four squares. But all four, when they lack their own sides, make 4 units; and when they receive in addition 1 unit, make 5 units. Thus, I am reduced to dividing 5 into four squares.[70] But 5 is divided into four squares, (namely, the squares) 9 25ths, 16 25ths, 64 25ths, and 36 25ths. I take their sides: they are 3 5ths, 4 5ths, 8 5ths, and 6 5ths. I add to each of them 2′ of a unit, and I find the sides to be 11 10ths, 13 10ths, 21 10ths, and 17 10ths. Thus, the sought-after squares will be 121 100ths, 169 100ths, 441 100ths, and 289 100ths.

31. To divide unit into two numbers, and add to each of them a given number, and make their product a square.

264

Let the unit be divided into two numbers, and add 3 units to the one, 5 units to the other, and make their product a square.

Let the first be assigned to be 1 Number. The second, therefore, will be 1 unit lacking 1 Number. And if 3 units are added to the first, it will be 1 Number, 3 units; while, if 5 units (are added) to the second, it will be 6 units lacking 1 Number. And their product becomes 3 Numbers, 18 units lacking 1 Power, (which are) equal to a square, say to 4 Powers. And let the lacking be added in common. 3 Numbers, 18 units become equal to 5 Powers, and the (solution to the) equation is not expressible. But the 5, (the multitude of) Powers,[71] is a square together with 1 unit. These, multiplied by the 18 units, and receiving in addition the square of half of the 3, (the multitude of) Numbers,[72] that is 2 4′, must produce a square. Well then, for this reason I am reduced to looking for a square ⟨which⟩,[73] if it receives in addition 1 unit, and has been taken 18 times, and it receives in addition 2 4′ units, makes a square.

Let the square be 1 Power. This, together with 1 unit, and having been taken 18 times, and having received in addition 2 4′ units, ⟨makes⟩[74] 18 Powers, 20 4′ units, (which must be) equal to a square. Everything 4 times. 72 Powers, 81 units become equal to a square. So, I form the square from 8 Numbers, 9 units, and the Number becomes 18 units. To the hypostases: the square will be 324.

I return to the initial (problem), to set up the equation: 3 Numbers, 18 units lacking 1 Power are equal to a square. Now I assign (the square) to be 324 Powers, and the Number becomes 78 325ths, that is, 6 25ths.

266

To the hypostases: the first will be 6 (25ths), and the second 19 (25ths).

---

70 Here the text has the following phrase, which is clearly an interpolation: "I added to each one of the sides 2′ of a unit, and I found the sides of the sought-after squares".

71 Lit. "But the 5 Powers".

72 Lit. "the 3 Numbers".

73 Supplemented by Bachet.

74 Supplemented by Bachet.

Otherwise

To divide unit into two numbers, and add to each of them a given number, and make their product a square.

Now, let the unit be divided into two numbers, and add 3 units to the one, 5 units to the other, and make their product a square.

Let the first be assigned to be 1 Number and lacking 3 units[75] that it receives in addition. The remaining second, therefore, will be 4 units lacking 1 Number. And if 3 units are added to the first, it becomes 1 Number; while, if 5 units are added to the second, it becomes 9 units lacking 1 Number. And so their product is 9 Numbers lacking 1 Power, (which must be) equal to a square; let it be to 4 Powers, and the Number becomes 9 5ths. To the hypostases: but I cannot subtract 3 units from the Number.

Therefore, the Number must be greater than 3 units, but smaller than 4 units. But the Number was found by dividing 9 by 5, which is a square together with 1 unit. But if 9, when divided by a certain square together with 1 unit it makes 3 units, therefore, the (number) by which it is divided is, of course, 3. But the number which divides 9 is a square | ⟨together with⟩[76] a unit. Therefore, the square together with 1 unit ⟨is smaller than 3 units⟩.[77] Let the unit be removed. Therefore, the square is ⟨smaller than⟩[78] 2 units. Again, we want that when we divide the 9 by a square together with 1 unit, it makes 4 units. The (number) by which it is divided, therefore, ⟨is, of course, 2 4′ units. But the (number) which divides⟩[79] 9 is a square together with 1 unit. Therefore, the square together with a unit is greater than 2 4′ units. Let the unit be removed. Therefore, the square is greater than 1 4′ units. But it has been shown, also, that the square is smaller than 2. Thus, I am reduced to finding a certain square which is greater than 1 4′ units and smaller than 2. I reduce these (numbers) into square parts, (namely,) into 64ths, and they become 80 and 128 (64ths). But this is easy, and the square is 100 64ths, that is 25 16ths.

Thus, I return to the initial (problem). I was seeking 9 Numbers lacking 1 Power to be equal to a square, that is, equal to the 25 16ths Powers we found, and the Number becomes 144 41sts.

To the hypostases: the first will be 21 (41sts), and the second 20 (41sts).

32. To divide a given number into three numbers such that the product of the first and second, if it receives in addition the third, or if it lacks (it), makes a square.

Let the given (number) be 6.

---

75 **A** (102r) and **T** (101v) have "καὶ λεῖψις μονάδων Γ̄", lit. "and a lacking of 3 units".
76 "σύν", supplemented by Bachet.
77 Supplemented by Bachet.
78 Supplemented by Bachet.
79 Text restored initially by Bachet, then, more accurately, by Tannery.

Let the third be assigned to be 1 Number, and the second (a multitude) of units smaller than 6; | let it be 2 units. Therefore, the first will be 4 units lacking 1 Number. And so two (of the) proposed tasks remain: that the product of the first and second, if it receives in addition the third, or if it lacks (it), to make a square. And so a double equality arises: 8 units lacking 1 Number is equal to a square, and 8 units lacking 3 Numbers is equal to a square. But this is inexpressible, for the (multitudes of) Numbers do not have to each other the ratio that a square number has to a square number. But the 1 Number is smaller than the 2 by 1 unit, while the 3 Numbers are, similarly, 1 ⟨unit⟩[80] greater than the 2. Thus, I am reduced to finding a certain number, like (in this instance) the 2, so that the (number) one unit greater than it has to the (number) one unit ⟨smaller than it the ratio that a square number has to⟩[81] a square number.

Let the sought-after (number) be 1 Number. ⟨The⟩[82] number 1 unit greater than it will be 1 Number, 1 unit, while the (number) 1 unit smaller than it, 1 Number lacking 1 unit. So, we want that these (numbers) have to each other the ratio that a square number has to a square number. Let them have the (ratio) 4 to 1. Therefore, 1 Number lacking 1 unit (multiplied) by 4 units becomes 4 Numbers lacking 4 units, while 1 Number, 1 unit (multiplied) by 1 unit ⟨becomes 1 Number, 1 unit⟩.[83] And these are the numbers set out which have to each other the ratio that a square number has to a square number. Now, 4 Numbers lacking 4 units are equal to 1 Number, 1 unit, and the Number becomes 5 3rds units.

So, I assign the second to be 5 3rds units; now the third is 1 Number; therefore, the first will be 13 3rds units lacking 1 Number. | Then, the proposed task which remains is that the product of the first and second, if it receives in addition the third, must make a square, and if it lacks the third, it must (also) make a square. But the product of the first and second, when it receives in addition the third, makes 65 9ths units lacking 3″ of a Number, which (must be) equal to a square; while, lacking the third, makes 65 9ths units lacking 2 3″ Numbers, which (must be) equal to a square. And everything by 9. 65 units lacking 6 Numbers become equal to a square, and 65 units lacking 24 Numbers (become) equal to a square. And I equate the Numbers of the greater equality[84] by quadruplicating. We have 260 units lacking 24 Numbers equal to a square, and 65 units lacking 24 Numbers equal to a square. Now I take their difference, which is 195 units, and I set out two numbers whose product is 195 units: 15 and 13 are (such numbers). And half of their difference (multiplied) by itself is equal to the smaller square, and the Number becomes 8 3rds.

80 Supplemented by Bachet.
81 Supplemented by Bachet.
82 Supplemented by Tannery.
83 Supplemented by Tannery.
84 That is, the multitude of Numbers in "65 units lacking 6 Numbers" is made equal to the multitude of Numbers in "65 units lacking 24 Numbers".

To the hypostases: the first will be 5 (3rds), the second 5 (3rds), the third 8 (3rds), and the proof is obvious.

33. To find two numbers such that each of them, if it receives in addition from the other the same part, or the same parts, has to what remains from the giver a proposed ratio.

Now, let the first, if it receives in addition from the second a certain part, or parts, is the triple of the remainder, while the second, | if it receives from the first the same part, or the same parts, is the quintuple of the remainder.

Let the second be assigned to be 1 Number, 1 unit, and let its part, or parts, be 1 unit. Therefore, the first will be 3 Numbers lacking 1 unit. And the first, when it receives in addition a certain part, or parts, of the second, that is 1 unit, becomes the triple of the remainder. We want, also, the second, if it receives in addition the same part, or the same parts, ⟨of the first⟩,[85] to be the quintuple of the remainder. But since the two are 4 Numbers, and the second receives something while the first gives, and the produced (number) becomes the quintuple of the remainder, therefore, the sum of the two, the produced (number) and the remainder, will be 4 Numbers; thus, the remainder will be (found) if we take from 4 Numbers the sixth,[86] that is 3″ of a Number. If, therefore, from 3 Numbers lacking 1 unit we remove 3″ of a Number, we shall have the part, or parts, of the first. But when we remove (it), the remainder becomes 7 3rds Numbers lacking 1 unit; the second, which is 1 Number, 1 unit, having received from the first 7 3rds Numbers lacking 1 unit, becomes the quintuple of what remains from the first. It remains at this point to see if 7 3rds Numbers lacking 1 unit is the same part, or the same parts, of 3 Numbers lacking 1 unit, as 1 unit is a part, or parts, of 1 Number, 1 unit. But when you seek something like that, the product of 7 3rds Numbers lacking 1 unit by 1 Number, 1 unit is equal to the product of 3 Numbers lacking 1 unit by a unit, | that is, the parts are multiplied alternately; from which 7 3rds Powers, 4 3rds Numbers lacking 1 unit are equal to 3 Numbers lacking 1 unit, and the Number becomes 5 7ths.

To the hypostases: the first will be 8 7ths, and the second 12 7ths. But the parts of the second were 1 unit. We investigate: 1 unit is 7 12ths of the second; and I septuplicate the two numbers. The first will be 8 units, the second 12 units, and the parts 7 12ths. But since the first does not have a 12th part, I triplicate them in order to avoid fractions. The first will be 24, the second 36, their parts are 7 12ths, and the proof is obvious.

Lemma to the next

To find two indeterminate numbers such that their product together with their sum makes a given number.

---

85 Supplemented by Bachet.
86 That is, a sixth of the four Numbers.

274

276

Let it make 8 units.

Let the first be assigned to be 1 Number, and the second 3 units. Their product, together with their sum, is 4 Numbers, 3 units. These are equal to 8 units, and the Number becomes 5 4ths. To the hypostases: the first will be 5 4ths, and the second 3 units.

Now, I investigate how the Number came to be 5 4ths. (It came) from the 5, which had to be divided by 4, the (multitude of) Numbers. But the 5 comes from the excess | by which the 8 exceeds the 3, while 4, the (multitude of) Numbers, are the second (number) increased by one unit. If, therefore, we assign the second to be (any expression) whatever in terms of a Number, and I remove this from 8 units, and I divide the remainder by the second (number) increased by one unit, I shall have the first. For example, let the second be 1 Number lacking 1 unit. I remove these from 8 units. There remain 9 units lacking 1 Number. I divide these by the (second number) increased by 1 unit, that is, by 1 Number; it results in 9 inverse Numbers lacking 1 unit; (this) will be the first (number).

And it has been solved indeterminately so that their product together with the sum of the two makes 8 units. Indeed, (solving a problem) indeterminately is such that, for however many units someone might wish the Number to be, when he brings (it) to the hypostases, the problem is completed.

34. To find three numbers such that the product of any two, if it receives in addition their sum, makes a given number.[87] It is certainly necessary that the given (numbers) be squares except one unit.

Now, let the product of the first and second, together with their sum, make 8 units; the product of the second and third, together with their sum, make 15 units; and the product of the first and third, together with their sum, make 24 units.

So, since I want the product of the first and second, together with their sum, to make 8 units, if, therefore, I assign the second to be any (number) whatever, and I remove this from 8 units, and I divide (the remainder) by the (number which is) by a unit greater than the second, I shall have the first. | Let the second be assigned to be 1 Number lacking 1 unit. If I remove this[88] from 8 units, and I divide (the remainder) by the (number which is) 1 unit greater than the second, the first will be 9 inverse Numbers lacking 1 unit. Again, similarly, since I want the product of the second and third, together with their sum, to make 15 units, ⟨if from 15 units⟩[89] I subtract 1 Number lacking 1 unit and I divide (the remainder) by the (number which is) 1 unit greater than the second, that is, by 1 Number – (thus) they become 16 inverse Numbers lacking 1 unit –, I shall have the third. It remains for the product of the first and third, together with their sum – which makes 144 inverse Powers lacking 1 unit – (for) these to be equal to 24 units, and the Number becomes 12 5ths.

278

280

---

87 Lit. "given numbers".
88 Text has "these".
89 Supplemented by Auria.

To the hypostases: the first will be 33 12ths, the second 7 5ths, and the third 68 12ths. Everything in one part: the first becomes 165 60ths, the second, 84 60ths, and the third, 340 60ths.

Lemma to the next

To find two indeterminate numbers such that their product, if it lacks their sum, makes a given (number).

Let (it make) 8.

Let the first be assigned to be 1 Number, and the second 3 units. Their product, when it lacks their sum, makes 2 Numbers lacking 3 units, (which are) equal to 8 units, and the Number becomes 5 2′ units. To the hypostases: the first will be 5 2′ units, and the second 3 units.

282   So, I investigate again[90] how the Number came to be 5 2′ units: (it came) from the 11, which had to be divided by the 2. But the 11 is the given (number) together with the second; while the 2, (the multitude of) Numbers, is by a unit smaller than the second (number). If, then, I assign the second to be (any number) whatever, and we add this to the given (number) and we divide the result by the (number which is) 1 unit smaller than the second, we shall find the first. Let the second be 1 Number, 1 unit. These, together with 8 units, make 1 Number, 9 units. I divide these by the (number which is) 1 unit smaller than the second, that is, by 1 Number, and it results in 1 unit, 9 inverse Numbers.

And it has been solved indeterminately so that their product, if it lacks their sum, makes 8 units.

35. To find three numbers such that the product of any two, if it lacks their sum, makes a given (number).[91] It is certainly necessary that the given (numbers) be squares except a unit.

Now, let it be proposed that the product of the first and the second, if it lacks their sum, makes 8 units; the product of the second and the third, if it lacks their sum, makes 15 units; and the product of the third and the first, if it lacks their sum, makes 24 units.

Since I want the product of the first and the second, if it lacks their sum, to make 8 units, if, therefore, I assign the second to be (any number) whatever, and we add this to 8 units and I divide the result by the second decreased by one unit, I

284   shall have, according to the previously described lemma, the first. | Let the second be 1 Number, 1 unit. I add 8 units to it. It becomes 1 Number, 9 units. I divide these by the (number which is) one unit smaller than the second,[92] that is, by 1 Number. It becomes 1 unit, 9 inverse Numbers; this will be the first. Similarly,

---

90   That is, as in the lemma preceding Problem 34.
91   Lit. "given (numbers)".
92   Lit. "the first number which is smaller than the second".

the third will also be 1 unit, 16 inverse Numbers, and two (of the) proposed tasks have been fulfilled. Further, the product of the first and third, if it lacks their sum – (which) makes 144 inverse Powers lacking 1 unit – must be equal to 24 units; and the Number becomes 12 5ths.

To the hypostases: the first will be 57 12ths, the second 17 5ths, and the third 92 12ths. And if you wish them to be of one part, everything in 60ths, and ⟨the first⟩[93] will be 285 (60ths) the second 204 (60ths), the third 460 (60ths).

Lemma to the next

To find two indeterminate numbers such that their product has to their sum a given ratio.

Now, let it be proposed that their product be three times their sum.

Let the first be assigned to be 1 Number, and the second 5 units. So, their product is 5 Numbers. We want these to be thrice 1 Number, 5 units. Therefore, 3 Numbers, 15 units are equal to 5 Numbers, and the Number becomes 7 2′ units. To the hypostases: the first will be 7 2′ units, and the second 5 units.

So, I consider ⟨how⟩[94] the Number came to be 7 2′ units. (It came) from the 15, which had to be divided by 2, the (multitude of) Numbers. But the 15 is the second number multiplied by the ratio, while the 2 comes from the excess by which the second (number) exceeds the ratio. If, therefore, we assign the second to be whatever (multitude) of Numbers, and we multiply it by the ratio, it makes 3 Numbers. And if (this) is divided by the excess by which the second (number) exceeds the ratio, that is by 1 Number lacking 3 units, the first becomes 3 Numbers in a part of 1 Number lacking 3 units.

36. To find three numbers such that the product of any two has to their sum a given ratio.

Now, let it be proposed that the product of the first and second be three times their sum, the product of the second and third be four times their sum, and the product of the third and first be five times their sum.

Let the second be assigned to be 1 Number. So, by virtue of the lemma, the first will be 3 Numbers in a part of 1 Number lacking 3 units, and, similarly, the third, 4 Numbers in a part of 1 Number lacking 4 units. Then, the product of the first and the third must be five times their sum. But the product of the first and third is 12 Powers in a part of 1 Power, 12 units lacking 7 Numbers, while the sum of both the first and the third is 7 Powers lacking 24 Numbers of a part of 1 Power, 12 units lacking 7 Numbers.

286

---

93 Supplemented by Bachet.
94 Supplemented by Bachet and Auria.

288   (The explanation is) as follows: whenever it is required to add fractions, for example, 3 Numbers of a part of 1 Number lacking 3 units, and 4 Numbers of a part of 1 Number lacking 4 units, the Numbers of the part must be multiplied by the alternate parts, that is to say, 3 Numbers, by the parts of the other, that is by 1 Number lacking 4 units, and again 4 Numbers, by the parts of the other, by 1 Number lacking 3 units. In this way the addition produced 7 Powers lacking 24 Numbers of a part which is the product of the parts, that is 1 Power, 12 units lacking 7 Numbers.

But we also have the product of the first and third, 12 Powers of a part of 1 Power, 12 units lacking 7 Numbers. Thus, 12 Powers ⟨of a part of 1 Power, 12 units⟩[95] lacking 7 Numbers are the quintuple of the sum. 5 times the sum, therefore, becomes 35 Powers lacking 120 Numbers of a part of 1 Power, 12 units lacking 7 Numbers. And everything by their common part, (namely,) by 1 Power, 12 units lacking 7 Numbers. 12 Powers become equal to 35 Powers lacking 120 Numbers, and the Number becomes 120 23rds.

To the hypostases: well, the first was 3 Numbers of a part of 1 Number lacking 3 units, the second 1 Number, and the third 4 Numbers of a part of 1 Number lacking 4 units. But the Number was found to be 120 23rds. If you apply (this) to the first, to 3 Numbers, there will be 360 units; then, to the part, (that is,) 120 units to 1 Number lacking 3 units, they become 51 units; therefore, the first is 360 51sts.

290   And the | second, 120 23rds, for it did not have a numerical part. And the third, similarly, (if you apply) 120 to the 4 Numbers they become 480; and, similarly, to the part, (that is,) 120 to 1 Number lacking 4 units, they become 28 units; therefore, the third is 480 28ths. And the proof is obvious.

37. To find three numbers such that the product of any two has to the (number) composed of all three a given ratio.

Now, let it be proposed that the product of the first and the second be the triple of the three; the product of the second and the third be the quadruple of the three; and the product of the third and the first be the quintuple of the three.

So, since the product of any two has to the (number composed) of all three a given ratio, I first search for three numbers and an arbitrary (number) such that the product of any two has to the arbitrary (number) the proposed ratio. Let the arbitrary (number) be 5 units. And since the product of the first and the second is the triple of the arbitrary (number), that is, the 5, therefore, the product of the first and the second will be 15 units. Let the second be 1 Number. The first, therefore, will be 15 inverse Numbers. Again, since the product of the second and the third is the quadruple of the 5, the product of the second and third, therefore, will be 20 units. But the second is 1 Number. Therefore, the third will be 20 inverse Numbers. It remains, also, for the product of the third and the first, which is 300 inverse Powers, to be the quintuple of the 5. They become 300 inverse Powers, (which are) 292   equal to 25 units. | And if the species had to the species a ratio that a square has

95 Supplemented by Bachet.

to a square, the proposed task would have been fulfilled. But 300, (the multitude of) the inverse Powers, is the product of the 15 and the 20; and the 15 is the triple of the 5, while the 20 is the quadruple of the 5. So, we want the product of thrice the 5 by four times the 5 to have to five times the 5 a ratio that a square has to a square. But the 5 is arbitrary. Thus, I am reduced to seeking a certain number such that its triple, if multiplied by its quadruple, has to its quintuple a ratio that a square has to a square.

Let the sought-after (number) be 1 Number, and let its triple multiplied by its quadruple make 12 Powers. Then, this must have to its quintuple a ratio that a square has to a square. Therefore, we want 12 Powers to 5 Numbers to be in a ratio that a square number has to a square number. Therefore, their product will also be a square. Thus, 60 Cubes are equal to a square. But this is easy: (let them be) equal to 900 Powers, and the Number becomes 15 units. To the hypostases: the sought-after (number) will be 15 units.

Thus, I assign it to be 15 units. Therefore, the product of the first and the second will be 45 units. And the second is 1 Number. The first, therefore, will be 45 inverse Numbers, and similarly, the third will be 60 inverse Numbers. It remains (for us) to establish that the product of the first and third, that is 2700 inverse Powers, is the quintuple of the 15 units; 2700 inverse Powers are equal to 75 units, and the Number becomes 6 units. To the hypostases: the first will be 7 2′ units, the second 6 units, and the third 10 units.

And if the[96] (number) composed of all three were 15 units, the proposed task would have been fulfilled. So, I assign the (number) composed of all three to be 15 Powers, while the three themselves, as we found (them), to be in (terms of) Numbers, (namely,) the first 7 2′ Numbers, the second 6 Numbers, and the third 10 Numbers. Then, the three must be 15 Powers. But the three are 23 2′ Numbers. Thus, 23 2′ Numbers are equal to 15 Powers, and the Number becomes 47 30ths.

To the hypostases: the first will be 352 2′ (30ths), the second 282 (30ths), and the third 470 (30ths).

294

38. To find three numbers such that the (number) composed of all three, if multiplied by the first makes a triangular (number), by the second makes a square, and by the third makes a cube.

So, let the three be assigned to be 1 Power; the first, a triangular (multitude) of inverse Powers,[97] let it be 6 inverse Powers; the second, 4 inverse Powers; and the third, a cubic (multitude) of inverse Powers,[98] let it be 8 inverse Powers. And the 1 Power, when multiplied by the first, makes 6 units, which is a triangular (number); by the second, makes 4 units, which is a square; and by the third, makes 8 units, which is a cube. It remains for all three to be 1 Power. But the three | are

296

---

96 Tannery's text has the feminine article ἡ instead of the masculine ὁ (294.1).
97 Lit. "of triangular inverse Powers".
98 Lit. "of cubic inverse Powers".

18 inverse Powers,[99] (which are) equal to 1 Power. And everything by 1 Power. It becomes 1 Power-Power equal to 18 units. So, 18 must be a square, having a square as a side. But the 18 is the sum of a triangular (number) and a square and a cube. Thus, I am reduced to finding a square having a square as a side, and to divide (it) into a triangular (number) and a square and a cube.

Let the square be 1 Power-Power, 1 unit lacking 2 Powers. If, therefore, from 1 Power-Power I remove 1 Power-Power, 1 unit lacking 2 Powers, it leaves as a remainder 2 Powers lacking 1 unit. Again, these must be divided into a cube and a triangular (number). And let the cube be 8 units. Therefore, the remaining triangular (number, namely,) 2 Powers lacking 9 units, is equal to a triangular (number). But every triangular (number), taken 8 times, and receiving in addition 1 unit, becomes a square. Therefore, 16 Powers lacking 71 units are equal to a square. I form the square from 4 Numbers lacking 1 unit. The square becomes 16 Powers, 1 unit ⟨lacking 8 Numbers⟩,[100] and the Number becomes 9 units. To the hypostases: the triangular (number) will be 153 units, the square 6400 units, and the cube 8 units.

I return to the initial (problem) and I assign the square which is composed of all three to be 1 Power, the first to be 153 inverse Powers, since it must become a triangular (number), the second to be 6400 inverse Powers, since it must become a square, and the third to be 8 inverse Powers, since it must become a cube. And the 1 Power, being a square, when multiplied by whatever (of these numbers), makes the one a triangular (number), another a square, and the other a cube. | Now all three must be 1 Power. But (all three) are 6561 inverse Powers; (these are) equal to 1 Power. And everything by a Power: it results in 1 Power-Power equal to 6561 units, and the Number becomes 9 units.

To the hypostases: the first will be 153 81sts, the second 6400 81sts, and the third 8 81sts, and the proof is obvious.

39. To find three numbers such that the difference of the greater and the middle has to the difference of the middle and the smaller a given ratio, and, moreover, taken by twos, produce a square.

Now, let it be proposed that the difference of the greater and the middle is the triple of the difference of the middle and the least.

Since the sum of both the middle and the smaller makes a square, let it make 4 units. Therefore, the middle is greater than a dyad. Let it be 1 Number, 2 units. The least, therefore, will be 2 units lacking 1 Number. And since the difference of the greater and the middle ⟨is⟩[101] the triple of the difference of the middle and the least, and the difference of the middle and the least is 2 Numbers, therefore, the difference of the greater and the middle will be 6 Numbers; and thus, the

---

99 Tannery's text has, erroneously, "18 Powers".
100 Supplemented by Tannery.
101 Supplemented by Bachet.

greater will be 7 Numbers, 2 units. Two proposed tasks remain: that the sum of the two, ⟨the greater and the least, make a square, and that the greater⟩[102] and the middle make a square. And so, I am led to a double equality: 8 Numbers, 4 units are equal to a square, and 6 Numbers, 4 units are equal to a square. | And since the units are square (numbers), the setting-up of the equation is easy. I form two numbers such that their product is 2 Numbers, according to what we know for the double equality. So, let them be 2' of a Number and 4 units, and the Number becomes 112 units. Coming back to the hypostases, I cannot subtract 1 Number, that is 112 units, from 2 units. So, I want the Number to be found to be smaller than 2 units, so that 6 Numbers, 4 units will also be smaller than 16 units; for the dyad, multiplied by (the multitude) 6 (of) Numbers and receiving in addition 4 units makes 16 units. Accordingly, since I seek 8 Numbers, 4 units to be equal to a square, and 6 Numbers, 4 units to be equal to a square, and the square of the dyad, that is, 4 units, is also a square, three squares have been produced – 8 Numbers, 4 units; and 6 Numbers, 4 units; and 4 units – and the difference of the greater and the middle is a third part of the difference of the middle and the least.

Thus, I am reduced to finding ⟨three⟩[103] squares so that the difference of the greater and the middle is a third part of the difference of the middle and the least, and, moreover, the least is 4 units, and the middle (is) smaller than 16 units. Let the least be assigned to be 4 units, and the side of the middle, 1 Number, 2 units. The square itself, therefore, will be 1 Power, 4 Numbers, 4 units. So, since the difference of the greater and the middle is a third part of the difference of the middle and the least, and the difference of the middle and the least is 1 Power, 4 Numbers, then the difference of the greater and the middle will be a 3rd of a Power, 1 3' Numbers. But the middle is 1 Power, 4 Numbers, 4 units. Therefore, the greater will be 1 3' Powers, 5 3' Numbers, 4 units, (which is) equal to a square. | Everything 9 times. Thus, 12 Powers, 48 Numbers, 36 units are equal to a square. And (I take) their fourth: 3 Powers, 12 Numbers, 9 units are equal to a square. Further, I also want the middle square to be smaller than 16 units, that is to say, (I want) the side to be smaller than 4 units. But the side of the middle is 1 Number, 2 units; these are smaller than 4 units. And the 2 units being subtracted in common, the Number will be smaller than 2 units.

So, I am led to making 3 Powers, 12 Numbers, 9 units equal to a square. I form a certain square from 3 units lacking a certain (multitude of) Numbers. And the Number will be produced from a certain number taken six times and receiving in addition 12 – that is the 12 (of the) Numbers of the equation – and divided by the excess by which the square of the number exceeds 3, the Powers in the equation. So, I am reduced to finding a certain number which, taken 6 times, and receiving in addition 12 units, and divided by the excess by which its square exceeds a triad, makes the quotient smaller than 2 units.

300

302

---

102 Supplemented by Bachet.
103 Supplemented by Bachet.

Let the sought-after (number) be 1 Number. Thus, taken 6 times and receiving in addition 12 units, it makes 6 Numbers, 12 units, while its square lacking 3 units makes 1 Power lacking 3 units. So, I want 6 Numbers, 12 units to be divided by 1 Power lacking 3 units and make the quotient smaller than 2 units. But 2 divided by 1 unit makes the quotient 2. Therefore, 6 Numbers, 12 units have to 1 Power lack-

304   ing 3 units a ratio smaller than 2 to 1. | And the product is unequal to the product.[104] Therefore, the product of 6 Numbers, 12 units by 1 unit is smaller than the product of a dyad by 1 Power lacking 3 units; that is, 6 Numbers, 12 units are smaller than 2 Powers lacking 6 units. And let 6 units be added in common: 6 Numbers, 18 units are smaller than 2 Powers. But when we solve such an equation,[105] we multiply[106] half of the Numbers by itself; it becomes 9; and the 2 Powers, by the 18 units; they become 36; add (this) to 9; they become 45, of which (we take) the side; it is not smaller than 7 units; add half of the Numbers; ⟨it becomes not smaller than 10 units; and divide by the (multitude of) Powers;⟩[107] it becomes not smaller than 5 units. Thus, I am reduced to equating 3 Powers, 12 Numbers, 9 units to a square, (say) the one from a side of 3 units lacking 5 Numbers, and the Number becomes 42 22nds units, that is, 21 11ths. I have assigned the side of the middle square to be 1 Number, 2 units; the side of the square will be 43 11ths units, and the square itself 1849 121sts units.

Thus, I return to the initial (problem) and I assign 1849 121sts, which is a square, to be equal to 6 Numbers, 4 units. And everything by 121; the Number becomes 1365 726ths, and it is smaller than a dyad.

To the hypostases of the initial problem: we posited the middle to be 1 Number, 2 units, the least to be 2 units lacking 1 Number, and the greatest to be 7 Numbers,

306   2 units. The greatest will be | 11,007 (726ths), the second 2817 (726ths), and the least 87 (726ths). And now, since the part, being 726ths, is not a square, but a sixth of a square, if we take 121, which is a square,[108] then we take the sixth of everything, and similarly the first will be 1834 2′ 121sts, the second 469 2′ (121sts), and the third 14 2′ (121sts). And if you wish to have whole (numbers), without the presence of the 2′, restore by 4;[109] so the first will be 7338 484ths, the second 1878 484ths, the third 58 484ths, and the proof is obvious.

---

104 By "product" we translate the word χωρίον, lit. "area" or "surface".

105 "ὅταν δὲ τοιαύτην ἴσωσιν ἰσώσωμεν". Tannery's Latin translation is "Quando talem aequationem solvimus" (Diophantus 1893–95, II, 305), so he makes the verb "ἰσοῦν" mean "to solve"; yet in his Index Graecitatis apud Diophantum, *s.v.* ἰσοῦν he gives the principal meaning of the verb, that is, "aequare" (Diophantus 1893–95, II, 272). The French translations by Ver Eecke (Diophantus 1959, 178) and Allard (1980, 794) follow Tannery: "Lorsque nous résolvons une pareille équation", "Lorsque nous résolvons une équation de ce type", respectively. Heath also follows Tannery: "When we solve such an equation" (1910, 198). Xylander (1575, 115) and Bachet (Diophantus 1621, 277) translate the verb ἰσοῦν by "explico".

106 Lit. "we make".

107 Text restored first by Bachet, then, in a more satisfactory manner, by Tannery.

108 I.e., we want the denominators to be the square 121.

109 "εἰς τέσσαρα ἔμβαλε". The Arabic word *al-jabr*, "restoration", is used for the same purpose in arithmetic, such as to "restore" 3 to 10 one multiplies by $3\frac{1}{3}$. The Greek verb used here is the

40. To find three numbers such that the excess by which the square of the greatest exceeds the square of the middle has to the excess of the middle over the least a given ratio, and moreover, taken by twos, make a square.

Now, let the excess by which the square of the greatest exceeds the square of the middle be the triple of the excess that the middle exceeds over the least.

Since the greatest and the middle make a square, let them make 16 Powers. Therefore, the greatest will be greater than 8 Powers; let it be 8 Powers, 2 units. Since the sum of the greatest and the middle is greater than the sum of the greatest and the least, and the sum of the greatest and the middle is 16 Powers, then the sum of the greatest and the least will be smaller than 16 Powers, but greater than 8 Powers. So, let | the sum of the greatest and the least be 9 Powers. The greatest and the middle are 16 Powers, of which 8 Powers, 2 units is the greatest. Therefore, the middle will be 8 Powers lacking 2 units, and the third, 1 Power lacking 2 units. And since I want the excess by which the square of the greatest exceeds the square of the middle to be the triple of the excess of the middle over the least; and the excess by which the square of the greatest exceeds the square of the middle is 64 Powers, while the excess of the middle over the least is 7 Powers; so we want the 64 Powers to be the triple of the 7 Powers. And the 7 Powers, taken three times, make 21 Powers, while the (multitude) 64 of the Powers is produced by taking the 2 units 32 times. Thus, I am reduced to finding a certain number which taken 32 times makes 21 units. It is obviously the 21 32nds. I assign, therefore, the first to be 8 Powers, 21 32nds units, the middle, 8 Powers lacking 21 32nds units, and the third, 1 Power lacking 21 32nds units. And one (of the) proposed tasks remains, (namely,) that the sum of both the middle and the least must be a square. But the middle and the least are 9 Powers lacking 42 32nds units; this is equal to a square, (let it be the one) from a side of 3 Numbers lacking 6 units, and the Number becomes 597 576ths.

To the hypostases: the first will be 3,069,000 of a part of 331,776, the second 2,633,544 (of a part of 331,776), and the third 138,681 (of a part of 331,776).

308

---

verb ἐμβάλλω, one of the meanings of which is "*put into* place, *to set* a broken or dislocated limb", see LSJ (= Liddell, Scott, and Jones 1996), *s.v.* ἐμβάλλω, A.6. The corresponding noun is ἐμβολή, the first meaning of which (LSJ) is "putting into its place", "setting or reduction of a fracture or dislocated limb", "insertion (of a letter)"

Book V^G

310   **1.** To find three numbers in geometrical proportion such that, if each one of them lacks a given number, it makes a square.

Let the given (number) be 12 units.

Now, a proportion is geometrical when the (number produced) by the extremes has the middle as a side. First I look for a ⟨square⟩[1] which, if it lacks 12 units, ⟨makes a square⟩.[2] But this is easy[3] and it is 42 4'. ⟨Thus, I assign the first of the extremes to be 42 4' units⟩,[4] and the second, 1 Power. Therefore, the middle will be 6 2' Numbers. It remains for each of the others lacking 12 units to make a square. And so, 1 Power lacking 12 units is equal to a square, and 6 2' Numbers lacking 12 units are equal to a square. Their difference is 1 Power

312   lacking 6 2' Numbers. The measurement: | 1 Number measures it according to 1 Number lacking 6 2' units. Half of the difference (multiplied) by itself is 169 16ths. These are equal to the smaller, that is, to 6 2' Numbers lacking 12 units, and ⟨the Number⟩[5] becomes 361 104ths.

To the hypostases: the first will be 42 4' units, the second 2346 2' 104ths, and the third 130,321 10,816ths.

2. To find three numbers in geometrical proportion such that, if each of them receives in addition a given (number), it makes a square.

Well, say the (number) 20.

I look again for a square which, if it receives in addition 20 units, makes a square; and it is 16. Accordingly, I assign one of the extremes to be 16 units, and the second of the extremes, 1 Power. Therefore, the middle will be 4 Numbers. So, following the previous (proposition), it is reduced to requiring that 4 Numbers, 20 units (be made) equal to a square, and 1 Power, 20 units (be made)

---

1 Supplemented by Bachet.
2 Supplemented by Bachet.
3 See Problem II.10.
4 Supplemented by Bachet.
5 Supplemented by Bachet.

DOI: 10.4324/9781315171470-16

equal to a square. And their difference is 1 Power lacking 4 Numbers. Measurement: ⟨1 Number⟩⁶ measures it ⟨according to⟩⁷ 1 Number lacking 4 units. The half of the difference (multiplied) by itself makes 4 units, (which are) equal to the smaller, (namely, to) 4 Numbers, 20 units, which is absurd, for the 4 units should not be smaller than 20 units. But the 4 units are a fourth of the 16; and the 16 is not arbitrary, but it is a square which, when it receives in addition 20 units, makes a square.

Thus, I am reduced to looking for a square which has a fourth part | greater than 20 units, and, if it receives in addition 20 units, makes a square. And so, the square becomes greater than 80 units. Now 81 is a square greater than 80. If, therefore, we establish the side of the sought-after square as 1 Number, 9 units, then the square itself will be 1 Power, 18 Numbers, 81 units. This, together with 20 units, must be made a square. Therefore, 1 Power, 18 Numbers, 101 units (must be) equal to a square, let it be from a side of 1 Number lacking 11 units. Therefore, the square will be 1 Power, 121 units lacking 22 Numbers. These are equal to 1 Power, 18 Numbers, 101 units, and the Number becomes a half of a unit. But the side of the sought-after square was 1 Number, 9 units; therefore, the square will be 90 4′ units.

Now I return to the initial (problem) and assign one of the extremes to be 90 4′ units, while the third, 1 Power. Therefore, the middle will be 9 2′ Numbers. And so, I am reduced to require (that) 1 Power, 20 units (be made) equal to a square, and 9 2′ Numbers, 20 units (be made) equal to a square. And the difference is 1 Power lacking 9 2′ Numbers. (The measurement:) 1 Number measures it according to 1 Number lacking 9 2′ units. Half of the difference (multiplied) by itself is 361 16ths, (which are) equal to the smaller, that is, to 9 2′ Numbers, 20 units, and the Number becomes 41 152nds.

To the hypostases: the first will be 90 4′, the second 389 2′ 152nds, and the third 1681 23,104ths.

3. To associate⁸ three numbers with a given number so that each of them, and the product of any two, if it receives in addition a given number, makes a square.

Well, say the (number) 5.

And since we have in the *Porisms* that "if each of two numbers, as well as their product, together with the same given (number), make a square, they have been obtained from the squares of two consecutive (numbers)", I set out two squares of consecutive (numbers), the one from 1 Number, 3 units, the other from 1 Number, 4 units. And so, one of the squares becomes 1 Power, 6 Numbers, 9 units, and the other 1 Power, 8 Numbers, 16 units. I remove 5 units from each of them, and I assign the one to be 1 Power, 6 Numbers, 4 units, the other 1 Power, 8

---

6 Supplemented by Bachet.
7 Supplemented by Bachet.
8 The word προσθεῖναι is usually translated as "to add", but the range of meanings includes also "to associate", "to link", "to connect".

Numbers, 11 units, and the third twice their sum except one unit, that is, 4 Powers, 28 Numbers, 29 units. Then, it also remains for this, together with 5 units, to make a square. Therefore, 4 Powers, 28 Numbers, 34 units are equal to a square, (say) the one from a side of 2 Numbers lacking 6 units. And so, the square becomes 4 Powers, 36 units lacking 24 Numbers, (which is) equal to 4 Powers, 28 Numbers, 34 units, and the Number becomes one 26th of a unit.

318   To the hypostases: the first will be 2861 676ths, the second 7645 676ths, and the third 20,336 676ths.

4. Given one number, to find three numbers such that any one of them, as well as the product of any two, if it lacks a given (number), makes a square.

Let the given (number) be 6 units.

Well, again, I set out likewise two consecutive squares, the one 1 Power, the other 1 Power, 2 Numbers, 1 unit, and I add the given (number) to them, and I assign the first to be 1 Power, 6 units, the second 1 Power, 2 Numbers, 7 units, and the third, similarly, twice their sum except one unit, that is, 4 Powers, 4 Numbers, ⟨25 units. Then, it also remains for this, lacking 6 units, to make a square. Therefore, 4 Powers, 4 Numbers, 19 units are equal to a square, (say) the one from a side of 2 Numbers lacking 6 units. And the square becomes 4 Powers, 36 units lacking 24 Numbers, (which is) equal to 4 Powers, 4 Numbers, 19 units⟩,[9] and the Number becomes 17 28ths.

To the hypostases: the first will be 4993 784ths, the second 6729 784ths, and the third 22,660 784ths.

320   **5.** To find three square (numbers) such that the product of any two, if it receives in addition their sum, or the remaining one, makes a square.

And again, we have in the *Porisms* that "for any two squares from consecutive (numbers), another number can be found, namely the (number) by a dyad greater than twice the sum of the two (squares), which constitutes the greater number of three numbers such that the product of any two, if it receives in addition their sum, or the remaining one, makes a square". Accordingly, we assign the first of the three set-out squares to be 1 Power, 2 Numbers, 1 unit, and the other 1 Power, 4 Numbers, 4 units, and the third 4 Powers, 12 Numbers, ⟨12 units).[10] Further, it is required to make the third, that is, 4 Powers, 12 Numbers, ⟨12 units)[11] equal to a square. And (we take) the fourth part in common: it results in 1 Power, 3 Numbers, 3 units equal to a square. I form the square from 1 Number lacking 3 units. Therefore, the square itself becomes 1 Power, 9 units

9 Supplemented by Bachet.
10 Supplemented by Bachet.
11 Supplemented by Bachet.

lacking 6 Numbers, (which is) equal to 1 Power, 3 Numbers, 3 units, and the Number becomes 3″ of a unit.

To the hypostases: the first will be 25 9ths, the second 64 9ths, and the third 196 9ths.

6. To find three numbers such that each of them, if it lacks a dyad, makes a square, while the product of any two, if it lacks their sum or the remaining one, makes a square.

322

If to each one of the numbers found in the previous (problem)[12] I add a dyad, the (numbers) produced fulfill what was proposed; obviously, this is what is said (in the proposition). We assign one of the required (numbers) to be 1 Power, 2 units, the other 1 Power, 2 Numbers, 3 units, the third 4 Powers, 4 Numbers, 6 units, and the proposed tasks are fulfilled. It remains (for us) to equate 4 Powers, 4 Numbers, 4 units to a square. And (we take) the fourth (of it), so that 1 Power, 1 Number, 1 unit should also be equal to a square. And if we assign the side of the square to be from a difference, let it be from 1 Number lacking 2 units. The square becomes 1 Power, 4 units lacking 4 Numbers, (which is) equal to 1 Power, 1 Number, 1 unit, and the Number becomes 3 5ths.

To the hypostases: the first will be 59 25ths, the second 114 25ths, the third 246 25ths, and the proof is obvious.

Lemma to the next

To find two numbers such that their product, if it receives in addition the sum of their squares, makes a square.

Let the first be 1 Number, and the second however many units you wish; let it be 1 unit. And their product becomes 1 Number, while the (sum) of their squares makes 1 Power, 1 unit; together with the 1 Number, it becomes 1 Power, 1 Number, 1 unit, (which must be) equal to a square. So, let it be to the one from a side of 1 Number lacking 2 units. The square becomes 1 Power, 4 units lacking 4 Numbers, (which is) equal to 1 Power, 1 Number, 1 unit, and the Number becomes 3 5ths.

324

To the hypostases: the first will be 3 (5ths), and the second 5 (5ths). And by eliminating the part, the first will be 3 units, the second 5, and they fulfill what was proposed. Indeed the (sum) of their squares, together with their product, makes a square, while as many times as you might wish to make the numbers 3 and 5, the produced numbers will fulfill what was proposed.

Lemma to the next

To find three right-angled triangles having equal areas.

---

12 That is, to numbers assigned according to the porism mentioned in the previous problem.

First, we must search for two numbers such that their squares, together with their product,[13] make a ⟨square. But this has been shown,[14] and they are 3 and 5, whose squares, together with their product, make a square⟩[15] having 7 as a side. Now I assign three right-angled triangles to be (formed) from two numbers, 7 and 3,[16] and again from 7 and 5, and further from 7 and the sum of the found | numbers 3 and 5, that is 8, therefore from 7 and 8.

326

The triangles will be: 40, 42, 58, and 24, 70, 74, and 15, 112, 113, and the triangles are equal, having areas of 840 units.

7. To find three numbers such that the square of any one, if it receives in addition, or if it lacks, the (number) composed of all three, makes a square.

And since we seek that the square of the first, if it receives in addition, or if it lacks, the (number) composed of the three, to make a square; and in any right-angled triangle the square on the hypotenuse, if it receives in addition, or if it lacks, four times the area, makes a square; therefore, the three numbers will be hypotenuses of right-angled triangles, while the (number) composed of the three will be the quadruple of the area of the triangles, the hypotenuses of which are (these numbers). Thus, I am reduced to looking for three triangles ⟨having⟩[17] equal areas. But this has been shown before, and the triangles are 40, 42, 58, and 24, 70, 74, and 15, 112, 113. Now, returning to the initial (problem), I assign the three (sought-after numbers) to be the hypotenuses of the triangles (set) in (terms of) Numbers. And so, the first will be 58 Numbers, the second 74 Numbers, and the third 113 Numbers; while (I assign) the (number) composed of the three to be the quadruple of the area in (terms of a) Power. Thus, 3360 Powers are equal to 245 Numbers, and the Number becomes 7 96ths.

328

To the hypostases: the first will be 406 (96ths), the second 518 (96ths), and the third 791 (96ths).

Lemma to the next

With respect to three squares (formed) from given (numbers), it is possible to find three numbers such that the product of any two makes the given square numbers.

Indeed, if the given squares are 4, 9, and 16, and we assign one of the sought-after (numbers) to be 1 Number, (then) one of the remaining two will be 4 inverse Numbers, the other 9 inverse Numbers, and so it remains for the product of the second and third to make 16 units. But the product of the second and third is 36

---

13 Diophantus's highly condensed phrase is "such that the (τά, neuter article, plural) on them, together with the (τοῦ, masculine article, genitive, singular) by them".

14 See the previous lemma.

15 Supplemented by Bachet.

16 In general, one forms a right triangle from two numbers a and b ($a > b$) by making the perpendiculars $2ab$ and $a^2 - b^2$, and the hypotenuse $a^2 + b^2$.

17 Supplemented by Bachet.

inverse Powers, (which are) equal to the square (number) 16, and the Number becomes 1 2′ units. To the hypostases: the first will be 1 2′, the other 2 2′ 6′, and the third 6.

And in order that it be set out as a rule, I found 36 inverse Powers equal to 16 units. And everything by 1 Power: 16 Powers become equal to 36 units, and the Power (becomes) 36 16ths, whose side is 6 4ths. But the 6 is the product of the sides of the (squares) 4 and 9, that is, of the second and the third,[18] while the part, that is, the 4, is a side of the square 16. So, whenever it is proposed for you to find three numbers such that the product of any two makes given square (numbers), for example, the (square numbers) 4, 9, and 16, multiply the sides of 4 and 9, it becomes 6, divide these by the side of 16, and the first (number) becomes 6 4ths. | Now, (divide) again the square (number) 4 by 6 4ths, they become ⟨16 6ths, and also the square (number) 9 by 6 4ths, they become⟩[19] 6 units. Thus, the first will be 6 4ths, the second 16 6ths, and the third 6 units.

8. To find three numbers such that the product of any two, if it receives in addition the (number) composed of all three, or if it lacks (them), makes a square.

Again, we first look for three triangles ⟨having equal areas⟩,[20] and having found (them), we take the (squares) on the hypotenuses. The first is 3364, the second 5476, and the third 12,769.[21] And having (them), we find, as described before, three numbers such that the product of any two makes given squares; say the ones set out.[22] But we set out these (squares), because each one, if it receives in addition 3360 units, or if it lacks (it), makes a square. But 3360 units is two times twice[23] the area of each of the triangles. Accordingly, by virtue of this, I assign, in (terms of) Numbers, the first to be 4292 113rds Numbers, the second 3277 37ths (Numbers), and the third 4181 29ths (Numbers);[24] and so, the product of any two of them makes the above squares. | Further it is required to equate (the sum of) the three to 3360 Powers. And, in order to have one part, we make everything to be in a part of 121,249.[25] And so ⟨the first becomes 4,605,316 Numbers of a part of 121,249⟩,[26] the second 17,480,781 Numbers of the same part, and the third 10,738,729 Numbers of the same part.[27] And the three (together)

330

332

---

18 Bachet corrected this to "that is of the first and the second".
19 Supplemented first by Bachet, then, in a slightly modified way, by Tannery.
20 Supplemented by Bachet.
21 These are the squares of 58, 74, and 113, the hypotenuses found in the previous problem.
22 That is, set out in the rule of the preceding lemma.
23 In place of "two times twice", Tannery's text has ὁ δ$^{πλ.}$, "the quadruple". **T** (108r), **B** (228r) have δὶς δίς ("twice twice", or, as we translate, "two times twice").
24 Tannery has changed the last two numbers into 380,132 4292nds and 618,788 4292ths, respectively.
25 Tannery changed this number into 484,996.
26 Supplemented by Tannery, then modified by us.
27 Tannery changed the three numbers into "18,421,264 Numbers of a part of 484,996", "42,954,916 Numbers of the same part", and "69,923,044 Numbers of the same part".

become 32,824,806 Numbers of a part of 121,249,[28] (which are) equal to 3360 Powers. And everything by 121,249.[29] And so, 32,824,806 Numbers become equal to 407,396,640 Powers,[30] and the Number becomes 32,824,806 of a part of 407,396,640.[31] And by taking a common divisor,[32] the Number will be 781,543 of a part of 9,699,920.

To the hypostases: the first will be. . . .[33]

9. To divide unit into two parts, and add a given (number) to either part and make a square. It is certainly necessary that the given (number) not be odd, nor its double increased by one unit be | measured by any prime number which, if increased by one unit, has a fourth part.[34]

Now, let it be proposed to add 6 units to either part and make a square.

So, since we want to cut the unit and add 6 units to each part and make a square, therefore, the sum of the squares is 13 units. Thus, it is required to divide 13 into two squares such that each one is greater than 6 units. If, therefore, I divide 13 into two squares whose difference is smaller than 1 unit, I fulfill the requirement. I take the half of the 13, it becomes 6 2′, and I look for a part to add to 6 2′ units to make a square. And everything four times. Therefore, I look for a square part to add to 26 units and to make a square. Let the added part be 1 inverse Power. And so, 26 units, 1 inverse Power become equal to a square. And everything by a Power: 26 Powers, 1 unit become equal to a square. Let it be to the one from a side of 5 Numbers, 1 unit, and the Number becomes 10 units. Therefore, the Power is 100 units, and the inverse Power a 100th of a unit. So, the (part which is) added to 26 units is a 100th; (the part added) to 6 2′ units, therefore, becomes a 400th, and it makes a square, the one from a side of 51 20ths.

So, it is necessary, in dividing 13 into two squares, to establish the side of each of them as close as possible to 51 20ths.[35] So, I look for what the triad must lack, and the dyad must receive in addition, in order to make the same, that is, 51 20ths. | So, I assign two squares, one to be from (a side of) 11 Numbers, 2 units, the other from 3 units lacking 9 Numbers. The (number) composed of their squares

334

336

---

28 Tannery has "131,299,224 Numbers of a part of 484,996".

29 Tannery has 484,996.

30 Tannery has "131,299,224 Numbers become equal to 1,629,586,560 Powers".

31 Tannery has "131,299,224 of a part of 1,629,586,560".

32 The text contains at this point the following phrase: "which is impossible, for the numbers are prime to each other", which, of course, is wrong.

33 The end of the proposition is missing. According to Tannery's reconstruction, it continues as follows: "781,543 255,380ths, the second 781,543 109,520ths, and the third 781,543 67,280ths".

34 The text of the second half of this condition is corrupted. Our translation follows Tannery's reconstruction (332–33). Other tentative reconstructions have been proposed by Bachet, Nesselman, Hankel, and Allard. Cf. Heath (1910, 107–08) and Allard (1980, 848.17–19).

35 The method to do that is called by Diophantus in problems V.11 and V.14 below παρισότητος ἀγωγή, an expression which Heath renders as "method of approximation" (Heath 1910, 95–98). A more accurate translation would be "method of approximate equality", or "method of quasi-equality".

becomes 202 Powers, 13 units lacking 10 Numbers, (which are) equal to 13 units, and the Number becomes 5 101sts. Therefore, the side of the one square will be 257 101sts, and (the side) of the other 258 101sts. And if from the square of each of them we remove 6 units, (then) the one part of the unit will be 5358 10,201sts of a unit, the other 4843 10,201sts, and it is manifest that each one together with 6 units makes a square.

10. To cut a unit ⟨into two parts⟩[36] and add to either a different given number and make a square.[37]

Now, let it be proposed to cut a unit and add 2 units to the one (part), 6 units to the other, and make each of them a square.

Let a unit, AB, be set out, and let it be cut at Γ, and let a dyad, AΔ, be added to AΓ, and a hexad, BE, to ΓB. Therefore, each of ΓΔ, ΓE is a square. And since AB is 1 unit, while AΔ and BE together is an octad, therefore, ΔE as a whole becomes 9 units, and these need to be divided into two squares, the (squares) ΓΔ and ΓE. | But since one of the squares is greater than AΔ, that is, of a dyad, but smaller than ΔB, that is, of a triad, I am reduced to dividing a proposed square, as in the present case the 9, into two squares, ΔΓ and ΓE, so that one, ΓΔ, is in the interval between the dyad and the triad. Indeed, ΓΔ having been found, and the dyad AΔ being given, then the remainder AΓ is given. But AB is 1 unit; therefore, BΓ, the remainder, is given. Therefore, Γ, (the point) at which the unit is cut, is, also, given.

The method[38] is described below. Indeed, let one of the squares, (that) between the dyad and the triad, be 1 Power. Then the remainder will be 9 units lacking 1 Power. These are equal to a square. And making them equal to a square is easy, but it is necessary to find a Power between 2 and 3. We take two squares, one greater than 2, the other smaller than 3; 289 144ths and 361 144ths are (such squares). If, therefore, we establish the 1 Power to be in the interval between the two aforesaid squares, we will fulfill the requirement. Therefore, the side of 1 Power, that is 1 Number, must be greater than 17 12ths but smaller than 19 12ths. And so, seeking to equate 9 units lacking 1 Power to a square, we must find the Number to be greater than 17 12ths but smaller than 19 12ths. | And if we (wish to) make 9 units lacking 1 Power equal to a square, we form the side of the square from 3 units lacking a certain (multitude) of Numbers, and we find that the Number is produced from a certain number made six times and divided by its square increased by 1 unit. Therefore, we are reduced to finding a certain number which, if made

338

340

36 Supplemented by Auria.
37 Tannery followed Bachet in this problem by adding the diagram suggested by the text but which is not found in any manuscript:

38 The word ἀγωγή can mean "method", "working out", "procedure". Ancient rhetoric bears witness of the use of ἀγωγή with the meaning of "method". See, for example, Patillon in Aelius Théon (2002, 15) and Webb (2001, 309).

six times, and being divided by its square increased by 1 unit, makes the quotient greater than 17 12ths but smaller than 19 12ths.

Let the sought-after (number) be 1 Number.[39] I seek, according to the determination, for 6 Numbers in a part of 1 Power, 1 unit to be greater than 17 12ths but smaller than 19 12ths. But when 17 is divided by 12 it makes the quotient 17 12ths units. Therefore, 6 Numbers must have to 1 Power, 1 unit a ratio greater than 17 has to 12. Then the product of 6 Numbers and 12 units, that is 72 Numbers, should be greater than ⟨the product of 1 Power, 1 unit and 17 units, that is 17 Powers, 17 units⟩.[40] (Multiply) half the Numbers by itself; it becomes 1296. Subtract the Powers (multiplied) by the units, that is 289; so, the remainder is 1007. (Take) the side of these; it is not greater than 31.[41] Add the half of the Numbers; it becomes | not greater than 67. Divide by the multitude of the Powers; the Number becomes ⟨not greater than⟩[42] 67 17ths. And similarly, 6 Numbers should have to 1 Power, 1 unit a ratio smaller ⟨than 19 has to 12⟩.[43] We will find that the Number is not smaller than 66 19ths, but it is also not greater than 67 17ths.

342

Let it be 3 2' units. I form, therefore, the side of the square from 3 units lacking 3 2' Numbers. The square becomes 12 4' Powers, 9 units lacking 21 Numbers. These are equal to 9 units lacking 1 Power, from which the Number is 84 53rds, and the Power 7056 2809ths. And if from this we subtract the dyad, one part of the unit will be 1438 2809ths, and so the other will be 1371 2809ths. And the proposed task is fulfilled.

11. To divide a unit into three numbers and, first,[44] add to each of them the same given (number) and make each one a square. It is certainly necessary that the given number be neither a dyad nor any of the (numbers produced) from the dyad when increased consecutively by eight (units).[45]

Now, let it be proposed to divide a unit into three numbers, and add to each 3 units, and make each one a square.

344

Again, we must divide 10 into three squares in such a manner that each of them be greater than 3 units. Indeed, if we again divide 10 into three squares through the 'method of the approximate equality', each of them will be greater than a triad, and so we will be able, by subtracting 3 units from each of them, to have those into

---

39 **ABVT** have all "1 Number, 1 unit".

40 Supplemented initially by Bachet, then, in a slightly different way, by Tannery.

41 In fact, the square root of 1007 lies between 31 and 32. The solution proposed by Diophantus adopts 31 as an upper limit.

42 Supplemented by Tannery.

43 Supplemented by Bachet.

44 The conditions on the division of the unit in Problem 12 are a generalization of those in Problem 11. Thus, Diophantus writes here that we "first" (πρότερον) cover the simple case.

45 That is, the given number must not be 2 or, in modern notation, a number of the form $8k + 2$, $k = 1, 2, \ldots$

which the unit is divided. We take this time[46] the third of 10, it is 3 3′, and we look for a square part which, if we add it to 3 3′ units, we will make ⟨a square⟩.[47] Everything 9 times. And so, a square part must be added to 30, and make the whole a square. Let the added part be 1 inverse Power. And everything by a Power: 30 Powers, 1 unit become equal to a square, (say) to the one from a side of 5 Numbers, 1 unit. The square becomes 25 Powers, 10 Numbers, 1 unit, (which is) equal to 30 Powers, 1 unit, from which the Number (is found to be) 2 units, the Power 4 units, and the inverse Power a 4th of a unit. So, if to the 30 units is added a 4th of a unit, to the 3 3′ units will be added a 36th (of a unit), and it becomes 121 36ths. So, we must divide 10 into three squares in such a manner that the side of each square be approximatively equal[48] to 11 6ths units. But the 10 is composed of two squares, the 9 and the unit. We divide the unit into two squares, 9 25ths and 16 25ths, and so the 10 is composed of three squares, 9, | 9 25ths, and 16 25ths. Now, we must render each of the sides of these (squares) to be approximatively equal to 11 6ths. But their sides are 3 units, 4 5ths of a unit, and 3 5ths of a unit. And everything 30 times. They become 90 units, 24 units, and 18 units, while the 11 6ths become 55 units. So, we must establish each side (to be almost equal to) 55. We form the side of the one to be 3 units lacking 35 Numbers; then of the other 31 Numbers, 4 5ths units, and of the other 37 Numbers, 3 5ths units. The squares of these[49] (sides) become, (together,) 3555 Powers, 10 units lacking 116 Numbers; these are equal to 10 units, from which the Number is found to be 116 3555ths.

To the hypostases: the sides of the squares come out to be given, therefore, also (the squares) themselves. The rest is obvious.

12. To divide a unit into three numbers and add to each of them a different given (number) and make each one a square.

Let the given (numbers) be 2, 3, and 4.

And we are reduced, again, to dividing 10 into three squares, such that the first is greater than a dyad, the second is greater than a triad, and the third is greater than four units. So, bisecting 1 unit, if we add to (each of) the | given (numbers) a half of a unit, then we should look for one of the squares to be greater than a dyad but smaller than 2 2′ units, the other greater than 3 units but smaller than 3 2′ units, and the third greater than 4 units but smaller than 4 2′ units. And all together are reduced to dividing 10, which is composed of two squares, anew into two other squares in such a manner that one of them be greater than 2 units but smaller than 2 2′ units. And if we subtract a dyad from it, we will find one of the (parts) of the unit. And again, we divide the other of the squares anew into two other squares in such a manner that one of them be

346

348

---

46 In contrast with Problem V.9, in which we took half of 13.
47 Supplemented by Bachet.
48 "πάρισος", meaning "approximatively equal", "almost equal". For this term, see our Introduction, Section 4.2.8.
49 Lit. "The squares of the said".

greater than 3 units but smaller than 3 2' units. And again, if we subtract 3 units from it, we will find one of the sought-after (numbers). And so, similarly, we shall find the third.[50]

13. To divide a proposed number into three numbers such that, taken by twos, make a square.

Now, let it be proposed (to divide) the (number) 10.

And since, among the three sought-after numbers, the greater and the middle make a square, and, similarly, the middle together with the third make a square, and (also) the third together with the first, therefore, the three taken two times make three squares, each of which is smaller than 10 units. But twice the three make 20 units. Therefore, we must divide the 20 into three squares in such a manner that each is ⟨smaller⟩[51] than 10 units. But 20 is composed of two

350   squares, 16 and | 4. And if we assign one of the sought-after (squares) to be 4 units, it is required to divide the 16 into two squares, in such a manner that each of them is smaller than 10 units. And we learned how to divide a given square into two squares in such a manner that one of them is greater than 6 units and smaller than 10 units.[52] Let the sum of both be 16 units and let it have been divided into (two) squares such that each of them is smaller than 10 units. And if we subtract each from 10 units, we will find the remainders which, taken by twos, make a square.

14. To divide a given number into four numbers such that, taken by threes, make a square.

Now, let it be proposed (to divide) the (number) 10.

So, since the ⟨three taken⟩[53] in order from the first make a square but, also, the three from the second make the same, and the three from the third make the same, and (also) the three from the fourth, therefore, thrice all four make four squares. But thrice the four make 30 units. Thus, it is required to divide 30 units into four squares in such a manner that each is smaller than 10 units. And this will be found as follows: If, through the (method) of approximate equality, we assign each of them (to be approximately equal to) 7 2' units, and we subtract each square from 10 units, we shall find the sought-after (numbers). But if not,[54] I observe that 30 is

352   composed of 16, 9, 4, and 1 unit. | We assign the 4 and the 9 because each of them is smaller than 10 units. It is reduced to require that 17 units be divided into two squares such that each of them is smaller than 10 units. Accordingly, if we divide

---

50 Diophantus merely outlines the solution of this problem. Detailed reconstructions have been proposed by Wertheim (Diophantus 1890, 218–20 n.), Heath (1910, 209–10 n. 1), Ver Eecke (Diophantus 1959, 206), Czwalina (Diophantus 1952, 129), and Stamatis (1963, 500–04).

51 Supplemented by Bachet.

52 See Problem V^G.10.

53 Supplemented by Tannery.

54 That is, if one wants to follow a different path.

17, as we have learned,[55] into two squares in such a manner that one of them is greater than 8 2′ units but smaller than 10 units, each of them will be smaller than 10 units, and if we subtract each of them from 10 units we will find the remaining of the sought-after (numbers), [the one to be 6 units, the other 1 unit so that what was required will be fulfilled.][56]

15. To find three numbers such that the cube of the (number) composed of all three, if it receives in addition any one (of them), makes a cube.

Let the (number) composed of all three be assigned to be 1 Number, and each of the sought-after (numbers), 7 Cubes the one, another 26 Cubes, and the third 63 Cubes. And (what was proposed) is fulfilled since the cube of the (number) composed of all three, if it receives in addition any one of them, makes a cube. It remains (for us) to equate the three with 1 Number. But the three are 96 Cubes. Therefore, 96 Cubes are equal to 1 Number. And everything by a Number: 96 Powers are equal to 1 unit. And the unit is a square; if 96 units were also a square, the proposed task would have been fulfilled. Accordingly, I seek how the 96 arose. It is the sum of three numbers, any one of which, together with 1 unit, makes a cube. So, I am reduced to finding three numbers, such that any one of them, | together with 1 unit, makes a cube, and moreover the sum of the three is a square.

354

Let the side of the first (cube) be set out to be 1 Number, 1 unit, the (side) of the second, 2 units lacking 1 Number, and the[57] (side) of the third, 2 units. One of the cubes becomes 1 Cube, 3 Powers, 3 Numbers, 1 unit, another 6 Powers, 8 units lacking 1 Cube, 12 Numbers, and the third 8 units. I remove 1 unit from each one, and I assign the first to be 1 Cube, 3 Powers, 3 Numbers, the second 6 Powers, 7 units lacking 1 Cube, 12 Numbers, and the third 7 units. It remains for them, if added together, to make a square. So, 9 Powers, 14 units lacking 9 Numbers, become equal to a square, (say) the one from a side of 3 Numbers lacking 4 units, and the Number becomes 2 15ths. One of the sought-after (numbers) will be 1538 3375ths, the other 18,577 3375ths, and the third 7 units.

I return to the initial (problem) and we assign, again, the first of the three numbers to be 1538 3375ths Cubes, the second 18,577 3375ths Cubes, and the third 7 Cubes. And again, we assign all three to be 1 Number, and so 43,740 3375ths Cubes become equal to 1 Number. (We take) the 15th of everything and (we divide) by a Number. So, 2916 Powers become equal to 225 units, and the Number becomes 15 54ths.

To the hypostases, and (what was proposed) is fulfilled.

16. To find three numbers such that the cube of the (number) composed of all three, if it lacks any one (of them), makes a cube.

356

55 See Problem V^G.10.
56 According to Tannery the bracketed text is interpolated.
57 Tannery's text has here the masculine article ὁ instead of the feminine ἡ.

Again, let the three be assigned to be 1 Number, and among them, again, the one to be 7 8ths Cubes, another 26 27ths Cubes, and the other 63 64ths Cubes. It remains (for us) to equate all three with 1 Number. A certain multitude of Cubes[58] becomes equal to 1 Number. Everything by a Number: so a certain multitude of Powers becomes equal to 1 unit. And the unit is a square. Therefore, the (multitude of) Powers must also be a square. Where did the multitude of Powers come from? From subtracting from a triad three cubes, each one of which is smaller than 1 unit. And so we are reduced to finding three cubes such that each of them is smaller than 1 unit, and their sum, if removed from a triad, makes a square. And, moreover, we seek each cube among them to be smaller than 1 unit. If, therefore, we establish the three numbers (together) to be smaller than 1 unit, (then) each of them will be much smaller than 1 unit. Therefore, it is necessary for the square which is left to be greater than a dyad. Let the square which is left be assigned to be greater than a dyad, let it be 2 4′ units. So, it is necessary to divide 3 4ths into ⟨three⟩[59] cubes; or (to divide) their[60] multiples according to (a cube) divided into
358   certain cubes. | So, let it be according to 216. We must, therefore, divide 162 into three cubes. But 162 is composed of a cube, namely, 125, and a difference of two cubes, namely, 64 and 27. Now, we have in the *Porisms* that "the difference of any two cubes ⟨is the sum of two⟩[61] cubes".

We return to the initial (problem) and assign each (number) to be (of so many) Cubes as (the numbers) found, while the three (together) to be 1 Number. And so, it comes out that the cube of the (number) composed of all three, when it lacks each one, makes a cube. It remains (for us) to equate the three with 1 Number. But the three become 2 4′ Cubes. These are equal to 1 Number, from which the Number becomes 2 thirds.

To the hypostases.

17. To find three numbers such that the cube of the (number) composed of all three, if removed from each one, makes a cube.

Again, let the three be assigned to be 1 Number, and among the three, the one to be 2 Cubes, another 9 Cubes, and the other 28 Cubes. It remains (for us) to equate the three with 1 Number. But the three are 39 Cubes. Thus, 39 Cubes are equal to 1 Number. And (everything) by a Number: therefore, 39 Powers are equal to
360   1 unit. | And if the 39 Powers were ⟨a square, the proposed task would have been fulfilled. But 39 is⟩[62] the sum of three cubes, together with 3 units. Therefore, one must find three cubes whose sum, together with 3 units, makes a square.

So, let the side of the first cube be assigned to be 1 Number, the (side) of the second 3 units lacking 1 Number, and the remaining (side) a certain (multitude)

---

58 Lit. "a certain cubic multitude".
59 Supplemented by Bachet.
60 I.e., to the 3 and the 4 of the fraction 3 4ths.
61 Supplemented by Tannery.
62 Supplemented by Bachet.

of units; so, let it be 1 unit. And the sum of the three cubes becomes 9 Powers, 28 units ⟨lacking 27 Numbers⟩.[63] These, together with 3 units, become 9 Powers, 31 units lacking 27 Numbers. (These) ⟨are equal⟩[64] to a square, (say) the one from a side of 3 Numbers lacking 7 units, and the Number becomes 6 5ths. ⟨The side of the first will be 6 (5ths)⟩,[65] the (side) of the second 9 (5ths), and the (side) of the other 1 unit.

And I add 1 unit to the cube of each of them, and I return to the initial (problem). I assign each (number) to be of so many Cubes (as the numbers found), while (the sum of) the three is supposed to be 1 Number. It remains (for us) to equate the three with 1 Number. The three become 289 25ths Cubes. These are equal to 1 Number, and the Number becomes 5 17ths.

To the hypostases.

18. To find three numbers equal to a ⟨square⟩[66] such that the cube of the (number) composed of all three, if it receives in addition any one (of them), makes a square.

Let the (number) composed of all three, in order that it be a square, be assigned to be 1 Power, and, among the sought-after (numbers), the one to be 3 Cube-Cubes, another 8 Cube-Cubes, | and the third 15 Cube-Cubes. And it comes out that the cube of the (number) composed of all three, when it receives each one, makes a square. It remains (for us) to equate the three with 1 Power. But the three are 26 Cube-Cubes. These are equal to 1 Power. And everything by 1 Power: 26 Power-Powers become equal to 1 unit. And the 1 unit is a square (number) having a square side. Therefore, 26 Power-Powers must also be a square having a square side. And the said multitude of Power-Powers came from a certain three numbers, each of which, together with 1 unit, makes a square. ⟨Therefore, we are reduced to finding three numbers such that each (of them), together with 1 unit, makes a square⟩,[67] and moreover the (number) composed of all three is a square having a square side.

Let one of the sought-after (numbers) be assigned to be 1 Power-Power lacking 2 Powers, the other 1 Power, 2 Numbers, and the remaining one 1 Power lacking 2 Numbers. And it is fulfilled that each one of them, together with 1 unit, makes a square, and, moreover, that the three added together make a square, ⟨which has a square side⟩.[68] And so what was required has been solved in indeterminate numbers. So, let the Number be posited to be 3 units. Thus, one of the sought-after (numbers) will be 63 units, the second 15 units, and the third 3 units.

We return to the initial (problem) and assign, again, the three to be 1 Power, and, among the sought-after (numbers), one to be 63 Cube-Cubes, another 15

362

---

63 Supplemented by Bachet.
64 Supplemented by Bachet.
65 Supplemented initially by Auria, then by Bachet, and in a better way by Tannery.
66 Supplemented by Bachet.
67 Supplemented by Bachet.
68 Supplemented by Bachet.

Cube-Cubes, and the other 3 Cube-Cubes. It remains (for us) to equate the three with 1 Power, and so they result in 81 Cube-Cubes equal to 1 Power, and the Number becomes a 3rd.

The rest is obvious.

19. To find three numbers equal to a square such that the cube of the (number) composed of all three, if it lacks any one of them, makes a square.[69]

364

. . . .

And we are led, again, to divide 2, as before, and the cube of 2 is 8 units. So, we must subtract each one from 8 units, and make a square. Therefore, 22 should be divided into three squares such that each of them is greater than 6 units. And if we remove each of them from 8 units, we will find the three sought-after numbers. But this has been shown before,[70] namely, how 22 can be divided into three squares such that each of them is greater than 6 units.

20. To divide a given part of a unit into three parts such that each of them, if it lacks the cube of the (number) composed of all three, makes a square.

Let the given part be a 4th of a unit, and so we must let the 4′ be divided into three parts, as proposed. | Therefore, each one of them lacking a 64th of a unit should make a square. Therefore, the three lacking 3 64ths of a unit make three squares, and if we add a 64th to each one, we will find each of the sought-after (numbers). But this is easy. Indeed, we are reduced to dividing 13 64ths into three squares, which is easy.

366

21. To find three squares such that the solid (number formed) from the three, if it receives in addition any one (of them), makes a square.

Let the solid (number formed) from the three be assigned to be 1 Power. So, we seek three squares such that any one of them, together with 1 unit, makes a square. But this (can be obtained) from any right-angled triangle. I set-out the three right-angled triangles, and taking the square of one of the perpendiculars, I divide it by the square of the other perpendicular. And so, we will find that one of the squares is 9 16ths Powers, another 25 144ths Powers, the third 64 225ths Powers. And (what was proposed) is fulfilled since each one of them, together with 1 Power, makes a square. | It remains (for us) to equate the solid (number formed) from the

368

69 The solution of this problem is missing from the manuscripts. The excerpt that follows is the last part of a solution to a problem which is not Problem 19. In fact after Problem 19, for which only the enunciation is preserved, and before Problem 20, three problems are missing. These are referred to in the literature as 19a, 19b, and 19c. The preserved solution is the last part of the solution to Problem 19c. The missing Greek text of the three problems, as well as of the solution to Problem 19, have been reconstructed hypothetically by E. S. Stamatis, and we give our translation of his text in Appendix 1.

70 See Problem V.11.

three with 1 Power. But the solid (number formed) from the three becomes 14,400 518,400ths Cube-Cubes, and these are equal to 1 Power. And everything by a Power: it results in 14,400 518,400ths Power-Powers equal to 1 unit. And the side (is equal) to the side; it results in 120 720ths Powers equal to 1 unit. And the unit is a square; if 120 720ths Powers were also a square, then what was required would have been fulfilled. But it is not. Therefore, we are reduced to finding three right-angled triangles such that the solid (number formed) from the three perpendiculars, multiplied by the solid from their bases, makes a square.

Let (the square) have as a side the product of the (sides) about the right (angle) of one of the right-angled triangles. And if we divide everything by the product of the (sides) about the right (angle) of the said right-angled (triangle), it will become the product of the (sides) about the right (angle) of one of the triangles (multiplied) by the product ⟨of the sides⟩[71] about the right (angle) of the other of the triangles. And so, if we assign one of these (triangles) to be 3, 4, 5, we are reduced to finding two right-angled triangles such that the product of the (sides) about the right (angle of the one) is duodecuple of the product of the (sides) about the right (angle of the other). Therefore, the area will also be the duodecuple of the area. But if it is the duodecuple, it will be the triple as well. But this is easy, and ⟨the one⟩[72] (triangle) is similar to the (triangle) 9, 40, 41, and the other to the (triangle) 8, 15, 17.

Now, having the three | right-angled triangles, we return to the initial (problem) and assign one of the three sought-after squares to be 9 16ths (Powers), another 225 64ths (Powers), the other 81 1600ths (Powers). And so, if we equate the solid (number formed) from them with 1 Power, the Number will become expressible.

To the hypostases.

22. To find three squares such that the solid (number formed) from the three, if it lacks any one of them, makes a square.

Let the solid (number formed) from them be assigned to be 1 Power, and the three sought-after squares, once again (obtained) from the (three) right-angled triangles, be 16 25ths, 25 169ths, and 64 289ths. I assign them to be in (terms of) Powers, and it is established that the 1 Power, when it lacks any one of them, makes a square. It remains (for us) to equate the solid (number formed) from them with 1 Power. And the solid (formed) from them is 25,600 Cube-Cubes in a part of 1,221,025; these are equal to 1 Power. And everything by 1 Power: it results in 25,600 Power-Powers in a part of 1,221,025 equal to 1 unit. And the unit is a square having a square side. Therefore, 25,600 Power-Powers in a part of 1,221,025 should be a square ⟨having a square side⟩.[73] And, again, we are reduced to finding three right-angled triangles such that the solid (formed) from their | perpendiculars, if multiplied by the solid (formed) from the hypotenuses, makes a

370

372

71 Supplemented by Tannery.
72 Supplemented by Tannery.
73 Supplemented by Bachet.

square. And if we divide everything by the product of the hypotenuse and perpendicular of one of the right-angled (triangles), the product of the hypotenuse and perpendicular should be a multiple of the product of the hypotenuse and perpendicular, according to the product of the hypotenuse and perpendicular of a certain right-angled (triangle). Let one of the right-angled triangles be 3, 4, 5. Therefore, we are reduced to finding two right-angled triangles such that the product of the hypotenuse and perpendicular is twenty times the product of the hypotenuse and perpendicular. But if it is twenty times, it is five times as well. This is easy,[74] and the greater triangle is 5, 12, 13, while the smaller is 3, 4, 5. So, by means of these (triangles), two other (right-angled triangles) must be sought such that the product of the hypotenuse and perpendicular ⟨of the one⟩[75] is 6 units, ⟨of the other, 30 units⟩.[76] And the hypotenuse of the greater (triangle) is 6 2′ units, and the perpendicular 60 13ths; while the number for the hypotenuse of the smaller triangle is 2 2′ units, and the number for the side adjacent to the right angle,[77] 12 5ths. | And so, taking the least of the similar (triangles), we return to the initial (problem), and assign the solid (number formed) from the three to be 1 Power, and the squares themselves to be 16 25ths Powers, the one, another 576 625ths Powers, and the other 14,400 Powers in a part of 28,561. It remains (for us) to equate the solid (number formed) from the three with 1 Power. And everything by a Power. So, the side (will be equal) to the side, and the Number is found to be 65 48ths.

374

To the hypostases.

23. To find three squares such that the (solid number formed) from them, if subtracted from each of them, makes a square.

Again, let the solid (formed) from them be assigned to be 1 Power, and (the squares) themselves be (obtained) from any three right-angled (triangles we wish). And again, we are reduced, here too, to seeking the same as in the previous proposition. So, if in this (proposition,) too, we make use of the same right-angled (triangles) and assign one of the sought-after squares to be 25 16ths Powers, another 625 576ths Powers, and the other 28,561 14,400ths Powers, (then) it is again established that the solid (number formed) from the three, when removed from any one of them, makes a square. It remains, also, (for us) to equate the solid (formed) from them with 1 Power, from which the Number is found to be 48 65ths.

And (what was proposed) is fulfilled.

376   **24**. To find three squares such that the product of any two, if it receives in addition one unit, makes a square.

And since I seek the product of the first and second, together with 1 unit, to make a square, (I multiply) everything by the third, which is a square; therefore,

---

74 Tannery added here ⟨ἐπὶ τῶν ἐμβαδῶν⟩, that is, "with respect of the areas".
75 Supplemented by Tannery.
76 Supplemented by Tannery.
77 Lit. "about the right angle".

the product of the first and second, (multiplied) ⟨by the third⟩,[78] that is the solid (number formed) from the three, together with the third, must make a ⟨square⟩,[79] as well as together with the first, and with the second. But we have shown this before.[80] Therefore, these numbers fulfill this question, too.

25. To find three squares such that the product of any two, if it lacks one unit, makes a square.

(We multiply) everything by the third. Therefore, the product of the first and second, (multiplied) by the third, that is the solid (number formed) from the three, if it lacks the third, makes a square. As also, the solid from the three, if it lacks any one of the first and second, makes a square. But this has been shown before.[81] So, these numbers fulfill this (problem), too.

26. To find three squares such that the product of any two, if subtracted from one unit, makes a square.

Again, since we seek the product of any two, when removed from 1 unit, to make a square, if we multiply everything by the third, we are reduced again to finding three numbers such that the | solid (number formed) from them, when removed from any one (of them), makes a square. But we have shown this before.[82]

378

27. Given a number, to find three squares such that the squares, taken by twos, and receiving in addition the given number, make a square.

Let the given (number) be 15 units.

Let one of the sought-after (numbers) be 9 units. So, we must look for two other (squares) such that, on the one hand, either of them together with 24 units makes a square, and, on the other hand, the sum of the two together with 15 units makes a square. It is required, therefore, to look for two squares such that either of them together with 24 units makes a square. We take the numbers which measure 24 units.[83] ⟨Let (the measurement) be according to 4 inverse Numbers, the corresponding (number) being 6 Numbers. Half of the sum of the two becomes 2 inverse Numbers and 3 Numbers. Again⟩,[84] let (the measurement) be according to 3 inverse Numbers, the corresponding (number) being 8 Numbers. Half of the sum of the two becomes 1 2′ inverse Numbers

---

78 Supplemented by Bachet.

79 Supplemented by Bachet.

80 See Problem V^G.21.

81 See Problem V^G.22.

82 See Problem V^G.23.

83 "μετροῦντας", i.e., "divisors". The text has also the phrase "καὶ τριγώνου ὀρθογωνίου π^λ τὰς περὶ τὴν ὀρθήν", i.e., "and (we take) the sides about the right (angle) of a right-angled triangle". Heath remarks (1910, 222 n. 1), "I think these words must be a careless interpolation: they are not wanted and give no sense; nor do they occur in the corresponding place in the next problem".

84 Supplemented by Bachet.

and 4 Numbers. Let the side of the first (square) be from the difference of 2 inverse Numbers and 3 Numbers, ⟨and the (side) of the other from the difference of 1 2′ inverse Numbers and 4 Numbers⟩.[85] And it is fulfilled that each of them, together with 24 units, makes a square. | It remains, also, for the sum of the two together with 15 units to make a square. And it results in 6 4′ inverse Powers, 25 Powers lacking 9 units equal to a square; (say) equal to 25 units, and the Number becomes 5 6ths units.

380

To the hypostases.

28. Given a number to find three squares such that, taken by twos, and lacking the given (number), make a square.

Let the given (number) be 13 units.

Again, let one of the sought-after squares be assigned to be 25 units. ⟨So, we must look for two other (squares) such that⟩,[86] on the one hand, either of them together with 12 units makes a square, and on the other hand, the sum of the two lacking 13 units makes a square. We take, again, the measurement according to 3 Numbers and 4 inverse Numbers. The side of the first results from the difference of 1 2′ Numbers and 2 inverse Numbers, and (the side) of the other from the difference of 2 Numbers and 1 2′ inverse Numbers. And it is established that the square of either of them, together with 12 units, makes a square. It remains for the sum of the two lacking 13 units to make a square. And it results in 6 4′ inverse Powers, 6 4′ Powers lacking 25 units equal to a square; let it be equal to 6 4′ inverse Powers, and the Number becomes 2 units.

To the hypostases.

29. To find three squares such that the (number) composed of their squares makes a square.

382

Now, of the sought-after (squares), let one be assigned to be 1 Power, another 4 units, and the other 9 units; and so, the (number) composed of their squares becomes 1 Power-Power, 97 units, (which are) equal to a square; (say) to the one from a side of 1 Power lacking 10 units. And so, there remain 20 Powers equal to 3 units. And if each one were a square the proposed task would have been fulfilled. And so, we are reduced to finding two squares, and a certain number, ⟨such that⟩[87] its square, if decreased by the squares of the sought-after (squares), makes a certain ⟨number⟩[88] which has to the double of the initial number a ratio that a square number has to a square number.

---

85 Supplemented by Bachet.
86 Supplemented by Bachet.
87 Supplemented by Bachet.
88 Supplemented by Bachet.

Let one of the sought-after squares be assigned to be 1 Power, the other 4 units, ⟨while the arbitrary number, 1 Power, 4 units⟩.[89] And so, the square of the latter, decreased by (the sum of) their squares, leaves 8 Powers. We want these to have to twice 1 Power, 4 units, that is, to 2 Powers, 8 units, a ratio that a square has to a square. And the half of everything. Therefore, (we want) 4 Powers to also have to 1 Power, 4 units a ratio that a square has to a square (number). But 4 Powers is a square; therefore, 1 Power, 4 units are also equal to a square, (say) to the one from a side of 1 Number, 1 unit; therefore, the Number is 1 2' units. Of the sought-after squares, one will be 2 4' units, another 4 units, while the arbitrary one, 6 4' units. And everything four times. One becomes 9 units, another 16 units, and the arbitrary one 25 units.

We return to the initial (problem) and assign one of the three squares to be 1 Power, another 9 units, and the other 16 units. And so, | the (number) composed of their squares becomes 1 Power-Power, 337 units. These are equal to a square, (say) to the one from a side of 1 Power lacking 25 units; therefore, the Number is 12 5ths units.

And the rest is obvious.

30. "Obliged to do something useful for his sailing companions,
   someone mixed up eight-drachma[90] and five-drachma congii,[91]
   and paid for all a square (number),
   which, if it receives the proposed units, gives you again another square (number),
   having as a side the sum of the congii.
   So, child, tell apart[92] how many the eight-drachmas were,
   and, also, say the others, the five-drachmas".

What is meant by the epigram is the following: someone bought two qualities[93] of wine, a congius of the one being worth 8 drachmas, a congius of the other being worth 5 drachmas, and paid for all a square number, which, receiving 60 units, made a square (number) having as (its) side the multitude of the congii; tell apart the eight-drachmas and the five-drachmas.

Let the multitude of congii be 1 Number, so that the price will be 1 Power lacking 60 units. Then, 1 Power lacking 60 units must be made equal to a square, and so it is necessary to assign the side of the square to be (formed) from a side of 1 Number

384

---

89 Supplemented by Bachet.
90 "δραχμή", drachma: an Attic silver coin equivalent to 6 obols.
91 "χοῦς", a measure of capacity of wine or other liquids, equivalent to about 3.28 liters. A *chous* was equal to 12 *cotylae* (κοτύλαι).
92 The translation "tell apart" of the verb διάστειλον has been proposed by Acerbi (2013b, 711 n. 13).
93 We adopt Tannery's translation of the *hapax legomenon* in the Greek mathematical corpus term ἐνῆ, by "qualitates". See the discussion on this term by Acerbi (2013a); cf. (Tannery 1912j).

lacking whatever (multitude of) units (we wish).[94] But since 1 Power lacking 60 units is composed of a certain two numbers – the price of the eight-drachmas and the price

386 | of the five-drachmas – ⟨and the fifth of the price of the five-drachmas⟩[95] makes the multitude of the five-drachmas, while the eighth of the price of the eight-drachmas makes the multitude of the eight-drachmas, and since added together the multitudes of the congii make 1 Number, we are led to divide a certain (number), namely, 1 Power lacking 60 units, into two numbers such that a fifth of the one and an eighth of the other make 1 Number. But this is not always possible, unless the Number was established to be greater than the eighth of 1 Power lacking 60 units and smaller than the fifth of 1 Power lacking 60 units.

Let 1 Power lacking 60 units be greater than 5 Numbers but smaller than 8 Numbers. So, since 1 Power lacking 60 units is greater than 5 Numbers, let 60 units be added in common, so that 1 Power is greater than 5 Numbers, 60 units. Therefore, 1 Power ⟨is equal⟩[96] to 5 Numbers increased by a certain number greater than 60 units. Therefore, the Number must be not smaller than 11 units. Again, since 1 Power lacking 60 units is smaller than 8 Numbers, let 60 units be added in common, so that 1 Power is equal to 8 Numbers increased by a certain number smaller than 60 units, from which the Number must be found to be not greater than 12 units. But it was shown, also, to be not smaller than 11 units. Therefore, the Number must be found to be greater than 11 units but smaller than 12 units.

Now, if we seek 1 Power lacking 60 units to be equal a ⟨square⟩,[97] we form the side of the square from 1 Number lacking certain units, and so the Number arises from a certain number multiplied by itself and having received in addi-

388   tion 60 units, which has been divided by its double. | And we are reduced to finding a certain number such that its square, after having received in addition 60 units and having been divided by its double, makes the quotient greater than 11 units but smaller than 12 units.[98] And if we assign the sought-after (number) to be 1 Number, then it is required that, if one divides 1 Power, 60 units by 2 Numbers, (the quotient) should be made greater than 11 units, so that 1 Power, 60 units must be greater than 22 Numbers. Therefore, 22 Numbers are equal to 1 Power increased by a certain number smaller than 60 units; therefore, the Number should not be smaller than 19 units. Again, it is required that if one divides 1 Power, 60 units by 2 Numbers, one should find (the quotient to be) smaller than 12 units; therefore, 1 Power, 60 units are smaller than 24 Numbers. Therefore, 24 Numbers are equal to 1 Power increased by a certain number greater than 60 units; therefore, the Number should be smaller than 21 units, but, also, greater than 19

---

94 The word ὁσανδήποτε in Tannery's edition (p. 384.25) must be corrected to ὁσασδήποτε (A, f. 120v, line 7).
95 Supplemented by Bachet.
96 Supplemented by Bachet.
97 Supplemented by Bachet.
98 Tannery's text has the passage "And if we assign the sought-after (number) to be 1 Number, it is required that if one divides 1 Power, 60 units by 2 Numbers, the quotient should be made greater than 11 units but smaller than 12 units", which is superfluous since it is covered by what follows.

units. Let it be 20 units. Therefore, in making 1 Power lacking 60 units equal to a square, one must assign the side of the square to be from 1 Number lacking 20 units; from which the Number is found to be 11 2' units, and the square 132 4'. I remove 60 units; there remain 72 4' units.

So, we must divide the 72 4' units into two numbers such that a fifth of the first, together with an eighth of the second, make 11 4' units. Let the fifth of the first be 1 Number; therefore, an eighth of the second will be 11 2' units lacking 1 Number. Therefore, the numbers themselves | will be 5 Numbers, and 92 units lacking 8 Numbers. These are equal to 72 4' units; and thus, ⟨the Number⟩[99] will be 79 12ths units.

Thus, the multitude of the five-drachmas will be 6 congii, 7 cotylae, while, of the eight-drachmas, 4 congii, 11 cotylae. And the rest is obvious.

390

---

99 Supplemented by Bachet.

Book VI^G

1. To find a right-angled triangle such that the number for the hypotenuse,[1] if it lacks the number for each of the two perpendiculars, makes a cube.

Let the required triangle be formed from two numbers, and let one be 1 Number and the other 3 units. Therefore, the hypotenuse becomes 1 Power, 9 units, the perpendicular 6 Numbers, and the base 1 Power lacking 9 units.[2] But if it lacks the number for one of the perpendiculars, namely, 1 Power lacking 9 units, the hypotenuse becomes 18 units, and it is not a cube.

How did the 18 arise? It is the square of the 3 taken two times. Thus, we must find a certain number such that its square, taken two times, makes a cube. Let the required (amount) be 1 Number. So, it results in 2 Powers equal to a cube; let it be equal to 1 Cube, and the Number becomes 2 units.

Once more, I form the triangle from 1 Number and now not 3 units, but 2 units. So, the hypotenuse becomes 1 Power, 4 units, the perpendicular 4 Numbers, and 394 the base 1 Power lacking 4 units. And it is established that | the hypotenuse, if it lacks the number for the base, that is, 1 Power lacking 4 units, makes a cube. It remains (for us) ⟨to make it a cube when it lacks⟩[3] also the ⟨side⟩ which is 4 Numbers. So, it results in 1 Power, 4 units lacking 4 Numbers equal to a cube. But (this) is a square from a side of 1 Number lacking 2 units. Therefore, if we equate 1 Number lacking 2 units to a cube, we shall accomplish what is required. Let it be equal to 8 units, so the Number becomes 10 units.

And so, the triangle will be formed from 10 units and ⟨2⟩[4] units, and the hypotenuse becomes 104 units, the perpendicular 40 units, and the base 96 units. And (what was proposed) is fulfilled.

---

1 Diophantus's expression is "ὁ ἐν τῇ ὑποτεινούσῃ", lit. "the in the hypothenuse", which means "the number representing the hypotenuse". Similarly for the expressions "ὁ ἐν μιᾷ τῶν ὀρθῶν", "ὁ ἐν ἑκατέρᾳ τῶν ὀρθῶν", or "ὁ ἐν τῷ ἐμβαδῷ". He uses these formulations many times in this book.

2 In general, the hypotenuse and the two prependiculars formed from two numbers a and b ($a > b$) are $a^2 + b^2$, $2ab$, and $a^2 - b^2$, respectively.

3 Following Allard, who adds in brackets the phrase κάθετον λείψασαν ποιεῖν κύβον.

4 Supplemented by Bachet.

DOI: 10.4324/9781315171470-17

2. To find a right-angled triangle such that the number for the hypotenuse, if it receives in addition the number for each of the two perpendiculars, makes a cube.

If we form the required (triangle) from two numbers, as in the previous (problem), we are led to look for a certain square such that its double ⟨is⟩[5] a cube; and (such) is the square from a side of 2 units. So we form the required (triangle) from 1 Number and 2 units, and similarly, the hypotenuse becomes 1 Power, 4 units, one of the perpendiculars, 4 Numbers; the other, 4 units ⟨lacking 1 Power⟩.[6] It remains for the number for the hypotenuse, if it receives the number for the other of the two perpendiculars, to make a cube and, when we arrive at the hypostasis, to find the Power to be smaller than 4 units. Thus, the ⟨Number⟩[7] is smaller than 2 units, and we are reduced to finding | a cube smaller than 4 units but greater than 2 units; and (such a cube) is 27 8ths. So let 1 Number, 2 units be equal to 27 8ths, and the Number becomes 11 (8ths).

396

Thus, the hypotenuse will be 377 64ths, while one of the perpendiculars will be 135 64ths, the other 5 2' units. And (all) in 64ths. Thus, the triangle will be 377 and 135 and 352, and (what was proposed) is fulfilled.

3. To find a right-angled triangle such that the number for its area, if it receives in addition a given number, makes a square.

Let the given (number) be 5 units.

And let the triangle be assigned to be given in form: 3 Numbers, 4 Numbers, 5 Numbers. So, the number for the area, together with 5 units, becomes 6 Powers, 5 units (which are equal) to a square. Let it be equal to 9 Powers. And from the likes the likes. 3 Powers remain equal to 5 units. And the species must have to the species the ratio that a square number has to a square number. So, we are reduced to finding a right-angled triangle and a square number such that the square, if it lacks the number for the area of the triangle, makes a fifth of a square, since the given (number) is 5 units.

Let ⟨the triangle⟩ be formed ⟨from⟩ 1 Number ⟨and 1 inverse Number⟩.[8] So, the number for the area becomes 1 Power ⟨lacking 1 inverse Power⟩.[9] Let the | side of the square be 1 Number and as many inverse Numbers as the double of the given number is, that is 10 inverse Numbers. So, the square becomes 1 Power, 100 inverse Powers, 20 units. And if we remove from this the area, that is, 1 Power ⟨lacking 1 inverse Power⟩,[10] there remains 101 inverse Powers, 20 units. These 5 times: the square becomes equal to 505 inverse Powers, 100 units. And everything by 1 Power. 100 Powers, 505 units become equal ⟨to a square⟩.[11] Let it be equal

398

---

5 Supplemented by Bachet.
6 Supplemented by Bachet.
7 Supplemented by Bachet.
8 Restorations proposed by Bachet.
9 Supplemented by Bachet.
10 Supplemented by Bachet.
11 Supplemented by Bachet.

to the one from a side of 10 Numbers, 5 units, from which the Number is found to be 24 5ths.

To the hypostases: thus, the triangle is formed from 24 5ths and 5 24ths, while the side of the square from 413 60ths. So, if we assign the right-angled (triangle) to be in (terms of) Numbers, and we make its area together with 5 units equal to 170,569 3600ths Powers, the rest will be obvious to us.

4. To find a right-angled triangle such that the number for its area, if it lacks a given ⟨number⟩,[12] makes a square.

Let the given (number) be 6 units.

And let the triangle be assigned to be given in form,[13] and so, according to the supposition, 6 Powers lacking 6 units are equal to a square. Let them be equal to 4 Powers. So, once again, we are reduced to finding a right-angled triangle | and a square number such that if the square is removed from the (number) of the area, and the remainder is taken 6 times, it makes a square. Again, let the triangle be formed from 1 Number and 1 inverse Number, and the side of the square (from) 1 Number ⟨lacking as many inverse Numbers⟩[14] as half the multitude of the given number is, that is, 3 inverse Numbers: 6 units lacking 10 inverse Powers (result). And 6 times: It results in 36 Powers lacking 60 units, (which are) equal to a square; (say) to the one from a side of 6 Numbers lacking 2 units, from which the Number is found to be 8 3rds. Thus, the triangle is formed from 8 3rds and 3 8ths, while the ⟨side⟩[15] of the square from 37 24ths. And so, having found the triangle I assign (it) in (terms of) Numbers and proceeding according to the proposition, I will find the Number to be expressible, and (what was proposed) is fulfilled.

5. To find a right-angled triangle such that the number for its area, if subtracted from a given number, makes a square.

Let the given (number) be 10 units.

And once again let the triangle be assigned to be 3 Numbers, 4 Numbers, 5 Numbers. 10 units lacking 6 Powers become equal to a square. And if we make (it) equal to a square (multitude of) Powers, we are reduced, once again, to finding a right-angled triangle and a square number, such that the square, if it receives in addition the number for the area, makes a tenth of a square. | Let the triangle be formed from 1 Number and 1 inverse Number, and the side of the square (from) 1 inverse Number and 5 Numbers. And so, the (number) composed of the number for the area and the ⟨square⟩[16] becomes 26 Powers, 10 units. These ten times. It results in 260 Powers, 100 units equal to a square.

400

402

---

12 Supplemented by Bachet.
13 That is, as in the previous problem, its sides being 3 Numbers, 4 Numbers, 5 Numbers.
14 Supplemented by Bachet.
15 Supplemented by Bachet.
16 Supplemented by Bachet.

And the fourth: 65 Powers, 25 units become equal to a square; (say) to the one from a side of 5 units, 8 Numbers, from which the Number is found to be 80 units.

To the hypostases: so, similarly to the previous (problems) we will find (what we are looking for).

6. To find a right-angled triangle such that the number for the area, if it receives in addition the number for one of the perpendiculars, makes a given number.

Let the given (number) be 7 units.

Let once again the triangle be assigned to be given in form: 3 Numbers, 4 Numbers, 5 Numbers. So, 6 Powers, 3 Numbers become equal to 7 units. And it is necessary to add to half the Numbers multiplied by itself (the product of) the Powers ⟨by the units⟩,[17] and make a square. But it doesn't make (a square). Therefore, it will be necessary to find a right-angled triangle such that the square of half of one of the perpendiculars, if it receives in addition seven times the number for the area, makes a square.

Let the number for one of the perpendiculars be 1 Number, and the number for the other, 1 unit. So, they become 3 2′ Numbers (and) a 4th of a unit. And everything 4 times: 14 Numbers, ⟨1 unit⟩[18] become equal to a square. And in order to construct the right-angled (triangle) so that to be expressible, it is also necessary that 1 Power together with 1 unit be a square. | The difference is 1 Power lacking 14 Numbers. The measurement: 1 Number, according to 1 Number lacking 14 units. Half of the difference (multiplied) by itself is 49 units. (These are) equal to the smaller, and the Number becomes 24 7ths. To the hypostases.

I assign, therefore, one of the perpendiculars of the triangle to be 24 7ths, and the other 1 unit. And everything 7 times. One becomes 24, the other 7, and the hypotenuse 25. The number for the area, together with the second of the perpendiculars, becomes 84 Powers, 7 Numbers. These are equal to 7 units, from which the Number is found to be ⟨a 4th. Thus, the triangle will be⟩[19] 6 units, 7 4ths, 25 4ths, and (what was proposed) is fulfilled.

7. To find a right-angled triangle such that the number for the area, if it lacks the number for one of the perpendiculars, makes a given number.

Let the given (number) be 7 units.

Once again, if we assign it to be given in form, we are reduced to finding a right-angled triangle such that half of one perpendicular, multiplied by itself, and receiving in addition seven times the number for the area, makes a square. But it was found, and it is 7, 24, 25. | I assign, therefore, (the triangle) in (terms of) Numbers, so the area, when it lacks the number for one of the perpendiculars,

404

406

17 Supplemented by Tannery.
18 Supplemented by Bachet.
19 Supplemented by Tannery.

becomes 84 Powers lacking 7 Numbers. These are equal to 7 units, and the Number becomes a 3rd of a unit.

To the hypostases.

8. To find a right-angled triangle such that the number for its area, if it receives in addition the number for the sum of the two perpendiculars, makes a given (number).

Let the given (number) be 6 units.

Once again, let (the triangle) be assigned to be given in form, and again we are reduced to finding a right-angled triangle such that half the sum of the two perpendiculars (multiplied) by itself, together with six times the number for the area, makes a square. Once again, let ⟨one⟩[20] of the perpendiculars be posited to be 1 Number, and the other 1 unit. And we are led to require (that) a 4th of a Power, 3 2′ Numbers, a 4th of a unit be equal to a square. And everything 4 times: it results in 1 Power, 14 Numbers, 1 unit equal to a square. But 1 Power, 1 unit is equal ⟨to a square⟩.[21] The difference is 14 Numbers. The measurement: 2 Numbers, according to 7 units. Half their difference (multiplied) by itself becomes 1 Power, 12 4′ units lacking 7 Numbers, (which is) equal to 1 Power, 1 unit. And the Number becomes 45 28ths units. Thus, the triangle will be 45 28ths units, 1 unit, 53 28ths units. And everything | 28 times. Therefore, the triangle becomes 45 Numbers, 28 Numbers, 53 Numbers, and the area, together with the sum of the two perpendiculars, becomes 630 Powers, 73 Numbers, (which is) equal to 6 units, and the Number is expressible.

To the hypostases.

9. To find a right-angled triangle such that the number for its area, if it lacks the number for the sum of the two perpendiculars, makes a given number.

Let the given (number) be 6 units.

Once again, if we assign the required triangle to be given in form, we are led to look for a right-angled triangle such that half the sum of the two perpendiculars (multiplied) by itself, together with six times the number for the area, makes a square. But this was shown before, and it is 28, 45, 53. So, I assign these to be in (terms of) Numbers, and it results again in 630 Powers lacking 73 Numbers equal to 6 units, from which the Number is found to be 6 35ths of a unit.

To the hypostases.

10. To find a right-angled triangle such that the number for its area, if it receives in addition the number for the sum of both the hypotenuse and one of the perpendiculars, makes a given number.

---

20 Supplemented by Bachet.
21 Supplemented by Bachet.

Let the given (number) be 4 units.

Once again, let us assign it to be given in form. | We are reduced, again, to finding   410
a right-angled triangle such that half the sum of both the hypotenuse and one of the
perpendiculars (multiplied) by itself, ⟨together with the number for the area taken
four times, makes a square. Let the triangle be formed from 1 unit and 1 Number, 1
unit. And so, half of both the hypotenuse and one of the perpendiculars (multiplied)
by itself⟩[22] becomes 1 Power-Power, 4 Cubes, 6 Powers, 4 Numbers, 1 unit, while
four times the number for the area, 4 Cubes, 12 Powers, 8 Numbers. Thus, we should
require (to make) 1 Power-Power, 8 Cubes, 18 Powers, 12 Numbers, 1 unit equal to
a square; (say) to the one from a side of 6 Numbers, 1 unit lacking 1 Power, and the
Number becomes 4 5ths. Therefore, the triangle is formed from ⟨1 unit and⟩[23] 9 5ths.
And everything 5 times; again, the triangle will be formed from 9 and 5. And taking
the least among the similar (triangles) I assign it to be in (terms of) Numbers. It is 28
Numbers, 45 Numbers, 53 Numbers. And so, the number for the area, together with
the sum of both the hypotenuse and one of the perpendiculars, becomes 630 Powers,
81 Numbers, (which are) equal to 4 units, and the Number becomes 4 105ths.

To the hypostases.

11. To find a right-angled triangle such that the number for its area, if it lacks the
number for the sum of both the hypotenuse and one of the perpendiculars, makes
a given number.

Let the given (number) be 4 units.

Once again, let us assign it to be given in form. We are reduced, again, to find-   412
ing a right-angled triangle such that the number for its area taken four times, if it
receives in addition half the sum of both the hypotenuse and one of the perpendiculars (multiplied) by itself, makes a square. It can be shown that (the triangle)
is 28, 45, 53. I assign it to be in (terms of) Numbers, and they become 630 Powers
lacking 81 Numbers equal to 4 units, and the Number becomes a 6th of a unit.

To the hypostases.

Lemma to the next

To find a right-angled triangle such that ⟨the difference of the perpendiculars is a
square⟩,[24] and the number for the greater of the perpendiculars is a square, and further the number for its area, together with the smaller perpendicular, also makes
a square.

Let the triangle be formed from two numbers and let it be supposed that the
greater perpendicular is made from twice their product. Therefore, it is required
to find two numbers such that twice their product is a square, and the excess by

---

22 Text restored initially by Bachet then, with modifications, by Tannery.
23 Supplemented by Tannery.
24 Texte restored initially by Bachet then, with modifications, by Tannery.

which twice their product exceeds the difference of their squares makes a square. But this (is true) for any two numbers, when the greater is the double of the smaller. Further, we also require the area of the triangle, together with the smaller

414   of the perpendiculars, to make a square. But | the area of the triangle is the sextuple of the power-power[25] of the ⟨smaller⟩[26] number, while the number for the smaller of the perpendiculars is 3 of the squares of the smaller.[27] And everything by the square of the smaller. Therefore, we will look for a certain number such that 6 of its squares, together with 3 units, makes a square. But the unit is one (solution), as well as an unlimited multitude of other numbers. Thus, the required right-angled triangle will be formed from 1 unit and 2 units.

Second (lemma) needed for the same

Given two numbers the sum of which makes a square, an unlimited multitude of squares can be found such that any one of them, if multiplied by one of the given (numbers), ⟨and receiving in addition the other⟩,[28] makes a square.

Let the two given numbers be 3 and 6. It is required to find a square which, when multiplied, say by 3, and receiving in addition 6 units, makes a square.

Let the sought-after square be 1 Power, 2 Numbers, 1 unit. So, 3 Powers, 6 Numbers, 9 units become equal to a square. But (this) can be obtained in an unlimited number of ways, since the units is a square (number). So, let it be (equal) to a (square) from a side of 3 units lacking 3 Numbers, hence the Number becomes 4 units. And thus, the side of the square is 5 units. And an unlimited multitude of other (squares) can be found.

12. To find a right-angled triangle such that the number for its area, if it receives in addition the number for either of the perpendiculars, makes a square.

416   Let the triangle be assigned to be given in form: 5 Numbers, 12 Numbers, 13 Numbers. It results in 30 Powers, 12 Numbers equal to a square, and 30 Powers, 5 Numbers equal to a square;[29] let (the first of these) be equal to 36 Powers, so the Number becomes 2 units. And the Number being 2 units, 30 Powers, 5 Numbers, as well, should be a square. But it isn't. Therefore, we are reduced to finding a certain square which, if it lacks the 30 and the 12 is divided by the remainder and

---

25 Here Diophantus is using the term "power-power" in an arithmetical sense to mean the fourth power of a number.

26 Supplemented by Tannery.

27 If Diophantus had named the smaller number from which he formed the triangle "1 Number", the two terms together would have been "6 Power-Powers, 3 Powers". Because nothing has been named, the language belongs to arithmetic. He next divides both by the arithmetical "square of the smaller", mimicking the corresponding step that would be taken for the algebraic version.

28 Supplemented by Bachet.

29 Tannery considers the authenticity of the phrase "and 30 Powers, 5 Numbers equal to a square" doubtful.

the produced number (multiplied) by itself is taken 30 times and receives in addition five times the found number, makes a square number.[30]

Let the sought-after (number) make the square 1 Power. And ⟨if it lacks the 30, and the 12 is divided by the remainder⟩,[31] the number ⟨becomes⟩ 12 units in a part of 1 Power lacking 30 units. The square becomes 144 ⟨units⟩[32] in a part of 1 Power-Power, 900 units lacking 60 Powers.[33] These 30 times, together with its quintuple, produces 60 Powers, 2520 units in a part of 1 Power-Power, 900 units lacking 60 Powers.[34] And the part is a square (number), so 60 Powers, 2520 units should also be a square. But the Number results from a certain square. ⟨Therefore, we must look for a⟩[35] Power which, if taken 60 times and received in addition 2520 units, makes a square. Accordingly, if by modifying the right-angled (triangle), we establish the 60 together with the 2520 to make a square, we shall accomplish the proposed task. But the 60 results from the product of the sides about the right angle, while the 2520 is from the solid (number) contained by the greater of the | perpendiculars, and the difference of the perpendiculars, and the area. And we are reduced to finding a right-angled triangle such that the product of the sides about its right angle, if it receives in addition the solid (number) contained by the greater of the perpendiculars, and the difference of the perpendiculars, and its area, makes a square. But if we assign the greater of the perpendiculars to be a square, and if we divide everything by it, we shall require the number for the smaller of its perpendiculars, together with the product of the area and the difference of the perpendiculars, ⟨to make⟩[36] a square.

418

We are reduced to this: having found two numbers,[37] ⟨the product⟩[38] of the area and the difference of the perpendiculars, ⟨and the number for the smaller of the perpendiculars⟩,[39] to find after that a certain square which, multiplied by one of the givens, ⟨and receiving in addition the other⟩,[40] makes a square. But these lemmas have been shown before,[41] and the right-angled (triangle) is 3, 4, 5. I assign it to be in (terms of) Numbers, and I am led to require 6 Powers, 4 Numbers to be equal to a square, and 6 Powers, 3 Numbers to be equal to a square. And once

30 Heath describes the meaning of this sentence as follows. "I must find a square m^2x^2, to replace $36x^2$, such that $\frac{12}{m^2-30}$, the value of x obtained from the first equation, is real and satisfies the condition $30x^2 + 5x =$ a square" (1910, 233).
31 Supplemented by Bachet.
32 Supplemented by Tannery.
33 Tannery's Latin translation has $36x^2$ instead of $60x^2$.
34 Again, Tannery's translation has $36x^2$ instead of $60x^2$.
35 Supplemented by Tannery.
36 Supplemented by Tannery.
37 Tannery corrected the text "$\bar{\beta}$ ἀριθμοὺς ὄντας" (being two numbers) of the manuscripts into "δύο ἀριθμοὺς εὑρόντας" (having found two numbers).
38 Supplemented by Tannery.
39 Supplemented by Tannery.
40 Supplemented by Bachet.
41 Heath (1910, 234 n. 1) remarks that Tannery's restoration of the text does not correspond exactly to the phrasing of the enunciation of the first lemma.

again, if we resolve the greater equality, the Number becomes 4 units in a part of 1 Power lacking 6 units. The Power, therefore, becomes 16 units in a part of 1 Power-Power, 36 units lacking 12 Powers. And so, 6 Powers together with 3 Numbers become 12 Powers, 24 units in a part of 1 Power-Power, 36 units lacking 12 Powers. ⟨Accordingly, 12 units and⟩[42] 24 units must produce | a square which, if multiplied by the smaller of the givens and receiving in addition the greater, makes a square. And (this square) is 25. Therefore, the Power becomes 25 units, and so the Number will be 5 units.

420

So, seeking to equate 6 Powers, 4 Numbers, we make them equal to 25 Powers, and the Number becomes 4 19ths.

Thus, the triangle will be 12 (19ths), 16 (19ths), 20 (19ths), and (what was proposed) is fulfilled.

13. To find a right-angled triangle such that the number for its area, if it lacks the number for either of the perpendiculars, makes a square.

Once again, if we assign it to be given in form, similarly to the previous (problem), we are reduced to finding a right-angled triangle similar to the (triangle) 3, 4, 5. Then, let it be assigned to be in (terms of) Numbers; it becomes 3 Numbers, 4 Numbers, 5 Numbers. And so, 6 Powers lacking 4 Numbers are equal to a square. And let us assign the square to be smaller than 6 Powers. The Number becomes 4 units in a part of the excess by which the ⟨6⟩[43] exceeds a certain square. So, if we assign the square to be 1 Power, it comes – the Number being of such a size – to make 6 Powers lacking 3 Numbers equal to a square. And the 6 Powers (become) 96 units in a part of 1 Power-Power, 36 units lacking 12 Powers, while the triple of the side, 12 units in a part of 6 units lacking 1 Power, that is to say, 72 units lacking 12 Powers in the same part.[44] | And if we remove these from 96 units in the same part, there remain 12 Powers, ⟨24 units⟩[45] in a part of 1 Power-Power, 36 units lacking 12 Powers. And the part is a square (number), so 12 Powers, 24 units are also equal to a square; and the Number is 1 unit. Thus, I assign 6 Powers lacking 4 Numbers to be equal to 1 Power, and the Number becomes 4 5ths. Thus, the sides of the required right-angled (triangle) will be 12 5ths, 16 5ths, 4 units.

422

And if you don't wish to use the unit,[46] assign the smaller to be 1 Number, 1 unit. Therefore, the 3 Powers, 6 units are equivalent to 3 Powers, 6 Numbers, 9 units. And to make these equal to a square is easy, and the Number will be found to be no greater than 13 9ths. But the Number was 1 Number, 1 unit. Therefore,

---

42 Supplemented initially by Bachet then more satisfactorily by Tannery.
43 Supplemented by Bachet.
44 That is, the same as in the penultimate fraction.
45 Supplemented by Bachet.
46 That is, in the expression "12 Powers, 24 units", which had to be made a square, the solution employed before was 1. But we know, by virtue of the second lemma that precedes Problem 12, that the forthcoming equation "12 Powers, 24 units equal a square" has an unlimited number of solutions.

the Number will be no greater than 22 9ths, and its square, if removed from 6 units, makes the Number expressible.

14. To find a right-angled triangle such that the number for its area, if it lacks the number for either the hypotenuse or one of the perpendiculars, makes a square.

Let the triangle be given in form: 3 Numbers, 4 Numbers, 5 Numbers. And, once again, I am led to require 6 Powers lacking 5 Numbers to be equal to a square, and 6 Powers lacking 3 Numbers to be equal to a square. And if I equate 6 Powers lacking 3 Numbers to a square, the Number becomes 3 units in a part of 6 units lacking 1 Power. And (the Number) having been found to be such, the 6 Powers become 54 units in a part of 1 Power-Power, 36 units lacking 12 Powers. And it is required ⟨to subtract the 5 Numbers⟩[47] from 54 units in a part of 1 Power-Power, 36 units lacking 12 Powers – so there will be 90 units lacking 15 Powers in the same part – and to make the remainder equal to a square. But there remain 15 Powers lacking 36 units in a part of 1 Power-Power, 36 units lacking 12 Powers equal to a square. And the part is a square. Therefore, 15 Powers lacking 36 units are also a square. But this equality is impossible, since 15 cannot be divided into two squares. However, the initial (problem) is by no means impossible. Therefore, we must determine the triangle accurately. The 15 Powers came from a certain square smaller than the number for the area [6],[48] which had been multiplied by the product of the hypotenuse [5] and one of the perpendiculars [3]; while the lacking 36 units, from the solid (number) which is contained by the area [6] and one of the perpendiculars [3] and the excess by which the hypotenuse [5] exceeds the said perpendicular [3]. So, we are reduced to finding first a right-angled triangle, and a square number smaller than the area, such that the square, multiplied by the ⟨product of the⟩[49] hypotenuse and one of the perpendiculars, ⟨lacking⟩[50] the solid (number) which is contained by the area and the said perpendicular and the excess | by which the hypotenuse ⟨exceeds the said perpendicular, makes a square. And if we form the triangle from two numbers, and if we suppose⟩[51] that the said perpendicular stems from the double of their product, and if we divide everything by the square of their difference, ⟨that is, the difference⟩[52] of the hypotenuse and the aforesaid of the perpendiculars, we will look once again for a certain other square ⟨which⟩,[53] if multiplied by the product of the hypotenuse and one of the perpendiculars, exceeds over (the product of) the area and this (particular)

424

426

---

47 Restoration proposed with doubts by Tannery.
48 We put in square brackets the numbers to which Diophantus is referring.
49 ὑπὸ τῆς, lit. "by the" (meaning the product), supplemented by Tannery.
50 Supplemented by Bachet.
51 Text restored by Tannery.
52 Supplemented by Tannery.
53 Supplemented by Tannery.

perpendicular by a square. And if we assign the (numbers) which form the right-angled (triangle) to be similar plane (numbers),[54] we will solve the question.

Let the triangle be formed from 4 units and 1 unit, and let the square be 36 units, so that it is smaller than the area. And having formed the triangle, I set[55] it in (terms of Numbers:) 8 Numbers, 15 Numbers, 17 Numbers. So, the number for the area lacking the number for one of the perpendiculars becomes 60 Powers lacking 8 Numbers; these are equal to 36 Powers, and the Number becomes a 3rd of a unit.

To the hypostases: thus, the triangle will be 8 3rds, 15 3rds, 17 3rds, and (what was proposed) is fulfilled.

428   **Lemma to the next**

Given two numbers, if, when a certain square, after being multiplied by one of them and (the product) being decreased by the other, makes a square, another square greater than the aforesaid square can be found, making the same.

Let two numbers, (namely,) 3 and 11, be given, and a certain square, the one from 5, which, multiplied by the 3 and (the product) decreased by the 11, makes a square, (namely,) the one from a side of 8 units. Let another square be required, greater than the 25, which makes the same. Let the side of the square be 1 Number, 5 units. The square becomes 1 Power, 10 Numbers, 25 units. These (taken) 3 times, lacking 11 units, become 3 Powers, 30 Numbers, 64 units, which are equal to a square, (say) the one from a side of 8 units lacking 2 Numbers; and the Number becomes 62 units. Thus, the side will be 67 units, the square 4489, and the latter makes the proposed.

15. To find a right-angled triangle such that the number for its area, if it receives in addition the number for either the hypotenuse or one of the perpendiculars, makes a square.

If we assign it to be given in form, we are reduced, once again, to determine and look for a right-angled triangle and a square number greater than the number for the area, such that the square, when multiplied by the ⟨product of the⟩[56] hypotenuse and one
430   of the perpendiculars | of the required right-angled (triangle), ⟨lacking⟩[57] the solid (number) contained ⟨by the⟩[58] number for the area and the aforesaid perpendicular and the excess by which the hypotenuse exceeds over that one – an excess ⟨which

---

54 A "plane number" is a number of the form $p \cdot q$, and two plane numbers $p \cdot q$ and $r \cdot s$ are similar if $p : q :: r : s$. Diophantus intends here Proposition IX.1 of Euclid's *Elements*, which states that "If two similar plane numbers by multiplying one another make some number, the product will be square" (Heath 1926, II, 384).

55 Lit. "I form".

56 ὑπὸ τῆς, lit. "by the", supplemented by Bachet.

57 Supplemented by Bachet.

58 Supplemented by Bachet.

is⟩[59] a square – ⟨makes a square⟩.[60] So, let the triangle be formed from 4 units and 1 unit, and the square be 36 units. But it is not greater than the area. Thus, we have two numbers: one, the product of the hypotenuse and one of the perpendiculars, namely, 136 units, (and) the other, the solid (number) which is contained by the area and one of the perpendiculars and the difference of the hypotenuse and the aforesaid one of the perpendiculars, (namely), 4320. Therefore, since a certain square, the one of 36 units, if multiplied by the 136 and lacking the 4320, makes a square, and we require the square to be greater than 36, if, therefore, we assign (it) to be 1 Power, 12 Numbers, 36 units, and proceed according to the demonstration shown before,[61] we shall find an unlimited number of squares which make the problem, one of which is 676 units. Thus, we assign the right-angled (triangle) to be 8 Numbers, 15 Numbers, 17 Numbers, so 60 Powers, 8 Numbers become equal to 676 Powers, and the Number becomes a 77ths.

To the hypostases.

16. To find a right-angled triangle such that if ⟨one⟩[62] of the acute angles is bisected, the number for the line cutting the angle[63] is expressible.

Let the (line) bisecting the angle be assigned to be 5 Numbers, and one segment of the base, 3 Numbers; therefore, the perpendicular will be 4 Numbers. Now, let also the whole base[64] be assigned to be of however many units having a third part. Let it be 3 units. So, the other segment of the basis is 3 units lacking 3 Numbers. But since the angle was bisected, and the perpendicular is sesquitertian of the segment cut off; therefore, the hypotenuse is also sesquitertian of the remaining (segment) of the base. And the remaining segment of the base has been assigned to be 3 units lacking 3 Numbers. Therefore, the hypotenuse is 4 units lacking 4 Numbers. It remains (for us) to equate the square on them – that is, 16 Powers, 16 units lacking 32 Numbers – to the squares on the perpendiculars, that is, to 16 Powers, 9 units; so, the Number becomes 7 32nds. The rest is obvious. And if I take everything 32 times, then the perpendicular will be 28 units, the base 96 units, the hypotenuse 100 units, the (line) cutting the angle 35 units, and ⟨one of the segments of the base, 21 units, and the other, 75 units⟩.[65]

17. To find a right-angled triangle such that the number for its area, if it receives in addition the number for the hypotenuse, makes a square, while the number for its perimeter is a cube.

432

---

59 Supplemented by Tannery.
60 Supplemented by Bachet.
61 Diophantus refers to the lemma preceding this problem.
62 Supplemented by Tannery.
63 That is, the length of the line of bisection.
64 ἡ ἐξ ἀρχῆς βάσις, lit. "the base from the beginning". Ver Eecke translates this by "la base, prise à partir de son origine" (Diophantus 1959, 264).
65 Supplemented by Tannery.

Let the number for its area be assigned to be 1 Number, while the number for the hypotenuse a certain square (multitude of) units lacking 1 Number; let it be 16

434   units lacking 1 Number. But since we supposed that the number for its area is | 1 Number, therefore, the number for the product of the (sides) about the right angle of it becomes 2 Numbers. But the 2 Numbers are contained by 1 Number and 2 units. If, therefore, we assign one of the perpendiculars to be 2 units, the other will be 1 Number. And the perimeter becomes 18 units, and it is not a cube. But the 18 came from a certain square and 2 units. Therefore, a certain square should be found which, if it receives in addition 2 units, makes a cube, so that the cube exceeds the square by 2 units.

So, let the ⟨side⟩[66] of the square be assigned to be 1 Number, 1 unit, while that of the cube, 1 Number lacking 1 unit. The square becomes 1 Power, 2 Numbers, 1 unit, and the cube, ⟨1 Cube⟩,[67] 3 Numbers lacking 3 Powers, 1 unit. So, I wish the cube to exceed over the square by a dyad. Therefore, the square together with a dyad, that is, 1 Power, 2 Numbers, 3 units, is equal to 1 Cube, ⟨3⟩ Numbers ⟨lacking 3 Powers⟩, 1 ⟨unit⟩,[68] from which the Number is found to be 4 units. Thus, the side of the square will be 5 units, and that of the cube, 3 units. The square itself, therefore, is 25 units, and the cube 27 units.

So, I set up the right-angled (triangle) anew, and having assigned its area to be 1 Number, I assign the hypotenuse to be 25 units lacking 1 Number, while the base remains 2 units, and the perpendicular 1 Number. It remains for the square of the hypotenuse to be equal to the squares of the sides about the right angle. It becomes 1 Power, 625 units lacking 50 Numbers; this is equal to 1 Power, 4 units, from which the Number will be 621 50ths.

To the hypostases, and (what was proposed) is fulfilled.

436   **18.** To find a right-angled triangle such that the number for its area, if it receives in addition the number for the hypotenuse, makes a cube, while the number for its perimeter is a square.

Now if, similarly to the previous (problem), we assign the number for the area to be 1 Number, while the one for the hypotenuse to be a cubic (multitude of) units lacking 1 Number, we are reduced to looking for which cube, together with 2 units, makes a square.

Let the side of the cube be assigned to be 1 Number lacking 1 unit. The cube ⟨together with 2 units⟩[69] becomes 1 Cube, 3 Numbers, 1 unit lacking 3 Powers. It must be a square; let it be from a side of 1 2′ Numbers, 1 unit, and the Number becomes 21 4ths of a unit. Thus, the side of the cube will be 17 4ths, (and the cube) itself, therefore, will be 4913 64ths.

I assign once again the number for the area to be 1 Number, while the hypotenuse, 4913 64ths units lacking 1 Number. And we also have that the base is 2

---

66 Supplemented by Bachet.
67 Supplemented by Bachet.
68 Restorations proposed by Bachet.
69 Supplemented by Bachet.

units, and the perpendicular, 1 Number. So, if we equate the square of the hypotenuse to the squares of the sides about the right (angle), we will find the Number to be expressible.

19. To find a right-angled triangle such that the number for its area, if it receives in addition the number for one of the perpendiculars, makes a square, while the number for its perimeter is a cube.

Let the right-angled (triangle) be assigned to be (formed) from a certain indeterminate odd number,[70] let it be 2 Numbers, 1 unit. Then, the perpendicular will be 2 Numbers, 1 unit, the base 2 Powers, 2 Numbers, and the hypotenuse 2 Powers, 2 Numbers, 1 unit. It remains for its perimeter to be a cube, and the area, together with one of the perpendiculars, to make a square. And the perimeter becomes 4 Powers, 6 Numbers, 2 units, which are equal to a cube. But this is a composite number; for it is contained by 4 Numbers, 2 units and 1 Number, 1 unit. So, if we divide each side by 1 Number, 1 unit, we shall have that its perimeter is 4 Numbers, 2 units; this will be a cube. Then it remains for the number for its area, together with one of the perpendiculars, to make a square. But the number for its area becomes 2 Cubes, 3 Powers, 1 Number in a part of 1 Power, 2 Numbers, 1 unit, while one of the perpendiculars, 2 Numbers, 1 unit in a part of 1 Number, 1 unit. And if we make the two to be in the same part, they become 2 Cubes, 5 Powers, 4 Numbers, 1 unit.[71] And they have, as a common part, 1 Power, 2 Numbers, 1 unit. Therefore, the sum of the two makes 2 Numbers, 1 unit, (which are) equal to a square. But we also require that 4 Numbers, 2 units are equal to a cube. So, we are reduced to finding a cube which is the double of a square. (Such) is 8, (the double of) 4 units. Let 4 Numbers, 2 units be equal to 8 units, and the Number becomes 1 2′.

Thus, the right-angled (triangle) will be 8 5ths, 15 5ths, 17 5ths, and (what was proposed) is fulfilled.

438

20. To find a right-angled triangle such that the number for its area, if it receives in addition the number for one of the perpendiculars, makes a cube, while the number for its perimeter is a square.

440

Once again, if we make use of the same procedure as in the previous (proposition), we are reduced to making 4 Numbers, 2 units equal to a square, and 2 Numbers, 1 unit equal to a cube. So, we are led to look for a square which is the double of a cube; it is 16, and (it is the double of) 8. Again, we equate 16 units to 4 Numbers, 2 units, and the Number becomes 3 2′ units. Thus, the right-angled (triangle) will be 16 9ths, 63 9ths, 65 9ths.

---

70 That is, if m is an odd number, the sides of the right-angled triangle formed from it are m, $\frac{1}{2}\left(m^2-1\right)$, $\frac{1}{2}\left(m^2+1\right)$. Proclus (1873, 428.10) attributes this method of producing right-angled triangles from odd numbers to the Pythagoreans.

71 This is the numerator of the sum of the two fractions.

21. To find a right-angled triangle such that the number for its perimeter is a square, and if it receives in addition the number for its area, it makes a cube.

Let the right-angled (triangle) be formed from 1 Number, 1 unit. One of the perpendiculars becomes 2 Numbers, the other 1 Power lacking 1 unit, and the hypotenuse 1 Power, 1 unit. And we are led to require that 2 Powers, 2 Numbers be equal to a square, and 1 Cube, 2 Powers, 1 Number be equal to a cube. And it is easy to make the 2 Powers, 2 Numbers a square; for if you divide a dyad by a square except a dyad, you will find the Number.[72] But it is required for it to be found to be such that its cube[73] and two of its squares and itself, when added together, make a cube.

442   Well, the Number is made from a dyad divided by 1 Power lacking 2 units. The Cube becomes 8 units in a part of the ⟨cube⟩[74] of 1 Power lacking 2 units. And 2 of its squares become 8 units in a part of the square of 1 Power lacking 2 units. And (the Number) itself is 2 units in a part of 1 Power lacking 2 units. And everything in the same part; they become 2 Power-Powers in a part of the cube of 1 Power lacking 2 units. And the part is a cubic number. Let 2 Power-Powers be equal to a cube. Everything by 1 Cube: 2 Numbers become equal ⟨to a cube⟩.[75] And if we assign it to be equal to a cubic (multitude) of units, the Number will be found to be the half of a certain cube. Let the cube be 8 units; therefore, its half becomes 4 units, ⟨and its square, 16 units. If, therefore, we remove 2 units from 16 units, and we divide 2 units by the remainder, then the Number will be 2 14ths of a unit, that is a 7th, while its⟩[76] square becomes a 49th, from which we must remove 1 unit, because one of the perpendiculars is 1 Power lacking 1 unit.

And we are reduced to looking for a cube so that a fourth of its square is greater than 2 units, but smaller than 4 units. And if we assign the cube to be 1 Cube, we must require that a 4th of a Cube-Cube is greater than 2 units, but smaller than 4 units. Therefore, the Cube-Cube will be greater than 8 units, but smaller than 16 units. 729 64ths is (such a Cube-Cube), so the cube is 27 8ths.

Thus, I assign 2 Numbers to be equal to 27 8ths units, and the Number becomes
444   27 16ths, and the Power, 729 256ths. And if we divide the dyad by two units | smaller than it, we will find the Number to be 512 217ths of a unit, and we can remove 1 unit from its square.

---

72 The text is τὸν ἀριθμὸν ἕνα, lit. "the Number one", which is awkward. It is difficult to explain the presence of the arithmetical ἕνα, which occurs in all our main manuscripts, even if we adopt the translation "the one Number".

73 Tannery's text has K^Y, that is, the algebraic Cube. On the contrary, Allard's text has κύβον, that is, cube in a non technical sense. In **ATBV** the word is written in full.

74 Supplemented by Tannery.

75 Supplemented by Bachet.

76 Text restored by Allard. Tannery chose not to restore the lacuna. Bachet proposed the following restoration: "whose square is 16 units. I assign it to be in (terms of) Powers, and they become 16 Powers, equal to 2 Powers, 2 Numbers, and the Number becomes $\frac{1}{7}$".

22. To find a right-angled triangle such that the number for its perimeter is a cube, and if it receives in addition the number for its area, it makes a square.

First, we must examine (the following): given two numbers, to find a right-angled triangle so that the number for its perimeter is equal to one of the givens, and the number for its area (is equal) to the other. Let the two numbers be 12 and 7. And let it be proposed that the 12 is the number for its perimeter, and the 7 is the number for the area. Then, the product of the (numbers) about its right (angle) will be 14 units. So, if we assign one of its perpendiculars to be 1 inverse Number, its other will be 14 Numbers. But also, its perimeter is 12 units. Therefore, the hypotenuse will be 12 units lacking 1 inverse Number, 14 Numbers. It remains (for us) to equate its square – which is 1 inverse Power, 196 Powers, 172 units lacking 24 inverse Numbers, 336 Numbers – to the squares of the sides about the right (angle), that is to 1 inverse Power, 196 Powers.

Let the lacking be added in common and likes from likes, and everything by a Number: 172 Numbers become equal to 336 Powers, 24 units. But this is not always possible, unless half the Numbers (multiplied) by itself, when it lacks (the product of) the Powers by the units, makes a square. And | the (multitude of) Numbers comes from (the sum of the) square of the perimeter and the quadruple of the number for the area, while (the product of) the (multitude of) Powers by the (multitude of) units from eight times the square of the perimeter (multiplied) by the area. Accordingly, if the numbers are given such as these – and let the number for the area be 1 Number, while the one for the perimeter be a cube and a square simultaneously, namely, 64 units – then, in order to construct the triangle, we must subtract eight times the square of the perimeter (multiplied) by 1 Number from half (the sum of) the square of 64 units and 4 Numbers multiplied[77] ⟨by itself⟩,[78] and require the remainder to be equal to a square. They become 4 Powers, 4,194,304 units lacking 24,576 Numbers. And the fourth of everything: it results in 1 Power, 1,048,576 lacking 6144 Numbers equal to a square. But 1 Number, 64 units must also be equal to a square. And let the units be made equal. The difference; the measurement.

And the rest is obvious.

446

23. To find a right-angled triangle such that the square of the hypotenuse is (the sum of) another square and (its) side, ⟨and⟩[79] if divided by the number for one of the perpendiculars, it makes a cube and (its) side.

Let one of the perpendiculars be assigned to be 1 Number, and the other, 1 Power. And it is established that the square of the hypotenuse is (the sum) of a square and (its) side. It remains (for us) to equate 1 Power-Power, 1 Power with a square. And everything by | a Power: it results in 1 Power, 1 unit equal to a square, (say) to the

448

---

77 Lit. "made".
78 Supplemented by Bachet.
79 Supplemented by Bachet.

one from a side of 1 Number lacking 2 units, from which the Number becomes 3 4ths of a unit.

And the rest is obvious.

24. To find a right-angled triangle such that the number for one of the perpendiculars is a cube, the number for the other is a cube lacking (its) side, and the number for the hypotenuse is a cube and (its) side.

Let the number for the hypotenuse be assigned to be 1 Cube, 1 Number, and the number for one of the perpendiculars, 1 Cube lacking 1 Number. Then, the number for the other will be 2 Powers. It remains (for us) to equate 2 Powers with a cube, say, to equate it with 1 Cube, and the Number becomes 2 units.

To the hypostases: the triangle will be 6, 8, 10, and (what was proposed) is fulfilled.

Part III

Commentary

Writers on the *Arithmetica* from Heath to Sesiano and Rashed have invariably summarized the solutions to the problems of Diophantus using modern algebra. Indeed, until recently modern algebraic notation has been the only tool available for such a task since the conceptual and procedural differences between premodern and modern algebra had not yet been identified. Now that we have a good grasp of Diophantus's premodern approach, it is more appropriate to summarize his solutions in a format and with a notation that is compatible with his mode of thought.

The basic steps in a premodern algebraic solution to a problem consist of (i) setting up a polynomial equation, (ii) simplifying the equation to a standard form, and (iii) solving the equation (Sections 2.6, 4.1). While these last two steps are nearly always straightforward, the first can be particularly difficult and call for some creative thought. This is especially true for Diophantus, who strove to set up equations that simplify to two terms of consecutive powers to guarantee a positive rational solution or, when this was not possible, to ensure in some other way that he would obtain such a solution. We summarize the enunciations and solutions to each problem in the extant books with a particular focus on the setting-up of the equation. To this end we have adopted a modified version of the tabular presentations we read in the Michigan papyrus and in the commentary of Maximus Planudes. Our tables are divided into three columns, with the assignments of algebraic names on the left, operations on these names in the middle, and equations on the right.

Here, and also in other parts of this book, we employ a notation that reflects the nature of an algebraic term as a multitude-species pair, so that it is in accordance with the notations of algebraists writing in Greek, Arabic, Latin, and European vernaculars through the sixteenth century (Sections 2.5.1, 2.8). We abbreviate the species (powers) with the first letter of its English translation, whether from Greek or Arabic. The first-degree species is abbreviated as "N" for "Number" from the Greek books and "T" for "Thing" from the Arabic books. The second-degree species is shown as "P" for "Power" from the Greek and "M" for "*Māl*" from the Arabic. The third-degree species is "C" for "Cube" from both languages, and higher powers repeat the letters, so "PP" stands for "Power-Power", "MCC" for "*Māl* Cube Cube", etc. Units are abbreviated as "u". For each term we place

DOI: 10.4324/9781315171470-18

the multitude to the left of the species, always writing the "1" when there is one of them. Thus, "1N" stands for "1 Number", "5T" for "five Things", "$\frac{1}{8}$C" for "an eighth of a Cube", etc. These correspond to the modern x, $5x$, and $\frac{1}{8}x^3$.

It is because premodern algebraic expressions are aggregations of the powers that we avoid the symbols "+" and "−" (Section 2.5). As in the various premodern notations, existing terms are merely juxtaposed one after another, while lacking terms are preceded by a sign meaning "lacking" or "less", and which we show as "ℓ". In Greek this sign is ⋔, which is formed from the first two letters of the stem of the word λεῖψει that it signifies. In Arabic notation, which is absent in the Arabic translation of the *Arithmetica*, the sign is either the entire word *illā* or a truncated version ⋎ that looks like an upside-down "ℓ". The polynomial that we would translate into the modern $6x^2 + 7 - x^3 - 12x$, from Problem 15 of the Greek Book V, is shown as 6P 7u ℓ 1C 12N, and the polynomial that we write as $x^3 + 75x - 15x^2 - 125$, from Problem 8 of the Arabic Book V, is shown as 1C 75T ℓ 15M 125u.

Moreover, we apply the prime (') to indicate reciprocals of the powers as Diophantus does, so that 3N', for example, stands for "3 inverse Numbers", 18P' for "18 inverse Powers", and in Problem V^G.28 we write $2\frac{1}{4}$P 4P'ℓ6u for what in modern notation would be $2\frac{1}{4}x^2 + 4x^{(-2)} - 6$. Also, we use the notation for absolute value to indicate the difference between two terms in cases where we do not know which is greater, like $\left|1\frac{1}{2}\text{N}\,\ell\,2\text{N}'\right|$ for the difference between $1\frac{1}{2}$N and 2N' in Problem V^G.28. Sometimes Diophantus switches to a different unknown and its powers before he has finished with his first unknown. He does not distinguish between them in the text, but we abbreviate them in a different font, changing N and P first to $\mathcal{N}$ and $\mathcal{P}$, and where a third unknown is introduced, to $\mathbb{N}$ and $\mathbb{P}$. This happens in Problems IV^G.16, 17, and VI^G.12, 13, and 21.

Today we use the sign "=" with different meanings in different settings, including assignments (as in "Let $y = x^2$"), operations (like "$3(10 - x) = 30 - 3x$"), and equations (like "$x^2 + 3x = 4$"). In premodern algebra these three were consistently differentiated, so we adopt different notations for them in the three columns of our tables. For assignments we use "≔", to be further described later, and we reserve the equal sign "=" for the statements of equations. Operations in premodern arithmetic and algebra are performed in time with a particular outcome (Sections 2.2.2, 4.2.4), so instead of the modern symbols +, −, ·, ÷, we write the operations on algebraic expressions in words, and instead of "=" we use an arrow "→" to express the result. For example, the operation from Problem III.6 that we might translate as $(x^2 + 2x + 1) - (x^2 + 1 - 2x) = 4x$ is written as "Subtract 1P 1u ℓ 2N from 1P 2N 1u → 4N". Here only one operation is intended, while in modern notation we see three "+"s and two "−"s. In the same spirit, other operations are expressed by verbs like "multiply", "divide", etc., or by expressions like "a third of". In our tables we include not only the operations that Diophantus explicitly states but also those that are occasionally only implied.

Neither Diophantus nor any other premodern algebraist showed a notation for summarizing enunciations. This is because the unknowns in enunciations have not yet been given algebraic names. Any notation for capturing the meaning of enunciations that shows symbols for the unknowns is therefore misleading since the task of naming takes place in the solutions. But it would be inconvenient here, and especially in the conspectus of problems in Appendix 4, to simply restate the enunciations in rhetorical form. So, we cast aside any aspiration of rendering them in a historically responsible manner and adopt modern algebraic notation for them, which is as useful for us as it is alien to premodern practice. Accordingly, we write x, y, z, x', y', x'', etc. for the unnamed sought-after unknowns, a, b, c, etc. for given numbers, and m, n, p for given multitudes and ratios. When a number is prescribed to be a square or a cube, we show this by the shapes $\square$ and $\boxed{\boxdot}$, while the sides of such squares or cubes are shown as $\sqrt{\square}$ and $\sqrt[3]{\boxdot}$. For example, the enunciation to Problem IV.34 reads:

> We want to find two numbers, a square and a cube, such that the cube, if the square is added to it, amounts to a square number, and similarly, if the square is subtracted from it, leaves a square number.

We show this in our commentary and in Appendix 4 as:

$$\begin{cases} x^3 + y^2 = \square \\ x^3 - y^2 = \square' \end{cases}$$

We hope that this notation is distant enough from ancient practice that readers will understand that it neither fully nor entirely accurately represents the rhetorical statement that it encodes. The solution to this problem begins: "We assign the cube to be one Cube", which we represent in the assignments column of the table as " $x^3 := 1C$ ". This should be read as the assignment of the unnamed cube, which *we* have given the name x^3, as the algebraic "one Cube", or "1C".

Book I

I.1 $\begin{cases} x + y = a \\ x - y = b \end{cases}$ $\left(a = 100,\, b = 40 \right)$

The second assignment is designed to satisfy the second condition, and the equation is set up from the first condition.

Assignments	Operations	Equation
$y := 1N$		
	Add 1N and 40u $\rightarrow$ 1N 40u	
$x := 1N\ 40u$		
	Add 1N and 1N 40u $\rightarrow$ 2N 40u	
		2N 40u = 100u

Thus, 1N is found to be 30u, so the two sought-after numbers are $y = 30$ and $x = 70$

I.2 $\begin{cases} x + y = a \\ x = my \end{cases}$ $\left(a = 60,\, m = 3 \right)$

Again, the second assignment is designed to satisfy the second condition, and the equation is set up from the first condition.

Assignments	Operations	Equation
$x := 1N$		
	Three times 1N $\rightarrow$ 3N	
$y := 3N$		
	Add 1N and 3N $\rightarrow$ 4N	
		4N = 60u

Thus, 1N is 15u, so the two numbers are $x = 15$ and $y = 45$.

DOI: 10.4324/9781315171470-19

I.3 $\begin{cases} x+y=a \\ x=my+b \end{cases}$ $(a=80, b=4, m=3)$

The second assignment is designed to satisfy the second condition, and the equation is set up from the first condition.

Assignments	Operations	Equation
$x := 1N$		
	Add 4u to the triple of 1N→ 3N 4u	
$y := 3N\ 4u$		
	Add 1N and 3N 4u → 4N 4u	
		4N 4u = 80u

Thus, 1N is 19u, and the two numbers are $x=19$ and $y=61$.

I.4 $\begin{cases} x=my \\ x-y=a \end{cases}$ $(m=5, a=20)$

The second assignment is designed to satisfy the first condition, and the equation is set up from the second condition.

Assignments	Operations	Equation
$x \sim 1N$		
	Five times 1N → 5N	
$y := 5N$		
	Subtract 1N from 5N → 4N	
		4N = 20u

Thus, 1N is 5u, and the two numbers are $x=5$ and $y=25$.

I.5 $\begin{cases} x+y=a \\ \frac{1}{m}x+\frac{1}{n}y=b \end{cases}$ Determination (presuming without loss of generality $n>m$): $\frac{1}{n}a<b<\frac{1}{m}a$ $(a=100,\ m=3,\ n=5,\ b=30)$

For the determination, from the first condition we have $\frac{1}{m}x+\frac{1}{m}y=\frac{1}{m}a$ and $\frac{1}{n}x+\frac{1}{n}y=\frac{1}{n}a$. The first will be greater than b and the second will be smaller than b. The third assignment fulfills the second condition, and the equation is set up from the first condition.

Assignments	Operations	Equation
$\frac{1}{5}y := 1N$		
	Five times $1N \rightarrow 5N$	
$y := 5N$		
	Subtract $1N$ from $30u \rightarrow 30u \ell 1N$	
$\frac{1}{3}x := 30u \ell 1N$		
	Three times $30u \ell 1N \rightarrow 90u \ell 3N$	
$x := 90u \ell 3N$		
	Add $5N$ and $90u \ell 3N \rightarrow 2N 90u$	
		$2N 90u = 100u$

Thus, $1N$ is $5u$ and the two sought-after numbers are $y = 25$ and $x = 75$.

I.6 $\begin{cases} x + y = a \\ \frac{1}{m}x - \frac{1}{n}y = b \end{cases}$ Determination: $b < \frac{1}{m}a$

$(a = 100, m = 4, n = 6, b = 20)$

The determination follows by comparing $\frac{1}{m}x + \frac{1}{m}y = \frac{1}{m}a$, obtained from the first condition, with the second condition. The third assignment fulfills the second condition, and the equation is set up from the first condition.

Assignments	Operations	Equation
$\frac{1}{6}y := 1N$		
	Six times $1N \rightarrow 6N$	
$y := 6N$		
	Add $1N$ to $20u \rightarrow 1N 20u$	
$\frac{1}{4}x := 1N 20u$		
	Four times $1N 20u \rightarrow 4N 80u$	
$x := 4N 80u$		
	Add $6N$ and $4N 80u \rightarrow 10N 80u$	
		$10N 80u = 100u$

Thus, $1N$ is $2u$ and the two sought-after numbers are $y = 12$ and $x = 88$.

I.7 $x - a = m(x - b)$ $(a = 20, b = 100, m = 3)$

Assignments	Operations	Equation
$x := 1N$		
	Subtract 100u from 1N → 1N ℓ 100u	
	Subtract 20u from 1N → 1N ℓ 20u	
	Three times 1N ℓ 100u → 3N ℓ 300u	
		3N ℓ 300u = 1N ℓ 20u

Diophantus breaks down the simplification of the equation by first adding "the lacking" to get 3N = 1N 280u. Then he subtracts the 1N to get 2N = 280. Thus, 1N is 140u, so the sought-after number is $x = 140$.

I.8 $a + x = m(b + x)$ $(a = 100, b = 20, m = 3)$

Assignments	Operations	Equation
$x := 1N$		
	Add 1N to 100u → 1N 100u	
	Add 1N to 20u → 1N 20u	
	Three times 1N 20u → 3N 60u	
		3N 60u = 1N 100u

Thus, 1N is 20u, so the sought-after number is $x = 20$.

I.9 $a - x = m(b - x)$ Determination : $m > a : b$
$(a = 100, b = 20, m = 6)$

The determination can be found by tracing the origin of the value of 1N found in the solution.

Assignments	Operations	Equation
$x := 1N$		
	Subtract 1N from 100u → 100u ℓ 1N	
	Subtract 1N from 20u → 20u ℓ 1N	
	Six times 20u ℓ 1N → 120u ℓ 6N	
		120u ℓ 6N = 100u ℓ 1N

Thus, 1N is 4u, so the sought-after number is $x = 4$.

I.10 $a - x = m(b - x)$ $(a = 20, b = 100, m = 4)$

Assignments	Operations	Equation
$x := 1N$		
	Add 20u to 1N $\rightarrow$ 1N 20u	
	Subtract 1N from 100u $\rightarrow$ 100u ℓ 1N	
	Four times 100u ℓ 1N $\rightarrow$ 400u ℓ 4N	
		400u ℓ 4N = 1N 20u

Thus, 1N is 76u, so the sought-after number is $x = 76$.

I.11 $x + a = m(x - b)$ $(a = 20, b = 100, m = 3)$

Assignments	Operations	Equation
$x := 1N$		
	Add 20u to 1N $\rightarrow$ 1N 20u	
	Subtract 100u from 1N $\rightarrow$ 1N ℓ 100u	
	Three times 1N ℓ 100u $\rightarrow$ 3N ℓ 300u	
		3N ℓ 300u = 1N 20u

Thus, 1N is 160u; therefore, the sought-after number is $x = 160$.

I.12 $\begin{cases} x + y = x' + y' = a \\ x = my' \\ x' = ny \end{cases}$ $(a = 100, m = 2, n = 3)$

The second assignment satisfies the second condition, and the third assignment satisfies $x + y = a$. Then the fourth assignment satisfies the third condition, and the equation is set up from the condition $x' + y' = a$.

Assignments	Operations	Equation
$y' := 1N$		
	Two times 1N $\rightarrow$ 2N	
$x := 2N$		
	Subtract 2N from 100u $\rightarrow$ 100u ℓ 2N	
$y := 100u \ell 2N$		
	Three times 100u ℓ 2N $\rightarrow$ 300u ℓ 6N	

$x' \coloneqq 300u \ \ell \ 6N$

Add $300u \ \ell \ 6N$ and $1N \rightarrow 300u \ \ell \ 5N$

$300u \ \ell \ 5N = 100u$

Thus, 1N is 40u , hence the sought-after numbers are $x = 80$, $y = 20$, $x' = 60$, and $y' = 40$.

I.13 $\begin{cases} x + y = x' + y' = x'' + y'' = a \\ x = my' \\ x' = ny'' \\ x'' = py \end{cases}$ $\left(a = 100, \ m = 3, \ n = 2, \ p = 4 \right)$

The assignments follow the pattern in the previous problem, and the equation is set up from the condition $x'' + y'' = a$.

Assignments	Operations	Equation
$y'' \coloneqq 1N$		
	Two times $1N \rightarrow 2N$	
$x' \coloneqq 2N$		
	Subtract 2N from $100u \rightarrow 100u \ \ell \ 2N$	
$y' \coloneqq 100u \ \ell \ 2N$		
	Three times $100u \ \ell \ 2N \rightarrow 300u \ \ell \ 6N$	
$x \coloneqq 300u \ \ell \ 6N$		
	Subtract $300u \ \ell \ 6N$ from $100u \rightarrow$ 6N ℓ 200u	
$y \coloneqq 6N \ \ell \ 200u$		
	Four times $6N \ \ell \ 200u \rightarrow 24N \ \ell \ 800u$	
$x'' \coloneqq 24N \ \ell \ 800u$		
	Add $1N$ and $24N \ \ell \ 800u \rightarrow 25N \ \ell \ 800u$	
		$25N \ \ell \ 800u =$ 100u

Thus, 1N is 36u; therefore, the sought-after numbers are $x = 84$, $y = 16$, $x' = 72$, $y' = 28$, $x'' = 64$, and $y'' = 36$.

I.14 $x \cdot y = m(x + y)$ Determination: The value assigned to one unknown must be $> m$ $(m = 3)$

The determination is obtained by following Diophantus's solution. In this determination, he makes reference to one of the sought-after numbers.

Assignments	Operations	Equation
$x := 1N$		
$y := 12u$		
	Multiply 1N by 12u $\rightarrow$ 12N	
	Add 1N and 12u $\rightarrow$ 1N 12u	
	Three times 1N 12u $\rightarrow$ 3N 36u	
		3N 36u = 12N

Thus, 1N is 4u, so the sought-after numbers are $x = 4$ and $y = 12$.

I.15 $\begin{cases} x + a = m(y - a) \\ y + b = n(x - b) \end{cases}$ $(a = 30, b = 50, m = 2, n = 3)$

The second assignment satisfies the first condition, and the equation is set up from the second condition.

Assignments	Operations	Equation
$y := 1N\ 30u$		
$x := 2N\ \ell\ 30u$		
	Subtract 50u from 2N ℓ 30u $\rightarrow$ 2N ℓ 80u	
	Add 50u to 1N 30u $\rightarrow$ 1N 80u	
	Three times 2N ℓ 80u $\rightarrow$ 6N ℓ 240u	
		1N 80u = 6N ℓ 240u

Thus, 1N is 64u, so the two sought-after numbers are $x = 98$ and $y = 94$.

I.16 $\begin{cases} x + y = a \\ y + z = b \\ z + x = c \end{cases}$ Determination: $\frac{1}{2}(a + b + c) >$ each of a, b, and c

$(a = 20, b = 30, c = 40)$

The determination can be obtained by adding the three conditions. Using our modern equations, $x + y + z = \frac{1}{2}(a + b + c)$, so $x = \frac{1}{2}(a + b + c) - b$, $y = \frac{1}{2}(a + b + c) - c$, $z = \frac{1}{2}(a + b + c) - a$. Therefore, x, y, and z are positive if $\frac{1}{2}(a + b + c)$ is greater than each of a, b, and c. The assignments for z, x, and y satisfy the three conditions, and the equation is set up based on the first assignment.

Assignments	Operations	Equation
$x + y + z := 1N$		
	Subtract 20u from $1N \rightarrow 1N \,\ell\, 20u$	
$z := 1N \,\ell\, 20u$		
	Subtract 30u from $1N \rightarrow 1N \,\ell\, 30u$	
$x := 1N \,\ell\, 30u$		
	Subtract 40u from $1N \rightarrow 1N \,\ell\, 40u$	
$y := 1N \,\ell\, 40u$		
	Add $1N \,\ell\, 30u$, $1N \,\ell\, 40u$, $1N \,\ell\, 20u$ $\rightarrow 3N \,\ell\, 90u$	
		$3N \,\ell\, 90u = 1N$

Thus, 1N is 45u, hence the three sought-after numbers are $x = 15$, $y = 5$, and $z = 25$.

I.17
$$\begin{cases} w + x + y = a \\ x + y + z = b \\ y + z + w = c \\ z + w + x = d \end{cases}$$
Determination: $\frac{1}{2}(a + b + c + d) >$ each of a, b, c, and d

$(a = 20, b = 22, c = 24, d = 27)$

The determination, assignments, and equation are obtained by the same reasoning as in Problem I.16.

Assignments	Operations	Equation
$w + x + y + z := 1N$		
	Subtract 20u from $1N \rightarrow 1N \,\ell\, 20u$	
$z := 1N \,\ell\, 20u$		
	Subtract 22u from $1N \rightarrow 1N \,\ell\, 22u$	
$w := 1N \,\ell\, 22u$		
	Subtract 24u from $1N \rightarrow 1N \,\ell\, 24u$	
$x := 1N \,\ell\, 24u$		
	Subtract 27u from $1N \rightarrow 1N \,\ell\, 27u$	
$y := 1N \,\ell\, 27u$		
	Add $1N \,\ell\, 22u$, $1N \,\ell\, 24u$, $1N \,\ell\, 27u$, $1N \,\ell\, 20u \rightarrow 4N \,\ell\, 93u$	
		$4N \,\ell\, 93u = 1N$

Thus, 1N is 31u , so the four sought-after numbers are $w = 9$, $x = 7$, $y = 4$, and $z = 11$.

I.18 $\begin{cases} x+y=z+a \\ y+z=x+b \\ z+x=y+c \end{cases}$ $\left(a=20,\, b=30,\, c=40\right)$

First solution

This solution follows a variation on the reasoning in Problems I.15 and I.16. In modern notation, the assignment for z is found by adding z to both sides of $x + y = z + 20$ to get $x + y + z = 2z + 20$, from which we obtain $z = \frac{1}{2}(x+y+z)-10$. The others are found similarly.

Assignments	Operations	Equation
$x + y + z := 2N$		
	Subtract 20u from 2N → 2N ℓ 20u	
$2z := 2N\ \ell\ 20u$		
	Half of 2N ℓ 20u → 1N ℓ 10u	
$z := 1N\ \ell\ 10u$		
	Subtract 30u from 2N → 2N ℓ 30u	
$2x := 2N\ \ell\ 30u$		
	Half of 2N ℓ 30u → 1N ℓ 15u	
$x := 1N\ \ell\ 15u$		
	Subtract 40u from 2N → 2N ℓ 40u	
$2y := 2N\ \ell\ 40u$		
	Half of 2N ℓ 40u → 1N ℓ 20u	
$y := 1N\ \ell\ 20u$		
	Add 1N ℓ 10u, 1N ℓ 15u, 1N ℓ 20u → 3N ℓ 45u	
		3N ℓ 45u = 2N

Thus, 1N is 45u, so the three sought-after numbers are $x = 30$, $y = 25$, and $z = 35$.

Second solution

The assignment for y can be explained in modern notation as follows: adding $x + y = z + a$ and $y + z = x + b$ gives $x + 2y + z = x + z + (a + b)$, hence, $y = \frac{1}{2}(a + b)$.

Assignments	Operations	Equation
$z := 1N$		
	Add 1N and 20u → 1N 20u	
$x + y := 1N\ 20u$		
	Half of 20u and 30u → 25u	
$y := 25u$		
	Subtract 25u from 1N 20u → 1N ℓ 5u	
$x := 1N\ \ell\ 5u$		
	Add 1N ℓ 5u and 1N → 2N ℓ 5u	
	Add 25u and 40u → 65u	
		2N ℓ 5u = 65u

Thus, 1N is 35u, so $x = 30$, $y = 25$, and $z = 35$.

I.19
$$\begin{cases} w + x + y = z + a \\ x + y + z = w + b \\ y + z + w = x + c \\ z + w + x = y + d \end{cases}$$
Determination: $\frac{1}{2}(a + b + c + d) >$ each of $a, b, c,$ and d

$(a = 20, b = 30, c = 40, d = 50)$

The determination is obtained again by adding the conditions. Adding our four modern equations, we get $2(w + x + y + z) = a + b + c + d$. Adding w to both sides of the second equation and substituting for $w + x + y + z$ gives $w = \frac{1}{4}(a + b + c + d) - \frac{1}{2}b$, and similarly for the others. Thus, $w, x, y,$ and z are positive if $\frac{1}{2}(a + b + c + d)$ is greater than each of $a, b, c,$ and d. The assignments and equation follow the reasoning in the first solution to I.18.

Assignments	Operations	Equation
$w + x + y + z$ $:= 2N$		
	Subtract 20u from 2N → 2N ℓ 20u	
$2z := 2N\ \ell\ 20u$		
	Half of 2N ℓ 20u → 1N ℓ 10u	
$z := 1N\ \ell\ 10u$		
	Subtract 30u from 2N → 2N ℓ 30u	

$2w := 2N \, \ell \, 30u$		
	Half of 2N ℓ 30u → 1N ℓ 15u	
$w := 1N \, \ell \, 15u$		
	Subtract 40u from 2N → 2N ℓ 40u	
$2x := 2N \, \ell \, 40u$		
	Half of 2N ℓ 40u → 1N ℓ 20u	
$x := 1N \, \ell \, 20u$		
	Subtract 50u from 2N → 2N ℓ 50u	
$2y := 2N \, \ell \, 50u$		
	Half of 2N ℓ 50u → 1N ℓ 25u	
$y := 1N \, \ell \, 25u$		
	Add 1N ℓ 15u, 1N ℓ 20u, 1N ℓ 25u, 1N ℓ 10u → 4N ℓ 70u	
		4N ℓ 70u = 2N

Thus, 1N is 35u, so the sought-after numbers are $w = 20$, $x = 15$, $y = 10$, and $z = 25$.

Second solution

The assignments follow the pattern of the second solution to Problem I.18.

Assignments	Operations	Equation
$z := 1N$		
	Add 1N and 20u → 1N 20u	
$w + x + y := 1N$ 20u		
	Half the sum of 20u and 30u → 25u	
$x + y := 25u$		
	Subtract 25u from 1N 20u → 1N ℓ 5u	
$w := 1N \, \ell \, 5u$		
	Half the sum of 30u and 40u → 35u	
$y + z := 35u$		
	Subtract 1N from 35u → 35u ℓ 1N	
$y := 35u \, \ell \, 1N$		
	Subtract 35u ℓ 1N from 25u → 1N ℓ 10u	

$x := 1N \ \ell \ 10u$

Add 1N, 1N ℓ 5u, 1N ℓ 10u $\rightarrow$
3N ℓ 15u

Add 50u to 35u ℓ 1N $\rightarrow$ 85u ℓ 1N

85u ℓ 1N = 3N ℓ 15u

Thus, 1N is 25u, so the sought-after numbers are $w = 20$, $x = 15$, $y = 10$, and $z = 25$.

I.20 $\begin{cases} x+y+z = a \\ x+y = mz \\ y+z = nx \end{cases}$ $\left(a = 100, m = 3, n = 4 \right)$

Diophantus solves this problem in two stages, each one constituting a complete algebraic solution to part of the problem. First, he finds z and the sum $x + y$ from the first two conditions. Then, considering the first and the third conditions, and given that the value of z has been found, he determines the values of x and y individually.

Assignments	Operations	Equation
$z := 1N$		
	Three times 1N $\rightarrow$ 3N	
$x+y := 3N$		
	Add 1N and 3N $\rightarrow$ 4N	
$x+y+z := 4N$		
		4N = 100u

Thus, 1N is 25u, so $z = 25$ and $x + y = 75$. Diophantus next finds x and y.

Assignments	Operations	Equation
$x := 1N$		
	Four times 1N $\rightarrow$ 4N	
$y+z := 4N$		
	Add 1N and 4N $\rightarrow$ 5N	
$x+y+z := 5N$		
		5N = 100u

Thus, 1N is 20u; therefore, $x = 20$ and $y = 55$.

That both z and x are named 1N does not mean that Diophantus introduces a second unknown. Because the problem is worked out in two distinct solutions by algebra, one after the other, he did not need a different name in the second stage.

I.21
$$\begin{cases} x = y + \frac{1}{m}z \\ y = z + \frac{1}{n}x \\ z = \frac{1}{p}y + a \end{cases}$$
Determination: $np > n + p$

$(m = 3, n = 3, p = 3, a = 10)$

The determination can be obtained by tracing the origin of Diophantus's solution $1N = 12\frac{1}{2}u$ in terms of the given values. The $12\frac{1}{2}u$ comes from dividing (in modern notation) $na + \frac{a}{m}$ by $np - n - p - \frac{1}{m}$. The numerator is positive, so the denominator must be also. The fraction $\frac{1}{m}$ can be disregarded, so the requirement is that $np > n + p$. To decipher the statement of the determination, the part by which the middle exceeds the least is $\frac{1}{n}$, and its homonymous part is n. This is multiplied by the difference of the middle and the least according to the assignments Diophantus makes. The multitude of numbers in this amount will be $p - 1$, and this product, $np - n$, must be greater than p, the multitude of the Numbers in the assignment of y. Wertheim (1890, 27–28) pointed out that this determination refers to the particular assignments that Diophantus makes.

Maximus Planudes comments on this determination as follows:

> The determination of the 21st (problem) is, in this (particular) example, the following. The homonymous number with the part of the first number, which is the third, is 3. This, when multiplied by the excess of the middle over the least, that is by 2 Numbers less 10 units, gives more Numbers than (those) in the middle. Indeed, 6 Numbers less 30 units are more than 3 Numbers. And it seems to me that the above cannot happen otherwise.
>
> (Diophantus 1893–95, II, 182.15–21)

First solution

The first two assignments are made to satisfy the third condition. The assignment for x then follows from the second condition, and the equation is formed based on the first condition.

Assignments	Operations	Equation
$z := $ 1N 10u		
$y := $ 3N		
	Subtract 1N 10u from 3N →	
	2N ℓ 10u	

$\frac{1}{3}x := 2\text{N} \,\ell\, 10\text{u}$

	Three times 2N ℓ 10u $\rightarrow$ 6N ℓ 30u	
$x := 6\text{N} \,\ell\, 30\text{u}$		
	Subtract 3N from 6N ℓ 30u $\rightarrow$ 3N ℓ 30u	
	Three times 3N ℓ 30u $\rightarrow$ 9N ℓ 90u	
		9N ℓ 90u = 1N 10u

Thus, 1N is $12\frac{1}{2}$u , so the sought-after numbers are $x = 45$, $y = 37\frac{1}{2}$, and $z = 22\frac{1}{2}$.

Second solution

Here the first two assignments remain the same, but the assignment for x now comes from the first condition. The equation is set up from the second condition.

Assignments	Operations	Equation
$z := 1\text{N} \, 10\text{u}$		
$y := 3\text{N}$		
	A third of 1N 10u $\rightarrow \frac{1}{3}$N $3\frac{1}{3}$u	
	Add 3N and $\frac{1}{3}$N $3\frac{1}{3}$u $\rightarrow 3\frac{1}{3}$N $3\frac{1}{3}$u	
$x := 3\frac{1}{3}\text{N}\, 3\frac{1}{3}\text{u}$		
	A third of $3\frac{1}{3}$N $3\frac{1}{3}$u $\rightarrow 1\frac{1}{9}$N $1\frac{1}{9}$u	
	Add 1N 10u and $1\frac{1}{9}$N $1\frac{1}{9}$u $\rightarrow 2\frac{1}{9}$N $11\frac{1}{9}$u	
		$2\frac{1}{9}$N $11\frac{1}{9}$u = 3N

Thus, 1N is again $12\frac{1}{2}$u, so the sought-after numbers are $x = 45$, $y = 37\frac{1}{2}$, and $z = 22\frac{1}{2}$.

Tannery regards the second solution to be an ancient interpolation. The enunciation is restated, and the determination is entirely reformulated in a way that ultimately agrees with the first determination. To decipher the determination, "the given part of the greatest" is in our notation $\frac{1}{n}x$, and we should consider it in the form of its algebraic assignment in the solution. The multitude of Numbers in the assignment of x is $p + \frac{1}{m}$, so the multitude of Numbers in the "given part of the greatest" is $\frac{1}{n}\left(p + \frac{1}{m}\right)$. We add to this 1, the assigned multitude of Numbers of the least, z. This must be less than p, "the ones of the middle (y) that were assumed in the beginning." Altogether, $\frac{1}{n}\left(p + \frac{1}{m}\right) + 1 < p$, or again $p + \frac{1}{m} + n < np$, which gives $np > n + p$. And again, this way of formulating the determination derives from the particular solution that follows. The only real difference in the solutions is that in the first the assignment for *x* is derived from the second condition and the equation is set up from the first condition, while in the second solution the two are reversed. The solution given in al-Karajī's *al-Fakhrī*, his Problem III.32, follows the first solution.

This is how Planudes comments the determination of the alternative solution:

> And here I think that the determination is not possible to be other that the one he states. And it arose as follows: the third of the greatest and the least is 2 Numbers and a ninth of a Number and 11 units and a ninth of a unit, which are smaller than the middle, that is, the 3 Numbers. Indeed, 2 and $\frac{1}{9}$ Numbers are smaller than 3 Numbers.

> (Diophantus 1893–95, II, 183.8–13)

I.22 $x - \frac{1}{m}x + \frac{1}{p}z = y - \frac{1}{n}y + \frac{1}{m}x = z - \frac{1}{p}z + \frac{1}{n}y \ (m = 3, \ n = 4, \ p = 5)$

Assignments	Operations	Equation
$x := 3N$		
$y := 4u$		
	A fourth of 4u → 1u	
	A third of 3N → 1N	
	4u decreased by 1u and increased by 1N → 1N 3u	
	Subtract 1N from 3N → 2N	
	Subtract 2N from 1N 3u → 3u ℓ 1N	
$\frac{1}{5}z := 3u \ \ell \ 1N$		
	Five times 3u ℓ 1N → 15u ℓ 5N	
$z := 15u \ \ell \ 5N$		
	Subtract 3u ℓ 1N from 15u ℓ 5N → 12u ℓ 4N	
	Add 1u to 12u ℓ 4N → 13u ℓ 4N	
		13u ℓ 4N = 1N 3u

Thus, 1N is 2u, so the sought-after numbers are $x = 6$, $y = 4$, and $z = 5$.

This problem is an abstract version of the common problem type of two or more men who each gives a fraction of his money to one or more of the others. One example is Problem I.25 of al-Karajī's *al-Fakhrī*, whose enunciation reads:

Two men meet, and each of them has some (money). One of them says to the other, if you give me a third of what you have, and I give you a fourth of what I have, then your money and my money will be equal.

(Saidan 1986, 175.15)

Problems of this kind are indeterminate, which is why Diophantus can assign y to be four units and al-Karajī can make the money of the second "three dirhams". Al-Karajī then solves purely arithmetical variations on this problem in Problems I.27 and I.34. The enunciation to I.27 reads: "Two different quantities: subtracting from the first its fifth and adding it to the second, and subtracting from the second its ninth and adding it to the first, makes them equal" (Saidan 1986, 176.7).

I.23
$$w - \tfrac{1}{m}w + \tfrac{1}{q}z = x - \tfrac{1}{n}x + \tfrac{1}{m}w = y - \tfrac{1}{p}y + \tfrac{1}{n}x = z - \tfrac{1}{q}z + \tfrac{1}{p}y$$
$$(m = 3,\ n = 4,\ p = 5,\ q = 6)$$

Assignments	Operations	Equation
$w := 3N$		
$x := 4u$		
	A fourth of 4u → 1u	
	A third of 3N → 1N	
	4u decreased by 1u and increased by 1N → 1N 3u	
	Subtract 1N from 3N → 2N	
	Subtract 2N from 1N 3u → 3u ℓ 1N	
$\tfrac{1}{6}z := 3u\ \ell\ 1N$		
	6 times 3u ℓ 1N → 18u ℓ 6N	
$z := 18u\ \ell\ 6N$		
	Subtract 3u ℓ 1N from 18u ℓ 6N → 15u ℓ 5N	
	Subtract this from 1N 3u → 6N ℓ 12u	
$\tfrac{1}{5}y := 6N\ \ell\ 12u$		
	Five times 6N ℓ 12u → 30N ℓ 60u	
$y := 30N\ \ell\ 60u$		
	Subtract 6N ℓ 12u from 30N ℓ 60u → 24N ℓ 48u	
	Add 1u to it → 24N ℓ 47u	
		24N ℓ 47u = 1N 3u

The 1N is $\frac{50}{23}$ u; therefore, the sought-after numbers are $w = \frac{150}{23}$, $x = \frac{92}{23}$, $y = \frac{120}{23}$, and $z = \frac{114}{23}$, or in whole numbers, 150, 92, 120, and 114. This problem follows the same reasoning as the previous problem.

I.24 $x + \frac{1}{m}(y + z) = y + \frac{1}{n}(z + x) = z + \frac{1}{p}(x + y)$ $(m = 3, n = 4, p = 5)$

Assignments	Operations	Equation
$x \coloneqq 1\text{N}$		
$y + z \coloneqq 3\text{u}$		
	Add 1N and 3u $\rightarrow$ 1N 3u	
$x + y + z \coloneqq 1\text{N } 3\text{u}$		
	A third of 3u $\rightarrow$ 1u	
$x + \frac{1}{3}(y + z) \coloneqq 1\text{N } 1\text{u}$		
	4 times 1N 1u $\rightarrow$ 4N 4u	
$4y + (z + x) \coloneqq 4\text{N } 4\text{u}$		
	Subtract 1N 3u from 4N 4u $\rightarrow$ 3N 1u	
$3y \coloneqq 3\text{N } 1\text{u}$		
	A third of 3N 1u $\rightarrow$ 1N $\frac{1}{3}$ u	
$y \coloneqq 1\text{N } \frac{1}{3}\text{u}$		
$z + \frac{1}{5}(x + y) \coloneqq 1\text{N } 1\text{u}$		
	Five times 1N 1u $\rightarrow$ 5N 5u	
$5z + (x + y) \coloneqq 5\text{N } 5\text{u}$		
	Subtract 1N 3u from 5N 5u $\rightarrow$ 4N 2u	
$4z \coloneqq 4\text{N } 2\text{u}$		
	A fourth of 4N 2u $\rightarrow$ 1N $\frac{1}{2}$ u	
$z \coloneqq 1\text{N } \frac{1}{2}\text{u}$		
	Add 1N, 1N $\frac{1}{3}$ u, 1N $\frac{1}{2}$ u $\rightarrow$ 3N $\frac{1}{3}\frac{1}{2}$ u	
		3N $\frac{1}{3}\frac{1}{2}$ u = 1N 3u

Thus, 1N is $\frac{13}{12}$ u, hence the three required numbers are $x = \frac{13}{12}$, $y = \frac{17}{12}$, and $z = \frac{19}{12}$, or in whole numbers, 13, 17, and 19.

This problem is an abstract version of the common problem type from Arabic and later problem solving in which two or more men want to buy a horse. Here is the enunciation of one such problem from the *Trattato d'abaco* of Piero della Francesca (15th c.):

> Three men want to buy a horse which is worth 30 ducats, and no one by himself can buy it. The first says to the second and the third, give me half of your (money), so I can buy the horse. The second says to the first and the third, give me 1/3 of your (money), so I can buy the horse. The third says to the first and the second, give me 1/4 of your (money), so I can buy the horse. I ask how much money each one has.
>
> (Piero della Francesca 1970, 100)

Piero's problem is determinate because he states the value of the horse. Other authors solve indeterminate versions like Diophantus, who makes his problem determinate when he assigns $y + z$ to be three units. This is adjusted in the end by a factor of 12 to eliminate fractions.

I.25 $w + \frac{1}{m}(x + y + z) = x + \frac{1}{n}(y + z + w) = y + \frac{1}{p}(z + w + x) = z + \frac{1}{q}(w + x + y)$

$(m = 3, n = 4, p = 5, q = 6)$

Assignments	Operations	Equation
$w := 1N$		
$x + y + z := 3u$		
	Add 1N and 3u $\rightarrow$ 1N 3u	
$w + x + y + z$ $:= 1N\ 3u$		
	A third of 3u $\rightarrow$ 1u	
$w + \frac{1}{3}(x + y + z)$ $:= 1N\ 1u$		
	4 times 1N 1u $\rightarrow$ 4N 4u	
$4x + (y + z + w)$ $:= 4N\ 4u$		
	Subtract 1N 3u from 4N 4u $\rightarrow$ 3N 1u	
$3x := 3N\ 1u$		
	A third of 3N 1u $\rightarrow$ 1N $\frac{1}{3}$u	
$x := 1N\ \frac{1}{3}u$		
$y + \frac{1}{5}(z + w + x)$ $:= 1N\ 1u$		
	Five times 1N 1u $\rightarrow$ 5N 5u	

$5y + (z + w + x) :=$ 5N 5u		
	Subtract 1N 3u from 5N 5u → 4N 2u	
$4y := 4N\ 2u$		
	A fourth of 4N 2u → 1N $\frac{1}{2}$u	
$y := 1N\ \frac{1}{2}u$		
$z + \frac{1}{6}(w + x + y)$ $:= 1N\ 1u$		
	Six times 1N 1u → 6N 6u	
$6z + (w + x + y)$ $:= 6N\ 6u$		
	Subtract 1N 3u from 6N 6u → 5N 3u	
$5z := 5N\ 3u$		
	A fifth of 5N 3u → 1N $\frac{3}{5}$u	
$z := 1N\ \frac{3}{5}u$		
	Add 1N, 1N $\frac{1}{3}$u, 1N $\frac{1}{2}$u, 1N $\frac{3}{5}$u → 4N $\frac{43}{30}$u	
		4N $\frac{43}{30}$ u = 1N 3u

The 1N is $\frac{47}{90}$u, hence, the required numbers are $w = \frac{47}{90}$, $x = \frac{77}{90}$, $y = \frac{92}{90}$, and $z = \frac{101}{90}$. The text gives the numbers 47, 77, 92, and 101. It is not clear if the denominator 90 is implied, or whether the numbers result after multiplying by 90. This problem is like I.24, a purely arithmetical version of a "four men want to buy a horse" problem.

I.26 $\begin{cases} a \cdot x = \square \\ b \cdot x = \sqrt{\square} \end{cases}$ $(a = 200, b = 5)$

Assignments	**Operations**	**Equation**
$x := 1N$		
	Multiply 1N by 200u → 200N Multiply 1N by 5u → 5N Square 5N → 25P	
		25P = 200N

Thus, 1N is 8u; therefore, the sought-after number is $x = 8$. Sesiano questions the authenticity of this problem in his commentary to Problem IV.20 (Sesiano 1982, 195–96). Al-Samaw'al solves this problem, so if it is not original it had already been added by the ninth century (see Section 3.3.9).

I.27 $\begin{cases} x + y = a \\ x \cdot y = b \end{cases}$ Determination: $\left(\frac{1}{2}a\right)^2 - b$ must be a square

$\qquad\qquad\qquad\quad (a = 20, b = 96)$

The determination is found by following Diophantus's solution. Subtracting 96u (b) from 100u $\left(\frac{1}{2}a\right)^2$ must leave a square.

The assignment $x - y \coloneqq 2N$ allows Diophantus to make x and y the sum and difference of 10u and 1N, respectively, which satisfies the first condition. Their product, 100u ℓ 1P, then consists only of units and Powers, which leads to a two-term equation.

Assignments	Operations	Equation
$x - y \coloneqq 2N$		
$x \coloneqq 1N\ 10u$		
$y \coloneqq 10u\ \ell\ 1N$		
	Multiply 1N 10u by 10u ℓ 1N $\rightarrow$ 100u ℓ 1P	
		100u ℓ 1P = 96u

Thus, 1N is 2u, so the two sought-after numbers are $x = 12$ and $y = 8$.

I.28 $\begin{cases} x + y = a \\ x^2 + y^2 = b \end{cases}$ Determination : $2b - a^2$ must be a square

$\qquad\qquad\qquad\quad (a = 20, b = 208)$

The determination follows from the solution: subtracting a^2, or double 200u, from $2b$, or double 208u, leaves 4P, which must be a square.

The assignments here are similar to those in the last problem. By assigning $x - y \coloneqq 2N$, Diophantus can again make x and y the sum and difference of 10u and 1N. This time, it is the sum of their squares that has only two terms, which again leads to a two-term equation.

Assignments	Operations	Equation
$x - y \coloneqq 2N$		
$x \coloneqq 1N\ 10\ u$		

$y := 10u \, \ell \, 1N$

	Add the squares of 1N 10u and 10u ℓ 1N $\rightarrow$ 2P 200u	
		2P 200u = 208u

Thus, 1N is 2u; therefore, the two sought-after numbers are $x = 12$ and $y = 8$.

I.29 $\begin{cases} x+y=a \\ x^2 - y^2 = b \end{cases}$ $\quad (a = 20, \, b = 80)$

The assignments here are the same as in Problem I.28. This makes the difference of the squares of x and y the single-term amount 40N, which yields a two-term equation.

Assignments	Operations	Equation
$x - y := 2N$		
$x := 1N \, 10u$		
$y := 10u \, \ell \, 1N$		
	The excess of the square of 1N 10u over the square of 10u ℓ 1N $\rightarrow$ 40N	
		40N – 80u

Thus, 1N is 2u; therefore, the two numbers are $x = 12$ and $y = 8$.

I.30 $\begin{cases} x - y = a \\ x \cdot y = b \end{cases}$ $\quad$ Determination : $4b + a^2$ must be a square $\quad (a = 4, \, b = 96)$

The determination is found from the equation, where the sum of b (96), and $\left(\frac{1}{2}a\right)^2$ (4) is equal to 1P, so it must be a square.

As in Problem I.27, $x + y$ is assigned to be 2N, so that x and y can be made the sum and difference of two terms, this time 1N and 2u. Their product consists of units and Powers, so the equation will have two terms.

Assignments	Operations	Equation
$x + y := 2N$		
$x := 1N \, 2u$		
$y := 1N \, \ell \, 2u$		

> Multiply 1N 2u by 1N ℓ 2u $\to$ 1P ℓ 4u
>
> 1P ℓ 4u = 96u

Thus, 1N is 10u, so, the sought-after numbers are $x = 12$ and $y = 8$.

I.31 $\begin{cases} x = my \\ x^2 + y^2 = n(x+y) \end{cases}$ $(m = 3,\ n = 5)$

Assignments	Operations	Equation
$y := 1N$		
	Three times 1N $\to$ 3N	
$x := 3N$		
	Add the squares of 1N and 3N $\to$ 10P	
	Add 1N and 3N $\to$ 4N	
	Five times 4N $\to$ 20N	
		20N = 10P

Thus, 1N is 2u, so the required numbers are $x = 6$ and $y = 2$.

I.32 $\begin{cases} x = my \\ x^2 + y^2 = n(x-y) \end{cases}$ $(m = 3,\ n = 10)$

Assignments	Operations	Equation
$y := 1N$		
	Three times 1N $\to$ 3N	
$x := 3N$		
	Add the squares of 1N and 3N $\to$ 10P	
	Subtract 1N from 3N $\to$ 2N	
	Ten times 2N $\to$ 20N	
		10P = 20N

Thus, 1N is 2u, hence, the two numbers are $x = 6$ and $y = 2$. We show the statement "10 Powers are the decuple of 2 Numbers" in the table as the equation "10P = 20N", though the way it is worded, without the adjective ἴσος and without having yet found the 20N, suggests that it is not yet the equation. But immediately after, Diophantus simplifies it as if it were already an equation. There are rare instances of cases like this in premodern algebra in other languages, and they make little difference in the solution.

I.33
$$\begin{cases} x = my \\ x^2 - y^2 = n(x+y) \end{cases} \quad (m = 3, \, n = 6)$$

Assignments	Operations	Equation
$y \coloneqq 1N$		
	Three times $1N \to 3N$	
$x \coloneqq 3N$		
	The excess of the square of 3N over the square of 1N $\to$ 8P	
	Add 1N and 3N $\to$ 4N	
	Six times 4N $\to$ 24N	
		$24N = 8P$

Thus, 1N is 3u, so the sought-after numbers are $x = 9$ and $y = 3$.

I.34
$$\begin{cases} x = my \\ x^2 - y^2 = n(x-y) \end{cases} \quad (m = 3, \, n = 12)$$

Assignments	Operations	Equation
$y \coloneqq 1N$		
	Three times $1N \to 3N$	
$x \coloneqq 3N$		
	The excess of the square of 3N over the square of 1N $\to$ 8P	
	The excess of 3N over 1N $\to$ 2N	
	Twelve times 2N $\to$ 24N	
		$24N = 8P$

Thus, 1N is 3u , so the sought-after numbers are $x = 9$ and $y = 3$.

Diophantus then states two problems as corollaries, which can be solved by the same procedure:

I.34.1
$$\begin{cases} x = my \\ x \cdot y = n(x+y) \end{cases}$$

I.34.2 $\begin{cases} x = my \\ x \cdot y = n(x - y) \end{cases}$

Maximus Planudes has the following comment concerning these corollaries:

> What is said in the corollary is this: Let the greater be triple the smaller, and their product be the double-sesquiquartan ($2\frac{1}{4}$) of the sum of the two (numbers). Let the smaller be assigned to be 1 Number. The greater, therefore, will be 3 Numbers. Then, I want their product to be the double-sesquiquartan of the sum of the two. But their product is 3 Powers, while the sum of the two is 4 Numbers. Therefore, twice 4 Numbers and their fourth will be equal to 3 Powers. Thus, 9 Numbers are equal to 3 Powers. Therefore, 1 Power is 3 Numbers, and the Number, 3 units. The greater will be 9, the smaller 3, the sum of the two 12, their product 27, and 27 is the double-sesquiquartan of 12. Again, let the greater be triple the smaller, and their product be the quadruple-sesquialter ($4\frac{1}{2}$) of their difference; and it results in the same with 9 and 3. This is why, I believe, Diophantus did not examine these, because they cannot be demonstrated with multiple ratios, as the preceding, but only with multiple-superparticulars.

> (Diophantus 1893–95, II, 204.17–205.10)

I.35 $\begin{cases} x = my \\ y^2 = nx \end{cases}$ $(m = 3, n = 6)$

Assignments	Operations	Equation
$y := 1N$		
	Three times $1N \to 3N$	
$x := 3N$		
	Square $1N \to 1P$	
	Six times $3N \to 18N$	
		$18N = 1P$

Thus, $1N$ is $18u$, so the required numbers are $x = 54$ and $y = 18$.

I.36 $\begin{cases} x = my \\ y^2 = ny \end{cases}$ $(m = 3, n = 6)$

Assignments	Operations	Equation
$y := 1N$		
	Three times $1N \to 3N$	

$x := 3N$

Square $1N \rightarrow 1P$	
Six times $1N \rightarrow 6N$	
	$6N = 1P$

Thus, $1N$ is 6u, and the sought-after numbers are $x = 18$ and $y = 6$.

I.37 $\begin{cases} x = my \\ y^2 = n(x+y) \end{cases}$ $\quad (m = 3, n = 2)$

Assignments	Operations	Equation
$y := 1N$		
	Three times $1N \rightarrow 3N$	
$x := 3N$		
	Square $1N \rightarrow 1P$	
	Add $1N$ and $3N \rightarrow 4N$	
	Two times $4N \rightarrow 8N$	
		$8N = 1P$

Thus, $1N$ is 8u, so the sought-after numbers are $x = 24$ and $y = 8$.

I.38 $\begin{cases} x = my \\ y^2 = n(x-y) \end{cases}$ $\quad (m = 3, n = 6)$

Assignments	Operations	Equation
$y := 1N$		
	Three times $1N \rightarrow 3N$	
$x := 3N$		
	Square $1N \rightarrow 1P$	
	The excess of $3N$ over $1N \rightarrow 2N$	
	Six times $2N \rightarrow 12N$	
		$12N = 1P$

Thus, $1N$ is 12u; therefore, the sought-after numbers are $x = 36$ and $y = 12$.

Diophantus then states four problems as corollaries which can be solved by the same procedure:

I.38.1 $\begin{cases} x = my \\ x^2 = ny \end{cases}$

I.38.2 $\begin{cases} x = my \\ x^2 = nx \end{cases}$

I.38.3 $\begin{cases} x = my \\ x^2 = n(x+y) \end{cases}$

I.38.4 $\begin{cases} x = my \\ x^2 = n(x-y) \end{cases}$

Here is the commentary of Planudes:

> What is said in the corollary (Τὰ τοῦ πορίσματος) is this: According to the first proposition the greater becomes 6 units, the smaller is 2 units, and they are in a triple ratio; while the square of 6, which is 36, is eighteen times the 2. According to the second, the greater is again 6, the smaller is 2, in a triple ratio, while the square of 6, the 36, is six times this 6. According to the third, the greater is 12, the smaller is 4, in a triple ratio, while the square of 12, which is 144, is nine times the two together, that is 16. According to the fourth, the greater is 6, the smaller is 2, in a triple ratio; while the square of 6, which is 36, is nine times their excess, that is 4.
>
> (Diophantus 1893–95, II, 207.13–27)

I.39   The three numbers $(a + b)x$, $(b + x)a$, and $(x + a)b$, arranged in order, have equal differences
($a = 3$, $b = 5$)

Diophantus assigns the sought-after number x to be 1N. Therefore, given that $a = 3$ and $b = 5$, the numbers $(a + b)x$, $(a + x)b$, and $(b + x)a$ are given the algebraic names 8N, 5N 15u, and 3N 15u, respectively. Now, there are three possibilities regarding the ordering of these three numbers: (1) 5N 15u > 8N > 3N 15u, (2) 5N 15u > 3N 15u > 8N, or (3) 8N > 5N 15u > 3N 15u. For cases (1) and (3) Diophantus forms his equation from the fact that for any three numbers in equal difference, the sum of the greatest and the least is twice the middle. For case (2) he simply equates the differences:

1   The sum of 5N 15u and 3N 15u is 8N 30u, and twice 8N is 16N, hence 8N 30u = 16N; thus 1N is $\frac{15}{4}$ u.
2   The excess of 5N 15u over 3N 15u is 2N, and that of 3N 15u over 8N is 15u ℓ 5N, hence, 15u ℓ 5N = 2N, and so 1N is $\frac{15}{7}$ u.
3   The sum of 8N and 3N 15u is 11N 15u, and twice the 5N 15u is 10N 30u, hence 11N 15u = 10N 30u ; therefore 1N is 15u.

Book II

According to Tannery, Problems II.1–7 are later additions to the text of the *Arithmetica*. He observes that these determinate problems of the second degree with two unknowns constitute a class of problems already treated in the first book, and to this class

> doivent être rattachées les sept premières questions du livre II, qui y sont évidemment interpolées, ce livre ayant son commencement naturel à la question 8, c'est-à-dire au problème fondamental de l'analyse indéterminée du second degré:
>
> $$x^2 + y^2 = a^2$$
>
> Les questions II, 1 à 5 sont notamment des répétitions des questions I, 34 à 37,[1] et en particulier II, 3 donne la solution de deux énoncés contenus dans I, 33, mais non traités alors. Toutefois, dans l'énoncé de ces cinq premières questions du livre II, une des conditions a été supprimée à tort, en sorte qu'à première vue, les problèmes apparaissent comme indéterminés.
>
> (Tannery 1912b, 370; cf. Diophantus 1893–95, I, 83 n. 1)

Heath (1910, 143–44) and Ver Eecke (Diophantus 1959, xxvi) both agree. The fact that Problems II.1–7 are missing from the problems of the *Arithmetica* borrowed or adapted by al-Karajī gives additional weight to Tannery's view. On the other hand, the reason Rashed and Houzel find for supporting the authenticity of at least Problems II.1–5, by resorting to modern algebraic geometry to distinguish these problems from their counterparts of Book I, has no historical value (Rashed and Houzel 2013, 179–84).

II.1 $x + y = m\left(x^2 + y^2\right)$ $(m = \frac{1}{10})$

---

1 The numbering of problems refers to the edition of 1621 by Bachet. In Tannery's edition the problems are I.31, I.34, the corollaries of I.34, and Problems I.32 and I.33, in order.

DOI: 10.4324/9781315171470-20

Assignments	Operations	Equation
$x := 1N$		
$y := 2N$		
	Add 1N and 2N $\rightarrow$ 3N	
	Add the squares of 1N and 2N $\rightarrow$ 5P	
	Ten times 3N $\rightarrow$ 30N	
		$30N = 5P$

1N is 6u, so the two sought-after numbers are $x = 6$ and $y = 12$.

II.2 $x - y = m(x^2 - y^2)$ $(m = \frac{1}{6})$

Assignments	Operations	Equation
$y := 1N$		
$x := 2N$		
	Subtract 1N from 2N $\rightarrow$ 1N	
	Subtract the square of 1N from the square of 2N $\rightarrow$ 3P	
	Six times 1N $\rightarrow$ 6N	
		$6N = 3P$

Thus, 1N is 2u, and the sought-after numbers are $y = 2$ and $x = 4$.

II.3   (a) $x \cdot y = m(x + y)$ $(m = 6)$

Assignments	Operations	Equation
$x := 1N$		
$y := 2N$		
	Multiply 1N by 2N $\rightarrow$ 2P	
	Add 1N and 2N $\rightarrow$ 3N	
	Six times 3N $\rightarrow$ 18N	
		$18N = 2P$

Thus, 1N is 9u, and the two numbers are $x = 9$ and $y = 18$.

(b) $x \cdot y = m(x - y)$ $(m = 6)$

Assignments	Operations	Equation
$x := 2N$		
$y := 1N$		
	Multiply 1N by 2N $\rightarrow$ 2P	
	Subtract 1N from 2N $\rightarrow$ 1N	
	Six times 1N $\rightarrow$ 6N	
		$6N = 2P$

Thus, 1N is 3u, and the two numbers are $y = 3$ and $x = 6$.

II.4 $\quad x^2 + y^2 = m(x - y) \qquad (m = 10)$

Assignments	Operations	Equation
$y := 1N$		
$x := 2N$		
	Add the squares of 1N and 2N $\rightarrow$ 5P	
	Subtract 1N from 2N $\rightarrow$ 1N	
	Ten times 1N $\rightarrow$ 10N	
		$5P = 10N$

Thus, 1N is 2u, so, the two numbers are $y = 2$ and $x = 4$.

II.5 $\quad x^2 - y^2 = m(x + y) \qquad (m = 6)$

Assignments	Operations	Equation
$y := 1N$		
$x := 2N$		
	Subtract the square of 1N from the square of 2N $\rightarrow$ 3P	
	Add 1N and 2N $\rightarrow$ 3N	
	Six times 3N $\rightarrow$ 18N	
		$3P = 18N$

Thus, 1N is 6u, and the two numbers are $y = 6$ and $x = 12$.

After the fifth problem Planudes makes the following comment:

> The problems from the first up to the fifth included, seem to be the same with previous ones, namely,
>
> | the first | with the thirty-first ⟨of the first⟩ book, |
> |---|---|
> | the second | with the thirty-fourth, |

the third, being double,   with the twenty-seventh and the thirtieth,
the fourth   with the thirty-second,
and moreover the fifth   with the thirty-third.

But, compared with the latter, they are imperfect. For that which was prescribed in the latter is (prescribed) in these (problems) too, and, in addition,
the ratio of the sought-after numbers with each other. But in these (problems),
this is not prescribed at all. So, according to those (problems), the same is
manifest (here too). And so, he assigns in these (problems), as in the previous ones, the first to be 1 Number and the other to be 2 Numbers, without
discrimination. Indeed, it makes no difference if we assign either to be any
(multitude of) Numbers, provided that the one is greater than the other. For,
again, the problem is done.

(Diophantus 1893–95, II, 210.18–211.13)

The fact that Planudes makes Problem II.3 correspond with Problems I.27 and
I.30 must be attributed to the defective state of his manuscript of reference. Problem II.3 corresponds to corollaries 1 and 2 of Problem I.34.

II.6 $\begin{cases} x - y = a \\ x^2 - y^2 = (x-y) + b \end{cases}$ Determination : $a^2 < a + b$
$(a = 2, b = 20)$

The determination follows from the fact that the difference of two squares
(this difference being here $a + b$) is always greater than the square of the difference (a^2).

Assignments	Operations	Equation
$y := 1N$		
$x := 1N\ 2u$		
	The excess of 1N 2u over 1N $\rightarrow$ 2u	
	The excess of the square of 1N over the square of 1N 2u $\rightarrow$ 4N 4u	
	Add 2u and 20u $\rightarrow$ 22u	
		$4N\ 4u = 22u$

Thus, 1N is $4\frac{1}{2}$u, so the two sought-after numbers are $y = 4\frac{1}{2}$ and $x = 6\frac{1}{2}$.

II.7 $x^2 - y^2 = m(x - y) + a$ $(m = 3, a = 10)$
Determination: $(x - y)^2 < m(x - y) + a$

The determination follows from the fact that $(x - y)^2$ is always less than $x^2 - y^2$. This determination exhibits two irregularities: it is presented after the instantiation instead of before it, and it involves sought-after numbers. No other problem in the extant *Arithmetica* shows the first irregularity, and only Problem I.14 shows the second irregularity. Problem II.7 is clearly a follow-up to Problem II.6, and the irregularities can be eliminated by modeling it on II.6 and representing the enunciation like this:

$$\begin{cases} x - y = a \\ x^2 - y^2 = m(x-y) + b \end{cases} \quad \text{Determination: } a^2 < ma + b \quad (m = 3,\ a = 2,\ b = 10)$$

Our second condition can be reformulated as $x^2 - y^2 = ma + b$, so the determination follows by the same reasoning as in the previous problem.

Assignments	Operations	Equation
$x - y := 2u$		
$y := 1N$		
$x := 1N\ 2u$		
	The excess of the square of 1N 2u over the square of 1N $\rightarrow$ 4N 4u	
	Three times 2u increased by 10u $\rightarrow$ 16u	
		$16u = 4N\ 4u$

Thus, 1N is 3u; therefore, the two sought-after numbers are $y = 3$ and $x = 5$.

II.8 $x^2 + y^2 = a^2$ $(a = 16)$

First solution

Assignments	Operations	Equation
$x^2 := 1P$		
	Subtract 1P from 16u $\rightarrow$ 16u ℓ 1P	
$y^2 := 16u\ \ell\ 1P$		
		$16u\ \ell\ 1P = \square$
$\sqrt{\square} := 2N\ \ell\ 4u$		
	Square 2N ℓ 4u $\rightarrow$ 4P 16u ℓ 16N	
		$4P\ 16u\ \ell\ 16N = 16u\ \ell\ 1P$

Thus, 1N is $\frac{16}{5}$ u, hence the two squares are $x^2 = \frac{256}{25}$ and $y^2 = \frac{144}{25}$.

Second solution

Assignments	Operations	Equation
$x := 1\text{N}$		
$y := 2\text{N} \; \ell \; 4\text{u}$		
	Square 1N → 1P	
	Square 2N ℓ 4u → 4P 16u ℓ 16N	
	Add this to 1P → 5P 16u ℓ 16N	
		5P 16u ℓ 16N = 16u

Again, 1N is $\frac{16}{5}$ u, and the two squares are $x^2 = \frac{256}{25}$ and $y^2 = \frac{144}{25}$.

In the first solution three assignments take place. The first is direct, the second derivative, and the third is an assignment we find for the first time in the *Arithmetica*, which we call after the name by which Arabic algebraists designated it: *al-istiqrā'*. The 4u is chosen in this assignment so that the 16u on the two sides of the resulting equation will cancel to yield a two-term equation. The second solution is worked out without an *al-istiqrā'* assignment.

II.9 $x^2 + y^2 = a^2 + b^2$ ($a^2 = 4$, $b^2 = 9$)

Assignments	Operations	Equation
$x := 1\text{N} \; 2\text{u}$		
$y := 2\text{N} \; \ell \; 3\text{u}$		
	Square 1N 2u → 1P 4N 4u	
	Square 2N ℓ 3u → 4P 9u ℓ 12N	
	Add this to 1P 4N 4u → 5P 13u ℓ 8N	
		5P 13u ℓ 8N = 13u

Thus, 1N is $\frac{8}{5}$ u, so the sought-after squares are $x^2 = \frac{324}{25}$ and $y^2 = \frac{1}{25}$.

II.10 $x^2 - y^2 = a$ ($a = 60$)

Assignments	Operations	Equation
$y := 1\text{N}$		
$x := 1\text{N} \; 3\text{u}$		
	Square 1N → 1P	
	Square 1N 3u → 1P 6N 9u	
	Subtract 1P from 1P 6N 9u → 6N 9u	
		6N 9u = 60u

Thus, 1N is $8\frac{1}{2}$ u, so $y = 8\frac{1}{2}$ and $x = 11\frac{1}{2}$, and the two squares are $y^2 = 72\frac{1}{4}$ and $x^2 = 132\frac{1}{4}$.

II.11 $\begin{cases} a + x = \Box \\ b + x = \Box' \end{cases}$ $(a = 2, b = 3)$

First solution (by the double-equality)

Assignments	Operations	Equation
$x := 1$N		
	Add 1N and 2u $\rightarrow$ 1N 2u	
		1N 2u $= \Box$
	Add 1N and 3u $\rightarrow$ 1N 3u	
		1N 3u $= \Box'$

The expressions 1N 2u and 1N 3u must be both squares. "This kind," writes Diophantus, "is called 'double-equality'". To solve it, we observe that the difference of 1N 3u and 1N 2u, which is 1u, is equal to the difference between $\Box'$ and $\Box$. We can write the latter as $r'^2 - r^2$, where r' and r are the roots of $\Box'$ and $\Box$, respectively. This difference factors into $(r' - r)(r' + r)$, and Diophantus chooses to make $r' - r = \frac{1}{4}$ and $r' + r = 4$. From here it is easy to see that r' is "half the sum" of $\frac{1}{4}$ and 4, or $\frac{17}{8}$, and r is "half the difference," or $\frac{15}{8}$. So $\Box' = \frac{289}{64}$ and $\Box = \frac{225}{64}$. He can then set up the equation 1N 2u $= \frac{225}{64}$ u, and find that 1N is $\frac{97}{64}$ u. The sought-after number is then $x = \frac{97}{64}$. The same result can be obtained by equating the square of half the sum of 4u and $\frac{1}{4}$ u, which is $\frac{289}{64}$ u, with 1N 3u.

Second solution (without the double-equality)

To avoid the double-equality, Diophantus completes a forthcoming equation by *al-istiqrā'*. The first assignment satisfies the first condition, and 1N is chosen in the *al-istiqrā'* assignment so that the 1Ps will cancel in the equation.

Assignments	Operations	Equation
$x := 1$P ℓ 2u		
	Add 1P ℓ 2u and 3u $\rightarrow$ 1P 1u	
		1P 1u $= \Box'$
$\sqrt{\Box'} := 1$N ℓ 4u		
	Square 1N ℓ 4u $\rightarrow$ 1P 16u ℓ 8N	
		1P 16u ℓ 8N = 1P 1u

Thus, 1N is $\frac{15}{8}$ u, so the sought-after number is $x = \frac{97}{64}$.

II.12 $\begin{cases} a - x = \Box \\ b - x = \Box' \end{cases}$ $(a = 9, b = 21)$

Assignments	Operations	Equation
$x := 9u \; \ell \; 1P$		
	Subtract 9u ℓ 1P from 21u → 1P 12u	
		1P 12u = $\Box'$
$\sqrt{\Box'} := 1N \, \ell \, 4u$		
	Square 1N ℓ 4u → 1P 16u ℓ 8N	
		1P 16u ℓ 8N = 1P 12u

Thus, 1N is $\frac{4}{8}$u, hence the sought-after number is $x = \frac{576}{64}$. This solution again makes use of an *al-istiqrā'* assignment to complete the forthcoming equation.

Problems II.11–13 constitute a group of similar problems. In each of Problems 11 and 13 two procedures are followed, one which makes use of an *al-istiqrā'* assignment, and the other which applies the technique of the double-equality. In Problem 12, however, only the former procedure is used. It is likely that a solution by the double-equality was also included for Problem 12 in the original text of Diophantus and which disappeared in the process of textual transmission. Such a solution is preserved in al-Karajī's *al-Fakhrī* (Problem III.41). Next we tabulate al-Karajī's solution (Saidan 1986, 235.22):

$\begin{cases} a - x = \Box \\ b - x = \Box' \end{cases}$ $(a = 5, b = 3)$

Assignments	Operations	Equation
$x := 1T$		
	Subtract 1T from 5u → 5u ℓ 1T	
		5u ℓ 1T = $\Box$
	Subtract 1T from 3u → 3u ℓ 1T	
		3u ℓ 1T = $\Box'$

The difference of the squares is 2u, which al-Karajī factors into 2u by 1u. By equating the square of half the sum of the factors with 5u ℓ 1T we obtain the equation 5u ℓ 1T = $2\frac{1}{4}$ u, so 1T is $2\frac{3}{4}$ u. Or, by equating the square of half the difference of the factors with 3u ℓ 1T, we obtain the equation 3u ℓ 1T = $\frac{1}{4}$u, so again 1T is $2\frac{3}{4}$ u. Therefore, the sought-after number is $x = 2\frac{3}{4}$.

II.13 $\begin{cases} x - a = \square \\ x - b = \square' \end{cases}$ $(a = 6, b = 7)$

First solution (by the double-equality)

Assignments	Operations	Equation
$x := 1N$		
	Subtract 6u from $1N \rightarrow 1N \ell 6u$	
		$1N \ell 6u = \square$
	Subtract 7u from $1N \rightarrow 1N \ell 7u$	
		$1N \ell 7u = \square'$

Thus, a double-equality arises. The difference of the squares is 1u which Diophantus factors into 2u by $\frac{1}{2}$u. By equating the square of half the difference of the two factors with $1N \ell 7u$ we obtain the equation $1N \ell 7u = \frac{9}{16}u$, so $1N$ is $\frac{121}{16}u$, which is the sought-after number x.

Second solution (without the double-equality)

Assignments	Operations	Equation
$x := 1P\, 6u$		
	Subtract 7u from $1P\, 6u \rightarrow 1P \ell 1u$	
		$1P \ell 1u = \square'$
$\sqrt{\square'} := 1N \ell 2u$		
	Square $1N \ell 2u \rightarrow 1P\, 4u \ell 4N$	
		$1P\, 4u \ell 4N = 1P \ell 1u$

Thus, $1N$ is $\frac{5}{4}u$, which makes the sought-after number $x = \frac{121}{16}$. This time the equation is completed with an *al-istiqrā'* assignment.

II.14 $\begin{cases} x + y = a, \\ z^2 + x = \square \\ z^2 + y = \square' \end{cases}$ $(a = 20)$

Diophantus begins by taking two numbers such that the sum of their squares is less than 20 units, which are 2u and 3u. The assignments for $\square$ and $\square'$ are made with their squares, and with the idea that z^2 will be assigned to be $1P$.

Assignments	Operations	Equation
	Square the sum of $1N$ and $2u \rightarrow 1P\, 4N\, 4u$	
$\square := 1P\, 4N\, 4u$		

	Square the sum of 1N and 3u → 1P 6N 9u
□′ ≔ 1P 6N 9u	
z² ≔ 1P	
	Subtract 1P from 1P 4N 4u → 4N 4u
x ≔ 4N 4u	
	Subtract 1P from 1P 6N 9u → 6N 9u
y ≔ 6N 9u	
	Add 4N 4u and 6N 9u → 10N 13u
	10N 13u = 20u

Thus, 1N is $\frac{7}{10}$u, so, the sought-after numbers are $x = \frac{68}{10}$ and $y = \frac{132}{10}$, and the square is $z^2 = \frac{49}{100}$.

Because Diophantus assigns the sides of the two squares to be 1N and (instead of "lacking") two numbers, the sum of the squares of these two numbers, which become the 13u in the equation, must be less than 20 to ensure that the value of 1N is positive. Tannery noted that 1N could be negative and still yield positive x and y, and he proposed the modified condition that the two numbers α and β (the 2 and 3 in Diophantus) need only satisfy $\alpha > \beta$ and $\beta > \alpha - \frac{20}{\alpha}$ (1912b, 387). Of course, Diophantus would not have stated this condition because he solves for 1N and not 1P or 4N 4u in his solution.

II.15 $\begin{cases} x + y = a, \\ z^2 - x = \square \\ z^2 - y = \square' \end{cases}$ $(a = 20)$

Assignments	Operations	Equation
z ≔ 1N 2u		
	Square 1N 2u → 1P 4N 4u	
x ≔ 4N 4u		
y ≔ 2N 3u		
z² ≔ 1P 4N 4u		
	Add 4N 4u and 2N 3u → 6N 7u	
		6N 7u = 20u

Thus, 1N is $\frac{13}{6}$u, so the two sought-after numbers are $x = \frac{76}{6}$ and $y = \frac{44}{6}$, and the square is $z^2 = \frac{625}{36}$.

II.16 $\begin{cases} x = my \\ y + a^2 = \square \\ x + a^2 = \square' \end{cases}$ ($m = 3$, $a^2 = 9$)

The first assignment is designed to satisfy the second condition, the second assignment covers the first condition, and the equation is formed from the third condition.

Assignments	Operations	Equation
	Square 1N 3u → 1P 6N 9u	
$y := $ 1P 6N		
	Three times 1P 6N → 3P 18N	
$x := $ 3P 18N		
	Add 9u to 3P 18N → 3P 18N 9u	
		3P 18N 9u = $\square'$
$\sqrt{\square'} := $ 2N ℓ 3u		
	Square 2N ℓ 3u → 4P 9u ℓ 12N	
		3P 18N 9u = 4P 9u ℓ 12N

Thus, 1N is 30u, hence, the two sought-after numbers are $y = 1080$ and $x = 3240$.

II.17 $x - \left(\frac{1}{m}x + a\right) + \left(\frac{1}{p}z + c\right) = y - \left(\frac{1}{n}y + b\right) + \left(\frac{1}{m}x + a\right)$

$= z - \left(\frac{1}{p}z + c\right) + \left(\frac{1}{n}y + b\right)$

($m = 5$, $a = 6$, $n = 6$, $b = 7$, $p = 7$, $c = 8$)

This is a variation on the problem of two or more men who have some money. See Problem I.22.

First solution

This problem is indeterminate. It becomes determinate with the second assignment, which fixes the ratio of x to y.

Assignments	Operations	Equation
$x := $ 5N		
$y := $ 6N		

A fifth of 5N increased by 6u → 1N 6u

Add this to 6N → 7N 6u

A sixth of 6N increased by 7u → 1N 7u

Subtract this from 7N 6u → 6N ℓ 1u

Subtract 1N 6u from 5N → 4N ℓ 6u

Subtract this from 6N ℓ 1u → 2N 5u

Subtract 8u from this → 2N ℓ 3u

$\frac{1}{7}z := 2N\,\ell\,3u$

Seven times 2N ℓ 3u → 14N ℓ 21u

$z := 14N\,\ell\,21u$

Add 8u to 2N ℓ 3u → 2N 5u

Subtract it from 14N ℓ 21u →
12N ℓ 26u

Add to this 1N 7u → 13N ℓ 19u

13N ℓ 19u =
6N ℓ 1u

Thus, 1N is found to be $\frac{18}{7}$u; hence, the three sought-after numbers are $x = \frac{90}{7}$, $y = \frac{108}{7}$, and $z = \frac{105}{7}$.

Second solution[2]

Here the problem becomes determinate by assigning y to be a particular number.

Assignments	Operations	Equation
$x := 5N$		
$y := 12u$		
	Add a fifth of 5N and 6u → 1N 6u	
	Add 12u to this → 1N 18u	
	Add a sixth of 12u and 7u → 9u	
	Subtract this from 1N 18u → 1N 9u	
	Subtract 1N 6u from 5N → 4N ℓ 6u	
	Subtract this from 1N 9u → 15u ℓ 3N	
	Subtract 8u from this → 7u ℓ 3N	
$\frac{1}{7}z := 7u\,\ell\,3N$		

2 This solution is given after the enunciation of Problem II.18, whose solution is missing.

7 times 7u ℓ 3N $\rightarrow$ 49u ℓ 21N

$z := 49$u ℓ 21N

Add 9u to 49u ℓ 21N $\rightarrow$ 58u ℓ 21N

Subtract 15u ℓ 3N $\rightarrow$ 43u ℓ 18N

43u ℓ 18N =
1N 9u

Thus, 1N is $\frac{34}{19}$u, so the three sought-after numbers are $x = \frac{170}{19}$, $y = \frac{228}{19}$, and $z = \frac{217}{19}$.

II.18
$$\begin{cases} x + y + z = d \\ x - \left(\frac{1}{m}x + a\right) + \left(\frac{1}{p}z + c\right) = y - \left(\frac{1}{n}y + b\right) + \left(\frac{1}{m}x + a\right) \\ = z - \left(\frac{1}{p}z + c\right) + \left(\frac{1}{n}y + b\right) \end{cases}$$
$(d = 80, m = 5, a = 6, n = 6, b = 7, p = 7, c = 8)$

This is the same as the previous problem, with the addition of the first condition, which makes it determinate. The solution of this problem is missing. However, a version is preserved as Problem IV.40 in al-Karajī's *al-Fakhrī* (Saidan 1986, 264.12–266.9). Al-Karajī gives this problem right before his version of *Arithmetica* II.19, so we can be certain that he took it from Diophantus. Problems of this type are common in Arabic arithmetic books and are usually posed in terms of two or more men giving and taking money from each other. Al Karajī took some of his wording from this tradition, most notably where he writes, "So make the money (*māl*) of the first a Thing". If the problem were stated abstractly, then this assignment would have been written as "So make the first a Thing". But other parts are stated abstractly, like the enunciation, where the 50 dirhams are divided into three "parts" and not among three men. (The word "dirham" could mean either the denomination of silver coin or abstract units.) The problem runs as follows:

> If (someone) said, divide fifty dirhams into three parts such that if the first gives to the second its third and two dirhams, and the second gives to the third its fourth and three dirhams, and the third gives to the first its fifth and four dirhams, then they are equal after the taking and giving.
>
> So make the money (*māl*) of the first a Thing, and the money of the second eight dirhams. Then the second will have, if he gives its fourth and three dirhams and takes from the first its third and two dirhams, five dirhams and a third of a Thing.
>
> What remains with the first is two thirds of a Thing less two dirhams, and it is necessary that what the first has, by taking a fifth of the third and four dirhams, and also giving a third of what he has and two dirhams, is five dirhams and a

third of a Thing. But if you cast away what the first was left with from five dirhams and a third of a Thing, it leaves seven dirhams less a third of a Thing. This is necessarily a fifth of what the third has and four dirhams. So cast away four dirhams from it, leaving three dirhams less a third of a Thing. This is a fifth of what the third has. So the third will have, in total, fifteen dirhams less a Thing and two thirds of a Thing. He must give a fifth of what he has and four dirhams, leaving him with eight dirhams less a Thing and a third of a Thing. Adding to it what he takes from the second, which is five dirhams, gives thirteen dirhams less a Thing and a third of a Thing. That Equals five dirhams and a third of a Thing. So the one Thing Equals four dirhams and four fifths of a dirham. This is what the first has, and the second has eight dirhams, and the third has seven dirhams.

If you check this, you find that what each one will have, after the taking and giving, is six dirhams and three fifths of a dirham, and the total of what three have is nineteen dirhams and four fifths of a dirham. If this were fifty, then you would be right.

So start the problem again, and make what the first has a Thing, and what the second has twelve dirhams. So the second, if he gives a fourth of what he has and three dirhams, and he takes from the first a third of what he has and two dirhams, will have eight dirhams and a third of a Thing.

And the first is left with two thirds of a Thing less two dirhams. It should be that after this, the first, if he gives a third of what he has and two dirhams, and he takes from the third a fifth of what he has and four dirhams, will also be eight dirhams and a third of a Thing. So cast away what the first was left with from eight dirhams and a third of a Thing, leaving ten dirhams less a third of a Thing, and that is a fifth of the third and four dirhams. So if you cast away four dirhams from it, it leaves a fifth of the third, which is six dirhams less a third of a Thing. So the third has, in total, thirty dirhams less a Thing and two thirds of a Thing.

But the third, if he gives to the first its fifth and four dirhams, and takes from the second its fourth and three dirhams, will have twenty-six dirhams less a Thing and a third of a Thing, and that Equals eight and a third of a Thing. If you restore and confront, you find that the one Thing Equals ten and four fifths. This is the money of the first, and the money of the second is twelve dirhams, and the money of the third is also twelve dirhams.

If you check this, each one will have, after the taking and giving, eleven and three fifths, and they will have in total thirty-four dirhams and four fifths of a dirham.

The increase in the second is four dirhams, and the increase in the total number with the three is fifteen dirhams. So for each dirham, (the increase is to) three dirhams and a half and a fourth, and in place of fifteen, the increase should be thirty dirhams and a fifth, so that it gives fifty. Now, if for the one dirham, the increase is three and a half and a fourth, you see how many dirhams will increase to thirty and a fifth. So divide thirty and a fifth by three and a half and a fourth, resulting in eight and four parts of seventy-five parts of a unit.

So start the problem again, and make what the first has a Thing, and what the second has sixteen and four parts of seventy-five parts of a unit. And take from it its fourth and three dirhams, leaving it with nine and three parts of seventy-five parts of a unit. Add to it a third of the first and two dirhams, so it will be eleven dirhams and three parts of seventy-five parts of a unit and a third of a Thing. This is necessarily what each one of them has, after the taking and giving. So cast away from it what the first was left with, to get thirteen and three parts of seventy-five parts of a unit less a third of a Thing. This is a fifth of the third and four dirhams. And if we cast away four dirhams from it, and we multiply the remainder by five, the money of the third will be forty-five dirhams and fifteen parts of seventy-five parts of a unit less a Thing and two-thirds of a Thing. Cast away from it its fifth and four dirhams, leaving thirty-two dirhams and twelve parts of seventy-five parts of a unit less a Thing and a third of a Thing. Add to it what he took from the second, which is seven dirhams and a part of seventy-five parts of a unit, so he will have thirty-nine dirhams and thirteen parts of seventy-five parts of a unit ⟨less a Thing and a third of a Thing⟩. That Equals eleven dirhams and three parts of seventy-five parts of a unit and a third of a Thing. If you restore and confront and cast away common amounts, it leaves twenty-eight dirhams and ten parts of seventy-five parts, which Equals a Thing and two thirds of a Thing. So the one Thing Equals sixteen dirhams and sixty-six parts of seventy-five parts of a unit. This is the money of the first, and the money of the second is sixteen and four parts of seventy-five parts of a unit, and the money of the third is seventeen dirhams and five parts of seventy-five parts of a unit.

If you check (this), each one of them will have, after the taking and giving, sixteen dirhams and fifty parts of seventy-five parts of a unit, and the total money of the three is fifty dirhams, which is what was asked.

The problem can be summarized as:

$$
\begin{cases}
x + y + z = d \\
x - \left(\frac{1}{m}x + a\right) + \left(\frac{1}{p}z + c\right) = y - \left(\frac{1}{n}y + b\right) + \left(\frac{1}{m}x + a\right) \\
= z - \left(\frac{1}{p}z + c\right) + \left(\frac{1}{n}y + b\right)
\end{cases}
$$
$$(d = 50, \ m = 3, \ a = 2, \ n = 4, \ b = 3, \ p = 5, \ c = 4)$$

The solution unfolds in three stages. First, the first condition is disregarded and the indeterminate problem posed by the second condition is solved by making the arbitrary assignment $y := 8u$. The sum of x, y, and z in the solution is then $19\frac{4}{5}$, which is short of the required 50 by $30\frac{1}{5}$. The problem is then worked out again, this time making $y := 12u$, and the sum of x, y, and z in the solution is now $34\frac{4}{5}$. At this point Diophantus/al-Karajī is poised to apply a variation of double false position. Increasing y by $4u$ increases the sum of x, y, and z by 15, so increasing y by 1 increases the sum by $3\frac{3}{4}$. Thus, to increase the sum by $30\frac{1}{5}$, which would bring it to 50, he must make $y := 16\frac{4}{75}u$. The problem is then solved again the same way to get $x = 16\frac{66}{75}$, $y = 16\frac{4}{75}$, and $z = 17\frac{5}{75}$, which satisfy all the conditions of the problem.

The first stage of the solution is tabulated next. The other stages follow the same reasoning.

Assignments	Operations	Equation
$x \doteq 1\text{T}$		
$y \doteq 8\text{u}$		
	The number corresponding to $y - \left(\frac{1}{4}y + 3\right) + \left(\frac{1}{3}x + 2\right)$ after the operations is $5\text{u}\frac{1}{3}\text{T}$	
	The number corresponding to $x - \left(\frac{1}{3}x + 2\right)$ is $\frac{2}{3}\text{T}\,\ell\,2\text{u}$	
	Subtract $\frac{2}{3}\text{T}\,\ell\,2\text{u}$ from $5\frac{1}{3}\text{T} \rightarrow 7\text{u}\,\ell\,\frac{1}{3}\text{T}$.	
	This is the number corresponding to $\frac{1}{5}z + 4$	
	Subtract 4u from $7\text{u}\,\ell\,\frac{1}{3}\text{T} \rightarrow 3\text{u}\,\ell\,\frac{1}{3}\text{T}$.	
	Five times this $\rightarrow 15\text{u}\,\ell\,1\frac{2}{3}\text{T}$	
$z \doteq 15\text{u}\,\ell\,1\frac{2}{3}\text{T}$		
	Subtract from $15\text{u}\,\ell\,1\frac{2}{3}\text{T}$ its fifth and 4u $\rightarrow 8\text{u}\,\ell\,1\frac{1}{3}\text{T}$	
	Add 5u $\rightarrow 13\text{u}\,\ell\,1\frac{1}{3}\text{T}$	
		$13\text{u}\,\ell\,1\frac{1}{3}\text{T} = 5\text{u}\frac{1}{3}\text{T}$

Thus, 1T is $4\frac{4}{5}\text{u}$, so the three sought-after numbers are $x = 4\frac{4}{5}$, $y = 8$, and $z = 7$.

The answer to the problem as posed by Diophantus does not give nicer numbers: $x = 26\frac{2}{363}$, $y = 26\frac{348}{363}$, and $z = 27\frac{13}{363}$. Tannery (Diophantus 1893–95, I, 109 n. 1) considered that Problems II.17–18 belonged to an ancient commentary to Book I, from which they crept into the text of Diophantus.

II.19 $x^2 - y^2 = m(y^2 - z^2)$ $(m = 3)$

Assignments	Operations	Equation
$z^2 \doteq 1\text{P}$		
$y^2 \doteq 1\text{P 2N 1u}$		

	Subtract 1P from 1P 2N 1u → 2N 1u
	Three times 2N 1u → 6N 3u
	Add 6N 3u and 1P 2N 1u → 1P 8N 4u
$x^2 := $ 1P 8N 4u	

1P 8N 4u = □

$\sqrt{\square} := $ 1N 3u	
	Square 1N 3u → 1P 6N 9u

1P 8N 4u = 1P 6N 9u

Thus, 1N is $2\frac{1}{2}$u, hence, the three sought-after squares are $x^2 = 30\frac{1}{4}$, $z^2 = 6\frac{1}{4}$, and $y^2 = 12\frac{1}{4}$.

II.20 $\begin{cases} x^2 + y = \square \\ y^2 + x = \square' \end{cases}$

The first two assignments satisfy the first condition, and the second condition leads to the equation.

Assignments	Operations	Equation
$x := $ 1N		
$y - $ 1u 2N		
	Square 1u 2N → 4P 4N 1u	
	Add to this 1N → 4P 5N 1u	
		4P 5N 1u = □'
$\sqrt{\square'} := $ 2N ℓ 2u		
	Square 2N ℓ 2u → 4P 4u ℓ 8N	
		4P 5N 1u = 4P 4u ℓ 8N

Thus, 1N is $\frac{3}{13}$u; therefore, the two sought-after numbers are $x = \frac{3}{13}$ and $y = \frac{19}{13}$.

II.21 $\begin{cases} x^2 - y = \square \\ y^2 - x = \square' \end{cases}$

The assignment for y is direct, then the assignment for x is made with the implicit assignment of □' as 1P. The equation is formed from the first condition.

Assignments	Operations	Equation
$y := $ 1N 1u		
	Square 1N 1u → 1P 2N 1u	
	Subtract 1P from it → 2N 1u	
$x := $ 2N 1u		
	Square 2N 1u → 4P 4N 1u	
	Subtract 1N 1u from it → 4P 3N	
		4P 3N = □
$\sqrt{\square} := $ 3N		
	Square 3N → 9P	
		4P 3N = 9P

Thus, 1N is $\frac{3}{5}$ u; therefore, the two sought-after numbers are $y = \frac{8}{5}$ and $x = \frac{11}{5}$.

II.22 $\begin{cases} x^2 + (x+y) = \square \\ y^2 + (x+y) = \square' \end{cases}$

The first two assignments satisfy the second condition, and the equation is set up from the first condition.

Assignments	Operations	Equation
$y := $ 1N		
$x := $ 1N 1u		
	Square 1N 1u → 1P 2N 1u	
	Add to this the sum of 1N and 1N 1u→ 1P 4N 2u	
		1P 4N 2u = □
$\sqrt{\square} := $ 1N ℓ 2u		
	Square 1N ℓ 2u → 1P 4u ℓ 4N	
		1P 4N 2u = 1P 4u ℓ 4N

Thus, 1N is $\frac{2}{8}$ u, so, the two sought-after numbers are $y = \frac{2}{8}$ and $x = \frac{10}{8}$.

II.23 $\begin{cases} x^2 - (x+y) = \square \\ y^2 - (x+y) = \square' \end{cases}$

The first two assignments are designed to satisfy the first condition, then the equation is set up from the second condition. Note that in our notation "1P ℓ 2N 1u" means that everything to the right of the "ℓ" is lacking.

Assignments	Operations	Equation
$y := 1N$		
$x := 1N \ 1u$		
	Square $1N \rightarrow 1P$	
	Subtract from this $1N$ and $1N \ 1u \rightarrow$ $1P \ \ell \ 2N \ 1u$	
		$1P \ \ell \ 2N \ 1u = \square'$
$\sqrt{\square'} := 1N \ \ell \ 3u$		
	Square $1N \ \ell \ 3u \rightarrow 1P \ 9u \ \ell \ 6N$	
		$1P \ \ell \ 2N \ 1u = 1P$ $9u \ \ell \ 6N$

Thus, $1N$ is $2\frac{1}{2}u$, so the two sought-after numbers are $y = 2\frac{1}{2}$ and $x = 3\frac{1}{2}$.

II.24 $\begin{cases} (x+y)^2 + x = \square \\ (x+y)^2 + y = \square' \end{cases}$

The assignments are designed to satisfy the two conditions, and the equation is set up to make them agree for $x + y$.

Assignments	Operations	Equation
$x := 3P$		
$y := 8P$		
$(x+y)^2 := 1P$		
	Add $3P$ and $8P \rightarrow 11P$	
	Square this $\rightarrow 121PP$	
		$121PP = 1P$

The equation simplifies successively to $11P = 1N$, then to $11N = 1u$. Thus, $1N$ is $\frac{1}{11}u$, hence, the two numbers are $x = \frac{3}{121}$ and $y = \frac{8}{121}$.

II.25 $\begin{cases} (x+y)^2 - x = \square \\ (x+y)^2 - y = \square' \end{cases}$

The assignments are based on the arithmetical equalities $4^2 - 12 = 2^2$ and $4^2 - 7 = 3^2$, which are set "in terms of a Power", simulating the conditions of the problem, much like in the previous problem.

Assignments	Operations	Equation
$x := 12P$		
$y := 7P$		
$(x+y)^2 := 16P$		
	Square the sum of 12P and 7P $\rightarrow$ 361PP	
		$361PP = 16P$

The equation simplifies to $19P = 4N$, so 1N is $\frac{4}{19}$u, and the sought-after numbers are then $x = \frac{192}{361}$ and $y = \frac{112}{361}$.

$$\mathbf{II.26} \quad \begin{cases} x \cdot y + y = \square \\ x \cdot y + x = \square' \\ \sqrt{\square} + \sqrt{\square'} = a \end{cases} \quad (a = 6)$$

Diophantus begins by citing the identity that we write as $m(4m - 1) + m = (2m)^2$, which shares the structure of the first (and second) condition. This leads by simulation to the two assignments.

Assignments	Operations	Equation
$y := 1N$		
$x := 4N \, \ell \, 1u$		

Diophantus silently calculates $\sqrt{\square'}$ from the identity. He multiplies 4N ℓ 1u by 1N to get 4P ℓ 1N, then he adds 1N to get 4P, which is $\square$. Thus, $\sqrt{\square}$ is 2N, and subtracting this from 6u gives $\sqrt{\square'}$ as 6u ℓ 2N. The equation will then be set up from the second condition.

Assignments	Operations	Equation
	Add 4P ℓ 1N and 4N ℓ 1u $\rightarrow$ 4P 3N ℓ 1u	
	Square 6u ℓ 2N $\rightarrow$ 4P 36u ℓ 24N	
		4P 3N ℓ 1u = 4P 36u ℓ 24N

Thus, 1N is $\frac{37}{27}$u; therefore, the two sought-after numbers are $y = \frac{37}{27}$ and $x = \frac{121}{27}$.

$$\mathbf{II.27} \quad \begin{cases} x \cdot y - y = \square \\ x \cdot y - x = \square' \\ \sqrt{\square} + \sqrt{\square'} = a \end{cases} \quad (a = 5)$$

Diophantus begins by citing the identity $m(4m + 1) - m = (2m)^2$, which shares the structure of the first (and second) condition. This leads by simulation to the two assignments.

Assignments	Operations	Equation
$x := 4$N 1u		
$y := 1$N		

Like in the last problem, Diophantus silently calculates $\sqrt{\square}'$ from the identity. He multiplies 4N 1u by 1N to get 4P 1N, then he subtracts 1N to get 4P, which is $\square$. Thus, $\sqrt{\square}$ is 2N, and subtracting this from 5u gives $\sqrt{\square}'$ as 5u ℓ 2N. The equation is then set up from the second condition.

Assignments	Operations	Equation
	Subtract 4N 1u from 4P 1N → 4P ℓ 3N 1u	
	Square 5u ℓ 2N → 4P 25u ℓ 20N	
		4P ℓ 3N 1u =
		4P 25u ℓ 20N

Thus, 1N is $\frac{26}{17}$u, so the two sought-after numbers are $y = \frac{26}{17}$ and $x = \frac{121}{17}$.

II.28 $\begin{cases} x^2 \cdot y^2 + x^2 = \square \\ x^2 \cdot y^2 + y^2 = \square' \end{cases}$

Diophantus solves this problem in two stages. He begins by positing $y^2 := 1$P and $x^2 := 1$u. The latter is arbitrary for the first condition since the choice of x^2 will not affect the solution for y, which is found from the first condition by an *al-istiqrā'* argument to be $\frac{3}{4}$. The second condition then becomes $\frac{9}{16}x^2 + \frac{9}{16} = \square'$, so Diophantus makes the assignment $x^2 := 1$P, and another *al-istiqrā'* argument gives $x = \frac{7}{24}$.

Assignments	Operations	Equation
$y^2 := 1$P		
$x^2 := 1$u		
$x^2 \cdot y^2 := 1$P		
	Add 1P and 1u → 1P 1u	
		1P 1u = $\square$
$\sqrt{\square} := 1$N ℓ 2u		
	Square 1N ℓ 2u → 1P 4u ℓ 4N	
		1P 1u = 1P 4u ℓ 4N

Thus, 1N is $\frac{3}{4}$u, which is y.

Now the second condition becomes $\frac{9}{16}x^2 + \frac{9}{16} = \square'$, and a new solution for x begins:

Assignments	Operations	Equation
$x^2 := 1P$		
$\frac{9}{16}x^2 + \frac{9}{16} := \frac{9}{16}P\frac{9}{16}u$		
	Sixteen times $\frac{9}{16}P\frac{9}{16}u \rightarrow 9P\,9u$	
		$9P\,9u = \square''$
$\sqrt{\square''} := 3N\,\ell\,4u$		
	Square $3N\,\ell\,4u \rightarrow 9P\,16u\,\ell\,24N$	
		$9P\,9u = 9P$ $16u\,\ell\,24N$

Then 1N is $\frac{7}{24}$u, which is x. Therefore, the two sought-after squares are $y^2 = \frac{324}{576}$ and $x^2 = \frac{49}{576}$.

II.29 $\begin{cases} x^2 \cdot y^2 - x^2 = \square \\ x^2 \cdot y^2 - y^2 = \square' \end{cases}$

As in the previous problem, Diophantus begins by positing $y^2 := 1P$ and $x^2 := 1u$, the latter being arbitrary in the context of the first condition. From that condition he finds that $y^2 = \frac{25}{16}$. He does not explain how he got this, but it would have been by *al-istiqrā'*, equating $1P\,\ell\,1u$ to the square of $1N\,\ell\,2u$. The second condition then becomes $\frac{25}{16}x^2 - \frac{25}{16} = \square'$. Diophantus posits $x^2 := 1P$ and sets up a forthcoming equation, which he completes by *al-istiqrā'*.

Assignments	Operations	Equation
$y^2 := 1P$		
$x^2 := 1u$		
$x^2y^2 := 1P$		
	Subtract $1u$ from $1P \rightarrow 1P\,\ell\,1u$	
		$1P\,\ell\,1u = \square$
$\sqrt{\square} := 1N\,\ell\,2u$		
	Square $1N\,\ell\,2u \rightarrow 1P\,4u\,\ell\,4N$	
		$1P\,\ell\,1u = 1P\,4u\,\ell\,4N$

Thus, 1N is $\frac{5}{4}$u, so $y^2 = \frac{25}{16}$.

The second condition is now $\frac{25}{16}x^2 - \frac{25}{16} = \square'$, and a new solution for x begins:

Assignments	Operations	Equation
$x^2 := 1P$		
$\frac{25}{16}x^2 - \frac{25}{16} := \frac{25}{16}P\,\ell\,\frac{25}{16}u$		
	Sixteen times $\frac{25}{16}P\,\ell\,\frac{25}{16}u \to 25P\,\ell\,25u$	
	Take a 25th of this $\to 1P\,\ell\,1u$	
		$1P\,\ell\,1u = \square''$
$\sqrt{\square''} := 1N\,\ell\,4u$		
	Square $1N\,\ell\,4u \to 1P\,16u\,\ell\,8N$	
		$1P\,\ell\,1u = 1P$ $16u\,\ell\,8N$

Thus, 1N is $\frac{17}{8}$u, so the sought-after squares are $x^2 = \frac{289}{64}$ and $y^2 = \frac{100}{64}$.

II.30 $\begin{cases} x\cdot y + (x+y) = \square \\ x\cdot y - (x+y) = \square' \end{cases}$

Because the two conditions of the enunciation are of the form $m + n = \square$ and $m - n = \square'$, Diophantus begins by looking for two such determinate numbers m and n. He finds them by noting that for any numbers a and b, $(a^2 + b^2) + 2ab = \square$, and $(a^2 + b^2) - 2ab = \square'$. So he chooses a and b to be 2 and 3, from which he finds m and n to be 13 and 12. He can then assign $x \cdot y$ to be 13P and $x + y$ to be 12P. The factoring of 13P into 1N by 13N is arbitrary.

Assignments	Operations	Equation
$x \cdot y := 13P$		
$x := 1N$		
$y := 13N$		
$x + y := 12P$		
	Add 1N and 13N $\to 14N$	
		$14N = 12P$

Thus, 1N is $\frac{14}{12}$u $= \frac{7}{6}$u, so the sought-after numbers are $x = \frac{7}{6}$ and $y = \frac{91}{6}$.

II.31 $\begin{cases} x + y = \square \\ x\cdot y + (x+y) = \square' \\ x\cdot y - (x+y) = \square'' \end{cases}$

By the same reasoning as in the previous problem, Diophantus chooses a and b to be 2 and 4, from which he finds m and n to be 20 and 16. This way n is a square, which is required by the first condition.

Assignments	Operations	Equation
$x \cdot y := 20\text{P}$		
$x + y := 16\text{P}$		
$x := 2\text{N}$		
$y := 10\text{N}$		
	Add 2N and 10N $\rightarrow$ 12N	
		$16\text{P} = 12\text{N}$

Thus, 1N is $\frac{12}{16}\text{u} = \frac{3}{4}\text{u}$, so the sought-after numbers are $x = \frac{6}{4}$ and $y = \frac{30}{4}$.

II.32 $\begin{cases} x^2 + y = \square \\ y^2 + z = \square' \\ z^2 + x = \square'' \end{cases}$

The first two assignments are designed to satisfy the first condition, and the assignment for z is then designed to satisfy the second condition. The equation is formed from the third condition.

Assignments	Operations	Equation
$x := 1\text{N}$		
$y := 2\text{N } 1\text{u}$		
$z := 4\text{N } 3\text{u}$		
	Square 4N 3u $\rightarrow$ 16P 24N 9u	
	Add 1N $\rightarrow$ 16P 25N 9u	
		$16\text{P } 25\text{N } 9\text{u} = \square''$
$\sqrt{\square''} := 4\text{N } \ell\, 4\text{u}$		
	Square 4N ℓ 4u $\rightarrow$ 16P 16u ℓ 32N	
		$16\text{P } 25\text{N } 9\text{u} =$
		$16\text{P } 16\text{u } \ell\, 32\text{N}$

Thus, 1N is $\frac{7}{57}\text{u}$, hence, the three sought-after numbers are $x = \frac{7}{57}$, $y = \frac{71}{57}$, and $z = \frac{199}{57}$.

II.33 $\begin{cases} x^2 - y = \square \\ y^2 - z = \square' \\ z^2 - x = \square'' \end{cases}$

The assignments follow the pattern in the previous problem.

Assignments	Operations	Equation
$x := 1N\ 1u$		
$y := 2N\ 1u$		
$z := 4N\ 1u$		
	Square $4N\ 1u \rightarrow 16P\ 8N\ 1u$	
	Subtract $1N\ 1u \rightarrow 16P\ 7N$	
		$16P\ 7N = \square''$
$\sqrt{\square''} := 5N$		
	Square $5N \rightarrow 25P$	
		$16P\ 7N = 25P$

Thus, $1N$ is $\frac{7}{9}u$, so the three sought-after numbers are $x = \frac{16}{9}$, $y = \frac{23}{9}$, and $z = \frac{37}{9}$.

II.34
$$\begin{cases} x^2 + (x+y+z) = \square \\ y^2 + (x+y+z) = \square' \\ z^2 + (x+y+z) = \square'' \end{cases}$$

Diophantus begins by citing an identity that we write as $\left(\frac{m-n}{2}\right)^2 + mn = \left(\frac{m+n}{2}\right)^2$. This is structured like each of the conditions of the problem, where we identify $x + y + z$ with mn. In all three conditions this mn is the difference of two squares, which factors into the difference and sum of their roots, so mn should be expressible as a product in three different ways. Diophantus chooses mn to be 12, which is $12 \cdot 1$, $6 \cdot 2$, and $4 \cdot 3$. Thus, three equalities are generated from the identity, namely, $\left(5\frac{1}{2}\right)^2 + 12 = \left(6\frac{1}{2}\right)^2$, $2^2 + 12 = 4^2$, and $\left(\frac{1}{2}\right)^2 + 12 = \left(3\frac{1}{2}\right)^2$. By making these quantities multitudes of Numbers, he obtains the assignments.

Assignments	Operations	Equation
$x + y + z := 12P$		
$x := 5\frac{1}{2}N$		
$y := 2N$		
$z := \frac{1}{2}N$		
	Add $5\frac{1}{2}N$, $2N$, $\frac{1}{2}N \rightarrow 8N$	
		$8N = 12P$

Thus, $1N$ is $\frac{4}{6}u$, and so the three sought-after numbers are $x = \frac{22}{6}$, $y = \frac{8}{6}$, and $z = \frac{2}{6}$.

$$\textbf{II.35} \quad \begin{cases} x^2 - (x+y+z) = \square \\ y^2 - (x+y+z) = \square' \\ z^2 - (x+y+z) = \square'' \end{cases}$$

This time Diophantus begins by citing the identity $\left(\frac{m+n}{2}\right)^2 - mn = \left(\frac{m-n}{2}\right)^2$, which is structured like each of the conditions of the problem. Again, identifying $x + y + z$ with mn, and taking the same numbers as in the preceding problem, we find the equalities $\left(6\frac{1}{2}\right)^2 - 12 = \left(5\frac{1}{2}\right)^2$, $4^2 - 12 = 2^2$, $\left(3\frac{1}{2}\right)^2 - 12 = \left(\frac{1}{2}\right)^2$. These give the assignments.

Assignments	Operations	Equation
$x+y+z \coloneqq 12\text{P}$		
$x \coloneqq 6\frac{1}{2}\text{N}$		
$y \coloneqq 4\text{N}$		
$z \coloneqq 3\frac{1}{2}\text{N}$		
	Add $6\frac{1}{2}\text{N}, 4\text{N}, 3\frac{1}{2}\text{N} \rightarrow 14\text{N}$	
		$14\text{N} = 12\text{P}$

Thus, 1N is $\frac{7}{6}\text{u}$, and so the three sought-after numbers are $x = \frac{45\frac{1}{2}}{6} = \frac{91}{12}$, $y = \frac{28}{6}$, and $z = \frac{24\frac{1}{2}}{6} = \frac{49}{12}$.

Planudes makes the following comment on this problem:

> The 35th, as well as the 34th, require a lemma like this: if a number is measured by some number, and we add the one which measures it and the one according to which it measures, the square on half the sum, less the initial (number), makes a square. For example, 6 is measured by 2 according to 3, and *vice versa*. So, if we add the 2 and the 3, there becomes 5. Half this, $2\frac{1}{2}$. The square on this becomes $6\frac{1}{4}$. And if we subtract from this the initial (number), that is the 6, there remains $\frac{1}{4}$, which is a square from a side of half a unit. And he assigns the numbers according to the rule of this lemma, as he did also in the 34th.

> (Diophantus 1893–95, II, 254.8–19)

Book III

$$\text{III.1} \quad \begin{cases} (x+y+z)-x^2 = \square \\ (x+y+z)-y^2 = \square' \\ (x+y+z)-z^2 = \square'' \end{cases}$$

The first three assignments are designed to satisfy the first two conditions. The last assignment is found in the same manner, but now on the basis of the division of 5 into the sum of two other squares, $\left(\frac{2}{5}\right)^2$ and $\left(\frac{11}{5}\right)^2$, which can be found by the method of Problem II.9.

Assignments	Operations	Equation
$x + y + z := 5\text{P}$		
$x := 1\text{N}$		
$y := 2\text{N}$		
$z := \frac{2}{5}\text{N}$		
	Add 1N, 2N, and $\frac{2}{5}\text{N} \rightarrow 3\frac{2}{5}\text{N}$	
		$3\frac{2}{5}\text{N} = 5\text{P}$

Thus, 1N is $\frac{85}{125}$ u, so the three numbers are $x = \frac{85}{125}$, $y = \frac{170}{125}$, and $z = \frac{34}{125}$.

$$\text{III.2} \quad \begin{cases} (x+y+z)^2 + x = \square \\ (x+y+z)^2 + y = \square' \\ (x+y+z)^2 + z = \square'' \end{cases}$$

The first assignment guarantees that $(x + y + z)^2$ will be a square, and the next three are designed to satisfy the three conditions. The equation reconciles the first assignment with the other three.

DOI: 10.4324/9781315171470-21

Assignments	Operations	Equation
$(x+y+z)^2 := 1P$		
$x := 3P$		
$y := 8P$		
$z := 15P$		
	Add 3P, 8P, and 15P $\rightarrow$ 26P	
	The side of 1P is 1N	
		$26P = 1N$

Thus, 1N is $\frac{1}{26}$ u, so the three numbers are $x = \frac{3}{676}$, $y = \frac{8}{676}$, and $z = \frac{15}{676}$.

III.3
$$\begin{cases} (x+y+z)^2 - x = \square \\ (x+y+z)^2 - y = \square' \\ (x+y+z)^2 - z = \square'' \end{cases}$$

The assignments follow the reasoning of those in the previous problem.

Assignments	Operations	Equation
$x+y+z := 4N$		
	Square 4N $\rightarrow$ 16P	
$(x+y+z)^2 := 16P$		
$x := 7P$		
$y := 12P$		
$z := 15P$		
	Add 7P, 12P, and 15P $\rightarrow$ 34P	
		$34P = 4N$

Thus, 1N is $\frac{2}{17}$ u, so the three sought-after numbers are $x = \frac{28}{289}$, $y = \frac{48}{289}$, and $z = \frac{60}{289}$.

III.4
$$\begin{cases} x - (x+y+z)^2 = \square \\ y - (x+y+z)^2 = \square' \\ z - (x+y+z)^2 = \square'' \end{cases}$$

The assignments follow the reasoning of those in the previous two problems.

Assignments	Operations	Equation
$x+y+z := 1N$		
	Square 1N $\rightarrow$ 1P	
$(x+y+z)^2 := 1P$		

$$x := 2P$$
$$y := 5P$$
$$z := 10P$$

Add 2P, 5P, and 10P $\to$ 17P

$$17P = 1N$$

Thus, 1N is $\frac{1}{17}$ u, so the three sought-after numbers are $x = \frac{2}{289}$, $y = \frac{5}{289}$, and $z = \frac{10}{289}$.

III.5
$$\begin{cases} x + y + z = \Box \\ x + y = z + \Box' \\ y + z = x + \Box'' \\ z + x = y + \Box''' \end{cases}$$

First solution

The first assignment deliberately makes $x + y + z$ a three-term square, 1P 2N 1u. Since adding together $x + y - z$, $y + z - x$, and $z + x - y$ gives this sum $x + y + z$, or 1P 2N 1u, Diophantus makes $\Box'$ 1u and $\Box''$ 1P, which are both squares, so $\Box'''$, or $z + x - y$, must be 2N. He does not make this observation, so he has to calculate $z + x - y$. This 2N is made a square in the forthcoming equation, which is completed by *al-istiqrā'*.

Assignments	Operations	Equation
$x + y + z := $ 1P 2N 1u		
$x + y + z := $ 1u		
	Subtract 1u from 1P 2N 1u $\to$ 1P 2N	
	Half of 1P 2N $\to \frac{1}{2}$P 1N	
$z := \frac{1}{2}$P 1N		
$y + z - x := $ 1P		
	Subtract 1P from 1P 2N 1u $\to$ 2N 1u	
	Half of 2N 1u $\to$ 1N $\frac{1}{2}$ u	
$x := $ 1N $\frac{1}{2}$ u		
	Add $\frac{1}{2}$P 1N and 1N $\frac{1}{2}$ u $\to \frac{1}{2}$P 2N $\frac{1}{2}$ u	
	Subtract this from 1P 2N 1u $\to \frac{1}{2}$P $\frac{1}{2}$ u	
$y := \frac{1}{2}$P $\frac{1}{2}$ u		
	Subtract $\frac{1}{2}$P $\frac{1}{2}$ u from $\frac{1}{2}$P 2N $\frac{1}{2}$ u $\to$ 2N	
		$2N = \Box'''$
$\Box''' := $ 16u		
		$2N = 16u$

Thus, 1N is 8u, so the sought-after numbers are $x = 8\frac{1}{2}$, $y = 32\frac{1}{2}$, and $z = 40$.

Second solution

Again, by adding the last three equations in our modern algebraic rendering of the conditions, it is clear that $\square$ is equal to the sum of $\square'$, $\square''$, and $\square'''$. Diophantus thus looks for three squares that add up to a square. He posits $\square' := 4$ and $\square'' := 9$, and he then needs to find $\square'''$ such that when added together with 4 and 9 they give a square, namely, $\square$. This problem is equivalent to problem II.10, and Diophantus gives a solution, 36, without explanation. Thus, the four squares $\square$, $\square'$, $\square''$, $\square'''$ will be 49, 4, 9, and 36, respectively. The problem is then reduced to that of looking for three numbers such that the sum of each pair exceeds the remaining one by 4, 9,

and 36. In modern notation, $\begin{cases} x+y=z+4 \\ y+z=x+9 \\ z+x=y+36 \end{cases}$.

This problem has already been solved as Problem I.18, so Diophantus again gives the solution without explanation. The numbers we are seeking are $x = 20$, $y = 6\frac{1}{2}$, and $z = 22\frac{1}{2}$.

III.6 $\begin{cases} x+y+z = \square \\ x+y = \square' \\ y+z = \square'' \\ z+x = \square''' \end{cases}$

First solution

The first three conditions are fulfilled by the first four assignments, and the equation is formed from the last condition.

Assignments	Operations	Equation
$x+y+z := $ 1P 2N 1u		
$x+y := $ 1P		
$z := $ 2N 1u		
$y+z := $ 1P 1u ℓ 2N		
	Subtract 1P 1u ℓ 2N from 1P 2N 1u $\rightarrow$ 4N	
$x := $ 4N		
	Subtract 4N from 1P $\rightarrow$ 1P ℓ 4N	
	Add 4N and 2N 1u $\rightarrow$ 6N 1u	
		6N 1u $= \square''$
$\square'' := $ 121u		
		6N 1u $= $ 121u

Thus, 1N is 20u, so the sought-after numbers are $x = 80$, $y = 320$, and $z = 41$.

Second solution

The second solution is nearly identical to the first, the only difference being in the assignment of $\Box'''$, which is now taken to be 36u. In all likelihood this second solution is a later addition (Diophantus 1893–95, I, 149 n. 1; Heath 1910, 158).

III.7
$$\begin{cases} z - y = y - x \\ x + y = \Box \\ y + z = \Box' \\ z + x = \Box'' \end{cases}$$

Diophantus solves this problem in two stages. He first finds numerical values for the three squares $\Box$, $\Box'$, and $\Box''$ so that the first condition will be satisfied, after which he solves for x, y, and z.

By the first condition, the three squares will satisfy $\Box' - \Box'' = \Box'' - \Box$, and also half their sum will be greater than each of them individually. Diophantus assigns $\Box$ and $\Box''$ to be the squares 1P and 1P 2N 1u, respectively, and the equation is formed to make the derivative assignment of $\Box'$ a square. (It is true that our $\Box'$ and $\Box'''$ are the same square, but this is how it seems to be worded in the solution.)

Assignments	Operations	Equation
$\Box := $ 1P		
$\Box'' := $ 1P 2N 1u		
	Subtract 1P from 1P 2N 1u → 2N 1u	
$\Box' := $ 1P 4N 2u	Add 2N 1u and 1P 2N 1u → 1P 4N 2u	
		1P 4N 2u = $\Box'''$
$\sqrt{\Box'''} := $ 1N ℓ 8u		
	Square 1N ℓ 8u → 1P 64u ℓ 16N	
		1P 64u ℓ 16N = 1P 4N 2u

Thus, 1N is $\frac{31}{10}$ u; therefore, the three squares are $\Box = \frac{961}{100}$, $\Box'' = \frac{1681}{100}$, and $\Box' = \frac{2401}{100}$, and by removing the denominators, 961, 1681, and 2401.

Now, by identifying the three indeterminate squares of the original problem with the square numbers just found, the first condition is satisfied and the problem

becomes
$$\begin{cases} x + y = 961 \\ y + z = 1681 \cdot \\ z + x = 2401 \end{cases}$$

Assignments	Operations	Equation
$x + y + z := 1N$		
	Subtract 961u from $1N \rightarrow 1N \ \ell \ 961u$	
$z := 1N \ \ell \ 961u$		
	Subtract 2401u from $1N \rightarrow 1N \ \ell \ 2401u$	
$x := 1N \ \ell \ 2401u$		
	Subtract 1681u from $1N \rightarrow 1N \ \ell \ 1681u$	
$y := 1N \ \ell \ 1681u$		
	Add $1N \ \ell \ 961u$, $1N \ \ell \ 2401u$, $1N \ \ell \ 1681u \rightarrow 3N \ \ell \ 5043u$	
		$3N \ \ell \ 5043u = 1N$

Thus, $1N$ is $2521\frac{1}{2}$ u, hence the sought-after numbers are $x = 120\frac{1}{2}$, $y = 840\frac{1}{2}$, and $z = 1560\frac{1}{2}$.

Finding three squares of equal difference became a problem in its own right in Arabic arithmetic. This sub-problem is also solved in Problems VII.8–10. See Section 3.3.3.

III.8
$$\begin{cases} x + y + a = \square \\ y + z + a = \square' \\ z + x + a = \square'' \\ x + y + z + a = \square''' \end{cases} \qquad (a = 3)$$

The first three assignments are designed to satisfy the first, second, and fourth conditions. The equation is formed from the third condition.

Assignments	Operations	Equation
$x + y := 1P \ 4N \ 1u$		
$y + z := 1P \ 6N \ 6u$		
$x + y + z := 1P \ 8N \ 13u$		
	Subtract $1P \ 4N \ 1u$ from $1P \ 8N \ 13u \rightarrow 4N \ 12u$	
$z := 4N \ 12u$		
	Subtract $1P \ 6N \ 6u$ from $1P \ 8N \ 13u \rightarrow 2N \ 7u$	
$x := 2N \ 7u$		
	Subtract $2N \ 7u$ from $1P \ 4N \ 1u \rightarrow 1P \ 2N \ \ell \ 6u$	
$y := 1P \ 2N \ \ell \ 6u$		
	Add $4N \ 12u$, $2N \ 7u$, and $3u \rightarrow 6N \ 22u$	

		6N 22u = $\square''$
$\square'' := 100u$		
		6N 22u = 100u

Thus, 1N is 13u, so the three sought-after numbers are $x = 33$, $y = 189$, and $z = 64$.

III.9 $\begin{cases} x+y-a=\square \\ y+z-a=\square' \\ z+x-a=\square'' \\ x+y+z-a=\square''' \end{cases}$ $(a = 3)$

The assignments follow the reasoning of the previous problem.

Assignments	Operations	Equation
$x+y := 1P\ 3u$		
$y+z := 1P\ 2N\ 4u$		
$x+y+z := 1P\ 4N\ 7u$		
	Subtract 1P 3u from 1P 4N 7u → 4N 4u	
$z := 4N\ 4u$		
	Subtract 4N 4u from 1P 2N 4u → 1P ℓ 2N	
$y := 1P\ \ell\ 2N$		
	Subtract 1P ℓ 2N from 1P 3u → 2N 3u	
$x := 2N\ 3u$		
	Add 4N 4u and 2N 3u → 6N 7u	
	Subtract 3u from this → 6N 4u	
		6N 4u = $\square''$
$\square'' := 64u$		
		6N 4u = 64u

Thus, 1N is 10u, so the sought-after numbers are $x = 23$, $y = 80$, and $z = 44$.

III.10 $\begin{cases} x \cdot y + a = \square \\ y \cdot z + a = \square' \\ z \cdot x + a = \square'' \end{cases}$ $(a = 12)$

The first three assignments cover the first condition. The second condition is taken care of with the next two assignments, and the forthcoming equation is formed from the third condition.

Assignments	Operations	Equation
$\square := 25u$		
$x := 13N$		

$y := 1N'$

$\square' := 16u$

	Subtract 12u from 16u $\rightarrow$ 4u	
$y \cdot z := 4u$		
	Multiply 4u by 1N $\rightarrow$ 4N	
$z := 4N$		
	Multiply 13N and 4N $\rightarrow$ 52P	
	Add 52P and 12u $\rightarrow$ 52P 12u	
		52P 12u = $\square''$

If the 52 in the forthcoming equation were a square, it would be easy to make an *al-istiqrā'* assignment to solve it.[1] Therefore, the multitudes 4 and 13 in the expressions of z and x that gave us the 52 should be replaced by numbers whose product is a square and such that each of them, if increased by 12 units, makes a square. Now, if both numbers are squares then their product will be a square. Hence, we are led to the auxiliary problem of finding two squares, each of which together with 12 units makes a square, in order to satisfy the first two conditions. Diophantus says that this is easy, without giving any details. In fact he has already worked it out for the case of three squares in the course of solving Problem II.34. Here he chooses one square to be 4 and the other $\frac{1}{4}$, both taken to be multitudes of Numbers. He then begins the problem again, this time with the implicit reassignments $\square := 16u$ and $\square' := 12\frac{1}{4}u$. Again, the equation is formed from the third condition.

Assignments	Operations	Equation
$x := 4N$		
$y := 1N'$		
$z := \frac{1}{4}N$		
	Multiply 4N and $\frac{1}{4}N \rightarrow 1P$	
	Add 1P and 12u $\rightarrow$ 1P 12u	
		1P 12u = $\square''$
$\sqrt{\square''} := 1N\ 3u$		
	Square 1N 3u $\rightarrow$ 1P 6N 9u	
		1P 12u = 1P 6N 9u

Thus, 1N is $\frac{1}{2}$u; therefore, the sought-after numbers are $x = 2$, $y = 2$, and $z = \frac{1}{8}$.

1 Or equivalently, one could appeal to the solution given to Problem II.10. As Heath observes (1910, 69–70), since the sum 52 + 12 is a square, the forthcoming equation 52P 12u = $\square''$ can be solved following the procedure shown later in the *Arithmetica*, in the second lemma preceding Problem 12 of the Greek Book VI.

III.11 $\begin{cases} x \cdot y - a = \Box \\ y \cdot z - a = \Box' \\ z \cdot x - a = \Box'' \end{cases}$ $\quad (a = 10)$

The assignments follow the reasoning of the previous problem.

Assignments	Operations	Equation
$\Box := 4u$		
$x := 14N$		
$y := 1N'$		
$\Box' := 9u$		
$z := 19N$		
	Multiply 19N by 14N $\rightarrow$ 266P	
	Subtract 10u from 266P $\rightarrow$ 266P ℓ 10u	
		266P ℓ 10u $= \Box''$

As with the previous problem, if the 266 were a square it would be easy to solve the equation by *al-istiqrā'*.[2] This 266 is the product of 19 by 14 from the assignments of x and z, so we should look for two squares to replace them. If we let p^2 be the multitude of Numbers in the assignment for x, and q^2 the multitude of Numbers in the assignment for z, then we are faced with the pair of subsidiary problems $p^2 - 10 = \Box'''$ and $q^2 - 10 = \Box^{iv}$.

Unlike the solution to the previous problem, the text gives the full derivation of these squares, which led Tannery to suspect that the entire paragraph beginning "Indeed, you will find" was inserted by some editor. One of them, say p^2, is found to be $30\frac{1}{4}$ by applying the identity $\left(\frac{m-n}{2}\right)^2 + mn = \left(\frac{m+n}{2}\right)^2$ used in Problem II.34, regarding 10 as the product $10 \cdot 1$, and which makes $\Box''' = 20\frac{1}{4}u$. The second square could have been found to be $12\frac{1}{4}$ in the same way by expressing 10 as $5 \cdot 2$, but it is found instead by a separate algebraic solution.

Assignments	Operations	Equation
$q^2 := 1P$		
	Subtract 10u from 1P $\rightarrow$ 1P ℓ 10u	
		1P ℓ 10u $= \Box$

2 Like this stage in the solution to the previous problem, the forthcoming equation 266P ℓ 10u $= \Box''$ could be solved by the technique given in Problem II.10. And because 266 − 10 is a square, it could also be solved by the procedure in the second lemma to Problem VI^G.12. (Heath 1910, 160)

$\sqrt{\square} \coloneqq 1\mathrm{N}\,\ell\,2\mathrm{u}$

Square $1\mathrm{N}\,\ell\,2\mathrm{u} \rightarrow 1\mathrm{P}\,4\mathrm{u}\,\ell\,4\mathrm{N}$

$1\mathrm{P}\,\ell\,10\mathrm{u} = 1\mathrm{P}\,4\mathrm{u}\,\ell\,4\mathrm{N}$

From here $1\mathrm{N}$ becomes $3\frac{1}{2}\mathrm{u}$, so $q^2 = 12\frac{1}{2}$. The solution to the problem can then proceed.

Assignments	Operations	Equation
$x \coloneqq 30\frac{1}{4}\mathrm{N}$		
$y \coloneqq 1\mathrm{N}'$		
$z \coloneqq 12\frac{1}{4}\mathrm{N}$		
	Multiply $30\frac{1}{4}\mathrm{N}$ by $12\frac{1}{4}\mathrm{N} \rightarrow 370\frac{1}{2}\frac{1}{16}\mathrm{P}$	
	Subtract 10u from $370\frac{1}{2}\frac{1}{16}\mathrm{P} \rightarrow 370\frac{1}{2}\frac{1}{16}\mathrm{P}\,\ell\,10\mathrm{u}$	
		$370\frac{1}{2}\frac{1}{16}\mathrm{P}\,\ell\,10\mathrm{u} = \square$
	Multiply $370\frac{1}{2}\frac{1}{16}\mathrm{P}\,\ell\,10\mathrm{u}$ by $16\mathrm{u} \rightarrow 5929\mathrm{P}\,\ell\,160\mathrm{u}$	
		$5929\mathrm{P}\,\ell\,160\mathrm{u} = \square^v$
$\sqrt{\square^v} \coloneqq 77\mathrm{N}\,\ell\,2\mathrm{u}$	Square $77\mathrm{N}\,\ell\,2\mathrm{u} \rightarrow 5929\mathrm{P}\,4\mathrm{u}\,\ell\,308\mathrm{N}$	
		$5929\mathrm{P}\,\ell\,160\mathrm{u} = 5929\mathrm{P}\,4\mathrm{u}\,\ell\,308\mathrm{N}$

Thus, $1\mathrm{N}$ is $\frac{41}{77}\mathrm{u}$, so the sought-after numbers are $x = \frac{1240\frac{1}{4}}{77}$, $y = \frac{77}{41}$, and $z = \frac{520\frac{1}{4}}{77}$.

III.12 $\begin{cases} x \cdot y + z = \square \\ y \cdot z + x = \square' \\ z \cdot x + y = \square'' \end{cases}$

The assignments together satisfy the first condition, and the two equations are formed from the second and third conditions.

Assignments	Operations	Equation
$\sqrt{\square} \coloneqq 1\mathrm{N}\,3\mathrm{u}$		
	Square $1\mathrm{N}\,3\mathrm{u} \rightarrow 1\mathrm{P}\,6\mathrm{N}\,9\mathrm{u}$	
$z \coloneqq 9\mathrm{u}$		
	Subtract 9u from $1\mathrm{P}\,6\mathrm{N}\,9\mathrm{u} \rightarrow 1\mathrm{P}\,6\mathrm{N}$	
$x \coloneqq 1\mathrm{N}$		
$y \coloneqq 1\mathrm{N}\,6\mathrm{u}$		

	Multiply 1N 6u by 9u and add 1N → 10N 54u	
		10N 54u = □′
	Multiply 9u by 1N and add 1N 6u → 10N 6u	
		10N 6u = □″

A double-equality arises, in which the difference of the squares is 48 units. By making the difference and the sum of the roots of the squares 4 and 12, we get □′ = 64 and □″ = 16. Thus, 1N is 1u, and the sought-after numbers are $x = 1$, $y = 7$, and $z = 9$.

III.13 $\begin{cases} x \cdot y - z = \Box \\ y \cdot z - x = \Box' \\ z \cdot x - y = \Box'' \end{cases}$

The assignments satisfy the first condition, and the equations come from the other two conditions.

Assignments	Operations	Equation
$x := 1N$		
$y := 1N\ 4u$		
$z := 4N$		
	Multiply 1N 4u by 4N and subtract 1N → 4P 15N	
		4P 15N = □′
	Multiply 4N by 1N and subtract 1N 4u → 4P ℓ 1N 4u	
		4P ℓ 1N 4u = □″

Again, a double-equality arises, which Diophantus solves by factoring the difference 16N 4u into 4u and 4N 1u so that □′ is 4P 10N 6¼u and □″ is 4P 2¼u ℓ 6N. Thus, 1N is $\frac{25}{20}$ u, and the three sought-after numbers are $x = \frac{25}{20}$, $y = \frac{105}{20}$, and $z = \frac{100}{20}$.

III.14 $\begin{cases} x \cdot y + z^2 = \Box \\ y \cdot z + x^2 = \Box' \\ z \cdot x + y^2 = \Box'' \end{cases}$

The three assignments satisfy the first two conditions, taking advantage of the symmetry between the squares 4P 4N 1u and 1P 4N 4u, and the equation is set up from the third condition.

Assignments	Operations	Equation
$x := 1N$		
$y := 4N\ 4u$		
$z := 1u$		

Multiply 1u by 1N → 1N

Square 4N 4u → 16P 32N 16u

Add 1N to this → 16P 33N 16u

16P 33N 16u = □″

$\sqrt{\square}″ \coloneqq$ 4N ℓ 5u

Square 4N ℓ 5u → 16P 25u ℓ 40N

16P 33N 16u = 16P 25u ℓ 40N

Thus, 1N is $\frac{9}{73}$u, hence the sought-after numbers are $x = \frac{9}{73}$, $y = \frac{328}{73}$, and $z = \frac{73}{73}$.

III.15
$$\begin{cases} x \cdot y + (x+y) = \square \\ y \cdot z + (y+z) = \square' \\ z \cdot x + (z+x) = \square'' \end{cases}$$

First solution

The first two assignments are made to be consecutive squares since the product of consecutive squares together with their sum is always a square; and thus, they satisfy the first condition.

Assignments	Operations	Equation
$x \coloneqq 4$u		
$y \coloneqq 9$u		
$z \coloneqq 1$N		
	Multiply 9u by 1N → 9N	
	Add to this 9u and 1N → 10N 9u	
		10N 9u = □′
	Multiply 1N by 4u → 4N	
	Add to this 1N and 4u → 5N 4u	
		5N 4u = □″

A double-equality arises, which is solved by factoring the difference 5N 5u into 5u and 1N 1u so that □′ is $\frac{1}{4}$P 3N 9u and □″ is $\frac{1}{4}$P 4u ℓ 2N. Thus, 1N is 28u, so the required numbers are $x = 4$, $y = 9$, and $z = 28$.

Second solution

Diophantus works his second solution in two stages, by first finding x and y, satisfying the first condition, after which he solves for z from the second and third conditions. The numbers for x and y that he initially finds do not lead to a rational z, so he has to back up and add another condition on x and y and find them again.

Assignments	Operations	Equation
$x \coloneqq 1$N		

$y := 3u$

| Multiply 1N by 3u → 3N |
| Add to this 1N and 3u → 4N 3u |

$$4N\ 3u = \square$$

$\square := 25u$

$$4N\ 3u = 25u$$

Thus, 1N is $5\frac{1}{2}$u, so $x = 5\frac{1}{2}$ and $y = 3$. The first condition is satisfied, so the prob-

lem is reduced to the remaining two conditions: $\begin{cases} 5\frac{1}{2}z + \left(5\frac{1}{2} + z\right) = \square' \\ 3z + (z + 3) = \square'' \end{cases}$.

Assignments	Operations	Equation
$z := 1N$		
	Multiply 3u by 1N → 3N	
	Add to this 3u and 1N → 4N 3u	
		$4N\ 3u = \square'$
	Multiply 1N by $5\frac{1}{2}$u → $5\frac{1}{2}$N	
	Add to this 1N and $5\frac{1}{2}$u → $6\frac{1}{2}$N $5\frac{1}{2}$u	
		$6\frac{1}{2}N\ 5\frac{1}{2}u = \square''$

Thus, a double-equality arises. But since the multitude of the Numbers and the units in one expression are both greater than those in the other, and they do not have a ratio that a square has to a square, we arrive at a dead end. Hence, the numbers $5\frac{1}{2}$ and 3 must be replaced by two other numbers for x and y that satisfy not only the first condition but also that "the numbers one unit greater than them have to each other a ratio that a square has to a square". In modern notation, the condi-

tions for x and y become $\begin{cases} x \cdot y + (x + y) = \square \\ (x+1):(y+1) :: \square_1 : \square_2 \end{cases}$.

Diophantus chooses assignments that satisfy the second condition, and his equation is formed from the first condition.

Assignments	Operations	Equation
$x := 1N$		
$y := 4N\ 3u$		
	Multiply 1N by 4N 3u → 4P 3N	
	Add this to 1N and 4N 3u → 4P 8N 3u	
		$4P\ 8N\ 3u = \square$
$\sqrt{\square} := 2N\ \ell\ 3u$		
	Square 2N ℓ 3u → 4P 9u ℓ 12N	
		$4P\ 8N\ 3u = 4P\ 9u\ \ell\ 12N$

So, 1N is found to be $\frac{3}{10}$, and the required numbers are $x = \frac{3}{10}$ and $y = 4\frac{1}{5}$. The problem of finding z, in modern notation, becomes
$$\begin{cases} 4\frac{1}{5}z + \left(4\frac{1}{5}+z\right) = \square' \\ \frac{3}{10}z + \left(z+\frac{3}{10}\right) = \square'' \end{cases}.$$

Assignments	Operations	Equation
$z := 1N$		
	Multiply 1N by $4\frac{1}{5}u \rightarrow 4\frac{1}{5}N$	
	Add this to 1N and $4\frac{1}{5}u \rightarrow 5\frac{1}{5}N\,4\frac{1}{5}u$	
		$5\frac{1}{5}N\,4\frac{1}{5}u = \square'$
	Multiply 1N by $\frac{3}{10}u \rightarrow \frac{3}{10}N$	
	Add this to 1N and $\frac{3}{10}u \rightarrow \frac{13}{10}N\,\frac{3}{10}u$.	
		$\frac{13}{10}N\,\frac{3}{10}u = \square''$
	Multiply $5\frac{1}{5}N\,4\frac{1}{5}u$ by 25 $\rightarrow$ 130N 105u	
		130N 105u $= \square'''$
	Multiply $\frac{13}{10}N\,\frac{3}{10}u$ by 100u $\rightarrow$ 130N 30u	
		130N 30u $= \square^{iv}$

This time the double-equality is solvable. The difference of the two expressions is 75u, which is 3u by 25u. By following the standard procedure, 1N is found to be $\frac{7}{10}u$, so the three sought-after numbers are $x = \frac{3}{10}$, $y = \frac{42}{10}$, and $z = \frac{7}{10}$.

III.16 $\begin{cases} x \cdot y\ (x+y) = \square \\ y \cdot z - (y+z) = \square' \\ z \cdot x - (z+x) = \square'' \end{cases}$

Diophantus remarks that if x is assigned to be 1 Number and y any number of units whatever, as he initially did in the second solution of the preceding problem, then we will fall likewise into a dead end. So, the problem is reduced again to that of looking for two numbers such that their product, if it lacks their sum, makes a square, and, moreover, the numbers one unit less than them have to each other the ratio that a square number has to a square number. In our notation, $\begin{cases} x \cdot y - (x+y) = \square \\ (x-1):(y-1) :: \square_1 : \square_2 \end{cases}$.

Diophantus again chooses assignments for x and y that satisfy the second of these conditions, and the equation is formed from the first condition.

Assignments	Operations	Equation
$x := 4N\ 1u$		
$y := 1N\ 1u$		
	Multiply 4N 1u by 1N 1u $\rightarrow$ 4P 5N 1u	

	Subtract from this 4N 1u and 1N 1u → 4P ℓ 1u	
		4P ℓ 1u = □
$\sqrt{\Box} \coloneqq 2\mathrm{N}\,\ell\,2\mathrm{u}$	Square 2N ℓ 2u → 4P 4u ℓ 8N	
		4P ℓ 1u = 4P 4u ℓ 8N

1N is $\frac{5}{8}$u, hence $x = \frac{13}{8}$ and $y = \frac{28}{8}$. The value of z is then found from the remaining conditions, which are now $\begin{cases} 3\frac{1}{2}z - \left(3\frac{1}{2}+z\right) = \Box' \\ \frac{13}{8}z - \left(z + \frac{13}{8}\right) = \Box'' \end{cases}$.

Assignments	Operations	Equation
$z \coloneqq 1\mathrm{N}$		
	Multiply $3\frac{1}{2}$u by 1N → $3\frac{1}{2}$N	
	Subtract from this 1N and $3\frac{1}{2}$u → $2\frac{1}{2}$N ℓ $3\frac{1}{2}$u	
		$2\frac{1}{2}$N ℓ $3\frac{1}{2}$u = □′
	Four times $2\frac{1}{2}$N ℓ $3\frac{1}{2}$u → 10N ℓ 14u	
		10N ℓ 14u = □′′′
	Multiply 1N by $\frac{13}{8}$u → $\frac{13}{8}$N	
	Subtract from this 1N and $\frac{13}{8}$u → $\frac{5}{8}$N ℓ $\frac{13}{8}$u	
		$\frac{5}{8}$N ℓ $\frac{13}{8}$u = □′′
	Sixteen times $\frac{5}{8}$N ℓ $\frac{13}{8}$u → 10N ℓ 26u	
		10N ℓ 26u = □ⁱᵛ

As before, a double-equality arises. The difference of the two expressions is 12u, which factors into 2u by 6u. From there, 1N is found to be 3u. Hence, the three sought-after numbers are $x = \frac{13}{8}$, $y = 3\frac{1}{2}$ or $\frac{28}{8}$, and $z = 3$ or $\frac{24}{8}$.

III.17 $\begin{cases} x \cdot y + (x+y) = \Box \\ x \cdot y + x = \Box' \\ x \cdot y + y = \Box'' \end{cases}$

The assignments satisfy the second condition, and the two equations are formed from the third and first conditions.

Assignments	Operations	Equation
$x \coloneqq 1\mathrm{N}$		
$y \coloneqq 4\mathrm{N}\,\ell\,1\mathrm{u}$		
	Multiply 1N by 4N ℓ 1u → 4P ℓ 1N	
	Add to this 4N ℓ 1u → 4P 3N ℓ 1u	

$$4P\ 3N\ \ell\ 1u = \square''$$

Add 4P ℓ 1N, 1N, and 4N ℓ 1u $\rightarrow$ 4P 4N ℓ 1u

$$4P\ 4N\ \ell\ 1u = \square$$

In this double-equality the difference of the two expressions is 1N, which is $\frac{1}{4}$u by 4N. By the usual procedure, 1N is found to be $\frac{65}{224}$u. Hence, the two sought-after numbers are $x = \frac{65}{224}$ and $y = \frac{36}{224}$.

III.18 $\quad \begin{cases} x \cdot y - (x+y) = \square \\ x \cdot y - x = \square' \\ x \cdot y - y = \square'' \end{cases}$

The assignments satisfy the third condition, and the equations are formed from the other two.

Assignments	Operations	Equation
$x := 1N\ 1u$		
$y := 4N$		
	Multiply 1N 1u by 4N $\rightarrow$ 4P 4N	
	Subtract from this 1N 1u $\rightarrow$ 4P 3N ℓ 1u	
		$4P\ 3N\ \ell\ 1u = \square'$
	Subtract 1N 1u and 4N from 4P 4N $\rightarrow$ 4P ℓ 1N 1u	
		$4P\ \ell\ 1N\ 1u = \square$

A double-equality arises. The difference of the two expressions is 4N, which Diophantus factors into 4N by 1u, so 1N is found to be $1\frac{1}{4}$u. Hence, the two sought-after numbers are $x = 2\frac{1}{4}$ and $y = 5$.

III.19 $\quad \begin{cases} (w+x+y+z)^2 \pm w = \begin{cases} \square^{i} \\ \square^{ii} \end{cases} \\ (w+x+y+z)^2 \pm x = \begin{cases} \square^{iii} \\ \square^{iv} \end{cases} \\ (w+x+y+z)^2 \pm y = \begin{cases} \square^{v} \\ \square^{vi} \end{cases} \\ (w+x+y+z)^2 \pm z = \begin{cases} \square^{vii} \\ \square^{viii} \end{cases} \end{cases}$

Diophantus observes that in any right triangle (a, b, c) with $a < b < c$, $c^2 \pm 2ab = (b \pm a)^2$. This has the same structure as the conditions of the problem, so we are reduced to the problem of dividing a square into two squares in four different ways. Now, as in the text, "we learned how to divide a given square into two squares in an unlimited number of ways" in Problem II.8, though he does not follow the procedure indicated there. Instead, Diophantus first considers two right triangles in the smallest (i.e., relatively prime) numbers, (3, 4, 5) and (5, 12, 13), and he takes the triangles produced by multiplying the sides of the one by the hypotenuse of the other. Thus, two right triangles are formed with equal hypotenuses, (39, 52, 65) and (25, 60, 65). Accordingly, $65^2 = 39^2 + 52^2$, and $65^2 = 25^2 + 60^2$, so, 65^2 has been divided into two squares in two ways. "Further", as in the text, "the 65 is divided naturally into (two) squares in two ways: into 16 and 49, but, also, 64 and a unit. This is due to the fact that the number 65 is contained by the numbers 13 and 5, each of which is divided into two squares". By means of the numbers 7 and 4 Diophantus forms the right triangle ($7^2 - 4^2$, $2 \cdot 7 \cdot 4$, $7^2 + 4^2$), or (33, 56, 65), and similarly, by means of the numbers 8 and 1 forms the right triangle ($8^2 - 1^2$, $2 \cdot 8 \cdot 1$, $8^2 + 1^2$), or (63, 16, 65). Thus, four right triangles are created having 65 as the hypotenuse: $65^2 = 39^2 = 52^2$, $65^2 = 25^2 + 60^2$, $65^2 = 33^2 + 56^2$, and $65^2 = 63^2 + 16^2$. From these relations one obtains that the expressions $65^2 \pm 2 \cdot 39 \cdot 52$, $65^2 \pm 2 \cdot 25 \cdot 60$, $65^2 \pm 2 \cdot 33 \cdot 56$, and $65^2 \pm 2 \cdot 63 \cdot 16$ are all squares. These relations are applied in the assignments.

Assignments	Operations	Equation
$w + x + y + z := 65N$		
$w := 4 \cdot \frac{39 \cdot 52}{2} P = 4056P$		
$x := 4 \cdot \frac{25 \cdot 60}{2} P = 3000P$		
$y := 4 \cdot \frac{33 \cdot 56}{2} P = 3696P$		
$z := 4 \cdot \frac{63 \cdot 16}{2} P = 2016P$		
	Add 4056P, 3000P, 3696P, 2016P $\rightarrow$ 12,768P	
		$12{,}768P = 65N$

Thus, 1N is $\frac{65}{12,768}$ u, hence the four sought-after numbers are $w = \frac{17,136,600}{163.021.824}$, $x = \frac{12,675,000}{163.021.824}$, $y = \frac{15,615,600}{163,021,824}$, and $z = \frac{8,517,600}{163,021,824}$.

For the second division of 65 into two squares in two different ways, Diophantus resorts to a proposition that we can express in modern notation in the following manner. If $x = a^2 + b^2$ and $y = c^2 + d^2$, where $a \neq b$ and $c \neq d$, then xy can be expressed as the sum of two squares in two different ways, as $(ac + bd)^2 + (ad - bc)^2$ and $(ac - bd)^2 + (ad + bc)^2$. By the wording of the text, Diophantus seems to presume that his readers were already familiar with this proposition, perhaps from his *Porisms*, or from some other lost ancient book. The proposition resurfaces later in al-Khāzin (10th c.), Fibonacci (13th c.), and in other medieval authors (see Section 3.3.3).

Both Tannery and Heath suggest that the last two problems in Book III, 20 and 21, are later additions, since they are the same as Problems II.15 and 14, respectively (Diophantus 1893–95, I, 187 n. 1; Heath 1910, 167 n. 2).

III.20
$$\begin{cases} x+y=a \\ z^2 - x = \Box \\ z^2 - y = \Box' \end{cases} \quad (a=10)$$

The assignments satisfy the second and third conditions, and the equation is formed from the first condition.

Assignments	Operations	Equation
$z^2 := 1\text{P } 2\text{N } 1\text{u}$		
$x := 2\text{N } 1\text{u}$		
$y := 4\text{N}$		
	Add 2N 1u and 4N → 6N 1u	
		6N 1u = 10u

Thus, 1N is $1\frac{1}{2}$u, so the two sought-after numbers are $x = 4$ and $y = 6$, and the square is $z^2 = 6\frac{1}{4}$.

III.21
$$\begin{cases} x+y=a \\ z^2 + x = \Box \\ z^2 + y = \Box' \end{cases} \quad (a=20)$$

Again, the assignments satisfy the second and third conditions.

Assignments	Operations	Equation
$z^2 := 1\text{P } 2\text{N } 1\text{u}$		
$y := 2\text{N } 3\text{u}$		
$x := 4\text{N } 8\text{u}$		
	Add 2N 3u and 4N 8u → 6N 11u	
		6N 11u = 20u

Thus, 1N is $1\frac{1}{2}$u , and the two sought-after numbers are $x = 6$ and $y = 14$, and the

square is $z^2 = 6\frac{1}{4}$.

Book IV (Arabic)

IV.1 $x^3 + y^3 = \square$

Assignments	Operations	Equation
$y := 1T$		
	Cube $1T \to 1C$	
$x := 2T$		
	Cube $2T \to 8C$	
	Add $1C$ and $8C \to 9C$	
		$9C = \square$
$\sqrt{\square} := 6T$		
	Square $6T \to 36M$	
		$9C = 36M$

Thus, 1T is equal to 4u, so the two cubes are $y^3 = 64$ and $x^3 = 512$.

IV.2 $x^3 - y^3 = \square$

Assignments	Operations	Equation
$y := 1T$		
	Cube $1T \to 1C$	
$x := 2T$		
	Cube $2T \to 8C$	
	Subtract $1C$ from $8C \to 7C$	
		$7C = \square$
$\sqrt{\square} := 7T$		
	Square $7T \to 49M$	
		$7C = 49M$

DOI: 10.4324/9781315171470-22

Thus, 1T is 7u, so the two cubes are $y^3 = 343$ and $x^3 = 2744$.

IV.3 $x^2 + y^2 = ⬚$

Assignments	Operations	Equation
$y^2 := 1M$ $x^2 := 4M$		
	Add 1M and 4M $\to$ 5M	
		$5M = ⬚$
$\sqrt[3]{⬚} := 1T$		
	Cube 1T $\to$ 1C	
		$5M = 1C$

Thus, 1T = 5u, so the two squares are $y^2 = 25$ and $x^2 = 100$.

IV.4 $x^2 - y^2 = ⬚$

Assignments	Operations	Equation
$y := 1T$ $x := 5T$		
	Square 5T $\to$ 25M Square 1T $\to$ 1M Subtract 1M from 25M $\to$ 24M	
		$24M = ⬚$
$\sqrt[3]{⬚} := 2T$		
	Cube 2T $\to$ 8C	
		$24M = 8C$

Thus, 1T is 3u, so the two squares are $y^2 = 9$ and $x^2 = 225$.

IV.5 $x^2 \cdot y^2 = ⬚$

Assignments	Operations	Equation
$y^2 := 1M$ $x := 2T$		
	Square 2T $\to$ 4M Multiply 1M by 4M $\to$ 4MM	

		$4MM = \boxdot$
$\sqrt[3]{\boxdot} := 2T$		
	Cube 2T $\rightarrow$ 8C	
		$4MM = 8C$

Thus, $1T = 2u$, hence the two squares are $y^2 = 4$ and $x^2 = 16$.

IV.6 $\quad x^2 \cdot y^3 = \square$

Assignments	Operations	Equation
$x := 1T$		
	Square 1T $\rightarrow$ 1M	
$y := 2T$		
	Cube 2T $\rightarrow$ 8C	
	Multiply 1M by 8C $\rightarrow$ 8MC	
		$8MC = \square$
$\sqrt{\square} := 4M$		
	Square 4M $\rightarrow$ 16MM	
		$8MC = 16MM$

Thus, $1T$ is $2u$, so, the square is $x^2 = 4$ and the cube is $y^3 = 64$.

Before assigning $\sqrt{\square}$ Diophantus explains what would happen "if we were to assign a side of the square to be Things", or $\sqrt{\square} := Ts$, where "Ts" denotes the unspecified multitude of Things. The forthcoming equation would then be of the form MCs = Ms, which reduces to Cs = units. This is not desirable since we would need to ensure that the result of dividing the units by the multitude of Cubes is a cube to get a rational solution. Thus, he makes $\sqrt{\square} := Ms$, specifically 4M. He gives similar explanations for his choice of power in the next two problems.

IV.7 $\quad x^2 \cdot y^3 = \boxdot$

Assignments	Operations	Equation
$x := 1T$		
	Square 1T $\rightarrow$ 1M	
$y := 4T$		
	Cube 4T $\rightarrow$ 64C	
	Multiply 1M by 64C $\rightarrow$ 64CM	
		$64CM = \boxdot$

$\sqrt[3]{\boxdot} \coloneqq 2M$

	Cube 2M $\rightarrow$ 8CC	
		64MC = 8CC

Thus, 1T is 8u, so the sought-after square is $x^2 = 64$ and the sought-after cube is $y^3 = 32{,}768$.

Like in the previous problem, Diophantus shows what we would get if we were to make the assignment $\sqrt[3]{\boxdot} \coloneqq$ Ts. The forthcoming equation would simplify to Ms = units, and "we would need the units ⟨that Equal the one *Māl*⟩ to be a square". He then runs through the indeterminate calculations again for $\sqrt[3]{\boxdot} \coloneqq$ Ms and finds that it simplifies to "Things Equal units", which will give a rational solution.

IV.8 $x^3 \cdot y^3 = \square$

Diophantus begins with assignments that will not lead to a rational solution.

Assignments	Operations	Equation
$x \coloneqq 1\text{T}$		
	Cube 1T $\rightarrow$ 1C	
$y \coloneqq 2\text{T}$		
	Cube 2T $\rightarrow$ 8C	
	Multiply 1C by 8C $\rightarrow$ 8CC	
	8CC = $\square$	

At this point, Diophantus cannot assign a side of the square to be Things since the equation would simplify to 8MM equal units. So, he looks into assigning its side to be some multitude of *Māl*s without specifying the multitude:

$\sqrt{\square} \coloneqq \text{Ms}$		
	Square Ms $\rightarrow$ MMs	
	CCs = MMs	

The equation simplifies to Ms = units. (In this particular problem this equation would be 8M = units, which cannot have a rational solution since the units will necessarily be a square.) Thus, "we are led to look for a square (the 1M) and a cubic number (the cube of the multitude of Ts in a new assignment for y, which will replace the 8) containing a square number (the units)". This is Problem IV.6, and Diophantus gives the solution $2^2 \cdot 4^3 = 16^2$, or $4 \cdot 64 = 256$. These numbers will be applied in Problem IV.9, which is a continuation of the present problem.

IV.9 $x^3 \cdot y^3 = \square$

This is a continuation of Problem IV.8. The assignments are based on the values found at the end of the previous problem.

Assignments	Operations	Equation
$y := 4T$		
$x := 1T$		
	Cube $4T \to 64C$	
	Cube $1T \to 1C$	
	Multiply $64C$ by $1C \to 64CC$	
		$64CC = \square$
$\sqrt{\square} := 16M$		
	Square $16M \to 256MM$	
		$64CC = 256MM$

Thus, $1T$ is $2u$; therefore, the two sought-after cubes are $x^3 = 8$ and $y^3 = 512$.

It was not necessary for Diophantus to work through this solution because he had found at the end of the previous problem that $1M$ would be $4u$, so $1T$ is $2u$. The assignment for x remained unchanged, and the value of y is necessarily $4T$, which is $8u$. Perhaps Diophantus had obtained the solution directly from the numbers 4, 64, and 256 found at the end of IV.6, and a scholiast inserted the calculations of IV.9.

Diophantus then states two corollaries whose solutions are reduced to previous problems:

Corollary 9.1: The problem we write in modern notation as $x^3 \div y^3 = \square$ is reduced to $u^2 \cdot y^3 = \boxdot$, which is the same as Problem IV.7.

Corollary 9.2: The problem we write in modern notation as $x^2 \div y^2 = \boxdot$ is reduced to $y^2 \cdot u^3 = \square$, which is the same as Problem IV.6.

IV.10 $x^3 + mx^2 = \square$ $(m = 10)$

Assignments	Operations	Equation
$x := 1T$		
	Cube $1T \to 1C$	
	Square $1T \to 1M$	
	Add $1C$ and ten times $1M \to 1C\ 10M$	
		$1C\ 10M = \square$
$\sqrt{\square} := 4T$		
	Square $4T \to 16M$	
		$1C\ 10M = 16M$

Thus, $1T = 6u$, so the sought-after cube is $x^3 = 216$.

IV.11 $x^3 - mx^2 = \square$ $(m = 6)$

Assignments	Operations	Equation
$x := 1T$		
	Cube $1T \rightarrow 1C$	
	Square $1T \rightarrow 1M$	
	Subtract from $1C$ six times $1M$ $\rightarrow 1C \, \ell \, 6M$	
		$1C \, \ell \, 6M = \square$
$\sqrt{\square} := 2T$		
	Square $2T \rightarrow 4M$	
		$1C \, \ell \, 6M = 4M$

Thus, $1T = 10u$; therefore, the sought-after cube is $x^3 = 1000$.

IV.12 $x^3 + mx^2 = \boxdot$ $(m = 10)$

Assignments	Operations	Equation
$x := 1T$		
	Cube $1T \rightarrow 1C$	
	Square $1T \rightarrow 1M$	
	Add $1C$ and ten times $1M \rightarrow 1C \, 10M$	
		$1C \, 10M = \boxdot$
$\sqrt[3]{\boxdot} := 2T$		
	Cube $2T \rightarrow 8C$	
		$1C \, 10M = 8C$

Thus, $1T$ is $\frac{10}{7}u$, and the required cube is $x^3 = \frac{1000}{343}$.

IV.13 $x^3 - mx^2 = \boxdot$ $(m = 7)$

First solution

Assignments	Operations	Equation
$x := 1T$		
	Cube $1T \rightarrow 1C$	

	Square 1T → 1M	
	Subtract seven times 1M from 1C → 1C ℓ 7M	
		1C ℓ 7M = ⌸
$\sqrt[3]{⌸} \coloneqq \frac{1}{2}T$		
	Cube $\frac{1}{2}$T → $\frac{1}{8}$C	
		1C ℓ 7M = $\frac{1}{8}$C

Thus, 1T is 8u, and the required cube is $x^3 = 512$.

Second solution, interpreting the enunciation as asking $x^3 - mx^2 = y^3$ (line 262)

Assignments	Operations	Equation
$x \coloneqq 2T$		
	Cube 2T → 8C	
$y \coloneqq 1T$		
	Cube 1T → 1C Subtract 1C from 8C → 7C Square 2T → 4M Seven times 4M → 28M	
		28M = 7C

Thus, 1T is 4u; therefore, ⌸ = 64 and $x^3 = 512$.

In the first solution the enunciation is interpreted as asking for a single number, a cube, whose remainder after taking away a multiple of the square is a cubic number. In the second solution the enunciation is interpreted as asking for two cubes, which we would translate into modern notation as the condition $x^3 - mx^2 = y^3$. The text does not state the assignment for y, and overall this solution reads like a confused version of the first solution. Sesiano (1982, 185) observes that it is probably an interpolation, since it is not much different from the first solution and it is "rather carelessly done".

IV.14 $\begin{cases} a \cdot x = ⌸ \\ b \cdot x = □' \end{cases}$ $(a = 10, b = 5, \text{ or } a = 5, b = 10)$

The text shows three solutions each to two different problems. The first four solutions, (a)–(d), follow the same line of reasoning that begins by assigning the unknown x as 1T and by declaring the ratio between □ and □'. They differ by interchanging the values of a and b and by inverting the ratio. The last two solutions, (e) and (f), name the side of ▱ and □' as 1T, respectively, and then make an *al-istiqrā'* assignment for the other.

(a) First insltantiation: $\begin{cases} 10 \cdot x = ▱ \\ 5 \cdot x = □' \end{cases}$

Assignments	Operations	Equation
$x := 1T$		
	Multiply 1T by 5u → 5T	
	Multiply 1T by 10u → 10T	
		$10T = ▱$
		$5T = □'$
$□ := \frac{1}{4}□'$		
	Multiply 5T by $\frac{1}{4}$u → $1\frac{1}{4}$T	
$□ := 1\frac{1}{4}T$		
	Divide 10T by $1\frac{1}{4}$T → 8u	
$\sqrt[3]{▱} := 8u$		
	Cube 8u → 512u	
		$10T = 512u$

The 1T is $51\frac{1}{5}$u, and this is the value of the sought-after x. The second forthcoming equation, $5T = □'$ serves to make the substitution from the second to the third assignment, i.e., to replace $\frac{1}{4}□'$ with $1\frac{1}{4}$T. The $\frac{1}{4}$ of the second assignment must be a square because it is the ratio of a square to a square (□ to □'). This will also guarantee that the value of 1T, which derives from the second condition, will also satisfy the first condition.

(b) Second instantiation: $\begin{cases} 5 \cdot x = ▱ \\ 10 \cdot x = □' \end{cases}$　　　(line 305)

Assignments	Operations	Equation
$x := 1T$		
	Multiply 1T by 5u → 5T	
	Multiply 1T by 10u → 10T	
		$5T = ▱$
		$10T = □'$

$\square := \frac{1}{4}\square'$

	Multiply 10T by $\frac{1}{4}$u → $2\frac{1}{2}$T

$\square := 2\frac{1}{2}$T

	Divide 5T by $2\frac{1}{2}$T → 2u

$\sqrt[3]{\boxdot} := 2u$

	Cube 2u → 8u

		5T = 8u

Thus, 1T is $\frac{8}{5}$u, which is the value of the sought-after number x. The solution follows the same reasoning as in (a).

(c) First instantiation again: $\begin{cases} 10 \cdot x = \boxdot \\ 5 \cdot x = \square' \end{cases}$ (line 318)

Assignments	Operations	Equation
$x := 1$T		
	Multiply 1T by 10u → 10T	
	Multiply 1T by 5u → 5T	
		10T = $\boxdot$
		5T = $\square'$
$\square := 4\square'$		
	Multiply 5T by 4u → 20T	
$\square := 20$T		
	Divide 10T by 20T → $\frac{1}{2}$u	
$\sqrt[3]{\boxdot} := \frac{1}{2}$u		
	Cube $\frac{1}{2}$u → $\frac{1}{8}$u	
		10T = $\frac{1}{8}$u

Thus, 1T is $\frac{1}{80}$u, which is the value of the sought-after number x. The solution follows the same reasoning as the previous two.

(d) Second instantiation again: $\begin{cases} 5 \cdot x = \boxdot \\ 10 \cdot x = \square' \end{cases}$ (line 328)

Again, the argument begins with the second assignment.

Assignments	Operations	Equation
$x := 1T$		
	Multiply $1T$ by $5u \to 5T$	
	Multiply $1T$ by $10u \to 10T$	
		$5T = \square$
		$10T = \square'$
$\square := 4\square'$		
	Multiply $10T$ by $4u \to 40T$	
$\square := 40T$		
	Divide $5T$ by $40T \to \frac{1}{8}u$	
$\sqrt[3]{\square} := \frac{1}{8}u$		
	Cube $\frac{1}{8}u \to \frac{1}{512}u$	
		$5T = \frac{1}{512}u$

Thus, $1T = \frac{1}{2560}u$, which is the value of the sought-after x. The same reasoning is applied again.

(e)   Another technique, for $\begin{cases} 10 \cdot x = \square \\ 5 \cdot x = \square' \end{cases}$ (line 343)

Assignments	Operations	Equation
$\sqrt[3]{\square} := 1T$		
	Cube $1T \to 1C$	
$x := \frac{1}{10}C$		
	Multiply $\frac{1}{10}C$ by $5u \to \frac{1}{2}C$	
		$\frac{1}{2}C = \square'$
$\sqrt{\square'} := 2T$		
	Square $2T \to 4M$	
		$\frac{1}{2}C = 4M$

Thus, $1T$ is $8u$, so the sought-after number is $x = 51\frac{1}{5}$.

(f)   Similarly, again for $\begin{cases} 10 \cdot x = \square \\ 5 \cdot x = \square' \end{cases}$ (line 358)

Assignments	Operations	Equation
$\sqrt{\square'} := 1T$		
	Square $1T \rightarrow 1M$	
$x := \frac{1}{5}M$		
	Multiply $\frac{1}{5}M$ by $10u \rightarrow 2M$	
		$2M = \boxdot$
$\sqrt[3]{\boxdot} := 1T$		
	Cube $1T \rightarrow 1C$	
		$2M = 1C$

Thus, $1T = 2u$, so the sought-after number is $x = \frac{4}{5}$.

Sesiano writes, "There is in fact no substantial difference between the resolution of (a)–(d) and that of (e) and (f), since (e) and (f) ultimately amount to setting a proportion between u^2 [our $\square'$] and v^2 [the square on the side of $\boxdot$], as was done before" (Sesiano 1982, 190). The solutions in (a)–(d) are different, however, from those in (e) and (f) from the perspective of procedure. In the former, the unknown number x is assigned to be $1T$, the proportion between the squares is explicitly assigned, and the equation is then formed by a derivative assignment of a side of the cube. In (e) and (f), instead, a side of $\boxdot$ or $\square'$ is assigned to be $1T$ to satisfy one of the conditions, while the second assignment is an *al-istiqrā'* assignment that will turn the other condition into an equation. That assignment implicitly gives the proportion. We agree with Sesiano that some scholiast has added solutions to this problem.

IV.15 $\begin{cases} a \cdot x = \boxdot \\ b \cdot x = \square' \end{cases}$ $(a = 10, b = 4)$

This problem is determinate.

First solution

Diophantus follows "the rule from the previous problem", which is the procedure to Problem IV.14(a).

Assignments	Operations	Equation
$x := 1T$		
	Multiply $1T$ by $10u \rightarrow 10T$	
	Multiply $1T$ by $4u \rightarrow 4T$	

$\sqrt[3]{\boxdot} := 2\frac{1}{2}\mathrm{u}$

Divide 10T by 4T $\rightarrow 2\frac{1}{2}\mathrm{u}$

Square $2\frac{1}{2}\mathrm{u} \rightarrow 6\frac{1}{4}\mathrm{u}$

$4\mathrm{T} = 6\frac{1}{4}\mathrm{u}$

Thus, 1T is $\frac{25}{16}\mathrm{u}$, which is the value of the sought-after x.

Second solution   (line 388)

Here Diophantus follows "the method in the second problem", referring to the solution to Problem IV.14(c).

Assignments	Operations	Equation
$\sqrt[3]{\boxdot} := 1\mathrm{T}$		
	Cube 1T $\rightarrow$ 1C	
$x := \frac{1}{10}\mathrm{C}$		
	Multiply $\frac{1}{10}\mathrm{C}$ by 4u $\rightarrow \frac{4}{10}\mathrm{C}$	
		$\frac{4}{10}\mathrm{C} = \square$
	Square 1T $\rightarrow$ 1M	
		$\frac{4}{10}\mathrm{C} = 1\mathrm{M}$

Thus, $1\mathrm{T} = 2\frac{1}{2}\mathrm{u}$; therefore, the sought-after number is $x = \frac{25}{26}$.

Corollary: $x^3 : y^2 = a : b \quad (a : b = 3 : 1)$

The procedure to solve this is as follows: we take two numbers in the given ratio, and we look for a number z which if multiplied by one of them makes a cube and if multiplied by the other makes a square. In modern notation, $\begin{cases} a \cdot z = \boxdot \\ b \cdot z = \square' \end{cases}$, which is Problem IV.14. Then $x = \boxdot$ and $y = \square'$.

IV.16 $\begin{cases} a \cdot x = \boxdot \\ a \cdot y = \sqrt[3]{\boxdot} \end{cases}$ $\quad (a = 10)$

Assignments	Operations	Equation
$y := 1\mathrm{T}$		
	Multiply 1T by 10u $\rightarrow$ 10T	

$\sqrt[3]{\boxed{}} := 10\text{T}$

Cube 10T $\rightarrow$ 1000C

$x := 300\text{M}$

Multiply 300M by 10u $\rightarrow$ 3000T

1000C = 3000M

Thus, 1T is 3u, so the sought-after numbers are $y = 3$ and $x = 2700$. The third assignment is made in terms of *Māls* to guarantee a rational solution.

IV.17 $\begin{cases} x = my \\ a \cdot x^2 = \boxed{} \\ a \cdot y^2 = \sqrt[3]{\boxed{}} \end{cases}$ Determination: $m \cdot a$ must be a square

$(a = 5,\ m = 20)$

If one follows the calculations, the value of 1M will be the result of dividing m by a. This accounts for the determination.

Assignments	Operations	Equation
$y := 1\text{T}$		
	Square 1T $\rightarrow$ 1M	
	Multiply 1T by 20u $\rightarrow$ 20T	
$x := 20\text{T}$		
	Square 20T $\rightarrow$ 400M	
	Multiply 400M by 5u $\rightarrow$ 2000M	
	Multiply 1M by 5u $\rightarrow$ 5M	
	Cube 5M $\rightarrow$ 125CC	
		125CC = 2000M

Thus, 1T = 2u. The two squares we are looking for arc $y^2 = 4$ and $x^2 = 1600$.

IV.18 $\begin{cases} x = my \\ a \cdot x^3 = \square \\ a \cdot y^3 = \sqrt{\square} \end{cases}$ Determination: a must be a cube

$(a = 8,\ m = 3)$

The determination follows from tracing the calculations made with 8u in the solution.

Assignments	Operations	Equation
$y := 1\text{T}$		
	Cube 1T $\rightarrow$ 1C	

$x := 3T$

Three times $1T \to 3T$

Cube $3T \to 27C$
Multiply $27C$ by $8u \to 216C$
Multiply $1C$ by $8u \to 8C$
Square $8C \to 64CC$

$216C = 64CC$

Thus, $1C$ is $3\frac{3}{8}u$, so $1T = 1\frac{1}{2}u$. Hence, the two cubes are $y^3 = 3\frac{3}{8}$ and $x^3 = 91\frac{1}{8}$.

IV.19 $\begin{cases} a \cdot x = \boxed{} & \text{Determination: } a \cdot b \text{ must be a square} \\ b \cdot x = \sqrt[3]{\boxed{}} & (a = 20, \ b = 5) \end{cases}$

The determination follows from the calculations. Since $a \div b$ must be a square, $a \cdot b$ must also be a square.

Assignments	Operations	Equation
$x := 1T$		
	Multiply $1T$ by $5u \to 5T$	
	Multiply $1T$ by $20u \to 20T$	
	Divide $20T$ by $5T \to 4u$	
	A side of $4u \to 2u$	
		$5T = 2u$

Thus, $1T$ is $\frac{2}{5}u$, so the sought-after number is $x = \frac{2}{5}$.

IV.20 $\begin{cases} a \cdot x^3 = \square & \text{Determination: } a \div b^2 \text{ must be a cube} \\ b \cdot x^3 = \sqrt{\square} & (a = 200, \ b = 5) \end{cases}$

The determination follows from the calculations. The 200 (a) is divided by the 5 (b) twice, and the result must be a cube. A simpler way to state the determination would be to require $a \cdot b$ to be a cube, but Diophantus generally found his determinations from the particular calculations in his solutions, and this problem is not an exception.

Assignments	Operations	Equation
$x := 1T$		
	Cube $1T \to 1C$	
	Multiply $1C$ by $200u \to 200C$	

Multiply 1C by 5u → 5C
Divide 200C by 5C → 40u

5C = 40u

Thus, 1T is 2u, and the sought-after cube is $x^3 = 8$.

IV.21 $\begin{cases} a \cdot x^2 = \boxed{} \\ b \cdot x^2 = \sqrt[3]{\boxed{}} \end{cases}$ Determination: $a \cdot b$ must be the square of a square $\left(a = 40\frac{1}{2}, \ b = 2 \right)$

Assignments	Operations	Equation
$x^2 := 1M$		
	Multiply 1M by $40\frac{1}{2}$u → $40\frac{1}{2}$M	
	Multiply 1M by 2u → 2M	
	Divide $40\frac{1}{2}$M by 2M → $20\frac{1}{4}$u	
	Square 2M → 4MM	
		$4MM = 20\frac{1}{4}$u

Diophantus calculated the 4MM but did not state it. Taking the square roots of both sides and dividing by 2u gives $1M = 2\frac{1}{4}$u. Then, taking the square roots again we obtain 1T is $1\frac{1}{2}$u. So, the sought-after square is $x^2 = 2\frac{1}{4}$.

To explain the determination, Diophantus follows his calculations based on the assignment $x^2 := 1M$ but without specifying the values of a and b. Dividing a $M\bar{a}l$s by b $M\bar{a}l$s gives the square of the side of the cube, so it must be a square. Thus, $a \cdot b$ must also be a square. Now, $\sqrt{\frac{a}{b}}$ is equal to a side of the cube, which by the second condition is b $M\bar{a}l$s, so dividing $\sqrt{\frac{a}{b}}$ by b leaves 1M, which is a square. This means that the product of $\sqrt{\frac{a}{b}}$ by b will also be a square. And this, Diophantus writes, means us that $a \cdot b$ is "a square with a square side". The repeated use of the algebraic "Equal" shows that Diophantus (or at least Qusṭā) is thinking about the equations associated with the calculations. It is easier for us to see this by shifting entirely to modern notation. Since $\frac{a}{b}$ is $(bx^2)^2 = b^2x^4$, then multiplying both sides by b^2 gives $ab = b^4x^4$.

IV.22 $\begin{cases} a \cdot x^3 = \boxed{} \\ b \cdot x^3 = \sqrt[3]{\boxed{}} \end{cases}$ Determination : $a \div b$ must be a square and $\sqrt{a \div b} \div b$ must be a cube $\left(a = 91\frac{1}{8}, \ b = 2 \right)$

Diophantus begins by explaining the determination much like he just described the determination of the previous problem. With the assignment $x^3 := 1C$, he follows calculations like those given previously, again without specifying a and b.

Dividing a Cubes by b Cubes gives the square of the side of ⬚, so it must be a square. Thus, $a \div b$ must be a square. Now, $\sqrt{\frac{a}{b}}$ is equal to a side of that cube, which by the second condition is b Cubes. Therefore, dividing $\sqrt{\frac{a}{b}}$ by b leaves 1C, which is a cube, so we write $\sqrt{a \div b} \div b$ must be a cube. Again, the repeated use of the algebraic "Equal" shows that Diophantus (or Qusṭā) is thinking about the equations associated with the calculations.

We can use our modern notation to make simpler conditions. The first condition implies that a is a cube. Then, since $\frac{a}{b}$ is the square of a side of the cube, it must be a square, so $a \cdot b$ is also a square. So, the conditions could have been stated more simply as: "the number that multiplies to give a cube must be a cube, and the product of the two numbers must be a square", or, a must be a cube and $a \cdot b$ must be a square. But the conditions given by Diophantus, which follow the calculations, are perhaps easier to understand for the premodern reader.

Diophantus then sets about finding two numbers a and b satisfying these conditions. He first makes $b := 2$. Because $\sqrt{a \div 2} \div 2$ should be a cube, he makes $\sqrt{a \div 2}$ the double of a cube, which he chooses to be the double of $3\frac{3}{8}$. From there he finds $a := 91\frac{1}{8}$. After making the assignment $x^3 := 1C$, the text reads "we work it out like we did in the previous problems". We reconstruct the solution accordingly:

Assignments	Operations	Equation
$x^3 := 1C$		
	Multiply 1C by $91\frac{1}{8}$u $\rightarrow 91\frac{1}{8}$C	
	Multiply the 1C by 2u $\rightarrow$ 2C	
	Divide $91\frac{1}{8}$C by 2C $\rightarrow 45\frac{9}{16}$u	
	Square 2C $\rightarrow$ 4CC	
		$4CC = 45\frac{9}{16}$u

Taking square roots and dividing by 2 gives $1C = 3\frac{3}{8}$u, so 1T is $1\frac{1}{2}$u. Thus, the sought-after x^3 is $3\frac{3}{8}$.

IV.23 $(x^2)^2 + (y^2)^2 = $ ⬚

Assignments	Operations	Equation
$y^2 := 1M$		
$x := 2T$		
	Square 2T $\rightarrow$ 4M	
$x^2 := 4M$		
	Square 1M $\rightarrow$ 1MM	
	Square 4M $\rightarrow$ 16MM	

	Add 1MM and 16MM → 17MM	
		17MM = ▱
$\sqrt[3]{▱} := 3T$	Cube 3T → 27C	
		17MM = 27C

After dividing by 1C and then by 17, 1T is found to be $\frac{27}{17}$ u. The two squares are then $y^2 = \frac{729}{289}$ and $x^2 = \frac{2916}{289}$.

IV.24 $(x^2)^2 - (y^2)^2 = ▱$

Assignments	Operations	Equation
$y := 1T$		
$x := 2T$		
	Square 1T → 1M	
	Square 2T → 4M	
	Square 1M → 1MM	
	Square 4M → 16MM	
	Subtract 1MM from 16MM → 15MM	
		15MM = ▱
$\sqrt[3]{▱} := 5T$		
	Divide 15MM by 5T → 3C	
	Square 5T → 25M	
		3C = 25M

Therefore, $1T = 8\frac{1}{3}$ u, so the two squares are $y^2 = 69\frac{4}{9}$ and $x^2 = 277\frac{7}{9}$. The solution takes a different path from the usual after the side of the cube is assigned. Instead of cubing the 5T and equating it to 15MM, the latter is divided by 5T, and this is equated to the square of 5T. The version of this problem in al-Karajī's *al-Fakhrī*, Problem V.22, is worked out the same way. See also Problems IV.28 and IV.31.

IV.25 $(x^3)^2 + (y^2)^2 = \square$

Assignments	Operations	Equation
$x := 1T$		
	Cube 1T → 1C	
$y := 2T$		
	Square 2T → 4M	

Square 1C → 1CC
Square 4M → 16MM
Add 1CC and 16MM → 1CC 16MM

1CC 16MM = □

Diophantus wishes to make a side of the □ some multitude of *Māl*s, so that the □ itself is *Māl Māl*s. The multitude of MMs must be chosen so that when 16 is subtracted from it, it leaves a square. Such a multitude is 25, which can be found by the method in Problem II.10. The solution continues:

Assignments	Operations	Equation
□ := 25MM		
		1CC 16MM = 25MM

Thus, 1T is 3u, so the sought-after cube is $x^3 = 27$ and the sought-after square is $y^2 = 36$.

IV.26   (a) $(x^3)^2 - (y^2)^2 = □$

Assignments	Operations	Equation
$x^3 := 1C$		
$y^2 := 4M$		
	Square 1C → 1CC	
	Square 4M → 16MM	
	Subtract 16MM from 1CC → 1CC ℓ 16MM	
		1CC ℓ 16MM = □

As in the last problem, Diophantus wants to make a side of the □ a multitude of *Māl*s, so that the □ itself is *Māl Māl*s, which, with the 16MM added to it, becomes a square. This is again Problem II.10, and Diophantus chooses the multitude to be 3.

Assignments	Operations	Equation
$\sqrt{□} := 3M$		
	Square 3M → 9MM	
		1CC ℓ 16MM = 9MM

Thus, 1T is 5u, so the sought-after cube is $x^3 = 125$, and the sought-after square is $y^2 = 100$.

(b) $(y^2)^2 - (x^3)^2 = \square$ (line 730)

Assignments	Operations	Equation
$x^3 := 1C$		
$y := 5T$		
	Square 5T → 25M	
	Square 1C → 1CC	
	Square 25M → 625MM	
	Subtract 1CC from 625MM → 625MM ℓ 1CC	
		625MM ℓ 1CC = $\square$

Again, Diophantus wants to make a side of the $\square$ a multitude of *Māl*s, so that the $\square$ itself is *Māl Māl*s, and subtracting a square of this number from 625 should give a square. This is an instance of Problem II.8, and Diophantus chooses the square of the multitude to be 225.

Assignments	Operations	Equation
$\square := 255MM$		
		625MM ℓ 1CC = 225MM

Thus, 1T is 20u, so the sought-after cube is $x^3 = 8000$, and the sought-after square is $y^3 = 10{,}0000$.

IV.27 $(x^3)^2 + my^2 = \square$ $(m = 5)$

Assignments	Operations	Equation
$x^3 := 1C$		
	Multiply 1C by 1C → 1CC	
$y := 2M$		
	Square 2M → 4MM	
	Multiply 4MM by 5u → 20MM	
	Add it to 1CC → 1CC 20MM	
		1CC 20MM = $\square$

The multitude of *Māl Māl*s to assign the $\square$ should be a square that is 20 greater than a square. This is Problem II.10, and Diophantus chooses 36.

Assignments	Operations	Equation
$\square := 36\text{MM}$		
		$1\text{CC } 20\text{MM} = 36\text{MM}$

So $1\text{T} = 4\text{u}$; therefore, the cube is $x^3 = 64$ and the square is $y^2 = 1024$.

IV.28 $(x^2)^2 + my^3 = \square$ $(m = 10)$

Assignments	Operations	Equation
$y^3 := 1\text{C}$		
	Multiply 1C by $10\text{u} \rightarrow 10\text{C}$	
$x := 2\text{T}$		
	Square $2\text{T} \rightarrow 4\text{M}$	
	Square $4\text{M} \rightarrow 16\text{MM}$	
	Add that to $10\text{C} \rightarrow 16\text{MM } 10\text{C}$	
		$16\text{MM } 10\text{C} = \square$
$\sqrt{\square} := 6\text{M}$		
	Divide $16\text{MM } 10\text{C}$ by $6\text{M} \rightarrow 2\tfrac{2}{3}\text{M}1\,\tfrac{2}{3}\text{T}$	
		$2\tfrac{2}{3}\text{M } 1\tfrac{2}{3}\text{T} = 6\text{M}$

Therefore, $1\text{T} = \tfrac{1}{2}\text{u}$, so the sought-after cube is $y^3 = \tfrac{1}{8}$, while the sought-after square is $x^2 = 1$. The calculations after the last assignment are unusual. It would have been easier to set up the equation $16\text{MM } 10\text{C} = 36\text{MM}$, which gives $10\text{C} = 20\text{MM}$, and from which 1T is found to be $\tfrac{1}{2}\text{u}$. Instead, the text follows the reasoning of Problem IV.24 by dividing $16\text{MM } 10\text{C}$ by 6M, the side of the square, and then equating the result with that side.

IV.29 $(x^3)^3 + (y^2)^2 = \square$

Assignments	Operations	Equation
$x^3 := 1\text{C}$		
	Cube $1\text{C} \rightarrow 1\text{CCC}$	
$y := 2\text{M}$		

	Square 2M → 4MM	
	Square 4MM → 16CCM	
	Add 1CCC and 16CCM → 1CCC 16CCM	
		1CCC 16CCM = □
$\sqrt{□} := 6M$		
	Multiply 6MM by 6MM → 36CCM	
		1CCC 16CCM = 36CCM

Thus, $1T = 20u$, so the sought-after cube is $x^3 = 8000$ and the sought-after square is $y^2 = 640{,}000$.

IV.30 $(x^3)^3 - (y^2)^2 = □$

Assignments	Operations	Equation
$x^3 := 1C$		
	Cube 1C → 1CCC	
$y := 2M$		
	Square 2M → 4MM	
	Square 4MM → 16CCM	
	Subtract 16CCM from 1CCC → 1CCC ℓ 16CCM	
		1CCC ℓ 16CCM = □
$\sqrt{□} := 2MM$		
	Square 2MM → 4CCM	
		1CCC ℓ 16CCM = 4CCM

Thus, $1T = 20u$, so the sought-after cube is $x^3 = 8000$ and the sought-after square is $y^2 = 640{,}000$.

At the end of the solution of this problem Diophantus adds the following remark regarding this and the previous problem:

> Thus, it is clear that we have also found two numbers, a cube and a square, such that a cube of the cube, if a square of the square is added to it, amounts to a square number, and if a square of the square is subtracted from a cube of the cube, then the remainder is a square number.

Later, in Problem IV.42(a), Diophantus will explain how to find numbers satisfying the conditions of both problems simultaneously.

IV.31 $(y^2)^2 - (x^3)^3 = \square$

Assignments	Operations	Equation
$x^3 := 1C$		
	Cube $1C \rightarrow 1CCC$	
$y := 2M$		
	Square $2M \rightarrow 4MM$	
	Square $4MM \rightarrow 16CCM$	
	Subtract $1CCC$ from $16CCM$ $\rightarrow 16CCM \; \ell \; 1CCC$	
		$16CCM \; \ell \; 1CCC = \square$
$\sqrt{\square} := 2MM$		
	Divide $16CCM \; \ell \; 1CCC$ by $2MM \rightarrow 8MM \; \ell \; \frac{1}{2}MC$	
		$8MM \; \ell \; \frac{1}{2}MC = 2MM$

Therefore, $1T = 12u$, so the sought-after cube is $x^3 = 1728$, and the sought-after square is $y^2 = 82{,}944$. As in Problems IV.24 and IV.28, the solution by *al-istiqrā'* is a variation in which the roots are equated.

IV.32 $(x^3)^3 + m(y^2 \cdot x^3) = \square$ $(m = 5)$

Assignments	Operations	Equation
$x^3 := 1C$		
	Cube $1C \rightarrow 1CCC$	
$y := 2C$		
	Square $2C \rightarrow 4CC$	
	Multiply $4CC$ by $1C \rightarrow 4CCC$	
	Five times $4CCC \rightarrow 20CCC$	
	Add it to $1CCC \rightarrow 21CCC$	
		$21CCC = \square$
$\sqrt{\square} := 7MM$		
	Square $7MM \rightarrow 49CCM$	
		$21CCC = 49CCM$

Thus, $1T = 2\frac{1}{3}u$, hence the sought-after cube is $x^3 = \frac{343}{27}$, while the sought-after square is $y^2 = \frac{470,596}{729}$.

IV.33 $(x^3)^3 = m(y^2 \cdot x^3) + \square$ $(m = 3)$

Assignments	Operations	Equation
$x^3 := 1C$		
	Cube $1C \rightarrow 1CCC$	
$y := \frac{1}{2}C$		
	Square $\frac{1}{2}C \rightarrow \frac{1}{4}CC$	
	Multiply that by $1C \rightarrow \frac{1}{4}CCC$	
	Three times $\frac{1}{4}CCC \rightarrow \frac{3}{4}CCC$	
	Subtract that from $1CCC \rightarrow \frac{1}{4}CCC$	
		$\frac{1}{4}CCC = \square$
$\sqrt{\square} := 1MM$		
	Square $1MM \rightarrow 1CCM$	
		$\frac{1}{4}CCC = 1CCM$

Thus, $1T = 4u$; therefore, the sought-after cube is $x^3 = 64$ and the square is $y^2 = 1024$.

After the solution the text states two corollaries which are variations of Problem IV.33, "and also vice versa, and whatever is similar".

Corollary IV.33.1: $(y^2)^2 + m(y^2 \cdot x^3) = \square$
Corollary IV.33.2: $(y^2)^3 + m(y^2 \cdot x^3) = \square$

Their solutions require three assignments of the same type as those of Problem IV.33.

IV.34 $\begin{cases} x^3 + y^2 = \square \\ x^3 - y^2 = \square' \end{cases}$

Assignments	Operations	Equation
$x^3 := 1C$		
$y^2 := 4M$		
	Add $1C$ and $4M \rightarrow 1C\ 4M$	

> Subtract 4M from 1C → 1C ℓ 4M
>
> 1C 4M = □
> 1C ℓ 4M = □′

From this point Diophantus works out the problem in two different ways.

First solution (by the double-equality)   (line 960)

The difference of the squares is 8M, which Diophantus factors into 2T by 4T. He equates the square of half the difference of the two factors with 1C ℓ 4M, and the square of half their sum with 1C 4M, to get the equations 1M = 1C ℓ 4M and 9M = 1C 4M, which both simplify to 1C = 5M. Thus, 1T = 5u, the sought-after cube is $x^3 = 125$, and the sought-after square is $y^2 = 100$.

Second solution (by equating the corresponding sides)   (line 977)

There is another way that Diophantus solves indeterminate problems from a pair of forthcoming equations and which we encounter here for the first time. Instead of factoring the difference of the squares, he states his intention to assign each of □ and □′ to be square multitudes of *Māl*s, and he works through the solutions of these equations for 1T without specifying these multitudes, after which he equates the indeterminate solutions. The equation 1C 4M = □ will be of the form 1C 4M = Ms (we write "Ms" for the unspecified *Māl*s, whose multitude is necessarily a square), which, after subtracting the 4M from both sides, becomes 1C = Ms. Dividing both sides by 1M, the value of 1T is found to be the multitude of *Māl*s remaining after subtracting the 4 from the square multitude. Similarly, from the other equation, 1T must also be the multitude of *Māl*s obtained from adding 4 to the square multitude of the assignment for □′. Thus, the remainder and the sum must be equal. In modern notation, writing m^2 for the multitude of *Māl*s in the assignment for □ and m'^2 for the multitude of *Māl*s in □′, then $m^2 - 4 = m'^2 + 4$. So, we need to find two squares m^2 and m'^2 whose difference is 8. This is Problem II.10, and Diophantus finds them to be $12\frac{1}{4}$ and $20\frac{1}{4}$, but without showing the work. The solution continues:

Assignments	Operations	Equation
□ = $20\frac{1}{4}$M		1C4M = $20\frac{1}{4}$M
□′ = $12\frac{1}{4}$M		1C ℓ 4M = $12\frac{1}{4}$M

Both equations simplify to 1C = $16\frac{1}{4}$M. Thus, 1T = $16\frac{1}{4}$u. Hence the sought-after cube is $x^3 = 4291\frac{1}{64}$ and the sought-after square is $y^2 = 1056\frac{1}{4}$.

It is curious that Diophantus goes to so much trouble to find that the differ-ence of the square multitudes is 8 when it is immediately apparent from the two forthcoming equations, 1C 4M = □ and 1C ℓ 4M = □′, that the difference of the squares is 8M. He will do this in other problems solved by this method, too.

Remarks on the word *ṭarafīn* and on our name for the method of the second solution

Once the forthcoming equations are set up, the first solution begins: "We solve this by working out the double-equality", and after that, the second solution begins: "We can also work this out without the ⟨double⟩ equality". So even if the second method also works from a pair of forthcoming equations, it was regarded as being distinct from the one Diophantus calls "the double-equality". The method of the second solution will be the main method applied in the rest of Book IV, and in three solutions it is given a description. In Problem IV.42(a) this is: "And if we wish, we can work this out by seeking equality in the *ṭarafīn* (the dual form of *ṭaraf*) of the Equation, as we explained in the preceding problems" (line 1363). The other instances are worded similarly.[1] Understanding just what "equality in the *ṭarafīn* of the Equation" means is not a simple matter. We have already encountered this word in the first solution to IV.34. After arriving at the simplified equation 1C = 5M, the text has: "So both *ṭarafīn* of the equation (*muʿādala*) are equal, each of them turning out to be a Cube Equals five *Māls*" (line 969). The first part of the phrase seems to imply that the *ṭarafīn* are the two sides of the equation, but the latter part suggests that they are the pair of equations themselves, which will both reduce to 1C = 5M. Sesiano wrote regarding this passage (1982, 111 n. 64):

> The text has two distinct words which can be rendered by 'equation': one (*ṭaraf*) designates a given expression in x [our 1T] equal to 'a square' (hence *ṭarafān*, as a dual noun, is the system to be solved), the other (*muʿādala^h*) is the resulting equation in x.[2]

He translates the passage as: "Thus the (resulting) equation for the two equations (of the proposed system) turned out to be the same, ending in each one with x^3 equal to $5x^2$" (Sesiano 1982, 110–11). Rashed, for his part, translates *ṭarafīn* in this passage as "deux cas": "L'équation est donc la même dans les deux cas et aboutit dans chacun d'eux à un cube égal à cinq carrés" (Diophantus 1984, 3, 59.1). Lane (1863–93, 1843) gives the meanings of *ṭaraf* as "extremity", "end", "side", "adjacent part", etc. The word is commonly used in geometry to mean an endpoint of a line, and in arithmetic and algebra it frequently refers to the first and last

---

1 They are: "Then we take up the search for something that makes the *ṭarafīn* of the Equation equal, as we did before" (Problem IV.40, line 1240) and "We work this out by looking for equality of the Equation in both *ṭarafīn*, as we explained above for this kind of problem" (Problem IV.42(b), line 1417).

2 The two terms are elided in Rashed's translation: "L'équation est donc la même dans les deux cas et aboutit dans chacun d'eux à un cube égal à cinq carrés" (Diophantus 1984, 3, 59.1).

terms in a sequence, whether a sequence of numbers in proportion (as in 2, 4, 8), a sequence of numbers to be added (like $1 + 2 +. . .+10$), a sequence of algebraic powers (like Things, *Māls*, Cubes) or the terms in a polynomial arranged according to power (like 1CC 4MC 4MM 6C 12T 9u). The word was also sometimes, though not commonly, used to designate the "sides" of an equation. For example, having arrived at 2T 20 = 100, al-Fārisī writes: "cast away the twenty from the two sides (*ṭarafīn*), leaving two Things Equal to eighty" (al-Fārisī 1994, 530.10). Other people who used *ṭaraf* to mean a side of an equation are Ibn al-Hā'im, Bahā al-Dīn al-'Āmilī (1547–1622) and Ḥājjī Khalīfa (1609–57).[3]

It would thus be alien to the core meaning of the word to translate *ṭarafīn* as either "two equations (of the proposed system)" or "deux cas". Sesiano's interpretation is made more difficult by the fact that in the three descriptions of the method he has to insert "(resulting from)" or "(resulting)" to make it work. For example, in Problem IV.40 (line 1240) he writes (1982, 117): "Next, we begin to search for what will make one and the same the equation (resulting from) the two (proposed) equalities, in the way we did before". Here Rashed translates *al-ṭarafīn* as "des seconds membres des deux équations", with a footnote explaining that the text literally reads "des deux membres (sous-entendu des seconds membres)" (Diophantus 1984, 3, 76). So he interprets the *ṭarafīn* as the "right" side of the equations after simplification, which, in the case of Problem IV.34, are both 5M.

Al-Karajī does not copy any of the phrases with *ṭarafīn*, but in two problems he gives a name to this method that is not found in our text. After setting up the forthcoming equations and stating the difference of the squares in Problem V.37 (his version of *Arithmetica* Problem IV.36) he writes: "And one can work it out by the equality of the *mathnāt*",[4] and in Problem V.38 (*Arithmetica* IV.37), after setting up the forthcoming equations, he has: "So if you wish, you work it out by the equality of the *mathnāt*" (Saidan 1986, 303.21). The word *mathnāt* means something that is "repeated, or recited twice" (Lane 1863-93, 360). From the context, this something could be the equation or one side of the equation. But taking into account the *ṭarafīn* from Qusṭā's translation, the latter is more likely.

We agree with Rashed, then, that the word *ṭarafīn* intends the second sides of the equation. We translate the passage from the present problem as: "So both sides of the equation are equal, each of them turning out to be a Cube Equals five *Māls*", where we are expected to understand that these "sides" are the 5M from the simplified versions of the two equations and not the entire equation 1C = 5M. The passage from Problem IV.42(a) that we quoted at the start of this discussion then becomes "And if we wish, we can work this out by seeking equality in the two sides of the Equation, as we explained in the preceding problems".

---

3 For other examples, see al-Fārisī (1994, 514.1, 531.5, 559–60), Ibn al-Hā'im (2003, 68.1, 99.1), Ḥājjī-Khālifa (1837, II, 583), and Al-'Āmilī (1976, 107.5,6). Sesiano already noted that these last two used the word this way (1982, 445–46).

4 (Saidan 1986, 303.10; MS Paris 2459 f. 106r12). Saidan's edition has the negation *la-* preceding *yumkin* so that it means "one cannot", which makes no sense. The prefix is not in the Paris MS.

There is support for this interpretation in the one other passage with *ṭarafīn*, given at the end of the third solution to Problem IV.42(a). This solution is worked out by a variation of the new method, in which Diophantus divides each side of his forthcoming equations by the same square monomial to reduce them to equations in Things and units, from which he solves for the value of 1T directly. He writes: "Most problems of the self-same sides (*dhawāt al-ṭarafīn*) presented above can be worked out by the method we (just) explained" (line 1410). From the larger context of the solutions in this last part of Book IV, it seems that "the self-same sides" refers to the new method introduced in the second solution to Problem IV.34 and applied at least once in every problem through IV.43. The "self-same" (*dhawāt*) indicates that he intends not the opposite sides of one equation but the corresponding sides of the two equations. Thus, we call this the method of "equating the corresponding sides". Of course, the method of the double-equality also seeks to make these two sides equal, and the first instance of *ṭarafīn* occurs in a solution by that method. Both methods work with a double-equality, and both methods make the two sides equal.

IV.35 $\begin{cases} y^2 + x^3 = \Box \\ y^2 - x^3 = \Box' \end{cases}$

Assignments	Operations	Equation
$x^3 := 1C$ $y^2 := 4M$		
	Add 4M and 1C → 4M 1C Subtract 1C from 4M → 4M ℓ 1C	
		4M 1C = $\Box$ 4M ℓ 1C = $\Box'$

Unlike the pair of forthcoming equations in the previous problem, this pair cannot be solved by "the double-equality". Thus, Diophantus works it out by "equating the corresponding sides", like he did in the second solution to IV.34. Diophantus wants to assign each square to be some square multitude of *Māl*s. Solving each of the forthcoming equations for 1T, he finds that "the (multitude of) *Māl*s remaining in the first equation is a square number (with) four subtracted from it, and the (multitude of) *Māl*s remaining in the second equation is a square number subtracted from four".

Diophantus then takes an approach to simplify this sub-problem that he will repeat again in Problems IV.37–41 and 43: he rephrases the equality as if it were an equation. "Thus, we say: a square less four units Equals four units less another square". This statement, as with those in the other problems, only mimics the form of an equation since the unknown amounts, here the two squares, remain unnamed. The phrase "Thus, we say" marks the transition to algebraic language, with a nod to the fact that it is not an equation in the full sense. Algebraic features of this language include the use of the verb "Equals" (from *'adala*) relating the two amounts, and the expression of the difference with "less" (*illā*) and not with the operation of subtraction. Further, the "equation" is simplified just as any equation would be. Here it is stated as: "we add the four units excluded from the first square in common to both sides", with similar wording in the other problems. The phrasing of the simplification in each case comes not from Arabic but is translated from Greek. The "equation" in the present problem becomes "two (different) squares, and they Equal eight units". The switch to algebraic language is helpful specifically because it allows for the articulation of this simplification. Al-Karajī also switches to algebraic language in his solution of this problem (his V.36), and in his case he simplifies the quasi-equation with the Arabic phrase "you restore and confront" (from *al-jabr wa-l-muqābala*).[5] And like we saw the second solution to the previous problem, it seems that it would have been much easier for Diophantus to simply add the two forthcoming equations together to find that the sum of □ and □′ is 8M. Diophantus was able to subtract the expressions of one equation from those of another, so it was certainly within his ability to add them, too.

So, Diophantus looks for two squares whose sum is 8. Because 8 is already the sum of two equal squares, "we should divide the eight into two other square numbers as we explained in the second Book", in Problem II.9. He finds the solution $\frac{4}{25}$u and $7\frac{21}{25}$u. Returning now to the original problem:

Assignments	Operations	Equation
$\square := 7\frac{21}{25}\,\mathrm{M}$		
		$4\mathrm{M}\ 1\mathrm{C} = 7\frac{21}{25}\,\mathrm{M}$
$\square' := \frac{4}{25}\,\mathrm{M}$		
		$4\mathrm{M}\ \ell\ 1\mathrm{C} = \frac{4}{25}\,\mathrm{M}$

Both equations simplify to $3\frac{21}{25}\,\mathrm{M} = 1\mathrm{C}$. Thus, $1\mathrm{T} = 3\frac{21}{25}$u, so the sought-after cube is $x^3 = \frac{884,736}{15,625}$ and the sought-after square is $y^2 = \frac{36,864}{625}$.

IV.36 $\begin{cases} x^3 + mx^2 = \square \\ x^3 - nx^2 = \square' \end{cases}$ $(m = 4, n = 5)$

Assignments	Operations	Equation
$x^3 := 1\mathrm{C}$		
$x^2 := 1\mathrm{M}$		

---

5 (Saidan 1986, 302.10). Al-Karajī also switches to algebraic language in his Problems V.38, V.40, and V.41 (IV.37, IV.39, and IV.40 in Diophantus), but in those problems he uses the non-algebraic *mithl* for "equal".

The solution is again by "equating the corresponding sides" (see the two preceding problems). After making the initial assignments, Diophantus wishes for both $\square$ and $\square'$ to be assigned as square multitudes of *Māl*s. But he looks for these multitudes in advance, before setting up the forthcoming equations. He seeks m^2 and m'^2 such that $m^2 - 4 = m'^2 + 5$, and the reason will only become clear afterward, when the forthcoming equations are found. Finding two squares whose difference is 9 is solved in Problem II.10, and he finds the squares 25 and 16. He then continues the solution:

Assignments	Operations	Equation
	Four times 1M → 4M	
	Add 1C and 4M → 1C 4M	
		1C 4M = $\square$
$\square := 25$M		1C 4M = 25M
	Subtract 4M from both sides	
		1C = 21M
	Five times 1M → 5M	
	Subtract 5M from 1C → 1C ℓ 5M	
		1C ℓ 5M = $\square'$
$\square' := 16$M		
		1C ℓ 5M = 16M
	Add 5M to both sides	
		1C = 21M

Both equations give 1T = 21u, so the sought-after cube is $x^3 = 9261$.

The text then notes that if the roles of the given 4 and 5 are reversed, then 1T will be 20u, and the sought-after cube $x^3 = 8000$. As Sesiano writes (1982, 209–10), this was probably inserted by some scholiast. We add that it may also be the scholiast who rearranged the solution by putting the finding of the multitudes m^2 and m'^2 before the forthcoming equations were set up. In al-Karajī's solution to the corresponding Problem V.37 the forthcoming equations are set up first.

IV.37 $\begin{cases} x^3 + mx^2 = \square \\ x^3 + nx^2 = \square' \end{cases}$ $(m = 5, n = 10)$

Assignments	Operations	Equation
$x^3 := 1$C		
$x^2 := 1$M		
	Multiply 1M by 5u → 5M	

Multiply 1M by 10u → 10M

Add 1C and 5M → 1C 5M

$$1C\ 5M = \square$$

Add 1C and 10M → 1C 10M

$$1C\ 10M = \square'$$

This problem is worked out again by "equating the corresponding sides", much like Problem IV.35. Once the two forthcoming equations are found, Diophantus writes that each square will be some square multitude of *Māl*s. Then he solves the equations for 1T and re-expresses the relationship between the multitudes as if it were an algebraic equation: "Thus, we say, a square less five units Equals the other square less ten units", which he simplifies to "a square and five units Equal a square". Again, it would have been simpler for him to subtract the first forthcoming equation from the second to arrive at $5M = \square'\ \ell\ \square$. There is also the condition that the smaller square must be greater than 5 units. Again, this is Problem II.10, and the two squares are found to be $53\frac{7}{9}$ and $58\frac{7}{9}$. The solution continues:

Assignments	Operations	Equation
$\square := 53\frac{7}{9}M$		
		$1C\ 5M = 53M\,\frac{7}{9}M$
$\square' := 58\frac{7}{9}M$		
		$1C\ 10M = 58\frac{7}{9}M$

Both equations simplify to $1C = 48\frac{7}{9}M$. Thus, $1T = 48\frac{7}{9}u$, so the sought-after cube is $x^3 = \frac{84,604,519}{729}$.

IV.38 $\begin{cases} x^3 - mx^2 = \square \\ x^3 - nx^2 = \square' \end{cases}$ $(m = 5, n = 10)$

Assignments	Operations	Equation
$x^3 := 1C$		
$x^2 := 1M$		
	Multiply 1M by 5u → 5M	
	Multiply 1M by 10u → 10M	
		$1C\ \ell\ 5M = \square$
		$1C\ \ell\ 10M = \square'$

The solution proceeds just as in Problems IV.35 and 37, again writing the relationship between the multitudes as if it were an algebraic equation in order to simplify it. This time Diophantus seeks any two squares whose difference is five units. He finds the squares 4 and 9, and the solution continues:

Assignments	Operations	Equation
$\square \doteq 9M$		
		$1C\ \ell\ 5M = 9M$
$\square' \doteq 4M$		
		$1C\ \ell\ 10M = 4M$

Both equations simplify to $1C = 14M$. Now $1T = 14u$, so the sought-after cube is $x^3 = 2744$.

IV.39 $\begin{cases} mx^2 - x^3 = \square \\ nx^2 - x^3 = \square' \end{cases}$ $(m = 3, n = 7)$

Assignments	Operations	Equation
$x^3 \doteq 1C$		
$x^2 \doteq 1M$		
	Multiply 1M by 3u → 3M	
	Multiply 1M by 7u → 7M	
	Subtract 1C from 3M → 3M ℓ 1C	
	Subtract 1C from 7M → 7M ℓ 1C	
		$3M\ \ell\ 1C = \square$
		$7M\ \ell\ 1C = \square'$

Again, by "equating the corresponding sides", Diophantus will assign the two squares to be square multitudes of *Māl*s, and he runs through the simplifications of the two forthcoming equations to find a relation between the multitudes. And again, the relationship between the multitudes is stated as if it were an algebraic equation in order to simplify it. This time he seeks two squares whose difference is 4 and so that the smaller is less than 3. This is again Problem II.10, and he finds $2\frac{1}{4}$ and $6\frac{1}{4}$. The solution continues:

Assignments	Operations	Equation
$\square \doteq 2\frac{1}{4}M$		
		$3M\ \ell\ 1C = 2\frac{1}{4}M$

$\square' := 6\frac{1}{4}M$

$7M\,\ell\,1C = 6\frac{1}{4}M$

Both equations simplify to $1C = \frac{3}{4}M$. Thus, $1T$ is $\frac{3}{4}u$ and the sought-after cube is $x^3 = \frac{27}{64}$.

IV.40 $\begin{cases} \left(x^2\right)^2 + y^3 = \square \\ \left(x^2\right)^2 - y^3 = \square' \end{cases}$

Assignments	Operations	Equation
$x := 2T$		
	Square $2T \to 4M$	
	Square $4M \to 16MM$	
$y := 4T$		
	Cube $4T \to 64C$	
	Add this to $16MM \to 16MM\ 64C$	
		$16MM\ 64C = \square$
	Subtract $64C$ from $16MM$ $\to 16MM\ \ell\ 64C$	
		$16MM\ \ell\ 64C = \square'$

The two forthcoming equations are solved again by "equating the corresponding sides" (see in particular Problems IV.34 and 35). Diophantus writes that the two squares will be assigned to be square multitudes of *Māl Māl*s, and he solves both equations for $1T$ and equates the results. Again, as he first did in Problem IV.35, he expresses this equality as if it were an algebraic equation: "a great square number less sixteen Equals sixteen less a small square number", which he simplifies to "a great square and a small square Equal thirty-two units". Such squares can be found by Problem II.9, since 32 is already the sum of 4^2 and 4^2. He finds, without showing his work, $\frac{16}{25}$ and $31\frac{9}{25}$. The solution continues:

Assignments	Operations	Equation
$\square' := 31\frac{9}{25}MM$		
		$16MM\ 64C = 31\frac{9}{25}MM$
$\square := \frac{16}{25}MM$		
		$16MM\ \ell\ 64C = \frac{16}{25}MM$

Both equations simplify to $64C = 15\frac{9}{25}$ MM, from which 1T is $4\frac{1}{6}$ u, so the sought-after square is $x^2 = 69\frac{4}{9}$ and the sought-after cube is $y^3 = 4629\frac{17}{27}$.

IV.41 $\begin{cases} y^3 + (x^2)^2 = \square \\ y^3 - (x^2)^2 = \square' \end{cases}$

As Sesiano suggests (1982, 118 n. 81), this problem was most likely the second part of Problem IV.40. Diophantus begins with the forthcoming equations that arise from the same assignments as in that problem.

Assignments	Operations	Equation
		64C 16MM = $\square$
		64C ℓ 16MM = $\square'$

Again by "equating the corresponding sides", Diophantus wishes to assign the squares to be square multitudes of *Māl Māls*. He then solves each equation for 1T and reframes the equality that relates the two square multitudes as if it were an algebraic equation (see Problem IV.35). After simplification of this quasi-equation, he needs to find two squares that differ by 32. This is Problem II.10, and without showing the process, Diophantus finds the squares 4 and 36. The solution continues:

Assignments	Operations	Equation
$\square := 36$MM		
		64C 16MM = 36MM
$\square' := 4$MM		
		64C ℓ 16MM = 4MM

Both equations simplify to $64C = 20$MM. Thus, 1T is $3\frac{1}{5}$u. Hence, the sought-after cube $y^3 = 2097\frac{95}{625}$, and the sought-after square $x^2 = 40\frac{24}{25}$.

IV.42 (a) $\begin{cases} (x^3)^3 + (y^2)^2 = \square \\ (x^3)^3 - (y^2)^2 = \square' \end{cases}$

Assignments	Operations	Equation
$x := 2$T		
	Cube 2T $\to$ 8C	

$y := 4M$

Cube 8C → 512CCC

Square 4M → 16MM
Square 16MM → 256CCM
Add 512CCC and 256CCM
→ 512CCC 256CCM
Subtract 256CCM from 512CCC
→ 512CCC ℓ 256CCM

512CCC 256CCM = □
512CCC ℓ 256CCM = □'

First solution (by the double-equality)   (line 1348)

Diophantus gives only the instructions for solving the problem. The difference of the squares is 512CCM, so one finds two multitudes of MMs that multiply to give 512CCM. Then a root of the greater square is half the sum of the factors, and a root of the smaller square is half the difference of the factors. Either equation simplifies to give the value of 1T. Then the vales of x^3 and y^2 are calculated.

Second solution (by equating the corresponding sides)   (line 1363)

Here, too, only the process of finding the solution is given. The two squares are to be assigned to be square multitudes of Cube Cube *Māl*s, and the forthcoming equations are solved indeterminately. Diophantus finds that he needs two squares whose sum is 512. These will be the multitudes of CCMs, and the forthcoming equations can be simplified and solved.

Third solution (by a variation on equating the corresponding sides)   (line 1391)

Instead of declaring that the two squares will be made multiples of CCMs, solving for 1T, and then setting up an equation to find the multitudes, Diophantus first divides every term by a square multitude of CCMs to reduce the degree of the forthcoming equations to the level of Things and units, and then he searches for the value of 1T directly.

Assignments	Operations	Equation
	Divide 512CCC 256CCM by 16CCM → 32T 16u	
		32T 16u = □''

Divide 512CCC ℓ 256CCM by 16CCM
$\rightarrow$ 32T ℓ 16u

32T ℓ 16u = $\square'''$

The two forthcoming equations are then read as the problem of finding a number (32T) such that adding 16 to it gives a square, subtracting 16 from it gives a
square. This can be solved by Problem II.10, and the value of 1T can then easily be found. He does not finish the problem. If, for example, he had found the
squares to be 36 and 4, so that 32T = 20u, then 1T = $\frac{5}{8}$u. It is this reduction of
the powers (and multitudes) and the subsequent solution for 1T that Diophantus
refers to when he writes "Most problems of 'the self-same sides' (like the ones)
presented above can be worked out by the method just described". He later applies
this variation by reduction of powers in the three parts of Problem IV.44. See
Problems V.4, V.5, and V.6 for another variation in which the final forthcoming
equations are solved differently.

(b) $\begin{cases} \left(y^2\right)^2 + (x^3)^3 = \square \\ \left(y^2\right)^2 - (x^3)^3 = \square' \end{cases}$ (line 1412)

Now, with the roles of the square and the cube reversed in the second condition, Diophantus proposes the same assignments as before so that the forthcoming
equations are:

Assignments	Operations	Equation
		256CCM 512CCC = $\square$
		256CCM ℓ 512CCC = $\square'$

The solution is by "equating the corresponding sides". He skips over declaring
the squares to be square multitudes of CCMs, the simplification of the equations,
and the simplification of the relation of the multitudes, straight to the problem of
dividing double the 256, or 512, into two squares. This is Problem II.9 again, and
Diophantus finds the squares $10\frac{6}{25}$ and $501\frac{19}{25}$ without showing the derivation. He
then continues to sketch the solution. Had he stated the equations, they would have
been 256CCM 512CCC = $501\frac{19}{25}$ CCM and 256CCM ℓ 512CCC = $10\frac{6}{25}$ CCM,
which both simplify to 512CCC = $245\frac{19}{25}$ CCM. Thus, 1T is $\frac{12}{25}$u, so the cube
$x^3 = \frac{13,824}{15,625}$ and the square $y^2 = \frac{331,776}{390,625}$.

Analysis and synthesis (by J. Oaks)

In Problems IV.37, IV.42a, and IV.43 we find phrases mentioning "analysis"
(تجليل, *taḥlīl*) and "synthesis" (تركيب, *tarkīb*). My coauthor holds that these are

Arabic interpolations and do not originate with Diophantus. He makes his arguments in (Christianidis 2021b). I hold that what we read in the Arabic translation is perfectly consistent with what Diophantus himself could have written, and this view is argued presently. I begin by giving my interpretation of the phrases in which the words appear and why they only occur in these problems. I then address analysis and synthesis generally and conclude with some comments on the interpretations of other historians.

Al-Samaw'al's explanation that algebra belongs to analysis is translated in Section 4.1.4. In the next subsection, 4.1.5, we list the parts of a Diophantine problem. I repeat that list here and identify which parts belong to analysis and which to synthesis in a manner consistent with both al-Samaw'al and the Greek definitions of the terms:

1   Enunciation
2   Determination
3   Setting out, or assignment of given numbers
4   Definition of goal or instantiated enunciation

Analysis
5   Stage 1: Naming unknowns, performing operations, establishing an equation
6   Stages 2–3: Simplifying and solving the equation for the value of 1 Number/ one Thing

Synthesis
7   Stage 4: Calculating the desired numbers
8   Proof

9   Conclusion

The first instance of the word "analysis" is encountered in the solution to Problem IV.37. This problem is solved by what we call the "method of equating the corresponding sides", which is a way of extracting a solution from a pair of forthcoming equations that is applied in Problems IV.34 (second solution), 35–41, 42a (second solution), 42b, and 43. See our commentary to Problem IV.34 for a description. For this method Diophantus needs to refer to the value of the Thing before actually finding it, one time for each forthcoming equation. In Problem IV.34, for example, he finds from the first forthcoming equation, "the number that was assigned in the problem to be one Thing is equal to the number of the remaining *Māl*s", and from the second equation, "the number that was assigned in the problem to be a Thing is equal to the number of the *Māl*s added together." Among the problems solved by this method, Diophantus refers to the value of one Thing by long phrases like these, usually "the number assigned[6] in the problem to be a

---

6 The text literally reads "the assigned number", but we write it as "the number assigned" to make the reading easier.

Thing", 13 times. He writes merely "the Thing" five times, and in two problems, IV.36 and 42b, he does not express this step.

In 11 of those 13 instances the text reads "in the/this problem", but one of the two instances in IV.37 shows "in this analysis", and one of the two in IV.43 shows "in this solution". The complete phrases are "the number assigned in this analysis to be a Thing" and "the number assigned in this solution to be a Thing". It makes no difference whether one writes of the value of the Thing as assigned "in this problem", "in this solution", or "in this analysis". The assignment takes place in the analysis, the analysis is part of the solution, and the solution is part of the problem.

The long phrase reappears soon after the last of this string of problems, but not in connection with a forthcoming equation. In Problem IV.44b the text announces the value of one Thing at the end of stage 3 not with the usual "the one Thing is" or something similar but with "this number is the (number) assigned in the solution of the problem to be a Thing, and it is one hundred ninety-two". And then again, in IV.44c, it appears as "forty-seven, and that is the number assigned to be a Thing in the solution of this problem". Long phrases like these to designate the value of the Thing are not found in the remainder of the work. The long version must have been echoing in the mind of the author from the previous problems when he came to this part in the solutions to IV.44b and 44c.

The three solutions to Problem IV.42a are the only ones in the extant *Arithmetica* in which, after making initial assignments and performing some calculations, Diophantus merely describes how to complete the solution to the problem, without making the further assignments that would be needed to produce a particular answer. In describing how to proceed, he distinguishes between the analytical part, which begins with the initial assignments, and the synthetic part, where the sought-after numbers are calculated. For example, after setting up two forthcoming equations $32T\ 16u = \square''$ and $32T\ \ell\ 16u = \square'''$ in the third solution to IV.42a, the text reads (beginning at line 1405):

> Let us look for a number such that, if we add a given number to it, which is sixteen, it gives a square, and if we subtract a given number from it, which is sixteen, it gives a square. Once we have found that number, we divide it by thirty-two, so what results from the division is the Thing. Once we know it, we return to synthesize the problem according to the way we set it up in the analysis.

When the value of one Thing has been found, the reader is instructed to "synthesize the problem", that is, to calculate the values of the sought-after cube and square, "according to the way we set it up in the analysis", that is, according to the assignments for the cube and the square made in earlier in the solution. The corresponding instruction in the first solution is: "After knowing the Thing, we can synthesize everything in the problem" (line 1361), and for the second solution it is: "and from that we can find the Thing whose value we are looking for. Then we return to perform the synthesis of the problem" (line 1389).

The next Problem, IV.43, is worked out completely. Perhaps it is because of the repeated instructions in IV.42a that we read, after the value of the Thing is found to be 16 units, "We will now perform the synthesis of the problem in the way we did the analysis" (line 1493). The sought-after cube and square are then calculated to be 4096 and 262,144, respectively. Here, too, the longer phrase from IV.42a seems to have been echoing in the mind of the author when writing this solution.

The uses of "analysis" and "synthesis" in Book IV are thus consistent with the definitions given to the terms in Pappus, Hero, the two anonymous Greek scholiasts, and with al-Samaw'al, all of whom characterize the terms only by the direction of the argument. The occurrences of the terms in precisely these problems have to do with the ways the problems are solved, either because the "method of equating the corresponding sides" entails reference to the value of one Thing before actually finding it (IV.37), or because the solution is merely sketched out (IV.42a, from which IV.43). Diophantus, like most authors, did not often have a need to refer to the value of the Thing or to particular parts of his solutions, and when he did, there was no rigid formula for how it should be done. Repetition with minor variations of an unusual phrase in consecutive problems is to be expected because the wording from the previous problem will still be resonating in the author's mind as he writes the next one. Then, many problems later, another wording might be devised.

Setting aside the *Arithmetica*, all extant examples of analysis in Greek mathematics belong to geometry.[7] And because analysis is applied much more often to problems than to theorems in our extant texts, the structure of problematical analysis in geometry might be mistaken to represent analysis generally. Recall that the two parts of an analysis in Greek problematic geometry identified by Hankel (1874) are the transformation and the resolution. Hintikka and Remes (1974) understandably call the transformation the "analysis proper", since it is there that the desired construction is assumed, and the argument proceeds to something independently constructible, or given. The resolution, which includes the chain of givens, serves an auxiliary role to verify that given object found at the end of the transformation will lead to the desired construction. The resolution thus guarantees that the synthesis will be successful, and it necessarily follows the logical direction of the synthesis, beginning with something given and proceeding to what is desired. The chain of givens itself, then, is synthetic in its direction.

The structure of theorematic analysis is different. In particular, there is no need for a chain of givens, so there is no resolution. Instead, Sidoli and Saito (2012) identify two stages of theorematic analysis as the "construction" and the "deduction", which are often followed by a "verification". They also describe a third type of analysis in Greek geometry which they call "comparative analysis",

---

7 By "analysis", I mean of course a line of reasoning that satisfies the ancient definitions. The chain of givens in Diophantus's Problem $V^G.10$ is a reduction (see Berggren and van Brummelen 2000 on reductions), and the chain of givens in Diophantus's *On Polygonal Numbers* serves as instructions for calculating a polygonal number given its side, based in a theorem that had already been proven (Diophantus 2011, 197.7–13, 211.1–7; Heath 1910, 253). Neither chain belongs to an analysis.

whose structure differs again from the other two. Further, we know from our ancient sources that analysis in geometry inspired its use outside of mathematics to problems in philosophy, medicine, and theology,[8] and we should expect that those arguments would have also exhibited their own structures that also differ from the transformation/resolution structure of problematic analysis in geometry.

Thus, analysis and synthesis as defined by Pappus, Hero, and the scholiasts is not a particular method of problem-solving or proof-writing with its own structure and vocabulary the way, say, that algebra and double false position are methods. Analysis and synthesis are simply ways of approaching problems or theorems, the former beginning with what is sought and the latter beginning with what is posited (Mahoney 1968, 319; Sidoli and Saito 2012, 2). In a particular setting, such as geometrical problems, analysis and synthesis will take on a particular structure, lending it the appearance of a method in that context. But analysis applied in other settings, such as in the problems in logistic that we find in Diophantus, will naturally differ in the ways the arguments unfold. One important difference is that in algebra, as in theorematic geometry, there is no need for a resolution. Knowing the value of 1 Number or one Thing found in the analysis, the synthesis consists of calculating the desired values from the assignments, and that is always a straightforward matter. Because there is no need to verify that the synthesis will work, there is no resolution, and thus no chain of givens, in the algebraic solutions in the *Arithmetica* or in those of any Arabic or later book.

The lack of any mention of analysis or synthesis by Diophantus in his introduction makes sense if one understands the two as ways of *approaching* problems or theorems, and not as methods. The *Arithmetica*, like many Arabic books and chapters on algebra, is a pedagogical work. His introduction presents the vocabulary and rules needed for solving problems by the method of algebra. Neither Diophantus nor the Arabic authors of introductory texts on algebra have any reason to situate the method in the context of analysis and synthesis.

There is no reason for us to regard the instances of "analysis" and "synthesis" in Book IV as Arabic interpolations. That does not mean that they are absolutely genuine but only that they are consistent with our Greek sources. In my view, we are fairly safe in presuming that they are authentic.

It is because the structure of premodern algebra has only recently been elucidated that previous historians have not fully understood the nature of analysis in Diophantus. Writing in 2011, Fabio Acerbi noted that although the method exhibited in the *Arithmetica* is apparently analytical, "[t]he *Arithmetica* are in the end a gigantic mass of reductive procedures".[9] This can be said only by someone

---

8 (Menn 2002; Byrne 1997, 3ff, 56–58; Hankinson 2009, 222–25; Havrda 2016, 92–95; 106–09).

9 "Gli Arithmetica sono insomma una gigantesca congerie di procedure riduttive" (Diophantus 2011, 50). This view is due in part to his misunderstanding of the species: "Diofanto introduce poi il termine δύναμις «potenza» come sinonimo di numero «quadrato»" (Diophantus 2011, 11), and "Diophantus settles the problem of separating particular square

who does not recognize the naming of unknowns and the building of equations from those names. Reviel Netz accepted the analytical nature of the solutions, but wrote: "In the Arabic Diophantus, besides the quick verification one also has a formal synthesis, repeating the argumentation of the analysis *backwards* so that one sees that, given the solution, the terms of a problem cannot fail to hold" (Netz 2012, 350, his emphasis). Of course, the synthesis does not repeat the analysis backwards. It refers only to the initial assignments, disregarding the formation of expressions and the setting-up, simplification, and solution of the equation. Roshdi Rashed and Christian Houzel (2013, 30) also see the synthesis as the strict reversal of the analysis. In reference to the instances of the words "analysis" and "synthesis" in the Arabic Problems IV.37, 42a, and 43, they claim that "Par le terme « synthèse » . . . Diophante entend la démarche inverse de l'analyse (il y a double implication stricte)." Jacques Sesiano, however, correctly identified the analytic and synthetic parts of a Diophantine solution, though he did not explain how the analysis unfolds (Sesiano 1982, 49).

And last, the philosopher Jacob Klein's remarks on analysis and synthesis in connection with Diophantus take place in the context of his search for the inspiration for François Viète's new algebra, dubbed *logistice speciosa* (see Section 3.5.6). Klein supposed without any clear evidence that, for Viète, "the 'analytical' manner of finding solutions in Diophantus and geometric ('problematical') analysis are understood as completely parallel procedures" (Klein 1968, 163). Then, because analysis in problematic geometry "does not need to use the 'given' magnitudes as unequivocally determinate *but only as having the character of being 'given'*" (p. 164, his emphasis), Klein looked for a parallel procedure in Diophantus where the givens, i.e., the multitudes of the powers, are not determined. He found it in Diophantus's solutions "in the indeterminate" presented, among other places, in Problem IV^G.19 and in the lemmas to Problems IV^G.34, 35, and 36.[10] It is the set-up and simplification of equations in these problems in particular that are, for Klein, "the properly 'analytic' procedure" (p. 164).[11] Viète himself never singled out these problems. His only remark on the method of Diophantus was to claim that he had also solved his problems by *logistice speciosa*, but that he presented his solutions numerically instead.[12] It is Klein who decided that to be "properly analytic" the givens must be undetermined. There is nothing in our

---

numbers from the species "square" by means of the opposition τετράγωνος/δύναμις" (Acerbi 2015, 905).

10 Some other problems, too, are mentioned on page 134.

11 Klein overlooks the fact that the algebraic manipulations in these solutions take place with determinate givens and that it is by tracing the origins of the particular solutions that they can be stated indeterminately at the end.

12 At the end of Chapter VI in his *In Artem Analyticem Isagoge* Viète wrote: "Diophantus in those books which concern arithmetic employed zetetics most subtly of all. But he presented it as if established by means of numbers and not also by species (which, nevertheless, he used), in order that his subtlety and skill might be the more admired" (translated by J. Winfree Smith, and published in Klein 1968, 345).

ancient or medieval texts to support this presumption, so from the perspective of analysis there is nothing special about the "solutions in the indeterminate". Klein's understanding of analysis in Diophantus is thus misguided.

IV.43
$$\begin{cases} \left(x^3\right)^3 + m(y^2)^2 = \square \\ \left(x^3\right)^3 - n(y^2)^2 = \square' \end{cases} \qquad (m = 1\tfrac{1}{4}, \; n = \tfrac{3}{4})$$

Assignments	Operations	Equation
$x^3 := 1C$		
	Cube 1C → 1CCC	
$y := 2M$		
	Square 2M → 4MM	
	Square 4MM → 16CCM	
	$1\tfrac{1}{4}$ of 16CCM → 20CCM	
	Add 1CCC and 20CCM → 1CCC 20CCM	
		1CCC 20CCM = □
	$\tfrac{3}{4}$ of 16CCM → 12CCM	
	Subtract 12CCM from 1CCC → 1CCC ℓ 12CCM	
		1CCC ℓ 12CCM = □′

The solution is by "equating the corresponding sides". The squares will be assigned to be square multiples of CCMs, and the forthcoming equations are solved for 1T and their indeterminate solutions are equated to determine the relationship of those multitudes. This is expressed again as a quasi-equation that simplifies to: "a small square and thirty-two Equal a great square". This is Problem II.10, and Diophantus finds the squares to be 4 and 36. The solution continues:

Assignments	Operations	Equation
□ := 36CCM		
		1CCC 20CCM = 36CCM
□′ := 4CCM		
		1CCC ℓ 12CCM = 4CCM

These equations both simplify to 1CCC = 16CCM. The solution is 1T = 16u, the sought-after cube is $x^3 = 4096$, and the sought-after square is $y^2 = 262,144$.

IV.44   (a) $\begin{cases} \left(x^3\right)^3 + m(y^2)^2 = \square \\ \left(x^3\right)^3 + n(y^2)^2 = \square' \end{cases}$ ($m = 3, n = 8$)

Assignments	Operations	Equation
$x := 1T$		
	Cube 1T → 1C	
	Cube 1C → 1CCC	
$y := 2M$		
	Square 2M → 4MM	
	Square 4MM → 16CCM	
	Multiply 16CCM by 3u → 48CCM	
	Multiply 16CCM by 8u → 128CCM	
	Add 48CCM to 1CCC → 1CCC 48CCM	
	Add 128CCM to 1CCC → 1CCC 128CCM	
		1CCC 48CCM = $\square$
		1CCC 128CCM = $\square'$
	Divide 1CCC 48CCM by 1CCM → 1T 48u	
	Divide 1CCC 128CCM by 1CCM → 1T 128u	
		1T 48u = $\square''$
		1T 128u = $\square'''$

The solution to this problem, and for (b) and (c) as well, is by the variation of "equating the corresponding sides" first shown in the third solution to Problem IV.42(a). Rather than relate the two squares, Diophantus looks for a number (the 1T) such that adding 48 and 128 give squares. This is Problem II.11, though Problem II.10 could also work, and Diophantus finds that 1T is 16u without showing the calculations. The sought-after cubic and square numbers are then $x^3 = 4096$ and $y^2 = 262,144$, the same numbers found in the previous problem.

(b) $\begin{cases}\left(x^3\right)^3 - m(y^2)^2 = \square \\ \left(x^3\right)^3 - n(y^2)^2 = \square'\end{cases}$ $(m = 3, n = 8)$ (line 1549)

Assignments	Operations	Equation
$x^3 \coloneqq 1C$		
	Cube 1C → 1CCC	
$y^2 \coloneqq 4MM$		
	Square 4MM → 16CCM	
	Multiply 16CCM by 3u → 48CCM	
	Multiply 16CCM by 8u → 128CCM	
	Subtract 48CCM from 1CCC → 1CCC ℓ 48CCM	
	Subtract 128CCM from 1CCC → 1CCC ℓ 128CCM	
		1CCC ℓ 48CCM = $\square$
		1CCC ℓ 128CCM = $\square'$
	Divide 1CCC ℓ 48CCM by 1CCM → 1T ℓ 48u	
	Divide 1CCC ℓ 128CCM by 1CCM 1T ℓ 128u	
		1T ℓ 48u = $\square''$
		1T ℓ 128u = $\square'''$

Again, Diophantus looks for a number (1T) satisfying these forthcoming equations. This is Problem II.13, though Problem II.10 could also have been applied, and 1T is found to be 192u. This time Diophantus does not calculate the values of x^3 and y^2 directly from the assignments. If he had done so, he would have found $x^3 = 7,077,888$ and $y^2 = 37,748,736$. Perhaps the problem with this solution is that the verification leads to unmanageably large numbers: $(x^3)^3 - 3(y^2)^2$ and $(x^3)^3 - 8(y^2)^2$ turn out to be on the order of 10^{20}. The greatest number stated in the extant *Arithmetica* is 2,641,807,540,224, in Problem IV.42, which is only of the order of 10^{12}. Similarly, in Problem VI.4 (line 2270) Diophantus chooses a certain remainder to be "a small number" to ensure that subsequent calculations do not wildly exceed the trillions.

Instead, Diophantus notes that the 1T = 192u of this problem is 12 times the 1T = 16u that was found in part (a). Since $x \coloneqq 1T$ and $y \coloneqq 2M$ in both problems,

the sought-after cube x^3 in part (b) is 12^3 (1728) times greater than 16^3 (4096), the sought-after cube of part (a), and the sought-after square y^2 in part (b) is $(144)^2 = 20{,}736$ times greater than $4 \cdot 16^4 = 262{,}144$, the square in part (a). As he notes, "a square of that square $((4 \cdot 16^4)^2)$ is equal to a cube of that cube $((16^3)^3)$" (line 1572), so this square amount can be factored out of both $(x^3)^3 - 3(y^2)^2$ and $(x^3)^3 - 8(y^2)^2$. Thus, $x^3 = 12^3$ (1728) and $y^2 = 144^2$ (20,736) is also a solution to the problem. Now $(x^3)^3$ and $(y^2)^2$ are only in the billions and hundreds of millions, respectively, and the solution can be more easily verified, and this verification is provided.

(c) $\quad \begin{cases} m(y^2)^2 - \left(x^3\right)^3 = \square \\ n(y^2)^2 - \left(x^3\right)^3 = \square' \end{cases} \qquad (m = 3, n = 8) \qquad \text{(line 1594)}$

The solution begins with the forthcoming equations, which come from the same assignments as parts (a) and (b).

Assignments	Operations	Equation
		48CCM ℓ 1CCC $= \square$
		128CCM ℓ 1CCC $= \square'$
	Divide 48CCM ℓ 1CCC by 1CCM $\rightarrow$ 48u ℓ 1T	
	Divide 128CCM ℓ 1CCC by 1CCM $\rightarrow$ 128u ℓ 1T	
		48u ℓ 1T $= \square''$
		128u ℓ 1T $= \square'''$

A solution to these equations can be found by Problem II.12, but again, II.10 could also work. Diophantus gives the solution 47u for the value of 1T. Thus, $x^3 = 103{,}823$ and $y^2 = 19{,}518{,}724$. Here the calculations to verify the solution are on the order of 10^{15}. Only the square roots of the two squares $\square$ and $\square'$ are stated, however.

Book V (Arabic)

$$V.1 \quad \begin{cases} \left(x^2\right)^2 + my^3 = \square \\ \left(x^2\right)^2 - ny^3 = \square' \end{cases} \qquad (m = 4, n = 3)$$

Problems V.1–3 are solved in an unusual way. Diophantus begins by establishing a pair of forthcoming equations in which one side is partially named in terms of the powers and the other is not named at all. From them then sets up a third forthcoming equation in which neither side has any named part, where both sides will be assigned different names for the same sought-after amount, y^3.

In the present problem Diophantus begins by assigning a side of the square x^2 as 1T, so the square is 1M and its square is 1MM. Thus, $\square$, 1MM, and $\square'$ are three descending squares whose successive differences are $4y^3$ and $3y^3$. The ratio of these differences is 4:3, so to model these squares he looks for three square numbers whose differences are in that ratio. These can be found by the method of Problem II.19, and they are given here as 81, 49, and 25. Diophantus makes the three squares $\square$, 1MM, and $\square'$ proportional to these numbers by regarding $\square$ as being 81 parts, 1MM as 49 parts, and $\square'$ as 25 parts, for some value he calls "a part". This makes $4y^3$ and $3y^3$, the differences between the squares, 32 parts and 24 parts, respectively. With these relations established, y^3 can be expressed in terms of MMs as $\frac{8}{49}$ MM. But y^3 is a cube, so Diophantus will also want to name it in terms of Cs.

It is these two different namings of the same amount that will form the equation from which 1T will be found. This equation is introduced near the beginning of the solution as the forthcoming equation "the cube (y^3) Equals a certain quantity (q)", where neither side has yet been named. Diophantus writes "the cube" for the side that will be assigned in terms of Cs, and "a certain quantity (*miqdār*)" for the side that will be assigned in terms of MMs. The "quantity" is found first, filling in one side of the forthcoming equation: "the sought-after cube Equals eight parts of forty-nine parts of a *Māl Māl*". (It is convenient that the latter amount can be read either as 8 of the 49 parts that make up the 1MM or as the common Arabic fraction $\frac{8}{49}$ of a MM.) The other side of the equation is then named via *al-istiqrā'*, so the equation is completed as $8C = \frac{8}{49}$ MM. From there 1T is found to be 49 units.

DOI: 10.4324/9781315171470-23

Assignments	Operations	Equation
$x := 1\text{T}$		
	Square $1\text{T} \rightarrow 1\text{M}$	
	Square $1\text{M} \rightarrow 1\text{MM}$	
		$1\text{MM} \ (4y^3) = \square$
		$1\text{MM} \ \ell \ (3y^3) = \square'$
		The cube $(y^3) = $ a quantity (q)

The only named terms in the first two forthcoming equations are the 1MMs. We have placed the second terms in parentheses to emphasize that they are not named in terms of the algebraic powers. The $4y^3$ in the first of these equations is spoken as "four times a certain cube", and similarly for the $3y^3$. These incomplete equations serve only to establish the next forthcoming equation, which says that y^3 can be expressed in terms of MMs.

Assignments	Operations	Equation
$q := \frac{8}{49}\text{MM}$		
		The cube $\left(y^3\right) = \frac{8}{49}\text{MM}$
$y := 2\text{T}$		
	Cube $2\text{T} \rightarrow 8\text{C}$	
		$8\text{C} = \frac{8}{49}\text{MM}$

After 1T is found to be 49u, Diophantus calculates $x^2 = 2401$ and $y^3 = 941{,}192$.

V.2
$$\begin{cases} \left(x^2\right)^2 + my^3 = \square \\ \left(x^2\right)^2 + ny^3 = \square' \end{cases} \quad (m = 12, \ n = 5)$$

Following the reasoning from the previous problem, Diophantus names a side of x^2 as 1T, making $(x^2)^2$ 1MM, and he then states the two forthcoming equations shown in the following table. This time he does not state the forthcoming equation "the cube Equals a certain quantity", but it is implied when he writes "Then let us look for the quantity in the known ratio with respect to the *Māl Māl*". This time $\square$, $\square'$, and 1MM are three descending squares whose successive differences are in the ratio of $12y^3 - 5y^3$ to $5y^3$, or 7 : 5. Following Problem II.19 can give the squares 16, 9, and 4, so Diophantus regards $\square$, $\square'$, and 1MM as being 16 parts, 9 parts, and 4 parts, respectively. This makes the "quantity" $\frac{1}{4}\text{MM}$, and the forthcoming equation becomes "a fourth of a *Māl Māl* Equals a cubic number". The assignment for y is then made by *al-istiqrā'*.

Assignments	Operations	Equation
$x := 1T$		
	Square $1T \to 1M$	
	Square $1M \to 1MM$	
		$1MM\,(12y^3) = \square$
		$1MM\,(5y^3) = \square'$
		a cube (y^3) = a quantity (q)
$q := \frac{1}{4}MM$		a cube $(y^3) = \frac{1}{4}MM$
$y := 2T$	Cube $2T \to 8C$	$8C = \frac{1}{4}MM$

Thus, $1T$ is $32u$, so $x^2 = 1024$ and $y^3 = 262{,}144$.

V.3
$$\begin{cases} \left(x^2\right)^2 - my^3 = \square \\ \left(x^2\right)^2 - ny^3 = \square' \end{cases} \quad (m = 12,\ n = 7)$$

The solution follows the same reasoning as the previous two problems.

Assignments	Operations	Equation
$x := 1T$		
	Square $11 \to 1M$	
	Square $1M \to 1MM$	
		$1MM\,\ell\,(12y^3) = \square$
		$1MM\,\ell\,(7y^3) = \square'$
		(a cube (y^3) = a quantity (q))

$1MM$, $\square'$, $\square$, are three decreasing squares with differences in the ratio 7:5. So also are 16, 9, 4, so make the trios proportional. By regarding $1MM$ to be 16 "parts", the differences are 7 parts and 5 parts, making q 1 part.

Assignments	Operations	Equation
$q := \frac{1}{16}MM$		
		a cube $(y^3) = \frac{1}{16}MM$
$y := \frac{1}{2}T$		

$$\text{Cube } \tfrac{1}{2}T \to \tfrac{1}{8}C$$

$$\tfrac{1}{8}C = \tfrac{1}{16}MM$$

Thus, 1T is 2u, so $x^2 = 16$ and $y^3 = 1$.

V.4
$$\begin{cases} \left(x^2\right)^2 + m\left(y^3\right)^3 = \square \\ \left(x^2\right)^2 - n\left(y^3\right)^3 = \square' \end{cases} \qquad (m = 5,\ n = 3)$$

This problem is solved by the variation on "equating the corresponding sides" that we first saw in the third solution to Problem IV.42(a) and then in Problem IV.44. But here, in Book V, the final forthcoming equations are worked out by a different method.

Assignments	Operations	Equation
$y := 1T$		
	Cube 1T → 1C	
	Cube 1C → 1CCC	
$x := 2M$		
	Square 2M → 4MM	
	Square 4MM → 16CCM	
		16CCM 5CCC = $\square$
		16CCM ℓ 3CCC = $\square'$
	Divide 16CCM 5CCC by	
	1CCM → 16u 5T	
	Divide 16CCM ℓ 3CCC by	
	1CCM → 16u ℓ 3T	
		16u 5T = $\square''$
		16u ℓ 3T = $\square'''$

It is very difficult to solve these equations by "the double-equality". Diophantus notes that "any square number increased by five times its fourth is a square, and decreased by three times its fourth is a square". The two forthcoming equations fit these identities, so he identifies 5T with five times a fourth of 16 and 3T with three times its fourth, which makes 1T equal to 4u. Therefore, $x^2 = 1024$ and $y^3 = 64$.

V.5
$$\begin{cases} \left(x^2\right)^2 + m\left(y^3\right)^3 = \square \\ \left(x^2\right)^2 + n\left(y^3\right)^3 = \square' \end{cases} \qquad (m = 12,\ n = 5)$$

Assignments	Operations	Equation
$y := 1T$		
	Cube 1T → 1C	
	Cube 1C → 1CCC	
$x := 2M$		
	Square 2M → 4MM	
	Square 4MM → 16CCM	
		16CCM 12CCC = □
		16CCM 5CCC = □′
	Divide 16CCM 12CCC by 1CCM → 16u 12T	
	Divide 16CCM 5CCC by 1CCM → 16u 5T	
		16u 12T = □″
		16u 5T = □‴

As in Problem V.4, Diophantus cites two identities that are simulated in the two forthcoming equations. Here they are: increasing a square by its $\frac{5}{4}$ gives a square, and increasing a square by its $\frac{12}{4}$ gives a square. By reading the 12T as being $\frac{12}{4}$ of the 16u, and the 5T as being $\frac{5}{4}$ of 16u, both lead to 1T being 4u. Then $y^3 = 64$ and $x^2 = 1024$.

V.6 $\quad \begin{cases} \left(x^2\right)^2 - m\left(y^3\right)^3 = □ \\ \left(x^2\right)^2 - n\left(y^3\right)^3 = □′ \end{cases} \quad (m = 7,\ n = 4)$

Assignments	Operations	Equation
$y := 1T$		
	Cube 1T → 1C	
	Cube 1C → 1CCC	
$x := 3M$		
	Square 3M → 9MM	
	Square 9MM → 81CCM	
		81CCM ℓ 7CCC = □
		81CCM ℓ 4CCC = □′
	Divide 81CCM ℓ 7CCC by 1CCM → 81u ℓ 7T	
	Divide 81CCM ℓ 4CCC by 1CCM → 81u ℓ 4T	
		81u ℓ 7T = □″
		81u ℓ 4T = □‴

In the previous two problem the numbers 5 and 3 (in V.4), and 12 and 5 (in V.5) in the final two forthcoming equations just happened to be present in arithmetical identities that are simulated in those equations. This time there are no obvious identities to match with the numbers 7 and 4, so Diophantus will derive them. In place of the "fourth" in the identities from Problem V.5 he needs a "determined quantity in any square such that if seven times it is subtracted from the square, it leaves a square, and if four times it is also subtracted from the square, it leaves a square." In modern notation, he will find a quantity q such that for any square z^2, $z^2 - 7qz^2$ and $z^2 - 4qz^2$ are both squares. In the present context z^2 is 81.

Because z^2, $z^2 - 4qz^2$, and $z^2 - 7qz^2$ are three squares whose successive differences are in the ratio 4:3, he can borrow the solution that he has already found in Problem V.1 to find q. Three squares with differences in this ratio are 81, 49, and 25. These will be proportional to 81u, 81u ℓ 4T, and 81u ℓ 7T. At this point Diophantus could have stated and solved the equation 81u ℓ 4T = 49u to get 1T = 8u, but he first finds that q is $\frac{8}{81}$, or, as it is stated in the text, "eight ninths of a ninth", and from there he works with the differences $4q \cdot 81$ and $3q \cdot 81$, which are 32 and 24, respectively, to get "the one Thing is eight ninths of a ninth of the eighty-one, which is eight units." Then $x^2 = 36{,}864$ and $y^3 = 512$.

V.7 $\begin{cases} x + y = a \\ x^3 + y^3 = b \end{cases}$ $\quad$ Determination: $\frac{4b - a^3}{3a} = \square$ and $\left(4b - a^3\right) \cdot \frac{3}{4} a = \square'$

$\quad\quad\quad\quad\quad (a = 20, b = 2240)$

The two conditions of the determination are equivalent. The first of these can be found by following the calculations in the solution and guarantee that 1M is a square. Sesiano (1982, 232) suggests that the second condition was likely a marginal note that crept into the text.

The first three assignments satisfy the first condition. These assignments make the calculations that follow more manageable than assigning x and y to be, say, 1T and 20 ℓ 1T since two terms in the sum of the cubes drop out.

Assignments	Operations	Equation
$x - y := 2$T		
$x := 10$u 1T		
$y := 10$u ℓ 1T		
	Cube 10u 1T → 1000u 1C 300T 30M	
	Cube 10u ℓ 1T → 1000u 30M ℓ 1C 300T	
	Add these two cubes → 2000u 60M	
		2000u 60M = 2240u

1T is then found to be 2u, so the sought-after numbers are $x = 12$ and $y = 8$.

Diophantus explains verbally the rules for cubing a sum and a difference before performing the calculations, beginning with a warning: "Whenever we want to form a cube from a side composed of two different species we should avoid errors due to the (presence of a) multitude of species".

The same problem, with $a = 10$, $b = 370$ and without the determination, appears as Problem IV^G.1. That problem is probably interpolated.

V.8 $\begin{cases} x - y = a \\ x^3 - y^3 = b \end{cases}$ Determination: $\frac{4b-a^3}{3a} = \square$ and $\left(4b - a^3\right) \cdot \frac{3}{4} a = \square'$

$(a = 10, b = 2170)$

As with the determination of the previous problem, the two conditions are equivalent. The other remarks there apply here, too.

The following three assignments satisfy the first condition, and as in Problem V.7, are designed to simplify the subsequent calculations.

Assignments	Operations	Equation
$x + y := 2T$		
$x := 1T\ 5u$		
$y := 1T\ \ell\ 5u$		
	Cube $1T\ 5u \rightarrow 1C\ 125u\ 15M\ 75T$	
	Cube $1T\ \ell\ 5u \rightarrow 1C\ 75T\ \ell\ 15M\ 125u$	
	Subtract that from $1C\ 125u\ 15M\ 75T$ $\rightarrow 250u\ 30M$	
		$250u\ 30M = 2170u$

$1T$ is found to be $8u$, so the sought-after numbers are $x = 13$ and $y = 3$.

The same problem, with $a = 6$, $b = 504$, and without the determination, appears as Problem IV^G.2. Like Problem IV^G.1, it is probably interpolated.

V.9 $\begin{cases} x + y = a \\ x^3 + y^3 = m(x-y)^2 \end{cases}$ Determination: $\left(m - \frac{3}{4} a\right) \cdot a^3$ must be a square

$(a = 20, m = 140)$

Following the calculations, $1M$ becomes the number calculated from $\frac{\frac{1}{4}a^3}{4m-3a}$, which must be a square. This is equivalent to the determination.

The assignments are the same as in Problem V.7.

Assignments	Operations	Equation
$x - y := 2T$		
$x := 10u\ 1T$		

$y := 10u \; \ell \; 1T$

Add the cubes of 10u 1T and 10u ℓ 1T $\rightarrow$ 2000u 60M	
Square the difference of 10u ℓ 1T and 10u 1T $\rightarrow$ 4M	
140 times 4M $\rightarrow$ 560M	
	2000u 60M = 560M

1T is 2u, so the two sought-after numbers are $x = 12$ and $y = 8$.

V.10
$$\begin{cases} x - y = a \\ x^3 - y^3 = m(x+y)^2 \end{cases} \quad \left(a = 10, m = 8\tfrac{1}{8}\right)$$

Determination: $\left(m - \tfrac{3}{4}a\right) \cdot a^3$ must be a square

As with the previous problem, 1M becomes the number calculated from $\frac{\frac{1}{4}a^3}{4m-3a}$, which must be a square. This is equivalent to the determination.

The assignments are the same as in Problem V.8.

Assignments	Operations	Equation
$x + y := 2T$		
$x := 1T \; 5u$		
$y := 1T \; \ell \; 5u$		
	Subtract the cube of 1T ℓ 5u from the cube of 1T 5u $\rightarrow$ 250u 30M	
	Square the sum of 1T 5u and 1T ℓ 5u $\rightarrow$ 4M	
	$8\tfrac{1}{8}$ times 4M $\rightarrow$ $32\tfrac{1}{2}$ M	
		250u 30M = $32\tfrac{1}{2}$ M

1T is 10u, so $x = 15$ and $y = 5$.

V.11
$$\begin{cases} x - y = a \\ x^3 + y^3 = m(x+y) \end{cases} \quad (a = 4, m = 28)$$

Determination: $m - \tfrac{3}{4}a^2$ must be a square

The determination follows from the calculations: 1M becomes the number calculated from $m - \tfrac{3}{4}a^2$.

Assignments	Operations	Equation
$x + y := 2T$		
$x := 1T\ 2u$		
$y := 1T\ \ell\ 2u$		
	Cube $1T\ 2u \rightarrow 1C\ 8u\ 6M\ 12T$	
	Cube $1T\ \ell\ 2u \rightarrow 1C\ 12T\ \ell\ 6M\ 8u$	
	Add $1C\ 8u\ 6M\ 12T$ and $1C\ 12T\ \ell\ 6M$ $8u \rightarrow 2C\ 24T$	
	28 times $2T \rightarrow 56T$	
		$2C\ 24T = 56T$

1T is 4u, so $x = 5$ and $y = 2$.

V.12 $\begin{cases} x + y = a \\ x^3 - y^3 = m(x - y) \end{cases}$ $\quad$ Determination: $m - \frac{3}{4}a^2$ must be a square
$\qquad\qquad\qquad\qquad\qquad (a = 8, m = 52)$

Again, the determination follows from the calculations: 1M becomes the number calculated from $m - \frac{3}{4}a^2$.

Assignments	Operations	Equation
$x - y := 2T$		
$x := 4u\ 1T$		
$y := 4u\ \ell\ 1T$		
	Cube $4u\ 1T \rightarrow 64u\ 1C\ 48T\ 12M$	
	Cube $4u\ \ell\ 1T \rightarrow 64u\ 12M\ \ell\ 1C\ 48T$	
	Subtract $64u\ 12M\ \ell\ 1C\ 48T$ from $64u\ 1C$ $48T\ 12M \rightarrow 2C\ 96T$	
	52 times $2T \rightarrow 104T$	
		$2C\ 96T = 104T$

1T is 2u, so $x = 6$ and $y = 2$.

Sesiano (1982, 238) observes that the determinations for Problems V.7–12 do not guarantee that the smaller of the two numbers will be positive.

V.13 $\begin{cases} mx^2 + a = y + z^1 \\ x^3 + y = \boxed{} \\ x^3 + z = \boxed{}' \end{cases}$ $\quad (m = 9, a = 30)$

1 In Problems V.13–16 the numbers denoted by y and z are not asked for.

Assignments	Operations	Equation
$x := 1T$		
	Cube 1T → 1C	
	Square 1T → 1M	
	Nine times 1M → 9M	
	Add 9M to 30u → 9M 30u	

At this point Diophantus would like to name the sides of ▦ and ▦' as 1T and some units. This way the cubes will be 1C together with Ms, Ts, and units. Because x^3 is 1C, y and z will be the remaining Ms, Ts, and units. Their sum, too, will consist of Ms, Ts, and units, and will be set equal to the 9M 30u just calculated. For the *al-istiqrā'* solution to work, the number of Ms in the expression for $y + z$ must be 9, and the units in the sum must be less than 30. (The latter condition is in fact satisfied for any two positive numbers whose sum is 3.) Thus, the sum of the units in the assignments for ▦ and ▦' must be 3 (they will in turn be multiplied by 3 in the calculation of the cubes). Diophantus chooses the units to be 1 and 2. The solution continues:

Assignments	Operations	Equation
$\sqrt[3]{▦} := 1T\,2u$		
	Cube 1T 2u → 1C 6M 12T 8u	
$\sqrt[3]{▦'} := 1T\,1u$		
	Cube 1T 1u → 1C 3M 3T 1u	

There is no need to perform a subtraction to remove the 1C (i.e., x^3) from 1C 6M 12T 8u to find the expression for our y since the aggregation is by its nature already partitioned. The first term is x^3, so the rest is y. Similarly, our z is 3M 3T 1u, the last three terms in the expression for the second cube. In the enunciation Diophantus does not ask for y or z, so it is no surprise that he does not make any assignments for them. Instead, he adds the two aggregations just found, with the 1Cs removed, and he equates the sum with 9M 30u:

Assignments	Operations	Equation
	Add 6M 12T 8u and 3M 3T 1u → 9M 15T 9u	
		9M 15T 9u = 9M 30u

1T is then $1\frac{2}{5}$u, so $x = 1\frac{2}{5}$ and $x^3 = 2\frac{93}{125}$.

After the solution the text states "this problem can be worked out this way whenever a cube of a third of the number of times is smaller than four times the given number", or $\left(\frac{1}{3}m\right)^3 < 4a$. Sesiano notes that the wording implies that this condition is necessary for the particular solution pursued by Diophantus, in which the second-degree terms are made to drop out in the forthcoming equation. In fact, he shows that the condition is not necessary for another way of solving the problem. He also shows how this condition can be obtained from the determination in Problem V.7 (Sesiano 1982, 239–40). Although this is the most likely way that Diophantus found the condition, it also could have come from an examination of inequalities involving polynomials. The derivation can be obtained easily in modern algebra, and given the competence Diophantus exhibits with manipulations on algebraic expressions with undetermined multitudes in other problems, it was surely within his reach to accomplish this, too. Let us call the units in the assignments for $\sqrt[3]{\boxed{c}}$ and $\sqrt[3]{\boxed{c}'}$ b and c, where $b > c$. Because $m = 3$ $(b + c)$ and a $> b^3 + c^3$, we should look for a relation between $b + c$ and $b^3 + c^3$. Now, $(b - c)^2 > 0$, so $b^2 - bc + c^2 > bc$. Multiplying both sides by $b + c$, we get $b^3 + c^3 > bc(b + c)$. Thus, $3(b^3 + c^3) > 3b^2c + 3bc^2$, which, after adding $b^3 + c^3$ to both sides, gives $4(b^3 + c^3) > (b + c)^3$. It then follows that $4a > \left(\frac{1}{3}m\right)^3$.

V.14
$$\begin{cases} mx^2 - a = y + z \\ x^3 - y = \boxed{c} \qquad (m = 9,\ a = 26) \\ x^3 - z = \boxed{c}' \end{cases}$$

Assignments	Operations	Equation
$x := 1T$		
	Cube 1T → 1C	
	Square 1T → 1M	
	Nine times 1M → 9M	
	Subtract 26u from 9M → 9M ℓ 26u	

The procedure is similar to that in the previous problem. This time Diophantus wants the sides of $\boxed{c}$ and $\boxed{c}'$ to be 1T less some units. Now the cubes will be 1C and Ts, less Ms and units. Because x^3 is 1C, y and z will be Ms and units less Ts. Their sum, too, will consist of Ms and units less Ts, and will be set equal to the 9M ℓ 26u just calculated. Again, for the *al-istiqrā'* solution to work, the number of Ms in the expression for $y + z$ must be 9, but this time there is no condition on the number of units in the sum. Thus, the sum of the units in the assignments for $\boxed{c}$ and $\boxed{c}'$ must be 3 (they will in turn be multiplied by 3 in the calculation of the cubes). Diophantus chooses the units to be 2 and 1. The solution continues:

Assignments	Operations	Equation
$\sqrt[3]{⊡} := 1\mathrm{T}\,\ell\,2\mathrm{u}$		
	Cube $1\mathrm{T}\,\ell\,2\mathrm{u} \to 1\mathrm{C}\,12\mathrm{T}\,\ell\,6\mathrm{M}\,8\mathrm{u}$	
$\sqrt[3]{⊡}' := 1\mathrm{T}\,\ell\,1\mathrm{u}$		
	Cube $1\mathrm{T}\,\ell\,1\mathrm{u} \to 1\mathrm{C}\,3\mathrm{T}\,\ell\,3\mathrm{M}\,1\mathrm{u}$	

The first cube is the result of subtracting 6M 8u ℓ 12 T from 1C, and the second cube is the result of subtracting 3M 1u ℓ 3T from 1C. Thus, 6M 8u ℓ 12 T is y and 3M 1u ℓ 3T is z.

Assignments	Operations	Equation
	Add 6M 8u ℓ 12T and 3M 1u ℓ 3T $\to$ 9M 9u ℓ 15T	
		9M 9u ℓ 15T = 9M ℓ 26u

1T is found to be $2\frac{1}{3}$ u, so $x = 2\frac{1}{3}$ and $x^3 = 12\frac{19}{27}$.

V.15 $\quad \begin{cases} mx^2 - a = y + z \\ x^3 - y = ⊡ \\ x^3 - z = ⊡' \end{cases} \quad (m = 9,\, a = 18)$

Assignments	Operations	Equation
$x := 1\mathrm{T}$		
	Cube $1\mathrm{T} \to 1\mathrm{C}$	
	Square $1\mathrm{T} \to 1\mathrm{M}$	
	Nine times $1\mathrm{M} \to 9\mathrm{M}$	
	Subtract 18u from 9M $\to$ 9M ℓ 18u	

Diophantus makes a side of $⊡$ to be 1T and some units, and a side of $⊡'$ to be 1T less some units. He chooses the latter units to be greater that the former units, which guarantees that he will not have to posit a condition like the one given at the end of Problem V.13. The solution continues like those in the last two problems.

Assignments	Operations	Equation
$\sqrt[3]{⊡}' := 1\mathrm{T}\,\ell\,2\mathrm{u}$		
$\sqrt[3]{⊡} := 1\mathrm{T}\,1\mathrm{u}$		

$$\text{Cube } 1T \; \ell \; 2u \to 1C \; 12T \; \ell \; 8u \; 6M$$
$$\text{Cube } 1T \; 1u \to 1C \; 3M \; 3T \; 1u$$

Diophantus notes that if 6M 8u ℓ 12T is subtracted from 1C, then the remainder is a cube ($\boxdot'$), and if 3M 1u 3T is added to 1C, the sum is a cube ($\boxdot$). These are our z and y, so he adds them and forms the equation:

Assignments	Operations	Equation
	Add 6M 8u ℓ 12T and 3M 1u 3T $\to$ 9M 9u ℓ 9T	
		9M 9u ℓ 9T = 9M ℓ 18u

1T is found to be 3u, so $x = 3$ and $x^3 = 27$. The techniques used for the assignments are the same as in the last two problems.

V.16
$$\begin{cases} mx^2 - a = y + z \\ x^3 - y = \boxdot \qquad (m = 9, \, a = 16) \\ z - x^3 = \boxdot' \end{cases}$$

It is implicit after the first assignment that $x := 1T$ and $x^2 := 1M$.

Assignments	Operations	Equation
$x^3 := 1C$		
	Nine times 1M $\to$ 9M	
	Subtract 16u from 9M $\to$ 9M ℓ 16u	

Here he makes a side of $\boxdot$ to be 1T less some units, and a side of $\boxdot'$ to be some units less 1T, and the calculations follow those in the last three problems.

Assignments	Operations	Equation
$\sqrt[3]{\boxdot} := 1T \; \ell \; 1u$		
	Cube 1T ℓ 1u $\to$ 1C 3T ℓ 3M 1u	
$\sqrt[3]{\boxdot'} := 2u \; \ell \; 1T$		
	Cube 1T ℓ 2u $\to$ 8u 6M ℓ 1C 12T	

Diophantus notes that subtracting 3M 1u ℓ 3T from 1C leaves a cube ($\boxdot$) and subtracting 1C from 6M 8u ℓ 12T leaves a cube ($\boxdot'$). These trinomials are our y and z, and their sum will be equated to 9M ℓ 16u.

Assignments	Operations	Equation
	Add 3M 1u ℓ 3T and 6M 8u ℓ 12T $\to$ 9M 9u ℓ 15T	
		9M 9u ℓ 15T = 9M ℓ 16u

1T is found to be $1\frac{2}{3}$ u, so $x = 1\frac{2}{3}$ and $x^3 = 4\frac{17}{27}$. Again, the techniques used for the assignments are the same as in the last three problems.

Book VI (Arabic)

The first three problems of Book VI coincide with Problems IV.25, 26(a), and 26(b), except for the condition $x = my$ which is not stated in the Book IV versions. Moreover, in both groups the assignments are made by the same techniques, including, for the *al-istiqrā'* assignment, an appeal to either to Problem II.8 or II.10. Thus, we have here a similar situation with that of the first five problems of Book II and the analogous problems of Book I (see our remarks at the start of Book II). As with Problems II.1–5, we agree with Sesiano that Problems VI.1–3 must be considered as interpolations. Sesiano writes, referring to VI.1, "This problem is the first of a large set of interpolated problems which occupies almost half of Book VI" (Sesiano 1982, 244).

VI.1 $\begin{cases} x = my \\ \left(x^3\right)^2 \mid \left(y^2\right)^2 = \square \end{cases}$ $(m = 2)$

Assignments	Operations	Equation
$y \coloneqq 1T$		
	Square $1T \to 1M$	
	Two times $1T \to 2T$	
$x \coloneqq 2T$		
	Cube $2T \to 8C$	
	Add the squares of 8C and 1M $\to$64CC 1MM	
		$64CC\ 1MM = \square$

Both 64 and 1 are squares, so Diophantus can name the $\square$ as a multitude of either power. He chooses to name it in terms of Cube Cubes, and since the powers CC and MM differ by two, the remainder from subtracting 64 from this multitude must be a square. This is easily found by the procedure in Problem II.10, and he chooses 100.

Assignments	Operations	Equation
$\square \coloneqq 100CC$		
		$64CC\ 1MM = 100CC$

DOI: 10.4324/9781315171470-24

1T is found to be $\frac{1}{6}$u, so $x^3 = \frac{8}{216}$ and $y^2 = \frac{1}{36}$.

VI.2 $\begin{cases} x = my \\ \left(x^3\right)^2 - \left(y^2\right)^2 = \square \quad (m = 2) \end{cases}$

Assignments	Operations	Equation
$y := 1T$		
	Two times $1T \rightarrow 2T$	
$x := 2T$		
	Square $1T \rightarrow 1M$	
	Square $1M \rightarrow 1MM$	
	Cube $2T \rightarrow 8C$	
	Square $8C \rightarrow 64CC$	
	Subtract $1MM$ from $64CC \rightarrow 64CC \,\ell\, 1MM$	
		$64CC \,\ell\, 1MM = \square$

As in Problem VI.1, Diophantus wishes to make the $\square$ a multitude of CCs, so subtracting this multitude from 64 should leave a square. This is Problem II.8, and he finds $40\frac{24}{25}$.

Assignments	Operations	Equation
$\square := 40\frac{24}{25}CC$		
		$64CC\, 1MM = 40\frac{24}{25}CC$

1T is then $\frac{5}{24}$u, so $x^2 = \frac{125}{1728}$ and $y^2 = \frac{75}{1728}$.

VI.3 $\begin{cases} x = my \\ \left(y^2\right)^2 - \left(x^3\right)^2 = \square \end{cases} \quad (m = 2)$

Assignments	Operations	Equation
$y := 1T$		
	Two times $1T \rightarrow 2T$	
$x := 2T$		
	Cube $2T \rightarrow 8C$	
	Square $8C \rightarrow 64CC$	
	Square $1T \rightarrow 1M$	
	Square $1M \rightarrow 1MM$	
	Subtract $64CC$ from $1MM \rightarrow 1MM \,\ell\, 64CC$	
		$1MM \,\ell\, 64CC = \square$

As in Problem VI.1, Diophantus wishes to make the □ a multitude of CCs. This multitude together with the 64 should be a square, so again Problem II.10 can give the 36 that Diophantus found.

Assignments	Operations	Equation
$\sqrt{\square} := 6C$		
	Square 6C → 36 CC	
		1MM ℓ 64CC = 36CC

1T is then $\frac{1}{10}$u, so $x^3 = \frac{8}{1000}$ and $y^2 = \frac{10}{1000}$.

VI.4 $\begin{cases} x = my \\ x^3 \cdot y^2 + \left(x^3\right)^2 = \square \end{cases} \quad (m = 5)$

Assignments	Operations	Equation
$y := 1T$		
	Square 1T → 1M	
	Five times 1T → 5T	
$x := 5T$		
	Cube 5T → 125C	
	Multiply 125C by 1M → 125CM[1]	
	Square 125C → 15,625CC	
	Add 15,625CC to 125CM → 15,625CC 125CM	
		15,625CC 125CM = □

Diophantus will assign the □ to be CCs, the only restriction being that the square multitude should be greater than $125^2 = 15,625$. To keep the numbers in the calculations from becoming too large, he wants the difference between the square multitude and 125^2 to be small, so he chooses 126^2.

Assignments	Operations	Equation
$\sqrt{\square} := 126C$		
	Square 126C → 15,876CC	
		15,625CC 125CM = 15,876CC

1T is found to be $\frac{125}{251}$u, so $y^2 = \frac{15,625}{63,001}$ and $x^3 = \frac{244,140,625}{15,813,251}$. The text erroneously shows the number 2,563,001 for the denominator of the cube. Sesiano, and after

---

1 For CM the text has "Cube(s) multiplied by a *Māl*" here and in every instance through Problem VI.7.

him Rashed, trace the error to a corruption in the Greek original (Sesiano 1982, 246–47; Rashed in Diophantus 1984, 4, 122–23).

VI.5 $\begin{cases} x = y \\ x^3 \cdot y^2 + (y^2)^2 = \Box \end{cases}$

Assignments	Operations	Equation
$y := 1T$		
	Square $1T \to 1M$	
$x := 1T$		
	Cube $1T \to 1C$	
	Multiply $1C$ by $1M \to 1CM$	
	Square $1M \to 1MM$	
	Add $1MM$ to $1CM \to 1CM\ 1MM$	
		$1CM\ 1MM = \Box$
$\sqrt{\Box} := 2M$		
	Square $2M \to 4MM$	
		$1CM\ 1MM = 4MM$

$1T$ is found to be 3u, so $x^3 = 27$ and $y^2 = 9$.

VI.6 $\begin{cases} x = y \\ x^3 \cdot y^2 - (x^3)^2 = \Box \end{cases}$

Assignments	Operations	Equation
$y := 1T$		
	Square $1T \to 1M$	
$x := 1T$		
	Cube $1T \to 1C$	
	Multiply $1C$ by $1M \to 1CM$	
	Square $1C \to 1CC$	
	Subtract $1CC$ from $1CM \to 1CM\ \ell\ 1CC$	
		$1CM\ \ell\ 1CC = \Box$
$\sqrt{\Box} := 1C$		
	Square $1C \to 1CC$	
		$1CM\ \ell\ 1CC = 1CC$

$1T$ is found to be $\frac{1}{2}$u, so $y^2 = \frac{1}{4}$ and $x^3 = \frac{1}{8}$.

VI.7 $\begin{cases} x = y \\ x^3 \cdot y^2 - (y^2)^2 = \Box \end{cases}$

Assignments	Operations	Equation
$y := 1T$		
	Square $1T \to 1M$	
$x := 1T$		
	Cube $1T \to 1C$	
	Multiply $1C$ by $1M \to 1CM$	
	Square $1M \to 1MM$	
	Subtract $1MM$ from $1CM \to 1CM \ell 1MM$	
		$1CM \ell 1MM = \square$
$\sqrt{\square} := 1M$		
	Square $1M \to 1MM$	
		$1CM \ell 1MM = 1MM$

$1T$ is found to be 2u, so $y^2 = 4$ and $x^3 = 8$.

VI.8 $x^3 \cdot y^2 + \sqrt{x^3 \cdot y^2} = \square$

Assignments	Operations	Equation
$x^3 := 64u$		
$y^2 := 1M$		
	Multiply $64u$ by $1M \to 64M$	
	The side of $64M \to 8T$	
	Add $8T$ to $64M \to 64M\ 8T$	
		$64M\ 8T = \square$
$\sqrt{\square} := 10T$		
	Square $10T \to 100M$	
		$64M\ 8T = 100M$

$1T$ is $\frac{2}{9}u$, so $y^2 = \frac{4}{81}$ and $x^3 = 64$.

VI.9 $x^3 \cdot y^2 + \sqrt{x^3 \cdot y^2} = \square$

Assignments	Operations	Equation
$x^3 := 64u$		
$y^2 := 1M$		
	Multiply $64u$ by $1M \to 64M$	
	The side of $64M \to 8T$	
	Subtract $8T$ from $64M \to 64M \ell 8T$	

$$\sqrt{\square} := 7\mathrm{T}$$

	$64\mathrm{M}\ \ell\ 8\mathrm{T} = \square$
Square $7\mathrm{T} \to 49\mathrm{M}$	
	$64\mathrm{M}\ \ell\ 8\mathrm{T} = 49\mathrm{M}$

1T is $\frac{8}{15}$ u, so $y^2 = \frac{64}{225}$ and $x^3 = 64$.

VI.10 $\sqrt{x^3 \cdot y^2} - x^3 \cdot y^2 = \square$

Assignments	Operations	Equation
$x^3 := 64\mathrm{u}$		
$y^2 := 1\mathrm{M}$		
	Multiply $64\mathrm{u}$ by $1\mathrm{M} \to 64\mathrm{M}$	
	The side of $64\mathrm{M} \to 8\mathrm{T}$	
	Subtract $64\mathrm{M}$ from $8\mathrm{T} \to 8\mathrm{T}\ \ell\ 64\mathrm{M}$	
		$8\mathrm{T}\ \ell\ 64\mathrm{M} = \square$
$\sqrt{\square} := 4\mathrm{T}$		
	Square $4\mathrm{T} \to 16\mathrm{M}$	
		$8\mathrm{T}\ \ell\ 64\mathrm{M} = 16\mathrm{M}$

1T is $\frac{1}{10}$ u, so $y^2 = \frac{1}{100}$ and $x^3 = 64$.

VI.11 $\left(x^3\right)^2 + x^3 = \square$

Assignments	Operations	Equation
$x := 1\mathrm{T}$		
	Cube $1\mathrm{T} \to 1\mathrm{C}$	
	Square $1\mathrm{C} \to 1\mathrm{CC}$	
	Add $1\mathrm{CC}$ to $1\mathrm{C} \to 1\mathrm{CC}\ 1\mathrm{C}$	
		$1\mathrm{CC}\ 1\mathrm{C} = \square$
$\sqrt{\square} := 3\mathrm{C}$		
	Square $3\mathrm{C} \to 9\mathrm{CC}$	
		$1\mathrm{CC}\ 1\mathrm{C} = 9\mathrm{CC}$

1T is found to be $\frac{1}{2}$ u, so $x^3 = \frac{1}{8}$.

The *al-istiqrā'* assignment is stated in the manuscript in an awkward way: "Let us assign its side to be from a number giving (*yakūn*) three Cubes", so Sesiano suggested the following reconstruction: "Let us assign its side to be from a number (of Cubes such that if we subtract one Cube Cube from its square, then the remainder is a cube, say) it gives three Cubes". This condition is indeed necessary for obtaining a rational answer. In fact, the choice of 3 for the number of cubes is the only positive rational number that will give a valid solution.[2] Also, the remark a few lines after that, "the problem can be solved and working it out is not impossible", suggests to Sesiano that the missing passage might have contained more than just the condition, though he could find anything plausible for it. "This problem is odd", he wrote. "I am convinced that it must be an interpolation" (Sesiano 1982, 249).

VI.12 $\begin{cases} x^2 + x^2/y^2 = \square \\ y^2 + x^2/y^2 = \square' \end{cases}$ $x > y$

The first two assignments satisfy the second condition, and the equation is set up from the first condition.

Assignments	Operations	Equation
$y^2 := 1\mathrm{M}$		
$\dfrac{x^2}{y^2} := \dfrac{9}{16}\mathrm{M}$		
	Multiply $1\mathrm{M}$ by $\dfrac{9}{16}\mathrm{M} \to \dfrac{9}{16}\mathrm{MM}$	
$x^2 := \dfrac{9}{16}\mathrm{MM}$		
	Add $\dfrac{9}{16}\mathrm{M}$ to $\dfrac{9}{16}\mathrm{MM} \to \dfrac{9}{16}\mathrm{MM}\dfrac{9}{16}\mathrm{M}$	
		$\dfrac{9}{16}\mathrm{MM}\dfrac{9}{16}\mathrm{M} = \square$

Diophantus chooses to name the $\square$ as a square multitude of *Māl Māls*, so two conditions must be met for this multitude: if it is diminished by $\frac{9}{16}$ the remainder must be a square, and this remainder must be less than $\left(\frac{9}{16}\right)^2$ so that x^2 will be greater than y^2. This can be solved by the method of Problem II.10, and Diophantus finds the multitude $\frac{169}{256}$, which is the square of $\frac{13}{16}$.

Assignments	Operations	Equation
$\sqrt{\square} := \dfrac{13}{16}\mathrm{M}$		
	Square $\dfrac{13}{16}\mathrm{M} \to \dfrac{169}{256}\mathrm{MM}$	
		$\dfrac{9}{16}\mathrm{MM}\dfrac{9}{16}\mathrm{M} = \dfrac{169}{256}\mathrm{MM}$

Thus, $1\mathrm{T}$ is $\frac{12}{5}\mathrm{u}$, so $y^2 = \frac{144}{25}$ and $x^2 = \frac{11{,}664}{625}$.

---

2 Fueter (1930, 70–72) has shown that the only rational solutions of the equation $u^2 - 1 = v^3$ are $u = \pm 1$, $v = 0$; $u = 0$, $v = -1$; and $u = \pm 3$, $v = 2$. Cf. Rashed and Houzel (2013, 322–23).

VI.13 $\begin{cases} x^2 - x^2/y^2 = \square \\ y^2 - x^2/y^2 = \square' \end{cases}$ $\quad x > y$

First solution

The assignments for y and $\frac{x^2}{y^2}$ satisfy the second condition, and the equation is set up from the first condition.

Assignments	Operations	Equation
$y := 1T$		
	Square $1T \rightarrow 1M$	
$y^2 := 1M$		
$\frac{x^2}{y^2} := \frac{9}{25}M$		
	Multiply $1M$ by $\frac{9}{25}M \rightarrow \frac{9}{25}MM$	
$x^2 := \frac{9}{25}MM$		
	Subtract $\frac{9}{25}M$ from $\frac{9}{25}MM \rightarrow$ $\frac{9}{25}MM \,\ell\, \frac{9}{25}M$	
		$\frac{9}{25}MM \,\ell\, \frac{9}{25}M = \square$

Diophantus chooses to name the $\square$ as a square multitude of *Māl Māl*s, so to make the solution work, subtracting that multitude from $\frac{9}{25}$ should leave a square. This can be solved by the method of Problem II.8, and he finds $\frac{81}{625} = \left(\frac{9}{25}\right)^2$.

Assignments	Operations	Equation
$\sqrt{\square} := \frac{9}{25}M$		
	Square $\frac{9}{25}MM \rightarrow \frac{81}{625}MM$	
		$\frac{9}{25}MM \,\ell\, \frac{9}{25}M = \frac{81}{625}MM$

Diophantus does not solve for $1T$ but stops at finding $1M$ to be $1\frac{9}{16}$ u. Thus, $y^2 = 1\frac{9}{16} = \frac{25}{16}$ and $x^2 = \frac{225}{256}$.

In this solution $x^2 < y^2$, but the problem calls for $y^2 < x^2$. Nevertheless, the solution works for the problem in which the divisor is the smaller, which may be why Diophantus gave this solution a formal termination, with the usual "We have found two numbers according to the restriction that we restricted". He then proceeds to solve it again, this time finding a solution in which the divisor is the greater.

Second solution (line 2544)

The assignments for y and x satisfy the first condition, and the equation is set up from the second condition.

Assignments	Operations	Equation
$y := 1\frac{2}{3}$ u		
	Square $1\frac{2}{3}$ u $\rightarrow 2\frac{7}{9}$ u	
$y^2 := 2\frac{7}{9}$ u		
$x := 1$T		
	Square 1T $\rightarrow 1$M	
$x^2 := 1$M		
	Divide 1M by $2\frac{7}{9}$ u $\rightarrow \frac{9}{25}$ M	
	Subtract $\frac{9}{25}$ M from $2\frac{7}{9}$ u $\rightarrow$ $2\frac{7}{9}$ u $\ell \frac{9}{25}$ M	
		$2\frac{7}{9}$ u $\ell \frac{9}{25}$ M $= \square$
$\sqrt{\square'} := 1\frac{2}{3}$ u $\ell 1\frac{1}{5}$ T		
	Square $1\frac{2}{3}$ u $\ell 1\frac{1}{5}$ T $\rightarrow 2\frac{7}{9}$ u $1\frac{11}{25}$ M ℓ 4T	
		$2\frac{7}{9}$ u $\ell \frac{9}{25}$ M
		$= 2\frac{7}{9}$ u $1\frac{11}{25}$ M ℓ 4T

1T is found to be $2\frac{2}{9}$u, so $x^2 = \frac{400}{81}$ and $y^2 = \frac{25}{9}$. Diophantus does not mention it, but the multitude of Things in the *al-istiqrā'* assignment must lie between $\frac{1}{5}$ and $\frac{9}{5}$ so that $x > y$ (Sesiano 1982, 253 n. 10).

VI.14 $\begin{cases} x^2 / y^2 - x^2 = \square \\ x^2 / y^2 - y^2 = \square' \end{cases}$ $\quad x > y$

The assignments for y and x satisfy the first condition, and the equation is set up from the second condition.

Assignments	Operations	Equation
$x := 1$T		
	Square 1T $\rightarrow 1$M	
$x^2 := 1$M		
$y := \frac{4}{5}$ u		
	Square $\frac{4}{5}$ u $\rightarrow \frac{16}{25}$ u	
$y^2 := \frac{16}{25}$ u		
	Divide 1M by $\frac{16}{25}$ u $\rightarrow 1\frac{9}{16}$ M	

Subtract $\frac{16}{25}$u from $1\frac{9}{16}$M $\rightarrow \frac{25}{16}$M ℓ $\frac{16}{25}$u

$$\frac{25}{16}\text{M } \ell \ \frac{16}{25}\text{u} = \square'$$

$\sqrt{\square'} \coloneqq 1\frac{1}{4}\text{T} \ell \text{ 2u}$

Square $1\frac{1}{4}$T ℓ 2u $\rightarrow \frac{25}{16}$M 4u ℓ 5T

$$\frac{25}{16}\text{M } \ell \ \frac{16}{25}\text{u}$$
$$= \frac{25}{16}\text{M 4u } \ell \text{ 5T}$$

1T is thus $\frac{1}{5}$ of $4\frac{16}{25}$u, i.e., $\frac{116}{125}$u, so $x^2 = \frac{13,456}{15,625}$ and $y^2 = \frac{16}{25}$. Diophantus does not mention it, but in order for x to be greater than y, the units in the *al-istiqrā'* assignment can be any positive value except those between $\frac{2}{5}$ and $\frac{8}{5}$ inclusive (Sesiano 1982, 254 n. 11).

VI.15 $\begin{cases} x^2 + \left(x^2 - y^2\right) = \square \\ y^2 + \left(x^2 - y^2\right) = \square' \end{cases}$ $\quad x > y$

The assignments for x and $x^2 - y^2$ satisfies the first condition, and the second condition is automatically satisfied, though Diophantus does not mention it. The equation is set up with an *al-istiqrā'* assignment to ensure that y^2 is a square.

Assignments	Operations	Equation
$x \coloneqq 1\text{T}$		
	Square 1T $\rightarrow$ 1M	
$x^2 \coloneqq 1\text{M}$		
$x^2 - y^2 \coloneqq 2\text{T 1u}$		
	Subtract 2T 1u from 1M $\rightarrow$ 1M ℓ 2T 1u	
$y^2 \coloneqq 1\text{M} \ \ell \ 2\text{T 1u}$		
		1M ℓ 2T 1u $= y^2$
$y \coloneqq 1\text{T} \ \ell \ 2\text{u}$		
	Square 1T ℓ 2u $\rightarrow$ 1M 4u ℓ 4T	
		1M ℓ 2T 1u = 1M 4u ℓ 4T

1T is found to be $2\frac{1}{2}$u, so $x^2 = 6\frac{1}{4}$ and $y^2 = \frac{1}{4}$. The equation in the solution is formed from two ways to name y^2, much like we saw for the two ways to name y^3 in Problems V.1–3.

VI.16 $\begin{cases} x^2 - \left(x^2 - y^2\right) = \square \\ y^2 - \left(x^2 - y^2\right) = \square' \end{cases}$ $\quad x > y$

The first condition is automatically satisfied, and the assignments for x and $x^2 - y^2$ make the derivative assignment for y^2 a square. The equation is formed from the second condition.

Assignments	Operations	Equation
$x \coloneqq 1$T		
	Square 1T → 1M	
$x^2 \coloneqq 1$M		
$x^2 - y^2 \coloneqq 2$T ℓ 1u		
	Subtract 2T ℓ 1u from 1M → 1M 1u ℓ 2T	
$y^2 \coloneqq 1$M 1u ℓ 2T		
	Subtract 2T ℓ 1u from 1M 1u ℓ 2T → 1M 2u ℓ 4T	
		1M 2u ℓ 4T = $\square'$
$\sqrt{\square'} \coloneqq 1$T ℓ 4u		
	Square 1T ℓ 4u → 1M 16u ℓ 8T	
		1M 2u ℓ 4T = 1M 16u ℓ 8T

1T is $3\frac{1}{2}$u, so $x^2 = 12\frac{1}{4}$ and $y^2 = 6\frac{1}{4}$.

VI.17
$$\begin{cases} x^2 + y^2 + z^2 = \square \\ x^2 = y \\ y^2 = z \end{cases}$$

The first three assignments satisfy the last two conditions, and the equation is formed from the first condition. In the last assignment, the 1MM is necessary to cancel the 1MMMM in the forthcoming equation, and the $\frac{1}{2}$ is necessary to cancel the 1MM, leaving $1M = \frac{1}{4}$. Because no choice is made, this is not an *al-istiqrā'* assignment in the way that al-Karajī understood the term. Diophantus does not give a name to the technique, so we do not know if he saw the assignment in the present problem as being of a different kind from the others.

Assignments	Operations	Equation
$x^2 \coloneqq 1$M		
	Square 1M → 1MM	
$y^2 \coloneqq 1$MM		
	Square 1MM → 1MMMM	
$z^2 \coloneqq 1$MMMM		

Add 1MMMM, 1MM,

1M → 1MMMM 1MM 1M

1MMMM 1MM 1M

= □

$\sqrt{\square} := 1MM\frac{1}{2}u$

Square 1MM $\frac{1}{2}$u → 1MMMM 1MM $\frac{1}{4}$u

1MMMM 1MM 1M

=1MMMM1MM $\frac{1}{4}$u

1M is found to be $\frac{1}{4}$u, so $x^2 = \frac{1}{4}$, $y^2 = \frac{1}{16}$, and $z^2 = \frac{1}{256}$.

VI.18 $(x^2 \cdot y^2) \cdot z^2 + (x^2 + y^2 + z^2) = \square$

The numerical assignments for the first two squares are made so that their product increased by 1u is a square number since he wants the multitude of *Māl*s in the forthcoming equation to be a square.

Assignments	Operations	Equation
$x^2 := 1u$		
$y^2 := \frac{9}{16}u$		
$z^2 := 1M$		
	Multiply 1u by $\frac{9}{16}u \to \frac{9}{16}u$	
	Multiply $\frac{9}{16}u$ by 1M → $\frac{9}{16}M$	
	Add 1u, $\frac{9}{16}u$, and 1M → 1M $\frac{25}{16}u$	
	Add $\frac{9}{16}M$ and 1M $\frac{25}{16}u$ → $\frac{25}{16}M \frac{25}{16}u$	
		$\frac{25}{16}M \frac{25}{16}u = \square$
$\sqrt{\square} := 1\frac{1}{4}T\frac{1}{4}u$		
	Square $1\frac{1}{4}T \frac{1}{4}u$ → $\frac{25}{16}M \frac{10}{16}T \frac{1}{16}u$	
		$\frac{25}{16}M \frac{25}{16}u = \frac{25}{16}M \frac{10}{16}T \frac{1}{16}u$

1T is $2\frac{2}{5}$u, so $x^2 = 1$, $y^2 = \frac{9}{16}$, and $z^2 = \frac{144}{25}$.

VI.19 $(x^2 \cdot y^2) \cdot z^2 - (x^2 + y^2 + z^2) = \square$

The numerical assignments for the first two squares are made so that their product decreased by 1u is a square number, since he wants the multitude of *Māl*s in the forthcoming equation to be a square.

Assignments	Operations	Equation
$x^2 := 1\mathrm{u}$		
$y^2 := 1\frac{9}{16}\mathrm{u}$		
$z^2 := 1\mathrm{M}$	Multiply $1\mathrm{u}$ by $1\frac{9}{16}\mathrm{u} \to 1\frac{9}{16}\mathrm{u}$	
	Multiply $1\frac{9}{16}\mathrm{u}$ by $1\mathrm{M} \to 1\frac{9}{16}\mathrm{M}$	
	Add $1\mathrm{u}, 1\frac{9}{16}\mathrm{u}$, and $1\mathrm{M} \to 1\mathrm{M}\,2\frac{9}{16}\mathrm{u}$	
	Subtract $1\mathrm{M}\,2\frac{9}{16}\mathrm{u}$ from $1\frac{9}{16}\mathrm{M} \to \frac{9}{16}\mathrm{M}\,\ell\,2\frac{9}{16}\mathrm{u}$	
		$\frac{9}{16}\mathrm{M}\,\ell\,2\frac{9}{16}\mathrm{u} = \square$
$\sqrt{\square} := \frac{3}{4}\mathrm{T}\,\ell\,\frac{1}{4}\mathrm{u}$		
	Square $\frac{3}{4}\mathrm{T}\,\ell\,\frac{1}{4}\mathrm{u} \to \frac{9}{16}\mathrm{M}\,\frac{1}{16}\mathrm{u}\,\ell\,\frac{6}{16}\mathrm{T}$	
		$\frac{9}{16}\mathrm{M}\,\ell\,2\frac{9}{16}\mathrm{u} =$ $\frac{9}{16}\mathrm{M}\,\frac{1}{16}\mathrm{u}\,\ell\,\frac{6}{16}\mathrm{T}$

$1\mathrm{T}$ is $7\mathrm{u}$, so $x^2 = 1$, $y^2 = 1\frac{9}{16}$, and $z^2 = 49$.

VI.20 $\left(x^2 + y^2 + z^2\right) - (x^2 \cdot y^2)\cdot z^2 = \square$

The numerical assignments for the first two squares are made so that their product, if subtracted from $1\mathrm{u}$, gives a square number since he wants the multitude of *Māl*s in the forthcoming equation to be a square.

Assignments	Operations	Equation
$x^2 := 4\mathrm{u}$		
$y^2 := \frac{4}{25}\mathrm{u}$		
$z^2 := 1\mathrm{M}$	Multiply $4\mathrm{u}$ by $\frac{4}{25}\mathrm{u} \to \frac{16}{25}\mathrm{u}$	
	Multiply $\frac{16}{25}\mathrm{u}$ by $1\mathrm{M} \to \frac{16}{25}\mathrm{M}$	
	Add $4\mathrm{u}, \frac{4}{25}\mathrm{u}$, and $1\mathrm{M} \to 1\mathrm{M}\,4\frac{4}{25}\mathrm{u}$	
	Subtract $\frac{16}{25}\mathrm{M}$ from $1\mathrm{M}\,4\frac{4}{25}\mathrm{u} \to \frac{9}{25}\mathrm{M}\,4\frac{4}{25}\mathrm{u}$	
		$\frac{9}{25}\mathrm{M}\,4\frac{4}{25}\mathrm{u} = \square$
$\sqrt{\square} := \frac{3}{5}\mathrm{T}\,1\mathrm{u}$		
	Square $\frac{3}{5}\mathrm{T}\,1\mathrm{u} \to \frac{9}{25}\mathrm{M}\,1\frac{1}{5}\mathrm{T}\,1\mathrm{u}$	
		$\frac{9}{25}\mathrm{M}\,4\frac{4}{25}\mathrm{u} = \frac{9}{25}\mathrm{M}\,1\frac{1}{5}\mathrm{T}\,1\mathrm{u}$

1T is $\frac{79}{30}$u, so $x^2 = 4$, $y^2 = \frac{4}{25}$, and $z^2 = \frac{6241}{900}$.

VI.21 $\begin{cases} (x^2)^2 + (x^2 + y^2) = \square \\ (y^2)^2 + (x^2 + y^2) = \square' \end{cases}$

The first two assignments satisfy the second condition of the problem, and the equation is formed from the first condition.

Assignments	Operations	Equation
$y^2 := 1$M		
$x^2 := \frac{1}{4}$u		
	Square $\frac{1}{4}$u $\to \frac{1}{16}$u	
	Add 1M and $\frac{1}{4}$u $\to$ 1M $\frac{1}{4}$u	
	Add 1M $\frac{1}{4}$u to $\frac{1}{16}$u $\to$ 1M $\frac{5}{16}$u	
		1M $\frac{5}{16}$u $= \square$
$\sqrt{\square} := 1$T $\frac{1}{2}$u		
	Square 1T $\frac{1}{2}$u $\to$ 1M 1T $\frac{1}{4}$u	
		1M $\frac{5}{16}$u $=$ 1M1 T$\frac{1}{4}$u

1T is $\frac{1}{16}$u, so $x^2 = \frac{1}{4}$ and $y^2 = \frac{1}{256}$.

VI.22 $\begin{cases} x^2 + y^2 = \square \\ x^2 \cdot y^2 = \boxed{\square}' \end{cases}$.

First solution

The first two assignments satisfy the second condition, and the equation is set up from the first condition. Here the $1\frac{1}{4}$ is chosen so that its square is 1 more than a square, which ensures a rational solution (Sesiano 1982, 257 n. 15).

Assignments	Operations	Equation
	Square 1M $\to$ 1MM	
$x^2 := 1$M		
$y^2 := 1$MM		
	Add 1M and 1MM $\to$ 1MM 1M	
		1MM 1M $= \square$
$\sqrt{\square} := 1\frac{1}{4}$M		
	Square $1\frac{1}{4}$M $\to 1\frac{9}{16}$MM	
		1MM 1M $= 1\frac{9}{16}$MM

Thus, 1M is $\frac{16}{9}$ u, so $x^2 = \frac{16}{9}$ and $y^2 = \frac{256}{81}$.

Second solution (line 2842)

The first two assignments are found by taking a Pythagorean triple and putting the numbers in terms of a *Māl* to satisfy the first condition. The equation is formed from the second condition.

Assignments	Operations	Equation
$x^2 := 16M$		
$y^2 := 9M$		
	Multiply 16M by 9M → 144MM	
		144MM = ⌼′
⌼′ := 8C		
		144MM = 8C

1T is $\frac{1}{18}$ u, so $x^2 = \frac{4}{81}$ and $y^2 = \frac{1}{36}$.

VI.23 $\begin{cases} a^2/x^2 + a^2/y^2 = \square \\ x^2 + y^2 + a^2 = \square' \end{cases}$ $\left(a^2 = 9 \right)$

To satisfy the first condition Diophantus notes that whenever $m^2 + n^2$ is a square, then $\frac{a^2}{m^2} + \frac{a^2}{n^2}$ is also a square. So, he finds two square numbers that add to a square, which are $\frac{9}{25} + \frac{16}{25} = 1$. Any Pythagorean triple would do, but this particular choice is convenient because the sum is 1, which makes the multitude of Ms 1 in the forthcoming equation. That equation is formed from the second condition.

Assignments	Operations	Equation
$x^2 := \frac{9}{25}M$		
$y^2 := \frac{16}{25}M$		
	Add $\frac{9}{25}M$, $\frac{16}{25}M$, and 9u → 1M 9u	
		1M 9u = $\square'$
$\sqrt{\square'} := 1T\ 1u$		
	Square 1T 1u → 1M 2T 1u	
		1M 9u = 1M 2T 1u

1T is found to be 4u, so $x^2 = \frac{144}{25}$ and $y^2 = \frac{256}{25}$.

Book VII (Arabic)

VII.1 $\begin{cases} x = my \\ y = mz \\ (x^3 \cdot y^3) \cdot z^3 = \square \end{cases}$ $\qquad (m = 2)$

The assignments for x, y, and z fulfill the first two conditions, and the equation is formed from the third condition.

Assignments	Operations	Equation
$z := 1T$		
	Cube $1T \rightarrow 1C$	
$z^3 := 1C$		
	Two times $1T \rightarrow 2T$	
$y := 2T$		
	Cube $2T \rightarrow 8C$	
$y^3 := 8C$		
	Two times $2T \rightarrow 4T$	
$x := 4T$		
	Cube $4T \rightarrow 64C$	
$x^3 := 64C$		
	Multiply $64C$ by $8C \rightarrow 512CC$	
	Multiply $512CC$ by $1C \rightarrow 512CCC$	
		$512CCC = \square$
$\sqrt{\square} := 32MM$		
	Square $32MM \rightarrow 1024MMMM$	
		$512CCC =$ $1024MMMM$

$1T$ is $2u$, so $z^3 = 8$, $y^3 = 64$, and $x^3 = 512$.

DOI: 10.4324/9781315171470-25

VII.2 $(x^6 \cdot y^6) \cdot z^6 = \square^2$

In the enunciation Diophantus asks for "three cubic numbers which are also squares", which earlier we transformed into modern notation as x^6, y^6 and z^6. In Problem VII.4 he asks for "a cubic number with a square side", which is more satisfactorily transformed into $(w^2)^3$.

Assignments	Operations	Equation
$x^6 := \frac{1}{64}u$ $y^6 := 64u$ $z^6 := 1CC$		
	Multiply $\frac{1}{64}u$ by $64u \rightarrow 1u$ Multiply $1u$ by $1CC \rightarrow 1CC$ Take the side of $1CC \rightarrow 1C$	
		$1C = \square$
$\square := 4M$		
		$1C = 4M$

$1T = 4u$, so $x^6 = \frac{1}{64}$, $y^6 = 64$, and $z^6 = 4096$. Sesiano points out (1982, 264) that a solution to this problem could also be obtained by squaring the results in the preceding problem.

VII.3 $(w^2)^2 = x^3 + y^3 + z^3$

Assignments	Operations	Equation
$w^2 := 1M$		
	Square $1M \rightarrow 1MM$	
$(w^2)^2 := 1MM$ $x^3 := 1C$ $y^3 := 8C$ $z^3 := 64C$		
	Add $1C$, $8C$, and $64C \rightarrow 73C$	
		$73C = 1MM$

$1T$ is $73u$, so $x^3 = 389{,}017$, $y^3 = 3{,}112{,}136$, $z^3 = 24{,}897{,}088$, and $(w^2)^2 = 28{,}398{,}241$.

VII.4 $(w^2)^3 = x^2 + y^2 + z^2$

The last three assignments are made in such a way that they are three squares that add up to a square. To find their multitudes Diophantus recalls Problem III.5.

Assignments	Operations	Equation
$w^2 := 1M$		
	Cube $1M \to 1CC$	
$(w^2)^3 := 1CC$		
$x^2 := 1MM$		
$y^2 := 4MM$		
$z^2 := \frac{4}{9}MM$		
	Add $1MM$, $4MM$, and $\frac{4}{9}MM \to \frac{49}{9}MM$	
		$\frac{49}{9}MM = 1CC$

$1M$ is $\frac{49}{9}$u. Therefore, $(w^2)^3 = \frac{117,649}{729}$, $x^2 = \frac{21,609}{729}$, $y^2 = \frac{86,436}{729}$ and $z^2 = \frac{9604}{729}$. The text has instead the values of x^2, y^2, and z^2 as if they were *Māl*s instead of *Māl Māl*s: $x^2 = \frac{49}{9}$, $y^2 = \frac{196}{9}$, and $z^2 = \frac{196}{729}$. And even here, the last denominator is taken from the correct value. See footnote 5 in page 418.

VII.5 $\left(x^3\right)^3 \cdot y^3 + \left(x^3\right)^3 \cdot z^2 = \square$

Assignments	Operations	Equation
$x^3 := 8u$		
	Cube $8u \to 512u$	

Diophantus next restates the enunciation with this value: $512 \cdot y^3 + 512 \cdot z^2 = \square$. He continues:

Assignments	Operations	Equation
$y^3 := 1C$		
$z^2 := 1M$		
	Multiply $1C$ by $512u \to 512C$	
	Multiply $1M$ by $512u \to 512M$	
	Add $512C$ and $512M \to 512C\ 512M$	
		$512C\ 512M = \square$
$\sqrt{\square} := 64T$		
	Square $64T \to 4096M$	
		$512C\ 512M$ $= 4096M$

$1T$ is found to be $7u$, so $x^3 = 8$, $y^3 = 343$, and $z^2 = 49$.

VII.6 $\begin{cases} x^2 + y^2 = \square \\ x^2 \cdot y^2 = m\left(x^2 + y^2\right) \quad (m = 9) \end{cases}$ Determination : m must be a square

The first three assignments satisfy the first condition, and the equation is formed from the second condition.

Assignments	Operations	Equation
$x^2 + y^2 := 1M$		
$x^2 := \frac{16}{25}M$		
$y^2 := \frac{9}{25}M$	Multiply $\frac{16}{25}M$ by $\frac{9}{25}M \to \frac{144}{625}MM$	$\frac{144}{625}MM = 9M$

1M is then $39\frac{1}{16}$u, so $x^2 = 25$ and $y^2 = 14\frac{1}{16}$.

According to Sesiano, among the first six problems of Book VII "only VII.4 (not considering the miscomputations) would deserve to figure among the problems of the *Arithmetica*" (1982, 265), suggesting that the other problems of this group are later interpolations.

VII.7 $\begin{cases} \left(w^3\right)^2 = x + y + z \\ x + y = \square \\ y + z = \square' \\ z + x = \square'' \end{cases}$

First solution

Diophantus first assigns $(w^3)^2$ to be 1CC. Then he solves the problem for x, y, and z without requiring the side of their sum to be a cube, which is precisely Problem III.6. Those numbers can then be made *Māl Māl*s, and the solution then falls out.

Assignments	Operations	Equation
$w^3 := 1C$		
	Square 1C $\to$ 1CC	
$(w^3)^2 := 1CC$		
$x := 80MM$		
$y := 320MM$		
$z := 41MM$		
	Add 80MM, 320MM, and 41MM $\to$ 441MM	
		441MM = 1CC

$1M = 441u$, so $x = 15{,}558{,}480$, $y = 62{,}233{,}920$, $z = 7{,}973{,}721$, and $(w^3)^2 = 85{,}766{,}121$.

Second solution (line 3142)

We assign $(w^3)^2$ to be 64u, i.e., $(2^3)^2$, and once again the problem is reduced to III.6. Diophantus takes the same solution as before, 320, 80, and 41, but in a different order. Their sum is 441, but according to the assignment we adopted the sum must be 64. Therefore, we multiply each number by 64 and then divide each result by 441, obtaining the solution $x = \frac{20{,}480}{441}$, $y = \frac{5120}{441}$, $z = \frac{2624}{441}$, and $(w^3)^2 = 64$.

The only place algebra is applied in the second solution is in working out Problem III.6. The method used in this solution is single false position (see Section 2.3). In the present book of the *Arithmetica* this method is also used in Problems 8–11 and 15.

VII.8
$$\begin{cases} \left(x^3\right)^2 + y = \square \\ \left(x^3\right)^2 + 2y = \square' \end{cases}$$

As in the second solution of the preceding problem, Diophantus assigns $(x^3)^2$ to be 64u, i.e., $(2^3)^2$, so the problem becomes $\begin{cases} 64 + y = \square \\ 64 + 2y = \square' \end{cases}$. He considers this a special case of the more general problem $\begin{cases} u^2 + v = \square'' \\ u^2 + 2v = \square''' \end{cases}$, whose solution he works out.

The first two assignments satisfy the first condition, and the equation is formed from the second condition.

Assignments	Operations	Equation
$u^2 = 1M$		
$v = 2T\ 1u$		
	Twice 2T 1u → 4T 2u	
	Add 4T 2u to 1M → 1M 4T 2u	
		1M 4T 2u = $\square'''$
$\sqrt{\square'''} = 1T \ell 2u$		
	Square 1T ℓ 2u → 1M 4u ℓ 4T	
		1M 4T 2u = 1M 4u ℓ 4T

1T is found to be $\frac{1}{4}$u, so $u^2 = \frac{1}{16}$ and $v = 1\frac{1}{2}$. Diophantus multiplies them by 16, a square, to get the whole numbers 1 and 24. Then a solution to the original problem results by multiplying these numbers by 64. Thus, $(x^3)^2 = 64$ and $y = 1536$.

VII.9 $\quad \begin{cases} \left(x^3\right)^2 - y = \square \\ \left(x^3\right)^2 - 2y = \square' \end{cases}$

The solution is similar to that of the preceding problem. We assign $(x^3)^2$ to be 64u, so the problem becomes $\begin{cases} 64 - y = \square \\ 64 - 2y = \square' \end{cases}$, which we consider a special case of the more general problem $\begin{cases} u^2 - v = \square'' \\ u^2 - 2v = \square''' \end{cases}$.

The first two assignments fulfill the first condition, and the equation is formed from the second condition.

Assignments	Operations	Equation
$u^2 := 1M$		
$v := 2T \ell 1u$		
	Twice $2T \ell 1u \to 4T \ell 2u$	
	Subtract $4T \ell 2u$ from $1M \to$ $1M\ 2u \ell 4T$	
		$1M\ 2u \ell 4T = \square'''$
$\sqrt{\square'''} := 1T \ell 3$		
	Square $1T \ell 3u \to 1M\ 9u \ell 6T$	
		$1M\ 2u \ell 4T =$ $1M\ 9u \ell 6T$

Thus, 1T is found to be $3\frac{1}{2}$u, so $u^2 = 12\frac{1}{4}$ and $v = 6$. Multiplying them by 4, we get the whole numbers 49 and 24. But the 49 should be 64. Since 64 is $1\frac{15}{49}$ of 49, i.e., 49 and $\frac{15}{49}$ of 49, Diophantus mentions that to the 24 and its double 48 we should add $\frac{15}{49}$ of them. The calculation is made differently, however. He multiplies them by $\frac{64}{49}$ to get $y = \frac{1536}{49}$ and $2y = \frac{3072}{49}$.

VII.10 $\quad \begin{cases} \left(x^3\right)^2 + y = \square \\ \left(x^3\right)^2 - 2y = \square' \end{cases}$

Along the same lines as the previous two problems, Diophantus adopts the assignment $(x^3)^2 := 64u$ and solves the more general problem $\begin{cases} u^2 + v = \square'' \\ u^2 - v = \square''' \end{cases}$.

Here the first two assignments satisfy the second condition, and the equation is set up from the first condition.

Assignments	Operations	Equation
$u^2 \coloneqq 1M$		
$v \coloneqq 2T \ell \, 1u$		
	Add $2T \ell \, 1u$ to $1M \rightarrow 1M \, 2T \ell \, 1u$	
		$1M \, 2T \ell \, 1u = \square''$
$\sqrt{\square''} \coloneqq 1T \ell \, 3u$		
	Square $1T \ell \, 3u \rightarrow 1M \, 9u \, \ell \, 6T$	
		$1M \, 2T \ell \, 1u =$
		$1M \, 9u \, \ell \, 6T$

Thus, $1T = 1\frac{1}{4}u$. So $u^2 = \frac{25}{16}$ and $v = \frac{24}{16}$, or, after multiplying them by 16, 25 and 24. A solution for y in the original problem results by multiplying the value for v by $\frac{64}{25}$, which gives $y = \frac{1536}{25}$.

VII.11 $\begin{cases} a^2 = x + y \\ a^2 + x = \square \\ a^2 - y = \square' \end{cases}$ $(a^2 = 25)$

Diophantus first solves the more general problem $\begin{cases} u^2 = v + w \\ u^2 + v = \square'' \\ u^2 - w = \square''' \end{cases}$. The assignments

satisfy the last two conditions, and the equation is formed from the first condition.

Assignments	Operations	Equation
$u^2 \coloneqq 1M$		
$v \coloneqq 2T \, 1u$		
$w \coloneqq 2T \ell \, 1u$		
	Add $2T \, 1u$, and $2T \ell \, 1u \rightarrow 4T$	
		$1M = 4T$

Thus, $1T = 4u$, so $u^2 = 16$, $v = 9$, and $w = 7$. Multiplying u and v each by $\frac{25}{16}$, we obtain the sought-after numbers from the original problem, $x = \frac{225}{16}$ and $y = \frac{175}{16}$.

After the solution, Diophantus adds a remark stating that the problem $\begin{cases} z^2 = x + y \\ z^2 + x = \square \\ z^2 + y = \square' \end{cases}$ is impossible. Adding the second and third equations, and substituting $x + y$ with z^2, gives $3z^2 = \square + \square'$. It is easy to show via modular arithmetic that this equation has no solution in rational numbers, though we do not know how Diophantus found it.

VII.12
$$\begin{cases} a^2 = x + y \\ a^2 - x = \Box \\ a^2 - y = \Box' \end{cases} \quad (a^2 = 25)$$

Diophantus observes that if x and y are square numbers then the problem is reduced to that of dividing a given square into two squares. But this is Problem II.8, so he gives directly the solution 9 and 16.

VII.13
$$\begin{cases} a^2 = x + y + z \\ a^2 + x = \Box \\ a^2 + y = \Box' \\ a^2 + z = \Box'' \end{cases} \quad (a^2 = 25)$$

Adding the last three conditions and applying the first gives $4a^2 = \Box + \Box' + \Box''$. So, the problem is reduced to that of dividing 100u into the sum of three squares, each of which is greater than 25u. This can be done by two iterations of Problem II.8 since 100u is a square. Without working out the computations he gives 36, $33\frac{471}{841}$, and $30\frac{370}{841}$ for the squares $\Box$, $\Box'$, and $\Box''$, and by subtracting 25 from each he finds the sought-after numbers to be $x = 11$, $y = 8\frac{471}{841}$, and $z = 5\frac{370}{841}$.

VII.14
$$\begin{cases} a^2 = x + y + z \\ a^2 - x = \Box \\ a^2 - y = \Box' \\ a^2 - z = \Box'' \end{cases} \quad (a^2 = 25)$$

This time adding the last three conditions and applying the first gives $2a^2 = \Box + \Box' + \Box''$. So, the problem is reduced to that of dividing 50 into the sum of three squares, each of which is less than 25u. Because 50 is the sum of two (equal) squares, Problem II.9 can be applied, and after that II.8 can be applied to divide one of the square parts into two squares (Sesiano 1982, 174). Diophantus finds the squares to be 16, $22\frac{3}{169}$, and $11\frac{166}{169}$, and by subtracting each one of them from 25 he finds $x = 9$, $y = 2\frac{166}{169}$, and $z = 13\frac{3}{169}$.

VII.15
$$\begin{cases} a^2 = w + x + y + z \\ a^2 - w = \Box \\ a^2 - x = \Box' \\ a^2 + y = \Box'' \\ a^2 + z = \Box''' \end{cases} \quad (a^2 = 25)$$

As in Problems VII.8–11, Diophantus considers the problem as special case of the

more general problem $\begin{cases} v^2 = w + x + y + z \\ v^2 - w = \square \\ v^2 - x = \square' \\ v^2 + y = \square'' \\ v^2 + z = \square''' \end{cases}$, which he solves and then corrects by

applying the technique of single false position.

The assignments satisfy the last four conditions, and the equation is formed from the first condition.

Assignments	Operations	Equation
$v^2 := 1M$		
$y := 2T\ 1u$		
$z := 4T\ 4u$		
	Add 2T 1u and 4T 4u $\rightarrow$ 6T 5u	
$w := 2T\ \ell\ 1u$		
$x := 4T\ \ell\ 4u$		
	Add 2T ℓ 1u and 4T ℓ 4u $\rightarrow$ 6T ℓ 5u	
	Add 6T 5u and 6T ℓ 5u $\rightarrow$ 12T	
		1M = 12T

Then 1T is 12u, so $v^2 = 114$, $w = 23$, $x = 44$, $y = 25$, and $z = 52$. Now, by multiplying these numbers by $\frac{25}{144}$ we obtain a solution of the original problem, which is $w = \frac{575}{144}$, $x = \frac{1100}{144}$, $y = \frac{625}{144}$, and $z = \frac{1300}{144}$.

After the solution Diophantus remarks that in the same manner one can solve the same problem, but dividing the square into eight parts, adding the first four and subtracting the last four. The assignments for this expanded problem can be modelled on squares such as those of the form $m^2 \pm 2m + 1$, $m^2 \pm 4m + 4$, $m^2 \pm 6m + 9$, and $m^2 \pm 8m + 16$.

VII.16 $\begin{cases} x^2 : y^2 :: y^2 : z^2 \\ y^2 - x^2 = \square \\ z^2 - y^2 = \square' \end{cases}$

First Diophantus states the lemma that if $x^2 : y^2 :: y^2 : z^2$ and $y^2 - x^2$ is a square, then $z^2 - y^2$ is also a square. Sesiano (1982, 277) explains that a proof can be obtained by applying Propositions V.17 and VIII.24 of Euclid's *Elements*. $x^2 : y^2 :: y^2 : z^2$ is identical to $y^2 : x^2 :: z^2 : y^2$, and by V.17 we get $(y^2 - x^2) : x^2 :: (z^2 - y^2) : y^2$. Then, by VIII.24, if $y^2 - x^2$ is a square, then $z^2 - y^2$ is also a square. Applying these propositions in the present instance might be seen as problematic because Book V of the *Elements* deals with continuous geometric magnitudes and Book VIII with discrete numbers, neither of which correspond to the numbers in the *Arithmetica*. But these issues may not have bothered Diophantus, since the propositions hold also for the positive rational numbers that he worked with. The next two problems give similar lemmas, and as we explain below for Problem VII.18, they may have been proven in a lost book of his.

The first three assignments satisfy the first condition, so the remaining two conditions are equivalent. The equation is formed from the second condition.

Assignments	Operations	Equation
$x^2 := 1u$		
$z^2 := 1MM$		
$y^2 := 1M$		
	Subtract $1u$ from $1M \to 1M \ell 1u$	
		$1M \ell 1u = \square$
$\sqrt{\square} := 1T \ell 2u$		
	Square $1T \ell 2u \to 1M\ 4u \ell 4T$	
		$1M \ell 1u = 1M\ 4u \ell 4T$

$1T$ is found to be $1\frac{1}{4}u$, so $x^2 = 1$, $y^2 = \frac{25}{16}$, and $z^2 = \frac{625}{256}$.

VII.17 $\begin{cases} w^2 : x^2 :: y^2 : z^2 \\ w^2 + x^2 + y^2 + z^2 = \square \end{cases}$

Here Diophantus states the lemma that if $m : n :: p : q$ then $m \cdot q = p \cdot n$. This is proven twice in Euclid's *Elements*, first in Proposition VI.16 in the context of lines, where the formation of a rectangle takes the place of multiplication, and again in Proposition VII.19 for discrete numbers.

The assignments for x^2 and z^2 are chosen, so they add to a square multitude of *Māl*s in order that the *al-istiqrā'* assignment will work. Those for w^2 and y^2 are chosen using the lemma to satisfy the first condition. The equation is set up from the second condition.

Assignments	Operations	Equation
$w^2 := 1u$		
$z^2 := 16M$		
$x^2 := 9M$		
$y^2 := 1\frac{7}{9}u$		
	Add $1u$, $16M$, $9M$, and $1\frac{7}{9}u \to 25M\ 2\frac{7}{9}u$	
		$25M\ 2\frac{7}{9}u = \square$
$\sqrt{\square} := 5T\frac{1}{3}u$		

$$\left| \begin{array}{l} \text{Square } 5\text{T}\,\tfrac{1}{3}\text{u} \rightarrow 25\text{M }3\tfrac{1}{3}\text{T}\,\tfrac{1}{9}\text{u} \\[2mm] \qquad\qquad 25\text{M }2\tfrac{7}{9}\text{u} = 25\text{M }3\tfrac{1}{3}\text{T}\,\tfrac{1}{9}\text{u} \end{array} \right.$$

1T is $\tfrac{8}{10}$u, so $w^2 = 1$, $x^2 = \tfrac{576}{100}$, $y^2 = 1\tfrac{7}{9}$, and $z^2 = \tfrac{1024}{100}$.

VII.18 $\quad \begin{cases} w^2 : x^2 :: y^2 : z^2 \\ x^2 - w^2 = \square \\ y^2 - x^2 = \square' \\ z^2 - y^2 = \square'' \end{cases}$

Diophantus again begins with a lemma, that if $w^2 : x^2 :: y^2 : z^2$, and $x^2 - w^2$ is a square, then $z^2 - y^2$ is also a square. This is a more general form of the lemma stated in Problem VII.16, and its proof can be obtained by a similar argument from the one Sesiano proposed (1982, 277). By the way Diophantus presents it, "We have found that", suggests that perhaps this lemma, and also the lemmas to the previous two problems, were part of some lost book of Diophantus, perhaps the *Porisms*. The fact that the numbers in the *Arithmetica*, which can be any positive whole or fractional number, are not covered in Euclid's *Elements* at all, lends support to this suggestion.

The first two assignments are chosen to satisfy the second condition. The assignment for z^2 is direct, and that for y^2 derives from the fact that $w^2 \cdot z^2 = x^2 \cdot y^2$, a consequence of the first condition. The equation is then formed from the third condition, which by the lemma will also satisfy the fourth condition.

Assignments	Operations	Equation
$w^2 := 9\text{u}$		
$x^2 := 25\text{u}$		
$z^2 := 1\text{M}$		
	Multiply 9u by 1M $\rightarrow$ 9M	
	Divide 9M by 25u $\rightarrow \tfrac{9}{25}$M	
$y^2 := \tfrac{9}{25}\text{M}$		
	Subtract 25u from $\tfrac{9}{25}$M $\rightarrow$ $\tfrac{9}{25}$M ℓ 25u	
		$\tfrac{9}{25}$M ℓ 25u $= \square'$
$\sqrt{\square'} := \tfrac{3}{5}\text{T}\,\ell\,1\text{u}$		
	Square $\tfrac{3}{5}$T ℓ 1u $\rightarrow \tfrac{9}{25}$M 1u ℓ $\tfrac{6}{5}$T	
		$\tfrac{9}{25}$M ℓ 25u $= \tfrac{9}{25}$M 1u ℓ $\tfrac{6}{5}$T

1T is found to be $\tfrac{130}{6}$u, so $w^2 = 9$, $x^2 = 25$, $y^2 = \tfrac{6084}{36}$, and $z^2 = \tfrac{16{,}900}{36}$.

Book IV^G

IV^G.1 $\begin{cases} x+y=a \\ x^3+y^3=b \end{cases}$ $\quad (a=10, b=370)$

The assignments are designed to satisfy the first condition in such a way that the expression for $x^3 + y^3$ in the second condition has only two terms.

Assignments	Operations	Equation
$x := 1N\ 5u$		
$y := 5u\ \ell\ 1N$		
	Add the cubes of 1N 5u and 5u ℓ 1N → 30P 250u	
		30P 250u = 370u

1N is found to be 2u, hence, $x = 7$ and $y = 3$.

    This is the same problem as V.7, where $a = 20$ and $b = 2240$ and the determination $\frac{4b-a^3}{3a} = \square$ is given. Undoubtedly the present problem, as well as the next one, which coincides with Problem V.8, are later interpolations (Sesiano 1982, 4).

IV^G.2 $\begin{cases} x-y=a \\ x^3-y^3=b \end{cases}$ $\quad (a=6, b=504)$

The assignments are designed to satisfy the first condition so that the expression for x^3-y^3 in the second condition has only two terms.

Assignments	Operations	Equation
$x := 1N\ 3u$		
$y := 1N\ \ell\ 3u$		

DOI: 10.4324/9781315171470-26

> Subtract the cube of 1N ℓ 3u from the
> cube of 1N 3u $\rightarrow$ 18P 54u
>
> 189 54u = 504u

1N is found to be 5u, so $x = 8$ and $y = 2$. This problem has already been solved in Problem V.8, where the determination $\frac{4b-a^3}{3a} = \square$ is also given.

IV^G.3 $\begin{cases} x^2 \cdot y = \sqrt[3]{\boxdot} \\ x \cdot y = \boxdot \end{cases}$

The assignments are designed to satisfy the second condition, and the equation is formed from the first condition.

Assignments	Operations	Equation
$x^2 := 1P$		
$x := 1N$		
$y := 8N'$		
	Multiply 8N′ by 1P $\rightarrow$ 8N	
	Multiply 8N′ by 1N $\rightarrow$ 8u	
		2u = 8N

Thus, 1N is $\frac{2}{8}$u , so $x = \frac{2}{8}$ or $\frac{1}{4}$, while $y = 32$.

IV^G.4 $\begin{cases} x^2 + y = \square \\ x + y = \sqrt{\square} \end{cases}$

The assignments are designed to satisfy the first condition, and the equation is formed from the second condition.

Assignments	Operations	Equation
$x^2 := 1P$		
$x := 1N$		
$y := 3P$		
	Add 3P and 1P $\rightarrow$ 4P	
	Add 3P and 1N $\rightarrow$ 3P 1N	
	Take the square root of 4P $\rightarrow$ 2N	
		3P 1N = 2N

Thus, 1N is $\frac{1}{3}$u , so $x^2 = \frac{1}{9}$ and $y = \frac{3}{9}$.

IV^G.5 $\begin{cases} x^2 + y = \sqrt{\Box} \\ x + y = \Box \end{cases}$

The assignments are designed to satisfy the second condition, and the equation is formed from the first condition.

Assignments	Operations	Equation
$x^2 := 1P$		
$x := 1N$		
$y := 4P \, \ell \, 1N$		
	Add $4P \, \ell \, 1N$ to $1N \to 4P$	
	Add $4P \, \ell \, 1N$ to $1P \to 5P \, \ell \, 1N$	
	Take the square root of $4P \to 2N$	
		$5P \, \ell \, 1N = 2N$

Thus, $1N$ is found to be $\frac{3}{5}u$, so $x^2 = \frac{9}{25}$ and $y = \frac{21}{25}$.

IV^G.6 $\begin{cases} x^3 + z^2 = \boxed{\boxtimes} \\ y^2 + z^2 = \Box' \end{cases}$

The second and third assignments satisfy the second condition based in the identity $pq + \left(\frac{p-q}{2}\right)^2 = \left(\frac{p+q}{2}\right)^2$, where pq is 9, which factors into 1 by 9. This gives $16 + 9 = 25$, which are made multiples of Powers in the assignments. The equation is formed from the first condition.

Assignments	Operations	Equation
$x^3 := 1C$		
$y^2 := 9P$		
$z^2 := 16P$		
	Add $1C$ and $16P \to 1C \, 16P$	
		$1C \, 16P = \boxed{\boxtimes}$
$\boxed{\boxtimes} := 8C$		
		$1C \, 16P = 8C$

Thus, $1N$ is found to be $\frac{16}{7}u$, so $x^3 = \frac{4096}{343}$, $y^2 = \frac{2304}{49}$, and $z^2 = \frac{4096}{49}$.

IV^G.7 $\begin{cases} x^3 + z^2 = \Box \\ y^2 + z^2 = \boxed{\boxtimes}' \end{cases}$

First solution

Diophantus wishes to make x^3 equal to $⌸'$, so the conditions can be restated as $\begin{cases} x^3 + z^2 = \square \\ x^3 - z^2 = y^2 \end{cases}$. He models the two conditions on the identity $p^2 + q^2 \pm 2pq = (p \pm q)^2$, associating $p^2 + q^2$ with x^3, $2pq$ with z^2, $(p+q)^2$ with $\square$, and $(p-q)^2$ with y^2. In order for $2pq$ to be a square, he assigns p and q to be 2N and 1N, respectively. This gives the first three assignments. The equation is formed from the first condition.

Assignments	Operations	Equation
$x^3 \coloneqq 5P$		
$z^2 \coloneqq 4P$		
	Subtract 4P from 5P → 1P	
$y^2 \coloneqq 1P$		
		$5P = ⌸'$
$⌸ \coloneqq 1C$		
		$5P = 1C$

Thus, 1N is 5u, so $x^3 = 125$, $y^2 = 25$, and $z^2 = 100$.

Second solution

Again, Diophantus wishes to make x^3 equal to $⌸'$. Since $⌸'$ is the sum of two squares, and one of them (z^2) is added back to it in the first condition to make a square, Diophantus looks for numbers u and v such that $(u^2 + v^2) + u^2 = \square_1$.

Assignments	Operations	Equation
$u^2 \coloneqq 1P$		
$v^2 \coloneqq 4u$		
	Add 1P and 4u → 1P 4u	
	Add 1P 4u and 1P → 2P 4u	
		$2P\ 4u = \square_1$
$\sqrt{\square_1} \coloneqq 2N\ \ell\ 2u$		
	Square 2N ℓ 2u → 4P 4u ℓ 8N	$2P\ 4u = 4P\ 4u\ \ell\ 8N$

Thus, 1N is 4u, so $u^2 = 16$ and $v^2 = 4$. Now x^3, y^2, and z^2 will be assigned to be multitudes of Powers, where these multitudes are $16 + 4$, 4, and 16, respectively.

Assignments	Operations	Equation
$z^2 := 16P$		
$y^2 := 4P$		
$x^3 := 20P$		
		$20P = \square\,'$
$\square\,' := 1C$		
		$20P = 1C$

Thus, 1N is 20u, so $x^3 = 8000$, $y^2 = 1600$, and $z^2 = 6400$.

IV^G**.8** $\begin{cases} x^3 + y = \square \\ x + y = \sqrt[3]{\square} \end{cases}$

Assignments	Operations	Equation
$y := 1N$		
$x := 2N$		
	Cube 2N → 8C	
$x^3 := 8C$		
	Add 1N to 2N → 3N	
	Add 1N to 8C → 8C 1N	
	Cube 3N → 27C	
		8C 1N = 27C

The equation is formed from the first condition. It leads to $19P = 1u$, but because 19 is not a square, 1N is not rational. Now, 19 arises from $3^3 - 2^3$, that is from the difference of the cubes of two consecutive numbers. Thus, the problem is reduced to that of finding two consecutive numbers whose cubes differ by a square. In modern notation, $\begin{cases} u - v = 1 \\ u^3 - v^3 = \square \end{cases}$ $(u > v)$.

The first two assignments satisfy the first condition, and the equation is formed from the second condition.

Assignments	Operations	Equation
$v := 1N$		
$u := 1N\ 1u$		
	The difference of the cubes of 1N 1u and 1N → 3P 3N 1u	

$3P\ 3N\ 1u = \square$

$\sqrt{\square} := 1u\ \ell\ 2N$

Square $1u\ \ell\ 2N \rightarrow 4P\ 1u\ \ell\ 4N$

$3P\ 3N\ 1u = 4P\ 1u\ \ell\ 4N$

1N is found to be 7u, so $v = 7$ and $u = 8$. He then returns to the original problem.

Assignments	Operations	Equation
$y := 1N$		
$x := 7N$		
	Cube $7N \rightarrow 343C$	
$x^3 := 343C$		
	Add $1N$ to $7N \rightarrow 8N$	
	Add $1N$ to $343C \rightarrow 343C\ 1N$	
$\sqrt[3]{\boxplus} := 8N$		
	Cube $8N \rightarrow 512C$	
		$512C = 343C\ 1N$

Thus, 1N is $\frac{1}{13}$u, so $x^3 = \frac{343}{2197}$, its side is $x = \frac{7}{13}$, and $y = \frac{1}{13}$.

IV^G.9 $\begin{cases} x^3 + y = \sqrt[3]{\boxplus} \\ x + y = \boxplus \end{cases}$

The assignments satisfy the second condition, and the equation is formed from the first condition.

Assignments	Operations	Equation
$x^3 := 8C$		
$x := 2N$		
$y := 27C\ \ell\ 2N$		
	Add $27C\ \ell\ 2N$ to $2N \rightarrow 27C$	
$\boxplus := 27C$		
	Add $27C\ \ell\ 2N$ to $8C \rightarrow 35C\ \ell\ 2N$	
		$35C\ \ell\ 2N = 3N$

The equation simplifies to $35P = 5u$. The solution is not rational because the ratio of 35 to 5 is not a square. Now, the 35 arises from the addition of two cubes, 27 and 8, while the 5 arises from the addition of the sides of these cubes. Thus, the problem is reduced to that of finding two cubes to replace the 27 and 8, such that their sum has to the sum of their sides the ratio of a square number to a square number. In our notation, $(u^3 + v^3) : (u + v) :: \square : \square'$.

Assignments	Operations	Equation
$u + v = 2u$		
$u = 1N$		
$v = 2u \ell 1N$		
	Add the cubes of 1N and 2u ℓ 1N $\rightarrow$ 6P 8u ℓ 12N	

We want 6P 8u ℓ 12N : 2u to be the ratio of a square to a square. Since 2u is double a square, 6P 8u ℓ 12N should also be double a square.

Assignments	Operations	Equation
	Take half of 6P 8u ℓ 12N $\rightarrow$ 3P 4u ℓ 6N	3P 4u ℓ 6N = $\square''$
$\sqrt{\square''} = 2u \ell 4N$	Square 2u ℓ 4N $\rightarrow$ 4u 16P ℓ 16N	3P 4u ℓ 6N = 4u 16P ℓ 16N

Hence, 1N is $\frac{10}{13}$ u, so $u = \frac{10}{13}$ and $v = \frac{16}{13}$. Eliminating the denominators and taking half, they become 5 and 8. Diophantus then returns to the original problem, adjusting his assignments according to these numbers.

Assignments	Operations	Equation
$x = 5N$		
$x^3 = 125C$		
$y = 512C \ell 5N$		
	Add 512C ℓ 5N to 125C $\rightarrow$ 637C ℓ 5N	
		8N = 637C ℓ 5N

1N is found to be $\frac{1}{7}$ u, so $x^3 = \frac{125}{343}$, $x = \frac{5}{7}$, and $y = \frac{267}{343}$.

IV^G.10 $x^3 + y^3 = x + y$

Assignments	Operations	Equation
$x = 2N$		
$y = 3N$		
	Add the cubes of 2N and 3N $\rightarrow$ 35C	
	Add 2N and 3N $\rightarrow$ 5N	
		35C = 5N

The solution to this equation is not rational, again because the ratio of 35 to 5 is not a square. Therefore, as in the previous problem, we must find two numbers such that the sum of their cubes has to the sum of the sides the ratio that a square number has to a square number. This has already been solved, and the numbers are 5 and 8.

Assignments	Operations	Equation
$x := 5N$		
$y := 8N$		
	Add the cubes of 8N and 5N $\rightarrow$ 637C	
	Add 8N and 5N $\rightarrow$ 13N	
		$637C = 13N$

1N is found to be $\frac{1}{7}$ u , so $x = \frac{5}{7}$ and $y = \frac{8}{7}$, which makes $x^3 = \frac{125}{343}$ and $y^3 = \frac{512}{343}$.

IV^G.11 $x^3 - y^3 = x - y$

Assignments	Operations	Equation
$y := 2N$		
$x := 3N$		
	The difference of the cubes of 3N and 2N $\rightarrow$ 19C	
	The difference of 3N and 2N $\rightarrow$ 1N	
		$1N = 19C$

Again, the solution to the equation is not rational. Therefore, we must find two cubes whose difference has to the difference of their sides the ratio of a square number to a square number. In our notation, $(u^3 - v^3) : (u - v) :: \square : \square'$, $u > v$.

Assignments	Operations	Equation
$v := 1N$		
$u := 1N1u$		
	Subtract the cube of 1N from the cube of 1N 1u $\rightarrow$ 1N 1U $\rightarrow$ 3P 3N 1u	

Since 3P 3N 1u : 1u is the ratio of a square to a square, 3P 3N 1u must be a square.

Assignments	Operations	Equation
		$3P\ 3N\ 1u = \square''$
$\sqrt{\square''} := 1u\ \ell\ 2N$		

Square 1u ℓ 2N → 1u 4P ℓ 4N

3P 3N 1u = 1u 4P ℓ 4N

Thus, 1N is found to be 7u, so $v = 7$ and $u = 8$. The original problem is then begun again.

Assignments	Operations	Equation
$y := 7N$		
$x := 8N$		
	The difference of the cubes of 8N and 7N → 169C	
	The difference of 8N and 7N → 1N	
		169C = 1N

1N is then $\frac{1}{13}$u, so $x = \frac{8}{13}$ and $y = \frac{7}{13}$.

IV^G.12 $x^3 + y = y^3 + x$

Assignments	Operations	Equation
$y := 2N$		
$x := 3N$		
	The cube of 3N increased by 2N → 27C 2N	
	The cube of 2N increased by 3N → 8C 3N	
		8C 3N = 27C 2N

This leads to 19P − 1u, whose solution is not rational. Now, 19u arises from the difference of two cubes, while the 1u is the difference of the sides of the cubes. Therefore, as in the last problem, we need to find two numbers such that the difference of their cubes has to the difference of the numbers the ratio that a square has to a square. Such numbers have been found, and they are 7 and 8. The solution of the original problem is started afresh.

Assignments	Operations	Equation
$y := 7N$		
$x := 8N$		
	The cube of 8N increased by 7N → 512C 7N	

> The cube of 7N increased by
> 8N → 343C 8N
>
> 343C 8N = 512C 7N

1N is found to be $\frac{1}{13}$u , so $x = \frac{8}{13}$ and $y = \frac{7}{13}$

IV^G.13 $\begin{cases} x+1=\square \\ y+1=\square' \\ (x+y)+1=\square'' \\ (y-x)+1=\square''' \end{cases}$

The first assignment is made to satisfy the first condition, based on the square of 3N 1u.

Assignments	Operations	Equation
$x := $ 9P 6N		

From the second and third conditions, $y + 1$ must be a square, and adding this to 9P 6N must also make a square. He models the assignment of y on the identity $pq = \left(\frac{p+q}{2}\right)^2 - \left(\frac{p-q}{2}\right)^2$, associating pq with 9P 6N, and the right side with $\square''$ – $\square'$. By making p to be 9N 6u and q to be 1N, he finds $\square'$ to be the square of 4N 3u, or 16P 24N 9u, which is $y + 1$. The solution continues with the assignment for y, which satisfies the second and third conditions, after which the equation is formed from the last condition.

Assignments	Operations	Equation
$y := $ 16P 24N 8u		
	Subtract 9P 6N from 16P 24N 8u → 7P 18N 8u	
	Add 1u to that → 7P 18N 9u	
		7P 18N 9u = $\square'''$
$\sqrt{\square'''} := $ 3u ℓ 3N		
	Square 3u ℓ 3N → 9u 9P ℓ 18N	
		7P 18N 9u = 9u 9P ℓ 18N

1N is found to be 18u, so $x = 3024$ and $y = 5624$.

IV^G.14 $x^2 + y^2 + z^2 = (x^2 - y^2) + (y^2 - z^2) + (x^2 - z^2)$

Diophantus begins by making the observation that the condition is equivalent to $x^2 + y^2 + z^2 = 2(x^2 - z^2)$. Both z^2 and x^2 are assigned to be squares, and the equation is formed to make the derivative assignment of y^2 a square.

Assignments	Operations	Equation
$z^2 \doteq 1u$		
$x^2 \doteq 1P\ 2N\ 1u$		
	Twice the difference of 1u and 1P 2N 1u → 2P 4N	
	Add 1P 2N 1u and 1u → 1P 2N 2u	
	Subtract 1P 2N 2u from 2P 4N → 1P 2N ℓ 2u	
$y^2 \doteq 1P\ 2N\ \ell\ 2u$		
		1P 2N ℓ 2u = $\square$
$\sqrt{\square} \doteq 1N\ \ell\ 4$		
	Square 1N ℓ 4u → 1P 16u ℓ 8N	
		1P 16u ℓ 8N = 1P 2N ℓ 2u

1N is then $\frac{9}{5}$u, so $x^2 = \frac{196}{25}$, $y^2 = \frac{121}{25}$, and $z^2 = 1$. Or, multiplying by 25, $x^2 = 196$, $y^2 = 121$, and $z^2 = 25$.

IV^G.15 $\begin{cases} (x+y)\cdot z = a \\ (y+z)\cdot x = b \\ (z+x)\cdot y = c \end{cases}$ $\left(a = 35, b = 27, c = 32 \right)$

The first two assignments fulfill the first condition. Then x and y are assigned whatever convenient values that make their sum 35N', so two equations can be formed from the other two conditions.

Assignments	Operations	Equation
$z \doteq 1N$		
$x + y \doteq 35N'$		
$x \doteq 10N'$		
$y \doteq 25N'$		
	Multiply the sum of 1N and 25N' by 10N' → 10u 250P'	
		10u 250P' = 27u
	Multiply the sum of 1N and 10N' by 25N' → 25u 250P'	
		25u 250P' = 32u

Diophantus observes that 5u, the difference of 32u and 27u, is not equal to 15u, the difference of 25u 250P' and 10u 250P'. Therefore, we need to replace the numbers 25 and 10 in the assignments for x and y with other numbers whose sum is 35 and whose difference is 5. The new values are 15 and 20:

Assignments	Operations	Equation
$x \coloneqq 15N'$		
$y \coloneqq 20N'$		
	Multiply the sum of 1N and 20N' by 15N' $\rightarrow$ 15u 300P'	
		15u 300P' = 27u
	Multiply the sum of 1N and 15N' by 20N' $\rightarrow$ 20u 300P'	
		20u 300P' = 32u

The two equations give 1N equal to 5u, so the sought-after numbers are $x = 3$, $y = 4$, and $z = 5$.

IV^G.16
$$\begin{cases} x+y+z = \square \\ x^2 + y = \square' \\ y^2 + z = \square'' \\ z^2 + x = \square''' \end{cases}$$

Diophantus begins by solving the first three conditions of the problem indeterminately. The first two assignments are made with the help of the identity $\left(\frac{p-q}{2}\right)^2 + pq = \left(\frac{p+q}{2}\right)^2$, where p is 2N and q is 2u, to satisfy the second condition. He then works on the third condition for the assignment for z, and he sets up the forthcoming equation 13N = $\square$ based on the first condition.

Assignments	Operations	Equation
$y \coloneqq 4N$		
$x \coloneqq 1N \ell 1u$		
	Square 4N $\rightarrow$ 16P	
$\sqrt{\square''} \coloneqq 4N\ 1u$		
	Square 4N 1u $\rightarrow$ 16P 8N 1u	
	Subtract 16P from 16P 8N 1u $\rightarrow$ 8N 1u	
$z \coloneqq 8N\ 1u$		
	Add 1N ℓ 1u, 4N, 8N 1u $\rightarrow$ 13N	
		13N = $\square$

Because the last condition is still not satisfied, this equation will be solved indeterminately by making the □ a square number of Powers of a second, independent
unknown, which we show in notation as $\mathcal{P}$. Diophantus also calls this $\mathcal{P}$ a δύναμις
(Power).

Assignments	Operations	Equation
		$13N = \square$
$\square := 169\mathcal{P}$		
		$13N = 169\mathcal{P}$

Thus, 1N is $13\mathcal{P}$. Replacing 1N with $13\mathcal{P}$ in the expressions for x, y, and z, we
obtain the new names given here, which satisfy the first three conditions of the
problem indeterminately. Then the equation can be formed from the last condition.

Assignments	Operations	Equation
$x := 13\mathcal{P} \ \ell$ 1u		
$y := 52\mathcal{P}$	Square $104\mathcal{P}$ 1u $\rightarrow 10,816\mathcal{PP}$ $208\mathcal{P}$ 1u	
$z := 104\mathcal{P}$ 1u	Add this to $13\mathcal{P} \ \ell$ 1u $\rightarrow 10,816\mathcal{PP}$ $221\mathcal{P}$	$10,816\mathcal{PP}$ $221\mathcal{P} = \square'''$
	Divide $10,816\mathcal{PP}$ $221\mathcal{P}$ by $1\mathcal{P} \rightarrow 10,816\mathcal{P}$ 221u	$10,816\mathcal{P}$ 221u $= \square^{iv}$
$\sqrt{\square^{iv}} := 104\mathcal{N}$ 1u	Square $104\mathcal{N}$ 1u $\rightarrow 10,816\mathcal{P}$ $208\mathcal{N}$ 1u	$10,816\mathcal{P}$ 221u $=$ $10,816\mathcal{P}$ $208\mathcal{N}$ 1u

Thus, $1\mathcal{N}$ is $\frac{55}{52}$u . This makes $x = \frac{36,621}{2704}$, $y = \frac{157,300}{2704}$, and $z = \frac{317,304}{2704}$.

Problem IV^G.16 is the first problem in the preserved part of the *Arithmetica*
in which Diophantus solves a problem "in the indeterminate". He explains such
solutions at the end of Problem IV^G.19: "to seek indeterminately means that the
hypostasis (of the unknown) be such that whatever one might wish the Number to
be, when he brings (it) to the hypostases he will fulfill the proposed task". Or, in a
similar remark, in the lemma to Problem IV^G.34: "(solving a problem) indeterminately is such that, of however many units someone might wish the Number to be,
when he brings (it) to the hypostases, the problem is completed". Indeterminate
solutions are used in Problems IV^G.16, 17, 19, 21, V^G.18, in the lemmas to Problems IV^G.34–36, and in the first lemma to Problem V^G.7. As Jacob Klein noted,
such indeterminate solutions "have a merely auxiliary character" and serve as a
means of producing determinate solutions to problems (1968, 134).

Problem IV^G.16 is also remarkable for another reason. It is the first problem in
the preserved part of the *Arithmetica* in which Diophantus makes use of a second,
independent unknown. The equation we write as $13N = 169\mathcal{P}$ would be written
in modern algebra as $13x = y^2$. Similarly, we write $\mathcal{N}$ for the first-degree second
unknown and $\mathcal{C}$ for its cube. Diophantus has no other set of terms to designate

the powers of this second, independent unknown, so he writes ἀριθμός for both N and $\mathcal{N}$ and δύναμις for both P and $\mathcal{P}$. Which unknown is intended is clear by the context. In later problems we sometimes use this notation for the second unknown, and even $\mathbb{N}$, $\mathbb{P}$, and $\mathbb{C}$ for powers of a third unknown, to make it clear which unknown Diophantus intends.

$$\text{IV}^{\text{G}}.17 \quad \begin{cases} x+y+z = \square \\ x^2 - y = \square' \\ y^2 - z = \square'' \\ z^2 - x = \square''' \end{cases}$$

The solution follows the same reasoning as in the previous problem.

Assignments	Operations	Equation
$y := 4\text{N}$		
$x := 1\text{N } 1\text{u}$		
	Square $4\text{N} \rightarrow 16\text{P}$	
$\sqrt{\square''} := 4\text{N } \ell\, 1\text{u}$		
	Square $4\text{N } \ell\, 1\text{u} \rightarrow 16\text{P } 1\text{u } \ell\, 8\text{N}$	
	Subtract $16\text{P } 1\text{u } \ell\, 8\text{N}$ from $16\text{P} \rightarrow 8\text{N } \ell\, 1\text{u}$	
$z := 8\text{N } \ell\, 1\text{u}$		
	Add $1\text{N } 1\text{u}, 4\text{N}, 8\text{N } \ell\, 1\text{u} \rightarrow 13\text{N}$	
		$13\text{N} = \square$
$\square := 169\mathcal{P}$		
		$13\text{N} = 169\mathcal{P}$

Thus, 1N is $13\mathcal{P}$. Replacing 1N with $13\mathcal{P}$ in the expressions for x, y, and z, we obtain the new names given here, which satisfy the first three conditions of the problem indeterminately.

Assignments	Operations	Equation
$x := 13\mathcal{P} \text{ lu}$		
$y := 52\mathcal{P}$		
$z := 104\mathcal{P} \ell\, 1\text{u}$		
	Square $104\mathcal{P} \ell\, 1\text{u} \rightarrow$ $10{,}816\mathcal{PP} \text{ 1u } \ell\, 208\mathcal{P}$	
	Subtract from this $13\mathcal{P} \text{ lu} \rightarrow 10{,}816\mathcal{PP} \ell\, 221\mathcal{P}$	
		$10{,}816\mathcal{PP} \ell\, 221\mathcal{P} = \square$

Divide $10,816\mathcal{PP}\,\ell\,221\mathcal{P}$ by
$1\mathcal{P} \rightarrow 10,816\mathcal{P}\,\ell\,221\mathrm{u}$

$10,816\mathcal{P}\,\ell\,221\mathrm{u} = \square^{iv}$

$\sqrt{\square^{iv}} := 104\mathcal{N}\,\ell\,1\mathrm{u}$

Square $104\mathcal{N}\,\ell\,1\mathrm{u} \rightarrow$
$10,816\mathcal{P}\,1\mathrm{u}\,\ell\,208\mathcal{N}$

$10,816\mathcal{P}\,\ell\,221\mathrm{u} =$
$10,816\mathcal{P}\,1\mathrm{u}\,\ell\,208\mathcal{N}$

Thus, $1\mathcal{N}$ is $\frac{111}{104}$ u, so $x = \frac{170,989}{10,816}$, $y = \frac{640,692}{10,816}$, and $z = \frac{1,270,568}{10,816}$.

IV^G.18 $\begin{cases} x^3 + y = \boxed{\mathbin{\!/\mkern-5mu/\!}} \\ y^2 + x = \square' \end{cases}$

The first two assignments are designed to satisfy the first condition, and the equation is then set up based on the second condition.

Assignments	Operations	Equation
$x := 1\mathrm{N}$		
$y := 8\mathrm{u}\,\ell\,1\mathrm{C}$		
	Square $8\mathrm{u}\,\ell\,1\mathrm{C} \rightarrow 1\mathrm{CC}\,64\mathrm{u}\,\ell\,16\mathrm{C}$	
	Add $1\mathrm{CC}\,64\mathrm{u}\,\ell\,16\mathrm{C}\,1\mathrm{N}$ ﹐ $1\mathrm{CC}\,1\mathrm{N}\,64\mathrm{u}\,\ell\,16\mathrm{C}$	
		$1\mathrm{CC}\,1\mathrm{N}\,64\mathrm{u}\,\ell\,16\mathrm{C} = \square'$
$\sqrt{\square'} := 1\mathrm{C}\,8\mathrm{u}$		
	Square $1\mathrm{C}\,8\mathrm{u} \rightarrow 1\mathrm{CC}\,16\mathrm{C}\,64\mathrm{u}$	
		$1\mathrm{CC}\,1\mathrm{N}\,64\mathrm{u}\,\ell\,16\mathrm{C} = 1\mathrm{CC}\,16\mathrm{C}\,\ell\,64\mathrm{u}$

The equation simplifies to $32\mathrm{P} = 1\mathrm{u}$, whose solution is not rational. Therefore, the 32 should be replaced with a square number. Now, 32u come from twice 16u, while 16u is twice 8u, the multitude of units in the assignment of y. Therefore, 32u is four times 8u. So, we must replace 8u by a cube whose quadruple is a square. This is a new problem, which we write in modern notation as $4u^3 = \square''$.

Assignments	Operations	Equation
$u^3 := 1\mathrm{C}$		

	Four times 1C → 4C	
		$4C = \square''$
$\square'' := 16P$		
		$16P = 4C$

1N is found to be 4u, so $u^3 = 64$. Starting the solution of the original problem from the assignment of y, the procedure runs as follows:

Assignments	Operations	Equation
$y := 64u\ \ell\ 1C$		
	Square 64u ℓ 1C → 1CC 4096u ℓ 128C	
	Add 1CC 4096u ℓ 128C and	
	1N → 1CC 4096u 1N ℓ 128C	
		1CC 4096u 1N ℓ
		$128C = \square'$
$\sqrt{\square'} := 1C\ 64u$		
	Square 1C 64u → 1CC 128C 4096u	
		1CC 1u 4096N ℓ 128C
		= 1CC 128C 4096u

Now the *al-istiqrā* assignment for $\sqrt{\square'}$ leads to a rational solution. 1N is found to be $\frac{1}{16}$u, so $x = \frac{1}{16}$ and $y = \frac{262{,}143}{4096}$.

$$\text{IV}^\text{G}.19 \quad \begin{cases} x \cdot y + 1 = \square \\ y \cdot z + 1 = \square' \\ z \cdot x + 1 = \square'' \end{cases}$$

In this problem Diophantus looks for indeterminate solutions for x, y, and z that will become assignments to unknowns in Problem $\text{IV}^\text{G}.20$.

The first two assignments satisfy the first condition, then 1P 2N is factored into 1N by 1N 2u. The assignment for $\sqrt{\square'}$ leads to an assignment for z based on the second condition, and the equation is formed from the third condition.

Assignments	Operations	Equation
$\sqrt{\square} := 1N\ 1u$		
	Square 1N 1u → 1P 2N 1u	
$x \cdot y := 1P\ 2N$		

$y := 1N$

$x := 1N\ 2u$

$\sqrt{\square'} := 3N\ 1u$

	Square 3N 1u → 9P 6N 1u
$y \cdot z := 9P\ 6N$	
$z := 9N\ 6u$	
	Multiply 1N 2u by 9N 6u → 9P 24N 12u
	Add 1u to that → 9P 24N 13u
	$\quad$ 9P 24N 13u = $\square''$

Diophantus would like 9P 24N 13u to be the square of 3N and some multitude of units. That is, he would like a square number in place of the 13, and for twice its root multiplied by 3 to be the multitude of Numbers. Looking into how he arrived at the 24 and the 13, he determines that the multitudes of Numbers in the assignments of $\sqrt{\square}$ and $\sqrt{\square'}$ should be consecutive. He then makes his assignments again.

Assignments	Operations	Equation
$\sqrt{\square} := 1N\ 1u$		
$\sqrt{\square'} := 2N\ 1u$		
	Square 1N 1u → 1P 2N 1u	
$x \cdot y := 1P\ 2N$		
$y := 1N$		
$x := 1N\ 2u$		
	Square 2N 1u → 4P 4N 1u	
$y \cdot z := 4N\ 4N$		
$z := 4N\ 4u$		

With these assignments all conditions, including the last, are satisfied. Therefore, the indeterminate solution is $x = 1N\ 2u$, $y = 1N$, $z = 4N\ 4u$.

IV^G.20 $\begin{cases} w \cdot x + 1 = \square \\ w \cdot y + 1 = \square' \\ w \cdot z + 1 = \square'' \\ x \cdot y + 1 = \square'' \\ x \cdot z + 1 = \square^{iv} \\ y \cdot z + 1 = \square^{v} \end{cases}$

The assignments for x, y, and z found at the end of the previous problem become the assignments for x, w, and y in the present problem. The assignment for z is then derived from the third condition. The last condition is satisfied by virtue of a modification of the last part of the solution to the previous problem. Replacing $\sqrt{\square} \mathrel{\vcenter{:}}= 1\mathrm{N}\ 1\mathrm{u}$ and $\sqrt{\square'} \mathrel{\vcenter{:}}= 2\mathrm{N}\ 1\mathrm{u}$ with $\sqrt{\square'} \mathrel{\vcenter{:}}= 2\mathrm{N}\ 1\mathrm{u}$ and $\sqrt{\square''} \mathrel{\vcenter{:}}= 3\mathrm{N}\ 1\mathrm{u}$ (associated with wy and wz), then $yz + 1$ will be a square. Only the fifth condition remains, which is solved with an *al-istiqrāʾ* assignment.

Assignments	Operations	Equation
$\sqrt{\square} \mathrel{\vcenter{:}}= 1\mathrm{N}\ 1\mathrm{u}$		
	Square $1\mathrm{N}\ 1\mathrm{u} \rightarrow 1\mathrm{P}\ 2\mathrm{N}\ 1\mathrm{u}$	
$w \cdot x \mathrel{\vcenter{:}}= 1\mathrm{P}\ 2\mathrm{N}$		
$w \mathrel{\vcenter{:}}= 1\mathrm{N}$		
$x \mathrel{\vcenter{:}}= 1\mathrm{N}\ 2\mathrm{u}$		
$\sqrt{\square'} \mathrel{\vcenter{:}}= 2\mathrm{N}\ 1\mathrm{u}$		
	Square $2\mathrm{N}\ 1\mathrm{u} \rightarrow 4\mathrm{P}\ 4\mathrm{N}\ 1\mathrm{u}$	
$w \cdot y \mathrel{\vcenter{:}}= 4\mathrm{P}\ 4\mathrm{N}$		
$y \mathrel{\vcenter{:}}= 4\mathrm{P}\ 4\mathrm{u}$		
$\sqrt{\square''} \mathrel{\vcenter{:}}= 3\mathrm{N}\ 1\mathrm{u}$		
	Square $3\mathrm{N}\ 1\mathrm{u} \rightarrow 9\mathrm{P}\ 6\mathrm{N}\ 1\mathrm{u}$	
$w \cdot z \mathrel{\vcenter{:}}= 9\mathrm{P}\ 6\mathrm{N}$		
$z \mathrel{\vcenter{:}}= 9\mathrm{N}\ 6\mathrm{u}$		
	The product of $1\mathrm{N}\ 2\mathrm{u}$ by $9\mathrm{N}\ 6\mathrm{u}$, increased by $1\mathrm{u} \rightarrow 9\mathrm{P}\ 24\mathrm{N}\ 13\mathrm{u}$	
		$9\mathrm{P}\ 24\mathrm{N}\ 13\mathrm{u} = \square^{iv}$
$\sqrt{\square^{iv}} \mathrel{\vcenter{:}}= 3\mathrm{N}\ \ell\ 4\mathrm{u}$		
	Square $3\mathrm{N}\ \ell\ 4\mathrm{u} \rightarrow 9\mathrm{P}\ 16\mathrm{u}\ \ell\ 24\mathrm{N}$	
		$9\mathrm{P}\ 24\mathrm{N}\ 13\mathrm{u} = 9\mathrm{P}\ 6\mathrm{u}\ \ell\ 24\mathrm{N}$

$1\mathrm{N}$ is then $\frac{1}{16}\mathrm{u}$, so $w = \frac{1}{16}$, $x = \frac{33}{16}$, $y = \frac{68}{16}$, and $z = \frac{105}{16}$.

IV^{G}.21
$$\begin{cases} x : y :: y : z \\ x - y = \square \\ y - z = \square' \\ x - z = \square'' \end{cases}$$

Assignments	Operations	Equation
$z := 1N$		
$y := 1N\ 4u$		
$x := 1N\ 13u$		

The assignments satisfy the second and third conditions, but the fourth does not hold, since 13 is not a square. But 13 is the sum of the squares 4 and 9. Thus, we are led to the problem of finding a square which is the sum of two squares. Any Pythagorean triple will do, so Diophantus chooses the triple (3, 4, 5) and begins the solution again, and the equation is formed from the first condition.

Assignments	Operations	Equation
$z := 1N$		
$y := 1N\ 9u$		
$x := 1N\ 25u$		
	Multiply $1N$ by $1N\ 25u \rightarrow 1P\ 25N$	
	Square $1N\ 9u \rightarrow 1P\ 18N\ 81u$	
		$1P\ 25N = 1P\ 18N\ 81u$

Thus, $1N$ is $\frac{81}{7}u$, so $x = \frac{256}{7}$, $y = \frac{144}{7}$, and $z = \frac{81}{7}$.

$$\text{IV}^G.22 \quad \begin{cases} x \cdot y \cdot z + x = \square \\ x \cdot y \cdot z + y = \square' \\ x \cdot y \cdot z + z = \square'' \end{cases}$$

The first two assignments satisfy the first condition.

Assignments	Operations	Equation
$x \cdot y \cdot z := 1P\ 2N$		
$x := 1u$		
$\sqrt{\square'} := 1N\ 3u$		
	Square $1N\ 3u \rightarrow 1P\ 6N\ 9u$	
	Subtract $1P\ 2N$ from $1P\ 6N\ 9u \rightarrow 4N\ 9u$	
$y := 4N\ 9u$		

Now, xy is also $4N\ 9u$, and since xyz is $1P\ 2N$, Diophantus would like to divide $1P\ 2N$ by $4N\ 9u$ to find z. But if he were to assign $z := \frac{1P\ 2N}{4N\ 9u}$, then the equation resulting from the third condition would become $16PP\ 108C\ 242P\ 180N = \square'''$, which cannot be solved by *al-istiqrā'*. So, he seeks a reassignment of y so that the Numbers and units, currently 4 and 9, will be in the ratio of 1 : 2, as are the multitudes in 1P 2u. Now, 4 comes from the excess of 6N over 2N, the 6N being twice the product of 1N by 3u, while 9 comes from the square of 3u. Therefore, we are led to the problem of finding a number to replace 3u, which if doubled and decreased by a dyad, will be half of its square. In modern notation, $2u - 2 = \frac{1}{2}u^2$.

Assignments	Operations	Equation
$u := 1N$		
	Two times 1N less 2u $\rightarrow$ 2N ℓ 2u	
	Square 1N $\rightarrow$ 1P	
		$1P = 4N \, \ell \, 4u$

The solution of the quadratic equation is 2u. We now return to the original problem and rework the solution beginning with the assignment of $\sqrt{\square}'$.

Assignments	Operations	Equation
$\sqrt{\square}' := 1N \, 2u$		
	Square 1N 2u $\rightarrow$ 1P 4N 4u	
	Subtract 1P 2N from 1P 4N 4u $\rightarrow$ 2N 4u	
$y := 2N \, 4u$		
	Multiply 1u by 2N 4u $\rightarrow$ 2N 4u	
	Divide 1P 2N by 2N 4u $\rightarrow \frac{1}{2}N$	
$z := \frac{1}{2}N$		
	Add 1P 2N and $\frac{1}{2}N \rightarrow 1P \, 2\frac{1}{2}N$	
		$1P \, 2\frac{1}{2}N = \square''$
$\square'' := 4P$		
		$1P \, 2\frac{1}{2}N = 4P$

1N is $\frac{5}{6}u$, so $x = \frac{6}{6}$, $y = \frac{34}{6}$, and $z = \frac{2\frac{1}{2}}{6}$.

IV^G.23 $\begin{cases} x \cdot y \cdot z - x = \square \\ x \cdot y \cdot z - y = \square' \\ x \cdot y \cdot z - z = \square'' \end{cases}$

The first two assignments satisfy the first condition. The derivative assignment for yz is then factored, and a pair of forthcoming equations are formed from the second and third conditions.

Assignments	Operations	Equation
$x := 1N$		
$x \cdot y \cdot z := 1P\ 1N$		
$y \cdot z := 1N\ 1u$		
$y := 1u$		
$z := 1N\ 1u$		
	Subtract $1u$ from $1P\ 1N \rightarrow 1P\ 1N\ \ell\ 1u$	
		$1P\ 1N\ \ell\ 1u = \square'$
	Subtract $1N\ 1u$ from $1P\ 1N \rightarrow 1P\ \ell\ 1u$	
		$1P\ \ell\ 1u = \square''$

To solve the double-equality, the difference $1N$ is factored into $\frac{1}{2}u$ by $2N$. Following the procedure of Problem II.11, the square of half the difference between $2N$ and $\frac{1}{2}u$ equals $1P\ \ell\ 1u$, and the square of half the sum of $2N$ and $\frac{1}{2}u$ equals $1P\ 1N\ \ell\ 1u$. Both equations lead to the same solution, $1N = \frac{17}{8}u$, from which we obtain $x = \frac{17}{8}$, $y = 1$, and $z = \frac{25}{8}$.

IV^G.24 $\begin{cases} x + y = a \\ x \cdot y = \square - \sqrt[3]{\square} \end{cases}$ $(a = 6)$

The first two assignments satisfy the first condition. Then, the side $\sqrt[3]{\square}$ is given a direct assignment that will not work out and will need to be adjusted.

Assignments	Operations	Equation
$x := 1N$		
$y := 6u\ \ell\ 1N$		
	Multiply $1N$ by $6u\ \ell\ 1N \rightarrow 6N\ \ell\ 1P$	
$\sqrt[3]{\square} := 2N\ \ell\ 1u$		
	Cube $2N\ \ell\ 1u \rightarrow 8C\ 6N\ \ell\ 12P\ 1u$	
	Subtract from this $2N\ \ell\ 1u \rightarrow 8C\ 4N\ \ell\ 12P$	
		$8C\ 4N\ \ell\ 12P$ $= 6N\ \ell\ 1P$

The solution to this equation is not rational. However, in Diophantus's words, "if the (multitudes of) Numbers in either part of the equation were equal, it would be reduced to equate Cubes equal to Powers, and the Number would be expressible". Now 4, the multitude of Numbers in the first part of the equation, comes from 3 times 2, the multitude of Numbers in the assignment of the side of the cube, decreased by the same 2; while the 6 in the other part of the equation is the given number. We have, therefore, to replace 2 by another number such that its triple, decreased by this same number, equals 6, or, in modern notation, $3v - v = 6$. The solution to this subsidiary problem is 3. So, we start again from the assignment of $\sqrt[3]{⌧}$:

Assignments	Operations	Equation
$\sqrt[3]{⌧} \coloneqq 3\text{N}\,\ell\,1\text{u}$		
	Cube $3\text{N}\,\ell\,1\text{u} \rightarrow 27\text{C}\,9\text{N}\,\ell\,27\text{P}\,1\text{u}$	
	Subtract from this	
	$3\text{N}\,\ell\,1\text{u} \rightarrow 27\text{C}\,6\text{N}\,\ell\,27\text{P}$	
		$27\text{C}\,6\text{N}\,\ell\,27\text{P} = 6\text{N}\,\ell\,1\text{P}$

This time 1N is rational and is $\frac{26}{27}$u, so the sought-after numbers are $x = \frac{26}{27}$ and $y = \frac{136}{27}$.

$$\textbf{IV}^{\text{G}}\textbf{.25} \quad \begin{cases} x + y + z = a \\ x \cdot y \cdot z = ⌧ \\ (y - x) + (z - y) + (z - x) = \sqrt[3]{⌧} \end{cases} \quad (a = 4)$$

The second condition is satisfied first.

Assignments	Operations	
$x \cdot y \cdot z \coloneqq 8\text{C}$		
	The side of $8\text{C} \rightarrow 2\text{N}$	

Diophantus then moves to the third condition. He notes that $(y - x) + (z - y) + (z - x)$ is equal to $2(z - x)$, which is 2N. Thus, the next, derivative assignment:

Assignments	Operations	Equation
$z - x \coloneqq 1\text{N}$		
$x \coloneqq 2\text{N}$		
$z \coloneqq 3\text{N}$		
	Multiply 2N by $3\text{N} \rightarrow 6\text{P}$	
	Divide 8C by $6\text{P} \rightarrow 1\frac{1}{3}\text{N}$	
$y \coloneqq 1\frac{1}{3}\text{N}$		

We should have $x < y$, but instead, $y < x$. Looking into the origin of the problem, the name for y was found by dividing 8 by the product of 2 by 3, where 2 and 3 were chosen to be two consecutive numbers. So, we are reduced to the problem of finding two consecutive numbers such that dividing 8 by their product results in a number lying between the two numbers. In modern notation, $u - v = 1; v < \frac{8}{u \cdot v} < u$.

Assignments	Operations	
$v := 1N$		
$u := 1N\ 1u$		
	Multiply $1N$ by $1N\ 1u \rightarrow 1P\ 1N$	
	Divide $8u$ by $1P\ 1N \rightarrow \frac{8u}{1P\ 1N}$	

We want $1N\ 1u > \frac{8u}{1P\ 1N} > 1N$. The first inequality simplifies to $8u > 1C\ 1P$. Diophantus then finds a cube whose first two terms are $1C\ 1P$, and this is the cube of $1N\ \frac{1}{3}$ u. Equating this with $2u$, the cube root of $8u$, $1N$ becomes $\frac{5}{3}$ u. This makes $v = \frac{5}{3}$ and $u = \frac{8}{3}$, and the middle number $\frac{8}{u \cdot v} = \frac{9}{5}$. Multiplying by 15 to eliminate the fractions, these become $v = 25$, $u = 40$, and the middle number is 27. Diophantus does not mention it, but this adjustment means that the first assignment is changed to $x \cdot y \cdot z := 27{,}000C$, where the multitude 27,000 is $15^3 \cdot 8$. These numbers satisfy the second and third conditions of the original problem, but not yet the first. The solution is begun again with new assignments.

Assignments	Operations	Equation
$z := 40N$		
$y := 27N$		
$x := 25N$		
	Add $25N$, $27N$, and $40N \rightarrow 92N$	
		$92N = 4u$

$1N$ is $\frac{1}{23}$ u, so $x = \frac{25}{23}$, $y = \frac{27}{23}$, and $z = \frac{40}{23}$.

IV^G.26 $\begin{cases} x \cdot y + x = a \\ x \cdot y + y = \boxdot' \end{cases}$

The first condition is taken care of in the first two assignments, and the equation is formed from the second condition.

Assignments	Operations	Equation
$x := 8N$		
$y := 1P\ \ell\ 1u$		
	Multiply $8N$ by $1P\ \ell\ 1u \rightarrow 8C\ \ell\ 8N$	
	Add $8C\ \ell\ 8N$ and $1P\ \ell\ 1u \rightarrow 8C\ 1P\ \ell\ 8N\ 1u$	

Assignments	Operations	Equation
	8C 1P ℓ 8N 1u = ▱′	
∛▱′ ≔ 2N ℓ 1u		
	Cube 2N ℓ 1u → 8C 6N ℓ 12P 1u	
		8C 1P ℓ 8N 1u = 8C 6N ℓ 12P 1u

1N is $\frac{14}{13}$ u, so $x = \frac{112}{13}$, and $y = \frac{27}{169}$.

IV^G.27 $\begin{cases} x\cdot y - x = ▱ \\ x\cdot y - y = ▱' \end{cases}$

Like in the last problem, the first condition is satisfied with the first two assignments. The second condition gives rise to the forthcoming equation.

Assignments	Operations	Equation
$x ≔ 8N$		
$y ≔ 1P\ 1u$		
	Multiply 8N by 1P 1u → 8C 8N	
	Subtract 1P 1u from 8C 8N → 8C 8N ℓ 1P 1u	
		8C 8N ℓ 1P 1u = ▱′

This equation, Diophantus writes, is impossible (see the following explanation), so he begins the solution again, changing his first two assignments so that the second condition is satisfied:

Assignments	Operations	Equation
$x ≔ 8N\ 1u$		
$y ≔ 1P$		
	Multiply 1P by 8N 1u → 8C 1P	
	Subtract 8N 1u from 8C 1P → 8C 1P ℓ 8N 1u	
		8C 1P ℓ 8N 1u = ▱
∛▱ ≔ 2N ℓ 1u		

Cube 2N ℓ 1u $\rightarrow$ 8C 6N ℓ 12P 1u

8C 1P ℓ 8N 1u =
8C 6N ℓ 12P 1u

1N is found to be $\frac{14}{13}$ u, so $x = \frac{125}{13}$, and $y = \frac{196}{169}$.

Heath has the following note concerning the statement of Diophantus that the equation which corresponds to our $x^3 + 8x - x^2 - 1 = y^2$, "is impossible":

> Diophantus means that, if we are to get rid of the third power and the absolute term, we can only put the expression equal to $(2x - 1)^2$, which gives a negative and therefore "impossible" value for x. But the equation is not really impossible, for we can get rid of the terms in x^3 and x^2 by putting
>
> $$8x^3 + 8x - x^2 - 1 = \left(2x - \tfrac{1}{12}\right)^2, \text{ whence } x = \tfrac{1727}{13752},$$
>
> or we can make the term in x and the absolute term disappear by putting
>
> $$8x^3 + 8x - x^2 - 1 = \left(\tfrac{8}{3}x - 1\right)^2, \text{ whence } x = \tfrac{549}{296}.$$
>
> Diophantus has actually shown us how to do the former in IV.25 just preceding.
>
> (Heath 1910, 186 n. 1)

IV^G.28 $\begin{cases} x \cdot y + (x + y) = \boxdot \\ x \cdot y - (x + y) = \boxdot' \end{cases}$

First solution

Diophantus begins by assigning the cubes as cubic numbers and then performs calculations based on the fact that $x + y$ is half the difference of the cubes and $x \cdot y$ is half of their sum.

Assignments	Operations	Equation
$\boxdot := 64u$		
$\boxdot' := 8u$		
	Subtract 8u from 64u $\rightarrow$ 56u	
	Half of 56u $\rightarrow$ 28u	
$x + y := 28u$		
	Subtract 28u from 64u $\rightarrow$ 36u	
$x \cdot y := 36u$		

The problem is then reduced to finding x and y such that $\begin{cases} x+y=28 \\ x \cdot y = 36 \end{cases}$. As Diophantus will mention later in this solution, this problem has already been solved, in Problem I.27. But he must work out the steps because it will lead to a dead end, and he will need to trace the cause of the problem back to the initial assignments.

Assignments	Operations	Equation
$x := 1$N 14u		
$y := 14$u ℓ 1N		
	Multiply 1N 14u by 14u ℓ 1N → 196u ℓ 1P	
		196u ℓ 1P = 36u

The equation simplifies to 1P = 160u. If 160u were a square the equation would have a rational solution. This 160 results from the subtraction of 36 from 196, the latter being the square of half of 28, which is the square of a fourth of 56, that is, the difference of the two cubes. The 36, on the other hand, are half the sum of the two cubes. Thus, we are led to the subsidiary problem of finding two cubes such that a fourth of their difference multiplied by itself and lacking half their sum, makes a square. In modern notation: $\left(\frac{1}{4}\left(u^3 - v^3\right)\right)^2 - \frac{1}{2}\left(u^3 + v^3\right) = \square$. Assignments for u and v are chosen so that $u^3 - v^3$ and $u^3 + v^3$ each have two terms.

Assignments	Operations	Equation
$u := 1$N 1u		
$v := 1$N ℓ 1u		
	Cube 1N 1u → 1C 3P 3N 1u	1N 1u → 1C
	Cube 1N ℓ 1u → 1C 3N ℓ 3P 1u	3P 3N 1u
	A fourth of their difference → $1\frac{1}{2}$P $\frac{1}{2}$u	
	Square $1\frac{1}{2}$P $\frac{1}{2}$u → $2\frac{1}{4}$PP $1\frac{1}{2}$P $\frac{1}{4}$u	
	Half the sum of the cubes → 1C 3N	
	Subtract 1C 3N from	
	$2\frac{1}{4}$PP $1\frac{1}{2}$P $\frac{1}{4}$u → $2\frac{1}{4}$PP $1\frac{1}{2}$P $\frac{1}{4}$u ℓ 1C 3N	
		$2\frac{1}{4}$PP $1\frac{1}{2}$P $\frac{1}{4}$u ℓ
		1C 3N = $\square$.
	Four times $2\frac{1}{4}$PP $1\frac{1}{2}$P $\frac{1}{4}$u ℓ 1C 3N →	
	9PP 6P 1u ℓ 4C 12N	

$$9PP\ 6P\ 1u\ \ell\ 4C$$
$$12N = \square'$$

$\sqrt{\square'} := 3P\ 1u\ \ell\ 6N$

Square $3P\ 1u\ \ell\ 6N \rightarrow 9PP\ 42P\ 1u\ \ell\ 36C\ 12N$

$$9PP\ 6P\ 1u\ \ell\ 4C$$
$$12N = 9PP\ 42P$$
$$1u\ \ell\ 36C\ 12N$$

From there 1N is found to be $\frac{9}{8}$u, so $u^3 = \frac{4913}{512}$ and $v^3 = \frac{1}{512}$.

We then return to the original problem and replace the old assignments for $\square$ and $\square'$ with the values we found for u^3 and v^3. The same procedure is then applied.

Assignments	Operations	Equation
$\square := \frac{4913}{512}$u		
$\square' := \frac{1}{512}$u		
	Subtract $\frac{1}{512}$u from $\frac{4913}{512}$u $\rightarrow \frac{4912}{512}$u	
	Half of $\frac{4913}{512}$u $\rightarrow \frac{2456}{512}$u	
$x + y := \frac{2456}{512}$u		
	Subtract $\frac{2456}{512}$u from $\frac{4913}{512}$u $\rightarrow \frac{2457}{512}$u	
$x \cdot y := \frac{2457}{512}$u		

The problem is then reduced to finding x and y such that $\begin{cases} x + y = \frac{2456}{512} \\ x \cdot y = \frac{2457}{512} \end{cases}$. Again,

this is Problem I.27, this time with a rational solution. Diophantus works it out "for the sake of the (present) problem".

Assignments	Operations	Equation
$x := 1N\ \frac{1228}{512}$u		
$y := \frac{1228}{512}$u $\ell\ 1N$		

Assignments	Operations	Equation
	Multiply 1N $\frac{1228}{512}$ u by $\frac{1228}{512}$ u ℓ 1N $\rightarrow \frac{1,507,984}{262,144}$ u ℓ 1P	
		$\frac{1,507,984}{262,144}$ u ℓ 1P $= \frac{2457}{512}$ u

1N is then $\frac{500}{512}$ u , so $x = \frac{1728}{512}$ and $y = \frac{728}{512}$.

Second solution

The solution begins with the arithmetical identity that we can supplement with modern notation as follows: "if any square number (p^2) is partitioned into its side and the remainder (p and p^2-p), (then) it makes their product (p^3-p^2), together with the sum of the two (p^2), a cube (p^3)." This is the motivation for the first two assignments, which will satisfy the first condition.

Assignments	Operations	Equation
$x := 1N$		
$y := 1P \ell 1N$		
	Multiply 1N by 1P ℓ 1N $\rightarrow$ 1C ℓ 1P	
	Subtract 1P from 1C ℓ 1P $\rightarrow$ 1C ℓ 2P	
		1C ℓ 2P $= \boxdot'$
$\boxdot' := \frac{1}{8}C$		
		1C ℓ 2P $= \frac{1}{8}C$

1N becomes $\frac{16}{7}$ u, so $x = \frac{16}{7}$ and $y = \frac{144}{49}$.

IV^G.29 $w^2 + x^2 + y^2 + z^2 + (w + x + y + z) = a$ $(a = 12)$

From the identity $p^2 + p + \frac{1}{4} = \left(p + \frac{1}{2}\right)^2$ Diophantus finds that

$$\left(w^2 + w\right) + \left(x^2 + x\right) + \left(y^2 + y\right) + \left(z^2 + z\right) + 1 = \left(w + \frac{1}{2}\right)^2 + \left(x + \frac{1}{2}\right)^2 + \left(y + \frac{1}{2}\right)^2 + \left(z + \frac{1}{2}\right)^2$$

Therefore, $\left(w + \frac{1}{2}\right)^2 + \left(x + \frac{1}{2}\right)^2 + \left(y + \frac{1}{2}\right)^2 + \left(z + \frac{1}{2}\right)^2 = 13$. Thus, 13 must be divided into four squares. It is already the sum of two squares, 4 and 9, so by Problem II.8, these in turn can each be divided into two squares. Diophantus finds the four squares to be $\frac{64}{25}$, $\frac{36}{25}$, $\frac{144}{25}$, and $\frac{81}{25}$. These are, respectively, the squares of $w + \frac{1}{2}$, $x + \frac{1}{2}$, $y + \frac{1}{2}$, and $z + \frac{1}{2}$. Subtracting $\frac{1}{2}$ from the roots of the squares, he gets $w = \frac{11}{10}$, $x = \frac{7}{10}$, $y = \frac{19}{10}$, and $z = \frac{13}{10}$, so the squares are $w^2 = \frac{121}{100}$, $x^2 = \frac{49}{100}$, $y^2 = \frac{361}{100}$, and $z^2 = \frac{169}{100}$. Algebra enters into the solution only in the two applications of Problem II.8, which he does not show.

IV^G.30 $\quad w^2 + x^2 + y^2 + z^2 - (w + x + y + z) = a \qquad (a = 4)$

From the identity $p^2 - p + \frac{1}{4} = \left(p - \frac{1}{2}\right)^2$ Diophantus finds that

$$\left(w^2 - w\right) + \left(x^2 - x\right) + \left(y^2 - y\right) + \left(z^2 - z\right) + 1$$
$$= \left(w - \frac{1}{2}\right)^2 + \left(x - \frac{1}{2}\right)^2 + \left(y - \frac{1}{2}\right)^2 + \left(z - \frac{1}{2}\right)^2$$

Therefore, $\left(w - \frac{1}{2}\right)^2 + \left(x - \frac{1}{2}\right)^2 + \left(y - \frac{1}{2}\right)^2 + \left(z - \frac{1}{2}\right)^2 = 5$. So, the problem becomes one of dividing 5 into four squares. Because 5 is already the sum of two squares, 1 and 4, Problem II.8 can again be applied to divide each of them into two squares. He finds them to be $\frac{9}{25}$, $\frac{16}{25}$, $\frac{64}{25}$, and $\frac{36}{25}$. Adding $\frac{1}{2}$ to each of their roots he gets $w = \frac{11}{10}$, $x = \frac{13}{10}$, $y = \frac{21}{10}$, and $z = \frac{17}{10}$, so the squares are $w^2 = \frac{121}{100}$, $x^2 = \frac{169}{100}$, $y^2 = \frac{441}{100}$, and $z^2 = \frac{289}{100}$. As in the previous problem, algebra enters into the solution only in the two applications of Problem II.8, which he does not show.

IV^G.31 $\quad \begin{cases} x + y = 1 \\ (x + a) \cdot (y + b) = \square \end{cases} \qquad (a = 3, b = 5)$

First solution

The first two assignments are designed to satisfy the first condition, and the equation is set up from the second condition.

Assignments	Operations	Equation
$x := 1$N		
$y := 1$u ℓ 1N		
	Add 3u to 1N → 1N 3u	
	Add 5u to 1u ℓ 1N → 6u ℓ 1N	
	Multiply 1N 3u by 6u ℓ 1N → 3N 18u ℓ 1P	
		3N 18u ℓ 1P = $\square$
$\square := 4$P		
		3N 18u ℓ 1P = 4P

This simplifies to 3N 18u = 5P, but its solution is not rational because the discriminant $5 \cdot 18 + \left(\frac{3}{2}\right)^2$ is not a perfect square. But the 5, as 5P, comes from adding 1P to 4P. We should find a square number to replace the 4, such that if the number 1 greater is multiplied by 18 and receives in addition the square of $\frac{3}{2}$, that is, $2\frac{1}{4}$, makes a square. In modern notation: $18\left(u^2 + 1\right) + 2\frac{1}{4} \to '$.

Assignments	Operations	Equation
$u^2 := 1$P		
	Add 1u to 1P→1P 1u	
	Eighteen times 1P1u→18P 18u	
	Add $2\frac{1}{4}$u to 18P 18u →18P $20\frac{1}{4}$u	
		18P $20\frac{1}{4}$u $= \square'$
	Four times 18P $20\frac{1}{4}$u → 72P 81u	
		72P 81u $= \square''$
$\sqrt{\square''} := 8$N 9u		
	Square 8N 9u → 64P 144N 81u	
		72P 81u = 64P 144N 81u

1N is found to be 18u; and thus, $u^2 = 324$ We then return to the original problem and change the last assignment into $\square := 324$P:

Assignments	Operations	Equation
		3N 18u ℓ 1P $= \square$
$\square := 324$P		
		3N 18u ℓ 1P $= 324$P

1N is $\frac{78}{325}$u , or simplified, $\frac{6}{25}$u . Therefore, $x = \frac{6}{25}$ and $y = \frac{19}{25}$. Diophantus does not state the simplified form of the last equation, which is 3N 18u = 325P, nor does he show how he worked out the solution. His procedure would have been to calculate the discriminant as he had done for 3N 18u = 5P, and then its root would have been added to half of the 3 to get 78 and then divided by 325.

Second solution

Here the first two assignments again satisfy the first condition, but now the assignment for x makes the x + 3 in the second condition simply 1N.

Assignments	Operations	Equation
$x := 1$N ℓ 3u		

$y := 4N \ \ell \ 4N$

Add 3u to 1N ℓ 3u $\rightarrow$ 1N

Add 5u to 4u ℓ 1N $\rightarrow$ 9u ℓ 1N

Multiply 1N by 9u ℓ 1N $\rightarrow$ 9N ℓ 1P

9N ℓ 1P = $\square$

$\square := 4P$

9N ℓ 1P = 4P

1N is found to be $\frac{9}{5}$ u. But we cannot subtract 3 from $\frac{9}{5}$ to find x. From the assignments for x and y, 1N must be greater than 3u and less than 4u. By examining the origin of $\frac{9}{5}$, Diophantus determines that we need to replace the 4, the multiple of Powers in the assignment for $\square$, with a square that is greater than $1\frac{1}{4}$ but less than 2. He finds $\frac{25}{16}$. The last part of the solution then becomes:

Assignments	Operations	Equation
		9N ℓ 1P = $\square$
$\square := \frac{25}{16}$ P		
		9N ℓ 1P = $\frac{25}{16}$ P

This time 1N is $\frac{144}{41}$ u, whence $x = \frac{21}{41}$ and $y = \frac{20}{41}$.

IV^G.32 $\begin{cases} x + y + z = a \\ x \cdot y + z = \square \\ x \cdot y - z = \square' \end{cases}$ $(a = 6)$

The first three assignments satisfy the first condition, and the other two conditions lead to two forthcoming equations.

Assignments	Operations	Equation
$z := 1N$		
$y := 2u$		
$x := 4u \ \ell \ 1N$		
	Multiply 4u ℓ 1N by 2u $\rightarrow$ 8u ℓ 2N	
	Add 1N to 8u ℓ 2N $\rightarrow$ 8u ℓ 1N	
	Subtract 1N from 8u ℓ 2N $\rightarrow$ 8u ℓ 3N	
		8u ℓ 1N = $\square$
		8u ℓ 3N = $\square'$

The method of the double-equality does not lead to a rational solution, for, in Diophantus's words, "the (multitudes of) Numbers do not have to each other the ratio that a square number has to a square number". But these multitudes are, respectively, 1 unit smaller and 1 unit greater than 2. So, we are led to the problem of finding a number such that the number one unit greater than it has to the number one unit smaller the ratio that a square number has to a square number. Suppose the ratio is 4 to 1. In modern notation the problem is $(u + 1) : (u - 1) :: 4 : 1$.

Assignments	Operations	Equation
$u := 1\text{N}$		
	Add 1u to 1N → 1N 1u	
	Subtract 1u from 1N → 1N ℓ 1u	
	Multiply 1N ℓ 1u by 4u → 4N ℓ 4u	
	Multiply 1N 1u by 1u → 1N 1u	
		4N ℓ 4u = 1N 1u

Thus, 1N is $\frac{5}{3}$u . So, we return to the original problem, starting again with the assignment of y.

Assignments	Operations	Equation
$y := \frac{5}{3}\text{u}$		
$z := 1\text{N}$		
$x := \frac{13}{3}\text{u} \, \ell \, 1\text{N}$		
	Multiply $\frac{13}{3}$u ℓ 1N by $\frac{5}{3}$u → $\frac{65}{9}$u ℓ $\frac{5}{3}$N	
	Add 1N to $\frac{65}{9}$u ℓ $\frac{5}{3}$N → $\frac{65}{9}$u ℓ $\frac{2}{3}$N	
		$\frac{65}{9}$u ℓ $\frac{2}{3}$N = □
	Subtract 1N from $\frac{65}{9}$u ℓ $\frac{5}{3}$N → $\frac{65}{9}$u ℓ $2\frac{2}{3}$N	
		$\frac{65}{9}$u ℓ $2\frac{2}{3}$N = □'
	Multiply everything in both equations by 9	
		65u ℓ 6N = □"
		65u ℓ 24N = □‴
	Multiply 65u ℓ 6N by 4u	
		260u ℓ 24N = □^{iv}

The last two equations are solved by the method of the double-equality. The difference of the expressions, 195u, is the difference of the squares. Writing 195 as

15 by 13, Diophantus finds 1N to be $\frac{8}{3}$u, so the sought-after numbers are $z = \frac{5}{3}$, $y = \frac{5}{3}$, and $x = \frac{8}{3}$ (Diophantus has reversed the roles of the "first" and "third").

IV^G.33
$$\begin{cases} x + \dfrac{m}{n}y = p\left(y - \dfrac{m}{n}y\right) \\[2ex] y + \dfrac{m}{n}x = q\left(x - \dfrac{m}{n}x\right) \end{cases}$$
   ($p = 3$, $q = 5$, posits $\frac{m}{n}y = 1$ in solution)

The assignments are designed to satisfy the first condition. The equation then comes from the second condition.

Assignments	Operations	Equation
$y := 1\text{N } 1\text{u}$		
$\frac{m}{n}y := 1\text{u}$ $x := 3\text{N } \ell \text{ 1u}$		

Since $y + \frac{m}{n}x = 5\left(x - \frac{m}{n}x\right)$, $y + x = 6\left(x - \frac{m}{n}x\right)$. Then, since $y + x := 4\text{N}$, $x - \frac{m}{n}x := \frac{2}{3}\text{N}$, so $\frac{m}{n}x := \frac{7}{3}\text{N } \ell \text{ 1u}$. The proportion $\frac{m}{n}x : x :: \frac{m}{n}y : y$ thus corresponds to $\left(\frac{7}{3}\text{N } \ell \text{ 1u}\right) : (3\text{N } \ell \text{ 1u}) :: 1\text{u} : (1\text{N } 1\text{u})$, so by cross multiplication, the product of $\frac{7}{3}\text{N } \ell \text{ 1u}$ by 1N 1u should equal 3N ℓ 1u by 1u.

Assignments	Operations	Equation
		$\frac{7}{3}\text{P } \frac{4}{3}\text{N } \ell \text{ 1u} = 3\text{N } \ell \text{ 1u}$

Thus, 1N is found to be $\frac{5}{7}$u, so $x = \frac{8}{7}$, $y = \frac{12}{7}$, and $\frac{m}{n} = \frac{7}{12}$. Multiplying by 7 in order to obtain integer numbers, and then by 3 to obtain numbers having a 12th part, the sought-after numbers become $x = 24$ and $y = 36$.

Lemma to Problem IV^G.34 $x \cdot y + (x + y) = a$ $(a = 8)$

Diophantus finds a particular solution, and then he traces the origin of the value of 1N to find an indeterminate solution.

Assignments	Operations	Equation
$x := 1\text{N}$ $y := 3\text{u}$		
	Multiply 1N by 3u $\rightarrow$ 3N	
	Add 1N and 3u $\rightarrow$ 1N 3u	

> Add 3N and 1N 3u → 4N 3u
>
> 4N 3u = 8u

1N is $\frac{5}{4}$u , so $x = \frac{5}{4}$ and $y = 3$. The 5 comes from the excess of 8 over 3, while 4 is one unit greater than the multitude of units in the assignment of y. Thus, we assign y to be any expression whatever involving N, say, $y := 1N \ell 1u$. Then the excess of 8u over $1N \ell 1u$ is $9u \ell 1N$, while the amount one unit greater than 1N ℓ 1u is 1N. So, $x = \frac{9u \ell 1N}{1N}$, or $9N' \ell 1u$, and $y = 1N \ell 1u$.

IV^G.34 $\begin{cases} x \cdot y + (x + y) = a \\ y \cdot z + (y + z) = b \\ z \cdot x + (z + x) = c \end{cases}$ Determination: a, b, and c are each 1 unit less than a square ($a = 8$, $b = 15$, $c = 24$)

Following the calculations here, 1P will be the quantity $\frac{(a+1)(b+1)}{c+1}$, so it must be a square. As Heath observed (1910, 192 n. 2), it is not necessary that each of $a + 1$, $b + 1$, and $c + 1$ be a square.

After y is assigned, the procedure described in the previous lemma is applied to calculate the assignment for x from the first condition, and then again to find the assignment for z from the second condition. The equation then derives from the third condition.

Assignments	Operations	Equation
$y := 1N \ell 1u$		
	Subtract $1N \ell 1u$ from 8u → $9u \ell 1N$	
	Add 1u to $1N \ell 1u$ → 1N	
	Divide $9u \ell 1N$ by 1N → $9N' \ell 1u$	
$x := 9N' \ell 1u$		
	Subtract $1N \ell 1u$ from 15u → $16u \ell 1N$	
	Divide $16u \ell 1N$ by 1N → $16N' \ell 1u$	
$z := 16N' \ell 1u$		
	Multiply $9N' \ell 1u$ by $16N' \ell 1u$ → $144P' 1u \ell 25N'$	
	Add $9N' \ell 1u$ and $16N' \ell 1u$ → $25N' \ell 2u$	
	Add $144P' 1u \ell 25N'$ and $25N' \ell 2u$ → $144P' \ell 1u$	
		$144P' \ell 1u = 24u$

1N is $\frac{12}{5}$u, so $x = \frac{33}{12}$, $y = \frac{7}{5}$, and $z = \frac{68}{12}$, or, with the same denominator, $x = \frac{165}{60}$, $y = \frac{84}{60}$, and $z = \frac{340}{60}$.

Lemma to Problem IV^G.35 $x \cdot y - (x + y) = a$ $(a = 8)$

Assignments	Operations	Equation
$x := 1N$		
$y := 3u$		
	Multiply 1N by 3u → 3N	
	Add 1N and 3u → 1N 3u	
	Subtract 1N 3u from 3N → 2N ℓ 3u	
		2N ℓ 3u = 8u

1N is $5\frac{1}{2}$u, so $x = 5\frac{1}{2}$ and $y = 3$. Now, $5\frac{1}{2}$ come from the division of the sum of 8 and 3 by the number which is one unit less than the multitude of units in the assignment of y. Thus, we assign y to be any expression whatever involving N, say, $y := 1N\ 1u$. Then the sum of 8u and 1N 1u yields 1N 9u, while the amount one unit less than 1N 1u is 1N. So, $x = \frac{1N\ 9u}{1N}$, or 1u 9N', and $y = 1N\ 1u$

$$\text{IV}^G.35 \quad \begin{cases} x \cdot y - (x + y) = a \\ y \cdot z - (y + z) = b \\ z \cdot x - (z + x) = c \end{cases}$$ Determination: a, b, and c are each 1 unit less than a square ($a = 8$, $b = 15$, $c = 24$)

Following the calculations here, 1P will be the quantity $\frac{(a+1)(b+1)}{c+1}$, just as it was in the previous problem, so it must be a square. Again, it is not necessary that each of $a + 1$, $b + 1$, and $c + 1$ be a square (Heath 1910, 193 n. 2).

Like in Problem IV^G.34, after y is assigned, the procedure described in the lemma is applied to calculate the assignment for x from the first condition and then again to find the assignment for z from the second condition. The equation then derives from the third condition.

Assignments	Operations	Equation
$y := 1N\ 1u$		
	Add 8u and 1N 1u → 1N 9u	
	Subtract 1u from 1N 1u → 1N	
	Divide 1N 9u by 1N → 1u 9N'	
$x := 1u\ 9N'$		
	Add 15u and 1N 1u → 1N 16u	
	Divide 1N 16u by 1N → 1u 16N'	
$z := 16N'\ 1u$		
	Multiply 1u 9N' by 1u 16N' → 144P' 25N' 1u	
	Add 1u 9N' and 1u 16N' → 25N' 2u	

> Subtract 25N' 2u from
> 144P' 25N' 1u → 144P' ℓ 1u

$$144P'\, \ell\, 1u = 24u$$

1N is $\frac{12}{5}$ u, so $x = \frac{57}{12}$, $y = \frac{17}{5}$, and $z = \frac{92}{12}$, or, with the same denominator, $x = \frac{285}{60}$, $y = \frac{204}{60}$, and $z = \frac{460}{60}$.

Lemma to Problem IV^G.36 $x \cdot y = m(x + y)$ $(m = 3)$

As in the previous two lemmas, Diophantus finds a particular solution, and then he traces the origin of the value of 1N to find a general solution.

Assignments	Operations	Equation
$x := 1N$		
$y := 5u$		
	Multiply 1N by 5u → 5N	
	Add 1N and 5u → 1N 5u	
	Three times 1N 5u → 3N 15u	
		$3N\ 15u = 5N$

1N is $7\frac{1}{2}$ u, so $x = 7\frac{1}{2}$ and $y = 5$. Now, $7\frac{1}{2}$ comes from the division of 15 by 2, where 15 comes from the multiplication of the second number by the given ratio, and 2 is the excess of the second number over the ratio. Thus, if we assign the second number to be any number of Numbers, say 1N, then the multiplication by the ratio yields 3N, the excess over the ratio yields 1N ℓ 3u. Thus, $x = \frac{3N}{1N\,\ell\,3u}$ and $y = 1N$.

IV^G.36 $\begin{cases} x \cdot y = m(x+y) \\ y \cdot z = n(y+z) \qquad (m = 3, n = 4, p = 5) \\ z \cdot x = p(z+x) \end{cases}$

The second and third assignments apply the previous lemma to the first condition to find the assignment for x, and to the second condition for the assignment of z. The equation is formed from the third condition.

Assignments	Operations	Equation
$y := 1N$		
$x := \dfrac{3N}{1N\,\ell\,3u}$		
$z := \dfrac{4N}{1N\,\ell\,4u}$		

Multiply $\dfrac{3N}{1N\,\ell\,3u}$ by $\dfrac{4N}{1N\,\ell\,4u}$ $\rightarrow$ $\dfrac{12P}{1P\,12u\,\ell\,7N}$

Add $\dfrac{3N}{1N\,\ell\,3u}$ and $\dfrac{4N}{1N\,\ell\,4u}$ $\rightarrow$ $\dfrac{7P\,\ell\,24N}{1P\,12u\,\ell\,7N}$

Five times $\dfrac{7P\,\ell\,24N}{1P\,12u\,\ell\,7N}$ $\rightarrow$ $\dfrac{35P\,\ell\,120N}{1P\,12u\,\ell\,7N}$

Multiply $\dfrac{12P}{1P\,12u\,\ell\,7N}$ by $1P\;12u\;\ell\;7N \rightarrow 12P$

Multiply $\dfrac{35P\,\ell\,120N}{1P\,12u\,\ell\,7N}$ by

$1P\;12u\;\ell\;7N \rightarrow 35P\;\ell\;120N$

$12P = 35P\;\ell\;120N$

1N is $\frac{120}{23}$ u, so $x = \frac{360}{51}$, $y = \frac{120}{23}$, and $z = \frac{480}{28}$.

IV^G.37 $\begin{cases} x\cdot y = m(x+y+z) \\ y\cdot z = n(x+y+z) \\ z\cdot x = p(x+y+z) \end{cases}$ $\quad (m=3, n=4, p=5)$

Diophantus first solves the problem of finding x, y, z, and an arbitrary number that he will substitute for $x+y+z$ in the three conditions. He first picks 5 for the arbitrary number and solves the problem $\begin{cases} x\cdot y = 3\cdot5 \\ y\cdot z = 4\cdot5 \\ z\cdot x = 5\cdot5 \end{cases}$. The second assignment provided next satisfies the first condition, and the third assignment satisfies the second condition. The equation is formed from the third condition.

Assignments	Operations	Equation
$y := 1N$		
	Divide 15u by 1N $\rightarrow$ 15N′	
$x := 15N'$		
	Divide 20u by 1N $\rightarrow$ 20N′	
$z := 20N'$		
	Multiply 15N′ by 20N′ $\rightarrow$ 300P′	
		$300P' = 25u$

Because the ratio of 300 to 25 is not a square, the solution is not rational. But 300 is the product of 15 by 20, 25 is the product of 5 by 5, and 15 and 20 are triple and quadruple the given 5 units, respectively. We are led, therefore, to the problem of finding a number to replace the 5 such that its triple, multiplied by its quadruple, has to its quintuple the ratio of a square to a square. In modern notation, $(3p \cdot 4p) : 5p :: \square : \square'$.

Assignments	Operations	Equation
$p \coloneqq 1N$		
	Multiply triple 1N by quadruple 1N $\rightarrow$ 12P	
	Five times 1N $\rightarrow$ 5N	

We now have the proportion 12P : 5N :: $\square$: $\square'$. Thus, the product of 12P by 5N must be a square.

Assignments	Operations	Equation
	Multiply 12P by 5N $\rightarrow$ 60C	
		$60C = \square''$
$\square'' \coloneqq 900P$		
		$60C = 900P$

Thus, 1N is 15u, so $p = 15$. Now we can replace the arbitrarily chosen 5 with 15.

That problem now becomes $\begin{cases} x \cdot y = 3 \cdot 15 \\ y \cdot z = 4 \cdot 15 \\ z \cdot x = 5 \cdot 15 \end{cases}$

Assignments	Operations	Equation
$y \coloneqq 1N$		
	Divide 45u by 1N $\rightarrow$ 45N′	
$x \coloneqq 45N'$		
	Divide 60u by 1N $\rightarrow$ 60N′	
$z \coloneqq 60N'$		
	Multiply 45N′ by 60N′ $\rightarrow$ 2700P′	
		$2700P' = 75u$

1N is found to be 6u, so $x = 7\frac{1}{2}$, $y = 6$, and $z = 10$. Now, if the sum of these three numbers were 15, the problem would have been solved. So, we adjust them to make that happen. We assign them to be in terms of a Number, and the 15 to be 15 Powers.

Assignments	Operations	Equation
$x = 7\frac{1}{2}N$		
$y \coloneqq 6N$		
$z \coloneqq 10N$		
	Add $7\frac{1}{2}N$, 6N, and 10N $\rightarrow 23\frac{1}{2}N$	
		$23\frac{1}{2}N = 15P$

1N becomes $\frac{47}{30}$u, so $x = \frac{352\frac{1}{2}}{30}$, $y = \frac{282}{30}$, and $z = \frac{470}{30}$.

IV^G.38
$$\begin{cases} (x+y+z) \cdot x = \Delta \\ (x+y+z) \cdot y = \square' \\ (x+y+z) \cdot z = \text{⌘}'' \end{cases}$$

The four assignments satisfy all the conditions, but the equation to reconcile the two ways of naming $x + y + z$ will not work.

Assignments	Operations	Equation
$x + y + z := 1$P		
$x := 6$P' (6 being triangular)		
$y := 4$P' (4 being a square)		
$z := 8$P' (8 being a cube)		
	Add 6P', 4P', and 8P' $\rightarrow$ 18P'	
		18P' = 1P

The equation simplifies to 1PP = 18u, so the 18 should be replaced. In Diophantus's words, "I am reduced to finding a square having a square as a side, and to divide (it) into a triangular number and a square and a cube". Writing $\square\square$ to indicate the square of a square, in modern notation the problem is
$$\begin{cases} u+v+w = \square\square''' \\ u = \Delta^{iv} \\ v = \square^{v} \\ w = \text{⌘}^{vi} \end{cases}$$

Assignments	Operations	Equation
$u + v + w := 1$PP		
$v := 1$PP 1u ℓ 2P		
	Subtract 1PP 1u ℓ 2P from 1PP $\rightarrow$ 2P ℓ 1u	
$u + w := 2$P ℓ 1u		
$w := 8$u		
	Subtract 8u from 2P ℓ 1u $\rightarrow$ 2P ℓ 9u	
$u := 2$P ℓ 9u		

Diophantus then notes that eight times any triangular number, with 1 unit added, becomes a square.

Assignments	Operations	Equation
	Multiply 2P ℓ 9u by 8u $\rightarrow$ 16P ℓ 72u	

$$\sqrt{\square^{vii}} := 4N \,\ell\, 1u$$

Add 1u to this → 16P ℓ 71u	
16P ℓ 71u = □^{vii}	
Square 4N ℓ 1u → 16P 1u ℓ 8N	
	16P ℓ 71u =
	16P 1u ℓ 8N

1N is then 9u. Thus, the triangular number u = 153, the square v = 6400, and the cube w = 8. So, we return to the original problem and we modify the assignments according to these values.

Assignments	Operations	Equation
$x + y + z := 1P$		
$x := 153P'$ (153 being triangular)		
$y := 6400P'$ (6400 being a square)		
$z := 8P'$ (8 being a cube)		
	Add 153P′, 6400P′, and 8P′ → 6561P′	
		6561P′ = 1P

Now the value of 1N becomes 9u. The triangular number is $x = \frac{153}{81}$, the square number is $y = \frac{6400}{81}$, and the cubic number is $z = \frac{8}{81}$.

IV^G.39
$$\begin{cases} (x - y) = m(y - z) \\ x + y = \square \\ y + z = \square' \\ z + x = \square'' \end{cases} \quad (m = 3)$$

The first three assignments satisfy the third condition, and then x is calculated from the first condition. The two forthcoming equations arise from the other two conditions.

Assignments	Operations	Equation
$\square' := 4u$		
$y := 1N\,2u$		
	Subtract 1N 2u from 4u → 2u ℓ 1N	

$z = 2u \ell 1N$

	Subtract 2u ℓ 1N from 1N 2u $\rightarrow$ 2N	
	Three times 2N $\rightarrow$ 6N	
	Add 6N to 1N 2u $\rightarrow$ 7N 2u	
$x = 7N\ 2u$		
	Add 1N 2u and 7N 2u $\rightarrow$ 8N 4u	
	Add 7N 2u and 2u ℓ 1N $\rightarrow$ 6N 4u	
		8N 4u = $\square$
		6N 4u = $\square$"

Diophantus solves the two equations by the double-equality. The difference 2N is factored into 4u by $\frac{1}{2}$N , and the eater square is equated with the square of half the sum of the two factors. This gives the equation $\frac{1}{16}$P 1N 4u = 8N 4u , from which 1N is found to be 112u. This value cannot be accepted, for it makes z negative. Therefore, 1N must be found less than 2u, so 6N 4u must be a square less than 16u.

Diophantus then turns to finding values for the squares $\square$, $\square$", and $\square$' to make this work. The conditions are that $\square < \square" < \square$, $\square' = 4$ (from the first assignment), $\square" < 16$ (from the condition just derived to make z positive), and $\square - \square" = \frac{1}{3}(\square" - \square')$. This last equality comes from the facts that $\square - \square" = y - z$ and $\square" - \square' = x - y$.

Assignments	**Operations**	**Equation**
$\square' := 4u$		
$\sqrt{\square"} := 1N\ 2u$		
	Square 1N 2u $\rightarrow$ 1P 4N 4u	
$\square" := 1P\ 4N\ 4u$		
	Subtract 4u from 1P 4N 4u $\rightarrow$ 1P 4N	
$\square" - \square' := 1P\ 4N$		
	A third of 1P 4N $\rightarrow \frac{1}{3}$P $1\frac{1}{3}$N	
$\square - \square" := \frac{1}{3}$P $1\frac{1}{3}$		
	Add $\frac{1}{3}$P $1\frac{1}{3}$N to	
	1P 4N 4u $\rightarrow 1\frac{1}{3}$P $5\frac{1}{3}$N 4u	
$\square := 1\frac{1}{3}$P $5\frac{1}{3}$N 4u		
	Nine times $1\frac{1}{3}$P $5\frac{1}{3}$N 4u $\rightarrow$ 12P 48N 36u	
	A fourth of 12P 48N 36u $\rightarrow$ 3P 12N 9u	
		3P 12N 9u = $\square^{iv}$

The *al-istiqrā'* assignment for the side of $\square^{iv}$ should be 3u lacking some multitude of Numbers. Its square will be of the form 9u Ps ℓ Ns, where the multitude of Ns is six times the multitude for the side, and the multitude of Ps is the square of that multitude. Confronting this square with 3P 12N 9u, the value of 1N will be 12 more than six times the chosen multitude, divided by the excess of the square of the chosen multitude over 3. This value must be less than 2 in order for $\square'' < 16$. In modern notation, we must find a number p that satisfies $\frac{12+6p}{p^2-3} < 2$.

Assignments	Operations	Equation
$p := 1\text{N}$		
	Six times 1N → 6N	
	Add 6N and 12u → 6N 12u	
	Square 1N→1P	
	Subtract 3u from 1P → 1P ℓ 3u	

Dividing 6N 12u by 1P ℓ 3u should give a quotient less than 2. Thus, the ratio 6N 12u : 1P ℓ 3u should be less than 2:1. Cross-multiplying, we get 6N 12u < 2P ℓ 6u. This simplifies to 6N 18u < 2P. Diophantus then runs through the steps of solving this inequality as if it were an equation. Contorted into modern notation, his rule for solving an equation of the type $ax^2 = bx + c$ is $x = \frac{\sqrt{\left(\frac{1}{2}b\right)^2 + ac} + \frac{1}{2}b}{a}$. Multiplying half the 6 by itself gives 9; 2 multiplied by 18 gives 36. Adding 9 and 36 gives 45. Because he wants the 2P to be greater than 6N 18u, he wants a value for 1N to be greater than the 1N obtained by solving the corresponding equation. This makes $\sqrt{45}$ a lower bound, so he writes, "it is not smaller than 7". Continuing with the rule for solving the equation, he adds half of 6 to the 7, and then he divides it by 2 to get 1N = 5. This is the multitude of Numbers in the *al-istiqrā'* assignment.

Returning now to the problem of finding $\square$, $\square'$, and $\square''$, the equation can be completed and solved.

Assignments	Operations	Equation
	3P 12N 9u = $\square^{iv}$	
$\sqrt{\square^{iv}} := 3\text{u} \ell 5\text{N}$		
	Square 3u ℓ 5N → 9u 25P ℓ 30N	
		3P 12N 9u = 9u 25P ℓ 30N

1N is then $\frac{21}{11}$ u, so $\square'' = \frac{1849}{121}$ and $\square = \frac{2304}{121}$.

Turning now to the original problem for finding x, y, and z, the forthcoming equation 6N 4u = $\square''$ can now be completed and the equation can be solved.

Assignments	Operations	Equation
		6N 4u = $\square''$
$\square'' := \frac{1849}{121}$ u		
		6N 4u = $\frac{1849}{121}$ u

Now 1N is found to be $\frac{1356}{726}$ u, which is less than 2 units. Thus, $x = \frac{7338}{484}$, $y = \frac{1878}{484}$, and $z = \frac{58}{484}$.

IV^G.40
$$\begin{cases} (x^2 - y^2) = m(y - z) \\ x + y = \square \\ y + z = \square' \\ z + x = \square'' \end{cases} \quad (m = 3)$$

The first assignment satisfies the second condition, and the assignment for x is direct. The fourth condition is satisfied with the third assignment. The remaining calculations provided next work toward the first condition.

Assignments	Operations	Equation
$x + y := 16P$		
$x := 8P\ 2u$		
$x + z := 9P$		
	Subtract 8P 2u from 16P $\rightarrow$ 8P ℓ 2u	
$y := 8P\ \ell\ 2u$		
	Subtract 8P 2u from 9P $\rightarrow$ 1P ℓ 2u	
$z := 1P\ \ell\ 2u$		
	Subtract the square of 8P ℓ 2u from the square of 8P 2u $\rightarrow$ 64P	
	Subtract 1P ℓ 2u from 8P ℓ 2u $\rightarrow$ 7P	

According to the first condition, 64P has to be three times 7P, or 21P, but it is not. Now the 64 results from taking 2 units 32 times. Thus, we are reduced to the problem of finding a number to replace the 2 units such that when taken 32 times it makes 21 units. In modern notation, $32u = 21$. It "is obviously" $\frac{21}{32}$.

Returning to the original problem, the assignment for x is modified so that the units are $\frac{21}{32}$ instead of 2.

Assignments	Operations	Equation
$x + y \coloneqq 16\text{P}$		
$x \coloneqq 8\text{P}\,\frac{21}{32}\text{u}$		
$x + z \coloneqq 9\text{P}$		
	Subtract $8\text{P}\,\frac{21}{32}\text{u}$ from $16\text{P} \to 8\text{P}\,\ell\,\frac{21}{32}\text{u}$	
$y \coloneqq 8\text{P}\,\ell\,\frac{21}{32}\text{u}$		
	Subtract $8\text{P}\,\frac{21}{32}\text{u}$ from $9\text{P} \to 1\text{P}\,\ell\,\frac{21}{32}\text{u}$	
$z \coloneqq 1\text{P}\,\ell\,\frac{21}{32}\text{u}$		
	Subtract the square of $8\text{P}\,\ell\,\frac{21}{32}\text{u}$ from the square of $8\text{P}\,\frac{21}{32}\text{u} \to 21\text{P}$	
	Subtract $1\text{P}\,\ell\,\frac{21}{32}\text{u}$ from $8\text{P}\,\ell\,\frac{21}{32}\text{u} \to 7\text{P}$	

Now all conditions have been fulfilled except the third, $y + z = \square\,'$.

Assignments	Operations	Equation
	Add $8\text{P}\,\ell\,\frac{21}{32}\text{u}$ and $1\text{P}\,\ell\,\frac{21}{32}\text{u} \to 9\text{P}\,\ell\,\frac{42}{32}\text{u}$	
		$9\text{P}\,\ell\,\frac{42}{32}\text{u} = \square\,'$
$\sqrt{\square\,'} = 3\text{N}\,\ell\,6\text{u}$		
	Square $3\text{N}\,\ell\,6\text{u} \to 9\text{P}\,36\text{u}\,\ell\,36\text{N}$	
		$9\text{P}\,\ell\,\frac{42}{32}\text{u} = 9\text{P}\,36\text{u}\,\ell\,36\text{N}$

1N is found to be $\frac{597}{576}\text{u}$. Therefore, $x = \frac{3{,}069{,}000}{331{,}776}$, $y = \frac{2{,}633{,}544}{331{,}776}$, and $z = \frac{138{,}681}{331{,}776}$.

Book V^G

$$\text{V}^G.1 \quad \begin{cases} x:y::y:z \\ x-a=\square \\ y-a=\square' \\ z-a=\square'' \end{cases} \quad (a=12)$$

The second condition, conceived as $\left(\sqrt{x}\right)^2 - \square = 12$, can be solved by the method of Problem II.10. Diophantus finds the square x to be $42\frac{1}{4}$ without showing the solution, and this is the first assignment. Then, with x a square, z is assigned to be a square, which then determines a rational assignment for y based on the first condition. The two forthcoming equations are then formed from the last two conditions.

Assignments	Operations	Equation
$x := 42\frac{1}{4}\mathrm{u}$ $z := 1\mathrm{P}$		
	Multiply $42\frac{1}{4}\mathrm{u}$ by $1\mathrm{P} \rightarrow 42\frac{1}{4}\mathrm{P}$	
	Take the square root of $42\frac{1}{4}\mathrm{P} \rightarrow 6\frac{1}{2}\mathrm{N}$	
$y := 6\frac{1}{2}\mathrm{N}$		
	Subtract 12u from $1\mathrm{P} \rightarrow 1\mathrm{P}\,\ell\,12\mathrm{u}$	
		$1\mathrm{P}\,\ell\,12\mathrm{u} = \square''$
	Subtract 12u from $6\frac{1}{2}\mathrm{N} \rightarrow 6\frac{1}{2}\mathrm{N}\,\ell\,12\mathrm{u}$	
		$6\frac{1}{2}\mathrm{N}\,\ell\,12\mathrm{u} = \square$

The resulting double-equality is solved by the standard procedure. The difference $1\mathrm{P}\,\ell\,6\frac{1}{2}\mathrm{N}$ is factored into $1\mathrm{N}$ by $1\mathrm{N}\,\ell\,6\frac{1}{2}\mathrm{u}$, so the square of half the difference of the two factors is the smaller square. This gives $6\frac{1}{2}\mathrm{N}\,\ell\,12\mathrm{u} = \frac{169}{16}\mathrm{u}$, from which $1\mathrm{N}$ is found to be $\frac{361}{104}\mathrm{u}$. Therefore, $x = 42\frac{1}{4}$, $y = \frac{2346\frac{1}{2}}{104}$, and $z = \frac{130,321}{10,816}$.

DOI: 10.4324/9781315171470-27

$$\mathbf{V^G.2} \quad \begin{cases} x:y::y:z \\ x+a=\square \\ y+a=\square' \\ z+a=\square'' \end{cases} \quad (a=12)$$

Again, Diophantus would like x to be a square multitude of units and z to be a square multitude of Powers so that y is a rational multitude of Numbers. Problem II.10 can be invoked to find an x, and in this case he finds it to be 16.

Assignments	Operations	Equation
$x := 16u$		
$z := 1P$		
	Multiply 16u by 1P $\rightarrow$ 16P	
	Take the square root of 16P $\rightarrow$ 4N	
$y := 4N$		
	Add 1P and 20u $\rightarrow$ 1P 20u	
		1P 20u = $\square''$
	Add 4N and 20u $\rightarrow$ 4N 20u	
		4N 20u = $\square'$

The solution by the double-equality does not work because factoring the difference into 1N by 1N ℓ 4u makes the square of half the difference 4u so that the solution to the equation 4N 20u = 4u is not positive. Tracing the origin of this 4u, Diophantus finds that it is a fourth of the 16u that was assigned to be x. Therefore, the problem is reduced to finding a square number to replace 16u which has a fourth part greater than 20u, and, if it receives in addition 20u, makes a square. In our notation: $u^2 + 20 = \square'''$; $u^2 > 80$. The first assignment satisfies the second condition, and the resulting forthcoming equation is solved by *al-istiqrā'*.

Assignments	Operations	Equation
$u := 1N\ 9u$		
	Square 1N 9u $\rightarrow$ 1P 18N 81u	
	Add 20u to 1P 18N 81u $\rightarrow$ 1P 18N 101u	
		1P 18N 101u = $\square'''$
$\sqrt{\square'''} := 1N\ \ell\ 11u$		
	Square 1N ℓ 11u $\rightarrow$ 1P 121u ℓ 22N	
		1P 18N 101u = 1P 121u ℓ 22N

1N becomes $\frac{1}{2}$u, so $u^2 = 90\frac{1}{4}$. This will be the new assignment for x.

Assignments	Operations	Equation
$x := 90\frac{1}{4}u$ $z := 1P$		
	Multiply $90\frac{1}{4}u$ by $1P \rightarrow 90\frac{1}{4}P$	
	Take the square root of $90\frac{1}{4}P \rightarrow 9\frac{1}{2}N$	
$y := 9\frac{1}{2}N$		
	Add $1P$ and $20u \rightarrow 1P\ 20u$	
		$1P\ 20u = \square''$
	Add $9\frac{1}{2}N$ and $20u \rightarrow 9\frac{1}{2}N\ 20u$	
		$9\frac{1}{2}N\ 20u = \square'$

The difference $1P \ \ell \ 9\frac{1}{2}N$ is factored into $1N$ by $1N \ \ell \ 9\frac{1}{2}u$. The square of half the difference is $\frac{361}{16}$, which gives the equation $9\frac{1}{2}N\ 20u = \frac{361}{16}u$. $1N$ is then $\frac{41}{152}u$, so $x = 90\frac{1}{4}$, $y = \frac{389\frac{1}{2}}{152}$, and $z = \frac{1681}{23,104}$.

$$\mathbf{V^G.3} \quad \begin{cases} x + a = \square \\ y + a = \square' \\ z + a = \square'' \\ x \cdot y + a = \square''' \\ y \cdot z + a = \square^{iv} \\ z \cdot x + a = \square^{v} \end{cases} \quad (a = 5)$$

Diophantus begins the solution by citing a theorem from his *Porisms*,[1] the first such reference in the extant books of the *Arithmetica*. The theorem states that if x, y, and a satisfy the first, second, and fourth conditions provided earlier, then the square roots of $\square$ and $\square'$ differ by one. This is misstated. Instead, if the square roots of $\square$ and $\square'$ differ by one and the first two conditions are satisfied, then the fourth condition will be satisfied as well. In fact, as Heath notes (1910, 99), Diophantus's own technique can be used to find examples in which the three conditions are satisfied and the difference of the roots is not one.[2]

---

1 See Section I.2.
2 One such example is $x = 4$, $y = 4\frac{67}{121}$, and $a = 5$, where the difference of the roots of $\square$ and $\square'$ is $\frac{1}{11}$.

The first two assignments and the two derivative assignments that follow satisfy the first, second, and fourth conditions. The assignment for z is made to be the same as $2\,(x+y) - 1$. Diophantus does not explain why, but this satisfies the last two conditions. The forthcoming equation is formed from the third condition and is solved by *al-istiqrā'*.

Assignments	Operations	Equation
$\sqrt{\square} := $ 1N 3u		
$\sqrt{\square}' := $ 1N 4u		
	Square 1N 3u → 1P 6N 9u	
	Square 1N 4u → 1P 8N 16u	
	Subtract 5u from 1P 6N 9u → 1P 6N 4u	
$x := $ 1P 6N 4u		
	Subtract 5u from 1P 8N 16u → 1P 8N 11u	
$y := $ 1P 8N 11u		
	Twice the sum of 1P 6N 4u, 1P 8N 11u, less 1u → 4P 28N 29u	
$z := $ 4P 28N 29u		
	Add 5u to 4P 28N 29u → 4P 28N 34u	
		4P 28N 34u = $\square''$
$\sqrt{\square}'' := $ 2N ℓ 6u		
	Square 2N ℓ 6u → 4P 36u ℓ 24N	
		4P 28N 34u = 4P 36u ℓ 24N

Thus, 1N is $\frac{1}{26}$u, so $x = \frac{2861}{676}$, $y = \frac{7645}{676}$, and $z = \frac{20{,}336}{676}$.

$$\mathrm{V^G.4} \quad \begin{cases} x - a = \square \\ y - a = \square' \\ z - a = \square'' \\ x \cdot y - a = \square''' \\ y \cdot z - a = \square^{iv} \\ z \cdot x - a = \square^{v} \end{cases} \quad (a = 6)$$

The solution follows the same reasoning as the previous problem.

Assignments	Operations	Equation
$\square := 1P$		
$\square' := 1P\ 2N\ 1u$		
	Add 6u to 1P → 1P 6u	
$x := 1P\ 6u$		
	Add 6u to 1P 2N 1u → 1P 2N 7u	
$y := 1P\ 2N\ 7u$		
	Twice the sum of 1P 6u and 1P 2N 7u, less 1u → 4P 4N 25u	
$z := 4P\ 4N\ 25u$		
	Subtract 6u from 4P 4N 25u → 4P 4N 19u	
		4P 4N 19u = $\square''$
$\sqrt{\square''} := 2N\ \ell\ 6u$		
	Square 2N ℓ 6u → 4P 36u ℓ 24N	
		4P 4N 19u = 4P 36u ℓ 24N

Thus, 1N is $\frac{17}{28}$u, so $x = \frac{4993}{784}$, $y = \frac{6729}{784}$, and $z = \frac{22{,}660}{784}$.

$$V^G.5 \begin{cases} x^2 \cdot y^2 + \left(x^2 + y^2\right) = \square \\ y^2 \cdot z^2 + \left(y^2 + z^2\right) = \square' \\ z^2 \cdot x^2 + \left(z^2 + x^2\right) = \square'' \\ x^2 \cdot y^2 + z^2 = \square''' \\ y^2 \cdot z^2 + x^2 = \square^{iv} \\ z^2 \cdot x^2 + y^2 = \square^{v} \end{cases}$$

In this problem Diophantus refers for a second time to his *Porisms*, invoking a theorem which states, in modern notation, that for three numbers of the form p^2, $(p + 1)^2$, and $2(p^2 + (p + 1)^2) + 2$, "the product of any two, if it receives in addition their sum, or the remaining one, makes a square". This fits the conditions of the problem and accounts for the assignments of x^2, y^2, and z^2. It then remains to make the assignment for z^2 to be a square. The resulting forthcoming equation is solved by *al-istiqrā'*.

Assignments	Operations	Equation
$x^2 := 1P\ 2N\ 1u$		
$y^2 := 1P\ 4N\ 4u$		

	Twice the sum of 1P 2N 1u, 1P 4N 4u, increased by 2u → 4P 12N 12u	
$z^2 := 4\text{P } 12\text{N } 12\text{u}$		
		4P 12N 12u $= \square_1$
	Take a fourth of 4P 12N 12u → 1P 3N 3u	
		$1\text{P } 3\text{N } 3\text{u} = \square_2$
$\sqrt{\square_2} := 1\text{N } \ell \; 3\text{u}$		
	Square 1N ℓ 3u → 1P 9u ℓ 6N	
		1P 3N 3u = 1P 9u ℓ 6N

1N is then $\frac{2}{3}$u, so $x^2 = \frac{25}{9}$, $y^2 = \frac{64}{9}$, and $z^2 = \frac{196}{9}$.

$$\mathbf{V^G.6} \begin{cases} x - 2 = \square \\ y - 2 = \square' \\ z - 2 = \square'' \\ x \cdot y - (x + y) = \square''' \\ y \cdot z - (y + z) = \square^{iv} \\ z \cdot x - (z + x) = \square^{v} \\ x \cdot y - z = \square^{vi} \\ y \cdot z - x = \square^{vii} \\ z \cdot x - y = \square^{viii} \end{cases}$$

The solution is again based on the theorem from the *Porisms* stated in the preceding problem, by virtue of the identity $(a + 2) \cdot (b + 2) - ((a + 2) + (b + 2)) = a \cdot b + (a + b)$. Diophantus makes three assignments according to the porism, with 2u added to each of them. The assignments for x and y, and the derivative assignment for z, satisfy all conditions except $z - 2 = \square''$, from which the forthcoming equation is set up. This is then solved by *al-istiqrā'*.

Assignments	Operations	Equation
$x := 1\text{P } 2\text{u}$		
$y := 1\text{P } 2\text{N } 3\text{u}$		
	Twice the sum of 1P and 1P 2N 1u, increased by 2u → 4P 4N 4u	

$z := $ 4P 4N 6u	Add 2u to 4P 4N 4u → 4P 4N 6u	
	Subtract 2u from 4P 4N 6u → 4P 4N 4u	
		4P 4N 4u = □″
	Take a fourth of 4P 4N 4u → 1P 1N 1u	
		1P 1N 1u = □₁
$\sqrt{□_1} := $ 1N ℓ 2u		
	Square 1N ℓ 2u → 1P 4u ℓ 4N	
		1P 1N 1u=1P 4u ℓ 4N

1N is $\frac{3}{5}$u, so $x = \frac{59}{25}$, $y = \frac{114}{25}$, and $z = \frac{246}{25}$.

First lemma to Problem V^G.7. $x \cdot y + (x^2 + y^2) = □$

Assignments	Operations	Equation
$x := $ 1N		
$y := $ 1u		
	Multiply 1N by 1u → 1N	
	Add the squares of 1N and 1u → 1P 1u	
	Add 1N and 1P 1u → 1P 1N 1u	
		1P 1N 1u = □
$\sqrt{□} := $ 1N ℓ 2u		
	Square 1N ℓ 2u → 1P 4u ℓ 4N	
		1P 1N 1u = 1P 4u 4N

1N is $\frac{3}{5}$u, so $x = \frac{3}{5}$ and $y = 1$, or, by eliminating the denominators, 3 and 5, respectively.

Second lemma to Problem V^G.7. To find numerically three right-angled triangles having equal areas.

Diophantus tacitly assumes that if there are three numbers a, b, c such that $a^2 + b^2 + ab = c^2$, then three right triangles having equal areas can be formed from the pairs (c, a), (c, b), and $(c, a + b)$.[3] He adopts the numbers 3, 5, and 7 from the previous lemma, and he forms three right triangles from the pairs $(7, 3)$, $(7, 5)$, and

3 On this proposition see Heath's extensive note (1910, 204–05).

(7, 3 + 5) (7, 3 + 5). The sides of the first triangle are $7^2 - 3^2 = 40$, $2 \cdot 7 \cdot 3 = 42$, $7^2 + 3^2 = 58$; the sides of the second triangle, similarly, $7^2 - 5^2 = 24$, $2 \cdot 7 \cdot 5 = 70$, $7^2 + 5^2 = 74$; and of the third triangle, $8^2 - 7^2 = 15$, $2 \cdot 8 \cdot 7 = 112$, $8^2 + 7^2 = 113$. The area of each triangle is 840 units. This way of forming right triangles from a pair of numbers will be applied in many problems in Book VI.

$$\text{V}^\text{G}.7 \quad \begin{cases} x^2 + (x+y+z) = \square \\ x^2 - (x+y+z) = \square' \\ y^2 + (x+y+z) = \square'' \\ y^2 - (x+y+z) = \square''' \\ z^2 + (x+y+z) = \square^{iv} \\ z^2 - (x+y+z) = \square^{v} \end{cases}$$

In any right triangle with perpendiculars p and q, the square on the hypotenuse ($p^2 + q^2$) increased or decreased by four times the area ($2pq$) will be a square. Thus, the conditions of the problem will be satisfied if we can find three right triangles with hypotenuses $x, y,$ and z, and whose areas are all one-fourth of $x + y + z$. By the preceding lemma we have found three right triangles with equal areas, so it remains to scale the triangles so that the sum of the hypotenuses is four times the area.

Assignments	Operations	Equation
$x := 58\text{N}$		
$y := 74\text{N}$		
$z := 113\text{N}$		
	Add 58N, 74N, and 113N → 245N	
	Four times the area 840u → 3360u	
		$3360\text{P} = 245\text{N}$

1N is $\frac{7}{96}$ u; therefore, $x = \frac{406}{96}$, $y = \frac{518}{96}$, and $z = \frac{791}{96}$.

Lemma to Problem V$^\text{G}$.8. $\quad \begin{cases} x \cdot y = a^2 \\ y \cdot z = b^2 \\ z \cdot x = c^2 \end{cases} \quad (a = 2, b = 3, c = 4)$

Assignments	Operations	Equation
$y := 1\text{N}$		
	Divide 4u by 1N → 4N′	
$x := 4\text{N}'$		
	Divide 9u by 1N → 9N′	

$z := 9N'$

	Multiply 9N′ by 4N′ → 36P′	
		$36P' = 16u$

Thus, 1N is found to be $1\frac{1}{2}$ u, so $y = 1\frac{1}{2}$, $x = 2\frac{1}{2}\frac{1}{6}$ (or $2\frac{2}{3}$), and $z = 6$.

Observing that the 6 in $y = \frac{6}{4}$ is the product of 2 by 3, the sides of two of the given square numbers, and 4 is the side of the third given square number, Diophantus formulates a general rule for finding three numbers satisfying the conditions of the problem for any three given squares: select two of the squares, and divide the product of their roots by a root of the third square, repeating for each pair of roots. These will be the three numbers x, y, and z.

$$\mathbf{V^G.8} \quad \begin{cases} x \cdot y + (x + y + z) = \square \\ x \cdot y - (x + y + z) = \square' \\ y \cdot z + (x + y + z) = \square'' \\ y \cdot z - (x + y + z) = \square''' \\ z \cdot x + (x + y + z) = \square^{iv} \\ z \cdot x - (x + y + z) = \square^{v} \end{cases}$$

From the second lemma to Problem V.7, (40, 42, 58), (24, 70, 74), and (15, 112, 113) are three right triangles with the same area, 840 units. Now, according to the proposition implicitly assumed in Problem V.7, the squares of the three hypotenuses – which are 3364, 5476, and 12,769, respectively – increased or decreased by four times the area, make squares. But, according to the rule stated in the lemma preceding this problem, we can find three numbers such that their products taken by twos produce the squares 3364, 5476, and 12,769. These numbers are $\frac{4292}{113}$, $\frac{3277}{37}\left(=\frac{380,132}{4292}\right)$, and $\frac{4181}{29}\left(=\frac{618,788}{4292}\right)$. Thus, the products of the three pairs formed from these numbers, increased or decreased by four times the common area of the triangles, produce square numbers. These relations have a similar structure with the conditions of the problem, once the sum of the three numbers is assumed to be quadruple of the common area. So, Diophantus assigns the sought-after numbers to be these numbers set in terms of a Number, while their sum to be the quadruple of the common area set in terms of a Power.

Assignments	Operations	Equation
$x := \frac{4292}{113} N$		
$y := \frac{3277}{37} N$		
$z := \frac{4181}{29} N$		

Add $\frac{4292}{113}$ N, $\frac{3277}{37}$ N, and

$\frac{4181}{29}$ N $\rightarrow \frac{32,824,806}{121,249}$ N

Four times the common area

840u $\rightarrow$ 3360u

$\frac{32,824,806}{121,249}$ N = 3360P

1N is found to be $\frac{32,824,806}{407,396,640}$u, or $\frac{781,543}{9,699,920}$u, so $x = \frac{781,543}{255,380}$, $y = \frac{781,543}{109,520}$, and $z = \frac{781,543}{67,280}$.

V^G.9 $\begin{cases} x+y=1 \\ x+a=\square \\ y+a=\square' \end{cases}$ Determination:[4] *a* must not be odd, and
$2a+1$ should not be divisible by a prime of the form $4n-1$.
$(a=6)$

Adding the second and third conditions, 13 must be divided into two squares, each of which is greater than 6, or equivalently, the difference of which is < 1. Because their difference is small, the squares are both close to $6\frac{1}{2}$. Next, Diophantus will need a rational number whose square is close to $6\frac{1}{2}$, so he looks for a small number to add to $6\frac{1}{2}$ to make a square. This number must be a square for the following *al-istiqrā'* assignment to work out, so Diophantus calls for this small number to be a "square part", i.e., a number of the form $\frac{1}{u^2}$ for some whole number *u*. Thus, $6\frac{1}{2}+\frac{1}{u^2}=\square''$, or, multiplying by 4, $26+4\frac{1}{u^2}=\square'''$.

Assignments	Operations	Equation
$4\frac{1}{u^2} := 1P'$		
	Add 26u and 1P' $\rightarrow$ 26u 1P'	
		26u 1P' = $\square'''$
	Multiply 26u 1P' by 1P $\rightarrow$ 26P 1u	
		26P 1u = $\square^{iv}$
$\sqrt{\square^{iv}} := 5$N 1u		

---

4 The text of the determination is highly corrupted, to the point that, as Heath points out, "There is room for any number of conjectures as to what may have been Diophantus' words" (1910, 107). The true necessary and sufficient condition on *a* has been given by Fermat in his Observation XXVI on Diophantus: The given number must be even, and the greatest square which measures $2a+1$ must not be divisible by a prime number of the form $4n-1$. See Fermat (1891, 313–14); cf. Heath (1910, 107) and Rashed and Houzel (2013, 474).

$$\text{Square 5N 1u} \rightarrow \text{25P 10N 1u}$$

$$\text{26P 1u} = \text{25P 10N 1u}$$

The choice of 5 for the multitude of Numbers in the *al-istiqrā'* assignment for $\sqrt{\square^{iv}}$ makes u a whole number, though this is not necessary for the solution. It is also the best integer to choose to make $\frac{1}{u^2}$ small. 1N is found to be 10u, so $\frac{1}{u^2} = \frac{1}{400}$ and $\sqrt{\square''} = \frac{51}{20}$.

In dividing 13 into two squares, we should establish the side of each of them as close as possible to $\frac{51}{20}$. Now, $13 = 3^2 + 2^2$, so, in Diophantus's words, "I look for what the triad must lack, and the dyad must receive in addition, in order to make the same, that is, 51 20ths". He finds that $3 - \frac{9}{20} = \frac{51}{20}$, and $2 + \frac{11}{20} = \frac{51}{20}$. Because the sum of the squares is not equal to 13, we need to substitute $\frac{1}{20}$ in the expressions of the sides of the two squares with another fraction. In modern notation, we want to find v such that $(2 + 11v)^2 + (3 - 9v)^2 = 13$.

Assignments	Operations	Equation
$v := 1\text{N}$		
	Square 11N 2u $\rightarrow$ 121P 44N 4u	
	Square 3u ℓ 9N $\rightarrow$ 9u 81P ℓ 54N	
	Add 121P 44N 4u and 9u 81P ℓ 54N $\rightarrow$ 202P ℓ 3u 1 10N	
		202P 13u ℓ
		10N = 13u

From here 1N is found to be $\frac{5}{101}$u, so $13 = \left(\frac{257}{101}\right)^2 + \left(\frac{258}{101}\right)^2$. Returning now to the original problem, we assign the two indeterminate squares to be $\left(\frac{257}{101}\right)^2$ and $\left(\frac{258}{101}\right)^2$, and subtracting 6 from each we obtain the two sought-after numbers, $x = \frac{4843}{10,201}$ and $y = \frac{5358}{10,201}$, whose sum is 1 unit.

In this problem we find for the first time Diophantus's particular method for dividing a given number into the sum of two or three square numbers which are nearly equal to one another. He calls this παρισότητος ἀγωγή, which may be rendered as "method of approximate equality", or "method of quasi-equality". See Heath (1910, 95–98) for a description of how it functions. Diophantus will apply it again in Problem V^G.11. A variation is also applied in Problems V^G.10, 12, and 14, where the squares are subject to different conditions, one of them being that their sum must be a given number. The word παρισότης and its cognate παρίσωσις originate in classical rhetoric, and they are used to mean a succession of a number of coordinate clauses which have approximately the same length, based on their number of words or syllables (Rowe 1997, 137). Neither word appears in any other Greek mathematical treatise.

V^G.10 $\begin{cases} x+y=1 \\ x+a=\square \\ y+b=\square' \end{cases}$ $\quad (a=2, b=6)$

Diophantus begins with a description of the representation of the numbers as lines, which can be redrawn like this:

$\vdash\!\!\!\!\!\!\!\!\!\!\!-\!\!\!-\!\!\!-\!\!\!-\!\!\!-\!\!\!-\!\!\!-\!\!\!-\!\!\!-\!\!\!-\!\!\!-\!\!\!-\!\!\!-\!\!\!-\!\!\!-\!\!\!-\dashv$

Δ   A   Γ   B   E

Line AB is the unit that is cut at Γ, so that AΓ is x and BΓ is y. AΔ is the added 2 units, and BE is the added 6 units. Then ΔΓ will be $\square$ and ΓE will be $\square'$. The language is similar to that of Euclid's *Data*, in that Diophantus indicates successively which numbers will be given. (Like premodern diagrams generally, and unlike Heath [1910, 207], we have not drawn this to scale.) From the diagram, the square $\square$ is some value between 2 and 3, and the entire line ΔE is 9 units. Diophantus does not mention that one should be able to divide the number one more than the sum of the given numbers, here 9, into two squares. This is possible here because 9 is a square.

Assignments	Operations	Equation
$\square := 1P$		
	Subtract 1P from 9u → 9u ℓ 1P	
		9u ℓ 1P $= \square'$

The solution of this forthcoming equation, Diophantus says, is easy. Indeed, we simply have to assign the side of $\square'$ to be an expression of the form 3u ℓ Ns, where the multitude of Ns is any positive rational number. But 1P must lie between 2u and 3u. Diophantus takes two squares that lie between 2 and 3: $2 \le \frac{289}{144} < \frac{261}{144} \le 3$, and he sets out to find a value for 1N that lies between the square roots of these two squares, that is, between $\frac{17}{12}$ and $\frac{19}{12}$.

Working through the equation without specifying this multitude, Diophantus discovers that 1N will be found to be the result of dividing six times the multitude by one more than its square. That is, he seeks a multitude m such that $\frac{17}{12} < \frac{6m}{m^2+1} < \frac{19}{12}$. To solve the first inequality as a subsidiary problem he assigns $m := 1N$ and he derives the inequality 17P 17u < 72N. Then, to find a value of 1N that works, he considers it as if it were an equation. This particular equation type has two positive roots, and Diophantus knows that the inequality will hold for values of 1N that are a little less than the greater root, so he follows the rule for solving for that root. Squaring half of the 72 gives 1296, and subtracting the product of the 17 (Powers) by the 17 (units), or 289, gives 1007. Because he wants the 17P 17u to be less than the 72N, he wants a value less than $\sqrt{1007}$, so he picks 31, whose square is 961, as an upper bound. Adding half of the 72 gives an upper bound of 67, and dividing by the 17 (Powers), he finds that the desired Number is not greater than $\frac{67}{17}$. Then, repeating all this for the inequality $\frac{6p}{p^2+1} < \frac{19}{12}$ via 72N < 19P 19u, he finds that the desired number is not less than $\frac{66}{19}$. So, Diophantus chooses the multitude $m = 3\frac{1}{2}$, which lies between $\frac{66}{19}$ and $\frac{67}{17}$. The *al-istiqrā'* assignment can then be made and the solution can be completed.

Assignments	Operations	Equation
		9u ℓ 1P = ▱
$\sqrt{\square}' \coloneqq 3u\,\ell\,3\frac{1}{2}\,N$		
	Square $3u\,\ell\,3\frac{1}{2}\,N \rightarrow$ $12\frac{1}{4}P\;9u\,\ell\,21N$	
		9u ℓ 1P $= 12\frac{1}{4}$P 9u ℓ 21N

1N then $\frac{84}{53}$u, so $\square = \frac{7056}{2809}$. Subtracting 2, we find $x = \frac{1438}{2809}$, and subtracting that from 1 gives $y = \frac{1371}{2809}$.

V^G.11 $\begin{cases} x+y+z=1 \\ x+a=\square \\ y+a=\square' \\ z+a=\square'' \end{cases}$ Determination:[5] a must not be 2 or any multiple of 8 increased by 2. ($a = 3$)

Adding the last three conditions gives $10 = \square + \square' + \square''$, and each square must be greater than 3. A third of 10 is $3\frac{1}{3}$, so following the "method of approximate equality" as in Problem V^G.9, Diophantus looks for a square whole number whose reciprocal added to $3\frac{1}{3}$ makes a square. He poses, therefore, a subsidiary problem of the type $3\frac{1}{3}+\frac{1}{u^2}=\square'''$, or, multiplying by 9, $30+9\frac{1}{u^2}=\square^{iv}$.

Assignments	Operations	Equation
$9\frac{1}{u^2} \coloneqq 1P'$		
	Add 1P' to 30u $\rightarrow$ 30u 1P'	
		30u 1P' $= \square^{iv}$

5 The problem asks to divide a number of the form $3a + 1$ into three squares, and the determination stated by Diophantus says that this cannot be the case if $a = 2$ or if $a = 3(8k + 2) + 1 = 24k + 7$. As pointed out by Heath (1910, 109), this condition does not include *all* numbers which cannot be the sum of three squares. Legendre proved that numbers of the form $24k + 7$ are the only *odd* numbers which are not divisible into three squares, while Fermat (1891, 314–15) gave the correct sufficient condition to which a must be subject, namely, that it must not be of the form $4^n(24k + 7)$ and hence not $4^n(8k + 7)$; cf. (Dickson 1971, 259–61).

$$\sqrt{\square^v} := 5N\ 1u$$

Multiply 30u 1P' by 1P → 30P 1u

$$30P\ 1u = \square^v$$

Square 5N 1u → 25P 10N 1u

$$30P\ 1u = 25P\ 10N\ 1u$$

As in Problem V^G.9, the choice of 5 for the multitude of Numbers in the *al-istiqrā'* assignment for $\sqrt{\square^v}$ guarantees that u a whole number, and also that $\frac{1}{u^2}$ is small. 1N is then 2u, so $\frac{1}{u^2} = \frac{1}{36}$, and $\square''' = \left(\frac{11}{6}\right)^2$. It is thus necessary to divide the 10 into three squares so that each of their sides is as close as possible to $\frac{11}{6}$. Now, 10 can be divided into three squares as $3^2 + \left(\frac{3}{5}\right)^2 + \left(\frac{4}{5}\right)^2$. To adjust these squares so that their sides are closer to $\frac{11}{6}$, we multiply each of the three roots by 30 to get 90, 18, and 24, and the $\frac{11}{6}$ by 30 to get 55. Now, $90 = 55 + 31$, $24 = 55 - 31$, and $18 = 55 - 37$, therefore, $\frac{55}{30} = 3 - \frac{35}{30} = \frac{4}{5} + \frac{31}{30} = \frac{3}{5} + \frac{37}{30}$. But we cannot put $3 - 35 \cdot \frac{1}{30}$ units for the side of $\square$, $\frac{4}{5} + 31 \cdot \frac{1}{30}$ units for the side of $\square'$, and $\frac{3}{5} + 37 \cdot \frac{1}{30}$ units for the side of $\square''$, for then the sum of the three squares would be greater than 10. So, the problem is reduced to that of finding an appropriate number to substitute $\frac{1}{30}$ in the expressions of the sides of the three squares. In modern notation: $10 = \left(3 - 35v\right)^2 + \left(\frac{4}{5} + 31v\right)^2 + \left(\frac{3}{5} + 37v\right)^2$.

Assignments	Operations	Equation
$v := 1N$		
	Square 3u ℓ 35N → 9u 1225P ℓ 210N	
	Square 31N $\frac{4}{5}$u → 961P $\frac{248}{5}$N $\frac{16}{25}$u	
	Square 37N $\frac{3}{5}$u → 1369P $\frac{222}{5}$N $\frac{9}{25}$u	
	Add the three results → 3555P 10u ℓ 116N	
		3555P 10u ℓ 116N = 10u

1N is found to be $\frac{116}{3555}$u, so the three squares are $\square = \left(\frac{1,321}{711}\right)^2$, $\square' = \left(\frac{1288}{711}\right)^2$, and $\square'' = \left(\frac{1285}{711}\right)^2$. Thus, $x = \frac{228,478}{505,521}$, $y = \frac{142,381}{505,521}$, and $z = \frac{134,662}{505,521}$.

V^G.12 $\quad \begin{cases} x + y + z = 1 \\ x + a = \square \\ y + b = \square' \\ z + c = \square'' \end{cases} \quad (a = 2, b = 3, c = 4)$

Diophantus merely outlines the process of solving this problem and never states values for the sought-after numbers. Gustav Wertheim proposed a reconstruction of the solution in Diophantus (1890, 216–20), repeated in Heath (1910, 209–10 n. 1), Ver Eecke (1959, 205–6), Czwalina (1952, 129ff.), Stamatis (1963, 500–04), and others. We adapt Wertheim's reconstruction to the format of this commentary.

Adding the last three conditions gives $10 = \square + \square' + \square''$. So, we are reduced to dividing 10 into three squares such that $\square > 2$, $\square' > 3$, and $\square'' > 4$. The task will be broken into two stages. First, we will divide 10 into two squares, one of which lies between 2 and 3, and then we will divide the remainder into two squares such that the smaller lies between 3 and 4.

The square that lies between 2 and 3 should be close to $2\frac{1}{2}$, so we look for a square whole number such that if we add its reciprocal to $2\frac{1}{2}$, it makes a square. In modern notation, $2\frac{1}{2} + \frac{1}{u^2} = \square'''$, or, taking four times each term, $10 + 4\frac{1}{u^2} = \square^{iv}$.

Assignments	Operations	Equation
$4\frac{1}{u^2} := 1P'$		
	Add $10u$ and $1P' \to 10u\ 1P'$	
		$10u\ 1P' = \square^{iv}$
	Multiply $10u\ 1P'$ by $1P \to 10P\ 1u$	
		$10P\ 1u = \square^{v}$
$\sqrt{\square^v} := 3N\,1u$		
	Square $3N\ 1u \to 9P\ 6N\ 1u$	
		$10P\ 1u = 9P$ $6N\ 1u$

$1N$ is then $6u$. Therefore, $u^2 = 144$, and $2\frac{1}{2} + \frac{1}{144} = \frac{361}{144} = \left(\frac{19}{12}\right)^2$. Now, since one of the two squares whose sum is 10 is close to $2\frac{1}{2}$, the other must be close to $7\frac{1}{2}$. So, in the same manner we look for a square whole number such that if its reciprocal is added to $7\frac{1}{2}$, it makes a square. This is a second subsidiary problem, of the type $7\frac{1}{2} + \frac{1}{v^2} = \square^{vi}$, or, taking four times everything, $30 + 4\frac{1}{v^2} = \square^{vii}$.

Assignments	Operations	Equation
$4\frac{1}{v^2} := 1P'$		
	Add $30u$ and $1P' \to 30u\ 1P'$	
		$30u\ 1P' = \square^{vii}$
	Multiply $30u\ 1P'$ by $1P \to 30P\ 1u$	
		$30P\ 1u = \square^{viii}$

$\sqrt{\square^{viii}} := 5\text{N}\,1\text{u}$

Square 5N 1u $\to$ 25P 10N 1u

30P 1u = 25P 10N 1u

1N is 2u, so $v^2 = 16$, and $7\frac{1}{2} + \frac{1}{16} = \frac{121}{16} = \left(\frac{11}{4}\right)^2 = \left(\frac{33}{12}\right)^2$.

Adding $\left(\frac{19}{12}\right)^2$ and $\left(\frac{33}{12}\right)^2$ is close to, but not exactly, 10. Comparing the roots respectively with 1 and 3, the roots of 1 and 9, we see that $\frac{19}{12} = 1 + 7 \cdot \frac{1}{12}$, and $\frac{33}{12} = 3 - 3 \cdot \frac{1}{12}$. So, as in Problem V.9, we must find a number to substitute $\frac{1}{12}$ in the expressions of the sides of the two squares. In modern notation, $10 = (1 + 7w)^2 + (3 - 3w)^2$.

Assignments	Operations	Equation
$w := 1\text{N}$		
	Square 1u 7N $\to$ 1u 14N 49P	
	Square 3u ℓ 3N $\to$ 9u 9P ℓ 18N	
	Add 1u 14N 49P and 9u 9P ℓ 18N $\to$ 58P 10uℓ 4N	
		58P 10uℓ 4N = 10u

1N is then $\frac{2}{29}$u, so the two squares are $\left(1 + 7 \cdot \frac{2}{29}\right)^2 = \left(\frac{43}{29}\right)^2 = \frac{1849}{841}$, and $\left(3 - 3 \cdot \frac{2}{29}\right)^2 = \left(\frac{81}{29}\right)^2 = \frac{6561}{841}$. The first of these squares lies between 2 and $2\frac{1}{2}$, so we adopt it as the value for $\square$.

We next need to divide the other square, $\frac{6561}{841}$, into two squares, one of which lies between 3 and 4. In our notation, $\frac{6561}{841} = \square' + \square''$, with $3 < \square' < 4$.

Assignments	Operations	Equation
$\square' := 1\text{P}$		
	Subtract 1P from $\frac{6561}{841}$u $\to$ $\frac{6561}{841}$u ℓ 1P	
		$\frac{6561}{841}$u ℓ 1P = $\square''$

We apply the procedure employed in Problem V.10 and we assign the side of $\square''$ to be an expression of the form $\frac{81}{29}$u ℓ Ns, where the multitude of Ns is to be chosen in such a manner that 1N, the solution to the resulting equation, lies between the roots of two squares that in turn lie between 3 and 4. We choose $\frac{7}{4}$ and $\frac{8}{4}$, which satisfy $3 \leq \left(\frac{7}{4}\right)^2 < \left(\frac{8}{4}\right)^2 \leq 4$. Working through the equation without specifying this multitude, 1N is found to be the result of dividing 162 times the multitude by 29 times one more than its square. That is, we seek a multitude m such that $\frac{7}{4} < \frac{162m}{29(m^2+1)} < \frac{8}{4}$. Working with the first inequality, and assigning $m := 1\text{N}$, we arrive at 203P 203u < 648N. Again, as in Problem V^G.10, we solve this as if it were an equation. The square of half of the 648 is 104,976. Subtracting the product of 203 (Powers) by 203 (units) gives 63,767, so the square root that we want is not greater than 252. Adding half of the 648 and then dividing by 203 gives $\frac{576}{203}$. Thus, $m \leq \frac{576}{203}$. For the other inequality, $\frac{162p}{29(p^2+1)} < \frac{8}{4}$, we again assign $m := 1\text{N}$, and arrive at 232P 232u > 648N. The discriminant this time is 51,152, whose square root is less than 227. Thus, we want $m \geq \frac{551}{232}$. Choosing $m = 2\frac{1}{2}$ for the multitude, the solution can be completed.

Assignments	Operations	Equation
		$\frac{6561}{841}$ u ℓ 1P $= \square''$
$\sqrt{\square''} := \frac{81}{29}$ u ℓ $2\frac{1}{2}$ N		
	Square $\frac{81}{29}$ u ℓ $\frac{81}{29}$ u ℓ $2\frac{1}{2}$ N	
	$\rightarrow \frac{6561}{841}$ u $6\frac{1}{4}$ P ℓ $\frac{405}{29}$ N	
		$\frac{6561}{841}$ u ℓ 1P $= \frac{6561}{841}$ u $6\frac{1}{4}$ P ℓ $\frac{405}{29}$ N

Thus, 1N is $\frac{1620}{841}$ u, whose square $\square' = \frac{2{,}624{,}400}{707{,}281}$ is a little over 3.71. Subtracting this from $\frac{6561}{841}$ gives $\square'' = \frac{2{,}893{,}401}{707{,}281} = \left(\frac{1701}{841}\right)^2$, and we had already found $\square = \frac{1849}{841}$. Subtracting 2, 3, and 4 from $\square$, $\square'$, and $\square''$ respectively gives $x = \frac{140{,}447}{707{,}281}$, $y = \frac{502{,}557}{707{,}281}$, and $z = \frac{64{,}277}{707{,}281}$.

$$V^G.13 \quad \begin{cases} x+y+z = a \\ x+y = \square \\ y+z = \square' \\ z+x = \square'' \end{cases} \quad (a=10)$$

Adding the last three conditions gives $\square + \square' + \square'' = 20$, and each square must be less than 10. The sum 20 can be divided into two squares as $4 + 16$. If we assign $\square := 4\text{u}$, we will need to divide the remaining 16 into two squares such that each of them is less than 10. How to do this was shown in Problem V^G.10. In Diophantus's words, "we learned how to divide a given square into two squares, in such a manner that one of them be greater than 6 units and smaller than 10 units". The text does not pursue the solution. The procedure we describe next is based on Wertheim's reconstruction (Diophantus 1890, 219–21).

We look for two squares, $\square'$ and $\square''$, such that $16 = \square' + \square''$, with $6 < \square' < 10$.

Assignments	Operations	Equation
$\square' := 1\mathrm{P}$		
	Subtract 1P from $1\mathrm{P} \to 16\mathrm{u}\,\ell\,1\mathrm{P}$	
		$16\mathrm{u}\,\ell\,1\mathrm{P} = \square''$

As before, we choose two squares that lie between our bounds of 6 and $10\colon 6 \le \left(\frac{5}{2}\right)^2 < 3^2 \le 10$. We assign the side of the $\square''$ to be of the form $4\mathrm{u}\,\ell\,N\mathrm{s}$, with the multitude of Ns chosen so that the solution 1N lies between $\frac{5}{2}$ and 3. If we call the multitude m, then we will find that $\frac{5}{2} < \frac{8m}{m^2+1} < 3$. Solving these two inequalities separately, we find that m lies between $\frac{7}{3}$ and $\frac{14}{5}$. Picking the multitude to be $2\frac{1}{2}$, the solution can be completed.

Assignments	Operations	Equation
		$16\mathrm{u}\,\ell\,1\mathrm{P} = \square''$
$\sqrt{\square''} := 4\mathrm{u}\,\ell\,2\frac{1}{2}\mathrm{N}$		
	Square $4\mathrm{u}\,\ell\,2\frac{1}{2}\mathrm{N} \to$ $16\mathrm{u}\,6\frac{1}{4}\mathrm{P}\,\ell\,20\mathrm{N}$	
		$16\mathrm{u}\,\ell\,1\mathrm{P} = 16\mathrm{u}\,6\frac{1}{4}\mathrm{P}\,\ell\,20\mathrm{N}$

1N is then $\frac{80}{29}\mathrm{u}$, so $\square' = \frac{6400}{841}$ and $\square'' = 16 - \frac{6400}{841} = \frac{705}{841}$. Subtracting these, and $\square = 4$, from 10, we get $x = 10 - \frac{6400}{841} = \frac{2010}{841}$, $y = 10 - \frac{7056}{841} = \frac{1354}{841}$, and $z = 10 - 4 = 6$.

$$\mathbf{V^G.14} \quad \begin{cases} w + x + y + z = a \\ w + x + y = \square \\ x + y + z = \square' \quad (a = 10) \\ y + z + w = \square'' \\ z + w + x = \square''' \end{cases}$$

Adding the last four conditions, we find that the sum of the four squares is 30, and each square must be less than 10. Diophantus proposes two solutions of this problem without pursuing them in detail. The first solution consists in finding four squares each of which is close to $7\frac{1}{2}$, and which can be done by following the method of quasi-equality illustrated in Problem $V^G.11$, and then subtracting each square from 10. In the second solution, considering the relation $30 = 4^2 + 3^2 + 2^2 + 1^2$, Diophantus assigns two of the squares to be 9 and 4 and he looks for two other squares, each of which is less than 10, and whose sum is 17. The problem is then reduced to that of dividing 17 into two squares, one of which lies between $8\frac{1}{2}$ and 10.

We follow Wertheim's procedure for the solution of finding two such squares in the second solution (Diophantus 1890, 221–23; Heath 1910, 212 n. 1). We look for a square whole number such that its reciprocal added to $8\frac{1}{2}$ makes a square. This is a subsidiary problem, written in modern notation as $8\frac{1}{2}+\frac{1}{u^2}=\square^{iv}$, or, taking four times everything, $34+4\frac{1}{u^2}=\square^v$.

Assignments	Operations	Equation
$4\frac{1}{u^2}:=1P'$		
	Add 34u and 1P' $\rightarrow$ 34u 1P'	
		34u 1P' $=\square^v$
	Multiply 34u 1P' by 1P $\rightarrow$ 34P 1u	
		34P 1u $=\square^{vi}$
$\sqrt{\square^{vi}}:=6N\,\ell\,1u$		
	Square 6N ℓ 1u $\rightarrow$ 36P 1u ℓ 12N	
		34P 1u = 36P 1u ℓ 12N

1N is found to be 6u, so $u^2 = 144$. Thus, we have $8\frac{1}{2}+\frac{1}{144}=\frac{1225}{144}=\left(\frac{35}{12}\right)^2$, which is a square close to $8\frac{1}{2}$. We then want to divide 17 into two squares such that their sides are close to $\frac{35}{12}$. Since $17 = 1^2 + 4^2$, we can substitute the two sides, 1 and 4, with two other sides. Now, $\frac{35}{12}=1+23\cdot\frac{1}{12}=4-13\cdot\frac{1}{12}$, so we should find a number to replace $\frac{1}{12}$ so that the sum of the squares is 17 (they are just a little off). This is a new subsidiary problem, which we write in modern notation as $17 = (1+23v)^2 + (4 - 13v)^2$.

Assignments	Operations	Equation
$v:=1N$		
	Add 1u and 23N $\rightarrow$ 1u 23N	
	Square 1u 23N $\rightarrow$ 1u 529P 46N	
	Subtract 13N from 4u $\rightarrow$ 4u ℓ 13N	
	Square 4u ℓ 13N $\rightarrow$ 16u 169P ℓ 104N	
	Add 1u 529P 46N and	
	16u 169P ℓ 104N $\rightarrow$ 698P 17u ℓ 58N	
		698P 17u ℓ 58N = 17u

1N is $\frac{29}{349}$u; therefore, $1+23\cdot\frac{29}{349}=\frac{1016}{349}$, whose square is $\frac{1,032,256}{121,801}$, and $4-13\cdot\frac{29}{349}=\frac{1019}{349}$, whose square is $\frac{1,038,361}{121,801}$. Thus, $2^2+3^2+\left(\frac{1016}{349}\right)^2+\left(\frac{1019}{349}\right)^2=30$, and by subtracting each of these squares from 10 we obtain $w=6, x=1, y=\frac{185,754}{121,801}$, and $z=\frac{179,649}{121,801}$.

$$\mathbf{V^G.15} \quad \begin{cases} (x+y+z)^3 + x = \boxdot \\ (x+y+z)^3 + y = \boxdot' \\ (x+y+z)^3 + z = \boxdot'' \end{cases}$$

The assignments are designed to satisfy the three conditions, and the equation is set up to reconcile the two ways of representing $x + y + z$.

Assignments	Operations	Equation
$x + y + z := 1N$		
$x := 7C$		
$y := 26C$		
$z := 63C$		
	Add 7C, 26C, and 63C $\to$ 96C	
		96C = 1N

The equation simplifies to 96P = 1u, and since 96 is not a square, the solution is not rational. So, the 96 must be replaced by a square number. Because the 96 arose as the sum of three numbers, any one of which increased by 1 unit makes a cube, the problem is reduced to finding three such numbers, and such that their sum is a square. In modern notation, $u + 1 = \boxdot'''$, $v + 1 = \boxdot^{iv}$, $w + 1 = \boxdot^{v}$, $u + v + w = \Box$. As Heath remarked (1910, 213 n. 1), this problem is solved with assignments for the sides of these cubes such that the Cube term in their sum vanishes, and the Power term is a square. The resulting forthcoming equation is then solved by *al-istiqrā'*.

Assignments	Operations	Equation
$\sqrt[3]{\boxdot'''} := 1N\,1u$		
$\sqrt[3]{\boxdot^{iv}} := 2u\,\ell\,1N$		
$\sqrt[3]{\boxdot^{v}} := 2u$		
	Cube 1N 1u $\to$ 1C 3P 3N 1u	
	Cube 2u ℓ 1N $\to$ 6P 8u ℓ 1C 12N	
	Cube 2u $\to$ 8u	
	Subtract 1u from 1C 3P 3N 1u $\to$ 1C 3P 3N	
	Subtract 1u from 6P 8u ℓ 1C 12N $\to$ 6P 7u ℓ 1C 12N	
	Subtract 1u from 8u $\to$ 7u	
$u := 1C$ 3P 3N		

$v := 6P\ 7u\ \ell\ 1C\ 12N$

$w := 7u$

Add 1C 3P 3N, 6P 7u ℓ 1C 12N, and 7u → 9P 14u ℓ 9N	
	9P 14u ℓ 9N = □

$\sqrt{\square} := 3N\ \ell\ 4u$

Square 3N ℓ 4u → 9P 16u ℓ 24N	
	9P 14u ℓ 9N = 9P 16u ℓ 24N

Thus, 1N is $\frac{2}{15}$ u, so $u = \frac{1538}{3375}$, $v = \frac{18,577}{3375}$, and $w = 7$. Returning to the original problem, we start the solution again with modified assignments.

Assignments	Operations	Equation
$x + y + z := 1N$		
$x := \frac{1538}{3375} C$		
$y := \frac{18,577}{3375} C$		
$z := 7C$		
	Add $\frac{1538}{3375} C, \frac{18,577}{3375} C, 7C \to \frac{43,740}{3375} C$	
		$\frac{43,740}{3375} C = 1N$

The equation simplifies to $2916P = 225u$, from which 1N is $\frac{15}{54}$ u. Then three sought-after numbers are $x = \frac{1538}{157,464}$, $y = \frac{18,577}{157,464}$, and $z = \frac{23,625}{157,464}$.

V^G.16
$$\begin{cases} (x+y+z)^3 - x = \boxdot \\ (x+y+z)^3 - y = \boxdot' \\ (x+y+z)^3 - z = \boxdot'' \end{cases}$$

The four assignments given next are again designed to satisfy the three conditions, and the equation is set up to reconcile the two ways of representing x + y + z.

Assignments	Operations	Equation
$x + y + z := 1N$		
$x := \frac{7}{8} C$		
$y := \frac{26}{27} C$		

$z := \frac{63}{64}C$

$\qquad$ Add $\frac{7}{8}C$, $\frac{26}{27}C$, and $\frac{63}{64}C \rightarrow \frac{4877}{1728}C$

$\qquad\qquad\qquad\qquad\qquad$ $\frac{4877}{1728}C = 1N$

The equation simplifies to $\frac{4877}{1728}P = 1u$. But since $\frac{4877}{1728}$ is not a square, we should look for a square number to replace it. As in the previous problem, we see where this number came from. It is the remainder from subtracting three cubes from three units, where each cube is less than one unit. The remainder, then, must be a square. Diophantus then writes that if we make the sum of the three numbers (chosen before to be $\frac{1}{8}$, $\frac{1}{27}$, and $\frac{1}{64}$) to be less than 1, then each of them individually will be much less than 1, and the square obtained by subtracting this sum from 3 will be greater than 2. While it is not true that each of them will be much less than 1 (as the solution below shows), it is true that subtracting their sum from 3 will give a remainder between 2 and 3.

Diophantus assigns the square to be $2\frac{1}{4}$. Thus, we look for three cubes whose sum is $3 - 2\frac{1}{4} = \frac{1}{4}$. He re-expresses this fraction as $\frac{162}{216}$ so that its denominator is a cube, and he then looks for three cubes that add up to 162. Knowing that 162 is the sum of 53 and 43 − 33, he cites a theorem in the *Porisms* that the difference of two cubes is also the sum of two cubes. Diophantus does not give these two cubes, and he only outlines the rest of the solution. François Viète gave a general solution to this problem in Zetetic IV.18 (Viète 1646, 74–75), and Heath (1910, 214 n. 2) completes Diophantus's problem with the numbers obtained from Viète's solution: $4^3 - 3^3 = \left(\frac{303}{91}\right)^3 + \left(\frac{40}{91}\right)^3$. We have then $\frac{3}{4} = \frac{162}{216} = \left(\frac{5}{6}\right)^3 + \left(\frac{101}{182}\right)^3 + \left(\frac{20}{273}\right)^3$. Subtracting each of these cubes from 1 gives $\frac{91}{216}$, $\frac{4,998,267}{6,028,265}$, and $\frac{20,338,417}{20,346,417}$, which are then applied to the assignments for x, y, and z.

Assignments	Operations	Equation
$x + y + z := 1N$		
$x := \frac{91}{216}C$		
$y := \frac{4,998,267}{6,028,568}C$		
$z := \frac{20,338,417}{20,346,417}C$		
	Add $\frac{91}{216}C$, $\frac{4,998,267}{6,028,568}C$, and $\frac{20,338,417}{20,346,417}C \rightarrow 2\frac{1}{4}C$	
		$2\frac{1}{4}C = 1N$

1N is $\frac{2}{3}u$, so 1C is $\frac{8}{27}u$, from which we find $x = \frac{91}{729}$, $y = \frac{185,121}{753,571}$, and $z = \frac{162,707,336}{549,353,259}$.

$$\mathbf{V^G.17} \quad \begin{cases} x - (x+y+z)^3 = \boxdot \\ y - (x+y+z)^3 = \boxdot' \\ z - (x+y+z)^3 = \boxdot'' \end{cases}$$

As in the last two problems, the first set of assignments is designed to satisfy the three conditions, and the equation is formed from the two ways of naming $x + y + z$.

Assignments	Operations	Equation
$x + y + z := 1N$		
$x := 2C$		
$y := 9C$		
$z := 28C$		
	Add 2C, 9C, and 28C $\rightarrow$ 39C	
		$39C = 1N$

The equation simplifies to 39P = 1u, but its solution is not rational because 39 is not a square. This 39 is the result of adding 3 to three cubes, so we need to find three other cubes that will give a square after adding 3 to their sum: $u^3 + v^3 + w^3 + 3 = \square$. The following assignments are designed in such a way that the Cubes vanish and the Powers are a square, and the equation is completed with an al-istiqrā' assignment.

Assignments	Operations	Equation
$u := 1N$		
$v := 3u \ell 1N$		
$w := 1u$		
	Add the cubes of 1N, $3u \ell 1N$, $1u \rightarrow 9P\ 28u \ell\ 27N$	
	Add 3u to 9P 28u ℓ 27N $\rightarrow$ 9P 31u ℓ 27N	
		9P 31u ℓ 27N $= \square$
$\sqrt{\square} := 3N \ell 7u$		
	Square $3N \ell 7u \rightarrow$ 9P 49u ℓ 42N	
		9P 31u ℓ 27N $=$ 9P 49u ℓ 42N

1N is $\frac{6}{5}$u, so $u = \frac{6}{5}$, $v = \frac{9}{5}$, and w = 1, and their cubes are $\frac{216}{125}$, $\frac{729}{125}$, and 1, respectively. We return to the original problem, making the number of Cubes in the assignments for x, y, and z one more than these cubes.

Assignments	Operations	Equation
$x + y + z := 1N$		
$x := \frac{341}{125}C$		
$y := \frac{854}{125}C$		

$$z := 2C = \tfrac{250}{125}\,C$$

Add $\tfrac{341}{125}\,C$, $\tfrac{854}{125}\,C$, and $\tfrac{250}{125}\,C \to \tfrac{289}{25}\,C$

$\tfrac{289}{25}\,C = 1N$

1N is then $\tfrac{5}{17}\,u$, so 1C is $\tfrac{125}{4913}$, and $x = \tfrac{341}{4913}$, $y = \tfrac{854}{4913}$, and $z = \tfrac{250}{4913}$.

V^{G}.18
$$\begin{cases} x + y + z = \square \\ (x+y+z)^3 + x = \square' \\ (x+y+z)^3 + y = \square'' \\ (x+y+z)^3 + z = \square''' \end{cases}$$

The following four assignments satisfy the four conditions, and again the equation is set up to reconcile the two ways of representing $x + y + z$.

Assignments	Operations	Equation
$\square := 1P$		
$x := 3CC$		
$y := 8CC$		
$z := 15CC$		
	Add 3CC, 8CC, and 15CC $\to$ 26CC	
		26CC = 1P

The equation simplifies to 26PP = 1u, but because 26 is not a perfect fourth power, the solution is not rational. So, we need to find three numbers, call them u, v, and w, such that their sum is a square with a square side, and adding 1 to any of them makes a square. Diophantus adopts the assignments $u := 1PP\ \ell\ 2P$, $v := 1P\ 2N$, $w := 1P\ \ell\ 2N$, whose sum is 1PP, and adding 1u to any of them makes a square. Thus, the subsidiary problem is solved indeterminately. By assigning 1N to be 3u, he gets $u = 63$, $v = 15$, and $w = 3$. These numbers are now applied to the assignments for x, y, and z.

Assignments	Operations	Equation
$\square := 1P$		
$x := 63CC$		
$y := 15CC$		
$z := 3CC$		
	Add 63CC, 15CC, and 3CC $\to$ 81CC	
		81CC = 1P

Now 1N is found to be $\frac{1}{3}$u, so $x = \frac{63}{729}$, $y = \frac{15}{729}$, and $z = \frac{3}{729}$.

$$V^G.19 \quad \begin{cases} x+y+z = \square \\ (x+y+z)^3 - x = \square' \\ (x+y+z)^3 - y = \square'' \\ (x+y+z)^3 - z = \square''' \end{cases}$$

The solution of this problem is missing, but it can be reconstructed. For our translation of the reconstruction of the text by E. S. Stamatis (1961, 1963), see Appendix 1. Our commentary is based on that solution. The assignments follow the
pattern of the previous few problems.

Assignments	Operations	Equation
$x+y+z := 1P$		
$x := \frac{3}{4}CC$		
$y := \frac{8}{9}CC$		
$z := \frac{15}{16}CC$		
	Add $\frac{3}{4}CC$, $\frac{8}{9}CC$, and $\frac{15}{16}CC \rightarrow \frac{371}{144}CC$	
		$\frac{371}{144}CC = 1P$

The equation simplifies to $\frac{371}{144}PP = 1u$, but its solution is not rational because $\frac{371}{144}$
is not a square having a square side. So, we need to find three squares such that
removing their sum from 3 units leaves a perfect fourth power. If we require
that the squares together to be less than 1, then the remainder will lie between
2 and 3. We assign the square which has a square side to be $\left(\frac{6}{5}\right)^4 = \frac{1296}{625}$, and we
look for three squares whose sum is $3 - \frac{1296}{625} = \frac{579}{625}$. We find that $\frac{529}{625}$, $\frac{49}{625}$, and $\frac{1}{625}$ are
three such squares. We subtract each of these from 1 to get $\frac{96}{625}$, $\frac{576}{625}$, and $\frac{624}{625}$, and
we return to the initial problem substituting these numbers in the assignments
for x, y, and z.

Assignments	Operations	Equation
$x+y+z := 1P$		
$x := \frac{96}{625}CC$		
$y := \frac{576}{625}CC$		
$z := \frac{624}{625}CC$		
	Add $\frac{96}{625}CC$, $\frac{576}{625}CC$, and $\frac{624}{625}CC \rightarrow \frac{1296}{625}CC$	
		$\frac{1296}{625}CC = 1P$

Now 1N is found to be $\frac{5}{6}$u, so 1CC is $\frac{15,625}{46,656}$u. Therefore, $x = \frac{25}{486}$, $y = \frac{25}{81}$, and $z = \frac{325}{972}$.

Problems V^G.19a and V^G.19b are missing, and only the last part of V^G.19c is preserved. For our translation of the reconstruction of the text by E. S. Stamatis, see Appendix 1. Next we outline these solutions.

$$V^G.19a \quad \begin{cases} x+y+z=\square \\ x-(x+y+z)^3 =\square' \\ y-(x+y+z)^3 =\square'' \\ z-(x+y+z)^3 =\square''' \end{cases}$$

The following four assignments satisfy the four conditions, and the equation is formed based in the two ways of naming x + y + z.

Assignments	Operations	Equation
$x+y+z := 1P$		
$x := 2CC$		
$y := 5CC$		
$z := 10CC$		
	Add 2CC, 5CC, and 10CC → 17CC	
		17CC = 1P

The equation simplifies to 17PP = 1u, and again the 17 should be a square with a square side in order to get a rational solution. The 17 comes from the sum of three squares with three units. We assign the square with a square side to be 16, so we look for three squares whose sum is 16 − 3 = 13. Such squares are 9, $\frac{36}{25}$, and $\frac{64}{25}$. Adding 1 to each gives 10, $\frac{61}{25}$, and $\frac{89}{25}$. These numbers will replace those in the first set of assignments.

Assignments	Operations	Equation
$x+y+z := 1P$		
$x := 10CC$		
$y := \frac{61}{25}CC$		
$z := \frac{89}{25}CC$		
	Add 10CC, $\frac{61}{25}$CC, and $\frac{89}{25}$CC → 16CC	
		16CC = 1P

The equation simplifies to 16PP = 1u, so 1N is $\frac{1}{2}$u and 1CC is $\frac{1}{64}$u. Thus, $x = \frac{5}{32}$, $y = \frac{61}{1600}$, and $z = \frac{89}{1600}$.

$$\mathbf{V^G.19b} \quad \begin{cases} x+y+z=a \\ (x+y+z)^3 + x = \square \\ (x+y+z)^3 + y = \square' \\ (x+y+z)^3 + z = \square'' \end{cases} \quad (a=2)$$

Since the cube of 2 is 8, adding the last three conditions gives $26 = \square + \square' + \square''$, each square being greater than 8. If we assign $\square$ to be 9u, we should then divide the difference $26 - 9 = 17$ into the sum of the squares $\square'$ and $\square''$. According to the technique employed in Problem V.9 we take the number $8\frac{1}{2}$u, half of 17, and we look for a square whole number such that if its reciprocal is added to this number it makes a square. In our notation, $8\frac{1}{2}u + \frac{1}{u^2} = \square'$, or, multiplying by four, $34 + 4\frac{1}{u^2} = \square'$.

Assignments	Operations	Equation
$4\frac{1}{u^2} := 1P'$		
	Add 34u and 1P' → 34u 1P'	
		$34u\ 1P' = \square^{iv}$
	Multiply 34u 1P' by 1P → 34P 1u	
		$34P\ 1u = \square^{v}$
$\sqrt{\square^{v}} := 1u\ \ell\ 6N$		
	Square 1u ℓ 6N → 1u 36P ℓ 12N	
		$34P\ 1u = 36P$
		$1u\ \ell\ 12N$

1N is 6u, so $\frac{1}{u^2} = \frac{1}{144}$, and $\square''' = \frac{1225}{144} = \left(\frac{35}{12}\right)^2$. So, as in Problem V.9, we should divide 17 into squares $\square'$ and $\square''$ so that the sides of each of them are as close as possible to $\frac{35}{12}$. Now, $17 = 42 + 12$, so we have $4 - 13 \cdot \frac{1}{12} = \frac{35}{12}$ units, and $1 + 23 \cdot \frac{1}{12} = \frac{35}{12}$ units. The problem is then reduced to finding a number to replace $\frac{1}{12}$ in the expressions of the sides of the two squares. In our notation: $17 = (4 - 13v)^2 + (1 + 23v)^2$.

Assignments	Operations	Equation
$v := 1N$		
	Square 4u ℓ 13N → 16u 169P ℓ 104N	
	Square 1u 23N → 1u 529P 46N	
	Add these two squares → 698P 17u ℓ 58N	
		$698P\ 17u\ \ell$
		$58N = 17u$

1N is found to be $\frac{29}{349}$ u, so $\square' = \left(\frac{1019}{349}\right)^2 = \frac{1,038,361}{121,801}$, $\square'' = \left(\frac{1016}{349}\right)^2 = \frac{1,032,256}{121,801}$, while $\square = 9$.

Subtracting 8 from each of these, we get $x = \frac{63,953}{121,801}$, $y = \frac{57,848}{121,801}$, and $z = 1$.

V^G.19c
$$\begin{cases} x+y+z = a \\ \left(x+y+z\right)^3 - x = \square \\ \left(x+y+z\right)^3 - y = \square' \\ \left(x+y+z\right)^3 - z = \square'' \end{cases} \quad (a = 2)$$

Adding the last three conditions gives $22 = \square + \square' + \square''$, with each square being less than 8. Diophantus observes that this "22 should be divided into three squares such that each of them is greater than 6 units". He does not find such squares, but only explains how to find the solution from them. We have solved the problem following Wertheim, as related in (Heath 1910, 216 n. 1). In this solution the squares are each greater than 7.

We take the number $7\frac{1}{3}$, a third of 22, and we look for a square whole number such that if we add its reciprocal to $7\frac{1}{3}$ it makes a square. In our notation, $7\frac{1}{3} + \frac{1}{u^2} = \square'''$, or, multiplying by 9, $66 + 9\frac{1}{u^2} = \square^{iv}$.

Assignments	Operations	Equation
$9\frac{1}{u^2} := 1P'$		
	Add 66u and 1P' → 66u 1P'	
		$66u\ 1P' = \square^{iv}$
	Multiply 66u 1P' by 1P → 66P 1u	
		$66P\ 1u = \square^v$
$\sqrt{\square^v} := 1u\,8N$		
	Square 1u 8N → 1u 64P 16N	
		$66P\ 1u = 64P$ $16N\ 1u$

1N is 8 units, so $\frac{1}{u^2} = \frac{1}{576}$, and $\square'''' = \frac{4225}{576} = \left(\frac{65}{24}\right)^2$. We would like the three squares $\square$, $\square'$, and $\square''$ to be close to $\frac{4225}{576}$, and that they add up to 22. We can partition 22 into the sum $2^2 + \left(\frac{3}{5}\right)^2 + \left(\frac{21}{5}\right)^2$, so we look to replace these three squares with three other squares whose sides are close to $\frac{65}{24}$. But $\frac{65}{24} = 2 + 85 \cdot \frac{1}{120}$, $\frac{56}{24} = \frac{3}{5} + 253 \cdot \frac{1}{120}$, and $\frac{64}{24} = \frac{21}{5} - 179 \cdot \frac{1}{120}$. We have, therefore, to replace the fraction $\frac{1}{120}$ by another number, say v, such that $22 = \left(2 + 85v\right)^2 + \left(\frac{3}{5} + 253v\right)^2 + \left(\frac{21}{5} - 179v\right)^2$.

Assignments	Operations	Equation
$v := 1N$		
	Square 2u 85N → 4u 7225P 340N	

Square $\frac{3}{5}$u 253N $\rightarrow$ $\frac{9}{25}$u 64,009P $\frac{1518}{5}$ N

Square $\frac{21}{5}$u ℓ 179N $\rightarrow$ $\frac{441}{25}$u 32,041P ℓ $\frac{7518}{5}$ N

Add the three squares $\rightarrow$ 22u 103,275P ℓ 860N

22u 103,275P
ℓ 860N = 22u

1N is $\frac{172}{20,655}$u, making $\square = \left(\frac{55,930}{20,655}\right)^2 = \frac{3,128,164,900}{426,629,025}$, $\square' = \left(\frac{55,909}{20,655}\right)^2 = \frac{3,125,816,281}{426,629,025}$, and $\square = \left(\frac{55,963}{20,655}\right)^2 = \frac{3,131,857,369}{426,629,025}$. Subtracting these numbers from 8, we get $x = \frac{284,867,300}{426,629,025}$, $y = \frac{287,215,919}{426,629,025}$, and $z = \frac{281,174,831}{426,629,025}$.

V^G.20 $\begin{cases} x + y + z = \frac{1}{a} \\ x - \left(x + y + z\right)^3 = \square \\ y - \left(x + y + z\right)^3 = \square' \\ z - \left(x + y + z\right)^3 = \square'' \end{cases}$ $(a = 4)$

Adding the last three conditions of the problem gives $\frac{1}{4} - \frac{3}{64} = \frac{13}{64} = \square + \square' + \square''$. We know that $\frac{13}{64}$ can be written as the sum of two squares $\frac{9}{64} + \frac{4}{64}$, so we need to divide one of these squares into the sum of two other squares. Following the solution to Problem II.8, we divide $\frac{4}{64}$ into $\frac{1}{25} + \frac{9}{400}$. Therefore, $\frac{13}{64} = \frac{9}{64} + \frac{1}{25} + \frac{9}{400}$. By adding $\frac{1}{64}$ to each one of these numbers we find $x = \frac{10}{64}$, $y = \frac{89}{1600}$, and $z = \frac{61}{1600}$.

V^G.21 $\begin{cases} x^2 \cdot y^2 \cdot z^2 + x^2 = \square \\ x^2 \cdot y^2 \cdot z^2 + y^2 = \square' \\ x^2 \cdot y^2 \cdot z^2 + z^2 = \square'' \end{cases}$

Diophantus considers the auxiliary problem of finding three squares such that any one of them increased by 1 unit makes a square. This problem can be solved by means of the Pythagorean theorem, since $a^2 + b^2 = c^2$ implies that $\frac{a^2}{b^2} + 1 = \frac{c^2}{b^2}$. From the right triangles (3, 4, 5), (5, 12, 13), (8, 15, 17) we get $\frac{9}{16} + 1 = \frac{25}{16}$, $\frac{25}{144} + 1 = \frac{169}{144}$, and $\frac{64}{225} + 1 = \frac{289}{225}$. If we make these numbers multitudes of a Power, we get $\frac{9}{16}P + 1P = \frac{25}{16}P$, $\frac{25}{144}P + 1P = \frac{169}{144}P$, and $\frac{64}{225}P + 1P = \frac{289}{225}P$. On the basis of these relations Diophantus makes the assignments for the original problem.

Assignments	Operations	Equation
$x^2 \cdot y^2 \cdot z^2 := 1P$		
$x^2 := \frac{9}{16}P$		
$y^2 := \frac{25}{144}P$		
$z^2 := \frac{64}{255}P$		
	Multiply $\frac{9}{16}P$, $\frac{25}{144}P$, and $\frac{64}{225}P \rightarrow \frac{14,400}{518,400}CC$	
		$\frac{14,400}{518,400}CC = 1P$

From here Diophantus divides by 1P and then takes the square roots of both sides to get $\frac{120}{720}P = 1u$. If $\frac{120}{720}$ were a square, the problem would be solved. It is not, so we must replace it with a square number. But $\frac{120}{720} = \frac{3}{4} \cdot \frac{5}{12} \cdot \frac{8}{15}$, where 3, 5, 8 and 4, 12, 15 are, respectively, the perpendiculars and the bases of three right triangles. Thus, if b_is represent the bases and p_is the perpendiculars, we would like to find three triangles such that $\frac{b_1}{p_1} \cdot \frac{b_2}{p_2} \cdot \frac{b_3}{p_3}$ is a square, which implies also that the product $b_1 b_2 b_3 \cdot p_1 p_2 p_3$ will be a square. If one of the triangles is assumed to be the triangle (3, 4, 5), then we should make the number $12 b_2 b_3 p_2 p_3$ a square, and this will be true if $b_3 p_3$ is 12 times $b_2 p_2$. It is enough, then, that the area of one triangle be triple the area of the other. This, says Diophantus, is easy, though Heath's footnote describing Bachet's and Fermat's solutions occupy nearly a page (Heath 1910, 218 n. 1). Diophantus proposes the triangles (9, 40, 41) and (8, 15, 17), or any other triangles similar to them. Indeed, $3\left(\frac{1}{2}9 \cdot 40\right)\left(\frac{1}{2}8 \cdot 15\right) = 180^2$. So, we have three right triangles that give $\frac{3}{4} \cdot \frac{9}{40} \cdot \frac{15}{8} = \left(\frac{9}{16}\right)^2$ (he reverses the order of the two legs on the last triangle). This gives new assignments for the problem.

Assignments	Operations	Equation
$x^2 \cdot y^2 \cdot z^2 := 1P$		
$x^2 := \frac{9}{16}P$		
$y^2 := \frac{225}{64}P$		
$z^2 := \frac{81}{1600}P$		
	Multiply $\frac{9}{16}P$, $\frac{225}{64}P$, $\frac{81}{1600}P \rightarrow \frac{6561}{65,536}CC$	
		$\frac{6561}{65,536}CC = 1P$

Now 1N can be found, and it is $\frac{16}{9}$ u. Therefore, $x^2 = \frac{16}{9}$, $y^2 = \frac{100}{9}$, and $z^2 = \frac{4}{25}$.

V^G.22
$$\begin{cases} x^2 \cdot y^2 \cdot z^2 - x^2 = \square \\ x^2 \cdot y^2 \cdot z^2 - y^2 = \square' \\ x^2 \cdot y^2 \cdot z^2 - z^2 = \square'' \end{cases}$$

Following the same procedure as in the last problem, Diophantus considers three right triangles, (3, 4, 5), (5, 12, 13), (8, 15, 17), from which we get $1 - \left(\frac{4}{5}\right)^2 = \left(\frac{3}{5}\right)^2$, $1 - \left(\frac{5}{13}\right)^2 = \left(\frac{12}{13}\right)^2$, and $1 - \left(\frac{8}{17}\right)^2 = \left(\frac{15}{17}\right)^2$. Writing these relations in terms of a Power, $1P - \frac{16}{25}P = \frac{9}{25}P$, $1P - \frac{25}{169}P = \frac{144}{169}P$, and $1P - \frac{64}{289}P = \frac{225}{289}P$. Diophantus makes his assignments on the basis of these relations, which are structured like the conditions of the problem.

Assignments	Operations	Equation
$x^2 \cdot y^2 \cdot z^2 := 1P$		
$x^2 := \frac{16}{25}P$		
$y^2 := \frac{25}{169}P$		
$z^2 := \frac{64}{289}P$		
	Multiply $\frac{16}{25}P$, $\frac{25}{169}P$, $\frac{64}{289}P \rightarrow \frac{25,600}{1,221,025}$ CC	
		$\frac{25,600}{1,221,025}$ CC $= 1P$

The equation simplifies to $\frac{160}{1105}P = 1u$. If $\frac{160}{1105}$ were a square number, the problem would be solved. Since it is not, we must replace it with a square number. This $\frac{160}{1105}$ comes from $\frac{4}{5} \cdot \frac{5}{13} \cdot \frac{8}{17}$, where 4, 5, 8 and 5, 13, 17 are, respectively, the perpendiculars and the hypotenuses of three right triangles. If p_is represent the heights and h_is the hypotenuses, $\frac{p_1}{h_1} \cdot \frac{p_2}{h_2} \cdot \frac{p_3}{h_3}$ must be a square, therefore, the product $p_1p_2p_3 \cdot h_1h_2h_3$ must be a square. If one of the triangles is assumed to be the triangle (3, 4, 5), then we need $20p_2p_3h_2h_3$ to be a square. So, we are led to the problem of finding two right triangles such that $5 \cdot p_2h_2 \cdot p_3h_3$ is a square number. Diophantus writes that finding two such triangles "is easy", and they can be found from the two triangles (5, 12, 13) and (3, 4, 5). The reader is evidently expected to know the process, which may have been given in the *Porisms*, or perhaps in an earlier, lost book of the *Arithmetica*. Heath summarizes the solution, first given by Otto Schulz (1822, 546–51):

> With a view to this we have first (cf. the last proposition) to find two right-angled triangles such that, if x_1, y_1 are the two *perpendiculars* in one and x_2, y_2 the two *perpendiculars* in the other, $x_1y_1 = 5x_2y_2$. From such a pair of triangles we can form two more right-angled triangles such that the product of

the *hypotenuse* and *one perpendicular* in one is five times the product of the *hypotenuse* and *one perpendicular* in the other.

Since the triangles found satisfying the relation $x_1y_1 = 5x_2y_2$ are (5, 12, 13) and (3, 4, 5) respectively, we have in fact to find two new right-angled triangles from them, namely the triangles (h_1, p_1, b_1) and (h_2, p_2, b_2), such that $h_1p_1 = 30$ and $h_2p_2 = 6$, the numbers 30 and 6 being the areas of the two triangles mentioned.

These triangles are $\left(6\frac{1}{2}, \frac{60}{13}, \left[\frac{119}{26}\right]\right)$ and $\left(2\frac{1}{2}, \frac{12}{5}, \left[\frac{7}{10}\right]\right)$ respectively.

(Heath 1910, 219–20 (his emphases))

Thus, we have the "least[6] of the similar triangles" (3, 4, 5), (119, 120, 169), (7, 24, 25), from which we get $1-\left(\frac{4}{5}\right)^2 = \left(\frac{3}{5}\right)^2$, $1-\left(\frac{120}{169}\right)^2 = \left(\frac{119}{169}\right)^2$, and $1-\left(\frac{24}{25}\right)^2 = \left(\frac{7}{25}\right)^2$, so $1P - \frac{16}{25}P = \frac{9}{25}P$, $1P - \frac{14{,}400}{28{,}561}P = \frac{14{,}161}{28{,}561}P$, and $1P - \frac{576}{625}P = \frac{49}{625}P$. These give the new assignments.

Assignments	Operations		Equation
$x^2 \cdot y^2 \cdot z^2$ $:= 1P$			
$x^2 := \frac{16}{25}P$			
$y^2 := \frac{14{,}400}{28{,}651}P$			
$z^2 := \frac{576}{625}P$			
	Multiply $\frac{15}{25}P$, $\frac{14{,}400}{28{,}561}P$, and $\frac{576}{625}P \rightarrow$	$\frac{132{,}710{,}400}{446{,}265{,}625}$ CC	
			$\frac{132{,}710{,}400}{446{,}265{,}625}$ CC $= 1P$

1N is $\frac{65}{48}$ u, so $x^2 = \frac{169}{144}$, $y^2 = \frac{625}{676}$, and $z^2 = \frac{169}{100}$.

V^G.23 $\quad \begin{cases} x^2 - x^2 \cdot y^2 \cdot z^2 = \square \\ y^2 - x^2 \cdot y^2 \cdot z^2 = \square' \\ z^2 - x^2 \cdot y^2 \cdot z^2 = \square'' \end{cases}$

The solution of this problem is similar to that of the previous problem. Given $a^2 + b^2 = c^2$, then $\frac{c^2}{b^2} - 1 = \frac{a^2}{b^2}$. Using the same right triangles as before, we get $\left(\frac{5}{4}\right)^2 - 1 = \left(\frac{3}{4}\right)^2$, $\left(\frac{25}{24}\right)^2 - 1 = \left(\frac{7}{24}\right)^2$, and $\left(\frac{169}{120}\right)^2 - 1 = \left(\frac{119}{120}\right)^2$. These give the assignments for x^2, y^2, and z^2.

6 By "least" he means that the sides are relatively prime whole numbers.

Assignments	Operations	Equation
$x^2 \cdot y^2 \cdot z^2$ $:= 1P$		
$x^2 := \frac{25}{16}P$		
$y^2 := \frac{625}{576}P$		
$z^2 := \frac{28,561}{14,400}P$		
	Multiply $\frac{25}{16}P$, $\frac{625}{576}P$, and $\frac{28,561}{14,400}P \to \frac{446,265,625}{132,710,400}$ CC	
		$\frac{446,265,625}{132,710,400}$ CC $= 1P$

1N is $\frac{48}{65}$ u, so $x^2 = \frac{144}{169}$, $y^2 = \frac{100}{169}$, and $z^2 = \frac{676}{625}$.

$V^G.24$ $\begin{cases} x^2 \cdot y^2 + 1 = \square \\ y^2 \cdot z^2 + 1 = \square' \\ z^2 \cdot x^2 + 1 = \square'' \end{cases}$

By multiplying the three conditions, respectively, by z^2, x^2, and y^2, the problem is reduced to Problem $V^G.21$. Therefore, the numbers found in that problem also satisfy this problem.

$V^G.25$ $\begin{cases} x^2 \cdot y^2 - 1 = \square \\ y^2 \cdot z^2 - 1 = \square' \\ z^2 \cdot x^2 - 1 = \square'' \end{cases}$

Similarly, this problem is reduced to Problem $V^G.22$.

$V^G.26$ $\begin{cases} 1 - x^2 \cdot y^2 = \square \\ 1 - y^2 \cdot z^2 = \square' \\ 1 - z^2 \cdot x^2 = \square'' \end{cases}$

Again, this problem is reduced to Problem $V^G.23$.

$V^G.27$ $\begin{cases} x^2 + y^2 + a = \square \\ y^2 + z^2 + a = \square' \quad (a = 15) \\ z^2 + x^2 + a = \square'' \end{cases}$

Diophantus posits $z^2 := 9u$, so the problem becomes $\begin{cases} x^2 + y^2 + 15 = \square \\ y^2 + 24 = \square' \\ x^2 + 24 = \square'' \end{cases}$.

In both the second and third conditions, the 24 is the difference of two squares, which in turn is the product of the sum and difference of their roots. So, Diophantus factors 24 into $4N' \cdot 6N$ and $3N'' \cdot 8N$. In the first case, the smaller root is either $2N' \ell 3N$ or $3N \ell 2N'$, depending whether $2N'$ or $3N$ is greater. Diophantus does not say, and it will not matter because he will square the difference anyway. So, we express "the difference" between the two roots as $|2N' \ell 3N|$, and this will be the assignment for y. Similarly, from $3N' \cdot 8N$, x will be assigned to be $|1\frac{1}{2}N' \ell 4N|$. The equation is established by *al-istiqrā'*.

Assignments	Operations	Equation
$z^2 \coloneqq 9u$		
$x \coloneqq |2N' \ell 3N|$		
$y \coloneqq |1\frac{1}{2}N' \ell 4N|$		
	Square $|2N' \ell 3N| \rightarrow 4P' 9P \ell 12u$	
	Square $|1\frac{1}{2}N' \ell 4N| \rightarrow 2\frac{1}{4}P' 16P \ell 12u$	
	Add the two squares $\rightarrow 6\frac{1}{4}P' 25P \ell 24u$	
	Add 15u to $6\frac{1}{4}P' 25P \ell 24u \rightarrow$ $6\frac{1}{4}P' 25P \ell 9u$	
		$6\frac{1}{4}P' 25P \ell 9u = \Box$
$\Box \coloneqq 25P$		
		$6\frac{1}{4}P' 25P \ell 9u = 25P$

1N is $\frac{5}{6}u$, so it turns out that 3N is greater than 2N', and 4N is greater than $1\frac{1}{2}N'$. The solution is $x^2 = \frac{1}{100}$, $y^2 = \frac{529}{225}$, and $z^2 = 9$.

$$\mathbf{V^G.28} \quad \begin{cases} x^2 + y^2 - a = \Box \\ y^2 + z^2 - a = \Box' \quad (a = 13) \\ z^2 + x^2 - a = \Box''' \end{cases}$$

Diophantus posits $z^2 \coloneqq 25u$, so the problem becomes $\begin{cases} x^2 + y^2 - 13 = \Box \\ y^2 + 12 = \Box' \\ x^2 + 12 = \Box'' \end{cases}$

Again, in both the second and the third conditions, 12 is the difference of two squares, which is the product of the difference by the sum of their roots. This time Diophantus chooses to factor the 12 into $3N \cdot 4N'$ and $4N \cdot 3N'$. From the first, he assigns y to be $|1\frac{1}{2}N \ell 2N'|$ and from the second, x is $|2N \ell 1\frac{1}{2}N'|$.

Assignments	Operations	Equation
$z^2 \coloneqq 25u$		
$y \coloneqq |1\frac{1}{2}N \ell 2N'|$		
$x \coloneqq |2N \ell \frac{3}{2}N'|$		

Square $\left|1\frac{1}{2}\,\text{N}\,\ell\,2\text{N}'\right| \rightarrow 2\frac{1}{4}\,\text{P4P}'\,\ell\,6\text{u}$

Square $\left|2\text{N}\,\ell\,1\frac{1}{2}\,\text{N}'\right| \rightarrow 4\text{P}2\frac{1}{4}\,\text{P}'\,\ell\,6\text{u}$

Add the two squares $\rightarrow 6\frac{1}{4}\,\text{P}6\frac{1}{4}\,\text{P}'\,\ell\,12\text{u}$

Subtract 13u from this

$\rightarrow 6\frac{1}{4}\,\text{P}6\frac{1}{4}\,\text{P}'\,\ell\,25\text{u}$

$6\frac{1}{4}\,\text{P}6\frac{1}{4}\,\text{P}'\,\ell\,25\text{u} = \square$

$\square := 6\frac{1}{4}\,\text{P}'$

$6\frac{1}{4}\,\text{P}6\frac{1}{4}\,\text{P}'\,\ell\,25\text{u} = 6\frac{1}{4}\,\text{P}$

1N is 2u; therefore, $x^2 = \frac{169}{16}$, $y^2 = 4$, and $z^2 = 25$.

V^G.29 $\left(x^2\right)^2 + \left(y^2\right)^2 + \left(z^2\right)^2 = \square$

Assignments	Operations	Equation
$x^2 := 1\text{P}$		
$y^2 := 4\text{u}$		
$z^2 := 9\text{u}$		
	Add the squares of 1P, 4u, 9u $\rightarrow$ 1PP 97u	
		1PP 97u $= \square$
$\sqrt{\square} := 1\text{P}\,\ell\,10\text{u}$		
	Square 1P ℓ 10u $\rightarrow$ 1PP 100u ℓ 20P	
		1PP 97u = 1PP 100u ℓ 20P

The equation simplifies to 20P = 3u. If the ratio of 3 to 20 were the ratio of a square number to a square number, then the equation would be solved. But the 3 resulted from $10^2 - [(2^2)^2 + (3^2)^2]$ while the 20 are twice the 10. Therefore, the problem is reduced to that of finding two squares v_2 and w_2, and a certain number u, such that $[u^2 - ((v^2)^2 + (w^2)^2)] : 2u :: \square' : \square''$.

Assignments	Operations	Equation
$v^2 := 1\text{P}$		
$w^2 := 4\text{u}$		
$u := 1\text{P}\,4\text{u}$		
	Add the squares of 1P and 4u $\rightarrow$ 1PP 16u	

Square 1P 4u → 1PP 8P 16u

Subtract 1PP 16u from 1PP 8P 16u → 8P

Twice 1P 4u → 2P 8u

$8P : (2P\,8u) :: \square' : \square''$

$4P : (1P\,4u) :: \square' : \square''$

$1P\,4u = \square'''$

$\sqrt{\square'''} := 1N\,1u$

Square 1N 1u → 1P 2N 1u

$1P\,4u = 1P\,2N\,1u$

1N is $1\frac{1}{2}$u, so $u = 6\frac{1}{4}$u, $v^2 = 2\frac{1}{4}$, and $w^2 = 4$. Multiplying them all by 4, we obtain 25, 9, and 16, respectively. Now, having solved the subsidiary problem, we return to the original problem.

Assignments	Operations	Equation
$x^2 := 1P$		
$y^2 := 9u$		
$z^2 := 16u$		
	Add the squares of 1P, 9u, 16u → 1PP 337u	
		$1PP\,337u = \square$
$\sqrt{\square} := 1P\,\ell\,25u$		
	Square $1P\,\ell\,25u$ → 1PP 625u ℓ 50P	
		$1PP\,337u$ $= 1PP\,625u\,\ell\,50P$

The equation simplifies to 50P = 288u, so 1N is $\frac{12}{5}$u, $x^2 = \frac{144}{25}$, $y^2 = 9$, and $z^2 = 16$.

$$\mathbf{V^G.30} \quad \begin{cases} x + y = \square \\ \left(\frac{1}{a}x + \frac{1}{b}y\right)^2 = \square + c \end{cases} \quad (a = 5,\, b = 8,\, c = 60)$$

This is the only problem in the extant books of the Arithmetica posed as an epigram about material units, here congii and drachmas. In our notation, x is the total cost of the 5-drachma wine, and y is the total cost of 8-drachma wine. Then $\frac{1}{5}x$ is the number of congii of 5-drachma wine and $\frac{1}{8}y$ is the number of congii of 8-drachma wine.

Assignments	Operations	Equation
$\frac{1}{5}x + \frac{1}{8}y := 1N$		
	Square 1N → 1P	
	Subtract 60u from 1P → 1P ℓ 60u	
$x + y := 1P \ \ell \ 60u$		
		1P ℓ 60u = □

This equation will be completed with an *al-istiqrā'* assignment of the form $\sqrt{□} := 1N$ less a certain multitude of units, for a suitable multitude. To determine that multitude, Diophantus first needs to find bounds on the value that 1N will assume in the solution. We can put bounds on the total cost by the inequality $\frac{1}{8}x + \frac{1}{8}y < \frac{1}{5}x + \frac{1}{8}y < \frac{1}{5}x + \frac{1}{5}y$. Since $\frac{1}{5}x + \frac{1}{8}y$ is 1N and $x + y$ is 1P ℓ 60u, 1N lies between an eighth of 1P ℓ 60u and a fifth of 1P ℓ 60u.

Because 5N < 1P ℓ 60u , 1P is equal to 5N and a number of units that is ≥ 60. Diophantus notes that 1N then must be greater than 11u. (The actual bound, from the quadratic formula, is that 1N must be greater than $\frac{5}{2} + \sqrt{66\frac{1}{4}}$, which is a little more than 10.6.) Similarly, from 1P ℓ 60u < 8N, Diophantus finds that 1N must not be greater than 12u. (Again, by the quadratic formula, the actual bound is that 1N should be less than $4 + \sqrt{76}$, or about 12.7.) Thus, he wants to find 1N to lie between 11u and 12u.

Returning to the forthcoming equation 1P ℓ 60u = □, Diophantus will assign $\sqrt{□}$ to be 1N less a multitude of units. Working through the solution without specifying these units, he finds that the value of 1N will be "a certain number multiplied by itself and having received in addition 60 units, which has been divided by its double". If we call this number m, the value of 1N will be $\frac{m^2 + 60}{2m}$, and we have the double inequality $11 < \frac{m^2 + 60}{2m} < 12$. He then embarks on another algebraic investigation for each inequality, first for $11 < \frac{m^2 + 60}{2m}$.

Assignments	Operations	Inequality
$m := 1N$		
	Square 1N → 1P	
	Add 60u to 1P → 1P 60u	
	Multiply 2N by 11u → 22N	
		1P 60u > 22N

From here he concludes that 1N must be greater than 19, where the actual bound is that 1N must be greater than $\sqrt{61} + 11$, or about 18.8. By a similar investigation of $\frac{m^2+60}{2m} < 12$, he finds that 1N should be less than 21, when the actual bound is $\sqrt{84}+12$, or about 21.2. There are two positive roots each to the corresponding equations just solved, and perhaps it is because Diophantus was looking for only one solution that he did not mention the other roots, which are $11-\sqrt{61}$ and $12-\sqrt{84}$ (Heath 1910, 62). Those two roots are closer together, being approximately 3.18975 and 2.83485. Choosing 1N, the number of units for the *al-istiqrā'* assignment, to be 20, the solution continues.

Assignments	Operations	Equation
		1P ℓ 60u = $\square$
$\sqrt{\square} := 1$N ℓ 20u		
	Square 1N ℓ 20u → 1P 400u ℓ 40N	
		1P ℓ 60u = 1P 400u ℓ 40N

1N turns out to be $11\frac{1}{2}$u , from which 1P ℓ 60u is found to be $72\frac{1}{4}$u.

We have so far that the total number of congii is $\frac{1}{5}x+\frac{1}{8}y=11\frac{1}{2}$ and the total cost is $x+y=72\frac{1}{4}$. This problem is then solved by the method of Problem I.5.

Assignments	Operations	Equation
$\frac{1}{5}x := 1$N		
	Subtract 1N from $11\frac{1}{2}$u → $11\frac{1}{2}$u ℓ 1N	
$\frac{1}{8}y := 11\frac{1}{2}$u ℓ 1N		
	Five times 1N → 5N	
	Eight times $11\frac{1}{2}$u ℓ 1N → 92u ℓ 8N	
	Add 5N and 92u ℓ 8N → 92u ℓ 3N	
		92u ℓ 3N = $72\frac{1}{4}$u

Thus, 1N is $\frac{79}{12}$u , so $\frac{1}{5}x$, the amount of 5-drachma wine, is $\frac{79}{12}$ congii, or 6 congii and 7 cotylae, while $\frac{1}{8}y$, the amount of 8-drachma wine, is $\frac{59}{12}$ congii, or 4 congii and 11 cotylae (12 cotylae make 1 congius).

Book VI^G

It is in the second lemma to Problem VI^G.7 that we first encountered the technique of forming Pythagorean triples from two given numbers. Supposing that $p > q$, the two legs of the right triangle formed from p and q are $p^2 - q^2$ and $2pq$, and the hypotenuse is $p^2 + q^2$. This technique is used in many problems in Book VI^G. A second method for forming a Pythagorean triple, this time with one odd number, is applied in Problem VI^G.19.

VI^G.1
$$\begin{cases} x^2 + y^2 = z^2 \\ z - x = \boxdot \\ z - y = \boxdot' \end{cases}$$

Diophantus forms the triangle (x, y, z) from 1N and 3u.

Assignments	Operations	Equation
$z := 1P\ 9u$		
$y := 6N$		
$x := 1P\ \ell\ 9u$		
	Subtract $1P\ \ell\ 9u$ from $1P\ \ell\ 9u \rightarrow 18u$	

The 18 should be a cube, but it is not. Looking into its origin, we see that it is twice the square of the 3u. Therefore, we must replace the 3u by another number whose square, taken two times, makes a cube. This is a subsidiary problem, written in our notation, $2u^2 = \boxdot\,''$, and it is solved with an *al-istiqrā'* assignment.

Assignments	Operations	Equation
$u := 1N$		
	Two times the square of $1N \rightarrow 2P$	
		$2P = \boxdot\,''$
$\boxdot\,'' := 1C$		
		$2P = 1C$

DOI: 10.4324/9781315171470-28

Thus, 1N is 2u, so we now form the triangle from 1N and 2u, and the first two conditions will be satisfied. That leaves the third condition.

Assignments	Operations	Equation
$z := $ 1P 4u		
$y := $ 4N		
$x := $ 1P ℓ 4u		
	Subtract 4N from 1P 4u $\rightarrow$ 1P 4u ℓ 4N	

Because 1P 4u ℓ 4N is the square from a side of 1N ℓ 2u, making the side a cube will make the square a cube. This is easily solved with an *al-istiqrā'* assignment.

Assignments	Operations	Equation
		1N ℓ 2u = ⬛'''
⬛'' := 8u		
		1N ℓ 2u = 8u

Thus, 1N is found to be 10u, so $x = 96$, $y = 40$, and $z = 104$

$$\mathbf{VI^G.2} \quad \begin{cases} x^2 + y^2 = z^2 \\ z + x = ⬛ \\ z + y = ⬛' \end{cases}$$

Diophantus again forms the triangle from 1N and a certain number of units, and as in the previous problem, he finds that the double of its square must be a cube. So, he again forms the triangle from 1N and 2u, which satisfies the first two conditions.

Assignments	Operations	Equation
$z := $ 1P 4u		
$y := $ 4N		
$x := $ 4u ℓ 1P		
	Add 4N and 1P 4u $\rightarrow$ 1P 4N 4u	

Here, too, 1P 4N 4u is a square, so it is enough to equate its side 1N 2u with a cube.

Assignments	Operations	Equation
		1N 2u = ⬛''

The unnamed cube is subject to two conditions: on the one hand, it should be greater than 2u; on the other hand, we have from the assignment of x that $4u > 1P$, hence $2u > 1N$, so $1N\ 2u < 4u$. Thus, the cube should also be less than 4u. Diophantus chooses the cube to be $\frac{27}{8}$u .

Assignments	Operations	Equation
		$1N\ 2u = \text{⬠}''$
$\text{⬠}'' := \frac{27}{8}$ u		
		$1N\ 2u = \frac{27}{8}$ u

$1N$ is found to be $\frac{11}{8}$u. Therefore, $x = \frac{135}{64}$, $y = 5\frac{1}{2}$, and $z = \frac{377}{64}$. Or, removing the denominators, $x = 135$, $y = 352$, and $z = 377$.

VI^G.3 $\begin{cases} x^2 + y^2 = z^2 \\ \frac{1}{2} x \cdot y + a = \square \end{cases}$ $(a = 5)$

To satisfy the first condition, Diophantus begins by assigning the triangle to be "given in form", as (3N, 4N, 5N).

Assignments	Operations	Equation
$x := 3N$		
$y := 4N$		
$z := 5N$		
	Half of 3N by 4N $\rightarrow$ 6P	
	Add 5u to 6P $\rightarrow$ 6P 5u	
		$6P\ 5u = \square$
$\square = 9P$		
		$6P\ 5u = 9P$

The equation simplifies to $3P = 5u$, but its solution is not rational because $3:5$ is not the ratio of a square to a square. So, we are led to the subsidiary problem of looking for a right triangle (u, v, w) to replace the $(3, 4, 5)$ in the assignments, and a square number m^2 to replace 9, such that $m^2 - \frac{1}{2} u \cdot v = \frac{1}{5}\square'$. The amount on the left will be a multitude of Powers whose ratio to 5 units should be a square. If it is a fifth of a square, then the ratio will be $\frac{1}{5}\square' : 5$.

Diophantus forms the triangle from the numbers 1N and 1N'.

Assignments	Operations	Equation
$u := 1P \ell 1P'$		
$v := 2u$		
	Half of $1P \ell 1P'$ by $2u \rightarrow 1P \ell 1P'$	
$m := 1N\ 10N'$		
	Square $1N\ 10N' \rightarrow 1P\ 100P'\ 20u$	
	Subtract $1P \ell 1P'$ from $1P\ 100P'\ 20u \rightarrow$ $101P'\ 20u$	
	Five times $101P'\ 20u \rightarrow 505P'\ 100u$	
	Multiply $505P'\ 100u$ by $1P \rightarrow 100P$ $505u$	
		$100P\ 505u =$ $\square'$
$\sqrt{\square'} := 10N\ 5u$		
	Square $10N\ 5u \rightarrow 100P\ 100N\ 25u$	$100P\ 505u =$ $100P\ 100N$ $25u$

$1N$ is found to be $\frac{24}{5}u$, so the two numbers from which the triangle is formed are $\frac{24}{5}u$ and $\frac{5}{24}u$, so $u = \frac{331{,}151}{14{,}400}$, $v = 2$, and $w = \frac{332{,}401}{14{,}400}$. Also, $m^2 = \left(\frac{413}{60}\right)^2 = \frac{170{,}569}{3600}u$.

Now, we start from the beginning the solution of the original problem and we assign the sides of the triangle using these values.

Assignments	Operations	Equation
$x := \frac{331{,}151}{14{,}400}N$		
$y := 2N$		
$z := \frac{332{,}401}{14{,}400}N$		
	Half of $\frac{331{,}151}{14{,}400}N$ by $2N \rightarrow \frac{331{,}151}{14{,}400}P$	
	Add $5u$ to $\frac{331{,}151}{14{,}400}P \rightarrow \frac{331{,}151}{14{,}400}P\ 5u$	
		$\frac{331{,}151}{14{,}400}P\ 5u = \square$
$\square := \frac{170{,}569}{3600}P$		
		$\frac{331{,}151}{14{,}400}P\ 5u = \frac{170{,}569}{3600}P$

$1N$ is then $\frac{24}{53}u$, so the sides of the triangle are $x = \frac{331{,}151}{31{,}800}$, $y = \frac{28{,}800}{31{,}800}$, and $z = \frac{332{,}401}{31{,}800}$.

VI^G**.4** $\begin{cases} x^2 + y^2 = z^2 \\ \frac{1}{2}x \cdot y - a = \square \end{cases}$ $(a = 6)$

Diophantus begins this problem with the same assignments as in the previous problem.

Assignments	Operations	Equation
$x := 3N$		
$y := 4N$		
$z := 5N$		
	Half of 3N by 4N → 6P	
	Subtract 6u from 6P → 6P ℓ 6u	
		6P ℓ 6u = $\square$
$\square := 4P$		
		6P ℓ 6u = 4P

As in the last problem, we need to replace the (3, 4, 5) triangle with another right triangle (u, v, w), and replace the 4 in the assignment of $\square$ with another square m^2, such that $6\left(\frac{1}{2}u \cdot v - m^2\right) = \square'$. The new triangle will again be formed from the numbers 1N and 1N'.

Assignments	Operations	Equation
$u := 1P \ell 1P'$		
$v := 2u$		
	Half of 1P ℓ 1P' by 2u –› 1P ℓ 1P'	
$m := 1N \ell 3N'$		
	Square 1N ℓ 3N' → 1P 9P' ℓ 6u	
	Subtract 1P 9P' ℓ 6u from 1P ℓ 1P' → 6u ℓ 10P'	
	Six times 6u ℓ 10P' → 36u ℓ 60P'	
	Multiply 36u ℓ 60P' by 1P → 36P ℓ 60u	
		36P ℓ 60u = $\square'$
$\sqrt{\square'} := 6N \ell 2u$		
	Square 6N ℓ 2u → 36P 4u ℓ 24N	
		36P ℓ 60u = 36P 4u ℓ 24N

1N is then $\frac{8}{3}$u, so the triangle is formed from $\frac{8}{3}$u and $\frac{3}{8}$u. Thus, $u = \frac{4015}{576}$, $v = 2$, $w = \frac{4177}{576}$, and $m = \frac{37}{24}$. The rest of the solution is only briefly described but would have been worked out something like this:

Assignments	Operations	Equation
$x := \frac{4015}{576} N$		
$y := 2N$		
$z := \frac{4177}{576} N$		
	Half of $\frac{4015}{576} N$ by $2N \to \frac{4015}{576} P$	
	Subtract 6u from $\frac{4015}{576} P \to \frac{4015}{576} P \, \ell \, 6u$	
		$\frac{4015}{576} P \, \ell \, 6u = \square$
$\square := \frac{1369}{576} P$		
		$\frac{4015}{576} P \, \ell \, 6u = \frac{1369}{576} P$

Thus, 1N is $\frac{8}{7}$u, so the sides of the triangle are $x = \frac{4015}{504}$, $y = \frac{1152}{504}$, and $z = \frac{4177}{504}$.

$$\mathbf{VI^{G}.5} \quad \begin{cases} x^2 + y^2 = z^2 \\ a - \frac{1}{2} x \cdot y = \square \end{cases} \quad (a = 10)$$

The procedure is the same as the last two problems.

Assignments	Operations	Equation
$x := 3N$		
$y := 4N$		
$z := 5N$		
	Half of 3N by $4N \to 6P$	
	Subtract 6P from $10u \to 10u \, \ell \, 6P$	
		$10u \, \ell \, 6P = \square$

Diophantus would like to assign the $\square$ to be a certain multitude of Powers, and he recognizes from the last two problems that the triangle (3, 4, 5) will need to be replaced by another right triangle (u, v, w) and a square multitude, call it m^2, that satisfy $\frac{1}{2} u \cdot v + m^2 = \frac{1}{10} \square'$.

We form a triangle by means of the numbers 1N and 1N' and proceed with the assignments.

Assignments	Operations	Equation
$u := 1P \ell 1P'$		
$v := 2u$		
	Half of $1P \ell 1P'$ by $2u \rightarrow 1P \ell 1P'$	
$m := 1N' 5N$		
	Square $1N' 5N \rightarrow 1P' 25P 10u$	
	Add $1P \ell 1P'$ and $1P' 25P 10u \rightarrow 26P 10u$	
	Ten times $26P 10u \rightarrow 260P 100u$	
		$260P 100u = \square'$
	A fourth of $260P 100u \rightarrow 65P 25u$	
		$65P 25u = \square''$
$\sqrt{\square'} := 5u 8N$		
	Square $5u 8N \rightarrow 25u 64P 80N$	
		$65P 25u = 25u 64P 80N$

$1N$ is therefore $80u$, so $m = \frac{32,001}{80}$. The triangle is formed from $80u$ and $\frac{1}{80}u$, so $u = \frac{40,959,999}{6400}$, $v = 2$, and $w = \frac{40,960,001}{6400}$. The solution can then begin again.

Assignments	Operations	Equation
$x := \frac{40,959,999}{6400} N$		
$y := 2N$		
$z := \frac{40,960,001}{6400} N$		
	Half of $\frac{40,960,001}{6400} N$ by $2N \rightarrow \frac{40,959,999}{6400} P$	
	Subtract $\frac{40,959,999}{6400} P$ from $10u \rightarrow 10u \ell \frac{40,959,999}{6400} P$	
		$10u \ell \frac{40,959,999}{6400} P = \square$
$\square := \frac{1,024,064,001}{6400} P$		
		$10u \ell \frac{40,959,999}{6400} P$ $= \frac{1,024,064,001}{6400} P$

$1N$ is $\frac{1}{129}u$, so the sides of the triangle are $x = \frac{40,959,999}{825,600}$, $y = \frac{12,800}{825,600}$, and $z = \frac{40,960,001}{825,600}$.

VI^G.6 $\begin{cases} x^2 + y^2 = z^2 \\ \frac{1}{2}x \cdot y + x = a \end{cases}$ $(a = 7)$

This solution also begins by assigning the sides of the triangle to be proportional to (3, 4, 5).

Assignments	Operations	Equation
$x := 3N$		
$y := 4N$		
$z := 5N$		
	Half of 3N by 4N $\rightarrow$ 6P	
	Add 6P and 3N $\rightarrow$ 6P 3N	
		6P 3N = 7u

For the solution to this composite equation to be rational, the discriminant must be a square. For this equation type, it is calculated by squaring half of the 3 and adding to it the product of 6 by 7, and that amounts to $\frac{177}{4}$, which is not a square. Thus, the problem is reduced to that of finding a right triangle (u, v, w) with perpendiculars u and v such that $\left(\frac{1}{2}u\right)^2 + 7 \cdot \frac{1}{2}uv = \square$

Assignments	Operations	Equation
$u := 1N$		
$v := 1u$		
	Square half of 1u $\rightarrow \frac{1}{4}$u	
	Seven times half of 1N by 1u $\rightarrow 3\frac{1}{2}$ N	
	Add $\frac{1}{4}$ u and $3\frac{1}{2}$ N $\rightarrow 3\frac{1}{2}$ N $\frac{1}{4}$u	
	Multiply $3\frac{1}{2}$ N $\frac{1}{4}$u by 4u $\rightarrow$ 1u 14N	
		1u 14N = $\square$
		1P 1u = $\square'$

The last equation is necessary for 1N and 1u to be the sides of a rational right triangle. These two forthcoming equations are solved by the method of the double-equality. The difference is 1P ℓ 14N, which is factored into 1N by 1N ℓ 14u. The square of half the difference of the factors is 49u, so 1u 14N = 49u, hence 1N is $\frac{24}{7}$ u. The auxiliary triangle (u, v, w) is then $(\frac{24}{7}, 1, \frac{25}{7})$, or, taking seven times everything, (24, 7, 25). These become multitudes in the assignments for x, y, and z.

Assignments	Operations	Equation
$x := 24N$		
$y := 7N$		
$z := 25N$		
	Half of 24N by 7N → 84P	
	Add 84P and 7N → 84P 7N	
		84P 7N = 7u

Solving this quadratic equation by the procedure of Diophantus, we square half of the 7 (Numbers) to get $\frac{49}{4}$. We add to it the product of 7 (units) by 84 (Powers), to get $\frac{2401}{4}$, whose square root is $\frac{49}{2}$. Subtracting half of the 7 (Numbers) gives $\frac{42}{2}$, and then dividing by the 84 gives $\frac{1}{4}$, which is the value of 1N. Thus, the sides of the triangle are $x = 6$, $y = \frac{7}{4}$, and $z = \frac{25}{4}$.

VI^G.7 $\begin{cases} x^2 + y^2 = z^2 \\ \frac{1}{2}x \cdot y - x = a \end{cases}$ $(a = 7)$

Diophantus skips the initial part where he makes the assignments $x := 3N$, $y := 4N$ and $z := 5N$ and then finds that the quadratic equation does not have a rational solution. That work would have proceeded like this:

Assignments	Operations	Equation
$x := 3N$		
$y := 4N$		
$z := 5N$		
	Half of 3N by 4N → 6P	
	Subtract 3N from 6P → 6P ℓ 3N	
		6P ℓ 3N = 7u

Here, again, the discriminant is not a square. To ensure that it is a square Diophantus writes that the right triangle (u, v, w) used in the assignments should satisfy $\left(\frac{1}{2}u\right)^2 + 7 \cdot \frac{1}{2}uv = \square$. That triangle is the same as the one found in the previous problem: (24, 7, 25).

1N is $\frac{1}{3}u$, so the sides of the triangle are $x = 8$, $y = \frac{7}{3}$, and $z = \frac{25}{3}$.

VI^G.8 $\begin{cases} x^2 + y^2 = z^2 \\ \frac{1}{2}x \cdot y + (x + y) = a \end{cases}$ $(a = 6)$

If we were to adopt the usual assignments, the calculations would proceed like this:

Assignments	Operations	Equation
$x := 3N$		
$y := 4N$		
$z := 5N$		
	Half of 3N by 4N → 6P	
	Add 6P, 3N, and 4N → 6P 7N	
		6P 7N = 6u

The discriminant of this equation is found by squaring half of the 7 and adding to it the product of the two 6s, which gives $\frac{193}{4}$. This is not a square, so we need to solve the subsidiary problem $\left(\frac{1}{2}(u+v)\right)^2 + 6 \cdot \frac{1}{2}uv = \square$.

Assignments	Operations	Equation
$u := 1N$		
$v := 1u$		
	Square half of 1N 1u → $\frac{1}{4}$P $\frac{1}{2}$N $\frac{1}{4}$u	
	Six times half of 1N by 1u → 3N	
	Add $\frac{1}{4}$P $\frac{1}{2}$N $\frac{1}{4}$u and 3N → $\frac{1}{4}$P $3\frac{1}{2}$N $\frac{1}{4}$u	
		$\frac{1}{4}$P $3\frac{1}{2}$N $\frac{1}{4}$u = $\square$
	Four times $\frac{1}{4}$P $3\frac{1}{2}$N $\frac{1}{4}$u → 1P 14N 1u	
		1P 14N 1u = $\square'$
		1P 1u = $\square''$

As in Problem VI^G.6, the last equation is necessary for 1N and 1u to be the sides of a rational right triangle. The solution is found by the method of the double-equality. The difference 14N is factored into 2N by 7u, and the square of half the difference of the factors is 1P $12\frac{1}{4}$u ℓ 7N. This equals 1P 1u, and 1N becomes $\frac{45}{28}$u. So the triangle (u, v, w) is $(\frac{45}{28}, 1, \frac{53}{28})$ or, multiplying all by 28, (45, 28, 53).

Assignments	Operations	Equation
$x := 45N$		
$y := 28N$		
	Half of 45N by 28N → 630P	
	Add 630P, 45N, and 28N → 630P 73N	
		630P 73N = 6u

1N is found to be $\frac{1}{18}$ u, so the sides of the triangle are $x = \frac{45}{18}$, $y = \frac{28}{18}$, and $z = \frac{53}{18}$.

VI^G.9 $\begin{cases} x^2 + y^2 = z^2 \\ \frac{1}{2}x \cdot y - (x+y) = a \end{cases}$ $\qquad (a = 6)$

The solution follows that of the previous problem and is only briefly described. After assigning x, y, and z "in form", such as 3N, 4N, and 5N, respectively, Diophantus again needs to solve the auxiliary problem of finding another triangle (u, v, w) such that $\left(\frac{1}{2}(u+v)\right)^2 + 6 \cdot \frac{1}{2}uv = \square$, so the solution $(45, 28, 53)$ found in the last problem will work here, too. After making the assignments $x \coloneqq 45$N, $y \coloneqq 28$N, and $z \coloneqq 53$N, the equation becomes 630P ℓ 73N = 6u, from which 1N is found to be $\frac{6}{35}$ u. The sides of the triangle are then $x = \frac{270}{35}$, $y = \frac{168}{35}$, and $z = \frac{318}{35}$.

VI^G.10 $\begin{cases} x^2 + y^2 = z^2 \\ \frac{1}{2}x \cdot y + (z+x) = a \end{cases}$ $\qquad (a = 4)$

Diophantus begins the solution with: "Once again, we will assign it to be given in form", but he does not say what that form is. He certainly had in mind the $(3, 4, 5)$ triangle, and the work would have been something like this:

Assignments	Operations	Equation
$x \coloneqq 3$N		
$y \coloneqq 4$N		
$z \coloneqq 5$N		
	Half of 3N by 4N $\rightarrow$ 6P	
	Add 6P, 3N, and 5N $\rightarrow$ 6P 8N	
		6P 8N = 4u

The discriminant here is 40, which is not a square. We thus need to find a right triangle (u, v, w) such that $\left(\frac{1}{2}(u+w)\right)^2 + 4 \cdot \frac{1}{2}uv = \square$. This time the triangle is formed from 1u and 1N 1u, so its sides are 1P 2N, 2N 2u, and 1P 2N 2u.

Assignments	Operations	Equation
$u \coloneqq 1$P 2N		
$v \coloneqq 2$N 2u		
$w \coloneqq 1$P 2N 2u		
	Square half the sum of 1P 2N and 1P 2N 2u $\rightarrow$ 1PP 4C 6P 4N 1u	

Four times half of 1P 2N by 2N
2u → 4C 12P 8N

Add 1PP 4C 6P 4N 1u and 4C
12P 8N → 1PP 8C 18P 12N 1u

1PP 8C 18P 12N 1u
= □

$\sqrt{\Box}$ ≔ 6N 1u ℓ 1P

Square 6N 1u ℓ 1P → 1PP 34P
12N 1u ℓ 12C

1PP 8C 18P 12N 1u =
1PP 34P 12N 1u ℓ 12C

The equation simplifies to 20C = 16P, whose solution is $\frac{4}{5}$u. So, the triangle is formed from the numbers 1 and $\frac{9}{5}$, or, taking the quintuples, from 5 and 9. So (u, v, w) is (56, 90, 106), or, taking half, (28, 45, 53).

Assignments	Operations	Equation
$x ≔ 28N$		
$y ≔ 45N$		
$z ≔ 53N$		
	Half of 28N by 45N → 630P	
	Add 630P, 28N, and 53N → 630P 81N	
		630P 81N = 4u

1N is found to be $\frac{4}{105}$ u, and the sides of the triangle are $x = \frac{112}{105}$, $y = \frac{180}{105}$, and $z = \frac{212}{105}$.

VI^G.11 $\begin{cases} x^2 + y^2 = z^2 \\ \frac{1}{2}x \cdot y - (z + x) = a \end{cases}$ $(a = 4)$

The solution is similar to that of the preceding problem and is only briefly described. Starting with the usual assignments $x ≔ 3N$, $y ≔ 4N$, $z ≔ 5N$, we obtain the equation 6P ℓ 8N = 4u, whose solution is not rational since the discriminant $4^2 + 6 \cdot 4$ is not a square. Thus, we are led to the same subsidiary problem as in the previous problem, whose solution, 28, 45, 53, set in terms of Numbers, provides the modified assignments which solve the original problem. The new assignments are $x ≔ 28N$, $y ≔ 45N$, $z ≔ 53N$, and the equation is 630P ℓ 81N = 4u. 1N is found to be $\frac{1}{6}$u, and sides we are looking for are $x = 4\frac{2}{3}$, $y = 7\frac{1}{2}$, and $z = 8\frac{5}{6}$.

Lemma I to Problem VI^G.12 $\begin{cases} x^2 + y^2 = z^2 \\ y - x = \square \\ y = \square' \\ \frac{1}{2} x \cdot y + x = \square'' \end{cases}$ $\qquad y > x$

The solution to this lemma never crosses into the realm of algebra, that is, into the language of Numbers, Powers, etc., and their combinations. The whole argument remains arithmetical like the enunciation. Diophantus would like to form the right triangle by means of two numbers, in which y is double their product and x is the difference of their squares. The second and third conditions will be fulfilled if one of these generating numbers is double the other.[1] The calculations he then expresses with arithmetical language parallel the following algebraic solution, in which we (not Diophantus) form the triangle from 1N and 2N:

Assignments	Operations	Equation
$y := 4P$		
$x := 3P$		
	Half of the product of 4P and 3P $\to$ 6PP	
	Add 6PP and 3P $\to$ 6PP 3P	
	Divide 6PP 3P by 1P $\to$ 6P 3u	
		6P 3u $= \square''$

Diophantus observes that 1u is a solution, "as well as an unlimited multitude of other numbers". So, the triangle has sides $x = 3$, $y = 4$, and $z = 5$. Heath (1910, 232 n. 1) notes that all triangles that solve the problem are similar to this triangle.

Because Diophantus does not assign algebraic names to either the generating numbers or to x and y, he has to express the value of the shorter leg x as "3 of the squares of the smaller" (γ̄ τῶν ἀπὸ ἐλάσσονος τετραγώνων), that is, of the smaller number from which the triangle is formed, and not as the algebraic "3 Powers". The word *dynamis* for the Power does not appear in the solution, and to express the fourth power he uses the term *dynamodynamis* with its arithmetical meaning, as in Hero of Alexandria (see the Introduction, Section 3.1.1). The area of the triangle is not "6 Power-Powers", but "sextuple the power-power of the ⟨smaller⟩ number". The algebraic fourth power, by contrast, is never expressed as the Power-Power *of* something.

The fact that Diophantus concludes with the arithmetical "a certain number such that 6 of its squares, together with 3 units, makes a square" and not the algebraic "6 Powers, 3 units equal a square" will be convenient for the next lemma.

---

1 For the reader still clinging to modern algebra for guidance, if p and q are the generating numbers, then $y = 2pq$ and $x = p^2 - q^2$. If $p = 2q$, then the second and third conditions are satisfied.

There the "square" will be named not as "1 Power", but as "1 Power, 2 Numbers, 1 unit". It is possible, if not likely, that Diophantus worked out this lemma by algebra on a wax tablet or other surface, after which he then composed the arithmetical version for his book.

This lemma is not actually applied in the next problem, even if Diophantus claims to refer back to it.

Lemma II to Problem VI^G.12: $\begin{cases} a+b = \square \\ a \cdot x^2 + b = \square' \end{cases}$ $(a = 3, \ b = 6)$

This lemma justifies the claim made at the end of the previous lemma that there is an unlimited number of solutions to the problem of finding "a certain number such that 6 of its squares, together with 3 units, makes a square", or for any similar problem in which the sum of the multitudes is a square.

Assignments	Operations	Equation
$x^2 := 1\text{P } 2\text{N } 1\text{u}$		
	Multiply 1P 2N 1u by 3u $\rightarrow$ 3P 6N 3u	
	Add 6u to 3P 6N 3u $\rightarrow$ 3P 6N 9u	
		3P 6N 9u = $\square'$
$\sqrt{\square'} := 3\text{u } \ell \text{ 3N}$		
	Square 3u ℓ 3N $\rightarrow$ 9u 9P ℓ 18N	
		3P 6N 9u = 9u 9P ℓ 18N

1N is then 4u, so $x^2 = 25$.

VI^G.12 $\begin{cases} x^2 + y^2 = z^2 \\ \frac{1}{2} x \cdot y + x = \square \\ \frac{1}{2} x \cdot y + y = \square' \end{cases}$

Diophantus begins by making assignments so that the sides are similar to 5, 12, and 13, and he forms two forthcoming equations from the second and third conditions. He then makes a tentative *al-istiqrā'* assignment for one of the squares.

Assignments	Operations	Equation
$x := 5\text{N}$		
$y := 12\text{N}$		

$z := 13N$

Half of 5N by 12N → 30P
Add 12N to 30P → 30P 12N

$30P\ 12N = \square'$

Add 5N to 30P → 30P 5N

$30P\ 5N = \square$

$\square' := 36P$

$30P\ 12N = 36P$

From this last equation, 1N is 2u. But then 30P 5N in the other equation will be 130u, which is not a square. Thus, we look into replacing the 36 with another square that will work. Diophantus traces the solution just found in terms of this unnamed square, and finds that "if it lacks 30, and the 12 is divided by the remainder", then the quotient, let us call it q, is the value of 1N, which then must give a square if applied to 30P 5N. Diophantus then proceeds to solve this subsidiary problem. In modern notation, if we call the multitude of Powers m^2, the problem becomes $q = \frac{12}{m^2-30}$; $30q^2 + 5q = \square$. We abbreviate the Power here with $\mathcal{P}$ rather than P to avoid confusion with the original P later on.

Assignments	Operations	Equation
$m^2 := 1\mathcal{P}$		
	Subtract 30u from $1\mathcal{P} \to 1\mathcal{P}\ \ell\ 30u$	
	Divide 12u by $1\mathcal{P}\ \ell\ 30u \to \frac{12u}{1\mathcal{P}\ \ell\ 30u}$	
	Square $\frac{12u}{1\mathcal{P}\ \ell\ 30u} \to \frac{144u}{1\mathcal{P}\mathcal{P}\ 900u\ \ell\ 60\mathcal{P}}$	
	Thirty times that $\to \frac{4320u}{1\mathcal{P}\mathcal{P}\ 900u\ \ell\ 60\mathcal{P}}$	
	Five times $\frac{12u}{1\mathcal{P}\ \ell\ 30u} \to \frac{60u}{1\mathcal{P}\ \ell\ 30u}$	
	Add $\frac{4320u}{1\mathcal{P}\mathcal{P}\ 900u\ \ell\ 60\mathcal{P}}$ and $\frac{60u}{1\mathcal{P}\ \ell\ 30u} \to \frac{60\mathcal{P}\ 2520u}{1\mathcal{P}\mathcal{P}\ 900u\ \ell\ 60\mathcal{P}}$	
		$\frac{60\mathcal{P}\ 2520u}{1\mathcal{P}\mathcal{P}\ 900u\ \ell\ 60\mathcal{P}} = \square$

Because the denominator is a square, the numerator $60\mathcal{P}$ 2520u must also be a square. We would be able to solve this by the preceding second lemma if the sum of 60 and 2520 were a square, but it is not. So, Diophantus traces the origin of these two numbers. Supposing we were working with a triangle with sides (u, v, w) instead of (5, 12, 13), the 60 would correspond to $u \cdot v$, and the 2520 to $v \cdot (v-u) \cdot \frac{1}{2}uv$. If we make v a square and divide both amounts by it, we need their sum $u + (v-u) \cdot \frac{1}{2}uv$ to be a square. So, the problem then becomes:

$$\begin{cases} u^2 + v^2 = w^2 \\ v = \Box''' \\ u + \frac{1}{2}uv \cdot (v - u) = \Box^{iv} \end{cases} \qquad v > u$$

For the solution to this problem the text refers back to the lemmas that "have been shown before". Heath (1910, 234 n. 1) points out that this reference is problematic because Lemma I speaks not of making v and $u + \frac{1}{2}uv \cdot (v - u)$ squares but of making $v - u$, v, and $\frac{1}{2}uv + u$ squares. In fact, if we follow the assignments that we (not Diophantus) made in the first lemma and form the triangle for the new auxiliary problem from 1N and 2N, the forthcoming equation becomes $6\mathbb{PP}\,3u = \Box'$ and not $6\mathbb{P}\,3u = \Box''$. But the solution 1N = 1u that Diophantus gives solves the former as well as the latter equation. However, given the techniques at his disposal, it is doubtful that Diophantus could have found a method of obtaining other solutions to $6\mathbb{PP}\,3u = \Box$.[2]

The triangle formed from the numbers 1 and 2 is $(u, v, w) = (3, 4, 5)$, and the solution is begun again.

Assignments	Operations	Equation
$x := 3N$		
$y := 4N$		
$z := 5N$		
	Half of 3N by 4N $\rightarrow$ 6P	
	Add 4N to 6P $\rightarrow$ 6P $\cdot$ 4N	
		$6P\,4N = \Box'$
	Add 3N to 6P $\rightarrow$ 6P $\cdot$ 3N	
		$6P\,3N = \Box$

As before, Diophantus would like to assign the $\Box'$ to be some square multitude of Powers. Following the procedure he tried earlier that gave him the expression $\frac{12u}{1P\,\ell\,30u}$ in the first auxiliary problem, he finds that the value of 1N will now be $\frac{4u}{1P\,\ell\,6u}$, where the sought-after multitude of Powers is assigned to be $1P$ in a new auxiliary problem whose details are not provided. Squaring this expression, 1P will then be $\frac{16u}{1PP36u\,\ell\,12N}$, so that 6P 3N is $\frac{12P\,24u}{1PP36u\,\ell\,12P}$, which must be a square. Again, the denominator is a square, so $12P\,24u$ must also be a square. This time $12 + 24$ is a square, so the second lemma guarantees an unlimited number of solutions,

2 Reassigning $1\mathbb{P}$ to be a general second-degree polynomial and making $\Box''$ a fourth-degree square that cancels two or three terms does not work in this case. Also, considering all positive rational numbers for which neither the numerator nor denominator exceeds 100, $1N = \frac{11}{23}u$ is the only other solution.

and Diophantus chooses $1\mathcal{P}$ to be 25u. This makes $\frac{4u}{1\mathcal{P}\,\ell\,6u}$, which is 1N, equal to $\frac{4}{19}$u. Therefore, the sides of the triangle we seek are $x = \frac{12}{19}$, $y = \frac{16}{19}$, and $z = \frac{20}{19}$.

VI^G.13
$$\begin{cases} x^2 + y^2 = z^2 \\ \frac{1}{2}x \cdot y - x = \square \\ \frac{1}{2}x \cdot y - y = \square' \end{cases}$$

Because this problem is similar to the previous one, its solution is only briefly described. Diophantus finds again that the triangle will be similar to the (3, 4, 5) triangle. Like before, the second and third conditions lead to two forthcoming equations.

Assignments	Operations	Equation
$x := 3N$		
$y := 4N$		
$z := 5N$		
	Half of 3N by 4N $\rightarrow$ 6P	
	Subtract 4N from 6P $\rightarrow$ 6P ℓ 4N	
		6P ℓ 4N $= \square'$
	Subtract 3N from 6P $\rightarrow$ 6P ℓ 3N	
		6P ℓ 3N $= \square$

Diophantus would like to assign $\square'$ to be a square multitude of Powers. He names this multitude, which we call m^2 for reference, as $1\mathcal{P}$, for a new unknown Power $\mathcal{P}$, and he solves for 1N in terms of it.

Assignments	Operations	Equation
$m^2 := 1\mathcal{P}$		
	Subtract $1\mathcal{P}$ from 6u $\rightarrow$ 6u ℓ $1\mathcal{P}$	
	Divide 4u by 6u ℓ $1\mathcal{P}$ $\rightarrow \frac{4u}{6u\,\ell\,1\mathcal{P}}$	

This quotient is the value of 1N, and it should also satisfy the second forthcoming equation, 6P ℓ 3N $= \square$:

Assignments	Operations	Equation
	Square $\frac{4u}{6u\,\ell\,1\mathcal{P}} \rightarrow \frac{16u}{1\mathcal{PP}\,36u\,\ell\,12\mathcal{P}}$	
	Multiply that by 6u $\rightarrow \frac{96u}{1\mathcal{PP}\,36u\,\ell\,12\mathcal{P}}$	

$$\text{Three times } \frac{4u}{6u\ \ell\ 1\mathcal{P}} \rightarrow \frac{72u\ \ell\ 12\mathcal{P}}{1\mathcal{PP}\ 36u\ \ell\ 12\mathcal{P}}$$

$$\text{Subtract this from } \frac{96u}{1\mathcal{PP}\ 36u\ \ell\ 12\mathcal{P}} \rightarrow \frac{12\mathcal{P}\ 24u}{1\mathcal{PP}\ 36u\ \ell\ 12\mathcal{P}}$$

$$\frac{12\mathcal{P}\ 24u}{1\mathcal{PP}\ 36u\ \ell\ 12\mathcal{P}} = \Box$$

The denominator is a square, so we need $12\mathcal{P}\ 24u = \Box''$. Since $12 + 24$ is a square, 1u is a solution. Now the *al-istiqrā'* assignment can be made. Returning to the unknowns in the first part of the solution, $\Box' := 1\text{P}$, so $6\text{P}\ \ell\ 4\text{N} = 1\text{P}$. Thus, 1N is $\frac{4}{5}$u, which makes $x = \frac{12}{5}$, $y = \frac{16}{5}$, and $z = 4$.

Diophantus adds that if we would like a solution to $12\mathcal{P}\ 24u = \Box''$ other than $1\mathcal{N} = 1$u, we can apply the method of the second lemma of the previous problem. While it may appear that Diophantus directly substitutes "1 Number" with "1 Number, 1 unit", he in fact first reinterprets the equation $12\mathcal{P}\ 24u = \Box''$ arithmetically as a question about two squares, as it is formulated in the enunciation to the lemma, before making his new assignment. This is evident by his instruction "assign the smaller (square) to be 1 Number, 1 unit". This is an algebraic assignment to the side of an unnamed square in the arithmetic problem that derives from the equation $12\mathcal{P}\ 24u = \Box''$. In modern notation, the problem is to find p such that $12p^2 + 24 = \Box''$. Here we write the new unknown as $\mathbb{N}$.

Assignments	Operations	Equation
$p := 1\mathbb{N}\ 1u$		
	Square $1\mathbb{N}\ 1u \rightarrow 1\mathbb{P}\ 2\mathbb{N}\ 1u$	
	Multiply $1\mathbb{P}\ 2\mathbb{N}\ 1u$ by $12u \rightarrow 12\mathbb{P}\ 24\mathbb{N}\ 12u$	
	Add $24u$ to this $\rightarrow 12\mathbb{P}\ 24\mathbb{N}\ 36u$	
		$12\mathbb{P}\ 24\mathbb{N}\ 36u = \Box''$
	Divide $2\mathbb{P}\ 24\mathbb{N}\ 36u$ by $4u \rightarrow 3\mathbb{P}\ 6\mathbb{N}\ 9u$	
		$3\mathbb{P}\ 6\mathbb{N}\ 9u = \Box'''$

This $3\mathbb{P}\ 6\mathbb{N}\ 9u$ is equivalent to $3\mathcal{P}\ 6u$, a fourth of the original $12\mathcal{P}\ 24u$. The word Diophantus uses for "equivalent" is ἰσχύουσι, which LSJ defines as "to be worth or equivalent to". It is not the word used to state equations.

Because the multitude p^2 must be less than 6 in order to obtain solutions to $6\text{P}\ \ell\ 3\text{N} = \Box$ and $6\text{P}\ \ell\ 4\text{N} = \Box'$, Diophantus notes that 1N cannot be greater than $\frac{13}{9}$u. We complete the solution here since Diophantus does not pursue it further.

Assignments	Operations	Equation
		$3\mathbb{P}\ 6\mathbb{N}\ 9u = \square'''$
$\sqrt{\square'''} := 6\mathbb{N}\ \ell\ 3u$		
	Square $6\mathbb{N}\ \ell\ 3u \rightarrow 36\mathbb{P}\ 9u\ \ell\ 36\mathbb{N}$	
		$3\mathbb{P}\ 6\mathbb{N}\ 9u = 36\mathbb{P}\ 9u\ \ell\ 36\mathbb{N}$

Thus, $1\mathbb{N}$ is $\frac{14}{11}u$, so the multitude p^2 is $\frac{625}{121}$. That completes the equation $6\mathbb{P}\ \ell\ 4\mathbb{N} = \frac{625}{121}\mathbb{P}$, from which we find $1\mathbb{N}$ to be $\frac{484}{101}u$. The triangle calculated from it has sides $x = \frac{1452}{101}$, $y = \frac{1936}{101}$, and $z = \frac{2420}{101}$.

$$\text{VI}^G.14 \quad \begin{cases} x^2 + y^2 = z^2 \\ \frac{1}{2}x \cdot y - z = \square \\ \frac{1}{2}x \cdot y - x = \square' \end{cases}$$

As before, the triangle is assumed to be proportional to the (3, 4, 5) triangle, and a pair of forthcoming equations is set up.

Assignments	Operations	Equation
$x := 3\mathbb{N}$		
$y := 4\mathbb{N}$		
$z := 5\mathbb{N}$		
	Half of $3\mathbb{N}$ by $4\mathbb{N} \rightarrow 6\mathbb{P}$	
	Subtract $5\mathbb{N}$ from $6\mathbb{P} \rightarrow 6\mathbb{P}\ \ell\ 5\mathbb{N}$	
		$6\mathbb{P}\ \ell\ 5\mathbb{N} = \square'$
	Subtract $3\mathbb{N}$ from $6\mathbb{P} \rightarrow 6\mathbb{P}\ \ell\ 3\mathbb{N}$	
		$6\mathbb{P}\ \ell\ 3\mathbb{N} = \square$

Positing that the multitude of Powers m^2 in an assignment for $\square$ to be $1\mathbb{P}$ for a new unknown $\mathbb{P}$, the value of $1\mathbb{N}$ is found to be $\frac{3u}{6u\ \ell\ 1\mathbb{P}}$. This must give a square if applied to $6\mathbb{P}\ \ell\ 5\mathbb{N}$.

Assignments	Operations	Equation
	Square $\frac{3u}{6u\ \ell\ 1\mathbb{P}} \rightarrow \frac{9u}{1\mathbb{PP}\ 36u\ \ell\ 12\mathbb{P}}$	
	Multiply that by $6u \rightarrow \frac{54u}{1\mathbb{PP}\ 36u\ \ell\ 12\mathbb{P}}$	

Five times $\frac{3u}{6u\,\ell\,1\mathcal{P}} \to \frac{90u\,\ell\,15\mathcal{P}}{1\mathcal{P}\mathcal{P}\,36u\,\ell\,12\mathcal{P}}$

Subtract this from $\frac{54u}{1\mathcal{P}\mathcal{P}\,36u\,\ell\,12\mathcal{P}} \to \frac{15u\mathcal{P}\,\ell\,36u}{1\mathcal{P}\mathcal{P}\,36u\,\ell\,12\mathcal{P}}$

$$\frac{15\mathcal{P}\,\ell\,36u}{1\mathcal{P}\mathcal{P}\,36u\,\ell\,12\mathcal{P}} = \square'$$

Because $1\mathcal{P}\mathcal{P}\,36u\,\ell\,12\mathcal{P}$ is a square, we also need $15\mathcal{P}\,\ell\,36u$ to be a square. But as Diophantus remarks, "this equality is impossible, since 15 cannot be divided into two squares." So, he looks into the origin of the 15 and the 36, with the idea that we will need to find a triangle other than $(u, v, w) = (3, 4, 5)$ in order to find a suitable multitude m^2. He traces the origin of these numbers:

The 15 Powers came from a certain square $m^2 = 1\mathcal{P}$ smaller than the number for the area $(\frac{1}{2}uv = 6)$ which had been multiplied by the product of the hypotenuse $(w = 5)$ and one of the perpendiculars $(u = 3)$, while the lacking 36 units, from the solid number which is contained by the area $(\frac{1}{2}uv = 6)$ and one of the perpendiculars $(u = 3)$ and the excess by which the hypotenuse $(w = 5)$ exceeds the aforementioned of the perpendiculars $(u = 3)$.

So, we should find another right triangle (u, v, w) and a square m^2 such that $m^2 < \frac{1}{2}uv$ and $m^2 wu - \frac{1}{2}uv \cdot u \cdot (w - u) = \square''$. Suppose the triangle is formed from two numbers p and q such that u is $2pq$. Then, $v = p^2 - q^2$, and $w = p^2 + q^2$. Also, $w - u = (p - q)^2$. If we divide both terms of $m^2 wu - \frac{1}{2}uv \cdot u \cdot (w - u)$ by the square $w - u$, we get $wu \frac{m^2}{w-u} - \frac{1}{2}uv \cdot u$. We should then look for a square $t = \frac{m^2}{w-u}$ such that $wut - \frac{1}{2}uv \cdot u$ is a square. Expressing this in terms of the numbers that form the triangle, it is $2pq(p^2 + q^2)t^2 - pq(p^2 - q^2)2pq$. This will be a square if p and q are similar plane numbers, which by Euclid's *Elements* IX.1 implies that their product is a square, and if we make $t^2 = pq$ The whole expression then simplifies to $4p^2q^4$, and since $t^2 = \frac{m^2}{(p-q)^2}$, we find that $m^2 = pq(p - q)^2$. Diophantus chooses to form the triangle from $p = 4$ and $q = 1$, so the triangle $(u, v, w) = (8, 15, 17)$ and $m^2 = 36$. It should be kept in mind that although Diophantus did not use modern notation to work this out, he probably had some way of indicating the operations on a wax tablet or other surface, from which he composed his rhetorical solution. What we read is surely not the result of a purely mental process, but we do not know how it was performed.

The solution can then be found using these numbers in the assignments.

Assignments	Operations	Equation
$x := 8N$		
$y := 15N$		
$z := 17N$		
	Half of 8N by 15N P $\to$ 60P	

Subtract 17N from 60P → 60P ℓ 17N	
	60P ℓ 17N = □′
Subtract 8N from 60P → 60P ℓ 8N	
	60P ℓ 8N = □
□ ≔ 36P	
	60P ℓ 8N = 36P

1N is then $\frac{1}{3}$u , so the sides of the required triangle are $x = \frac{8}{3}$, $y = \frac{15}{3}$, and $z = \frac{17}{3}$.

Lemma to Problem VI^G.15:   Given $a \cdot c^2 - b = \square$, find $x^2 > c^2$ such that $a \cdot x^2 - b = \square'$

The given numbers are $a = 3$, $b = 11$, and $c^2 = 5^2$, so that $3 \cdot 5^2 - 11 = 64 = 8^2$.

Assignments	Operations	Equation
$x ≔ 1N\ 5u$		
	Square 1N 5u → 1P 10N 25u	
	Three times 1P 10N 25u → 3P 30N 75u	
	Subtract 11u from 3P 30N 75u → 3P 30N 64u	
		3P 30N 64u = □′
$\sqrt{\square} ≔ 8u\ \ell\ 2N$		
	Square 8u ℓ 2N → 4P 64u ℓ 32N	
		3P 30N 64u = 4P 64u ℓ 32N

1N is found to be 62u, so $x = 67$ and $x^2 = 4{,}489$.

VI^G.15 $\begin{cases} x^2 + y^2 = z^2 \\ \frac{1}{2}x \cdot y + z = \square \\ \frac{1}{2}x \cdot y + x = \square' \end{cases}$

It is implied that we are to follow the procedure in Problem VI^G.14, beginning with the assignments 3N, 4N, 5N for x, y, and z, and trying a number like 36 for the multitude of Powers in the *al-istiqrā'* assignment. Through this process we are again faced with the subsidiary problem of finding a right triangle (u, v, w) and a square m^2 such that $m^2 < \frac{1}{2}uv$ and $m^2wu - \frac{1}{2}uv \cdot u \cdot (w - u) = \square''$. Diophantus again chooses to form the triangle from 1 and 4, which gives the triangle (8, 15, 17). But we cannot make the square multitude of Powers 36 because it should be greater than the area of the triangle, which is 60.

So, we look for another value for m^2. Keeping the triangle (8, 15, 17), the condition becomes $136m^2 - 4320 = \Box'''$. We know that this is satisfied by $m^2 = 63$, and we need a greater square. This is found by the preceding the lemma.

Assignments	Operations	Equation
$m \coloneqq$ 1N 6u		
	Square 1N 6u → 1P 12N 36u	
	Multiply that by 136 → 136P 1632N 4896u	
	Subtract 4320 from that → 136P 1632N 576u	
		136P 1632N 576u = $\Box'''$
$\sqrt{\Box'''} \coloneqq$ 16N ℓ 24u		
	Square 16N ℓ 24u → 256P 576u ℓ 768N	
		136P 1632N 576u = 256P 576u ℓ 768N

1N is 20u, so $m^2 = 676$. The solution can then be worked out.

Assignments	Operations	Equation
$x \coloneqq$ 8N		
$y \coloneqq$ 15N		
$z \coloneqq$ 17N		
	Half of 8N by 15N → 60P	
	Add 17N to 60P → 60P 17N	
		60P 17N = $\Box'$
	Add 8N to 60P → 60P 8N	
		60P 8N = $\Box$
$\Box \coloneqq$ 676N		
		60P 8N = 676P

1N is found to be $\frac{1}{77}$u, so the sides of the required triangle are $x = \frac{8}{77}$, $y = \frac{15}{77}$, and $z = \frac{17}{77}$.

VI^G.16 Find the sides of a right-angled triangle such that if one of the acute angles is bisected, then the bisecting line is rational.

Diophantus assigns the bisecting line to be 5N, and one of the segments of the base to be 3N. (He does not mention that this segment is the one attached at the right angle.) Therefore, the perpendicular will be 4N. He also assigns the whole base to be 3u. Therefore, the other segment of the base is 3u ℓ 3N. For this problem we show the assignments in a diagram rather than in a table.

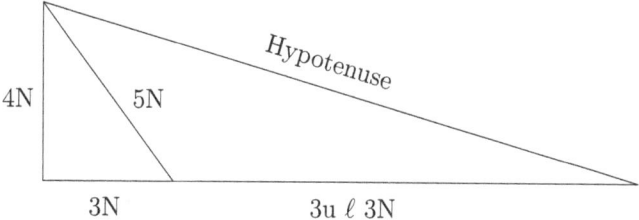

From Proposition VI.3 of Euclid's *Elements*, 3N: (3u ℓ 3N) :: 4N : Hypotenuse; and alternately, 3N : 4N :: (3u ℓ 3N) : Hypotenuse. Therefore, the number for the hypotenuse is sesquitertian ($1\frac{1}{3}$) of 3u ℓ 3N, that is, 4u ℓ 4N. The three sides of the large triangle are then 3u, 4N, and 4u ℓ 4N. By applying the Pythagorean theorem, we obtain the equation 16P 16u ℓ 32N = 16P 9u, from which 1N is found to be $\frac{7}{32}$u. Therefore, the base is 3, the altitude is $\frac{28}{32}$, and the hypotenuse is $\frac{100}{32}$. Or, if we multiply each length by 32, we get the triangle with sides 96, 28, and 100.

VI^G.17
$$\begin{cases} x^2 + y^2 = z^2 \\ \frac{1}{2}x \cdot y + z = \square \\ x + y + z = \boxed{\square}' \end{cases}$$

The solution begins with the assignments $\frac{1}{2}x \cdot y \coloneqq$ 1N and $z \coloneqq$ 16u ℓ 1N, so the second condition is satisfied. Diophantus factors $x \cdot y \coloneqq$ 2N into $x \coloneqq$ 1N and $y \coloneqq$ 2u, but then the perimeter $x + y + z$ is 18, which is not a cube. So, we need to find a square to replace the 16u so that if it received 2 units, it becomes a cube. This is the subsidiary problem $u^2 + 2 = v^3$.

Assignments	Operations	Equation
$u \coloneqq$ 1N 1u		
$v \coloneqq$ 1N ℓ 1u		
	Square 1N 1u → 1P 2N 1u	
	Cube 1N ℓ 1u → 1C 3N ℓ 3P 1u	
	Add 2u to 1P 2N 1u → 1P 2N 3u	
		1P 2N 3u = 1C 3N ℓ 3P 1u

The equation simplifies to 4P 4u = 1C 1N. Clearly both sides are divisible by 1P 1u, so it becomes 4u = 1N after the division. Therefore, $u = 5$ and $v = 3$, so the square and the cube are 25 and 27, respectively.

We return to the original problem, and we set up anew the right triangle. As before, the assignment of z fulfills the second condition.

Assignments	Operations	Equation
$\frac{1}{2}x \cdot y \doteq 1N$		
$z \doteq 25u \,\ell\, 1N$		
	Twice $1N \rightarrow 2N$	
$x \doteq 1N$		
$y \doteq 2u$		

The perimeter is now 27, which is a cube, so the third condition is fulfilled. That leaves the first condition.

Assignments	Operations	Equation
	Square $25u \,\ell\, 1N \rightarrow 1P\ 625u\,\ell\,50N$	
	Add the squares of $1N$ and $2u \rightarrow 1P\ 4u$	
		$1P\ 625u\,\ell\,50N = 1P\ 4u$

$1N$ is found to be $\frac{621}{50}u$. Hence, the sides of the triangle are $x = \frac{621}{50}, y = 2$, and $z = \frac{629}{50}$.

$$\text{VI}^{\text{G}}.18 \quad \begin{cases} x^2 + y^2 = z^2 \\ \frac{1}{2}x \cdot y + z = \boxdot \\ x + y + z = \square' \end{cases}$$

Again, Diophantus will assign $\frac{1}{2}x \cdot y$ to be $1N$ and z to be $1N$ less than a cubic number so that the second condition is satisfied. Then, if x and y are assigned to be $1N$ and $2u$, the perimeter will be $2u$ more than the cubic number. We are led, then, to a subsidiary problem, which in our notation is $u^3 + 2 = v^2$.

Assignments	Operations	Equation
$u \doteq 1N\,\ell\,1u$		
	Cube $1N\,\ell\,1u \rightarrow 1C\ 3N\,\ell\,3P\ 1u$	
	Add $2u$ to $1C\ 3N\,\ell\,3P\ 1u \rightarrow 1C$ $3N\ 1u\,\ell\,3P$	
		$1C\ 3N\,\ell\,3P\ 1u = v^2$
$v \doteq 1\frac{1}{2}N\ 1u$		
	Square $1\frac{1}{2}N\ 1u \rightarrow 2\frac{1}{4}P\ 3N\ 1u$	
		$2\frac{1}{4}P\ 3N\ 1u = 1C\ 3N\ 1u\,\ell\,3P$

This time the cubic equation simplifies to $5\frac{1}{4}P = 1C$, and 1N is found to be $\frac{21}{4}$u. Thus, $u^3 = \frac{4913}{64}$. The solution can now be worked out.

Assignments	Operations	Equation
$\frac{1}{2}x \cdot y := 1N$		
$z := \frac{4913}{64} u \, \ell \, 1N$		
$x := 2u$		
$y := 1N$		
	Add the squares of 2u and 1N $\rightarrow$ 4u 1P	
	Square $\frac{4913}{64} u \, \ell \, 1N \rightarrow \frac{24,137,569}{4096} u \, 1P \, \ell \, \frac{4913}{32} N$	
		$\frac{24,137,569}{4096} u \, 1P \, \ell \, \frac{4913}{32} N$
		$= 4u \, 1P$

So 1N is $\frac{24,121,185}{628,864}u$, and the sides of the triangle are $x = 2$, $y = \frac{24,121,185}{628,864}$, and $z = \frac{24,153,953}{628,864}$.

VI^G.19 $\begin{cases} x^2 + y^2 = z^2 \\ \frac{1}{2}x \cdot y + x = \square \\ x + y + z = \boxdot' \end{cases}$

Diophantus applies the method of forming a right triangle by means of an odd number, which Proclus attributes to Pythagoreans (Proclus 1873, 428.10–16; 1970, 340). According to this method, if a is an odd number then the sides of the right triangle are a, $\frac{1}{2}(a^2 - 1)$, $\frac{1}{2}(a^2 + 1)$. First, x is assigned to be an unknown odd number.

Assignments	Operations	Equation
$x := 2N \, 1u$		
	Square 2N 1u $\rightarrow$ 4P 4N 1u	
	From this, (a) subtract 1u, then take half $\rightarrow$ 2P 2N;	
	(b) add 1u, then take half $\rightarrow$ 2P 2N 1u	
$y := 2P \, 2N$		
$z := 2P \, 2N \, 1u$		
	Add 2N 1u, 2P 2N, and 2P 2N 1u $\rightarrow$ 4P 6N 2u	

The perimeter 4P 6N 2u must be a cube. Diophantus notes that it is composite (σύνθετος ἀριθμός), and he factors it into 4N 2u by 1N 1u. If we then divide each side by 1N 1u, then the perimeter will be simply 4N 2u. So, we reassign the sides:

Assignments	Operations	Equation
$x := \dfrac{2\text{N }1\text{u}}{1\text{N }1\text{u}}$		
$y := \dfrac{2\text{P }2\text{N}}{1\text{N }1\text{u}}$		
$z := \dfrac{2\text{P }2\text{N }1\text{u}}{1\text{N }1\text{u}}$		
	Add $\dfrac{2\text{N }1\text{u}}{1\text{N }1\text{u}}$, $\dfrac{2\text{P }2\text{N}}{1\text{N }1\text{u}}$, and $\dfrac{2\text{P }2\text{N }1\text{u}}{1\text{N }1\text{u}} \rightarrow 4\text{N }2\text{u}$	
		$4\text{N }2\text{u} = \boxdot'$
	Half of $\dfrac{2\text{N }1\text{u}}{1\text{N }1\text{u}}$ multiplied by $\dfrac{2\text{P }2\text{N}}{1\text{N }1\text{u}} \rightarrow \dfrac{2\text{C }3\text{P }1\text{N}}{1\text{P }2\text{N }1\text{u}}$	
	Add $\dfrac{2\text{C }3\text{P }1\text{N}}{1\text{P }2\text{N }1\text{u}}$ and $\dfrac{2\text{N }1\text{u}}{1\text{N }1\text{u}} \rightarrow \dfrac{2\text{C }5\text{P }4\text{N }1\text{u}}{1\text{P }2\text{N }1\text{u}}$	

Because 2C 5P 4N 1u is divisible by 1P 2N 1u, this last amount $\dfrac{2\text{C }5\text{P }4\text{N }1\text{u}}{1\text{P }2\text{N }1\text{u}}$ is just 2N 1u, so the forthcoming equation from the second condition is 2N 1u = □. Because 2N 1u is half of 4N 2u, we need to find a cube that is double a square, and 8 is a convenient choice. Now 4N 2u = 8u, so 1N is $1\frac{1}{2}$u. Thus, the sides of the required triangle are $x = \frac{8}{5}$, $y = \frac{15}{5}$, and $z = \frac{17}{5}$.

$$\text{VI}^{\text{G}}.20 \quad \begin{cases} x^2 + y^2 = z^2 \\ \frac{1}{2}x \cdot y + x = \boxdot \\ x + y + z = \square' \end{cases}$$

The solution follows the same procedure as the last problem, with the same assignments. This time the forthcoming equations are 4N 2u = □′ and 2N 1u = ⊡, so we need to find a square that is double a cube. Diophantus chooses 16 and 8. The equation becomes 4N 2u = 16u, so 1N is $3\frac{1}{2}$u, and the sides of the triangle are $x = \frac{16}{9}$, $y = \frac{63}{9}$, and $z = \frac{65}{9}$.

$$\text{VI}^{\text{G}}.21 \quad \begin{cases} x^2 + y^2 = z^2 \\ x + y + z = \square \\ (x + y + z) + \frac{1}{2}x \cdot y = \boxdot' \end{cases}$$

Diophantus forms the right triangle from 1N and 1u.

Assignments	Operations	Equation
$x := 2\text{N}$		
$y := 1\text{P }\ell\ 1\text{u}$		

Assignments	Operations	Equation
$z := 1P\ 1u$		
	Add 2N, $1P\ \ell\ 1u$, and $1P\ 1u \rightarrow 2P\ 2N$	
		$2P\ 2N = \square$
	Half of 2N multiplied by $1P\ \ell\ 1u \rightarrow 1C\ \ell\ 1N$	
	Add $2P\ 2N$ and $1C\ \ell\ 1N \rightarrow 1C\ 2P\ 1N$	
		$1C\ 2P\ 1N = \boxed{}\,'$

To solve $2P\ 2N = \square$ by *al-istiqrā'*, the square will be assigned to be a square multitude m^2 of Powers. The value of 1N found this way should also satisfy the other forthcoming equation, so m^2 should be chosen with that equation in mind. Diophantus next names this multitude m^2 as a new unknown $1P$, he finds 1N indeterminately as $\frac{2u}{1P\,\ell\,2u}$, and he then applies this expression to the polynomial $1C\ 2P\ 1N$.

Assignments	Operations	Equation
$m^2 := 1P$		
	Divide 2u by $1P\ \ell\ 2u \rightarrow \frac{2u}{2P\,\ell\,2u}$	
	Cube $\frac{2u}{1P\,\ell\,2u} \rightarrow \frac{8u}{(1P\,\ell\,2u)^3}$	
	Two squares of $\frac{2u}{1P\,\ell\,2u} \rightarrow \frac{8u}{(1P\,\ell\,2u)^2}$	

Now $1C\ 2P\ 1N$ is equivalent to $\frac{8u}{\text{cube of }1P\,\ell\,2u}\ \frac{8u}{\text{square of }1P\,\ell\,2u}\ \frac{2u}{1P\,\ell\,2u}$ [3] Diophantus has not expanded the denominators because it will not be necessary. They can remain unresolved because he will next make "everything in the same part", and then disregard the denominator because it is a cube.

Assignments	Operations	Equation
	Multiply 8u by $1P\ \ell\ 2u \rightarrow 8P\ \ell\ 16u$	
	Square $1P\ \ell\ 2u \rightarrow 1PP\ 4u\ \ell\ 4P$	
	Multiply $1PP\ 4u\ \ell\ 4P$ by $2u \rightarrow 2PP\ 8u\ \ell\ 8P$	
	Add 8u, $8P\ \ell\ 16u$, and $2PP\ 8u\ \ell\ 8P \rightarrow 2PP$	

3 This way of writing a fraction in notation with the semi-rhetorical "cube of $1P\ \ell\ 2u$" in the denominator is much like the same practice in some medieval Italian abacus books. Here is an example in a work by Antonio de' Mazzinghi (ca. 1380) as related by Benedetto da Firenze (mid-15th c.): " $\frac{72250\ p\ 4356\ cc.\ meno\ 102200\ c}{partiti\ per\ lo\ quadrato\ di\ 25\ meno\ 1\ c}$ " which, translated into the notation we have adopted for this commentary, is " $\frac{72{,}250u\ 4356PP\ \text{lacking}\ 102{,}200P}{\text{divided by the square of }25\ \text{lacking}\ 1P}$ " (Antonio de' Mazzinghi 1967, 60.15).

Combined into a single fraction, 1C 2P 1N is found to be equivalent to $\frac{2PP}{\text{cube of }1P\,\ell\,2u}$. Factoring away the denominator as well as $1C$ from the numerator, the forthcoming equation becomes $2\mathcal{N} = \boxplus''$. If we posit $\boxplus'' := 8u$, then the equation becomes $2\mathcal{N} = 8u$, so $1\,\mathcal{N}$ is 4u. Since 1N, equivalent to $\frac{2u}{1P\,\ell\,2u}$, is then $\frac{1}{7}$u, 1P is $\frac{1}{49}$u. But this value cannot be accepted because it makes the side of the triangle named 1P ℓ 1u negative. We need $\frac{2u}{1P\,\ell\,2u}$ to be both positive and greater than 1u, which means that $1P$ must lie between 2u and 4u. So, we return to the forthcoming equation $2\mathcal{N} = \boxplus'$. By squaring both sides and then dividing by 4u, we need to find a cube $\boxplus''$ such that a fourth of its square lies between 2u and 4u. Diophantus selects $\boxplus'' := \frac{27}{8}$u, which makes $1\mathcal{N}\,\frac{27}{16}$ u, and $1P\,\frac{729}{256}$; and thus, 1N is $\frac{512}{217}$u. The sides of the triangle are then $x = \frac{1024}{217}, y = \frac{215,055}{47,089}$, and $z = \frac{309,233}{47,089}$.

VI^G.22
$$\begin{cases} x^2 + y^2 = z^2 \\ x + y + z = \boxplus \\ (x+y+z) + \frac{1}{2}x \cdot y = \boxplus' \end{cases}$$

Diophantus approaches this problem by first finding values to assign to the perimeter and the area that guarantee rational values for x, y, and z in the first condition. For this he begins by investigating this alternate problem:
$$\begin{cases} x^2 + y^2 = z^2 \\ x + y + z = p \qquad (p = 12,\ a = 7). \\ \frac{1}{2}xy = a \end{cases}$$

The numbers 12 and 7 are chosen arbitrarily and serve only so that Diophantus can trace the calculations made with them to understand how to make a discriminant a square. The first two assignments given here satisfy the third condition, the third assignment derives from the second condition, and the equation comes from the first condition.

Assignments	Operations	Equation
$x := 1N'$		
$y := 14N$		
	Subtract 1N' and 14N from 12u $\rightarrow$ 12u ℓ 1N' 14N	
$z := 12u\ \ell\ 1N'\ 14N$		
	Square 12u ℓ 1N' 14N $\rightarrow$ 1P' 196P 172u ℓ 24N' 336N	

Add the squares of 1N′ and
14N → 1P′ 196P

1P′ 196P 172u ℓ 24N′
336N = 1P′ 196P

After simplification and multiplying both sides by 1N, the equation becomes
172N = 336P 24u. The discriminant $\left(\frac{1}{2}(172)\right)^{2} - 336 \cdot 24$ is not a square (it is
negative), so Diophantus traces the origin of these numbers. He determines that
the 172 comes from $p^2 + 4a$, and 336 · 24 comes from $8p^2a$. With this in mind he
searches for suitable values for p and a. He makes the number for the perimeter "a
cube and a square simultaneously, namely, 64 units". It must be a cube to satisfy
the second condition of the original problem, and next we will see why it also has
to be a square. The discriminant $\left(\frac{1}{2}(p^{2}+4a)\right)^{2} - 8p^{2}a$ is then calculated again,
beginning with an algebraic assignment for a.

Assignments	Operations	Equation
$a := 1N$		
$p := 64u$		
	The discriminant → 4P	
	4,194,304u ℓ 24,576N	
	Take a fourth of this →	
	1P 1,048,576u ℓ 6,144N	
		1P 1,048,576u ℓ 6,144N = □″

This forthcoming equation will be considered together with the one stemming
from the third condition but which must be adjusted so the units match.

Assignments	Operations	Equation
		1N 64u = □′
	Multiply 1N 64u by 16,384 →	
	16,384N 1,048,576u	
	16,384N 1,048,576u = □‴	

The multiplier 16,384 is $\frac{1}{16}p^{3}$, and because it must be a square, p must also be a
square.

The two forthcoming equations can now be solved by the method of the
double-equality. Diophantus concludes the problem with these words: "The

difference; the measurement. And the rest is obvious". We will finish it according to the reconstruction in Heath (1910, 245). The difference of the squares is 1P ℓ 22,528N, which can be factored into 1N and 1N ℓ 22,528u. Half of their difference is 11,264u, which is a root of $\square'''$. The equation becomes 16,384N 1,048,576u = 126,877,696u, from which we find 1N to be 7680u. But this value for the area of the triangle is impossible for a perimeter of 64. So, we should factor the 1P ℓ 22,528N differently. The only way to do it so that the units cancel in the final equation is to write 1P ℓ 22,528N as the product of 11N and $\frac{1}{11}$N ℓ 2048u. Again, taking half the difference of the factors, the equation becomes 16,384N 1,048,576u = $\frac{3600}{121}$P $\frac{122,880}{11}$N 1,048,576u, from which 1N is found to equal $\frac{39,424}{225}$u, which is the area $\frac{1}{2}xy$.

Now that $p = 64$ and $a = \frac{39,424}{225}$, we can find x, y and z. The procedure is the same as in the alternate problem given earlier. We give x and y assignments so that the area is $\frac{39,424}{225}$, and z is calculated from the equation for the perimeter.

Assignments	Operations	Equation
$x := 1N'$ $y := \frac{78,848}{225}N$ $z := 64u \ \ell \ 1N' \ \frac{78,848}{225}N$	 Subtract $1N'$ and $\frac{78,848}{225}N$ from $64u \rightarrow 64u \ \ell \ 1N' \ \frac{78,848}{225}N$	

To find 1N we apply these assignments to the first condition, the Pythagorean theorem.

Operations	Equation
Square $1N' \rightarrow 1P'$ Square $\frac{78,848}{225}N \rightarrow \frac{6,217,007,104}{50,625}P$ Add the two squares $\rightarrow 1P' \ \frac{6,217,007,104}{50,625}P$ Square $64u \ \ell \ 1N' \ \frac{78,848}{225}N \rightarrow \frac{6,217,007,104}{50,625}P$ $\frac{1,079,296}{225}u \ \ell \ \frac{10,029,544}{225}N \ 128N' \ 1P'$	 $1P' \ \frac{6,217,007,104}{50,625}P =$ $\frac{6,217,007,104}{50,625}P \ \frac{1,079,296}{225}u$ $1P' \ \ell \ \frac{10,029,544}{225}N \ 128N'$

The equation simplifies to $\frac{10,029,544}{225}$ N 128N′ $= \frac{1,079,296}{225}$ u. After multiplying each term by 1N, then by 225u, and dividing by 128u, we get $78,848$P 225u $= 8432$N, which has two solutions: 1N $= \frac{25}{448}$u and 1N $= \frac{9}{176}$u. Choosing the first of these, we get $x = \frac{448}{25}$, $y = \frac{176}{9}$, and $z = \frac{5968}{225}$. If we choose the other root, the numbers for x and y are reversed.

VI^G.23
$$\begin{cases} x^2 + y^2 = z^2 \\ z^2 = \boxdot + \sqrt{\boxdot} \\ z^2 \div x = \boxdot' + \sqrt[3]{\boxdot'} \end{cases}$$

The assignments for x, y and z^2 satisfy all three conditions. The equation comes from making the assignment for z^2 a square.

Assignments	Operations	Equation
$x \coloneqq 1$N		
$y \coloneqq 1$P		
	Add the squares of 1P and 1N → 1PP 1P	
$z^2 \coloneqq 1$PP 1P		
	Divide 1PP 1P by 1P → 1P 1u	
		1P 1u = $\Box'$
$\sqrt{\Box'} \coloneqq 1$N ℓ 2u		
	Square 1N ℓ 2u → 1P 4u ℓ 4N	
		1P 1u = 1P 4u ℓ 4N

1N is found to be $\frac{3}{4}$u, hence the sides of the required triangle are $x = \frac{3}{4}$, $y = \frac{9}{16}$, and $z = \frac{15}{16}$.

VI^G.24
$$\begin{cases} x^2 + y^2 = z^2 \\ x = \boxdot \\ y = \boxdot' + \sqrt[3]{\boxdot'} \\ z = \boxdot'' + \sqrt[3]{\boxdot''} \end{cases}$$

The first three assignments satisfy all conditions except the second.

Assignments	Operations	Equation
$z := 1C\ 1N$		
$y := 1C\ \ell\ 1N$		
	Subtract the square of $1C\ \ell\ 1N$ from the square of $1C\ 1N \rightarrow 4PP$	
	Take the square root of $4PP \rightarrow 2P$	
$x := 2P$		
		$2P = \boxdot$
$\boxdot = 1C$		
		$2P = 1C$

1N is 2u; therefore, the sides of the triangle are $x = 8$, $y = 6$, and $z = 10$.

Part IV

Appendices

Appendix 1

Reconstitution of the text of four missing problems from the fifth Greek book by Evangelos S. Stamatis

The text that follows the enunciation of problem V.19 in the Greek manuscripts clearly belongs to the solution of another problem. This is evident from the wording of the phrase immediately after the enunciation, which runs as follows: "And we are led, *again*, to divide 2, as *before*" (cf. Heath 1910, 215). There is therefore a lacuna after the enunciation to V.19 which covers at least the solution to that problem and the enunciation and first part of the solution to a different problem. In fact, even more than that is missing. As Bachet first proposed (1621, 329–30), the textual unit transmitted as "Problem V.19" covers in fact four problems whose enunciations and solutions were restored by Bachet in his Latin commentary. Later, Tannery gave the same restorations, in his own Latin wording, but without solutions (Diophantus 1893–95, I, 365 n. 1). E. S. Stamatis proposed a hypothetical restoration of the missing Greek text (1961, reprinted in Stamatis 1978, 45–59), and he included it in his 1963 Greek edition of the *Arithmetica*. Here we give the translation of the reconstructed by Stamatis text of the four problems, which are numbered as 19, 19a, 19b, and 19c.

19   To find three numbers equal to a square such that the cube of the (number) composed of all three, if it lacks any one of them, makes a square.

Again, let the three be assigned to be 1 Power, and among the sought-after (numbers, let) one be 3 4ths ⟨Cube-⟩Cubes,[1] another 8 9ths Cube-Cubes, and the third 15 16ths Cube-Cubes. And it comes out that the cube of the (number) composed of all three, if it lacks either one, makes a square. It remains (for us) to equate the three with 1 Power. But the three are 371 144ths Cube-Cubes, (which are) equal to 1 Power. And everything by 1 Power: 371 144ths Power-Powers become equal to 1 unit. And the 1 unit is a square having as a side a square. Therefore, 371 144ths Power-Powers must also be a square having as a side a square. And if the 371 144ths units were a square having as a side a square, the requirement would have been fulfilled. How did the multitude of Power-Powers arise? From subtracting from a triad three squares, each one of which is smaller than 1 unit. And we are

---

1 Omission due to a typographical error.

DOI: 10.4324/9781315171470-30

reduced to (the problem of) finding three squares such that each of them is smaller than 1 unit, while their sum, if removed from a triad, makes a square having as a side a square. Further, we require the square of each of them to be smaller than 1 unit. If, therefore, we construct the three numbers to be smaller than 1 unit, each of them will be much smaller than 1 unit. And so, the remaining square which has as a side a square must be greater than a dyad. Therefore, we must divide 371 144ths into three squares. Of the required (squares), one will be 529 625ths, another 49 625ths, and the third a 625th.

We return to the initial (problem) and assign again the three to be 1 Power, and among the sought-after (numbers), one to be 96 625ths Cube-Cubes, another 576 625ths ⟨Cube-Cubes,⟩² and the third 624 625ths Cube-Cubes. And it comes out that the cube of the sum of all three, if it lacks any one of them, makes a square. It remains (for us) to equate the three with ⟨1⟩³ Power. But the three become 1296 625ths Cube-Cubes. And everything by 1 Power: 1296 625ths Power-Powers become equal to 1 unit, and the Number becomes 5 6ths.

To the hypostases.

19a. To find three numbers equal to a square such that the cube of the (number) composed of all three, if subtracted from any one of them, makes a square.

Again, let the three be assigned to be 1 Power, and among the sought-after (numbers, let) one be 2 Cube-Cubes, another 5 Cube-Cubes, and the third 10 Cube-Cubes. And it comes out that each one, if it lacks the cube of the (number) composed of all three, makes a square. It remains (for us) to equate the three with 1 Power. But the three are 17 Cube-Cubes, and these are equal to 1 Power. And everything by 1 Power: 17 Power-Powers become equal to 1 unit. And the 1 unit is a square having as a side a square. If the 17 units were also a square having as a side a square, the requirement would have been fulfilled. But the 17 is the sum of three squares, together with 3 units. Therefore, three squares should be found whose sum, together with 3 units, makes a square having as a side a square; say 16 units. Therefore, we must divide 13 units into three squares. Of the required (squares), one will be 9, another 36 25ths, and the third 64 25ths. So, I add 1 unit to each of them and I return to the initial (problem). I assign each one to be these many Cube-Cubes, the three being assumed to be 1 Power. And one of them becomes 10 Cube-Cubes, another 61 25ths Cube-Cubes, and the third 89 25ths Cube-Cubes. And it will come out that each one, if it lacks the cube of the (number) composed of all three, makes a square. It remains (for us) to equate the three with 1 Power. But the three become 16 Cube-Cubes. And everything by 1 Power. They become ⟨16⟩⁴ Power-Powers equal to 1 unit, and the Number becomes a half.

---

2 Omission due to a typographical error.
3 Omission due to the fact that Stamatis was unaware of the difference between the modern x^2 and the premodern "1 Power".
4 Omission due to a typographical error.

19b. To divide a given number into three numbers such that the cube of the (number) composed of all three, if it receives in addition any one of them, makes a square.

Now let it be proposed (to divide) the (number) 2.

The cube of 2 is 8 units. Therefore, it is required to divide 2 into three numbers such that each of them, if it receives in addition 8 units, makes a square. Therefore, 26 should be divided into three squares such that each of them is greater than 8 units. I assign one of the sought-after squares to be 9 units, which is greater than 8 units. So, if I divide 17 into two squares such that each of them is greater than 8 units, I (will) fulfill what is required. I take the half of the 17; it becomes 8 2′; and I look for a square part to add to 8 2′ units so that it makes a square. And everything four times. Therefore, I look for a square fraction to add to 34 units so that it makes a square. Let the added fraction be 1 inverse Power. 34 units, 1 inverse Power become equal to a square. And everything by a Power: 34 Powers, 1 unit become equal to a square; let it be to the one from a side of 1 unit lacking 6 Numbers, and the Number becomes 6 units. Thus, the Power is 36 units, and the inverse Power a 36th of a unit. Therefore, the (square part) which is added to the 34 units will be a 36th. Therefore, the (square part) which is added to the 8 2′ units will be a 144th, and it makes a square, the one from a side of 35 12ths. Therefore, the side of each of the two squares to which 17 is divided must be as close as possible to 35 12ths. So, I look for a number, which if removed from a tetrad and added to a unit, makes the same, that is, 35 12ths. I assign two squares, one from (a side) of 23 Numbers, 1 unit, and the other from 4 units lacking 13 Numbers. And the number composed of their squares becomes 698 Powers, 17 units lacking 58 Numbers, (which is) equal to 17 units, and the Number becomes 29 349ths.

Thus, the side of one of the squares will be 1016 349ths, and (the side) of the other 1019 349ths. And if I remove 8 units from each square, I shall have the three sought-after (numbers).

19c. To divide a given number into three numbers such that the cube of the (number) composed of all three, if it lacks any one of them, makes a square.

Now let it be proposed (to divide) the (number) 2.

We are led, again, to divide 2, as before, and the cube of 2 is 8 units. So, we must subtract each one from 8 units and make a square. Therefore, 22 should be divided into three squares such that each of them is greater than 6 units. So, if we remove each of them from 8 units, we will find the three sought-after numbers. But this has been shown before, namely, how 22 can be divided into three squares such that each of them is greater than 6 units.

Appendix 2

Techniques of solving indeterminate problems by algebra

Recall that a forthcoming equation is an incomplete equation, in which one side, or rarely both sides, have not yet been named in terms of the names of the powers. One example from Problem II.22 is: "1 Power, 4 Numbers, 2 units. These are equal to a square", which we write in our notation as $1P\ 4N\ 2u = \square$. There are two main approaches that Diophantus took to solve forthcoming equations. Faced with a single forthcoming equation, he usually made an assignment by a technique that was called *al-istiqrā'* by al-Karajī and later Arabic algebraists. The other approach was to solve a pair of forthcoming equations simultaneously, usually by the technique he called "the double equality", but sometimes by some alternative.

Here we list forthcoming equations set up in the *Arithmetica*, first the single equations solved by *al-istiqrā'* and then the pairs of equations that are solved together. Some problems receive more than one solution, which we indicate with subscripts. So, $II.13_1$ means the first solution to Problem II.13, and $IV.42(a)_2$ means the second solution to subproblem (a) of Problem IV.42.

1 Single forthcoming equations in the *Arithmetica* solved by *al-istiqrā'*

The main approach Diophantus takes to complete forthcoming equations is to name the unnamed square or cube in such a way that the resulting equation simplifies to two terms of consecutive powers, thus guaranteeing a rational solution. Diophantus does not give a name to the technique by which he accomplishes this, nor is it named in the earlier Arabic books by Abū Kāmil and 'Alī al-Sulamī. Apparently al-Karajī was the first to call it *al-istiqrā'*, and several later Arabic algebraists did so as well. Recall that al-Karajī incorporated most problems from the first four books of the *Arithmetica* into the vast collection of solved problems in his algebra book *al-Fakhrī*. In the introduction he devoted a section to explaining *al-istiqrā'*, which begins:

> *Al-istiqrā'* in calculation is that you arrive at an expression formed from a type or two types or from three consecutive types in which that expression is not a square with regard to what is indicated in its formulation, but is a square in meaning and potentiality, and for which you want to know its root.
>
> (Saidan 1986, 165.20)

DOI: 10.4324/9781315171470-31

For example, "a *Māl* and four Things and four units" (1M 4T 4u) is an expression of three consecutive types (powers). It is a square with regard to what is indicated in its formulation because it is the square of 1T 2u. Thus, it will give a square number regardless what rational value is chosen for the Thing. On the other hand, "1 Power, four Numbers, two units" (1P 4N 2u, from Problem II.22) is not a square with regard to what is indicated in its formulation. For many values of the Number it will not be a square, but for some other values it will be. If we are presented with the forthcoming equation in which this expression is equal to a square, then the expression is "a square in meaning" because it must be a square by the conditions of the problem, and it is "a square in potentiality (*quwwa*)" because it has the capacity to be a square.

Al-Karajī goes on to explain how this is done. You want to name the square or cube in such a way that, after the equation is simplified, it becomes "one type equals one type, such as Things equal a number, or *Māl*s equal Things, or Cubes equal *Māl*s" (Saidan 1986, 166.9). He illustrates this through particular examples. If we have, say, 1M 4T = □, we can assign a side of the square to be any multitude of *Māl*s that is greater than one. He posits that the side is 4M, so the equation simplifies to 3M = 4T, and the Thing is found to be one and a third. But one can also assign a side of the square to be "a Thing less a dirham, or whatever you want of dirhams", for which the equation becomes (choosing one dirham) 1M 1u ℓ 2T = 1M 4T. This way the 1Ms cancel, and the equation simplifies to 4T = 1u. For the forthcoming equation 4M 16T 9u = □ we want either the *Māl*s or the units to cancel. Thus, we can assign a side of the square to be "two Things less whatever you wish in units, in which its square is greater than nine. So make it two Things less five dirhams" (Saidan 1986, 167.19). The equation becomes 4M 25u ℓ 20T = 4M 16T 9u, which simplifies to 36T = 16u. One can also make the units drop out by making a root of the square "three dirhams less whatever you wish in Things, so that if you multiply it by itself it is greater than four *Māl*s" (Saidan 1986, 168.3). Making its side 3u ℓ 3T, the equation becomes 9M 9u ℓ 18T = 4M 16T 9u, which simplifies to 5M = 34u. In every one of these assignments there is a free choice of coefficient, so that the technique can give an unlimited number of solutions.

But this technique will not work for every polynomial, such as those for which no appended power is a square, or for two non-square, non-consecutive powers. Al-Karajī explains:

> And this technique (*bāb*) is necessarily limited. Examples that do not allow the result of the root to be known are: ten Things less a *Māl* and (less) a dirham, and two *Māl*s and ten Things and ten units, and ten Things less a *Māl* and less five dirhams, and others like it.[1]

---

1 (Saidan 1986, 168.7), with adjustments from Paris MS BnF 2459, fol. 38b. "Examples" is singular in the text.

Later, in his *al-Badī'*, al-Karajī gave the most comprehensive extant account of *al-istiqrā'*,[2] in two parts. In the first part, consisting of three chapters, he classifies the different kinds of polynomials and how to name the square or cube to get a two-term equation. In the second part, consisting of two chapters beginning at p. 66.16, he explains techniques to solve some of those "limited" cases, and with limited success. Among the extant books, Diophantus encounters a problem covered in these later chapters only once, in Problem IV^G.31.

This particular meaning of *al-istiqrā'* to make an assignment that will ultimately give a two-term equation does not seem to agree with the usual meaning of the word, which in arithmetic is to examine or consider different numbers in succession. Rashed and Vahabzadeh (1999, 386) write of the corresponding verb, "« *istaqrā* » les pays, les choses ou les gens, c'est les considérer et les examiner successivement, un à un". Indeed, *istiqrā'* is often applied in arithmetic outside the context of solving forthcoming equations with just this meaning, sometimes to uncover arithmetical propositions, and other times to find solutions to problems by trial and error. Al-Karajī even characterizes the term this way in *al-Badī'*: "I say that *al-istiqrā'* is that you examine the quantities in succession in order to find what you are seeking" (al-Karajī 1964, 62.17). Jacques Sesiano found a way to reconcile this usual meaning with the technique for solving forthcoming equations explained in the first three chapters of *al-Badī'*. In modern notation, he considers the problem of making, for example, the three-term polynomial type $a_1^2 x^2 + bx + c$ equal to a square. The side of the square will be of the form $a_1 x \pm m$, and by examining the numbers one by one, you can find "un paramètre m tel que la solution x soit positive, et ceci se fera par tâtonnement (*istiqrā'*)" (Sesiano 1977a, 304). But, Diophantus and Arabic algebraists made the choice of their free term (like Sesiano's m) not by trial and error, but, as they explain, by considering whatever restrictions there might be on this value and then making their single choice based on it. This includes al-Karajī, whose practice does not match his definition. Instead, it seems that the word *istiqrā'* took on a new, technical meaning in the solutions to forthcoming equations, and that is what we intend by the term in relation to the technique of Diophantus.

Diophantus appears to be working with negative numbers in the solutions to some forthcoming equations by *al-istiqrā'* (Barner 2007). One example comes from his solution to the forthcoming equation 3P 4u ℓ 6N = □ in problem IV^G.9, where Diophantus has the option of assigning the side of the square to be of the form Ns 2u, Ns ℓ 2u, or 2u ℓ Ns, where by "Ns" we mean some multitude of Numbers. Squaring either of the latter two give the same polynomial, so there are really two forms to choose from. In this problem, the difference between choosing 2u ℓ 4N or 4N ℓ 2u for the side should lie with whether the answer is greater or less than $\frac{1}{2}$ u. Diophantus does not know ahead of time what the answer will be, so

---

2 (al-Karajī 1964, 62–72). It is translated into French in Sesiano (1977a). See Section 3.3.6 for a description.

he makes the choice arbitrarily. In this problem he chooses 2u ℓ 4N. Sometimes his answer agrees with the assignment of the side, and sometimes, as here, his answer will make that side negative: since 1N turns out to be $\frac{10}{13}$ u, his 2u ℓ 4N is a negative $\frac{14}{13}$ u. This does not mean that Diophantus worked with negative numbers. It just means that he did not bother to go back and change the assignment for the side of the square to be 4N ℓ 2u.

We infer from al-Karajī's polynomial types and from the accounts of *al-istiqrā*' in other Arabic books that the term intends assignments in which there is an open choice of coefficient to be made. An example where there is no possibility for a choice is in Problem IV^G.28, where Diophantus sets up the forthcoming equation 9PP 6P 1u ℓ 4C 12N = $\square'$. The only assignment that can lead to a two-term equation is $\sqrt{\square'} := 3P\ 1u\ \ell\ 6N$, so this is not an example of *al-istiqrā*'. To be an *al-istiqrā*' assignment an unlimited number of solutions should be attainable.

The technique of *al-istiqrā*' for obtaining two-term equations is applied 160 times in the extant books of the *Arithmetica*, and these are listed in the following table. Sometimes there are constraints for the assignment from other requirements of the problem. We place an asterisk (*) after them.

Problem	Forthcoming equation	Assignment
II.8$_1$	16u ℓ 1P = $\square$	$\sqrt{\square} := 2N\ \ell\ 4u$
II.11$_2$	1P 1u = $\square$	$\sqrt{\square} := 1N\ \ell\ 4u$
II.12	1P 12u = $\square$	$\sqrt{\square} := 1N\ \ell\ 4u$
II.13$_2$	1P ℓ 1u = $\square$	$\sqrt{\square} := 1N\ \ell\ 2u$
II.16	3P 18N 9u = $\square$	$\sqrt{\square} := 2N\ \ell\ 3u$
II.19	1P 8N 4u = $\square$	$\sqrt{\square} := 1N\ 3u$
II.20	4P 5N 1u = $\square$	$\sqrt{\square} := 2N\ \ell\ 2u$
II.21	4P 3N = $\square$	$\sqrt{\square} := 3N$
II.22	1P 4N 2u = $\square$	$\sqrt{\square} := 1N\ \ell\ 2u$
II.23	1P ℓ 2N 1u = $\square$	$\sqrt{\square} := 1N\ \ell\ 3u$
II.28	1P 1u = $\square$	$\sqrt{\square} := 1N\ \ell\ 2u$
	9P 9u = $\square$	$\sqrt{\square} := 3N\ \ell\ 4u$

II.29	1P ℓ 1u = □	√□ ≔ 1N ℓ 2u
	1P ℓ 1u = □	√□ ≔ 1N ℓ 4u
II.32	16P 25N 9u = □	√□ ≔ 4N ℓ 4u
II.33	16P 7N = □	√□ ≔ 5N
III.5₁	2N = □	□ ≔ 16u
III.6₁	6N 1u = □	□ ≔ 121u
III.6₂	6N 1u = □	□ ≔ 36u
III.7	1P 4N 2u = □	√□ ≔ 1N ℓ 8u
III.8	6N 22u = □	□ ≔ 100u
III.9	6N 4u = □	□ ≔ 64u
III.10	1P 12u = □	√□ ≔ 1N 3u
III.11	1P ℓ 10u = □	√□ ≔ 1N ℓ 2u
	5929P ℓ 160u = □	√□ ≔ 77N ℓ 2u
III.14	16P 33N 16u = □	√□ ≔ 4N ℓ 5u
III.15₂	4N 3u = □	□ ≔ 25u
	4P 8N 3u = □	√□ ≔ 2N ℓ 3u
III.16	4P ℓ 1u = □	√□ ≔ 2N ℓ 2u
IV.1	9C = □	√□ ≔ 6T
IV.2	7C = □	√□ ≔ 7T
IV.3	5M = ▱	∛▱ ≔ 1T
IV.4	24M = ▱	∛▱ ≔ 2T
IV.5	4MM = ▱	∛▱ ≔ 2T
IV.6	8MC = □	√□ ≔ 4M
IV.7	64MC = ▱	∛▱ ≔ 2M

IV.9	$64CC = \square$	$\sqrt{\square} \coloneqq 16M$
IV.10	$1C\ 10M = \square$	$\sqrt{\square} \coloneqq 4T$
IV.11	$1C\ \ell\ 6M = \square$	$\sqrt{\square} \coloneqq 2T$
IV.12	$1C\ 10M = ⧉$	$\sqrt[3]{⧉} \coloneqq 2T$
IV.13₁	$1C\ \ell\ 7M = ⧉$	$\sqrt[3]{⧉} \coloneqq \tfrac{1}{2}T$
IV.14(e)	$\tfrac{1}{2}C = \square$	$\sqrt{\square} \coloneqq 2T$
IV.14(f)	$2M = ⧉$	$\sqrt[3]{⧉} \coloneqq 1T$
IV.23	$17MM = ⧉$	$\sqrt[3]{⧉} \coloneqq 3T$
IV.24	$15MM = ⧉$	$\sqrt[3]{⧉} \coloneqq 5T$
IV.25	$1CC\ 16MM = \square$	$\square \coloneqq 25MM*$
IV.26a	$1CC\ \ell\ 16MM = \square$	$\sqrt{\square} \coloneqq 3M\ *$
IV.26b	$625MM\ \ell\ 1CC = \square$	$\square \coloneqq 225MM*$
IV.27	$1CC\ 20MM = \square$	$\square \coloneqq 36MM*$
IV.28	$16MM\ 10C = \square$	$\sqrt{\square} \coloneqq 6M$
IV.29	$1CCC\ 16CCM = \square$	$\sqrt{\square} \coloneqq 6MM$
IV.30	$1CCC\ \ell\ 16CCM = \square$	$\sqrt{\square} \coloneqq 2MM$
IV.31	$16CCM\ \ell\ 1CCC = \square$	$\sqrt{\square} \coloneqq 2MM$
IV.32	$21CCC = \square$	$\sqrt{\square} \coloneqq 7MM$
IV.33	$\tfrac{1}{4}CCC = \square$	$\sqrt{\square} \coloneqq 1MM$
V.1	$\tfrac{8}{49}MM = ⧉$	$\sqrt[3]{⧉} \coloneqq 2T$
V.2	$\tfrac{1}{4}MM = ⧉$	$\sqrt[3]{⧉} \coloneqq 2T$
V.3	$\tfrac{1}{16}MM = ⧉$	$\sqrt[3]{⧉} \coloneqq \tfrac{1}{2}T$
VI.1	$64CC\ 1MM = \square$	$\square \coloneqq 100CC*$
VI.2	$64CC\ \ell\ 1MM = \square$	$\square \coloneqq 40\tfrac{24}{25}CC*$

VI.3	1MM ℓ 64CC = □	□ := 36CC*
VI.4	15,625CC 125CM = □	$\sqrt{□}$:= 126C
VI.5	1CM 1MM = □	$\sqrt{□}$:= 2M
VI.6	1CM ℓ 1CC = □	$\sqrt{□}$:= 1C
VI.7	1CM ℓ 1MM = □	$\sqrt{□}$:= 1M
VI.8	64M 8T = □	$\sqrt{□}$:= 10T
VI.9	64M ℓ 8T = □	$\sqrt{□}$:= 7T
VI.10	8T ℓ 64M = □	$\sqrt{□}$:= 4T
VI.11	1CC 1C = □	$\sqrt{□}$:= 3C
VI.12	$\frac{9}{16}$ MM $\frac{9}{16}$ M = □	$\sqrt{□}$:= $\frac{13}{16}$ M*
VI.13$_1$	$\frac{9}{25}$ MM ℓ $\frac{9}{25}$ M = □	$\sqrt{□}$:= $\frac{9}{25}$ M*
VI.13$_2$	$2\frac{7}{9}$ u ℓ $\frac{9}{25}$ M = □	$\sqrt{□}$:= $1\frac{2}{3}$ u ℓ $1\frac{1}{5}$ T
VI.14	$\frac{25}{16}$ M ℓ $\frac{16}{25}$ u − □	$\sqrt{□}$:= $1\frac{1}{4}$ T ℓ 2u
VI.15	1M ℓ 2T 1u = □	$\sqrt{□}$:= 1T ℓ 2u
VI.16	1M 2u ℓ 4T = □	$\sqrt{□}$:= 1T ℓ 4u
VI.18	$\frac{25}{16}$ M $\frac{25}{16}$ u = □	$\sqrt{□}$:= $1\frac{1}{4}$ T $\frac{1}{4}$ u
VI.19	$\frac{9}{16}$ M ℓ $2\frac{9}{16}$ u = □	$\sqrt{□}$:= $\frac{3}{4}$ T ℓ $\frac{1}{4}$ u
VI.20	$\frac{9}{25}$ M $4\frac{4}{25}$ u = □	$\sqrt{□}$:= $\frac{3}{5}$ T 1u
VI.21	1M $\frac{5}{16}$ u = □	$\sqrt{□}$:= 1T $\frac{1}{2}$ u
VI.22$_1$	1MM 1M = □	$\sqrt{□}$:= $1\frac{1}{4}$ M
VI.22$_2$	144MM = ▱	▱ := 8C
VI.23	1M 9u = □	$\sqrt{□}$:= 1T 1u

VII.1	$512CCC = \square$	$\sqrt{\square} := 32MM$
VII.2	$1C = \square$	$\square := 4M$
VII.5	$512C\ 512M = \square$	$\sqrt{\square} := 64T$
VII.8	$1M\ 4T\ 2u = \square$	$\sqrt{\square} := 1T\ \ell\ 2u$
VII.9	$1M\ 2u\ \ell\ 4T = \square$	$\sqrt{\square} := 1T\ \ell\ 3u$
VII.10	$1M\ 2T\ \ell\ 1u = \square$	$\sqrt{\square} := 1T\ \ell\ 3u$
VII.16	$1M\ \ell\ 1u = \square$	$\sqrt{\square} := 1T\ \ell\ 2u$
VII.17	$25M\ 2\frac{7}{9}u = \square$	$\sqrt{\square} := 5T\ \frac{1}{3}u$
VII.18	$\frac{9}{25}M\ \ell\ 25u = \square$	$\sqrt{\square} := \frac{3}{5}T\ \ell\ 1u$
IV^G.6	$1C16P = \boxdot$	$\boxdot = 8C$
IV^G.7$_1$	$5P = \boxdot$	$\boxdot := 1C$
IV^G.7$_2$	$2P\ 4u = \square$	$\sqrt{\square} := 2N\ \ell\ 2u$
	$20P = \boxdot$	$\boxdot := 1C$
IV^G.8	$3P\ 3N\ 1u = \square$	$\sqrt{\blacksquare} := 1u\ \ell\ 2N$
IV^G.9	$3P\ 4u\ \ell\ 6N = \square$	$\sqrt{\square} := 2u\ \ell\ 4N$
IV^G.11	$3P\ 3N\ 1u = \square$	$\sqrt{\square} := 1u\ \ell\ 2N$
IV^G.13	$7P\ 18N\ 9u = \square$	$\sqrt{\square} := 3u\ \ell\ 3N$
IV^G.14	$1P\ 2N\ \ell\ 2u = \square$	$\sqrt{\square} := 1N\ \ell\ 4u$
IV^G.16	$13N = \square$	$\square := 169\mathcal{P}$ (indeterminate)
	$10,816\mathcal{P}\ 221u = \square$	$\sqrt{\square} := 104\mathcal{N}\ 1u$
IV^G.17	$13N = \square$	$\square := 169\mathcal{P}$ (indeterminate)
	$10,816\mathcal{P}\ \ell\ 221u = \square$	$\sqrt{\square} := 104\mathcal{N}\ \ell\ 1u$

IV^G.18	1CC 1N 64u ℓ 16C = □	$\sqrt{□} := $ 1C 8u
	4C = □	□ := 16P
	1CC 4096u 1N ℓ 108C = □	$\sqrt{□} := $ 1C 64u
IV^G.20	9P 24N 13u = □	$\sqrt{□} := $ 3N ℓ 4u
IV^G.22	1P2$\frac{1}{2}$N = □	□ := 4P
IV^G.26	8C1P ℓ 8N 1u = ⬛	$\sqrt[3]{⬛} := $ 2N ℓ 1u
IV^G.27	8C 1P ℓ 8N 1u = ⬛	$\sqrt[3]{⬛} := $ 2N ℓ 1u
IV^G.28$_1$	9PP 6P 1u ℓ 4C 12N = □	$\sqrt{□} := $ 3P 1u ℓ 6N
IV^G.28$_2$	1C ℓ 2P = ⬛	⬛ := $\frac{1}{8}$C
IV^G.31$_1$	72P 81u = □	$\sqrt{□} := $ 8N 9u
	3N 18u ℓ 1P = □	□ := 324P*
IV^G.31$_2$	9N ℓ 1P = □	□ := 4P
	9N ℓ 1P = □	□ := $\frac{25}{16}$P*
IV^G.37	60C = □	□ := 900P
IV^G.38	16P ℓ 71u = □	$\sqrt{□} := $ 4N ℓ 1u
IV^G.39	3P 12N 9u = □	$\sqrt{□} := $ 3u ℓ 5N *
IV^G.40	9P ℓ $\frac{42}{32}$u = □	$\sqrt{□} := $ 3N ℓ 6u
V^G.2	1P 18N 101u = □	$\sqrt{□} := $ 1N ℓ 11u
V^G.3	4P 28N 34u = □	$\sqrt{□} := $ 2N ℓ 6u
V^G.4	4P 4N 19u = □	$\sqrt{□} := $ 2N ℓ 6u
V^G.5	1P 3N 3u = □	$\sqrt{□} := $ 1N ℓ 3u
V^G.6	1P 1N 1u = □	$\sqrt{□} := $ 1N ℓ 2u
L1 V^G.7	1P 1N 1u = □	$\sqrt{□} := $ 1N ℓ 2u
V^G.9	26P 1u = □	$\sqrt{□} := $ 5N 1u

V^G.10	9u ℓ 1P = $\square$	$\sqrt{\square}$ ≔ 3u ℓ 3$\frac{1}{2}$N *
V^G.11	30P 1u = $\square$	$\sqrt{\square}$ ≔ 5N 1u
V^G.12	10P 1u = $\square$	$\sqrt{\square}$ ≔ 3N 1u
	30P 1u = $\square$	$\sqrt{\square}$ ≔ 5N 1u
	$\frac{6561}{841}$u ℓ 1P = $\square$	$\sqrt{\square}$ ≔ $\frac{81}{29}$u ℓ 2$\frac{1}{2}$N *
V^G.13	16u ℓ 1P = $\square$	$\sqrt{\square}$ ≔ 4u ℓ 2$\frac{1}{2}$N *
V^G.14	34P 1u = $\square$	$\sqrt{\square}$ ≔ 6N ℓ 1u
V^G.15	9P 14u ℓ 9N = $\square$	$\sqrt{\square}$ ≔ 3N ℓ 4u
V^G.17	9P 31u ℓ 27N = $\square$	$\sqrt{\square}$ ≔ 3N ℓ 7u
V^G.27	6$\frac{1}{4}$P′ 25P ℓ 9u = $\square$	$\square$ ≔ 25P
V^G.28	6$\frac{1}{4}$P 6$\frac{1}{4}$P′ ℓ 25u = $\square$	$\square$ ≔ 6$\frac{1}{4}$P′
V^G.29	1P 4u = $\square$	$\sqrt{\square}$ ≔ 1N 1u
V^G.30	1P ℓ 60u = $\square$	$\sqrt{\square}$ ≔ 1N ℓ 20u*
VI^G.1	2P = ▱	▱ ≔ 1C
	1N ℓ 2u = ▱	▱ ≔ 8u
VI^G.2	1N 2u = ▱	▱ ≔ $\frac{27}{8}$u *
VI^G.3	100P 505u = $\square$	$\sqrt{\square}$ ≔ 10N 5u
VI^G.4	36P ℓ 60u = $\square$	$\sqrt{\square}$ ≔ 6N ℓ 2u
VI^G.5	65P 25u = $\square$	$\sqrt{\square}$ ≔ 5u 8N
L2 VI^G.12	3P 6N 9u = $\square$	$\sqrt{\square}$ ≔ 3u ℓ 3N
VI^G.12	30P 12N = $\square$	$\square$ ≔ 36P
VI^G.13	6P ℓ 4N = $\square$	$\square$ ≔ 1P*
	3$\mathbb{P}$ 6$\mathbb{N}$ 9u = $\square$	$\sqrt{\square}$ ≔ 6$\mathbb{N}$ ℓ 3u

VI^G.14	60P ℓ 8N $= \square$	$\square \coloneqq$ 36P
L VI^G.15	3P 30N 64u $= \square$	$\sqrt{\square} \coloneqq$ 8u ℓ 2N
VI^G.15	136P 1632N 576u $= \square$	$\sqrt{\square} \coloneqq$ 16N ℓ 24u
	60P 8N $= \square$	$\square \coloneqq$ 676P
VI^G.19	4N 2u $= \boxtimes$	$\boxtimes \coloneqq$ 8u *
VI^G.20	4N 2u $= \square$	$\square \coloneqq$ 16u*
VI^G.21	2$\mathcal{N} = \boxtimes$	$\boxtimes \coloneqq$ 8u
VI^G.23	1P 1u $= \square$	$\sqrt{\square} \coloneqq$ 1N ℓ 2u
VI^G.24	2P $= \boxtimes$	$\boxtimes \coloneqq$ 1C

2 Pairs of forthcoming equations in the Arithmetica solved simultaneously

Often Diophantus sets up a pair of forthcoming equations from two conditions of the problem in which the unnamed portions are squares. The technique of *al-istiqrā'* cannot be applied because the solution to one equation might not work for the other. Diophantus usually solves these pairs of equations by what he calls "the double equality". Because the difference of two squares can be factored into the product of the sum and difference of their roots, it is possible to find a solution to both of them simultaneously. For example, in problem IV^G.39 he sets up the pair 8N 4u $= \square$ and 6N 4u $= \square$". Diophantus chooses to factor 2N, the difference of the squares, as 4u by $\frac{1}{2}$N (other choices will lead to different answers). One of these is the sum and the other is the difference of the roots of the unnamed squares. It is then easy to solve for the two roots: the greater root will be half the sum of 4u and $\frac{1}{2}$N, or 2u $\frac{1}{4}$N, and the smaller root will be half the difference, which is either 2u ℓ $\frac{1}{4}$N or $\frac{1}{4}$N ℓ 2u, depending whether 2u or $\frac{1}{4}$N is greater. But like the same choice in *al-istiqrā'*, it does not matter which one is greater because the expression is squared in the equation. The two forthcoming equations can then be completed: 8N 4u $=$ 4u $\frac{1}{16}$P 1N and 6N 4u $=$ 4u $\frac{1}{16}$P ℓ 1N, which both simplify to give 1N $=$ 112u.

There are some other techniques for solving a pair of forthcoming equations applied in the latter part of Book IV and the beginning of Book V. The technique we call "equating the corresponding sides" is first performed in the second solution to Problem IV.34. There, as in the other ten solutions in Book IV worked out by this technique, the completed sides of the forthcoming equations consist of two terms of consecutive powers, and their difference consists of only one term. In each case Diophantus wishes to assign the unnamed square to be some square multitude of the even-degree power in such a way that the two equations give the same solution. He runs through the simplification of these equations without specifying the multitudes and reduces the problem to finding a square that is either the sum or difference of two given squares. This subsidiary problem is easily solved by the method in one of the problems from the beginning of Book II. Such an alternative to "the double equality" is necessary since that method cannot work for the pair of forthcoming equations set up in Problems IV.35 and would be very difficult to apply in Problems IV.37 and IV.40. See our commentary to Problem IV.34₂ for an explanation of how "equating the corresponding sides" works.

A variation on "equating the corresponding sides" is applied in Problems IV.42(a)₃, IV.44(a), IV.44(b), and IV.44(c). Here Diophantus divides the two sides of the forthcoming equations by the least square power, so the completed parts of the equations are Things and units, and the multitudes of the Things happen to be the same. He then looks for a number (the Things) satisfying the conditions of the equations, which corresponds again to the problem solved in the beginning of Book II.

Then, Problems V.4, V.5, and V.6 exhibit a variation on this variation. Diophantus again divides by the least square power to get equations with Things and units, but now he appeals to an arithmetical identity to find the Things. See our commentaries to these problems for details.

For an explanation of the pairs of forthcoming equations in Problems V.1, V.2, and V.3, see our commentary. There one side of the equation is an unnamed square, and the other is not yet completely named. This technique is unrelated to the others just described.

In the following table, if the "Approach" is left blank the technique is "the double equality". We abbreviate "equating the corresponding sides" by "ecs". The entry for "Choice" shows the factoring of the difference for "the double equality", and the solution to the subsidiary problem for "equating the corresponding sides".

Problem	Forthcoming equations	Approach	Choice
II.11₁	1N 2u = □ 1N 3u = □′		1u = $\frac{1}{4}$u by 4u
II.12₂	5u ℓ 1T = □ (al-Karajī) 3u ℓ 1T = □′		2u = 2u by 1u
II.13₁	1N ℓ 6u = □ 1N ℓ 7u = □′		1u = 2u by $\frac{1}{2}$u
III.12	10N ℓ 54u = □′ 10N ℓ 6u = □″		48u = 4u by 12u
III.13	4P 15N = □′ 4P ℓ 1N 4u = □″		16N 4u = 4u by 4N 1u

III.15$_1$	10N 9u = $\square'$ 5N 4u = $\square''$		5N 5u = 5u by 1N 1u
III.15$_2$	4N 3u = $\square'$ $6\frac{1}{2}$N $5\frac{1}{2}$u = $\square''$		No solution
	130N 105u = $\square''$ 130N 30u = $\square^{iv}$		75u = 3u by 25u
III.16	10N ℓ 14u = $\square'''$ 10N ℓ 26u = $\square^{iv}$		12u = 2u by 6u
III.17	4P 3N ℓ 1u = $\square''$ 4P 4N ℓ 1u = $\square$		1N = $\frac{1}{4}$u by 4N
III.18	4P 3N ℓ 1u = $\square'$ 4P ℓ 1N 1u = $\square$		4N = 4N by 1u
IV.34$_1$	1C ℓ 4M = $\square$ 1C ℓ 4M = $\square'$		8M = 2T by 4T
IV.34$_2$	1C ℓ 4M = $\square$ 1C ℓ 4M = $\square'$	ecs	$8 = \left(4\frac{1}{2}\right)^2 - \left(3\frac{1}{2}\right)^2$
IV.35	4M 1C = $\square$ 4M ℓ 1C = $\square'$	ecs	$8 = \left(\frac{2}{5}\right)^2 + \left(2\frac{4}{5}\right)^2$
IV.36	1C 4M = $\square$ 1C ℓ 5M = $\square'$	ecs	$9 = 5^2 - 4^2$
IV.37	1C 5M = $\square$ 1C ℓ 10M = $\square'$	ecs	$5 = \left(2\frac{5}{9}\right)^2 - \left(2\frac{4}{9}\right)^2$
IV.38	1C 5M = $\square$ 1C ℓ 10M = $\square'$	ecs	$5 = 3^2 - 2^2$
IV.39	3M 1C = $\square$ 7M ℓ 1C = $\square'$	ecs	$4 = \left(2\frac{1}{2}\right)^2 - \left(1\frac{1}{2}\right)^2$
IV.40	16MM 64C = $\square$ 16MM ℓ 64C = $\square'$	ecs	$32 = \left(\frac{4}{5}\right)^2 + \left(5\frac{3}{5}\right)^2$
IV.41	64C 16MM = $\square$ 64C ℓ 16MM = $\square'$	ecs	$32 = 6^2 - 2^2$
IV.42(a)$_1$	512CCC 256CCM = $\square$ 512CCC ℓ 256CCM = $\square'$		Not worked out
IV.42(a)$_2$	512CCC 256CCM = $\square$ 512CCC ℓ 256CCM = $\square'$	ecs	Not worked out
IV.42(a)$_3$	512CCC 256CCM = $\square$ 512CCC ℓ 256CCM = $\square'$	ecs var$_1$	Not worked out
IV.42(b)	256CCC 512CCM = $\square$ 256CCC ℓ 512CCM = $\square'$	ecs	$512 = \left(3\frac{1}{5}\right)^2 + \left(22\frac{2}{5}\right)^2$

IV.43	1CCC 20CCM = □ 1CCC ℓ 12CCM = □′	ecs	$32 = 6^2 - 2^2$
IV.44(a)	1CCC 48CCM = □ 1CCC ℓ 128CCM = □′	ecs var$_1$	$16 + 48 = 8^2$; $16 + 128 = 12^2$
IV.44(b)	1CCC 48CCM = □ 1CCC ℓ 128CCM = □′	ecs var$_1$	$192 - 48 = 12^2$ $192 - 128 = 8^2$
IV.44(c)	48CCM ℓ 1CCC = □ 128CCM ℓ 1CCC = □′	ecs var$_1$	$48 - 47 = 1^2$ $128 - 47 = 9^2$
V.1	1MM ($4y^3$) = □ 1MM ℓ ($3y^3$) = □′	unnamed technique	See commentary
V.2	1MM ($12y^3$) = □ 1MM ($5y^3$) = □′	unnamed technique	See commentary to V.1
V.3	1MM ℓ ($12y^3$) = □ 1MM ℓ ($7y^3$) = □′	unnamed technique	See commentary to V.1
V.4	16CCM 5CCC = □ 16CCM ℓ 3CCC = □′	ecs var$_2$	Arithmetical identity
V.5	16CCM 12CCC = □ 16CCM 5CCC = □′	ecs var$_2$	Arithmetical identity
V.6	81CCM ℓ 7CCC = □′ 81CCM ℓ 4CCC = □′	ecs var$_2$	Arithmetical identity
IV^G.23	1P 1N ℓ 1u = □′ 1P ℓ 1u = ⊔″		$1N = \frac{1}{2}u$ by 2N
IV^G.32	8u ℓ 1N = ⊓ 8u ℓ 3N = □′		No solution
	65u ℓ 24N = □‴ 260u ℓ 24N = □^{iv}		$195u = 15u$ by $13u$
IV^G.39	8N 4u = □ 6N 4u = □″		$2N = 4u$ by $\frac{1}{2}N$
V^G.1	1P ℓ 12u = □″ $6\frac{1}{2}$N ℓ 12u = □′		$1P \ell 6\frac{1}{2}N = 1N$ by $1N \ell 6\frac{1}{2}u$
V^G.2	1P 20u = □″ 4N 20u = □′		No solution
	1P 20u = □″ $9\frac{1}{2}$N 20u = □′		$1P \ell 9\frac{1}{2}N = 1N$ by $1N \ell 9\frac{1}{2}u$
VI^G.6	1u 14N = □ 1P 1u = □′		$1P \ell 14N = 1N$ by $1N \ell 14u$

VI^G.8	1P 14N 1u = □′ 1P 1u = □″		14N = 2N by 7u
VI^G.22	1P 1,048,576u ℓ 6,144N = □″ 16,384N 1,048,576u = □‴		Not worked out

Appendix 3

Lexicon and conventions

In the translation of the Greek text we intervene in the text published by Tannery in several ways, which are the following:

All abbreviations of Tannery's text are resolved. Thus, the words behind the abbreviations are written in full. Also, the first letters of the names of the powers of the unknown are capitalized. For example, we write "number" for the common noun ἀριθμός, and "Number" for the technical term ἀριθμός, abbreviated in Tannery's text as ϛ, and transcribed in his translation as x.

Text in angle brackets is a proposed addition or restoration, either by Tannery or by some previous editor, usually by Bachet, or sometimes by Auria.

Words in parentheses are our addition, aiming to facilitate the reading of the text, which is often highly elliptic. In some instances the implied words are added without parentheses, as in the case of the participle ὁ μετρῶν, "the measuring", which is always given as "the measuring number". Also, expressions like ὁ ἐν τῇ ὑποτεινούσει, which appears in the sixth Greek book, meaning "the number representing the hypotenuse", are always given as "the number for the hypotenuse"; likewise, we write "the number for the area", "the number for the perpendicular", etc. Similarly, we write "the sides about the right angle" instead of "the (sides) about the right (angle)".

Constructions with genitive case, when a numerical value or a name is assigned, are transformed to constructions in nominative. Thus, the expression ἔσται/γίνεται μονάδων {n}, "it will be/it becomes of {n} units", is translated as "it will be/it becomes {n} units". Similarly, the expression τετάχθω ὁ ἐλάσσων ἀριθμοῦ ἑνός, "let the lesser be assigned to be of 1 Number", is translated as "let the lesser be assigned to be 1 Number".

Most often the product of two numbers is expressed by the formula ὁ ὑπ' αὐτῶν, that is, "the (number contained) by them". As a rule, throughout our translation it is translated by "their product" (cf. Section 4.2.4). Rarely we find the expression ὁ ἐκ τοῦ πολλαπλασιασμοῦ, that is, "the (number produced) from the multiplication". In our translation we write the complete phrase (cf. Section 4.2.4). Occasionally, when the multiplication is between sides of a right-angled triangle, as, for example, in the expression ὁ τῆς ὑποτεινούσης καὶ τῆς καθέτου, we translate it by "the product of the hypotenuse and the perpendicular". Similarly, for the expression ὁ ἐκ τῆς συνθέσεως, "the from the addition", a shortened form of the

DOI: 10.4324/9781315171470-32

phrase ὁ ἐκ τῆς συνθέσεως γενόμενος ἀριθμός, "the number produced from the addition", is translated as "their sum".

The square of a number is often expressed by the formula ὁ ἀπὸ τοῦ {*a*}, "the (square) on the {*a*}", occasionally the word τετράγωνος, "square", being added. We translated it as "the square on {*a*}" or "the square of {*a*}". Similarly, the expression οι ἀπ' αὐτῶν τετράγωνοι, lit. "the squares on them", is translated as "their squares".

What we call the "coefficient" of an algebraic term is rendered in the Greek text by πλῆθος, "multitude", as in Problem IV^G.8 where, having found the equation "19 Powers are equal to 1 unit", Diophantus says: "And the one unit is a square; if 19, the multitude of the Powers, were also a square, the equality would be solvable". However, often, expressions like "the multitude of Numbers" or "the multitude of Powers", etc. are shortened to simply "the Numbers", "the Powers". The same holds true also for the Arabic text. In such cases we sometimes add the word "multitude" in parentheses in our translation.

Apart from adding some clarifications in parentheses, translating the Arabic text does not call for interventions like these.

Lexica

A   English–Greek lexicon

A.1 Verbs

accomplish: λύειν
acquire: παραλαμβάνειν
    acquire practice: γεγυμνάσθαι
add: προστιθέναι; συντιθέναι
    add together: συντιθέναι
amount to: συνάγειν
apply: ἐπιτιθέναι; ποιεῖν
appears: δοκεῖν
arise: γίνεσθαι; συμβαίνειν
arrive: διέρχεσθαι; ἐξέρχεσθαι
ask: ζητεῖν
assign: ἐπιτάσσειν; τάσσειν
associate: προστιθέναι
assume: λαμβάνειν
avoid: μὴ ἐμπίπτειν
be: γίνεσθαι; εἶναι; ἐνυπάρχειν; καθιστάναι; ποιεῖν; τυγχάνειν
    be able: δύνασθαι
    be added: προσκεῖσθαι
    be amenable: ἐπιδέχεσθαι
    be built: συνιστάναι
    be composed: συγκεῖσθαι
    be constant: ἱστάναι

be contained: περιέχεσθαι
be convenient (to do): προσήκειν
be decreased: λείπειν
be due: συμβαίνειν
be equivalent: ἰσχύειν
be fulfilled: μένειν
be extant: ἐνυπάρχειν
be given: κτᾶσθαι
be greater: πλεονάζειν
be led: γίνεσθαι
be left: λοιπὸς/λοιπὸν εἶναι; καταλείπεσθαι; μένειν
    be left with: λοιπὸν/λοιποὺς ἔχειν
be made up: συγκεῖσθαι
be necessary: δεῖν; ὀφείλειν
be posited: ὑποκεῖσθαι
be possible: δύνασθαι; ἐνδέχεσθαι
be present: ἐπιτρέχειν
be produced: γίνεσθαι
be reduced: γίνεσθαι; ἔρχεσθαι
be required: δεῖν
be subtracted: λείπεσθαι
be supposed: ὑποκεῖσθαι
be unsuccessful: σχολάζειν
become: γίνεσθαι
begin: ἄρχεσθαι; ἐνάρχεσθαι
bisect: τέμνειν δίχα
    bisecting (line): τέμνουσα δίχα (εὐθεῖα)
bring (to the hypostases): ποιεῖν (ἐπὶ τὰς ὑποστάσεις)
buy: ἀγοράζειν
call: καλεῖν
can: δύνασθαι; ἔχειν
clarify: σαφηνίζειν
collect: συναθροίζειν
come: γίνεσθαι; ἔρχεσθαι
    come back: ἔρχεσθαι
    come out: συμβαίνειν; συνάγεσθαι
    come to be: γίνεσθαι
complete: περαίνειν
consider: βλέπειν; ὁρᾶν
constitute: ποιεῖν
construct: κατασκευάζειν; συνιστάναι
cut: τέμνειν
    cutting line: τέμνουσα (εὐθεῖα)
describe before/previously: προγράφειν
    describe below: ὑπογράφειν

determine: διορίζεσθαι
differ: διαφέρειν
distinguish: διαστέλλειν
divide: ἀναλύειν; διαιρεῖν; διελεῖν; μερίζειν; παραβάλλειν
　　divide again/anew: μεταδιαιρεῖν; μεταδιελεῖν
do: ποιεῖν; γίνεσθαι
eliminate: αἴρειν; περιαιρεῖν
equate: ἐξισοῦν; ἰσάζειν; ἰσοῦν; ποιεῖν ἴσον
establish: κατασκευάζειν; δοκιμάζειν
examine: ἐπισκέπτεσθαι
exceed: ὑπεραίρειν; ὑπερβάλλειν; ὑπερέχειν; πλεονάζειν
expose: ἐκτιθέναι
express: προδηλοῦν
fall: ἐμπίπτειν
fall short: ἐλλείπειν
find: εὑρίσκειν; προσευρίσκειν
　　is found to be: συνάγεται
follow: βαδίζειν; χωρεῖν
form: πλάσσειν; πλέκεσθαι
fulfill: μένειν; ποιεῖν; σώζειν
give: διδόναι; ποιεῖν
　　gives you . . . another square: ποιεῖν . . . ἕτερόν σε φέρειν τετράγωνον
happen: τυγχάνειν
have: ἔχειν; κτᾶσθαι
hold good: συνιστάναι
hypostasize: ὑφίστασθαι
increase: παραυξάνειν
investigate: ζητεῖν; σκέπτεσθαι
know: γιγνώσκειν; εἰδέναι
lack: λείπειν
learn: μανθάνειν
leave: καταλείπειν
look for: ζητεῖν
make: βάλλειν; γίνεσθαι; κατασκευάζειν; ποιεῖν
　　make equal: ἐξισοῦν
　　make use: χρῆσθαι
master: βεβαιοῦν
mean: λέγειν; σημαίνειν
measure: μετρεῖν
memorize: μνημονεύειν
mix up: μιγνύναι
multiply: γίνεσθαι; ποιεῖν; πολλαπλασιάζειν; πολυπλασιάζειν
must: δεῖν; ὀφείλειν
name: ἐπονομάζειν
observe: ὁρᾶν

obtain: εὑρίσκειν
    be obtained: γίνεσθαι
partition: διαιρεῖν
pay: ἀποδιδόναι
posit: ὑφίστασθαι
proceed: ἀκολουθεῖν; μεταβαίνειν
produce: γίνεσθαι; ποιεῖν
propose: ἐπιτάσσειν; προβάλλειν
receive: δέχεσθαι; λαμβάνειν
    receive in addition: προσλαμβάνειν
reduce: ἀναλύειν; ἀπάγειν; γίνεσθαι
remain: λοιπός/λοιπόν γίνεσθαι/δεῖν/εἶναι; μένειν; καταλείπεσθαι; περιλείπεσθαι;
    ὑπολείπεσθαι
remove: αἴρειν; ἀφαιρεῖν
render: παρασκευάζειν
require: δεῖν; ζητεῖν; δεῖ εὑρεῖν
    required: ζητούμενος
resolve: ἀπολύειν
restore: ἐμβάλλειν
result: γίνεσθαι
return: ἀνατρέχειν; ἔρχεσθαι
satisfy (the proposed task): συμφωνεῖν (τὸ ἐπίταγμα)
say: λέγειν; φάναι
    said: εἰρημένος
    aforesaid: προειρημένος
see: ζητεῖν
seek: ζητεῖν
    sought-after: ἐπιζητούμενος; ζητούμενος
septuplicate: ποιεῖν ἑπτάκις
set: τάσσειν
    set out: ἐκεῖσθαι; ἐκτιθέναι; κεῖσθαι
        set out before: προεκτιθέναι
    set up anew: μεθυφίσταμαι
    set up the equation: ἰσοῦν
should: δεῖ
show: δεικνύναι; ὑποδεικνύναι
    show before: προδεικνύναι
solve: διαλύειν; ἰσοῦν; λύειν
square: τετραγωνίζειν
stem: γίνεσθαι
subtract: ἀφαιρεῖν (ἀφελεῖν; ὑφαιρεῖν)
support: προσλαμβάνειν
suppose: ὑποτιθέναι
take: γίνεσθαι; εὑρίσκειν; λαμβάνειν
    take interchangeably: ἐναλλάσσειν

take the square: τετραγωνίζειν
tell apart: διαστέλλειν
the remainder is: λοιπός/λοιπόν γίνεσθαι/εἶναι
think worthwhile: δοκιμάζειν
try: πειρᾶσθαι
turn
    turn out to be: γίνεσθαι
    turn upside down: κάτω νεύειν
use: χρῆσθαι
want: βούλεσθαι; θέλειν
wish: θέλειν
work out with subtlety: φιλοτεχνεῖν
yield: συνάγειν

A.2 *Other terms*

abbreviated: συντομώτερος
absurd: ἄτοπος
acute: ὀξύς
addition: πρόσθεσις; σύνθεσις
angle: γωνία
arbitrary: τυχών
area: ἐμβαδόν
arising with the formation: πλασματικός
arithmetical: ἀριθμητικός
base: βάσις
    the whole base: ἡ ἐξ ἀρχῆς βάσις
beginning: ἀρχή
book: βιβλίον
characteristic: ἰδίωμα
child: παῖς
clear: φανερός
cogius: χοῦς
composite: σύνθετος
corollary: πόρισμα
corresponding: ἀντικείμενος
cotyla: κοτύλη
cube: κύβος
Cube: κύβος
Cube-Cube: κυβόκυβος
dead end: ἄπορον
denomination: ἐπωνυμία
demonstration: ἀπόδειξις
determinate: ὡρισμένος
determination: προσδιορισμός

difference: διαφορά; ὑπεροχή
difficult to be memorized: δυσμνημονευτός
division: διαίρεσις; μερισμός; παραβολή
double-equality or double equality: διπλῆ ἰσότης; διπλοϊσότης
double equation: διπλῆ ἴσωσις
drachma: δραχμή
dyad: δυάς
eagerness: προθυμία
easy to follow: εὐόδευτος
easy to grasp: εὐκατάληπτος
epigram: ἐπίγραμμα
equal: ἴσος
    approximatively equal: πάρισος
equality: ἰσότης
    approximate equality: παρισότης
equation: ἴσωσις
excess: ὑπεροχή
exchange: ἀντίδοσις
extant (species/term): ὕπαρξις
expressible: ῥητός
    inexpressible: οὐ ῥητός
extreme: ἄκρος
familiar: γνώριμος
formation: ὕπαρξις
foundation: θεμέλιον
fraction: μόριον
greater: μείζων
hexad: ἑξάς
homonymous: ὁμώνυμος
hypostasis: ὑπόστασις
hypotenuse: ὑποτείνουσα
immutable: ἀμετάθετος
impossible: ἀδύνατος
indeterminate: ἀόριστος
initial (number/square): ὁ ἐν ἀρχῇ (ἀριθμός/τετράγωνος)
interval: τόπος
inverse Cube: κυβοστόν
inverse Cube-Cube: κυβοκυβοστόν
inverse Number: ἀριθμοστόν
inverse Power: δυναμοστόν
inverse Power-Cube: δυναμοκυβοστόν
inverse Power-Power: δυναμοδυναμοστόν
just described: προκείμενος
keenness: ἐπιθυμία
knowledge: μάθησις

lacking (species/term): λεῖψις (appearing most often in the dative, λείψει,
   represented in Tannery's text by the sign ⋏1)
lemma: λῆμμα
lessening: ἐλάττωσις (spurious)
manifest: δῆλος
manner: τρόπος
mark: γραμμή
material: ὕλη
measurement: μέτρησις
method: ἀγωγή
most honourable: τιμιώτατος
multiplication: πολλαπλασιασμός; πολυπλασιασμός
multitude: πλῆθος
number: ἀριθμός
   initial (number): ὁ ἐν ἀρχῇ (ἀριθμός)
Number: ἀριθμὸς
numerical: ἀριθμητικός
numerical value: ὑπόστασις
obvious: καταφανής; φανερός
octad: ὀκτάς
of little hope: δυσέλπιστος
part: μέρος; μόριον; τμῆμα
partition: διαίρεσις
perimeter: περίμετρος
perpendicular: κάθετος; ὀρθή
physis: φύσις
plane (number): ἐπίπεδος (ἀριθμὸς)
porism: πόρισμα
possible: δυνατός
power: δύναμις
Power: δύναμις
Power-Cube: δυναμόκυβος
Power-Power: δυναμοδύναμις
price: τιμή
prime (number): πρῶτος (ἀριθμός)
problem: πρόβλημα
   initial (problem): τὸ ἐξ ἀρχῆς (πρόβλημα)
procedure: ἀγωγή
product: χωρίον; "product" is also the translation we employ for the expression
   ὁ ὑπ' αὐτῶν
proof: ἀπόδειξις
proportion: ἀναλογία
(what was) proposed: προκείμενος
   proposed task: ἐπίταγμα
question: ζήτημα; ζητούμενον

quotient: παραβολή
rather difficult: δυσχερέστερος
ratio: λόγος
remainder: λοιπός/λοιπόν
remaining: λοιπός/λοιπόν
right (angle): ὀρθή (γωνία)
right-angled (triangle): ὀρθογώνιον (τρίγωνον)
rule: μέθοδος
    to expose how: ὀργανῶσαι τὴν μέθοδον
sailing companion: ὁμόπλοος
segment: τμῆμα; τομή
    segment cut off: ἀποτομή
sesquitertian: ἐπίτριτος
setting-up of the equation: ἴσωσις
side: πλευρά
sign: σημεῖον
simpler: ἁπλούστερος
size: ὄγκος
slowly: βραδέως
smaller: ἐλάσσων
solid number: στερεὸς ἀριθμός
soul: ψυχή
species: εἶδος
square: τετραγωνικός; τετράγωνος
study: πραγματεία
subject: πρᾶγμα
subtraction: ἀφαίρεσις
successful accomplishment: κατόρθωσις
sum/sum of both/sum of the two/both together (as one)/the two together (as one):
    συναμφότερος; συναμφότεροι; σύνθεμα; σύνθεσις
supposition: ὑπόθεσις
teaching: ἀπόδειξις; διδαχή
tetrad: τετράς
treatment: πραγματεία
triad: τριάς
triangular: τριγωνικός; τρίγωνος
truncated: ἐλλιπής
unequal: ἄνισος
unit: μονάς
unknown: ἄδηλος
unlimited: ἄπειρος
unspoken: ἄλογος
way: ὁδός
wine: οἶνος
working out: ἀγωγή

B English–Arabic lexicon

See also Sesiano's Arabic index (1982, 431–60), which also gives references to the text and to corresponding Greek terms.

Note: the second and sometimes third words following "also" appear much less frequently than the first word.

to add: زاد; also ضاف
to add together: جمع
added: مزيد; مزاد
addition: زيادة
again: كذلك; also ايضا
all of that: جميع ذلك
also: ايضا
amounts to: اجتمع من
analysis: تحليل
any: كل; اى
appended: زائدة
as (for stating ratios): ك-
to assign: فرض
assigned: مفروض
because: من اجل
both sides (of an equation): كلى الناحيتين
to cast away: لقي
characteristic: خاصة
clear: ظاهر; بين
comes from: من; يكون من; الكائن من; يجنتمع من; مجتمع من
commensurable: مشاركة; قدر
common, in common: مشتركة
composed: مركب
condition: شريطة
to confront: قبل
confrontation: مقابلة
consist of: من
to contain: حاط
cube, cubic: مكعب
Cube (algebraic): كعب
to describe: وصف
description: صفة
to determine: حد
difference: تفاضل
difference between: فضل بين
different: مختلف
to dispense: غنى; استغنى

to divide: قسم

dividend: مقسوم

division: قسم

divisor: مقسوم عليه

double: ضعف

double equality: مساواة مثناة

to duplicate: ضعف

each of them: كل واحد منهما

equal: تساوى; استوى; مساو; مثل

to Equal, Equate: عدل

Equated: معادل

Equation: معادلة

everything: كل; جميع

to exceed: زاد

except: غير

excess: زيادة

excluded: مستثنى

to exhibit: ظهر

to explain: بين; وصف

to find: وجد

for, as for: اما

to form: عمل

formation (*plasmatikon*): مهينة

given: مفروض

goal: غرض

greater: أعظم; عظيم¹

greater than: اكثر من

however many: كم

hypostasis: اصول

to increase: زاد

known: معلوم

lacking: ناقص

larger: اكبر

to leave: بقي

less: الا

less than: اقل من

lesser in degree: اقعد

like terms: متشابهات

likewise: كذلك; also ايضا

look: لمس; also طلب

made up of: مؤلفة من

---

1 This form, which ordinarily means "great", appears at lines 1177, 1215, 1216, 1260, 1262, 1263, 1308, 1488, and 1490. See Sesiano (1982, 115 n. 75).

to make: جعل

Māl: مال

measure: عد

method: جهة

middle: اوسط

more than: اكثر من

multiple: امثال

multiplication: مضروب

to multiply: ضرب

must: وجب

namely: اعنى

it is necessary: وجب; also حاج; ينبغى

to need: حاج

number: عدد

number of times: امثال; مرات

once, (one) time: مرة

one: واحد

other than: غير

outcome: ما حصل; ما بلغ; بلغ

part (for stating fractions): جزء

part (divide a number into parts): قسم

partitioned: مقسوم

planar: مسطح

plasmatikon, formation: مهينة

portion: بعض

present (adj): زائدة

problem: مسئلة

procedure: عمل; also منفذ

product: مضروب

proportional: متناسبة

quantity: مقدار

quotient: قسم

ratio: نسبة

to remain: بقي

remainder: باقى

repetition: اعادة

to require: التماس; طلب; also حاج

restoration: جبر

to restore: جبر

to restrict: حدد

restriction: تحديد

to result in: خرج

reverse, reverse order, inverse, vice versa: جذذر

root/Root: جذر

rule: قياس

same: مثل

to say: قال

to search, to seek: طلب ;لمس

should: بغى ; also وجب ;حاج

to show: بين

side (of a square, cube, etc.): ضلع

side (of an equation): ناحية ; also طرف[2];جهة

similarly: ايضا ;كذلك

since: لأنا

size: مقدار

small: يسير

smaller/lesser: اصغر ;صغير[2]

smaller than: اصغر من

sought-after: مطلوب

species: نوع

square: مربع

to stipulate: شرط

stipulation: شرط

to subtract: نقص

subtracted: منقوص

sum: مجموع ;مجتمع ;جميع ;جملة

synthesis: تركيب

to synthesize: ركب

that is: اعنى

therefore, thus, etc.: لذلك

Thing: شيء

times (number of times): امثال

turn out to be: انتهى الى

unit: احد ;واحد

units: آحاد

value: مقدار

to want: راد

to wish: شاء

what is gathered: ما يجتمع ;الذى يجتمع

whatever: ما ; also اى

whole: كامل ;كل

whole number: عدد صحيح

with respect to (for expressing ratios): من ;عند

without: غير

work out: عمل

---

2 This form, which ordinarily means "small", appears at lines 1176, 1177, 1215, 1216, 1261 (twice), 1263, 1487, and 1488 (twice). See Sesiano (1982, 115 n. 75).

Appendix 4
Conspectus of problems

The purpose of a conspectus should be to give the reader a guide to the problems that can be read quickly and efficiently. It should reveal the mathematical structure of what is being asked and not be concerned with the particular way that a problem is worded. Any conspectus must make use of notation of some sort, and notation necessarily includes the naming of known and unknown quantities. In Diophantus naming is a task reserved for the solution, so by its very nature an enunciation cannot be faithfully rendered in notation. This means that no conspectus can be true to the enunciations on a conceptual level. For this reason, we have adopted unabashedly modern algebraic notation for our conspectus.

For example, an enunciation like "To add the same square number to a cube and to a square and make the same" in Problem IV^G.6 is not identical with our notational version $\begin{cases} x^3 + z^3 = \boxdot \\ y^2 + z^2 = \Box \end{cases}'$. To make the latter we have taken the steps of naming the "same square number" as z^2, and the cube and the square as x^3 and y^2, respectively. By contrast, when Diophantus names these unknowns, this latter square is not named "1 Power", but "9 Powers", and what we call z^2 is named "16 Powers". Where we have taken the step of setting up two modern indeterminate equations in three variables, in his solution Diophantus sets up a single determinate equation in one unknown: "1 Cube, 16 Powers . . . equal . . . 8 Cubes". As long as one keeps these distinctions in mind when reading over this list, it can be a very useful tool for investigating the *Arithmetica*.

We have adopted the form that Sesiano gives in his conspectus because it is clear and efficient. We have changed only a few things. Where Sesiano used a, b, c, and d for the sought-after numbers and k, l, etc. for givens, we use x, y, z, and w for the sought-after numbers and a, b, c, and d for the givens. As in Sesiano, multitudes and ratios are represented by m, n, and p in our list. When a number is to take a certain form such as a square or a cube, we show this form by the shapes $\Box$ and $\boxdot$. The sides of such squares or cubes are shown as $\sqrt{\Box}$ and $\sqrt[3]{\boxdot}$, respectively. Finally, the shape $\triangle$ denotes a triangular number.

DOI: 10.4324/9781315171470-33

Book I

I.1 $\begin{cases} x+y=a \\ x-y=b \end{cases}$ $(a=100, b=40)$

I.2 $\begin{cases} x+y=a \\ x=my \end{cases}$ $(a=60, m=3)$

I.3 $\begin{cases} x+y=a \\ x=my+b \end{cases}$ $(a=80, b=4, m=3)$

I.4 $\begin{cases} x=my \\ x-y=a \end{cases}$ $(m=5, a=20)$

I.5 $\begin{cases} x+y=a \\ \frac{1}{m}x+\frac{1}{n}y=b \end{cases}$ $(a=100, b=30, m=3, n=5)$

I.6 $\begin{cases} x+y=a \\ \frac{1}{m}x-\frac{1}{n}y=b \end{cases}$ $(a=100, b=20, m=4, n=6)$

I.7 $x-a=m(x-b)$ $(a=20, b=100, m=3)$

I.8 $x+a=m(x+b)$ $(a=100, b=20, m=3)$

I.9 $a-x=m(b-x)$ $(a=100, b=20, m=6)$

I.10 $a+x=m(b-x)$ $(a=20, b=100, m=4)$

I.11 $x+a=m(x-b)$ $(a=20, b=100, m=3)$

I.12 $\begin{cases} x+y=x'+y'=a \\ x=my' \\ x'=ny \end{cases}$ $(a=100, m=2, n=3)$

I.13 $\begin{cases} x+y=x'+y'=x''+y''=a \\ x=my' \\ x'=ny'' \\ x''=py \end{cases}$ $(a=100, m=3, n=2, p=4)$

I.14 $x \cdot y = m(x+y)$ $(m=3)$

I.15 $\begin{cases} x+a=m(y-a) \\ y+b=n(x-b) \end{cases}$ $(a=30, b=50, m=2, n=3)$

I.16 $\begin{cases} x+y=a \\ y+z=b \\ z+x=c \end{cases}$ $\qquad (a=20,\, b=30,\, c=40)$

I.17 $\begin{cases} w+x+y=a \\ x+y+z=b \\ y+z+w=c \\ z+w+x=d \end{cases}$ $\qquad \begin{array}{l}(a=20,\, b=22,\, c=24, \\ d=27)\end{array}$

I.18 $\begin{cases} x+y-z=a \\ y+z-x=b \\ z+x-y=c \end{cases}$ $\qquad (a=20,\, b=30,\, c=40)$

I.19 $\begin{cases} w+x+y-z=a \\ x+y+z-w=b \\ y+z+w-x=c \\ z+w+x-y=d \end{cases}$ $\qquad \begin{array}{l}(a=20,\, b=30,\, c=40, \\ d=50)\end{array}$

I.20 $\begin{cases} x+y+z=a \\ x+y=mz \\ y+z=nx \end{cases}$ $\qquad (a=100,\, m=3,\, n=4)$

I.21 $\begin{cases} x-y=\frac{1}{m}z \\ y-z=\frac{1}{n}x \\ z-a=\frac{1}{p}y \end{cases}$ $\qquad (a=10,\, m=3,\, n=3,\, p=3)$

I.22 $\begin{aligned} x-\tfrac{1}{m}x+\tfrac{1}{p}z &= y-\tfrac{1}{n}y+\tfrac{1}{m}x \\ &= z-\tfrac{1}{p}z+\tfrac{1}{n}y \end{aligned}$ $\qquad (m=3,\, n=4,\, p=5)$

I.23 $\begin{aligned} w-\tfrac{1}{m}w+\tfrac{1}{q}z &= x-\tfrac{1}{n}x+\tfrac{1}{m}w \\ &= y-\tfrac{1}{p}y+\tfrac{1}{n}x \\ &= z-\tfrac{1}{q}z+\tfrac{1}{p}y \end{aligned}$ $\qquad (m=3,\, n=4,\, p=5,\, q=6)$

I.24 $\begin{aligned} x+\tfrac{1}{m}(y+z) &= y+\tfrac{1}{n}(z+x) \\ &= z+\tfrac{1}{p}(x+y) \end{aligned}$ $\qquad (m=3,\, n=4,\, p=5)$

I.25 $\begin{aligned} w+\tfrac{1}{m}(x+y+z) &= x+\tfrac{1}{n}(y+z+w) \\ &= y+\tfrac{1}{p}(z+w+x) \\ &= z+\tfrac{1}{q}(w+x+y) \end{aligned}$ $\qquad (m=3,\, n=4,\, p=5,\, q=6)$

I.26 $\begin{cases} a \cdot x = \square \\ b \cdot x = \sqrt{\square} \end{cases}$ $\qquad$ $(a = 200, b = 5)$

I.27 $\begin{cases} x + y = a \\ x \cdot y = b \end{cases}$ $\qquad$ $(a = 20, b = 96)$

I.28 $\begin{cases} x + y = a \\ x^2 + y^2 = b \end{cases}$ $\qquad$ $(a = 20, b = 208)$

I.29 $\begin{cases} x + y = a \\ x^2 - y^2 = b \end{cases}$ $\qquad$ $(a = 20, b = 80)$

I.30 $\begin{cases} x - y = a \\ x \cdot y = b \end{cases}$ $\qquad$ $(a = 4, b = 96)$

I.31 $\begin{cases} x = my \\ x^2 + y^2 = n(x+y) \end{cases}$ $\qquad$ $(m = 3, n = 5)$

I.32 $\begin{cases} x = my \\ x^2 + y^2 = n(x-y) \end{cases}$ $\qquad$ $(m = 3, n = 10)$

I.33 $\begin{cases} x = my \\ x^2 - y^2 = n(x+y) \end{cases}$ $\qquad$ $(m = 3, n = 6)$

I.34 $\begin{cases} x = my \\ x^2 - y^2 = n(x-y) \end{cases}$ $\qquad$ $(m = 3, n = 12)$

Corollaries: I.34.1 $\begin{cases} x = my \\ x \cdot y = n(x+y) \end{cases}$

I.34.2 $\begin{cases} x = my \\ x \cdot y = n(x-y) \end{cases}$

I.35 $\begin{cases} x = my \\ y^2 = nx \end{cases}$ $\qquad$ $(m = 3, n = 6)$

I.36 $\begin{cases} x = my \\ y^2 = ny \end{cases}$ $\qquad$ $(m = 3, n = 6)$

I.37 $\begin{cases} x = my \\ y^2 = n(x+y) \end{cases}$ $(m = 3, n = 2)$

I.38 $\begin{cases} x = my \\ y^2 = n(x-y) \end{cases}$ $(m = 3, n = 6)$

Corollaries: I.38.1 $\begin{cases} x = my \\ x^2 = ny \end{cases}$

   I.38.2 $\begin{cases} x = my \\ x^2 = nx \end{cases}$

   I.38.3 $\begin{cases} x = my \\ x^2 = n(x+y) \end{cases}$

   I.38.4 $\begin{cases} x = my \\ x^2 = n(x-y) \end{cases}$

I.39   The 3 numbers $(a + b)x$, $(b + x)a$, and $(x + a)b$, arranged in order, have equal differences. $(a = 3, b = 5)$

Book II

II.1 $x + y = m\left(x^2 + y^2\right)$ $(m = \frac{1}{10})$

II.2 $x - y = m\left(x^2 - y^2\right)$ $(m = \frac{1}{6})$

II.3   (a) $x \cdot y = m(x + y)$ $(m = 6)$
       (b) $x \cdot y = m(x - y)$ $(m = 6)$

II.4 $x^2 + y^2 = m(x - y)$ $(m = 10)$

II.5 $x^2 - y^2 = m(x + y)$ $(m = 6)$

II.6 $\begin{cases} x - y = a \\ \left(x^2 - y^2\right) - (x - y) = b \end{cases}$ $(a = 2, b = 20)$

II.7 $(x^2 - y^2) - m(x - y) = a$ $(m = 3, a = 10)$

II.8 $x^2 + y^2 = a^2$ $(a = 16)$

II.9 $x^2 + y^2 = a^2 + b^2$ $(a^2 = 4, b^2 = 9)$

II.10 $x^2 - y^2 = a$ $(a = 60)$

II.11 $\begin{cases} x+a=\square \\ x+b=\square' \end{cases}$ $(a=2, k=3)$

II.12 $\begin{cases} a-x=\square \\ b-x=\square' \end{cases}$ $(a=9, b=21)$

II.13 $\begin{cases} x-a=\square \\ x-b=\square' \end{cases}$ $(a=6, b=7)$

II.14 $\begin{cases} x+y=a, \\ z^2+x=\square \\ z^2+y=\square' \end{cases}$ $(a=20)$

II.15 $\begin{cases} x+y=a, \\ z^2-x=\square \\ z^2-y=\square' \end{cases}$ $(a=20)$

II.16 $\begin{cases} x=my \\ x+a^2=\square \\ y+a^2=\square' \end{cases}$ $(m=3, a^2=9)$

II.17 $\begin{cases} x-\left(\frac{1}{m}x+a\right)+\left(\frac{1}{p}z+c\right)= \\ y-\left(\frac{1}{n}y+b\right)+\left(\frac{1}{m}x+a\right)= \\ z-\left(\frac{1}{p}z+c\right)+\left(\frac{1}{n}y+b\right) \end{cases}$ $(m=5, n=6, p=7, a=6, b=7, \\ c=8)$

II.18 $\begin{cases} x+y+z=d \\ x-\left(\frac{1}{m}x+a\right)+\left(\frac{1}{p}z+c\right) \\ \quad =y-\left(\frac{1}{n}y+b\right)+\left(\frac{1}{m}x+a\right) \\ \quad =z-\left(\frac{1}{p}z+c\right)+\left(\frac{1}{n}y+b\right) \end{cases}$ $(d=80, m=5, n=6, p=7, \\ a=6, b=7, c=8)$

II.19 $\quad x^2-y^2=m(y^2-z^2)$ $(m=3)$

II.20 $\begin{cases} x^2+y=\square \\ y^2+x=\square' \end{cases}$

II.21 $\begin{cases} x^2-y=\square \\ y^2-x=\square' \end{cases}$

II.22
$$\begin{cases} x^2 + (x+y) = \square \\ y^2 + (x+y) = \square' \end{cases}$$

II.23
$$\begin{cases} x^2 - (x+y) = \square \\ y^2 - (x+y) = \square' \end{cases}$$

II.24
$$\begin{cases} (x+y)^2 + x = \square \\ (x+y)^2 + y = \square' \end{cases}$$

II.25
$$\begin{cases} (x+y)^2 - x = \square \\ (x+y)^2 - y = \square' \end{cases}$$

II.26
$$\begin{cases} x \cdot y + x = \square \\ x \cdot y + y = \square' \\ \sqrt{\square} + \sqrt{\square'} = a \end{cases} \qquad (a = 6)$$

II.27
$$\begin{cases} x \cdot y - x = \square \\ x \cdot y - y = \square' \\ \sqrt{\square} + \sqrt{\square'} = a \end{cases} \qquad (a = 5)$$

II.28
$$\begin{cases} x^2 \cdot y^2 + x^2 = \square \\ x^2 \cdot y^2 + y^2 = \square' \end{cases}$$

II.29
$$\begin{cases} x^2 \cdot y^2 - x^2 = \square \\ x^2 \cdot y^2 - y^2 = \square' \end{cases}$$

II.30
$$\begin{cases} x \cdot y + (x+y) = \square \\ x \cdot y - (x+y) = \square' \end{cases}$$

II.31
$$\begin{cases} x + y = \square \\ x \cdot y + (x+y) = \square' \\ x \cdot y - (x+y) = \square'' \end{cases}$$

II.32
$$\begin{cases} x^2 + y = \square \\ y^2 + z = \square' \\ z^2 + x = \square'' \end{cases}$$

II.33
$$\begin{cases} x^2 - y = \square \\ y^2 - z = \square' \\ z^2 - x = \square'' \end{cases}$$

II.34
$$\begin{cases} x^2 + (x + y + z) = \square \\ y^2 + (x + y + z) = \square' \\ z^2 + (x + y + z) = \square'' \end{cases}$$

II.35
$$\begin{cases} x^2 - (x + y + z) = \square \\ y^2 - (x + y + z) = \square' \\ z^2 - (x + y + z) = \square'' \end{cases}$$

Book III

III.1
$$\begin{cases} (x + y + z) - x^2 = \square \\ (x + y + z) - y^2 = \square' \\ (x + y + z) - z^2 = \square'' \end{cases}$$

III.2
$$\begin{cases} (x + y + z)^2 + x = \square \\ (x + y + z)^2 + y = \square' \\ (x + y + z)^2 + z = \square'' \end{cases}$$

III.3
$$\begin{cases} (x + y + z)^2 - x = \square \\ (x + y + z)^2 - y = \square' \\ (x + y + z)^2 - z = \square'' \end{cases}$$

III.4
$$\begin{cases} x - (x + y + z)^2 = \square \\ y - (x + y + z)^2 = \square' \\ z - (x + y + z)^2 = \square'' \end{cases}$$

III.5
$$\begin{cases} x + y + z = \square \\ x + y - z = \square' \\ y + z - x = \square'' \\ z + x - y = \square''' \end{cases}$$

III.6
$$\begin{cases} x+y+z = \square \\ x+y = \square' \\ y+z = \square'' \\ z+x = \square''' \end{cases}$$

III.7
$$\begin{cases} z-y = y-x \\ x+y = \square \\ y+z = \square' \\ z+x = \square'' \end{cases}$$

III.8
$$\begin{cases} x+y+a = \square \\ y+z+a = \square' \\ z+x+a = \square'' \\ x+y+z+a = \square''' \end{cases}$$
$(a = 3)$

III.9
$$\begin{cases} x+y-a = \square \\ y+z-a = \square' \\ z+x-a = \square'' \\ x+y+z-a = \square''' \end{cases}$$
$(a = 3)$

III.10
$$\begin{cases} x \cdot y + a = \square \\ y \cdot z + a = \square' \\ z \cdot x + a = \square'' \end{cases}$$
$(a = 12)$

III.11
$$\begin{cases} x \cdot y - a = \square \\ y \cdot z - a = \square' \\ z \cdot x - a = \square'' \end{cases}$$
$(a = 10)$

III.12
$$\begin{cases} x \cdot y + z = \square \\ y \cdot z + x = \square' \\ z \cdot x + y = \square'' \end{cases}$$

III.13
$$\begin{cases} x \cdot y - z = \square \\ y \cdot z - x = \square' \\ z \cdot x - y = \square'' \end{cases}$$

III.14
$$\begin{cases} x \cdot y + z^2 = \square \\ y \cdot z + x^2 = \square' \\ z \cdot x + y^2 = \square'' \end{cases}$$

III.15 $\begin{cases} x \cdot y + (x + y) = \square \\ y \cdot z + (y + z) = \square' \\ z \cdot x + (z + x) = \square'' \end{cases}$

III.16 $\begin{cases} x \cdot y - (x + y) = \square \\ y \cdot z - (y + z) = \square' \\ z \cdot x - (z + x) = \square'' \end{cases}$

III.17 $\begin{cases} x \cdot y + (x + y) = \square \\ x \cdot y + x = \square' \\ x \cdot y + y = \square'' \end{cases}$

III.18 $\begin{cases} x \cdot y - (x + y) = \square \\ x \cdot y - x = \square' \\ x \cdot y - y = \square'' \end{cases}$

III.19   Eight conditions on w, x, y, z:
Adding or removing any one of them from
$(w + x + y + z)^2$ gives a $\square$.

III.20 $\begin{cases} x + y = a \\ z^2 - x = \square \\ z^2 - y = \square' \end{cases}$ $(a - 10)$

III.21 $\begin{cases} x + y = a \\ z^2 + x = \square \\ z^2 + y = \square' \end{cases}$ $(a = 20)$

Book IV (Arabic)

IV.1 $x^3 + y^3 = \square$

IV.2 $x^3 - y^3 = \square$

IV.3 $x^2 + y^2 = \boxed{\square}$

IV.4 $x^2 - y^2 = \boxed{\square}$

IV.5 $x^2 \cdot y^2 = \boxed{\square}$

IV.6 $x^2 \cdot y^3 = \square$

IV.7 $x^2 \cdot y^3 = \boxed{}$

IV.8 $x^3 \cdot y^3 = \square$

IV.9 $x^3 \cdot y^3 = \square$ (same as IV.8)

IV.9.1 $x^3 \div y^3 = \square$

IV.9.2 $x^2 \div y^2 = \boxed{}$

IV.10 $x^3 + mx^2 = \square$ $\qquad\qquad\qquad$ $(m = 10)$

IV.11 $x^3 - mx^2 = \square$ $\qquad\qquad\qquad$ $(m = 6)$

IV.12 $x^3 + mx^2 = \boxed{}$ $\qquad\qquad$ $(m = 10)$

IV.13 $x^3 - mx^2 = \boxed{}$ $\qquad\qquad$ $(m = 7)$

IV.14 $\begin{cases} a \cdot x = \boxed{} \\ b \cdot x = \square' \end{cases}$ $\qquad\qquad$ $(a = 10, b = 5)$
(then again, with $a = 5$, $b = 10$)

IV.15 $\begin{cases} a \cdot x = \boxed{} \\ b \cdot x = \square \end{cases}$ $\qquad\qquad$ $(a = 10, b = 5)$

IV.16 $\begin{cases} a \cdot x = \boxed{} \\ a \cdot y = \sqrt[3]{\boxed{}} \end{cases}$ $\qquad\qquad$ $(a = 10)$

IV.17 $\begin{cases} x = my \\ a \cdot x^2 = \boxed{} \\ a \cdot y^2 = \sqrt[3]{\boxed{}} \end{cases}$ $\qquad\qquad$ $(a = 5, b = 20)$

IV.18 $\begin{cases} x = my \\ a \cdot x^3 = \square \\ a \cdot y^3 = \sqrt{\square} \end{cases}$ $\qquad\qquad$ $(a = 8, m = 3)$

IV.19 $\begin{cases} a \cdot x = \boxed{} \\ b \cdot x = \sqrt[3]{\boxed{}} \end{cases}$ $\qquad\qquad$ $(a = 20, b=5)$

IV.20 $\begin{cases} a \cdot x^3 = \square \\ b \cdot x^3 = \sqrt{\square} \end{cases}$ $\qquad\qquad$ $(a = 200, b = 5)$

IV.21 $\begin{cases} a \cdot x^2 = \boxed{} \\ b \cdot x^2 = \sqrt[3]{\boxed{}} \end{cases}$ $(a = 40\frac{1}{2},\ b = 2)$

IV.22 $\begin{cases} a \cdot x^3 = \boxed{} \\ b \cdot x^3 = \sqrt[3]{\boxed{}} \end{cases}$ $(a = 91\frac{1}{8},\ b = 2)$

IV.23 $\left(x^2\right)^2 + \left(y^2\right)^2 = \boxed{}$

IV.24 $\left(x^2\right)^2 - \left(y^2\right)^2 = \boxed{}$

IV.25 $(x^3)^2 + (y^2)^2 = \square$

IV.26 (a) $(x^3)^2 - (y^2)^2 = \square$
(b) $(y^2)^2 - (x^3)^2 = \square$

IV.27 $(x^3)^2 + my^2 = \square$ $(m = 5)$

IV.28 $(x^2)^2 + my^3 = \square$ $(m = 10)$

IV.29 $(x^3)^3 + (y^2)^2 = \square$

IV.30 $(x^3)^3 - (y^2)^2 = \square$

IV.31 $(y^2)^2 - (x^3)^3 = \square$

IV.32 $(x^3)^3 + m(y^2 \cdot x^3) = \square$ $(m = 5)$

IV.33 $(x^3)^3 - m(y^2 \cdot x^3) = \square$ $(m = 3)$

Corollaries. $(x^3)^3 + m(y^2 \cdot x^3) = \square$
$(y^2)^2 + m(y^2 \cdot x^3) = \square$

IV.34 $\begin{cases} x^3 + y^2 = \square \\ x^3 - y^2 = \square' \end{cases}$

IV.35 $\begin{cases} y^2 + x^3 = \square \\ y^2 - x^3 = \square' \end{cases}$

IV.36 $\begin{cases} x^3 + mx^2 = \square \\ x^3 - nx^2 = \square' \end{cases}$ $(m = 4,\ n = 5)$

IV.37 $\begin{cases} x^3 + mx^2 = \square \\ x^3 + nx^2 = \square' \end{cases}$ $(m = 5,\ n = 10)$

IV.38 $\begin{cases} x^3 - mx^2 = \square \\ x^3 - nx^2 = \square' \end{cases}$ $(m = 5,\ n = 10)$

IV.39 $\begin{cases} mx^2 - x^3 = \square \\ nx^2 - x^3 = \square' \end{cases}$ $(m = 3, n = 7)$

IV.40 $\begin{cases} (x^2)^2 + y^3 = \square \\ (x^2)^2 - y^3 = \square' \end{cases}$

IV.41 $\begin{cases} x^3 + (y^2)^2 = \square \\ x^3 - (y^2)^2 = \square' \end{cases}$

IV.42

(a) $\begin{cases} (x^3)^3 + (y^2)^2 = \square \\ (x^3)^3 - (y^2)^2 = \square' \end{cases}$

(b) $\begin{cases} (x^3)^3 + (y^2)^2 = \square \\ (y^2)^2 - (x^3)^3 = \square' \end{cases}$

IV.43 $\begin{cases} (x^3)^3 + m(y^2)^2 = \square \\ (x^3)^3 - n(y^2)^2 = \square' \end{cases}$ $\left(m = 1\tfrac{1}{4}, n = \tfrac{3}{4} \right)$

IV.44

(a) $\begin{cases} (x^3)^3 + m(y^2)^2 = \square \\ (x^3)^3 + n(y^2)^2 = \square' \end{cases}$ $(m = 3, n = 8)$

(b) $\begin{cases} (x^3)^3 - m(y^2)^2 = \square \\ (x^3)^3 - n(y^2)^2 = \square' \end{cases}$ $(m = 3, n = 8)$

(c) $\begin{cases} m(y^2)^2 - (x^3)^3 = \square \\ n(y^2)^2 - (x^3)^3 = \square' \end{cases}$ $(m = 3, n = 8)$

Book V (Arabic)

V.1 $\begin{cases} (x^2)^2 + my^3 = \square \\ (x^2)^2 - ny^3 = \square' \end{cases}$ $(m = 4, n = 3)$

V.2 $\begin{cases} (x^2)^2 + my^3 = \square \\ (x^2)^2 + ny^3 = \square' \end{cases}$ $(m = 12, n = 5)$

V.3 $\begin{cases} (x^2)^2 - my^3 = \square \\ (x^2)^2 - ny^3 = \square' \end{cases}$ $(m = 12, n = 7)$

V.4 $\begin{cases} (x^2)^2 + m(y^3)^3 = \square \\ (x^2)^2 - n(y^3)^3 = \square' \end{cases}$ $(m = 5, n = 3)$

V.5 $\begin{cases} (x^2)^2 + m(y^3)^3 = \square \\ (x^2)^2 + n(y^3)^3 = \square' \end{cases}$ $(m = 12, n = 5)$

V.6 $\begin{cases} (x^2)^2 - m(y^3)^3 = \square \\ (x^2)^2 - n(y^3)^3 = \square' \end{cases}$ $(m = 7, n = 4)$

V.7 $\begin{cases} x + y = a \\ x^3 + y^3 = b \end{cases}$ $(a = 20, b = 2240)$

V.8 $\begin{cases} x - y = a \\ x^3 - y^3 = b \end{cases}$ $(a = 10, b = 2170)$

V.9 $\begin{cases} x + y = a \\ x^3 + y^3 = m(x - y)^2 \end{cases}$ $(a = 20, m = 140)$

V.10 $\begin{cases} x - y = a \\ x^3 - y^3 = m(x + y)^2 \end{cases}$ $(a = 10, m = 8\frac{1}{8})$

V.11 $\begin{cases} x - y = a \\ x^3 + y^3 = m(x + y) \end{cases}$ $(a = 4, m = 28)$

V.12 $\begin{cases} x + y = a \\ x^3 - y^3 = m(x - y) \end{cases}$ $(a = 8, m = 52)$

V.13 $\begin{cases} mx^2 + a = y + z^1 \\ x^3 + y = \square \\ x^3 + z = \square' \end{cases}$ $(m = 9, a = 30)$

1 In problems V.13–16 the numbers denoted by y and z are not asked for.

V.14 $\begin{cases} mx^2 - a = y + z \\ x^3 - y = \boxed{} \\ x^3 - z = \boxed{}' \end{cases}$ $(m = 9,\ a = 26)$

V.15 $\begin{cases} mx^2 - a = y + z \\ x^3 + y = \boxed{} \\ x^3 - z = \boxed{}' \end{cases}$ $(m = 9,\ a = 18)$

V.16 $\begin{cases} mx^2 - a = y + z \\ x^3 - y = \boxed{} \\ z - x^3 = \boxed{}' \end{cases}$ $(m = 9,\ a = 16)$

Book VI (Arabic)

VI.1 $\begin{cases} x = my \\ \left(x^3\right)^2 + (y^2)^2 = \square \end{cases}$ $(m = 2)$

VI.2 $\begin{cases} x = my \\ \left(x^3\right)^2 - (y^2)^2 = \square \end{cases}$ $(m = 2)$

VI.3 $\begin{cases} x = my \\ (y^2)^2 - \left(x^3\right)^2 = \square \end{cases}$ $(m = 2)$

VI.4 $\begin{cases} x = my \\ x^3 \cdot y^2 + \left(x^3\right)^2 = \square \end{cases}$ $(m = 5)$

VI.5 $\begin{cases} x = y \\ x^3 \cdot y^2 + (y^2)^2 = \square \end{cases}$

VI.6 $\begin{cases} x = y \\ x^3 \cdot y^2 - (x^3)^2 = \square \end{cases}$

VI.7 $\begin{cases} x = y \\ x^3 \cdot y^2 - (y^2)^2 = \square \end{cases}$

VI.8 $x^3 \cdot y^2 + \sqrt{x^3 \cdot y^2} = \square$

VI.9 $x^3 \cdot y^2 - \sqrt{x^3 \cdot y^2} = \square$

VI.10 $\sqrt{x^3 \cdot y^2} - x^3 \cdot y^2 = \square$

VI.11 $\left(x^3\right)^2 + x^3 = \square$

VI.12 $\begin{cases} x^2 + x^2 / y^2 = \square \\ y^2 + x^2 / y^2 = \square' \end{cases} x > y$

VI.13 $\begin{cases} x^2 - x^2 / y^2 = \square \\ y^2 - x^2 / y^2 = \square' \end{cases} x > y$

VI.14 $\begin{cases} x^2 / y^2 - x^2 = \square \\ x^2 / y^2 - y^2 = \square' \end{cases} x > y$

VI.15 $\begin{cases} x^2 + \left(x^2 - y^2\right) = \square \\ y^2 + \left(x^2 - y^2\right) = \square' \end{cases} x > y$

VI.16 $\begin{cases} x^2 - \left(x^2 - y^2\right) = \square \\ y^2 - \left(x^2 - y^2\right) = \square' \end{cases} x > y$

VI.17 $\begin{cases} x^2 + y^2 + z^2 = \square \\ x^2 = y \\ y^2 = z \end{cases}$

VI.18 $x^2 \cdot y^2 \cdot z^2 + \left(x^2 + y^2 + z^2\right) = \square$

VI.19 $x^2 \cdot y^2 \cdot z^2 - \left(x^2 + y^2 + z^2\right) = \square$

VI.20 $\left(x^2 + y^2 + z^2\right) - x^2 \cdot y^2 \cdot z^2 = \square$

VI.21 $\begin{cases} (x^2)^2 + \left(x^2 + y^2\right) = \square \\ (y^2)^2 + \left(x^2 + y^2\right) = \square' \end{cases}$

VI.22 $\begin{cases} x^2 + y^2 = \square \\ x^2 \cdot y^2 = \boxed{\square}' \end{cases}$

VI.23 $\begin{cases} a^2 / x^2 + a^2 / y^2 = \square \\ x^2 + y^2 + a^2 = \square' \end{cases}$ $(a^2 = 9)$

Book VII (Arabic)

VII.1 $\begin{cases} x = my \\ y = mz \\ x^3 \cdot y^3 \cdot z^3 = \square \end{cases}$ $(m = 2)$

VII.2 $x^6 \cdot y^6 \cdot z^6 = \square^2$

VII.3 $\left(w^2\right)^2 = x^3 + y^3 + z^3$

VII.4 $\left(w^2\right)^3 = x^2 + y^2 + z^2$

VII.5 $(x^3)^3 \cdot y^3 + (x^3)^3 \cdot z^2 = \square$

VII.6 $\begin{cases} x^2 + y^2 = \square \\ x^2 \cdot y^2 = m\left(x^2 + y^2\right) \end{cases}$ $(m = 9)$

VII.7 $\begin{cases} \left(w^3\right)^2 = x + y + z \\ x + y = \square \\ y + z = \square' \\ z + x = \square'' \end{cases}$

VII.8 $\begin{cases} \left(x^3\right)^2 + y = \square \\ \left(x^3\right)^2 + 2y = \square' \end{cases}$

VII.9 $\begin{cases} \left(x^3\right)^2 - y = \square \\ \left(x^3\right)^2 - 2y = \square' \end{cases}$

VII.10 $\begin{cases} \left(x^3\right)^2 + y = \square \\ \left(x^3\right)^2 - y = \square' \end{cases}$

VII.11 $\begin{cases} a^2 = x + y \\ a^2 + x = \square \\ a^2 - y = \square' \end{cases}$ $(a^2 = 25)$

Diophantus remarks that if the last condition is changed to an addition, the problem has no solution.

VII.12 $\begin{cases} a^2 = x + y \\ a^2 - x = \square \\ a^2 - y = \square' \end{cases}$ $(a^2 = 25)$

VII.13
$$\begin{cases} a^2 = x + y + z \\ a^2 + x = \square \\ a^2 + y = \square' \\ a^2 + z = \square'' \end{cases}$$
$\qquad (a^2 = 25)$

VII.14
$$\begin{cases} a^2 = x + y + z \\ a^2 - x = \square \\ a^2 - y = \square' \\ a^2 - z = \square'' \end{cases}$$
$\qquad (a^2 = 25)$

VII.15
$$\begin{cases} a^2 = w + x + y + z \\ a^2 - w = \square \\ a^2 - x = \square' \\ a^2 + y = \square'' \\ a^2 + z = \square''' \end{cases}$$
$\qquad (a^2 = 25)$

Diophantus then gives instructions for the same problem with eight unknowns, adding the first four and subtracting the last four.

VII.16
$$\begin{cases} x^2 : y^2 :: y^2 : z^2 \\ y^2 - x^2 = \square \\ z^2 - y^2 = \square' \end{cases}$$

VII.17
$$\begin{cases} w^2 : x^2 :: y^2 : z^2 \\ w^2 + x^2 + y^2 + z^2 = \square \end{cases}$$

VII.18
$$\begin{cases} w^2 : x^2 :: y^2 : z^2 \\ x^2 - w^2 = \square \\ y^2 - x^2 = \square' \\ z^2 - y^2 = \square'' \end{cases}$$

Book IV^G (Greek)

IV^G.1
$$\begin{cases} x + y = a \\ x^3 + y^3 = b \end{cases}$$
$\qquad (a = 10, b = 370)$

IV^G.2
$$\begin{cases} x - y = a \\ x^3 - y^3 = b \end{cases}$$
$\qquad (a = 6, b = 504)$

$\mathrm{IV^G}.3 \quad \begin{cases} x^2 \cdot y = \sqrt[3]{▱} \\ x \cdot y = ▱ \end{cases}$

$\mathrm{IV^G}.4 \quad \begin{cases} x^2 + y = \square \\ x + y = \sqrt{\square} \end{cases}$

$\mathrm{IV^G}.5 \quad \begin{cases} x^2 + y = \sqrt{\square} \\ x + y = \square \end{cases}$

$\mathrm{IV^G}.6 \quad \begin{cases} x^3 + z^2 = ▱ \\ y^2 + z^2 = \square' \end{cases}$

$\mathrm{IV^G}.7 \quad \begin{cases} x^3 + z^2 = \square \\ y^2 + z^2 = ▱' \end{cases}$

$\mathrm{IV^G}.8 \quad \begin{cases} x^3 + y = ▱ \\ x + y = \sqrt[3]{▱} \end{cases}$

$\mathrm{IV^G}.9 \quad \begin{cases} x^3 + y = \sqrt[3]{▱} \\ x + y = ▱ \end{cases}$

$\mathrm{IV^G}.10 \quad x^3 + y^3 = x + y$

$\mathrm{IV^G}.11 \quad x^3 - y^3 = x - y$

$\mathrm{IV^G}.12 \quad x^3 + y = y^3 + x$

$\mathrm{IV^G}.13 \quad \begin{cases} x + 1 = \square \\ y + 1 = \square' \\ (x + y) + 1 = \square'' \\ (y - x) + 1 = \square''' \end{cases}$

$\mathrm{IV^G}.14 \quad \begin{aligned} &x^2 + y^2 + z^2 = \\ &(x^2 - y^2) + (y^2 - z^2) + (x^2 - z^2) \end{aligned}$

$\mathrm{IV^G}.15 \quad \begin{cases} (x + y) \cdot z = a \\ (y + z) \cdot x = b \\ (z + x) \cdot y = c \end{cases} \qquad (a = 35,\ b = 27,\ c = 32)$

$$\text{IV}^{\text{G}}.16 \quad \begin{cases} x + y + z = \square \\ x^2 + y = \square' \\ y^2 + z = \square'' \\ z^2 + x = \square''' \end{cases}$$

$$\text{IV}^{\text{G}}.17 \quad \begin{cases} x + y + z = \square \\ x^2 - y = \square' \\ y^2 - z = \square'' \\ z^2 - x = \square''' \end{cases}$$

$$\text{IV}^{\text{G}}.18 \quad \begin{cases} x^3 + y = \boxed{\square} \\ y^2 + x = \square' \end{cases}$$

$$\text{IV}^{\text{G}}.19 \quad \begin{cases} x \cdot y + 1 = \square \\ y \cdot z + 1 = \square' \\ z \cdot x + 1 = \square'' \end{cases}$$

$$\text{IV}^{\text{G}}.20 \quad \begin{cases} w \cdot x + 1 = \square \\ w \cdot y + 1 = \square' \\ w \cdot z + 1 = \square'' \\ x \cdot y + 1 = \square''' \\ x \cdot z + 1 = \square^{iv} \\ y \cdot z + 1 = \square^{v} \end{cases}$$

$$\text{IV}^{\text{G}}.21 \quad \begin{cases} x : y :: y : z \\ x - y = \square \\ y - z = \square' \\ x - z = \square'' \end{cases}$$

$$\text{IV}^{\text{G}}.22 \quad \begin{cases} x \cdot y \cdot z + x = \square \\ x \cdot y \cdot z + y = \square' \\ x \cdot y \cdot z + z = \square'' \end{cases}$$

$$\text{IV}^{\text{G}}.23 \quad \begin{cases} x \cdot y \cdot z - x = \square \\ x \cdot y \cdot z - y = \square' \\ x \cdot y \cdot z - z = \square'' \end{cases}$$

$$\text{IV}^{\text{G}}.24 \quad \begin{cases} x + y = a \\ x \cdot y = \boxed{\square} - \sqrt[3]{\boxed{\square}} \end{cases} \qquad (a = 6)$$

$\text{IV}^{\text{G}}.25 \quad \begin{cases} x+y+z = a \\ x \cdot y \cdot z = \square \\ (y-x)+(z-y)+(z-x) = \sqrt[3]{\square} \end{cases} \qquad (a = 4)$

$\text{IV}^{\text{G}}.26 \quad \begin{cases} x \cdot y + x = \square \\ x \cdot y + y = \square' \end{cases}$

$\text{IV}^{\text{G}}.27 \quad \begin{cases} x \cdot y - x = \square \\ x \cdot y - y = \square' \end{cases}$

$\text{IV}^{\text{G}}.28 \quad \begin{cases} x \cdot y + (x+y) = \square \\ x \cdot y - (x+y) = \square' \end{cases}$

$\text{IV}^{\text{G}}.29 \quad \begin{aligned} & w^2 + x^2 + y^2 + z^2 + \\ & (w+x+y+z) = a \end{aligned} \qquad (a = 12)$

$\text{IV}^{\text{G}}.30 \quad \begin{aligned} & w^2 + x^2 + y^2 + z^2 - \\ & (w+x+y+z) = a \end{aligned} \qquad (a = 4)$

$\text{IV}^{\text{G}}.31 \quad \begin{cases} x+y = 1 \\ (x+a) \cdot (y+b) = \square \end{cases} \qquad (a = 4, b = 5)$

$\text{IV}^{\text{G}}.32 \quad \begin{cases} x+y+z = a \\ x \cdot y + z = \square \\ x \cdot y - z = \square' \end{cases} \qquad (a = 6)$

$\text{IV}^{\text{G}}.33 \quad \begin{cases} x + \frac{m}{n} y = p\left(y - \frac{m}{n} y\right) \\ y + \frac{m}{n} x = q\left(x - \frac{m}{n} x\right) \end{cases} \qquad \begin{aligned} & (p = 3, q = 5, \\ & \text{posits } \tfrac{m}{n} y = 1 \text{ in solution}) \end{aligned}$

Lemma: $\quad x \cdot y + (x+y) = a \qquad (a = 8)$

$\text{IV}^{\text{G}}.34 \quad \begin{cases} x \cdot y + (x+y) = a \\ y \cdot z + (y+z) = b \\ z \cdot x + (z+x) = c \end{cases} \qquad (a = 8, b = 15, c = 24)$

Lemma: $\quad x \cdot y - (x+y) = a \qquad (a = 8)$

$\text{IV}^{\text{G}}.35 \quad \begin{cases} x \cdot y - (x+y) = a \\ y \cdot z - (y+z) = b \\ z \cdot x - (z+x) = c \end{cases} \qquad (a = 8, b = 15, c = 24)$

Lemma: $x \cdot y = m\,(x + y)$ $(m = 3)$

IV^G.36 $\begin{cases} x \cdot y = m\left(x + y\right) \\ y \cdot z = n\left(y + z\right) \\ z \cdot x = p\left(z + x\right) \end{cases}$ $(m = 3, \, n = 4, \, p = 5)$

IV^G.37 $\begin{cases} x \cdot y = m\left(x + y + z\right) \\ y \cdot z = n\left(x + y + z\right) \\ z \cdot x = p\left(x + y + z\right) \end{cases}$ $(m = 3, \, n = 4, \, p = 5)$

IV^G.38 $\begin{cases} \left(x + y + z\right) \cdot x = \triangle \\ \left(x + y + z\right) \cdot y = \square' \\ \left(x + y + z\right) \cdot z = \boxed{\square}'' \end{cases}$

IV^G.39 $\begin{cases} \left(x - y\right) = m\left(y - z\right) \\ x + y = \square \\ y + z = \square' \\ z + x = \square'' \end{cases}$ $(m = 3)$

IV^G.40 $\begin{cases} \left(x^2 - y^2\right) = m\left(y - z\right) \\ x + y = \square \\ y + z = \square' \\ z + x = \square'' \end{cases}$ $(m = 3)$

Book V^G (Greek)

V^G.1 $\begin{cases} x : y :: y : z \\ x - a = \square \\ y - a = \square' \\ z - a = \square'' \end{cases}$ $(a = 12)$

V^G.2 $\begin{cases} x : y :: y : z \\ x + a = \square \\ y + a = \square' \\ z + a = \square'' \end{cases}$ $(a = 20)$

$V^G.3$

$$\begin{cases} x+a=\square \\ y+a=\square' \\ z+a=\square'' \\ x\cdot y+a=\square''' \\ y\cdot z+a=\square^{iv} \\ z\cdot x+a=\square^{v} \end{cases}$$

$(a=5)$

$V^G.4$

$$\begin{cases} x-a=\square \\ y-a=\square' \\ z-a=\square'' \\ x\cdot y-a=\square''' \\ y\cdot z-a=\square^{iv} \\ z\cdot x-a=\square^{v} \end{cases}$$

$(a=6)$

$V^G.5$

$$\begin{cases} x^2\cdot y^2+\left(x^2+y^2\right)=\square \\ y^2\cdot z^2+\left(y^2+z^2\right)=\square' \\ z^2\cdot x^2+\left(z^2+x^2\right)=\square'' \\ x^2\cdot y^2+z^2=\square''' \\ y^2\cdot z^2+x^2=\square^{iv} \\ z^2\cdot x^2+y^2=\square^{v} \end{cases}$$

$V^G.6$

$$\begin{cases} x-2=\square \\ y-2=\square' \\ z-2=\square'' \\ x\cdot y-(x+y)=\square''' \\ y\cdot z-(y+z)=\square^{iv} \\ z\cdot x-(z+x)=\square^{v} \\ x\cdot y-z=\square^{vi} \\ y\cdot z-x=\square^{vii} \\ z\cdot x-y=\square^{viii} \end{cases}$$

Lemma: $x\cdot y+(x^2+y^2)=\square$

Lemma: To find numerically three right tringles having equal areas

$$V^G.7 \quad \begin{cases} x^2 + (x+y+z) = \square \\ x^2 - (x+y+z) = \square' \\ y^2 + (x+y+z) = \square'' \\ y^2 - (x+y+z) = \square''' \\ z^2 + (x+y+z) = \square^{iv} \\ z^2 - (x+y+z) = \square^{v} \end{cases}$$

$$\text{Lemma:} \quad \begin{cases} x \cdot y = a^2 \\ y \cdot z = b^2 \\ z \cdot x = c^2 \end{cases} \qquad (a = 2,\ b = 3,\ c = 4)$$

$$V^G.8 \quad \begin{cases} x \cdot y + (x+y+z) = \square \\ x \cdot y - (x+y+z) = \square' \\ y \cdot z + (x+y+z) = \square'' \\ y \cdot z - (x+y+z) = \square''' \\ z \cdot x + (x+y+z) = \square^{iv} \\ z \cdot x - (x+y+z) = \square^{v} \end{cases}$$

$$V^G.9 \quad \begin{cases} x + y = 1 \\ x + a = \square \\ y + a = \square' \end{cases} \qquad (a = 6)$$

$$V^G.10 \quad \begin{cases} x + y = 1 \\ x + a = \square \\ y + b = \square' \end{cases} \qquad (a = 2,\ b = 6)$$

$$V^G.11 \quad \begin{cases} x + y + z = 1 \\ x + a = \square \\ y + a = \square' \\ z + a = \square'' \end{cases} \qquad (a = 3)$$

$$V^G.12 \quad \begin{cases} x + y + z = 1 \\ x + a = \square \\ y + b = \square' \\ z + c = \square'' \end{cases} \qquad (a = 6,\ b = 3,\ c = 4)$$

$$V^G.13 \quad \begin{cases} x+y+z=a \\ x+y=\square \\ y+z=\square' \\ z+x=\square'' \end{cases} \qquad (a=10)$$

$$V^G.14 \quad \begin{cases} w+x+y+z=a \\ w+x+y=\square \\ x+y+z=\square' \\ y+z+w=\square'' \\ z+w+x=\square''' \end{cases} \qquad (a=10)$$

$$V^G.15 \quad \begin{cases} (x+y+z)^3+x=\boxtimes \\ (x+y+z)^3+y=\boxtimes' \\ (x+y+z)^3+z=\boxtimes'' \end{cases}$$

$$V^G.16 \quad \begin{cases} (x+y+z)^3-x=\boxtimes \\ (x+y+z)^3-y=\boxtimes' \\ (x+y+z)^3-z=\boxtimes'' \end{cases}$$

$$V^G.17 \quad \begin{cases} x-(x+y+z)^3=\boxtimes \\ y-(x+y+z)^3=\boxtimes' \\ z-(x+y+z)^3=\boxtimes'' \end{cases}$$

$$V^G.18 \quad \begin{cases} x+y+z=\square \\ (x+y+z)^3+x=\square' \\ (x+y+z)^3+y=\square'' \\ (x+y+z)^3+z=\square''' \end{cases}$$

$$V^G.19 \quad \begin{cases} x+y+z=\square \\ (x+y+z)^3-x=\square' \\ (x+y+z)^3-y=\square'' \\ (x+y+z)^3-z=\square''' \end{cases}$$

Problems 19a–c are missing from the manuscripts. Their enunciations were reconstructed by Bachet from the symmetry of the triads of problems 18-19-19a and 19b-19c-20.

$$V^G.19a \quad \begin{cases} x+y+z = \Box \\ x-(x+y+z)^3 = \Box' \\ y-(x+y+z)^3 = \Box'' \\ z-(x+y+z)^3 = \Box''' \end{cases}$$

$$V^G.19b \quad \begin{cases} x+y+z = a \\ (x+y+z)^3 + x = \Box \\ (x+y+z)^3 + y = \Box' \\ (x+y+z)^3 + z = \Box'' \end{cases} \qquad (a=2)$$

$$V^G.19c \quad \begin{cases} x+y+z = a \\ (x+y+z)^3 - x = \Box \\ (x+y+z)^3 - y = \Box' \\ (x+y+z)^3 - z = \Box'' \end{cases} \qquad (a=2)$$

$$V^G.20 \quad \begin{cases} x+y+z = \frac{1}{a} \\ x-(x+y+z)^3 = \Box \\ y-(x+y+z)^3 = \Box' \\ z-(x+y+z)^3 = \Box'' \end{cases} \qquad (a=4)$$

$$V^G.21 \quad \begin{cases} x^2 \cdot y^2 \cdot z^2 + x^2 = \Box \\ x^2 \cdot y^2 \cdot z^2 + y^2 = \Box' \\ x^2 \cdot y^2 \cdot z^2 + z^2 = \Box'' \end{cases}$$

$$V^G.22 \quad \begin{cases} x^2 \cdot y^2 \cdot z^2 - x^2 = \Box \\ x^2 \cdot y^2 \cdot z^2 - y^2 = \Box' \\ x^2 \cdot y^2 \cdot z^2 - z^2 = \Box'' \end{cases}$$

$$V^G.23 \quad \begin{cases} x^2 - x^2 \cdot y^2 \cdot z^2 = \Box \\ y^2 - x^2 \cdot y^2 \cdot z^2 = \Box' \\ z^2 - x^2 \cdot y^2 \cdot z^2 = \Box'' \end{cases}$$

$$V^G.24 \quad \begin{cases} x^2 \cdot y^2 + 1 = \Box \\ y^2 \cdot z^2 + 1 = \Box' \\ z^2 \cdot x^2 + 1 = \Box'' \end{cases}$$

$$V^G.25 \quad \begin{cases} x^2 \cdot y^2 - 1 = \Box \\ y^2 \cdot z^2 - 1 = \Box' \\ z^2 \cdot x^2 - 1 = \Box'' \end{cases}$$

$\text{V}^{\text{G}}.26 \quad \begin{cases} 1 - x^2 \cdot y^2 = \square \\ 1 - y^2 \cdot z^2 = \square' \\ 1 - z^2 \cdot x^2 = \square'' \end{cases}$

$\text{V}^{\text{G}}.27 \quad \begin{cases} x^2 + y^2 + a = \square \\ y^2 + z^2 + a = \square' \\ z^2 + x^2 + a = \square'' \end{cases} \qquad (a = 15)$

$\text{V}^{\text{G}}.28 \quad \begin{cases} x^2 + y^2 - a = \square \\ y^2 + z^2 - a = \square' \\ z^2 + x^2 - a = \square'' \end{cases} \qquad (a = 13)$

$\text{V}^{\text{G}}.29 \quad \left(x^2\right)^2 + \left(y^2\right)^2 + \left(z^2\right)^2 = \square$

$\text{V}^{\text{G}}.30 \quad \begin{cases} a \cdot x + b \cdot y = \square \\ (x + y)^2 = \square + c \end{cases} \qquad (a = 8, \ b = 5, \ c = 60)$

Book VI^{G} (Greek)

$\text{VI}^{\text{G}}.1 \quad \begin{cases} x^2 + y^2 = z^2 \\ z - x = \boxdot \\ z - y = \boxdot' \end{cases}$

$\text{VI}^{\text{G}}.2 \quad \begin{cases} x^2 + y^2 = z^2 \\ z + x = \boxdot \\ z + y = \boxdot' \end{cases}$

$\text{VI}^{\text{G}}.3 \quad \begin{cases} x^2 + y^2 = z^2 \\ \frac{1}{2} x \cdot y + a = \square \end{cases} \qquad (a = 5)$

$\text{VI}^{\text{G}}.4 \quad \begin{cases} x^2 + y^2 = z^2 \\ \frac{1}{2} x \cdot y - a = \square \end{cases} \qquad (a = 6)$

$\text{VI}^{\text{G}}.5 \quad \begin{cases} x^2 + y^2 = z^2 \\ a - \frac{1}{2} x \cdot y = \square \end{cases} \qquad (a = 10)$

$\text{VI}^{\text{G}}.6 \quad \begin{cases} x^2 + y^2 = z^2 \\ \frac{1}{2} x \cdot y + x = a \end{cases} \qquad (a = 7)$

$\text{VI}^{\text{G}}.7 \quad \begin{cases} x^2 + y^2 = z^2 \\ \frac{1}{2} x \cdot y - x = a \end{cases} \qquad (a = 7)$

VI^G.8 $\begin{cases} x^2 + y^2 = z^2 \\ \frac{1}{2}x \cdot y + (x + y) = a \end{cases}$ $\qquad (a = 6)$

VI^G.9 $\begin{cases} x^2 + y^2 = z^2 \\ \frac{1}{2}x \cdot y - (x + y) = a \end{cases}$ $\qquad (a = 6)$

VI^G.10 $\begin{cases} x^2 + y^2 = z^2 \\ \frac{1}{2}x \cdot y + (z + x) = a \end{cases}$ $\qquad (a = 4)$

VI^G.11 $\begin{cases} x^2 + y^2 = z^2 \\ \frac{1}{2}x \cdot y - (z + x) = a \end{cases}$ $\qquad (a = 4)$

Lemma: $\begin{cases} x^2 + y^2 = z^2 \\ y - x = \square \\ y = \square' \\ \frac{1}{2}x \cdot y + x = \square' \end{cases}$ $\qquad y > x$

Lemma: $\begin{cases} a + b = \square \\ a \cdot x^2 + b = \square' \end{cases}$ $\qquad (a = 3,\ b = 6)$

VI^G.12 $\begin{cases} x^2 + y^2 = z^2 \\ \frac{1}{2}x \cdot y + x = \square \\ \frac{1}{2}x \cdot y + y = \square' \end{cases}$

VI^G.13 $\begin{cases} x^2 + y^2 = z^2 \\ \frac{1}{2}x \cdot y - x = \square \\ \frac{1}{2}x \cdot y - y = \square' \end{cases}$

VI^G.14 $\begin{cases} x^2 + y^2 = z^2 \\ \frac{1}{2}x \cdot y - z = \square \\ \frac{1}{2}x \cdot y - x = \square' \end{cases}$

Lemma: Given $a \cdot c^2 - b = d^2$, find $x^2 > c^2$ such that $a \cdot x^2 - b = \square$.

VI^G.15 $\begin{cases} x^2 + y^2 = z^2 \\ \frac{1}{2}x \cdot y + z = \square \\ \frac{1}{2}x \cdot y + x = \square' \end{cases}$

VI^G.16 Find $x^2 + y^2 = z^2$ such that if one of the acute angles is bisected, the bisecting line is rational.

$$\text{VI}^\text{G}.17 \quad \begin{cases} x^2 + y^2 = z^2 \\ \frac{1}{2}x \cdot y + z = \square \\ x + y + z = \boxdot' \end{cases}$$

$$\text{VI}^\text{G}.18 \quad \begin{cases} x^2 + y^2 = z^2 \\ \frac{1}{2}x \cdot y + z = \boxdot \\ x + y + z = \square' \end{cases}$$

$$\text{VI}^\text{G}.19 \quad \begin{cases} x^2 + y^2 = z^2 \\ \frac{1}{2}x \cdot y + x = \square \\ x + y + z = \boxdot' \end{cases}$$

$$\text{VI}^\text{G}.20 \quad \begin{cases} x^2 + y^2 = z^2 \\ \frac{1}{2}x \cdot y + x = \boxdot \\ x + y + z = \square' \end{cases}$$

$$\text{VI}^\text{G}.21 \quad \begin{cases} x^2 + y^2 = z^2 \\ x + y + z = \square \\ (x + y + z) + \frac{1}{2}x \cdot y = \boxdot' \end{cases}$$

$$\text{VI}^\text{G}.22 \quad \begin{cases} x^2 + y^2 = z^2 \\ x + y + z = \boxdot \\ (x + y + z) + \frac{1}{2}x \cdot y = \square' \end{cases}$$

$$\text{VI}^\text{G}.23 \quad \begin{cases} x^2 + y^2 = z^2 \\ z^2 = \square + \sqrt{\square} \\ z^2 / x = \boxdot' + \sqrt[3]{\boxdot'} \end{cases}$$

$$\text{VI}^\text{G}.24 \quad \begin{cases} x^2 + y^2 = z^2 \\ x = \boxdot \\ y = \boxdot' - \sqrt[3]{\boxdot'} \\ z = \boxdot'' + \sqrt[3]{\boxdot''} \end{cases}$$

Bibliography

I Manuscripts

A   For Diophantus's manuscripts, see Section 1.3

B   Greek manuscripts

Città del Vaticano, Biblioteca Apostolica Vaticana
  Vat. gr. 190. Facsimile available online: https://digi.vatlib.it/view/MSS_Vat.gr.190.pt.1,
    https://digi.vatlib.it/view/MSS_Vat.gr.190.pt.2
  Vat. gr. 1087. Facsimile available online: https://digi.vatlib.it/view/MSS_Vat.gr.1087
  Vat. gr. 1365.
  Vat. gr. 1584.
  Vat. Urb. gr. 78. Facsimile available online: https://digi.vatlib.it/view/MSS_Urb.gr.78
  Vat. Urb. gr. 80. Facsimile available online: https://digi.vatlib.it/view/MSS_Urb.gr.80
El Escorial, Real Biblioteca del Monasterio de S. Lorenzo
  Scorial. gr. Y.III.12
Firenze, Biblioteca Medicea Laurenziana5
  Laur. Plut. 28.11. Facsimile available online: http://mss.bmlonline.it/Catalogo.aspx?
    Shelfmark=Plut.28.11
  Laur. Plut. 28.18. Facsimile available online: http://mss.bmlonline.it/Catalogo.aspx?
    Shelfmark=Plut.28.18
  Laur. Plut. 58.29. Facsimile available online: http://mss.bmlonline.it/Catalogo.aspx?
    Shelfmark=Plut.58.29
  Laur. Plut. 86.3.
Heidelberg, Universitätsbibliothek
  Palatinus Heidelbergensis gr. 23. Facsimile available online: https://digi.ub.uni-heidel-
    berg.de/diglit/cpgraec23
Istanbul, Topkapi Sarayi Müzesi Kütüphanesi
  Constantinopolitanus Palatii veteris 1 (G.I.1)
Michigan, University of Michigan
  Papyrus Michigan 620. Facsimile available online: https://quod.lib.umich.edu/a/apis/x-
    2698/620R.TIF?lasttype=boolean;lastview=thumbnail;resnum=2;size=20;sort=a
    pis_inv;start=1;subview=detail;view=entry;rgn1=apis_inv;select1=regex;q1=620
Munich, Bayerische Staatsbibliothek
  Monac. gr. 384. Facsimile available online: https://daten.digitale-sammlungen.de/~db/
    0006/bsb00069139/images/index.html
Paris, Bibliothèque nationale de France

Paris. gr. 1630. Facsimile available online: https://gallica.bnf.fr/ark:/12148/btv1b1072 3587k/f23.item

Paris. gr. 2372. Facsimile available online: https://gallica.bnf.fr/ark:/12148/btv1b107236852

Paris. gr. 2448. Facsimile available online: https://gallica.bnf.fr/ark:/12148/btv1b1072 26329.r=Paris.%20gr.%202448?rk=21459;2

Paris. suppl. gr. 384. Facsimile available online: https://gallica.bnf.fr/ark:/12148/btv1b 8470199g/f8.item

C   Arabic manuscripts

Berlin, Staatsbibliothek
Ibn Fallūs, MS Landberg 199. Facsimile available online: https://digital.staatsbiblio-thek-berlin.de/werkansicht/?PPN=PPN646151142

Cairo, National Library (Dar al-kutub)
Sinān Ibn al-Fatḥ, Riyadāt 260/4

Città del Vaticano, Biblioteca Apostolica Vaticana
'Alī al-Sulamī, MS Sbath 5. Facsimile available online: https://digi.vatlib.it/view/ MSS_Sbath.5

Istanbul, Suleymaniye Library
al-Samaw'al, MS Esad Efendi 3155
al-Samaw'al, MS Aya Sofia 2718
Ibn al-Nadīm, Şehit Ali Paşa 1934

Leiden, Leiden University Libraries.
al-Nīsābūrī, MS Leiden Or. 780. Facsimile available online: http://hdl.handle.net/1887.1/ item:1573840

London, British Library
Ibn al-Majdī, Add MS 7469. Facsimile available online: www.qdl.qa/en/archive/81055/ vdc_100040139973.0x000001

Paris, Bibliothèque nationale de France
al-Karajī, MS arabe 2459. Facsimile available online : https://gallica.bnf.fr/ark:/12148/ btv1b11001788s
al-Khāzin, MS arabe 2457 . Facsimile available online: https://gallica.bnf.fr/ark:/12148/ btv1b11001636f
Ibn al-Khawwām, MS arabe 2470. Facsimile availavle online: https://gallica.bnf.fr/ ark:/12148/btv1b11001965c

Tehran, University of Tehran
al-Ḥilātī, MS 4409

University of Pennsylvania
al-Ḥaṣṣār, *Kitāb al-bayān wa'l-tadhkār fī ṣan'at 'amal al-ghubār.*
Lawrence J. Schoenberg collection, MS ljs 293, copied in Ṣafar 590 (Jan/Feb 1194 CE). Facsimile available online: http://openn.library.upenn.edu/Data/0001/html/ljs293.html
Ibn Qunfudh, *Ḥaṭṭ al-niqāb 'an wujūh a'māl al-ḥisāb.*
Lawrence J. Schoenberg collection, MS ljs 464, copied 11 Sha'bān 849 (November 1445). Facsimile available online: http://openn.library.upenn.edu/Data/0001/html/ ljs464.html

Washington, D.C., Library of Congress
Ibn Ghāzī, Mansuri Collection, 5-722. Facsimile available online: http://lccn.loc. gov/2008401022

D   Latin manuscripts

Città del Vaticano, Biblioteca Apostolica Vaticana
   Ottobon. lat. 3307. Facsimile available online: https://digi.vatlib.it/view/MSS_Ott.lat.3307
Firenze, Biblioteca Nazionale
   Palatino 573
Munich, Bayerische Staatsbibliothek
   CLM 14908. Facsimile available online: www.digitale-sammlungen.de/en/view/bsb00
      103422?page=,1
New York, Columbia University, Rare Book and Manuscript Library
   Plimpton 188

II Printed works

Abdeljaouad, Mahdi, ed., 2003. *Sharh al-urjūza al-Yāsmīnīya, de Ibn al-Hā'im.* Texte établi et commenté par Mahdi Abdeljaouad. Tunis: Association Tunisienne des Sciences Mathématiques.

Abdeljaouad, Mahdi, and Jeffrey Oaks, 2021. *Al-Hawārī's Essential Commentary: Arabic Arithmetic in the Fourteenth Century.* Berlin: Max-Planck-Gesellschaft zur Förderung der Wissenschaften.

Abū, Kāmil, 1986. *Kitāb fī al-Jabr wa'l-muqābala. A Facsimile Edition of MS Istanbul, Kara Mustafa Paşa 379, copied in 1253 C.E.* Edited by Jan P. Hogendijk. Frankfurt am Main: Institute for the History of Arabic-Islamic Science at the Johann Wolfgang Goethe University.

Abū, Kāmil, 2012. *Algèbre et analyse diophantienne.* Edition, traduction et commentaire par Roshdi Rashed. Berlin: Walter de Gruyter.

Acerbi, Fabio, 2003. "On the Shoulders of Hipparchus. A Reappraisal of Ancient Greek Combinatorics." *Archive for History of Exact Sciences* 57: 465–502.

Acerbi, Fabio, 2008. "Hypatia." In *New Dictionary of Scientific Biography*, edited by N. Koertge, 3.433–37. Detroit: Thomson Gale.

Acerbi, Fabio, 2009. "The Meaning of πλασματικόν in Diophantus' Arithmetica." *Archive for History of Exact Sciences* 63: 5–31.

Acerbi, Fabio, 2011. "Completing Diophantus, De polygonis numeris, prop. 5." *Historia Mathematica* 38: 548–60.

Acerbi, Fabio, 2012. "The Number of Endings of the Adjective Συναμφότερος." *Glotta* 88: 1–8.

Acerbi, Fabio, 2013a. "Ones." *Greek, Roman and Byzantine Studies* 53: 708–25.

Acerbi, Fabio, 2013b. "Why John Chortasmenos Sent Diophantus to the Devil." *Greek, Roman and Byzantine Studies* 53: 379–89.

Acerbi, Fabio, 2015. "Unaccountable Numbers." *Greek, Roman and Byzantine Studies* 55: 902–26.

Acerbi, Fabio, 2017. "I problemi aritmetici attribuiti a Demetrio Cidone e Isacco Argiro." *Estudios Bizantinos* 5: 131–206.

Acerbi, Fabio, Nicolas Vinel, and Bernard Vitrac, 2010. "Les prolégomènes à l'Almageste. Une édition à partir des manuscrits les plus anciens: Introduction générale – Parties I–III." *SCIAMVS* 11: 53–210.

Adler, Ada, ed., 1935. *Lexicographi Graeci. Vol. I: Suidae lexicon A–Ω. Index. Pars 4: Π–Ψ.* Leipzig: B. G. Teubner.

Aelius, Théon, 2002. *Progymnasmata*. Texte établi et traduit par Michel Patillon, avec l'assistance, pour l'arménien, de Giancarlo Bolognesi. Paris: Les Belles Lettres.

Agathias, 1975. *The Histories*. Translated with an introduction and short explanatory notes by Joseph D. Frendo. Berlin: Walter de Gruyter.

Aish, Seham Diab Almasry, 2016. "A Greek Mathematical Papyrus from the Cairo Museum." *Archiv Für Papyrusforschung Und Verwandte Gebiete* 62 (1): 43–56.

al-ʿĀmilī, Bahā al-Dīn, 1976. *Mathematical Works of Bahā al-Dīn al-ʿĀmilī, (953–1031 H.)/(1547–1622 A.D.)*. Edited by Jalāl Shawqī (in Arabic). Aleppo: Institute for the History of Arabic Science.

al-Baghdādī, 1985. *Kitāb al-takmila fī'l-ḥisāb*. Edited by Aḥmad Salīm Saʿīdān. Kuwait: Maʿhad al-Makhṭūṭāt al-ʿArabiyah.

Albiani, Maria Grazia, 2006. "Metrodorus." In *Brill's New Pauly: Encyclopedia of the Ancient World*, edited by H. Cancik and H. Schneider, VIII:839 (#9). Leiden: E. J. Brill.

al-Bīrūnī, 1934. *The Book of Instruction in the Elements of the Art of Astrology*. Facsimile edition of Brit. Mus. Ms. Or. 8349, with translation by R. Ramsay Wright. London: Luzac.

al-Fārābī, 2015. *Le recensement des sciences*. Texte, traduction critique et commentaire par Amor Cherni. [Beyrouth]/Paris: Albouraq/Diffusion Librairie de l'Orient.

al-Fārisī, 1994. *Asās al-qawāʿid fī uṣūl al-fawāʾid*. Edited by M. Mawāldī. Cairo: Maʿhad al-Makhṭūṭāt al-ʿArabīyah.

al-Karajī, 1964. *al-Badīʿ fī l-ḥisāb*. Edited by Adel Anbouba. Beirut: Manshūrāt al-Jāmiʿah al-Lubnānīah.

al-Karajī, 1986. *al-Kāfī fī l-ḥisāb*. Edited by S. Chalhoub. Aleppo: Jāmiʿat Ḥalab, Maʿhad al-Turāth al-ʿIlmī al-ʿArabī.

al-Khwārazmī, Muḥammad ibn Aḥmad, 1895. *Liber mafātīh al-olūm, explicans vocabula technica scientiarum tam Arabum quam peregrinorum, auctore Abū Abdallah Mohammed ibn Ahmed ibn Jūsof al-Kātib al-Khowarezmi*. Edited by G. van Vloten. Leiden: E. J. Brill.

al-Khwārazmī, Muhammad ibn Musa, 2009. *Al-Khwārizmī: the Beginnings of Algebra*. Edited, with translation and commentary by Roshdi Rashed. London: Saqi.

Allard, André, 1979. "L'*Ambrosianus et 157 sup.*, un manuscrit autographe de Maxime Planude." *Scriptorium* 33 (2): 219–34.

Allard, André, 1980. *Diophante d'Alexandrie, Les Arithmétiques. Histoire du texte grec, édition critique, traductions et scolies par A. Allard*, 2 vols. PhD thesis, Université Catholique de Louvain.

Allard, André, 1982–83. "La tradition du texte grec des *Arithmétiques* de Diophante d'Alexandrie." *Revue d'histoire des textes* 12–13: 57–138.

Allard, André, 1983a. "La tentative d'édition des *Arithmétiques* de Diophante d'Alexandrie par Joseph Auria." *Revue d'histoire des textes* 11 (1981): 99–122.

Allard, André, 1983b. "Les scholies aux *Arithmétiques* de Diophante d'Alexandrie dans le *Matritensis Bibl. Nat.* 4678 et les *Vaticani gr.* 191 et 304." *Byzantion* 53: 664–760.

Allard, André, 1984. "Un exemple de transmission d'un texte grec scientifique: le *Mediolanensis Ambrosianus* A 91 sup., un manuscrit de Jean-Vincent Pinelli prêté à Mathieu Macigno." *Les Études Classiques* 52: 317–31.

Allard, André, 1985. "Le manuscrit des *Arithmétiques* de Diophante d'Alexandrie et les lettres d'André Dudith dans le *Monacensis lat.* 10370." In *Mathemata: Festschrift für Helmuth Gericke*, edited by M. Folkerts and U. Lindgren, 297–315. Stuttgart: Franz Steiner Verlag.

Allard, André, 1987. "Chronique: À propos de plusieurs publications récentes sur les *Arithmétiques* de Diophante d'Alexandrie." *Revue des Questions Scientifiques* 158 (3): 309–16.

Allard, André, and Roshdi Rashed, 1984. "Chronique (review of Sesiano, 1982)." *Revue des questions scientifiques* 155 (3): 375–84.

Almási, Gábor, 2009. *The Uses of Humanism: Johannes Sambucus (1531–1584), Andreas Dudith (1533–1589), and the Republic of Letters in East Central Europe*. Leiden: Brill.

al-Qalaṣādī, 1999. *Sharḥ Talkhīs aʿmāl al-ḥisāb*. Edited by Farès Bentaleb. Beirut: Dār al-Gharb al-Islāmī.

Anbouba, Adel, 1978a, "Al-Samawʾal." In *Dictionary of Scientific Biography*, vol. 12, edited by C. C. Gillispie, 91–95. New York: Charles Scribner's Sons.

Anbouba, Adel, 1978b. "L'algèbre arabe aux IXe et Xe siècles. Aperçu général." *Journal for the History of Arabic Science* 2: 66–100.

Anbouba, Adel, 1979. "Un traité d'Abū Jaʿfar [Al-Khāzin] sur les triangles rectangles numériques." *Journal for the History of Arabic Science* 3: 134–78.

Anonimo, 1986. *Della radice deʾ numeri e metodo di trovarla: Trattatello di algebra e geometria dal Codice Ital. 578 della Biblioteca Estense di Modena*. Part I: Edited by Warren van Egmond; Part 2: Edited by Francesco Barbieri and Paola Lancellotti. Siena: Servizio Editoriale dell'Università di Siena.

Anonimo, 1988. *Il trattato d'algibra: Dal manoscritto Fond. Prin. II.V.152 della Biblioteca Nazionale di Firenze*. Edited by Rafaella Franci and Marisa Pancanti. Siena: Servizio Editoriale dell'Università di Siena.

Apollonius, 1566. *Apollonii Pergæi conicorum libri quattuor*. Translated by Federico Commandino. Bononiae: Ex Officina Alexandri Benatii.

Apollonius, 1990. *Conics, Books V to VII. The Arabic Translation of the Lost Greek Original in the Version of the Banū Mūsā*, 2 vols. Edited with translation and commentary by G. J. Toomer. New York: Springer-Verlag.

Archimedes, 2004. *The Works of Archimedes*. Translated into English, together with Eutocius' commentaries, with commentary, and critical edition of the diagrams by Reviel Netz, Vol I· *The Two Books on the Sphere and the Cylinder*. Cambridge: Cambridge University Press.

Arrighi, Gino, ed., 1967. *Antonio deʾ Mazzinghi, Trattato di fioretti nella trascelta a cura di Mº Benedetto secondo la lezione del Codice L.IV.21 (sec. XV) della Biblioteca degl'Intronati di Siena*. Siena: Domus Galilaeana.

Arrighi, Gino, 2004. "Nuovi contributi per la storia della matematica in Firenze nell'età di mezzo: Il codice Palatino 573 della Biblioteca Nazionale di Firenze." In *La matematica nell'età di mezzo. Scritti Scelti, a cura di F. Barberini, R. Franci & L. Toti Rigatelli*, 159–94. Pisa: Edizioni ETS.

Aubreton, Robert, 1968. "La tradition manuscrite des épigrammes de l'Anthologie Palatine." *Revue des études anciennes* 70: 32–81.

Aubreton, Robert, 1969. "Michel Psellos et l'Anthologie Palatine." *L'Antiquité Classique* 38: 459–62.

Aurel, Marco, 1552. *Libro primero de arithmetica algebratica*. Valencia: Joan de Mey Flandro.

Auzépy, Marie-France, 1994. "De la Palestine à Constantinople (VIIIe–IXe siècles): Étienne le Sabaïte et Jean Damascène." *Travaux et Mémoires* 12: 183–218.

Baillet, Jules, ed., 1892. *Le papyrus mathématique d'Akhmîm*. Paris: Ernest Leroux.

Balbus et al., 1996. *Corpus agrimensorum Romanorum II & III. II: Balbus, présentation systématique de toutes les figures. III: Podismus et textes connexes extraits d'Epaphrodite et de Vitruvius Rufus: La mesure des jugères*. Introduction, traduction et notes par J.-Y. Guillaumin. Napoli: Jovene.

Barhebraeus, 1663. *Historia compendiosa dynastiarum. Authore Gregorio Abul-Pharajio Malatiensi Medico, historiam complectens universalem, à mundo condito, usque ad tempora authoris, res orientalium accuratissimè describens. Arabice edita et Latine versa ab Eduardo Pocockio.* Oxford: Oxford University Press.

Barhebraeus, 1890. *Ta'rīkh mukhtaṣar al-duwal, lil-'allāma Ghrīghūriyūs al-Malaṭī al-Ma'rūf bi-Ibn al-'Ibrī.* Edited by A. Ṣāliḥānī. Bairūt: al-Maṭba'a al-kāthūlīkīya li-'l-ābā' al-yasū'īyīn.

Barner, Klaus, 2007. "Negative Größen bei Diophant?" *NTM: Nachrichtenblatt der Deutschen Gesellschaft für Geschichte der Medizin, Naturwissenschaften und Technik* 15: 18–49, 98–117.

Bashmakova, Isabella Grigoryevna, Evgenii Iosifovich Slavutin, and Boris A. Rozenfeld, 1978. "The Arabic Version of Diophantus's *Arithmetica* (in Russian)." *Istoriko-Matematicheskie Issledovaniya* 23: 192–225.

Bashmakova, Isabella Grigoryevna, Evgenii Iosifovich Slavutin, and Boris A. Rozenfeld, 1981. "The Arabic version of Diophantus' *Arithmetica*." In *Science and Technology, Humanism and Progress: Soviet Studies on the History of Science*, edited by P. N. Fedoseev, I. R. Grigulevich, and N. I. Maslova, 2: 151–61. Moscow: USSR Academy of Sciences.

Benedetto da Firenze, 1982. *La reghola de algebra amuchabale: Dal Codice L.IV.21 della Biblioteca Comunale di Siena.* A cura e con introduzione di Lucia Salomone. Siena: Servizio editoriale dell'Università di Siena.

Berggren, J. Lennart, and Glen Van Brummelen, 2000. "The Role and Development of Geometric Analysis and Synthesis in Ancient Greece and Medieval Islam." In *Ancient & Medieval Traditions in the Exact Sciences: Essays in Memory of Wilbur Knorr*, edited by P. Suppes, J. M. Moravcsik, and H. Mendell, 1–31. Stanford, CA: CSLI Publications.

Bernard, Alain, 2008. "Anatolios of Laodıkeıa (250–282 CE)." In *Encyclopedia of Ancient Natural Scientists: The Greek Tradition and Its Many Heirs*, edited by P. Keyser and G. Irby-Massie, 73. London: Routledge.

Bernard, Alain, and Jean Christianidis, 2012. "A New Analytical Framework for the Understanding of Diophantus's *Arithmetica I-III*." *Archive for History of Exact Sciences* 66 (1): 1–69.

Bertòla, Maria, 1942. *I due primi registri di prestito della Biblioteca Apostolica Vaticana: Codici Vaticani 3964, 3966.* Città del Vaticano: Biblioteca Apostolica Vaticana.

Biggs, Norman L., 1979. "The Roots of Combinatorics." *Historia Mathematica* 6: 109–36.

Blemmydes, 1896. *Nicephori Blemmydae curriculum vitae et carmina.* Nunc primum edidit Aug. Heisenberg: Praecedit dissertatio de vita et scriptis Nicephori Blemmydae. Leipzig: B. G. Teubner.

Bombelli, Rafael, 1572. *L'algebra, parte maggiore dell'arimetica divisa in tre libri.* Bologna: Giovanni Rossi.

Bombelli, Rafael, 1929. *L'algebra: opera di Rafael Bombelli da Bologna: Libri IV et V comprendenti « La parte geometrica » inedita, tratta dal manoscritto B. 1569, Biblioteca dell'Archginnasio di Bologna.* A cura di Ettore Bortolotti. Bologna: Nicola Zanichelli.

Bombelli, Rafael, 2017. *L'inedito terzo libro de L'Algebra di Rafael Bombelli.* A cura di Alessandra Fiocca ed Elisa Leone. Pisa: Edizioni della Normale.

Borrel, Jean, 1559. *Ioannis Buteonis Logistica, quæ & arithmetica vulgò dicitur in libros quinque digesta: Quorum index summatim habetur in tergo. Eiusdem, Ad locum Vitruuij corruptum restitutio, qui est de proportione lapidum mittendorum ad Balistæ Foramen, Libro Decimo.* Lyon: Apud Gulielmum Rouillium.

Bossut, Charles, 1802. *Histoire générale des mathématiques*, vol. I. Paris: Chez Louis, Libraire.

Brentjes, Sonja, 1990. "Sur quelques travaux mathématiques d'ibn Fallūs." *Archives Internationales d'Histoire des Sciences* 40: 239–57.

Bruins, Evert Marie, ed., 1963. *Codex Constantinopolitanus, palatii veteris No 1*, 3 vols. Leiden: E. J. Brill.

Buffière, Félix, ed., 1970. *Anthologie Grecque*. Première partie: *Anthologie Palatine*. Tome XII: *Livres XIII–XV*. Texte établi et traduit par Félix Buffière. Paris: Les Belles Lettres.

[Bulmer-]Thomas, Ivor, 1939. *Selections Illustrating the History of Greek Mathematics*, vol. I. Cambridge, MA: Harvard University Press.

Bulmer-Thomas, Ivor, 1981. "Hypsicles of Alexandria." In *Dictionary of Scientific Biography*, vol. VI, edited by C. C. Gillispie, 616–17. New York: Charles Scribner's Sons.

Bulmer-Thomas, Ivor, 1985. "Diophantus Arabus." *The Classical Review* 35 (2): 255–58.

Burnyeat, Myles Fredric, 1978. "The Philosophical Sense of Thaetetus' Mathematics." *Isis* 69: 489–513.

Burnyeat, Myles Fredric, 1979. "Methodology, Philology, and Philosophy: Reply to W. R. Knorr." *Isis* 70 (4): 569–70.

Byrne, James Steven, 2006. "A Humanist History of Mathematics? Regiomontanus's Padua Oration in Context." *Journal of the History of Ideas* 67 (1): 41–61.

Byrne, Patrick H., 1997. *Analysis and Science in Aristotle*. Albany: State University of New York.

Camerarius, Joachim, 1557. *De Graecis Latinisque numerorum notis*. Norimbergae: Heller.

Camerarius, Joachim, 1569. *De Graecis Latinisque numerorum notis*. Wittenberg: Krafft.

Camerarius, Joachim, 1583. *Epistolarum familiarum libri VI*. Francofurti: Andr. Wecheli.

Cameron, Alan, 1970. "Michael Psellus and the Date of the Palatine Anthology," *Greek, Roman and Byzantine Studies* 11: 339–50.

Cameron, Alan, 1990. "Isidore of Miletus and Hypatia: On the Editing of Mathematical Texts." *Greek, Roman and Byzantine Studies* 31: 103–27.

Cameron, Alan, 1993. *The Greek Anthology from Meleager to Planudes*. Oxford: Clarendon Press.

Cameron, Alan, and Jacqueline Long, 1993. *Barbarians and Politics at the Court of Arcadius*. Berkeley: University of California Press.

Cantor, Moritz, 1907. *Vorlesungen über Geschichte der Mathematik. Erster Band: Von den ältesten Zeiten bis zum Jahre 1200 n. Chr.* Leipzig: B. G. Teubner.

Cardano, Girolamo, 1539. *Hieronimi C. Cardani medici Mediolanensis, practica arithmetice & mensurandi singularis : In qua que preter alias cõtinentur versa pagina demonstrabit*. Milan: Io. Antonins Castellioneus Mediolani imprimebat, impensis B. Calusci.

Cardano, Girolamo, 1545. *Hieronymi Cardani, præstantissimi mathematici, philosophi, ac medici, artis magnae, sive, de regulis algebraicis, lib. unus. qui & totius operis de arithmetica, quod opvs perfectvm inscripsit, est in ordine decimus*. Norimbergae: per Ioh. Petreium excusum.

Chemla, Karine, ed., 2012. *The History of Mathematical Proof in Ancient Traditions*. Cambridge: Cambridge University Press.

Chemla, Karine, Régis Morelon, and André Allard, 1986. "La tradition arabe de Diophante d'Alexandrie: À propos de quatre livres des *Arithmétiques* perdus en grec retrouvés en arabe." *L'Antiquité Classique* 55: 351–75.

Christianidis, Jean, 1991. "'Αριθμητικὴ Στοιχείωσις: Un traité perdu de Diophante d'Alexandrie?" *Historia Mathematica* 18 (3): 239–46.

Christianidis, Jean, 1996. "Maxime Planude sur le sens du terme diophantien *plasmatikón.*" *Historia Scientiarum* 6 (1): 37–41.

Christianidis, Jean, 2007. "The Way of Diophantus: Some Clarifications on Diophantus' Method of Solution." *Historia Mathematica* 34 (3): 289–305.

Christianidis, Jean, 2008. "Ἡ μέθοδος του Διοφάντου για την επίλυση των αριθμητικών προβλημάτων και η ρήξη με την παλαιά λογιστική παράδοση." *Neusis* 17: 183–234.

Christianidis, Jean, 2015. "The Meaning of Hypostasis in Diophantus' *Arithmetica.*" In *Relocating the History of Science. Essays in Honor of Kostas Gavroglu*, edited by T. Arabatzis, J. Renn, and A. Simões, 315–27. Heidelberg: Springer.

Christianidis, Jean, 2018a. "Diophantus and Premodern Algebra. New Light on an Old Image." In *Revolutions and Continuity in Greek Mathematics*, edited by M. Sialaros, 35–65. Berlin: Walter de Gruyter.

Christianidis, Jean, 2018b. "La démarche de Diophante démystifiée: L'algèbre prémoderne au service de la résolution de problèmes." In *Actes du 12ème Colloque Maghrébin d'Histoire des Mathématiques Arabes (Marrakech 26–28 Mai 2016)*, edited by E. Laabid, 41–69. Marrakech: Ecole Normale Supérieure.

Christianidis, Jean, 2021a. "Diophantus." In *Oxford Research Encyclopedia of Oxford Classical Dictionary*. Oxford: Oxford University Press. https://doi.org/10.1093/acrefore/9780199381135.013.2229.

Christianidis, Jean, 2021b. "Analytical Reasoning and the Problem-solving in Diophantus's *Arithmetica*: Two Different Styles of Reasoning in Greek Mathematics." *Philosophia Scientiæ* 25 (3): 103–30.

Christianidis, Jean, and Athanasia Megremi, 2019. "Tracing the Early History of Algebra: Testimonies on Diophantus in the Greek-speaking World (4th–7th Century CE)." *Historia Mathematica* 47: 16–38. "Corrigendum" in *Historia Mathematica* 49 (2019): 115.

Christianidis, Jean, and Jeffrey A. Oaks, 2013. "Practicing Algebra in Late Antiquity: The Problem-solving of Diophantus of Alexandria." *Historia Mathematica* 40 (2): 127–63.

Christianidis, Jean, and Michalis Sialaros, 2022. "Rhetoric and History of Mathematics: The Case of Diophantus of Alexandria." In *Brill's Companion to the Reception of Ancient Rhetoric*, edited by S. Papaioannou, A. Serafim, and M. Edwards, 617–42. Leiden: E. J. Brill.

Christianidis, Jean, and Ioanna Skoura, 2013. "Solving Problems by Algebra in Late Antiquity: New Evidence from an Unpublished Fragment of Theon's Commentary on the Almagest." *SCIAMVS* 14: 41–57.

Chuquet, Nicolas, 1881. *Le Triparty en la science des nombres par maistre Nicolas Chuquet Parisien, publié d'apres le manuscrit fonds français No. 1346 de La Bibliothèque nationale de Paris et précédé d'une Notice par M. Aristide Marre*. Rome: L'Imprimerie des Sciences Mathématiques et Physiques.

Cifoletti, Giovanna, 1992. *Mathematics and Rhetoric: Peletier and Gosselin and the Making of the French Algebraic Tradition*. PhD thesis, Princeton University.

Clavius, Christoph, 1608. *Algebra*. Rome: Bartholomaeum Zannettum.

Collet, Claude-Georges, and Jean Itard, 1947. "Un mathématicien humaniste Claude-Gaspar Bachet de Méziriac (1581–1638)." *Revue d'histoire des sciences* 1: 26–50.

Constantinides, Costas N., 1982. *Higher Education in Byzantium in the Thirteenth and Early Fourteenth Centuries (1204–ca. 1310)*. Nicosia: Cyprus Research Centre.

Constantinides, Costas N., and Robert Browning, 1993. *Dated Greek Manuscripts from Cyprus to the Year 1570*. Washington, DC/Nicosia: Dumbarton Oaks Research Library and Collection/Cyprus Research Centre.

Cornford, Francis Macdonald, 1937. *Plato's Cosmology: The Timaeus of Plato*. London/ New York: Routledge & Kegan Paul/Humanities Press.

Cossali, Pietro, 1797. *Origine, trasporto in Italia, primi progressi in essa dell'algebra*, vol. I. Parma: Dalla Reale Tipografia Parmense.

Costil, Pierre, 1935. *André Dudith, humaniste hongrois, 1533–1589: Sa vie, son oeuvre et ses manuscrits grecs*. Paris: Les Belles Letters.

Coxe, Henry Octavius, 1969. *Greek Manuscripts. Reprinted with Corrections from the Edition of 1853*. Oxford: Bodleian Library.

Criscuolo, Ugo, 1977. "Due epistole inedite di Manuele Karanteno o Saranteno." *Bollettino della Badia Greca di Grottaferrata*, N.S. 31: 103–19.

Cufalo, Domenico, ed., 2007. *Scholia Graeca in Platonem, I: Scholia ad dialogos tetralogiarum I–VII continens*. Roma: Edizioni di Storia e Letteratura.

Curtze, Maximilian, 1895. "Ein Beitrag zur Geschichte der Algebra in Deutschland im 15. Jahrhundert." *Abhandlungen zur Geschichte der Mathematik* 7: 31–74.

Curtze, Maximilian, 1902. "Der Briefwechsel Regiomontan's mit Giovanni Bianchini, Jacob von Speier und Christian Rodet." *Abhandlungen zur Geschichte der Mathematischen Wissenschaften* 12 (1): 187–336.

Cydonès, Démétrius, 1956. *Correspondence*, 2 vols. Edited by Raymond-Joseph Loenertz. Città del Vaticano: Biblioteca Apostolica Vaticana.

Dasypodius, Conrad, 1570. *Mathematicum, complectens praecepta mathematica, astronomia, logistica, una cum typis et tabulis, ad explicationem eorundem necessariis*. Argentorati [Strasbourg]: Excudebat Josias Rihelius.

Dasypodius, Conrad, 1573. *ΛΕΞΙΚΟΝ seu dictionarium mathematicum, in quo definitiones, & divisiones continentur scientiarum mathematicarum arithmeticae, logisticae, geometriae, geodaesiae, astronomiae, harmonicae*. Argentorati [Strasbourg]: Excudebat Nicolaus Vvyriot.

De Biase, Emanuele, 2011. *Ricerche sul ms. Vaticano Greco 191: Scienza ed erudizione in età Paleologa*. PhD thesis, Università degli Studi Roma Tre.

De Biase, Emanuele, 2016. "The *Vat. gr.* 191, Manuel Bryennius, and the Circle of Scholars Alternative to that of Maximus Planudes." *Scriptorium* 70: 349–63.

Delatte, Armand, 1915. *Études sur la littérature pythagoricienne*. Paris: Libraire Ancienne Honoré Champion.

Dēmētrakos, Dēmētrios, 1958. *Μέγα λεξικὸν ὅλης τῆς Ἑλληνικῆς γλώσσης*. Athens: Ekdoseis Dēm. Dēmētrakos/Dēm. Revezikas.

Descartes, René, 1637. *La Geometrie*. Leiden: Ian Maire.

Devreesse, Robert, 1965. *Le fonds grec de la Bibliothèque Vaticane des origines à Paul V*. Città del Vaticano: Biblioteca Apostolica Vaticana.

Dickson, Leonard E., 1971. *History of the Theory of Numbers*, vol. II: *Diophantine Analysis*. New York: Chelsea.

Diogenes, Laertius, 2013. *Lives of Eminent Philosophers*. Edited by Tiziano Dorandi. Cambridge: Cambridge University Press.

Diophantus, 1575. *Diophanti Alexandrini rerum arithmeticarum libri sex*. Translated and commented by Xylander. Basileae: Eusebium Episcopium & Nicolai Fr. haeredes.

Diophantus, 1621. *Diophanti Alexandrini Arithmeticorum libri sex, et De numeris multangulis liber unus*. Nunc primùm gracè et latinè editi, atque absolutissimis commentariis illustrati. Auctore Claudio Gaspare Bacheto Meziriaco Sebusiano. Lutetiae Parisiorum: Hieronymi Drovart.

Diophantus, 1670. *Diophanti Alexandrini Arithmeticorum libri sex, et De numeris multangulis liber unus*. Cum commentariis C. G. Bacheti V. C. et obseruationibus D. P. de Fermat senatoris Tolosani. Accessit doctrinae analyticae inventum nouum, collectum ex

850   Bibliography

variis eiusdem D. de Fermat epistolis. Tolosae: Excudebat Bernardus Bosc, è Regione Collegii Societatis Jesu.

Diophantus, 1822. *Diophantus von Alexandria Arithmetische Aufgaben nebst dessen Schrift über die Polygon-Zahlen*. Aus dem Griechischen übersetzt und mit Anmerkungen begleitet von Otto Schulz. Berlin: Schlesingeschen Buch- und Musikhandlung.

Diophantus, 1890. *Die Arithmetik und die Schrift über Polygonalzahlen des Diophantus von Alexandria*. Übersetzt und mit Anmerkungen begleitet von Gustav Wertheim. Leipzig: B. G. Teubner.

Diophantus, 1893–95. *Diophanti Alexandrini opera omnia cum graeciis commentarii*. Edidit et latine interpretatus est P. Tannery. Vol. I (1893): *Diophanti quae exstant omnia continens*; Vol. II (1895): *Continens Pseudepigrapha, testimonia veterum, Pachymerae paraphrasin, Planudis commentarium, scholia vetera, omnia fere adhuc inedita, com prolegomenis et indicibus*. Leipzig: B. G. Teubner.

Diophantus, 1952. *Arithmetik des Diophantos aus Alexandria*. Aus dem Griechschen übertragen und erklärt von Arthur Czwalina. Göttingen: Vandenhoeck & Ruprecht.

Diophantus, 1959. *Diophante d'Alexandrie, les six livres arithmétiques et le livre des nombres polygones*. Œuvres traduites pour la première fois du grec en français, avec une introduction et des notes par Paul Ver Eecke. Bruges: Desclée de Brouwer.

Diophantus, 1975. *Şinā'at al-jabr (The Art of Algebra)*. Edited by Roshdi Rashed. Cairo: al-Hai'a al-Miṣrīya al-'Āmma lil-Kitāb.

Diophantus, 1984. *Les arithmétiques*, t. 3: *Livre IV*, t. 4: *Livres V–VII*. Texte établi et traduit par Roshdi Rashed. Paris: Les Belles Lettres.

Diophantus, 2007. *La Aritmética y el libro Sobre los números poligonales*, 2 vols. Translated by E. F. Moral, M. B. Muñoz, and M. Sánchez Benito. Madrid: Nivola.

Diophantus, 2011. *De polygonis numeris*. Introduzione, testo critico, traduzione italiana e commento di Fabio Acerbi. Pisa: Fabrizio Serra Editore.

Dold-Samplonius, Yvonne, 1970. "Al-Khāzin." In *Dictionary of Scientific Biography*, vol. 7, edited by C. C. Gillispie, 334–35. New York: Charles Scribner's Sons.

Dorandi, Tiziano, 2000. *Le stylet et la tablette: Dans le secret des auteurs antiques*. Paris: Les Belles Lettres.

Dudith, Andreas, 1995. *Epistulae. Pars II 1568–1573*. Editae Curantibus Lecho Szczucki et Tiburtio Szepessy. Budapest: Akadémiai Kiadó.

Euclid, 1896. *Euclidis Data, cum commentario Marini et scholiis antiquis*. Edited by H. Menge. Leipzig: Teubner.

Euclid, 1969–1977. *Euclides, Elementa*. Post I. L. Heiberg, edidit E. S. Stamatis (vol. I (1969): Libri I–IV cum appendicibus; vol. II (1970): Libri V–IX cum appendice; vol. III (1972): Liber X cum appendice; vol. IV (1973): Libri XI–XII cum appendice; vol. V,1 (1977): Prolegomena critica, Libri XIV–XV, Scholia in Libros I–V; vol. V,2 (1977): Scholia in Libros VI–XIII cum appendicibus). Leipzig: B. G. Teubner.

Eusebius, 1932. *The Ecclesiastical History II*. Translated by John Ernest Leonard Oulton. London/Cambridge, MA: Heinemann/Harvard University Press.

Evans, James, 2005. "Gnōmonikē Technē: The Dialer's Art and Its Meanings for the Ancient World." In *The New Astronomy: Opening the Electromagnetic Window and Expanding Our View of the Planet Earth (A Meeting to Honor Woody Sullivan on His 60th Birthday)*, edited by E. Orchiston, 273–92. Dordrecht: Springer.

Fathi-Chelhod, Jean, 2001. "L'origine du nom Bar 'Ebroyo: Une vieille histoire d'homonymes." *Hugoye: Journal of Syriac Studies* 4 (1): 7–43.

Fazzo, Silvia, 2008. "Nicolas, l'auteur du *Sommaire de la philosophie d'Aristote*: Doutes sur son identité, sa datation, son origine." *Revue des Études Grecques* 121 (1): 99–126.

Feliciano, Francesco, 1526. *Libro de arithmetica & geometria speculativa & practicale.* Vinegia: F. di A. Bindoni & M. Pasini.

Fermat, Pierre de, 1891. *Oeuvres de Fermat, t. I: Oeuvres mathématiques diverses; observations sur Diophante.* Edited by Paul Tannery and Charles Henry. Paris: Gauthier-Villars.

Fernández, Tomás, 2011. "Un fragment inédit attribué à Anatole d'Alexandrie." *Byzantion Nea Hellás* 30: 189–202.

Fibonacci, 1857. *Scritti di Leonardo Pisano, matematico del secolo decimoterzo.* Publicati da Baldassarre Boncompagni. Vol. I: *Il Liber Abbaci di Leonardo Pisano.* Roma: Tipografia delle Scienze Matematiche.

Fibonacci, 1987. *Book of Squares.* An annotated translation into modern English by Laurence Sigler. Boston: Academic Press.

Fibonacci, 2002. *Fibonacci's Liber Abaci.* A translation into modern English of Leonardo Pisano's book of calculation, by Laurence Sigler. New York: Springer.

Fitzpatrick, Richard, 2008. *Euclid's Elements of Geometry.* Independently Published.

Folkerts, Menso, 1992. "Mathematische Probleme im Corpus Agrimensorum." In *Die römische Feldmeßkunst. Interdisziplinäre Beiträge zu ihrer Bedeutung für die Zivilisationsgeschichte Roms*, 311–34. Göttingen: Vandenhoeck & Ruprecht.

Folkerts, Menso, 2002. "Regiomontanus' Role in the Transmission of Mathematical Problems." In *From China to Paris: 2000 Years Transmission of Mathematical Ideas*, edited by Y. Dold-Samplonius, J. W. Dauben, M. Folkerts, and B. Van Dalen, 411–28. Stuttgart: Franz Steiner Verlag.

Folkerts, Menso, 2006. "Diophantus." In *Brill's New Pauly* (antiquity volumes), edited by H. Cancik, H. Schneider, English Edition by C. F. Salazar. Leiden: Brill. http://doi.org/10.1163/1574-9347_bnp_e320610.

Fowler, David H., 1999. *The Mathematics of Plato's Academy. A New Reconstruction* (2nd ed.). Oxford: Clarendon Press.

Fowler, David H., and Eric G. Turner, 1983. "Hibeh Papyrus i 27: An Early Example of Greek Arithmetical Notation." *Historia Mathematica* 10 (3): 344–59.

Franci, Raffaella, and Marisa Pancanti, eds., 1988. *Anonimo (sec. XIV), Il trattato d'algibra dal manoscritto Fond. Prin. II. V. 152 della Biblioteca Nazionale di Firenze.* Siena: Servizio Editoriale dell'Università di Siena.

Fueter, Rudolf, 1930. "Ueber kubische diophantische Gleichungen." *Commentarii Mathematici Helvetici* 2: 69–89.

Galigai, Francesco, 1548. *Pratica d'arithmetica.* Florence: Appresso Bernardo Giunti.

Gaul, Niels, 2016. "All the Emperor's Men (and His Nephews): Paideia and Networking Strategies at the Court of Andronikos II Palaiologos (1290–1320)." *Dumbarton Oaks Papers* 70: 245–70.

Geminus, 1898. *Gemini elementa astronomiae.* Ad codicum fidem recensuit, germanica interpretatione et commentariis instruxit Carolus Manitius. Leipzig: B. G. Teubner.

Gerbert of Aurillac [Pope Sylvester II], 1899. *Gerberti, postea Silvestri II, papae, Opera mathematica (972–1003).* Accedunt aliorum opera ad Gerberti libellos aestimandos intelligendosque necessaria per septem appendices distributa. Collegit, ad fidem codicum manuscriptorum partim iterum, partim primum edidit, apparatu critico instruxit, commentario auxit, figuris illustravit Dr. Nicolaus Bubnov. Berlin: R. Friedländer.

Ghaligai, Francesco, 1521. *Summa de arithmetica.* Firenze: Per Bernardo Zuccheta.

Ghetaldi, Marino, 1630. *De resolutione & compositione mathematica libri quinque. Opus posthumum.* Romae: Typographia Reuerendae Camerae Apostolicae.

Gillings, Richard J., 1972. *Mathematics in the Time of the Pharaohs.* Cambridge, MA: The MIT Press.

Goldstein, Bernard R., 1964. "A Treatise on the Number Theory from a Tenth-century Arabic Source." *Centaurus* 10: 129–60.

Golitsis, Pantelis, 2007. "Georges Pachymère comme didascale. Essai pour une reconstitution de sa carrière et de son enseignement philosophique." *Jahrbuch der Österreichischen Byzantinistik* 58: 53–68.

Gori, Dionigi, 1984. *Libro e trattato della praticha d'alcibra: Dal Codice L. IV.22 della Biblioteca Comunale di Siena.* Edited by Laura Toti Rigatelli. Siena: Servizio Editoriale dell'Università di Siena.

Gosselin, Guillaume, 1577. *De arte magna, seu de occulta parte numerorum, quae et algebra et almucabala vulgo dicitur, libri qvatvor. In quibus explicantur aequationes Diophanti, regulae quantitatis simplicis et quantitatis surdae.* Paris: Apud Aegidium Beys.

Gosselin, Guillaume, 1578. *L'arithmetique de Nicolas Tartaglia Brescian, grand mathematicien, et prince des praticiens.* Paris: Gilles Beys.

Gosselin, Guillaume, 2016. *De arte magna libri IV: Traité d'algèbre; suivi de Praelectio: Leçon sur la mathématique.* Introduction, traduction et commentaire par Odile Le Guillou-Kouteynikoff. Paris: Les Belles Letters.

Goulet, Richard, 1994. "Anatolius." In *Dictionnaire des philosophes antiques*, edited by R. Goulet, I: 179–183. Paris: Éditions du Centre National de la Recherche Scientifique.

Grendler, Marcella, 1980. "A Greek Collection in Padua: The Library of Gian Vincenzo Pinelli (1535–1601)." *Renaissance Quarterly* 33: 386–416.

Griffith, Sidney H., 2001. "'Melkites', 'Jacobites' and the Christological Controversies in Arabic in Third/ninth-century Syria." In *Syrian Christians under Islam. The First Thousand Years*, edited by D. Thomas, 9–55. Leiden; Boston; Köln: E. J. Brill.

Griffith, Sidney H., 2016. "The Manṣūr Family and Saint John of Damascus: Christians and Muslims in Umayyad Times." In *Christians and Others in the Umayyad State*, edited by A. Borut and F. M. Donner, 29–51. Chicago: The Oriental Institute of the University of Chicago.

Gulchīn-i Maʿānī, A., 1973. *Fihrist-i kutub-i khaṭṭī-yi kitābkhāna-i Āstān-i Quds-i Riḍawī, VIII.* Meshed.

Gutas, Dimitri, 1998. *Greek Thought, Arabic Culture: The Graeco-Arabic Translation Movement in Baghdad and Early Abbasid Society (2nd–4th/8th–10th Centuries).* London: Routledge.

Habsieger, Laurent, Maxim Kazarian, and Sergei Lando, 1998. "On the Second Number of Plutarch." *American Mathematical Monthly* 105: 446.

Ḥājjī, Khālifa, 1837. *Lexicon bibliographicum et encyclopaedicum*, vol. II. Edidit latine vertit Gustavus Fluegel. Leipzig: Oriental Translation Fund of Great Britain and Ireland.

Halkin, François, ed., 1957. *Bibliotheca hagiographica Graeca*, 3 vols. Bruxelles: Société des Bollandistes.

Hankel, Hermann, 1874. *Zur Geschichte der Mathematik in Alterthum und Mittelalter.* Leipzig: Teubner.

Hankinson, R. J., 2009. "Galen on the Limitations of Knowledge". In *Galen and the World of Knowledge*, edited by Christopher Gill, Tim Whitmarsh, and John Wilkins, 206–42. Cambridge: Cambridge University Press.

Harlfinger, Dieter, 1971. *Die Textgeschichte der pseudo-aristotelischen Schrift Περὶ Ἀτόμων Γραμμῶν. Ein kodikologisch-kulturgeschichtlicher Beitrag zur Klärung der Überlieferungsverhältnisse im Corpus Aristotelicum.* Amsterdam: A. M. Hakkert.

Harriot, Thomas, 1631. *Artis Analyticae Praxis.* London: Apud Robertum Barker.

Havrda, Matyáš, 2016. *The So-called Eighth Stromateus by Clement of Alexandria: Early Christian Reception of Greek Scientific Methodology.* Leiden: Brill.

Hayashi, Takao, 1995. *The Bakhshālī Manuscript: An Ancient Indian Mathematical Treatise*. Groningen: Egbert Forsten.

Heath, Thomas L., 1885. *Diophantus of Alexandria. A Study in the History of Greek Algebra*. Cambridge: Cambridge University Press.

Heath, Thomas L., 1910. *Diophantus of Alexandria. A Study in the History of Greek Algebra* (2nd ed.). Cambridge: Cambridge University Press.

Heath, Thomas L., 1921. *A History of Greek Mathematics*, 2 vols. Oxford: Clarendon Press.

Heath, Thomas L., 1926. *The Thirteen Books of Euclid's Elements*, 3 vols. Oxford: Clarendon Press.

Heath, Thomas L., 1931. *A Manual of Greek Mathematics*. Oxford: Oxford University Press.

Hebesein, Christophe, 2009. *L'algèbre al-Badi d'al-Karagi*. PhD thesis, École polytechnique fédérale de Lausanne. https://doi.org/10.5075/epfl-thesis-4297.

Heeffer, Albrecht, 2010. "From the Second Unknown to the Symbolic Equation." In *Philosophical Aspects of Symbolic Reasoning in Early Modern Mathematics*, edited by A. Heeffer and M. Van Dyck, 57–101. London: Individual author and College Publications.

Heiberg, Johan Ludvig, 1901. "Anatolius, *Sur les dix premiers nombres*." In *Annales internationales d'histoire. Congrès de Paris 1900, Vᵉ section: Histoire des sciences*, 27–57. Paris: Librairie Armand Colin.

Henry, Charles, ed., 1879. *Opusculum de multiplicatione et divisione sexagesimalibus Diophanto vel Pappo attribuendum*. Primum edidit et notis illustrauit C. Henry. Halis Saxoniae: H. W. Schmidt.

Hero, 1864. *Heronis Alexandrini Geometricorum et Stereometricorum reliquiae. Accedunt Didymi Alexandrini Mensurae marmorum et anonymi variae collectiones ex Herone, Euclide, Gemino, Proclo, Anatolio aliisque*. Edidit Fridericus Hultsch. Berlin: Weidmann.

Hero, 1903. *Heronis Alexandrini opera quae supersunt omnia*. Vol. III: *Rationes dimetiendi et commentatio dioptrica*. Recensuit Hermannus Schoene. Leipzig: B. G. Teubner.

Hero, 1912. *Heronis Alexandrini opera quae supersunt omnia*. Vol. IV: *Heronis definitiones cum variis collectionibus. Heronis quae feruntur geometrica*. Copiis Guilelmi Schmidt usus. Edidit J. L. Heiberg. Leipzig: B. G. Teubner.

Hero, 1914. *Heronis Alexandrini opera quae supersunt omnia*. Vol. V: *Heronis quae feruntur stereometrica et de mensuris*. Copiis Guilelmi Schmidt usum; Edidit J. L. Heiberg. Leipzig: B. G. Teubner.

Hero, 2014. *Héron d'Alexandrie, Metrica*. Introduction, texte critique, traduction française et notes de commentaire par Fabio Acerbi et Bernard Vitrac. Pisa: Fabrizio Serra Editore.

Herrin, Judith, 2000. "Mathematical Mysteries in Byzantium: The Transmission of Fermat's Last Theorem." *Dialogos* 6: 22–42.

Hintikka, Jaakko, and Unto Remes, 1974. *The Method of Analysis: Its Geometrical Origins and its General Significance*. Dordrecht; Boston: Reidel.

Hogendijk, Jan P., 1985. "Review of (Sesiano, 1982)." *Historia Mathematica* 12: 82–90.

Hopkins, Burt C., 2011. *The Origin of the Logic of Symbolic Mathematics: Edmund Husserl and Jacob Klein*. Bloomington: Indiana University Press.

Høyrup, Jens, 1989. "Sub-scientific Mathematics: Observations on a Pre-modern Phenomenon." *History of Science* 27: 63–87.

Høyrup, Jens, 1990. "*Dýnamis*, the Babylonians, and Theaetetus 147c7–148d7." *Historia Mathematica* 17: 201–22.

Høyrup, Jens, 2002. *Lengths, Widths, Surfaces : A Portrait of Old Babylonian Algebra and Its Kin*. New York: Springer.

Høyrup, Jens, 2007. "Generosity: No Doubt, But at Times Excessive and Delusive." *Journal of Indian Philosophy* 35: 469–85.

Høyrup, Jens, 2019. *Selected Essays on Pre- and Early Modern Mathematical Practice.* Cham: Springer Nature Switzerland AG.

Hultsch, Friedrich, 1903. "Diophantus #18." In *Paulys Realencyclopädie der classischen Altertumswissenschaft,* edited by Georg Wissowa, V.1: 1051–73. Stuttgart: J. B. Metzler.

Hume, James, 1636. *Algebre de Viete d'une methode nouvelle, caire, et facile.* Paris: Chez Louis Boulenger.

Hunger, Herbert, 1978. *Die hochsprachliche profane Literatur der Byzantiner,* 2 vols. München: C. H. Beck.

Hunger, Herbert, and Kurt Vogel, eds., 1963. *Ein byzantinisches Rechenbuch des 15. Jahrhunderts.* Wien: Hermann Böhlaus.

Iamblichus, 1975. *Iamblichi, In Nicomachi arithmeticam introductionem liber. Ad fidem codicis Florentini edidit Hermenegildus Pistelli (1894).* Editionem addendis et corrigendis adiunctis curavit U. Klein. Stuttgart: Teubner.

Ibn al-Bannā', 1994. *Raf' al-ḥijāb 'an wujūh a'māl al-ḥisāb.* Edited by M. Aballagh. Fās: Jāmi'at Sīdī Muḥammad ibn 'Abd Allāh.

Ibn al-Hā'im, 2003. *Sharḥ al-urjūza al-Yāsmīnīya, de Ibn al-Hā'im.* Texte établi et commenté par Mahdi Abdeljaouad. Tunis: Publication de l'Association Tunisienne des Sciences Mathématiques.

Ibn al-Nadīm, 1871–72. *Kitâb al-fihrist,* 2 vols. Edited by Gustav Flügel. Leipzig: F. C. W. Vogel.

Ibn Ghāzī, 1983. *Bughyat al-ṭullāb fī sharḥ munyat al-ḥussāb.* Edited by M. Suwaysī. Ḥalab, Sūrīyah: Jāmi'at Ḥalab, Ma'had al-Turāth al-'Ilmī al-'Arabī.

Ifrah, Georges, 1994. *Histoire universelle des chiffres,* 2 vols. Paris: Robert Laffont.

Ignatios, 1998. *The Life of the Patriarch Tarasios by Ignatios the Deacon.* Introduction, text, translation and commentary by Stéphanos Efthymiadis. Aldershot: Ashgate.

Imhausen, Annette, 2003. *Ägyptische Algorithmen: Eine Untersuchung zu den mittelägyptischen mathematischen Aufgabentexten.* Wiesbaden: Harrassowitz Verlag.

Irigoin, Jean, 1997. *Tradition et critique des textes grecs.* Paris: Les Belles Lettres.

Italikos, Michel, 1972. *Lettres et discours.* Edités par Paul Gautier. Paris: Institut Français d'Etudes Byzantines.

Jaouiche, Khalil, 1987. "Review of (Rashed, 1984)." *Annals of Science* 44: 308–11.

Jaouiche, Khalil, 1990. "Correspondence: Réponse de M. K. Jaouiche." *Annals of Science* 47: 407–09.

Jayawardene, S. A., 1963. "Unpublished Documents Relating to Rafael Bombelli in the Archives of Bologna." *Isis* 54 (3): 391–95.

Jayawardene, S. A., 1965. "Rafael Bombelli, Engineer-architect: Some Unpublished Documents of the Apostolic Camera." *Isis* 56: 298–306.

Jayawardene, S. A., 1973. "The Influence of Practical Arithmetics on the Algebra of Rafael Bombelli." *Isis* 64 (4): 510–23.

Jones, Alexander. 2008. "Hipparchus." In *New Dictionary of Scientific Biography,* vol. 3, edited by Noretta Koertge, 320–22. New York: Thomson Gale.

Jones, Arnold Hugh Martin, John Robert Martindale, and John Morris, 1971. *The Prosopography of the Later Roman Empire,* vol. I. Cambridge: Cambridge University Press.

Kalvesmaki, Joel, 2013. *The Theology of Arithmetic: Number Symbolism in Platonism and Early Christianity.* Washington, DC: Center for Hellenic Studies.

Kaunzner, Wolfgang, and Karl Röttel, 2006. *Christoff Rudolff aus Jauer in Schlesien: zum 500. Geburtstag eines bedeutenden Cossisten Und Arithmetikers, der aus diesem seinerzeit hoheitlich zur Krone von Böhmen gehörenden Landesteil stammt.* Eichstätt: Polygon-Verlag.

Kazhdan, Alexander, and Stephen Gero, 1989. "Kosmas of Jerusalem: A More Critical Approach to His Biography." *Byzantinische Zeitschrift* 82: 122–32.

King, David A., 1988. "A Medieval Account of Algebra Before al-Khwārizmī." *Al-Masāq: Studia Arabo-Islamica Mediterranea* 1: 25–32.

Klaniczay, Tibor, 1973. "Contributi alle relazioni Padovane degli umanisti d'Ungheria: Nicasio Ellebodio e la sua attività filologica." In *Venezia e Ungheria nel rinascimento,* edited by Vittore Branca, 317–33. Firenze: Leo S. Olschki.

Klein, Jacob, 1968. *Greek Mathematical Thought and the Origin of Algebra.* Translated by Eva Brann. With an appendix containing Vieta's "Introduction to the analytical art", translated by the Reverend J. Winfree Smith. Cambridge, MA: The MIT Press.

Knorr, Wilbur Richard, 1975. *The Evolution of the Euclidean Elements.* Dordrecht: Reidel.

Knorr, Wilbur Richard, 1978. "Archimedes and the Elements: Proposal for a Revised Chronological Order of the Archimedean Corpus." *Archive for History of Exact Sciences* 19: 211–90.

Knorr, Wilbur Richard, 1979. "Methodology, Philology, and Philosophy." *Isis* 70 (4): 565–68.

Knorr, Wilbur Richard, 1985. "Review of Sesiano's *Books IV to VII of Diophantus' Arithmetica in the Arabic Translation Attributed to Qusṭā Ibn Lūqā.*" *The American Mathematical Monthly* 92 (2): 150–54.

Knorr, Wilbur Richard, 1989. *Textual Studies in Ancient and Medieval Geometry.* Boston: Birkhäuser.

Knorr, Wilbur Richard, 1993. "*Arithmêtikê stoicheiôsis*: On Diophantus and Hero of Alexandria." *Historia Mathematica* 20: 180–92.

Köhler, Franz, and Gustav Milchsack, 1913. *Die Handschriften der Herzoglichen Bibliothek zu Wolfenbüttel. Vierte Abteilung, die Gudischen Handschriften.* Wolfenbüttel: Julius Zwissler.

Kontouma, Vassa, 2000. "Jean Damascène." In *Dictionnaire des philosophes antiques,* edited by Richard Goulet, III: 989–1012. Paris: Éditions du Centre National de la Recherche Scientifique.

Kontouma, Vassa, 2015. "John of Damascus (c. 655–c. 745)." In *John of Damascus: New Studies on His Life and Works,* edited by Vassa Kontouma, 1–43. Farnham, Surrey: Ashgate Variorum.

Kunitzsch, Paul, 2003. "The Transmission of Hindu-Arabic Numerals Reconsidered." In *The Enterprise of Science in Islam: New Perspectives,* edited by J. P. Hogendijk and A. I. Sabra, 3–21. Cambridge, MA; London: The MIT Press.

Kwan, Alistair, 2014. "Hipparchus." In *Biographical Encyclopedia of Astronomers,* 2nd ed., edited by T. Hockey, 982–85. New York: Springer.

Labowsky, Lotte, 1979. *Bessarion's Library and the Bibliotheca Marciana.* Roma: Edizioni di Storia e Letteratura.

Lagrange, Joseph-Louis, 1877. *Œuvres de Lagrange.* Publiées par les soins de M. J. –A. Serret, tome VII. Paris: Gauthier-Villars.

Lane, Edward William, 1863–93. *An Arabic-English Lexicon.* London: Williams and Norgate.

Lattis, James M., 1994. *Between Copernicus and Galileo: Christoph Clavius and the Collapse of Ptolemaic Cosmology.* Chicago: University of Chicago.

Le Coz, Raymond, 1992. *Jean Damascène. Écrits sur l'Islam.* Paris: Les Éditions du Cerf.

Lehoux, Daryn, 2008. "Hipparkhos of Nikaia." In *Encyclopedia of Ancient Natural Scientists,* edited by P. T. Keyser and G. L. Irby-Massie, 397–99. London: Routledge.

L'Huillier, Ghislaine, 1990. *Le Quadripartitum numerorum de Jean de Murs: Introduction et édition critique*. Genève: Librairie Droz.

Liddell, Henry George, Robert Scott, and Henry Stuart Jones, 1996. *A Greek-English Lexicon*. Compiled by Henry George Liddell and Robert Scott; Revised and Augmented throughout by Sir Henry Stuart Jones; with the assistance of Roderick McKenzie and with the cooperation of many scholars (9th ed.). Oxford: Clarendon Press.

Lippert, Julius, ed., 1903. *Ibn al-Qifṭī's tā'rīkh al-ḥukamā'*. Leipzig: Dieterich'sche Verlagsbuchhandlung.

Litwa, David, tr., 2016. *Refutation of All Heresies*. Translated with an introduction and notes by M. David Litwa. Atlanta: SBL.

Lo Bello, Anthony, 2009. *The Commentary of al-Nayrizi on Books II–IV of Euclid's Elements of Geometry*. Leiden: Brill.

Louth, Andrew, 2002. *St John Damascene: Tradition and Originality in Byzantine Theology*. New York: Oxford University Press.

Macrobius, 1990. *Commentary on the Dream of Scipio*. Translated, with an introduction and notes by William Harris Stahl. New York: Columbia University.

Mahoney, Michael, 1968. "Another Look at Greek Geometrical Analysis." *Archive for History of Exact Sciences* 5: 318–48.

Martindale, John Robert, 1980. *The Prosopography of the Later Roman Empire*, vol. II. Cambridge: Cambridge University Press.

Martindale, John Robert, 1992. *The Prosopography of the Later Roman Empire*, vol. III. Cambridge: Cambridge University Press.

Masià, Ramon, 2015. "On Dating Hero of Alexandria." *Archive for History of Exact Sciences* 69: 231–55.

Matvievskaya, Galina Pavlovna, and Boris A. Rosenfeld, 1983. *Matematiki i astronomy Musulmanskogo srednevekovya i ikh trudy (VIII–XVII vv.)*, 3 vols. Moscow: Nauka.

McCarthy, Daniel P., 1996. "The Lunar and Paschal Tables of *De ratione paschali* Attributed to Anatolius of Laodicea." *Archive for History of Exact Sciences* 49: 285–320.

McCarthy, Daniel P., and Aidan Breen, eds., 2003. *The Ante-Nicene Christian Pasch: De ratione paschali: The Paschal Tract of Anatolius, Bishop of Laodicea*. Dublin: Four Courts Press.

Megremi, Athanasia, and Jean Christianidis, 2015. "Theory of Ratios in Nicomachus' Arithmetica and Series of Arithmetical Problems in Pachymeres' Quadrivium: Reflections about a Possible Relationship." *SHS Web of Conferences* 22: #00006. www.shs-conferences.org/articles/shsconf/abs/2015/09/contents/contents.html.

Menn, Stephen, 2002. "Plato and the Method of Analysis." *Phronesis* 47: 193–223.

Mennher, Valentin, 1556. *Arithmetiqve Seconde*. Anvers: Jan Loë.

Menninger, Karl, 1969. *Number Words and Number Symbols. A Cultural History of Numbers*. English translation by P. Broneer. Cambridge, MA: The MIT Press.

Mercati, Giovanni, 1926. *Scritti d'Isidoro Il cardinale ruteno, e codici a lui appartenuti che si conservano nella Biblioteca Apostolica Vaticana*. Città del Vaticano: Biblioteca Apostolica Vaticana.

Meskens, Ad, 2010. *Travelling Mathematics – The Fate of Diophantos' Arithmetic*. Basel: Birkhäuser.

Michael of Rhodes, 2009. *Book of Michael of Rhodes: A Fifteenth-century Maritime Manuscript. Vol. 2: Transcription and Translation*. Edited by Pamela O. Long, Dadid McGee, and Alan M. Stahl. Cambridge, MA: The MIT Press.

Montucla, Jean-Étienne, 1758. *Histoire des mathematiques*, vol. I. Paris: Ch. Ant. Jobert.

Moore, Paul, 2005. *Iter Psellianum: A Detailed Listing of Manuscript Sources for All Works Attributed to Michael Psellos, Including a Comprehensive Bibliography.* Toronto: Pontifical Institute of Mediaeval Studies.

Morelon, Régis, 1990. "Correspondence: Lettre de M. R. Morelon." *Annals of Science* 47: 407.

Morse, JoAnn S., 1981. *The Reception of Diophantus' Arithmetic in the Renaissance.* PhD thesis, Princeton University.

Mosshammer, Alden A., 2008. *The Easter Computus and the Origins of the Christian Era.* Oxford: Oxford University Press.

Mueller, Ian, 1981. *Philosophy of Mathematics and Deductive Structure in Euclid's Elements.* Cambridge, MA: The MIT Press.

Murr, Christoph Gottlieb von, 1786. *Memorabilia Bibliothecarum Publicarum Norimbergensium et Universitatis Altdorfinae.* Nuremberg: J. Hoesch.

Nagy, Gregory, 1996. *Poetry as Performance: Homer and Beyond.* Cambridge: Cambridge University Press.

Nasrallah, Joseph, 1950. *Saint Jean de Damas: Son époque, sa vie, son œuvre.* Harissa, Lebanon: Imp. St. Paul.

Nesselmann, Georg Heinrich Ferdinand, 1842. *Versuch einer kritischen Geschichte der Algebra, Theil 1: Die Algebra der Griechen.* Berlin: G. Reimer.

Nesseris, Ilias C., 2014. *Η παιδεία στην Κωνσταντινούπολη κατά τον 12ο αιώνα* (Higher Education in Constantinople in the 12th Century). PhD thesis, University of Ioannina.

Netz, Reviel, 1999a. *The Shaping of Deduction in Greeks Mathematics: A Study in Cognitive History.* Cambridge: Cambridge University Press.

Netz, Reviel, 1999b. "Proclus' Division of a Mathematical Proposition into Parts: How and Why Was It Formulated?" *Classical Quarterly* 49 (1 (n.s.)): 282–303.

Netz, Reviel, 2002. "It's Not That They Couldn't." *Revue d'histoire des mathématiques* 8: 263–89.

Netz, Reviel, 2012. "Reasoning and Symbolism in Diophantus: Preliminary Observations." In *The History of Mathematical Proof in Ancient Traditions*, edited by Karine Chemla, 327–61. Cambridge: Cambridge University Press.

Neugebauer, Otto, 1938. "Über eine Methode zur Distanzbestimmung Alexandria-Rom bei Heron I." *Det Kongelige Danske Videnskabernes Selskab* 26 (2): 3–26.

Nuñez, Pedro, 1567. *Libro de algebra en arithmetica y geometria.* Anvers: en casa de los herederos d'Arnoldo Birckman a la Gallina gorda.

Nunn, Thomas Percy, 1914. *The Teaching of Algebra (Including Trigonometry).* London: Longmans, Green and Co.

Oaks, Jeffrey A., 2009. "Polynomials and Equations in Arabic Algebra." *Archive for History of Exact Sciences* 63: 169–203.

Oaks, Jeffrey A., 2010a. "Equations and Equating in Arabic Mathematics." *Archives Internationales d'Histoire des Sciences* 60 (2): 265–98.

Oaks, Jeffrey A., 2010b. "Polynomials and Equations in Medieval Italian Algebra." *Bollettino di Storia delle Scienze Matematiche* 30 (1): 23–60.

Oaks, Jeffrey A., 2011. "Geometry and Proof in Abū Kāmil's Algebra." In *Actes du 10ème colloque Maghrébin sur l'histoire des mathématiques arabes (Tunis, 29-30-31 Mai 2010)*, 234–56. Tunis: L'Association Tunisienne des Sciences Mathématiques.

Oaks, Jeffrey A., 2012a. "Algebraic Symbolism in Medieval Arabic Algebra." *Philosophica* 87: 27–83.

Oaks, Jeffrey A., 2012b. "The Series of Problems in al-Khwārizmī's Algebra." https://problemata.hypotheses.org/157. (the file has been moved to: www.uindy.edu/cas/mathematics/oaks/files/oakscarnet.pdf).

Oaks, Jeffrey A., 2015. "Series of Problems in Arabic Algebra: The Example of ʿAlī al-Sulamī." *SHS Web of Conferences* 22: # 00005. https://doi.org/https://doi.org/10.1051/shsconf/20152200005.

Oaks, Jeffrey A., 2017. "Irrational 'Coefficients' in Renaissance Algebra." *Science in Context* 30: 141–72.

Oaks, Jeffrey A., 2018a. "Arithmetical Proofs in Arabic Algebra." In *Actes du 12ème colloque Maghrébin d'histoire des mathématiques arabes (Marrakech 26–28 Mai 2016)*, edited by E. Laabid, 215–38. Marrakech: Ecole Normale Supérieure.

Oaks, Jeffrey A., 2018b. "Diophantus, al-Karajī, and Quadratic Equations." In *Revolutions and Continuity in Greek Mathematics*, edited by M. Sialaros, 271–94. Berlin: Walter de Gruyter.

Oaks, Jeffrey A., 2018c. "François Viète's Revolution in Algebra." *Archive for History of Exact Sciences* 72 (3): 245–302.

Oaks, Jeffrey A., 2019. "Proofs and Algebra in al-Fārisī's Commentary." *Historia Mathematica* 47: 106–21.

Oaks, Jeffrey A., 2021. "Fermat and Descartes in Light of Premodern Algebra and Viète." In *Handbook of the History and Philosophy of Mathematical Practice*, edited by Bharath Sriraman, 36p. Heidelberg: Springer.

Oaks, Jeffrey A., forthcoming. "Algebra According to Early Arabic Authors." In *A Worldwide Approach to the Early History of Algebra*, edited by Karine Chemla and Tian Miao. Heidelberg: Springer.

Oaks, Jeffrey A., and Haitham M. Alkhateeb, 2005. "*Māl*, Enunciations, and the Prehistory of Arabic Algebra." *Historia Mathematica* 32: 400–25.

Oaks, Jeffrey A., and Haitham M. Alkhateeb, 2007. "Simplifying Equations in Arabic Algebra." *Historia Mathematica* 34: 45–61.

Oestermann, Günther, 2020. *The Astronomical Clock of Strasbourg Cathedral: Function and Significance.* English translation by Bruce W. Irwin. Leiden: E. J. Brill.

Olympiodorus, 1970. *Olympiodori in Platonis Gorgiam Commentaria.* Edited by Leendert Gerrit Westerink. Leipzig: B. G. Teubner.

Olympiodorus, 1998. *Commentary on Plato's Gorgias.* Translated with full notes by Robin Jackson, Kimon Lycos, and Harold Tarrant; introduction by Harold Tarrant. Leiden: E. J. Brill.

O'Meara, Dominic J., 1989. *Pythagoras Revived.* Oxford: Clarendon Press.

Omont, Henri, 1887. "Deux registres de prêts de manuscrits de la Bibliothèque de Saint-Marc à Venise (1545–1559)." *Bibliothèque de l'École des Chartes* 48: 651–86.

Omont, Henri, 1894. "Inventaire des manuscrits grecs at latins donnés à Saint-Marc de Venise par le Cardinal Bessarion (1468)." *Revue des Bibliothèques* 4: 129–87.

Orbán, Áron, 2021. "Nicasius Ellebodius." In *Companion to Humanism in East Central Europe. Vol. 1: Hungarian Humanism.* Edited by F. G. Kiss. Berlin: De Gruyter.

Oughtred, William, 1631. *Arithmeticae in numeris et speciebus institutio.* London: Apud Thomam Harperum.

Pachymeres, 1940. *Quadrivium de Georges Pachymère, Ου Σύνταγμα τῶν τεσσάρων μαθημάτων: ἀριθμητικῆς, μουσικῆς, γεωμετρίας καὶ ἀστρονομίας.* Edited by P. Tannery; texte revisé et établi par E. Stéphanou. Città del Vaticano: Biblioteca Apostolica Vaticana.

Pacioli, Luca, 1494. *Summa de aritmetica, geometria, proporzioni e proporzionalità.* Venice: Paganinus de Paganinis.

Pappus, 1876. *Pappi Alexandrini collectionis quae supersunt.* E libris manu scriptis edidit, latina interpretatione et commentariis instruxit Fridericus Hultsch, 3 vols. Berlin: Weidmann.

Pappus, 1986. *Pappus of Alexandria, Book 7 of the Collection.* Edited with translation and commentary by Alexander Jones. New York: Springer-Verlag.

Pappus, 2010. *Pappus of Alexandria, Book 4 of the Collection.* Edited with translation and commentary by Heike Sefrin-Weis. London: Springer.

Paton, William Roger, 1918. *The Greek Anthology,* with an English translation by W. R. Paton, vol. V. London/New York: William Heinemann/G. P. Putnam's Sons.

Peletier, Jacques, 1554. *L'Algebre.* Lyon: Ian de Tournes.

Pellat, Charles, 1977. *Textes arabes relatifs à la dactylonomie.* Paris: Maisonneuve et Larose.

Pérez de Moya, Juan, 1573. *Tratado de mathematicas en que se contienen cosas arithmetica, geometria, cosmographia, y philosophia natural.* Alcala de Henares: por Juan Gracian.

Pérez Martín, Inmaculada, 2006. "Maxime Planude et le Diophantus Matritensis (Madrid, Biblioteca Nacional, Ms. 4678): Un paradigme de la récupération des textes anciens dans la 'Renaissance Paléologue'." *Byzantion* 76: 433–62.

Pérez Martín, Inmaculada, 2010. "L'écriture de l'*hypatos* Jean Pothos Pédiasimos d'après ses scholies aux *Elementa* d'Euclide." *Scriptorium* 64: 109–19.

Peucer, Caspar, 1556. *Logistice astronomica hexacontadon et scrupulorum sexagesimorum.* Vitebergae: excvdebant haeredes Georgii Rhavv.

Pfeiffer, Rudolf, 1976. *History of Classical Scholarship: From 1300 to 1850.* Oxford: Clarendon Press.

Philo, 2001. *Philo of Alexandria: On the Creation of the Cosmos According to Moses.* Introduction, translation and commentary by David T. Runi. Leiden: E. J. Brill.

Philoponus, 1887. *Ioannis Philoponi in Aristotelis physicorum libros tres priores commentaria.* Edited by Girolamo Vitelli. Berlin: Georg Reimer.

Philoponus, 2014. *On Aristotle Physics 2.* Translated by A. R. Lacey. London; New Delhi; New York; Sydney: Bloomsbury.

Piero della Francesca, 1970. *Trattato d'abaco. Dal Codice Ashburnhamiano 280 (359*–291*) della Biblioteca Medicea Laurenziana di Firenze.* Edited by Gino Arrighi. Pisa: Domus Galilaeana.

Pingree, David, 1983. "The Byzantine tradition of Vettius Valens' Anthologies." *Harvard Ukrainian Studies* 7: 532–41.

Planudes, 1890. *Maximi Monachi Planudis epistulae.* Edited by Maximilian Treu. Wroslaw: Apud Guililmum Koobner.

Planudes, 1991. *Maximi Monachi Planudis epistulae.* Edited by Pietro Luigi Leone. Amsterdam: A. M. Hakkert.

Planudes, 2020. *Maxime Planoudès, Lettres.* Traduction et annotation par Jean Schneider. Leuven: Peeters.

Plato, 1926. *Cratylus. Parmenides. Greater Hippias. Lesser Hippias.* Translated by Harold North Fowler. Cambridge, MA/London: Harvard University Press/W. Heinemann.

Plato, 1955. *Charmides, Alcibiades I and II, Hipparchus, the Lovers, Theages, Minos, Epinomis.* Translated by Walter Rangeley Maitland Lamb. London/Cambridge, MA: W. Heinemann/Harvard University Press.

Plofker, Kim, 2009. *Mathematics in India.* Princeton: Princeton University Press.

Plutarch, 1976. *Moralia. Volume XIII,* Part 2: *Stoic Essays.* Edited by Harold Cherniss. Cambridge, MA: Harvard University Press.

Procissi, Angiolo, 1954. "I 'Ragionamenti d'algebra' di R. Canacci." *Bollettino dell'Unione Matematica Italiana (Serie 3)* 9: 300–26, 420–51.

Proclus, 1873. *Procli Diadochi in primum Euclidis Elementorum librum commentarii.* Ex Recognitione Godofredi Friedlein. Leipzig: B. G. Teubner.

Proclus, 1970. *A Commentary of the First Book of Euclid's Elements*. Translated with introduction and notes by Glenn R. Morrow. Princeton: Princeton University Press.

Ptolemy, 1898–1903. *Claudii Ptolemaei opera quae exstant omnia*. Vol I: *Syntaxis mathematica*. Edidit J. L. Heiberg, 1 vol. in 2 parts. Leipzig: B. G. Teubner.

Ptolemy, 1998. *Ptolemy's Almagest*. Translated and annotated by G. J. Toomer. Princeton: Princeton University Press.

Ramus, Petrus, 1560. *Algebra*. Paris: Apud Andream Wechelum.

Ramus, Petrus, 1569. *Mathematicarum libri unus et triginta*. Basileae: Eusebium Episcopium & Nicolai Fratris Haeredes.

Ramus, Petrus, 1586. *Arithmetices libri duo, et Algebrae totidem à Lazaro Schonero emendati & explicati*. Francofurdi: Andreae Wecheli.

Rashed, Marwan, 2013. "La connaissance de Diophante dans l'antiquité." In *Les 'Arithmétiques' de Diophante. Lecture historique et mathématique*, edited by R. Rashed and C. Houzel, 595–605. Berlin: Walter de Gruyter.

Rashed, Roshdi, 1979. "L'analyse diophantienne au Xe siècle: L'exemple d'al-Khāzin." *Revue d'histoire des sciences* 32: 193–222.

Rashed, Roshdi, 1983. "L'idée de l'algèbre selon al-Khwārizmī." *Fundamenta Scientiae* 4: 87–100.

Rashed, Roshdi, 1994a. "Notes sur la version arabe des trois premiers livres des *Arithmétiques* de Diophante, et sur le Problème I.39." *Historia Scientiarum* 4 (1): 39–46.

Rashed, Roshdi, 1994b. *The Development of Arabic Mathematics: Between Arithmetic and Algebra*. Dordrecht: Springer-Science+Business Media.

Rashed, Roshdi, 2021. *L'algèbre arithmétique au XIIᵉ siècle: Al-Bāhir d'al-Samaw'al*. Berlin: Walter De Gruyter.

Rashed, Roshdi, and Christian Houzel, 2013. *Les 'Arithmétiques' de Diophante. Lecture historique et mathématique*. Berlin: Walter de Gruyter.

Rashed, Roshdi, and Bijan Vahabzadeh, 1999. *Al-Khayyām mathématicien*. Paris: Albert Blanchard.

Regiomontanus, 1537. *Continentur in hoc libro, rudimenta astronomica Alfragrani*. Norimbergae: Apud Ioh. Petreium.

Reich, Karin, 1968. "Diophant, Cardano, Bombelli, Viète. Ein Vergleich ihrer Aufgaben." In *Rechenpfennige: Aufsätze zur Wissenschaftsgeschichte Kurt Vogel Zum 80. Geburtstag am 30. Sept. 1968*, edited by K. Elfering, 131–50. München: Forschungsinstitut des Deutschen Museums für die Geschichte der Naturwissehschaften und der Technik.

Reich, Karin, 2003. "Die Rezeption Diophants im 16. Jahrhundert." *NTM: Nachrichtenblatt der Deutschen Gesellschaft für Geschichte der Medizin, Naturwissenschaften und Technik* 11: 80–89.

Robbins, Frank Egleston, 1921. "The Tradition of Greek Arithmology." *Classical Philology* 16: 97–123.

Robbins, Frank Egleston, 1929. "P. Mich. 620: A Series of Arithmetical Problems." *Classical Philology* 24 (4): 321–29.

Roca, Antic, 1564. *Arithmetica*. Barcelona: Claudio Bornat.

Rome, Adolf, 1938. "Un manuscrit de la bibliothèque de Boniface VIII à la Médicéenne de Florence." *L'Antiquité Classique* 7 (2): 261–70.

Rommevaux, Sabine, 2012. "Qu'est-ce que l'algèbre pour Christoph Clavius?" In *Pluralité de l'algèbre à la Renaissance*, edited by Sabine Rommevaux, Maryvonne Speisser, and Maria Rosa Massa Esteve, 293–309. Paris: Honoré Champion.

Rose, Paul Lawrence, 1975. *The Italian Renaissance of Mathematics. Studies on Humanists and Mathematicians from Petrarch to Galileo*. Genève: Librairie Droz.

Rosenfeld, Boris A., and Ekmeleddin İhsanoğlu, 2003. *Mathematicians, Astronomers, and other Scholars of Islamic Civilization and Their Works (7th–19th C.)*. Istanbul: Research Center for Islamic History, Art and Culture (IRCICA).

Rowe, Galen O., 1997. "Style." In *Handbook of Classical Rhetoric in the Hellenistic Period 330 B.C. – A.D. 400*, edited by Stanley E. Porter, 121–57. Leiden: E. J. Brill.

Rudolff, Christoph, 1525. *Behend und Hübsch Rechnung durch die künstreichen regeln Algebre, so gemeimblich die Coß genannt werden*. Argentorati (Strasbourg): Vuolfius Cephaleus Ioanni Iung.

Ruska, Julius, 1917. *Zur ältesten arabischen Algebra und Rechenkunst*. Heidelberg: C. Winter.

Saidan, A. S., 1967. "Jāmi' al-ḥisāb bi'l-takht wa l-turāb, li Nasīr al-Dīn al-Ṭūsī." *Al-Abhath: Quarterly Journal of the American University* 20: 91–164; 213–92.

Saidan, A. S., 1971. *Tārīkh 'ilm al-ḥisāb al-'Arabī (History of the Science of Arithmetic in Islam)*. 'Amman: Jam'iyat 'Umāl al-Maṭābi' al-Ta'āwinīa.

Saidan, A. S., 1978. *The Arithmetic of al-Uqlīdisī. The Story of Hindu-Arabic Arithmetic as Told in Kitāb al-fuṣūl fī al-ḥisāb al-Hindī by Abū al-Ḥasan Aḥmad ibn Ibrāhīm al-Uqlīdisī*. Dordrecht; Boston: Reidel.

Saidan, A. S., ed., 1986. *Tārākh 'ilm al-jabr fī l-'ālam al-'Arabī (History of Algebra in Medieval Islam)*, 2 vols. Kuwait: l-Majlis al-Waṭanī lil-Thaqāfah wa'l-Funūn wa'l-Ādāb, Qism al-Turāth al-'Arabī.

Saito, Ken, and Nathan Sidoli, 2010. "The Function of Diorism in Ancient Greek Analysis." *Historia Mathematica* 37: 579–614.

Sammarchi, Eleonora, 2019. "Les collections de problèmes algébriques dans le *Qisṭās al-mu'ādala fī 'ilm al-jabr wa'l-muqābala* d'al-Zanjānī." *Médiévales* 77: 37–56.

Sasaki, Chikara, 2003. *Descartes's Mathematical Thought*. Dordrecht: Kluwer.

Savage-Smith, Emilie, Simon Swain, and Geert Jan van Gelder, eds., 2020. *A Literary History of Medicine*. Leiden: E. J. Brill.

Sayyid, Ayman Fu'ād, ed., 2009. *Ibn al-Nadīm, Kitāb al-fihrist*. London: Al-Furqan Islamic Heritage Foundation.

Schärlig, Alain, 2001. *Compter avec des cailloux. Le calcul élémentaire sur l'abaque chez les anciens grecs*. Lausanne: Presses Polytechniques et Universitaires Romandes.

Schärlig, Alain, 2006. *Compter du bout des doigts: cailloux, jetons et bouliers, de Périclès à nos jours*. Lausanne: Presses Polytechniques et Universitaires Romandes.

Scheibel, Johann Ephraim, 1775. *Einleitung zur mathematischen Bücherkenntnis. Zwenter Band: Welcher das siebente biß zwölfte Stück enthält*. Breßlau: Meyer.

Scheubel, Johann, 1550. *Evclidis Megarensis, Philosophi & Mathematici Excellentissimi, Sex libri priores, de geometricis principijs . . . algebrae porro regvlae. . . .* Basileae: Hervagius.

Schröder, Ernst, 1870. "Vier combinatorische Probleme." *Zeitschrift Für Mathematik Und Physik* 15: 361–76.

Sesiano, Jacques, 1977a. "Le traitement des équations indéterminées dans le *Badī' fī 'l-ḥisāb* d'Abū Bakr al-Karajī." *Archive for History of Exact Sciences* 17: 297–379.

Sesiano, Jacques, 1977b. "Les méthodes d'analyse indeterminée chez Abū Kāmil." *Centaurus* 21: 89–105.

Sesiano, Jacques, 1982. *Books IV to VII of Diophantus' Arithmetica in the Arabic Translation Attributed to Qusṭā ibn Lūqā*. New York: Springer-Verlag.

Sesiano, Jacques, 1993. "La version latine médiévale de l'algèbre d'Abū Kāmil." In *Vestigia Mathematica: Studies in Medieval and Early Modern Mathematics in Honour of H.L.L. Busard*, edited by M. Folkerts and J. P. Hogendijk, 315–452. Amsterdam: Rodopi.

Sesiano, Jacques, 2004. "Introduction to Part 4: 'Studies on Greek Algebra'." In *Classics in the History of Greek Mathematics*, edited by Jean Christianidis, 257–63. Dordrecht: Kluwer.

Sezgin, Fuat, 1974. *Geschichte des arabischen Schrifttums. 5, Mathematik bis ca. 430 H.* Leiden: E. J. Brill.

Sharaf al-Dīn al-Ṭūsī, 1986. *Algébre et géométrie au XIIe siècle.* Edited by Roshdi Rashed. Paris: Les Belles Lettres.

Sidoli, Nathan, 2011. "Heron of Alexandria's Date." *Centaurus* 53: 55–61.

Sidoli, Nathan, 2018a. "The Concept of Given in Greek Mathematics." *Archive for History of Exact Sciences* 72: 353–402.

Sidoli, Nathan, 2018b. "Uses of Construction in Problems and Theorems in Euclid's *Elements* I–VI." *Archive for History of Exact Sciences* 72: 403–52.

Sidoli, Nathan, and Yoichi Isahaya, 2018. *Thābit ibn Qurra's Restoration of Euclid's Data. Text, Translation, Commentary.* Springer International Publishing.

Sidoli, Nathan, and Ken Saito, 2012. "Comparative Analysis in Greek Geometry." *Historia Mathematica* 39: 1–33.

Skoura, Ioanna, 2016. *Τα Σχόλια του Θέωνα στη Μαθηματική σύνταξη του Πτολεμαίου.* PhD thesis, National and Kapodistrian University of Athens.

Skoura, Ioanna, 2019–20. "Μια ανέκδοτη επιστολή του Νικολάου Ραβδά για τους εκκλησιαστικούς λογαριασμούς." *Neusis* 27–28: 353–99.

Skoura, Ioanna, and Jean Christianidis, 2014. "Καθ' ἃ καὶ Διόφαντός φησιν: Θέων ο Αλεξανδρινός, ένας λόγιος αναγνώστης του Διοφάντου." *Neusis* 22: 5–52.

Smith, David Eugene, 1908. *Rara Arithmetica : A Catalogue of the Arithmetics Written Before the Year MDCI, with a Description of Those in the Library of George Arthur Plimpton, of New York.* Boston: Ginn and Company.

Smith, David Eugene, 1958. *History of Mathematics*, 2 vols. New York: Dover.

Spoerrl, Walter, 1980. "Die Edition der Aischylosscholien." *Museum Helveticum* 37 (1): 3–24.

Stamatis, E. S., 1961. "Ανακατασκευή του αρχαίου κειμένου τεσσάρων ελλειπόντων προβλημάτων του 5ου βιβλίου των *Αριθμητικών* του Διοφάντου." *ΠΛΑΤΩΝ* 13: 25–26.

Stamatis, E. S., ed., 1963. *Διοφάντου Αριθμητικά. Η Άλγεβρα των Αρχαίων Ελλήνων. Αρχαίον κείμενον – μετάφρασις – επεξηγήσεις Ευάγγελος Σ. Σταμάτης.* Athens: School Books Publishing Organization.

Stamatis, E. S., 1978. *Reprints.* Athens.

Stanley, Richard P., 1997. "Hipparchus, Plutarch, Schröder, and Hough." *American Mathematical Monthly* 104: 344–50.

Steinschneider, Moritz, 1897. *Die arabischen Uebersetzungen aus dem Griechischen.* Leipzig: Otto Harrassowitz.

Stevin, Simon, 1585. *L'arithmetique de Simon Stevin de Bruges: Contenant les computations des nombres arithmetiques ou vulgaires. Aussi l'algebre, avec les equations de cinc quantitez. Ensemble les quatre premiers livres d'algebre de Diophante d'Alexandrie, maintenant premierement traduits en François.* Leyde: De l'Imprimerie de Christophle Plantin.

Stewart, Devin J., 2014. "Editing the *Fihrist* of Ibn al-Nadīm." *Journal of Abbasid Studies* 1 (2): 159–205.

Stewart, Devin J., 2016. "*Kitāb al-waṣāyā* (The Book of Legacies) and the Works of Mathematician Abū Kāmil Shujāʿ b. Aslam." *Journal of Abbasid Studies* 3 (2): 129–66.

Stifel, Michael, 1544. *Arithmetica Integra.* Norimbergae: Petreius.

Suter, Heinrich, 1892. "Das Mathematiker-Verzeichniss im Fihrist des Ibn Abî Jaʾḳûb an-Nadîm." *Abhandlungen zur Geschichte der Mathematik* 6: 1–87.

Suter, Heinrich, 1900. *Die Mathematiker und Astronomen der Araber und ihre Werke.* Leipzig: B. G. Teubner.

Swerdlow, Noel M., 1993. "The Recovery of the Exact Sciences of Antiquity: Mathematics, Astronomy, Geography." In *Rome Reborn: The Vatican Library and Renaissance Culture*, edited by A. Grafton, 125–67. Washington/New Haven: Library of Congress/ Yale University Press.

Szabó, Árpad, 1969. *Anfänge der griechischen Mathematik.* München & Wien/Budapest: R. Oldenbourg/Akadémiai Kiadó.

Szabó, Árpad, 1978. *The Beginnings of Greek Mathematics.* Dordrecht: Reidel.

Taisbak, Christian Marinus, 1980. "'Dynamis' og 'dynasthai'. Et forslag til tolkning af en betydningstuld geometrisk terminus i den græske lære om usammålelige linjestykker." *Museum Tusculanum* 40–43: 119–31.

Takahashi, Hidemi, 2005. *Barhebraeus: A Bio-bibliography.* Piscataway, NJ: Gorgias Press.

Takahashi, Hidemi, 2010. "Between Greek and Arabic: The Sciences in Syriac from Severus Sebokht to Barhebraeus." In *Transmission of Sciences: Greek, Syriac, Arabic and Latin*, edited by H. Kobayashi and M. Kato, 16–39. Tokyo: Organization for Islamic Area Studies, Waseda University.

Takahashi, Hidemi, 2014. "Barhebraeus." In *Encyclopedia of Islam* (3rd ed.), vol. 2, edited by K. Fleet, G. Kramer, D. Matringe, J. Nawas, and E. Rowson, 40–44. Leiden: Brill.

Tannery, Paul, 1912a. "À quelle époque vivait Diophante?" In *Mémoires scientifiques de Paul Tannery*, edited by J. L. Heiberg and H. G. Zeuthen, I: 62–73. Toulouse/Paris: E. Privat/Gauthier-Villars.

Tannery, Paul, 1912b. "Études sur Diophante." In *Mémoires scientifiques de Paul Tannery*, edited by J. L. Heiberg and H. G. Zeuthen, II: 367–99. Toulouse/Paris: E. Privat/ Gauthier-Villars.

Tannery, Paul, 1912c. "L'article de Suidas sur Hypatia." In *Mémoires scientifiques de Paul Tannery*, edited by J. L. Heiberg and H. G. Zeuthen, I: 74–79. Toulouse/Paris: E. Privat/ Gauthier-Villars.

Tannery, Paul, 1912d. "La perte de sept livres de Diophante." In *Mémoires scientifiques de Paul Tannery*, edited by J. L. Heiberg and H. G. Zeuthen, II: 73–90. Toulouse/Paris: E. Privat/Gauthier-Villars.

Tannery, Paul, 1912e. "Les manuscrits de Diophante à l'Escorial." In *Mémoires scientifiques de Paul Tannery*, edited by J. L. Heiberg and H. G. Zeuthen, II: 418–32. Toulouse/ Paris: E. Privat/Gauthier-Villars.

Tannery, Paul, 1912f. "Rapport sur une mission en Italie du 24 janvier au 24 février 1886." In *Mémoires scientifiques de Paul Tannery*, edited by J. L. Heiberg and H. G. Zeuthen, II: 269–331. Toulouse/Paris: E. Privat/Gauthier-Villars.

Tannery, Paul, 1912g. "Sur la religion des derniers mathématiciens de l'antiquité." In *Mémoires scientifiques de Paul Tannery*, edited by J. L. Heiberg and H. G. Zeuthen, II: 527–39. Toulouse/Paris: E. Privat/Gauthier-Villars.

Tannery, Paul, 1912h. "Sur les épigrammes arithmétiques de l'Anthologie palatine." In *Mémoires scientifiques de Paul Tannery*, edited by J. L. Heiberg and H. G. Zeuthen, II: 442–46. Toulouse/Paris: E. Privat/Gauthier-Villars.

Tannery, Paul, 1912i. "Sur les manuscrits de Diophante à Paris." In *Mémoires scientifiques de Paul Tannery*, edited by J. L. Heiberg and H. G. Zeuthen, II: 64–72. Toulouse/Paris: E. Privat/Gauthier-Villars.

Tannery, Paul, 1912j. "Sur une épigramme attribué à Diophante." In *Mémoires scientifiques de Paul Tannery*, edited by J. L. Heiberg and H. G. Zeuthen, II: 433–39. Toulouse/ Paris: E. Privat/Gauthier-Villars.

Tannery, Paul, 1915. "Anatolius sur la décade et les nombres qu'elle comprend." In *Mémoires scientifiques de Paul Tannery*, edited by J. L. Heiberg and H. G. Zeuthen, III: 12–28. Toulouse/Paris: E. Privat/Gauthier-Villars.

Tannery, Paul, 1920a. "Notice sur les deux lettres arithmétiques de Nicolas Rhabdas." In *Mémoires scientifiques de Paul Tannery*, edited by J. L. Heiberg, IV: 61–198. Toulouse/ Paris: E. Privat/Gauthier-Villars.

Tannery, Paul, 1920b. "Psellus sur Diophante." In *Mémoires Scientifiques de Paul Tannery*, edited by J. L. Heiberg, IV: 275–82. Toulouse/Paris: E. Privat/Gauthier-Villars.

Tartaglia, Niccolò, 1560. *La sesta parte del general trattato de' numeri, et misure*. Venice: Per Curtio Troiano.

Taub, Liba, 2017. *Science Writing in Greco-Roman Antiquity*. Cambridge: Cambridge University Press.

Terian, Abraham, 1984. "A Philonic Fragment on the Decad." In *Nourished with Peace: Studies in Hellenistic Judaism in Memory of Samuel Sandmel*, edited by Frederick E. Greenspahn, Earle Hilgert, and Burton L. Mack, 173–82. Chico, CA: Scholars Press.

Theon de Smyrne, 1892. *Théon de Smyrne, philosophe platonicien: Exposition des connaissances mathématiques utiles pour la lecture de Platon*. Traduite pour la première fois du grec en français par Jean Dupuis. Paris: Librairie Hachette.

Theon of Alexandria, 1936. *Commentaires de Pappus et de Théon d'Alexandrie sur l'Almageste*. Texte établi et annoté par Adolf Rome. Tome II: *Commentaire sur les Livres 1 et 2 de l'Almageste*. Città del Vaticano: Biblioteca Apostolica Vaticana.

Theon of Alexandria, 1943. *Commentaires de Pappus et de Théon d'Alexandrie sur l'Almageste*. Texte établi et annoté par Adolf Rome. Tome III: *Commentaire lur les Livres 3 et 4 de l'Almageste*. Città del Vaticano: Biblioteca Apostolica Vaticana.

Theon of Smyrna, 1979. *Mathematics useful for understanding Plato*. Translated from the 1892 Greek/French edition of J. Dupuis by Robert and Deborah Lawlor and edited and annotated by Christos Toulis and others. San Diego: Wizards Bookshelf.

Theophylacte d'Achrida, 1986. *Lettres*. Introduction, texte, traduction et notes par Paul Gautier. Thessalonique: Association de Recherches Byzantines.

Thesleff, Holger, 1961. *An Introduction to the Pythagorean Writings of the Hellenistic Period*. Åbo: Åbo Akademi.

Thesleff, Holger, 1965. *The Pythagorean Texts of the Hellenistic Period*. Edited by Holger Thesleff. Åbo: Åbo Akademi.

Thulin, Carl, 1911. *Die Handschriften des Corpus agrimensorum Romanorum*. Berlin: Abhandlungen der Königlichen Akademie der Wissenschaften zu Berlin.

Tihon, Anne, 1977. "Un traité astronomique chypriote du XIVe siècle." *Janus* 64: 279–308.

Toomer, Gerald J., 1970. "Hipparchus." In *Dictionary of Scientific Biography*, Vol. 15: *Supplement I*, edited by C. C. Gillispie, 207–24. New York: Charles Scribner's Sons.

Toomer, Gerald J., 1985. "Chronique II: Réponse concernant la chronique de A. Allard et R. Rashed sur la traduction de l'*Arithmétique* de Diophante." *Revue des Questions Scientifiques* 156 (2): 237–41.

Toomer, Gerald J., 1996. "Diophantus." In *The Oxford Classical Dictionary* (3rd ed.), edited by Simon Hornblower and Antony Spawforth, 483. Oxford: Oxford University Press.

Toomer, Gerald J., and Reviel Netz, 2012. "Diophantus." In *The Oxford Classical Dictionary* (4th ed.), edited by Simon Hornblower and Antony Spawforth, 465. Oxford: Oxford University Press.

Tropfke, Johannes, 1934. "Zur Geschichte der quadratischen Gleichungen über dreieinhalb Jahrtausend (Fortsentzung)." *Jahresbericht Der Deutschen Mathematiker-Vereinigung* 44: 26–47.

Tsiotras, V. I., 2006. *Ἡ ἐξηγητικὴ παράδοση τῆς Γεωγραφικῆς Ὑφηγήσεως τοῦ Κλαυδίου Πτολεμαίου. Οἱ ἐπώνυμοι σχολιαστές.* Athens: Cultural Institution of National Bank.

Turyn, Alexander, 1964. *Codices Graeci Vaticani saeculis XIII et XIV scripti annorumque notis instructi.* Città del Vaticano: Biblioteca Apostolica Vaticana.

Turyn, Alexander, 1972. *Dated Greek Manuscripts of the Thirteenth and Fourteenth Centuries in the Libraries of Italy*, 2 vols. Urbana: University of Illinois Press.

Van der Pas, Stéphanie, 2014. "The Normal Road to Geometry: Δή in Euclid's *Elements* and the Mathematical Competence of His Audience." *Classical Quarterly* 64 (2): 558–73.

Van der Waerden, Bartel Leendert, 1976. "Defence of a 'Shocking' Point of View." *Archive for History of Exact Sciences* 15 (3): 199–210.

Van Egmond, Warren, 1985. "A Catalog of François Viète's Printed and Manuscript Works." In *Mathemata: Festschrift für Helmuth Gericke*, edited by Menso Folkers and Uta Lindgren, 359–96. Stuttgart: F. Steiner Verlag Wiesbaden.

Viète, François, 1591. *In artem analyticem Isagoge.* Tours: Apud Iametium Mettayer.

Viète, François, 1593a. "Effectionum geometricarum canonica recensio." In *Variorum de rebus mathematicis responsorum, liber VIII.* Turonis: Apud Iamettium Mettayer.

Viète, François, 1593b. *Supplementum geometriae.* Turonis: Excudebat Iametius Mettayer.

Viète, François, 1593c. *Variorum de rebus mathematicis responsorum, liber VIII.* Turonis: Apud Iamettium Mettayer.

Viète, François, 1593d. *Zeteticorum.* Turonis: Apud Iamettium Mettayer.

Viète, François, 1595. *Ad problema quod omnibus mathematicis totius orbis construendum proposuit Adrianus Romanus.* Parisiis: Apud Iametium Mettayer.

Viète, François, 1615a. *Ad angularium sectionum analyticen.* Parisiis: Apud Oliverium de Varennes.

Viète, François, 1615b. *De æquationum recognitione et emendatione tractatus duo.* Parisiis: Ex Typographia Ioannes Laquehay.

Viète, François, 1630. *Les cinq livres des zetetiques de Francois Viette.* Translated and commented by Jean-Louis Vaulezard. Paris: Iulian Iacquin.

Viète, François, 1646. *Franciscii Vietæ opera mathematica, in unum volumen congesta, ac recognita.* Opera atque studio Francisci à Schooten Leydensis, matheseos professoris. Lugduni Batavorum: Ex Officinâ Bonaventurae & Abrahami Elzviriorum.

Viète, François, 1983. *The Analytic Art: Nine Studies in Algebra, Geometry, and Trigonometry from the Opus Restitutae Mathematicae Analyseos, seu, algebrâ novâ by François Viète.* Translated by T. Richard Witmer. Kent, Ohio: Kent State University Press.

Vitrac, Bernard, 2005a. "Les classifications des sciences mathématiques en Grèce ancienne." *Archives de Philosophie* 68: 269–301.

Vitrac, Bernard, 2005b. "Peut-on parler d'algèbre dans les mathématiques." *Ayene-ye Miras* 3: 1–44.

Vitrac, Bernard, 2008. "Les formules de la 'Puissance' (Δύναμις, Δύνασθαι) dans les mathématiques grecques et dans les dialogues de Platon." In *DYNAMIS. Autour de la Puissance chez Aristote*, edited by M. Crubellier, A. Jaulin, D. Lefèbvre, and P.-M. Morel, 73–148. Louvain-la-Neuve: Peeters.

Vitrac, Bernard, 2018. "Quand ? Comment ? Pourquoi les textes mathématiques grecs sont-ils parvenus en occident ? (Version Août 2016)." www.academia.edu/32567771/Quand_Comment_Pourquoi_les_textes_mathématiques_grecs_sont-ils_parvenus_en_Occident.

Vitrac, Bernard, and Ahmed Djebbar, 2011. "Le Livre XIV des *Éléments* d'Euclide: versions grecques et arabes (première partie)." *SCIAMVS* 12: 29–158.

Vogel, Kurt, 1930. "Die algebräischen Probleme des P. Mich. 620." *Classical Philology* 30 (4): 373–75.

Vogel, Kurt, 1967. "Byzantine Science." In *The Cambridge Medieval History*, vol. IV: *The Byzantine Empire;* Part II: *Government, Church, Civilisation*, edited by J. M. Hussey, 264–305. Cambridge: Cambridge University Press.

Vogel, Kurt, ed., 1968. *Ein byzantinisches Rechenbuch des 14. Jahrhunderts*. Wien: Hermann Böhlaus.

Vogel, Kurt, 1971. "Diophantus of Alexandria." In *Dictionary of Scientific Biography*, vol. IV, edited by C. C. Gillispie, 110–19. New York: Charles Scribner's Sons.

Waltz, Pierre, 1928. "L'anthologie Grecque." *Bulletin de l'Association Guillaume Budé* 20: 2–23.

Waltz, Pierre, ed. 1960. *Anthologie Grecque*. Première partie: *Anthologie Palatine*. Tome I: *Livres I–IV.* Texte établi et traduit par P. Waltz. Tome II: *Livre V.* Texte établie et traduit par P. Waltz, en collaboration avec J. Guillon. Paris: Les Belles Lettres.

Waterhouse, William C., 1993. "Harmonic Means and Diophantus I.39." *Historia Mathematica* 20: 89–91.

Watt, Robert, 1824. *Bibliotheca Brittanica or a General Index to British and Foreign Literature*, 2 vols, vol. 1: Authors. Edinburgh/London: Archibald Constable and Company/Longman, Hurst, Rees, Orme, Brown, & Green.

Webb, Ruth, 2001. "The Progymnasmata as Practice." In *Education in Greek and Roman Antiquity*, edited by Y. L. Too, 289–316. Leiden: E. J. Brill.

Wehr, Hans, 1979. *A Dictionary of Modern Written Arabic* (4th ed.). New York: J Milton Cowan.

Wendel, Carl, 1940. "Planudea." *Byzantinische Zeitschrift* 40 (2): 406–45.

Wertheim, Gustav, 1897. "Die Schlussaufgabe in Diophants Schrift über Polygonalzalhen." *Zeitschrift Für Mathematik Und Physik, Hist.-Lit. Abteilung* 42: 121–26.

Westerink, Leendert Gerrit, ed., 1967. *Pseudo-Elias (Pseudo-David), Lectures on Porphyry's Isagoge*. Introduction, text and indices by L. G. Westerink. Amsterdam: North-Holland.

Wildberg, Christian, 2016. "Elias." In *The Stanford Encyclopedia of Philosophy* (Fall 2016 ed.). https://plato.stanford.edu/archives/fall2016/entries/elias/.

Wilson, Nigel Guy, 1992. *From Byzantium to Italy. Greek Studies in the Italian Renaissance*. London: Duckworth.

Wilson, Nigel Guy, 1996. *Scholars of Byzantium* (revised ed.). London/Cambridge, MA: Duckworth/The Medieval Academy of America.

Winter, John Garrett, ed., 1936. *Papyri in the University of Michigan Collection: Miscellaneous Papyri* (Michigan papyri, vol. III). Ann Arbor: University of Michigan Press.

Wissowa, Georg, ed., 1894. *Paulys Realencyclopädie der classischen Altertumswissenschaft*, vol. 1, Part 2. Stuttgart: J. B. Metzler.

Woepcke, Franz, 1851. *L'algèbre d'Omar Alkhayyâmî*. Publiée, traduite et accompagnée d'extraits de manuscrits inédits par F. Woepcke. Paris: Benjamin Duprat.

Woepcke, Franz, 1853. *Extrait du Fakhrî. Traité d'Algèbre par Aboù Bekr Mohammed Ben Alhaçan Alkarkhî (Manuscrit 952, Supplément arabe de la Bibliothèque Impériale); Précédé d'un mémoire sur l'algèbre indéterminée chez les arabes*. Paris: L'Imprimerie Impériale.

Woepcke, Franz, 1855. *Recherches sur l'histoire des sciences mathématiques chez les orientaux, d'apres des traités inédits arabes et persans*. Paris: Imprimerie Impériale.

Wright, Wilmer Cave, ed., 1922. *Philostratus and Eunapius, The Lives of the Sophists*. With an English translation by W. C. Wright. London: William Heinemann.

Xylander, ed., 1575. *Diophanti Alexandrini rerum Arithmeticarum libri sex, quorum primi duo adiecta habent scholia, Maximi (ut coniectura est) Planudis. Item Liber de numeris polygonis seu multiangulis*. Basileae: per Eusebium Episcopium, & Nicolai Fr. haeredes.

Yadegari, Mohammed, 1980. "The Binomial Theorem: A Widespread Concept in Medieval Islamic Mathematics." *Historia Mathematica* 7: 401–06.

Zemouli, T., 1993. *Muʾallafāt Ibn al-Yāsamīn al-riyāḍiyya* (Mathematical Writings of Ibn al-Yāsamīn). M.Sc. thesis in History of Mathematics, E.N.S., Algiers.

Zeuthen, Hieronymus Georg, 1902. *Histoire des mathématiques dans l'antiquité et le moyen age*. Translated by J. Mascart. Paris: Gauthier-Villars.

Zhmud, Leonid, 2016. "Greek Arithmology: Pythagoras or Plato?" In *Pythagorean Knowledge from the Ancient to the Modern World: Askesis, Religion, Science*, edited by A.-B. Renger and A. Stavru, 321–46. Wiesbaden: Harrassowitz Verlag.

Zinner, Ernst, 1990. *Regiomontanus: His Life and Work*. Translated by Ezra Brown. Amsterdam: North-Holland.

Index